Features of the Companion Website

Hands-On Problems Solving Exercises: These inquiry-based exercises challenge you to think as a scientist and to analyze and interpret experimental data. Exercises include data manipulation questions based on real experiments, as well as problems involving simulations of model systems.

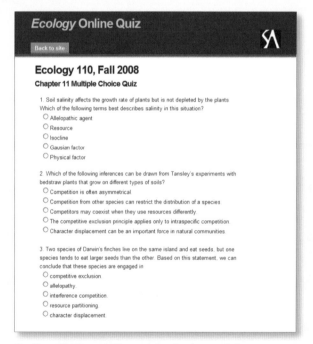

Online Quizzes: For each chapter of the textbook, the companion site includes a multiple-choice quiz that covers all the main topics presented in the chapter. Your instructor may assign these quizzes, or they may be made available to you as self-study tools. (Instructor registration is required for student access to the quizzes.)

Additional features of the Companion Website:

- **Chapter Summaries:** Concise overviews of the important concepts and topics covered in each chapter.
- **Chapter Outlines:** A convenient outline of each chapter's headings and numbered concepts.
- **Flashcards:** Flashcard activities help you master the many new terms introduced in each chapter of the textbook.
- **Web Extensions:** Expanded and additional coverage of selected topics introduced in the textbook. (These are referenced by number in the textbook.)
- **Web Stats Review:** A brief review of statistical methods and techniques mentioned in the textbook.
- **Glossary:** A complete online version of the glossary for quick access to definitions of important terms.

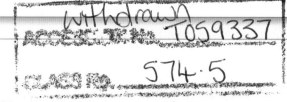

Hands-On Problem Solving Exercises

These exercises are referenced at the end of each chapter. All exercises are located on the Companion Website: www.sinauer.com/ecology.

1. Amphibian and Reptile Population Declines: This exercise explores declines in population sizes of amphibians and reptiles in Costa Rica, and suggests a climatic link. You will interpret data from a recent paper showing changes in population sizes of reptile and amphibian species, as well as changes in climatic conditions.

2. Effect of El Niño Events on Tuna Populations: This exercise shows the effect of the El Niño Southern Oscillation (ENSO) on tuna populations in the western Pacific. You will investigate whether changes in the locations of tuna populations covary with displacements of warm water that occur during ENSO events.

3. Losses of Coral Reefs: This exercise quantifies losses in coral reefs, a marine biome that is especially threatened. You will explore the rate, timing, and geographical extent of losses of coral cover in reefs in the Indo-Pacific region.

4. Thermal Adaptations in Urban Ants: This exercise demonstrates thermal adaptations in ants that live in cities. You will investigate whether adaptations to heat in urban leaf-cutter ants reduces tolerance to cold temperatures.

5. Optimal Foraging: This exercise illustrates optimal foraging. You will manipulate the foraging decision-rules of a predator to explore how different foraging strategies vary in the benefits they return.

6. Effects of Natural Selection and Genetic Drift: This exercise demonstrates how natural selection and genetic drift can alter the frequencies of alleles in populations. You will investigate the effects of manipulating population sizes (and thus the strength of genetic drift) and strengths of selection.

7. The Growth/Reproduction Tradeoff: This exercise explores the tradeoff that species must make between growth and reproduction. You will explore the effects of manipulating the set point at which fish start allocating resources into reproduction rather than further growth, under different levels of predation.

8. Abiotic and Biotic Factors: This exercise explores how the interaction of abiotic and biotic factors affects the abundance and distribution of lizards. You will interpret data to determine how well lizard populations are able to survive a hurricane on islands with predators and on islands without predators.

9. Density-dependent and Density-independent Factors: This exercise explores the effects of density-dependent and -independent factors on populations. You will analyze data from populations of arctic ground squirrels to determine

these effects on weaning success, over-wintering survival, and population growth rate.

10. Population Dynamics: This exercise explores how changes in population growth rates and the extent of delayed density dependence affect population dynamics. You will manipulate the values of these two factors, and run a simulation to determine their effects on a population.

11. Character Displacement: This exercise explores character displacement and phenotypic plasticity for a polyphenism in a spadefoot toad species. You will interpret data collected on different populations of toads and the results of common garden experiments to assess genetic effects.

12. Predation Alters Behavior: This exercise explores the effect of predation on lizard behavior. You will interpret data from a recent paper that demonstrated changes in the behaviors of *Anolis* lizards after the introduction of a predatory lizard. This behavioral shift was manifest as a change in habitat use.

13. Evolutionary Interactions of a Fly and Its Parasitoid: This exercise explores the evolutionary interactions between the fly *Drosophila melanogaster* and its wasp parasitoid. You will interpret data from a recent paper that examines the evolution of resistance of these flies to the parasitoids. Does resistance to parasitoids impose a cost?

14. From a Parasite to a Mutualist: This exercise explores the evolution of mutualism between bacteria that live inside *Drosophila* flies and their fly host. You will interpret data from a recent paper that demonstrates that, in the course of evolving with flies, the bacteria evolved from parasites into mutualists.

15. Measuring Marine Species Diversity: This exercise explores various methods of measuring species diversity. You will interpret data from a recent paper that measured marine species diversity, with emphasis on the benthic fauna of the continental shelf of Norway, using the Shannon index as well as other indices of diversity.

16. Soil Invertebrates and Succession: This exercise explores how invertebrates that live in the soil could affect patterns of succession. You will interpret data from a recent paper that shows the effects of soil invertebrates on the growth of early- and mid-succession plants.

17. Island Biogeography: This exercise explores the factors that determine the number of species that can live on different islands according to the theory of island biogeography. In a series of simulations, you will manipulate the size

of the island and the distance from the island to the mainland to demonstrate how these factors affect the equilibrium number of species on that island.

18. Periodic Disturbance and Its Effect on Species: This exercise demonstrates how periodic disturbance can maintain species in a community that otherwise could not coexist. You will manipulate the frequency and intensity of disturbances to investigate this effect.

19. Drought Reduces Productivity Across Europe: This exercise examines the effect of the 2003 European drought on primary productivity. You will interpret data from a recent paper that demonstrated that the drought did substantially reduce primary productivity at various sites in Europe.

20. Are Complex Food Webs More Stable? This exercise explores whether or not food webs that are more complex are also more stable. You will interpret data from a recent paper that experimentally addressed the relationship between species diversity in grassland food webs and the stability of the food web during the decade-long study.

21. Dry Decomposition: This exercise explores how plant litter decomposes in a dry climate. You will interpret data from a recent paper demonstrating the factors responsible for litter decomposition in a semi-arid ecosystem in Patagonia.

22. Evolution of an Invasive Species: This exercise explores hypotheses for why invasive plant species are often better competitors in their introduced ranges than in their native ranges. You will interpret data from a recent paper that was designed to test whether the success of an invasive tree species in its introduced range has resulted from evolutionary change.

23. Patch Movement: Crickets versus Cybercrickets: This exercise explores how organisms move across patches in the landscape. You will interpret data from a recent paper that simulated the movement of virtual organisms between patches across landscapes with different levels of connectivity, and then compared the simulation results to results from manipulation studies. Do real crickets move like cybercrickets?

24. Climate Change in the Arctic: This exercise explores regional variation in global warming. Different regions of Earth and the associated troposphere are warming at different rates. You will interpret data from a recent paper showing pronounced warming in the Arctic region at different levels in the atmosphere.

ECOLOGY

ECOLOGY

MICHAEL L. CAIN • *Bowdoin College*

WILLIAM D. BOWMAN • *University of Colorado*

SALLY D. HACKER • *Oregon State University*

Chapter 7, "Life History Analyses"
By **John Allen**, *Bowdoin College and*
Margaret Pizer, *The Nature Conservancy*

Chapter 22, "Conservation Biology" and
Chapter 23, "Landscape Ecology and Ecosystems Management"
By **Loraine Kohorn**, *Bowdoin College*

Sinauer Associates, Inc. • Sunderland, Massachusetts

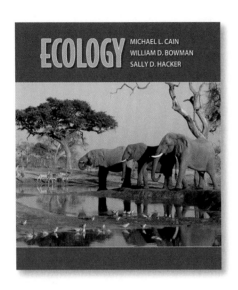

COVER PHOTOGRAPH

The community of organisms inhabiting Botswana's savanna includes African elephants (*Loxodonta africana*), impala (*Aepyceros melampus*), mourning doves (*Streptopelia decipiens*), and acacia trees (*Acacia* sp.).
© Frans Lanting/Minden Pictures.

Ecology

Address inquiries and orders to
Sinauer Associates, Inc.
23 Plumtree Road
Sunderland, MA 01375 U.S.A.

www.sinauer.com
FAX: 413-549-1118
orders@sinauer.com
publish@sinauer.com

Library of Congress Cataloging-in-Publication Data

Ecology / Michael L. Cain ... [et al.].
       p. cm.
   Includes bibliographical references and index.
   ISBN 978-0-87893-083-8
I. Cain, Michael L. (Michael Lee)

   QH541E31933 2008
   577—dc22

                                        2008007834

Printed in U.S.A.

5 4 3 2 1

For Debra and Hannah, with thanks and love.
MLC

For Jen, Gordon, and Miles and their unending patience,
and to my students for pushing me as much as I pushed them.
WDB

For my family and my students, whose gift of time has
made all the difference.
SDH

About the Authors

MICHAEL L. CAIN, (center in photo) having opted for an early retirement to focus full-time on writing, now brings his expertise to this new book. He received his Ph.D. in Ecology and Evolutionary Biology from Cornell University, taught at New Mexico State University and the Rose-Hulman Institute of Technology, and is currently affiliated with Bowdoin College. Dr. Cain is the senior author of *Discover Biology*, Third Edition, and a coauthor of Campbell's *Biology*, Eighth Edition. He has instructed students across a wide range of subjects, including introductory biology, ecology, field ecology, evolution, botany, mathematical biology, and biostatistics. His research interests include: plant population ecology; long-distance dispersal; ecological and evolutionary dynamics in hybrid zones; and search behavior in plants and animals. Dr. Cain has served on the Editorial Board of *Ecology* and *Ecological Monographs*, and is the recipient of numerous fellowships, grants, and awards, including the Pew Charitable Trust Teacher-Scholar Fellowship and research grants from the National Science Foundation.

WILLIAM D. BOWMAN is a Professor at the University of Colorado at Boulder, affiliated with both the Department of Ecology and Evolutionary Biology and the Institute of Arctic and Alpine Research. He earned his Ph.D. from Duke University. Dr. Bowman has taught courses in introductory ecology, plant ecology, plant–soil interactions, and ecosystems ecology, and for over a decade has directed undergraduate summer field research programs. He is coeditor of the 2001 book, *Structure and Function of an Alpine Ecosystem, Niwot Ridge, Colorado* (Oxford University Press). Dr. Bowman is editor for the journal *Arctic, Antarctic and Alpine Research*, and has directed the Mountain Research Station for 20 years. His research focuses on plant ecology, biogeochemistry, and community dynamics, and has been supported by the National Science Foundation, the Environmental Protection Agency, the National Park Service, and the Andrew W. Mellon Foundation.

SALLY D. HACKER is Associate Professor in the Department of Zoology at Oregon State University. She received her Ph.D. from the Department of Ecology and Evolutionary Biology at Brown University. She has taught courses in general ecology, community ecology, field ecology, and marine biology. As a community ecologist interested in natural and managed coastal, dune, and estuarine communities, Dr. Hacker's research explores species interactions and how they influence community formation, species diversity, and species invasions. Currently, her research is supported by grants from the National Science Foundation, the Environmental Protection Agency, and the National Oceanographic and Atmospheric Administration. She serves as a Receiving Editor for *Ecology Letters*, chairs the Science Advisor Board at the National Center for Ecological Analysis and Synthesis, and has received numerous fellowships and awards, including the Buell Award from the Ecological Society of America and the Young Investigator Prize from the American Society of Naturalists.

Brief Contents

Preface

This is an exciting and challenging time to study ecology. New discoveries are pouring in, revealing factors that affect local communities and link ecosystems to one another across broad geographic areas. The progress in these and other areas of ecology could not come at a better time: Ecologists are increasingly being asked to apply their knowledge towards efforts to solve current environmental problems and prevent future ones. In addition, as ecologists seek to answer basic and applied research questions, they have at their disposal a vast array of new tools, ranging from DNA sequencing machines to miniature radios that track the movements of small insects to improved satellite instruments that enable ecological systems to be studied at a global scale for the first time in history.

Developments such as these fuel the excitement that grips the field of ecology—but they also mean that what we know about ecology is increasing at an unprecedented rate. The current explosion of ecological information makes ecology a daunting subject, both to study and to teach. Students need to master a heady mix of abstract concepts, experimental reasoning, mathematical equations, and details about particular organisms and their habitats. Faced with the challenge of conveying the fundamental concepts, new discoveries, and the relevance and rigor of modern ecology, instructors have been calling for a textbook that offers a new balance in subject matter emphasis, clearly presented concepts, and engaging, fresh examples. Responding to that call, we have written this book for such instructors—and for the students who will use it.

What We Set Out to Achieve

This book is written for undergraduate students of ecology. We set out to introduce our readers to the beauty and importance of ecology, and to do so without boring them or overwhelming them with unnecessary detail. This is a tall order, and so when we began writing *Ecology*, we taped notes to our computers to remind ourselves of the two core principles that guided our writing: **"Teaching comes First!"** and **"Less is More!"**

Teaching truly does come first in *Ecology*—it motivates everything we did. We structured our chapters and selected their content primarily to make them good tools for teaching. For example, to introduce the material covered and capture student interest, each chapter begins with a story (a "Case Study," as described more fully below) about an applied problem or interesting bit of natural history. Once students are drawn in by the Case Study, the "storyline" that begins there is maintained throughout the rest of the chapter. We used a narrative writing style to link the sections of the chapter to one another, thus helping students keep the big picture in mind. In addition, the sections of the chapter are organized around a small number of Key Concepts (described more thoroughly below) that were carefully selected to summarize current knowledge and provide students with a clear overview of the subject at hand. Similarly, when designing the art, pedagogy came first: Many students are visual learners, so we worked very hard to ensure that each figure "tells a story" that can be understood on its own, without reference to the main body of the text.

As another way to help us achieve our primary goal of teaching students, we also embraced a "less is more" philosophy. Rather than write a tome that provided comprehensive coverage of all aspects of ecology, we were guided by the principle that if we covered less material—but presented it clearly and well—students would learn more. Hence, we chose to keep our chapters relatively short and to build them around a small number of Key Concepts (typically, three to five). We made these choices to prevent students from being overwhelmed by long, diffuse chapters, and to allow them to master the big ideas first, then fill in the details. In addition, as we worked on the many drafts of our chapters, we put our "less is more" philosophy into action by asking each other whether the text served one of the following purposes:

- Did it help to explain an essential concept?
- Did it show how the process of ecological inquiry works?
- Did it motivate readers by focusing on a key ecological application or a fascinating piece of natural history?

This approach made for some tough choices in our writing, but it enabled us to focus on teaching students what is currently known about ecology without overwhelming them with excess information.

Features of Ecology

The features of *Ecology* described here resulted largely from our core principles that "Teaching comes First!" and "Less is More!" But we also sought to address two other issues. First, we felt that none of the existing textbooks provided both the right emphasis (in terms of subject matter) and the right degree of difficulty. Second, we thought that there was room for improvement in how ecology textbooks integrate material across the levels at which ecology is studied (i.e., across an ecological hierarchy that goes from individuals to populations to communities to ecosystems). With these goals in mind, we turn now to the features of *Ecology*:

■ **Pedagogical Excellence**   Students taking their first course in ecology are exposed to a great deal of material, on a conceptual as well as individual-systems level. To help them manage this vast amount of information, each chapter of *Ecology* is organized around a small number of Key Concepts that provide up-to-date summaries of fundamental ecological principles. All of these Key Concepts are provided as a list on the book's back end papers. Chapter-opening vignettes (see "Case Studies") draw students in, while clear narrative, state-of-the-art illustrations, and links among chapter sections help them comprehend and digest chapter contents.

■ **Subject Matter**   This book differs from existing textbooks by placing comparatively greater emphasis on community ecology, ecosystem ecology, and applied aspects of ecology. We have devoted a greater number of chapters to these subjects, and include discussion of community/ecosystem topics in chapters that focus on individuals or populations. Two chapters of the book (Chapter 22, *Conservation Biology*, and Chapter 23, *Landscape Ecology and Ecosystem Management*) are devoted to issues related to the preservation of biological diversity, one of the most important challenges facing human societies today. Mechanisms that underlie communities and ecosystems—such as the impact of climate and the physical environment, evolution and ecology, population ecology, and interactions among organisms—are also discussed thoroughly.

■ **Case Studies**   Each chapter opens with an interesting vignette—a Case Study. By presenting an engaging story or interesting application, the Case Study captures the reader's attention while introducing the topic of the chapter. Later, the reader is brought full circle with the corresponding "Case Study Revisited" section at chapter's end. Each Case Study relates naturally to multiple levels of the ecological hierarchy, thereby providing a nice lead-in to the *Connections in Nature* feature described next.

■ **Connections in Nature**   In most ecology textbooks, connections among levels of the ecological hierarchy are discussed briefly, perhaps in the opening chapter. As a result, many opportunities are missed to highlight for students the fact that events in natural systems *really are* interconnected. To facilitate the ability of students to grasp how events in nature are interconnected, each chapter of *Ecology* closes with a section that discusses how the material covered in that chapter affects and is affected by interactions at other levels

of the ecological hierarchy. Where appropriate, these interconnections are also emphasized in the main body of the text.

■ **Ecological Inquiry**   Our understanding of ecology is constantly changing due to new observations and new results from ecological experiments and models. All chapters of the book emphasize the active, inquiry-based nature of what is known about ecology. In addition, *Ecology* includes hands-on interpretative and quantitative exercises, described next.

■ **Hands-on Problem Solving**   This unique component of the Companion Website is designed to sharpen scientific reasoning skills. Students are asked to manipulate data, to explore mathematical aspects of ecology in more detail through quantitative problems, to use simulations to analyze simple model systems, and to grapple with how to interpret results from real experiments. There are 24 *Hands-on Problem Solving* exercises, one for each chapter of the book. These inquiry exercises can be used in two important ways: as ready-to-go, five to ten minute problems for in-class use (e.g., with "clickers" or to stimulate class discussion), or as homework problems.

■ **Ecological Applications**   In recent years, ecologists have increasingly focused their attention on applied issues. Similarly, many students taking introductory ecology are very interested in applied aspects of ecology. Thus, ecological applications (including conservation biology) receive great attention in this book. Discussions of applied topics are woven into each chapter, helping to capture and retain student interest.

■ **Links to Evolution**   Evolution is a central unifying theme of all biology, and its connections with ecology are very strong. Yet, ecology textbooks typically present evolution almost as a separate subject. As an alternative to the standard approach, *Ecology*'s Chapter 6 is devoted to describing the joint effects of ecology and evolution. This chapter explores the ecology of evolution at both the population level and as documented in the sweeping history of life on Earth. Concepts or applications that relate to evolution are also described in many other chapters.

■ **Art Program**   Many of *Ecology*'s illustrations feature "balloon captions," which tell a story that can be understood at a glance, without relying on the accompanying text. The art program is available both as overhead transparencies and as part of the Instructor's Resource Library (see Media and Supplements section).

Ecology Is a Work in Progress

This book, like the subject we write about, does not consist of a set of unchanging ideas and fixed bits of information. Instead, the book will develop and change over time as we respond to new discoveries and new ways of teaching. As we roll up our sleeves to begin working on the next edition, we would love to hear from you—what you like about the book, what you don't like, and any questions or suggestions you may have for how we can improve the book. You can reach us individually or as a group by sending an email message to ecology@sinauer.com, or by writing us at *Ecology*, Sinauer Associates, 23 Plumtree Road, Sunderland, MA 01375.

Acknowledgments

We'd like to thank some of the many people who helped us turn our ideas into a book in print. First and foremost, thanks are due to three outstanding contributing authors, Loraine Kohorn of Bowdoin College, who wrote Chapters 22 and 23, and Jonathan Allen of Bowdoin College and Margaret Pizer of The Nature Conservancy, who coauthored Chapter 7. Special thanks to Ryan Bartlett at Duke, who read every chapter in page proofs and helped us catch lingering problems. We are grateful to all of our colleagues who generously critiqued the plan for the book or read one or more chapters in manuscript, who are listed on the following pages. Among the hundreds of people we contacted while researching this book, we also wish to thank the following individuals for their special efforts in providing guidance and generously sharing their time and expertise: Michael A. Bell, Eric Berlow, Robert Colwell, Megan Dethier, Jon Evans, Bruce Haines, Jenifer Hall-Bowman, Bob Jefferies, Bruce Menge, Richard B. Root, Richard Spellenberg, Debra VamVikites, and Jon Witman.

Finally, we would like to express our appreciation to some of our colleagues at Sinauer Associates, with whom we worked closely during the writing but especially during the book's production. Andy Sinauer supported the plan for the book from Day One. He enthusiastically participated in every phase along the way. Sydney Carroll, Kathaleen Emerson, and Julie HawkOwl did a terrific job guiding the book through its many stages of production. Norma Roche did a superb job of copyediting our manuscript, often encouraging us to expand certain topics or omit others to help us meet the goals outlined above. Elizabeth Morales provided the beautiful illustrations in the book and her queries always helped us sharpen the visual messages. David McIntyre, our photo editor, continually surprised us by coming up with exquisite images that enhance the information in the figures. Jefferson Johnson stayed with us through several rounds of interior design. We love his final version that you see here as well as his elegant cover design. Susan McGlew coordinated the more than 110 reviews of the plan and the manuscript with great perseverance. Dean Scudder masterminded the entire marketing effort and Marie Scavotto produced the attractive ads and mailings. Jason Dirks pulled together the impressive array of supplementary materials listed on page xiii. And Norman Johnson of UMass, Amherst, demonstrated his great skill and breadth in writing the 24 "Hands-On Problem Solving" exercises.

MICHAEL L. CAIN
mcain@bowdoin.edu

WILLIAM D. BOWMAN
william.bowman@colorado.edu

SALLY D. HACKER
hackers@science.oregonstate.edu

March 2008

Reviewers

Reviewers of the Plan for the Book

David Ackerly, Stanford University
Peter Alpert, University of Massachusetts, Amherst
Mark C. Belk, Brigham Young University
April Bouton, Villanova University
David D. Briske, Texas A&M University
Colin A. Chapman, University of Florida
John Faaborg, University of Missouri
William F. Fagan, University of Maryland
Thomas J. Givnish, University of Wisconsin
Jessica Gurevitch, Stony Brook University
Nelson Hairston, Cornell University
Mike Heithaus, Florida International University
Kringen Henein, Carleton University, Ontario
Sheila Lyons-Sobaski, Albion College
Richard Mack, Washington State University
Thomas E. Miller, Florida State University
Shahid Naeem, Columbia University
Shawn Nordell, Saint Louis University
Christopher Paradise, Davidson College
David M. Post, Yale University
Seth R. Reice, University of North Carolina
Heather Reynolds, Indiana University, Bloomington
Willem Roosenburg, Ohio University, Athens
Thomas Schoener, University of California, Davis
Frederick Singer, Radford University
Ethan Temeles, Amherst College
Kathleen Treseder, University of Pennsylvania
Jake F. Weltzin, University of Tennessee

Manuscript Reviewers

Gregory H. Adler, University of Wisconsin, Oshkosh
Stuart Allison, Knox College
Kama Almasi, University of Wisconsin, Stevens Point
Peter Alpert, University of Massachusetts, Amherst
David Armstrong, University of Colorado
James Barron, Montana State University
Christopher Beck, Emory University
Mark C. Belk, Brigham Young University
Eric Berlow, University of California, Merced
Charles Blem, Virginia Commonwealth University
Carl Bock, University of Colorado
Daniel Bolnick, University of Texas, Austin
Steve Brewer, University of Mississippi
David D. Briske, Texas A&M University

Judie Bronstein, University of Arizona
Romi Burks, Southwestern University
Aram Calhoun, University of Maine
Mary Anne Carletta, Georgetown College
Walter Carson, University of Pittsburgh
David D. Chalcraft, East Carolina University
Cory Cleveland, University of Montana
Rob Colwell, University of Connecticut
James Cronin, Louisiana State University
Todd Crowl, Utah State University
Anita Davelos Baines, University of Texas, Pan American
Andrew Derocher, University of Alberta
Megan Dethier, University of Washington
Jonathan Evans, University of the South
Rick Gillis, University of Wisconsin, LaCrosse
Elise Granek, Portland State University
Martha Groom, University of Washington, Bothell
Jack Grubaugh, University of Memphis
Jenifer Hall-Bowman, University of Colorado
Jason Hamilton, Ithaca College
Christopher Harley, University of British Columbia
Bradford Hawkins, University of California, Irvine
Nat Holland, Rice University
Stephen Howard, Middle Tennessee State University
Vicki Jackson, Central Missouri State University
John Jaenike, University of Rochester
Pieter Johnson, University of Colorado
Timothy Kittel, University of Colorado
Jeff Klahn, University of Iowa
Tom Langen, Clarkson University
Jack R. Layne, Jr., Slippery Rock University
Jeff Leips, University of Maryland, Baltimore County
Svata Louda, University of Nebraska
Lynn Mahaffy, University of Delaware
Michael Mazurkiewicz, University of Southern Maine
Andrew McCall, Denison University
Shannon McCauley, University of Michigan
Mark McPeek, Dartmouth College
Bruce Menge, Oregon State University
Sandra Mitchell, Western Wyoming College
Russell Monson, University of Colorado
Daniel Moon, University of North Florida
William F. Morris, Duke University
Kim Mouritsen, University of Aarhus
Courtney Murren, College of Charleston
Jason Neff, University of Colorado
Scott Newbold, Colorado State University

Timothy Nuttle, University of Pittsburgh
Matthew Parris, University of Memphis
William D. Pearson, University of Louisville
Jan Pechenik, Tufts University
Karen Pfennig, University of North Carolina
Joe Poston, Catawba College
Scott Ruhren, University of Rhode Island
Nathan Sanders, University of Tennessee
Dov Sax, Brown University
Maynard H. Schaus, Virginia Wesleyan College
Thomas Schoener, University of California, Davis

Janet Schwengber, SUNY Delhi
Erik P. Scully, Towson University
Dennis K. Shiozawa, Brigham Young University
John J. Stachowicz, University of California, Davis
Cheryl Swift, Whittier College
Michael Toliver, Eureka College
Bill Tonn, University of Alberta
Thomas Veblen, University of Colorado
Don Waller, University of Wisconsin
Carol Wessman, University of Colorado
Jon Witman, Brown University

Media & Supplements to accompany ECOLOGY

For the Student

Companion Website (www.sinauer.com/ecology)

Content by Norman Johnson, University of Massachusetts, Amherst and Julie Froehlig, University of California, Berkeley

The *Ecology* Companion Website offers students a wealth of study and review material, all available free of charge. The Hands-on Problem Solving exercises referenced at the end of each textbook chapter provide practical experience working with experimental data and interpreting results from simulations and models. The online quizzes (instructor registration required) are a great way for students to check their comprehension of the material covered in each chapter. And the flashcards encourage familiarity with the many new terms introduced in the ecology course.

The *Ecology* Companion Website includes:

- Chapter Outlines
- Chapter Summaries
- Hands-on Problem Solving Exercises
- Online Quizzes
- Web Extensions
- Flashcard Activities
- Web Stats Review
- Complete Glossary

(See the inside front cover for additional details.)

For the Instructor
(Available to qualified adopters)

Instructor's Resource Library

The *Ecology* Instructor's Resource Library includes a variety of resources to aid instructors in course planning, lecture development, and student assessment. The Resource Library includes:

- Figures & Tables: All of the line-art illustrations, photos, and tables from the textbook are provided as both high-resolution and low-resolution JPEGs, all optimized for use in presentation software (such as PowerPoint®).
- PowerPoint Resources: Two different PowerPoint presentations are provided for each chapter of the textbook:
  - All figures, photos, and tables
  - A complete lecture outline, including selected figures

- Hands-on Problem Solving Exercises: The exercises from the Companion Website are included in Microsoft Word® format, with suggested answers.

All of the resources included in the Instructor's Resource Library are also available to instructors online, via the instructor's side of the Companion Website. (Instructor login required.)

Test Bank

Content by Norman Johnson, University of Massachusetts, Amherst and Julie Froehlig, University of California, Berkeley

The *Ecology* Test Bank (included on the Instructor's Resource Library) includes a thorough set of multiple-choice questions for each chapter of the textbook. All important concepts are covered, and each question is referenced to a specific chapter heading, concept number, and page number. The Test Bank also includes key terms lists, for use in terminology quizzes, and all of the questions from the Companion Website online quizzes. The test bank is included in the Instructor's Resource Library in two formats:

- Microsoft Word
- Wimba Diploma® (software included): Diploma is a powerful, easy-to-use exam creation program that lets you quickly assemble exams using any combination of publisher-provided and custom questions.

Online Quizzing

The Companion Website includes online quizzes that can be assigned by instructors or used as self-review exercises. Quizzes can be customized with any combination of the default questions and an instructor's own questions, and can be assigned as desired. Results of the quizzes are stored in the online grade book. (*Note:* Instructors must register in order for their students to be able to take the quizzes.)

Overhead Transparencies

This set of transparencies includes 150 key figures selected from across the textbook. The figures are formatted and color-adjusted for excellent image quality when projected.

Contents

UNIT **1** Organisms and Their Environment

UNIT **2** Populations

UNIT **3** Interactions among Organisms

UNIT 4 Communities

UNIT **5** Ecosystems

UNIT **6 Applied and Large-Scale Ecology**

ECOLOGY

Introduction: The Web of Life

KEY CONCEPTS

CONCEPT 1.1 Events in the natural world are interconnected.

CONCEPT 1.2 Ecology is the scientific study of interactions between organisms and their environment.

CONCEPT 1.3 Ecologists evaluate competing hypotheses about natural systems with experiments, observations, and models.

CASE STUDY

Deformity and Decline in Amphibian Populations

In August of 1995, a group of elementary and middle school students from Henderson, Minnesota, made a gruesome discovery as they caught leopard frogs (*Rana pipiens*) for a summer science project: 11 of the 22 frogs they found were severely deformed. Some of the frogs had missing or extra limbs, others had legs that were shortened or bent in odd directions, and still others had legs projecting from their stomachs or bony growths coming out of their backs (**Figure 1.1**). The students reported their findings to the Minnesota Pollution Control Agency, which investigated the pond and found that overall, 30%–40% of the frogs there were deformed.

News of the students' discovery traveled fast, capturing public attention and spurring scientists to check for similar deformities in other parts of the country and in other amphibian species. It soon became apparent that the problem was widespread. In the United States, misshapen individuals were found in 46 states and in more than 60 species of frogs, salamanders, and toads. In some localities, more than 90% of the individuals were deformed. Deformed amphibians were also found in Europe, Asia, and Australia.

Nothing like this had ever been seen before. Previous reports of misshapen amphibians had been limited to a few individuals (less than 5%) and localities. Adding to the alarm caused by the gruesome deformities were observations, beginning in the late 1980s, of another disturbing trend: global amphibian populations seemed to be in decline. By 1993, over 500 populations of frogs and salamanders from around the world were reported to be decreasing in size or under threat of extinction. In some cases, entire species were in danger; across the globe, hundreds of species were extinct, missing, or threatened (**Figure 1.2**). Since 1980, at least 9 amphibian species have become extinct. An additional 113 species have not been seen since that time and are listed as "possibly extinct" (IUCN Global Amphibian Assessment 2004).

Species in other groups of organisms were also showing signs of decline, but scientists were especially worried about amphibians for three reasons. First, the decline appeared to have started recently across wide regions of the world. Second, some of the populations in decline were located in protected or pristine regions, seemingly far from the effects of human activities. Third, many scientists view amphibians as good "biological indicators" of environmental conditions for a number of reasons. Amphibians have permeable skin, with no hair, scales, or feathers to protect them, and their eggs lack shells or other protective coverings. Furthermore, most amphibians spend part of their lives in water and part on land, so they are exposed to a wide range of potential threats, including water and air pollution as well as changes in temperature and in the amount of ultraviolet (UV) light in their environment. Many amphibians never travel far from their birthplace, so the decline of a local population is likely to indicate a deterioration of local environmental conditions.

FIGURE 1.1 Deformed Leopard Frogs With their missing, misshapen, and extra legs, these individuals show several of the types of limb deformities that have become common in leopard frogs and other amphibian species.

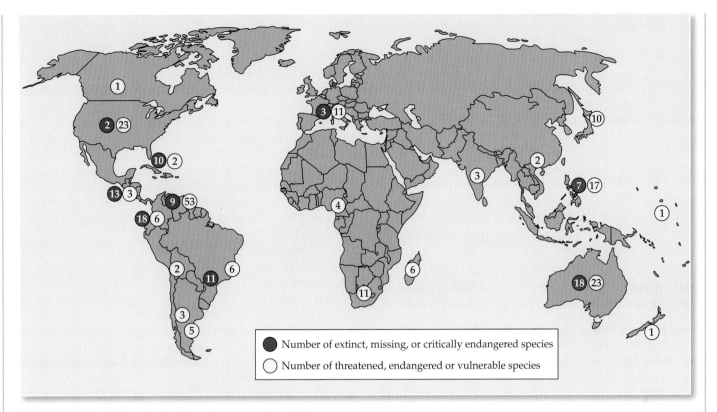

FIGURE 1.2 Amphibians in Decline In many regions of the world, amphibian species face increased risk of extinction.

Because amphibians worldwide were showing declining numbers and frequent deformities, scientists initially tried to find one or a few global causes that might explain these problems. However, as we'll see in this chapter, the story turned out to be more complicated than that: a single "smoking gun" has not emerged. What, then, has caused the global decline of amphibian populations?

We humans have an enormous impact on our planet. We have transformed nearly half of Earth's land surface, and we have altered the composition of the atmosphere, leading to global climate change. We have introduced many species to new regions, an action that can have severe negative effects on both native species and human economies. Even the oceans—seemingly so vast—show many signs of deterioration due to human activities, including declining fish stocks, the loss of once-spectacular coral reefs, and the formation of large "dead zones," regions where oxygen levels are too low to support marine life.

But people do not just affect the global environment—we are a part of it. Despite this, we have often taken actions that have affected our environment without considering how the natural systems of our environment work. Fortunately, we are beginning to realize that we must understand those systems so that we can anticipate the consequences of our actions and fix the problems we have already caused. Our growing realization that we must understand how natural systems work brings us to the subject of this book. Natural systems are driven by the ways in which organisms interact with one another and with their physical environment. Thus, to understand how natural systems work, we must understand *ecology*, the scientific study of how organisms affect—and are affected by—other organisms and their environment.

In this chapter, we'll introduce the study of ecology and its relevance for human societies. We'll begin by exploring a theme that runs throughout this book: connections in nature.

Connections in Nature

CONCEPT 1.1
Events in the natural world are interconnected.

From what you have read or observed about nature, can you think of examples that might illustrate the phrase *connections in nature*? In this book, we use that phrase to refer to the fact that events in the natural world can be linked or connected to one another. These connections occur as organisms interact with one another and with their environment. This does not necessarily mean that there are strong connections among all of the organisms that live in a given area. Two species may live in the same area but have little influence on each other. But all organisms are connected to features of their environment—they require food, space, and other resources, and they interact with other species and the physical environment as they pursue what they need to live. As a result, even species that do not interact directly with each other can be connected indirectly by shared features of their environment.

Connections in nature are revealed as ecologists ask questions about the natural world and examine what they've learned. To illustrate what this process can teach us about connections in nature, let's return to our discussion of amphibian deformities.

Early observations suggest that parasites cause amphibian deformities

Nine years before the Minnesota students made their startling discovery, Stephen Ruth was exploring ponds in northern California when he found Pacific tree frogs (*Hyla regilla*) and long-toed salamanders (*Ambystoma macrodactylum*) with extra limbs, missing limbs, and other deformities. He asked Stanley Sessions, an expert in amphibian limb development, to examine his grotesque specimens. Sessions found that the deformed amphibians all contained a parasite, now known to be *Ribeiroia ondatrae*, a trematode flatworm. Sessions and Ruth hypothesized that the parasite caused the deformities. They tried implanting small glass beads near the developing limb buds of tadpoles. These beads were meant to mimic the effects of *Ribeiroia* parasites, which often produce cysts close to the area where limbs form as a tadpole begins its metamorphosis into an adult frog. In a paper published in 1990, they reported that the beads caused deformities similar to those found by Ruth.

A laboratory experiment tests the role of parasites

When Ruth found deformed amphibians in the mid-1980s, he assumed (quite reasonably) that they were an isolated, local phenomenon. By 1996, Pieter Johnson, then an undergraduate at Stanford University, had learned of the Minnesota students' findings and of the paper by Sessions and Ruth. Although Sessions and Ruth provided indirect evidence that *Ribeiroia* could cause amphibian deformities, they did not infect *H. regilla* or *A. macrodactylum* with *Ribeiroia* and show that deformities resulted. In addition, the two amphibian species they used in their experiments (the African clawed frog, *Xenopus laevis*, and the axolotl salamander, *A. mexicanum*) were not known to have limb deformities in nature. Building on the work done by Sessions and Ruth (1990), Johnson and his colleagues set out to provide a more thorough test of whether *Ribeiroia* parasites can cause limb deformities in amphibians.

They began by surveying thirty-five ponds in Santa Clara County, California. They found Pacific tree frogs in thirteen ponds, four of which contained deformed frogs. Concentrating on two of these four ponds, they consistently found that 15%–45% of tadpoles in metamorphosis had extra limbs or other deformities (Johnson et al. 1999). One source of concern was that the deformities might be caused by pollutants, such as pesticides, polychlorinated biphenyls (PCBs), or heavy metals. However, none of these substances were found in water from the two ponds. In addition, the problem did not appear to be genetically based, since 200 *H. regilla* eggs collected from the two ponds developed normally when taken to the laboratory.

Johnson and his colleagues then turned their attention to other factors that might cause the deformities. Aware that Sessions and Ruth had hypothesized that parasites could be the cause, they noted that of the thirty-five ponds they had surveyed, the four ponds with abnormal frogs were the only ponds that contained both tree frogs and an aquatic snail, *Planorbella tenuis*. This snail is an intermediate host in the life cycle of the *Ribeiroia* parasite (**Figure 1.3**). In addition, dissections of abnormal frogs collected from the two ponds revealed *Ribeiroia* cysts in all the frogs with deformed limbs.

Like the findings of Sessions and Ruth, the results described in the previous paragraph provided indirect evidence that *Ribeiroia* caused deformities in Pacific tree frogs. Next, Johnson and his colleagues returned to the laboratory to test whether that idea really was true. They did this by using a standard scientific approach: they performed a **controlled experiment** in which an *experimental group* (that has the factor being tested) is compared with a *control group* (that lacks the factor being tested). Johnson and his colleagues collected *H. regilla* eggs from a region not known to have frog deformities, brought the eggs into the laboratory, and placed the tadpoles that hatched from them in individual 1-liter containers. Each tadpole was then assigned at random to one of four *treatments*, in which 0 (the control group), 16, 32, or 48 *Ribeiroia* parasites were placed in its container; these numbers were selected to match parasite levels that had been observed in the field.

Johnson and his colleagues found that as the number of parasites increased, fewer of the tadpoles survived to metamorphosis, and more of the survivors had deformities (**Figure 1.4**). In the control group (with zero *Ribeiroia*), however, 88% of the tadpoles survived, and none had deformities (Johnson et al. 1999). The link had been made: *Ribeiroia* could cause frog deformities. Furthermore, since exposure to *Ribeiroia* killed up to 60% of the tadpoles, the results also suggested that the parasites could contribute to amphibian declines.

A field experiment suggests that multiple factors influence frog deformities

Within a few years after Johnson and his colleagues published their research, scientists had shown that *Ribeiroia* parasites

1 In the snail, the parasite undergoes asexual reproduction to produce free-swimming cercariae.

6 Eggs in the bird's feces are released into water, where they hatch into free-swimming miracidia that infect the snail.

Miracidia

Eggs

Planorbella tenuis
(first intermediate host)

Cercariae

Adult parasite

5 The parasite matures to adulthood in the bird's body and reproduces sexually.

FIGURE 1.3 The Life Cycle of *Ribeiroia*   The life cycle of *Ribeiroia* utilizes three different kinds of hosts: snails, larval amphibians or fish, and birds or mammals. Many other parasites have similarly complex life cycles, and some of them are also known to alter the appearance or behavior of their second intermediate hosts, making them more vulnerable to predation by their final or definitive hosts.

2 The cercariae infect tadpoles and form cysts around their developing limb buds.

Amphibian
(second intermediate host)

Bird
(definitive host)

Metacercariae cyst

3 The cysts interfere with normal limb development, causing limb deformities.

4 The limb deformities make the amphibian more vulnerable to capture by predatory birds.

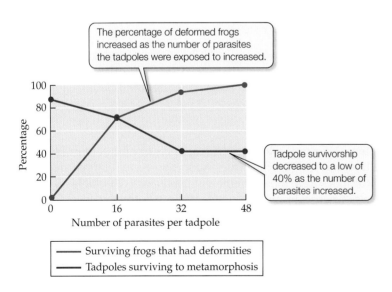

The percentage of deformed frogs increased as the number of parasites the tadpoles were exposed to increased.

Tadpole survivorship decreased to a low of 40% as the number of parasites increased.

— Surviving frogs that had deformities
— Tadpoles surviving to metamorphosis

FIGURE 1.4 Parasites Can Cause Amphibian Deformities   The graph shows the relationship between the number of *Ribeiroia* parasites that tadpoles were exposed to and their rates of survival and deformity. (After Johnson et al. 1999.)

could cause limb deformities in other amphibian species, including western toads (*Bufo boreas*), wood frogs (*Rana sylvatica*), and leopard frogs (*R. pipiens*, the species in which the Minnesota students had discovered deformities). While *Ribeiroia* clearly was important, researchers also suspected that other factors might play a role. Pesticides, for example, were known to contaminate some ponds in which deformed frogs were found. To examine the possible joint effects of parasites and pesticides, Joseph Kiesecker, then at Pennsylvania State University, conducted a field experiment in six ponds, all of which contained *Ribeiroia*, but only some of which contained pesticides (Kiesecker 2002).

Three of the ponds in Kiesecker's study were close to farm fields, and water tests indicated that each of these ponds contained detectable levels of pesticides; the other three ponds were not as close to farm fields, and none of them showed detectable levels of pesticides. In each of the six ponds, Kiesecker placed wood frog tadpoles in cages made with a mesh through which water could flow, but tadpoles could not escape. Six cages were placed in each pond; three of these cages had a mesh through which *Ribeiroia* parasites could pass, while the other three had a mesh too small for the parasites. Thus, in each pond, the tadpoles in three cages were

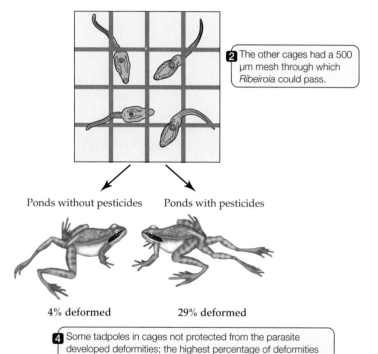

FIGURE 1.5 Do *Ribeiroia* and Pesticides Interact in Nature? To test the effects of *Ribeiroia* and pesticides on frog deformities in the field, screened cages were placed in six ponds. Three of the six ponds contained detectable levels of pesticides; the other three did not. (After Kiesecker 2002.)

1 Some of the cages in each pond had a 75 μm mesh that prevented the entry of *Ribeiroia*.

Ribeiroia cercariae

2 The other cages had a 500 μm mesh through which *Ribeiroia* could pass.

Ponds without pesticides   Ponds with pesticides   Ponds without pesticides   Ponds with pesticides

No deformities   No deformities   4% deformed   29% deformed

3 Tadpoles in cages that protected them from *Ribeiroia* did not develop deformities.

4 Some tadpoles in cages not protected from the parasite developed deformities; the highest percentage of deformities occurred in tadpoles exposed to both *Ribeiroia* and pesticides.

exposed to the parasites, while the tadpoles in the other three cages were not.

The results showed that *Ribeiroia* caused limb deformities in the field (**Figure 1.5**). No deformities were found in frogs raised in cages whose small mesh size (75 μm) prevented the entry of *Ribeiroia*, regardless of which pond the cages were in. Deformities were found in some of the frogs raised in cages whose larger mesh size (500 μm) allowed the entry of *Ribeiroia*. However, a greater percentage of frogs had deformities (29% vs. 4%) in ponds that contained pesticides than in ponds that did not contain pesticides. In addition, dissections revealed that every frog with a deformity was infected by *Ribeiroia*. Overall, the results of this experiment indicated

that (1) exposure to *Ribeiroia* was necessary for deformities to occur, and (2) when frogs were exposed to *Ribeiroia*, deformities were more common in ponds with detectable levels of pesticides than in ponds without detectable levels of pesticides.

Based on these results, Kiesecker hypothesized that pesticides might decrease the ability of frogs to resist infection by parasites. To examine whether pesticides had such an effect, Kiesecker (2002) brought wood frog tadpoles into the laboratory, where he reared some in an environment with pesticides and others in an environment without pesticides. The tadpoles exposed to pesticides had fewer white blood cells (indicating a suppressed immune system) and a higher

(A) Eosinophils

Tadpoles exposed to the pesticide had fewer eosinophils ...

(B) *Ribeiroia*

... and more *Ribeiroia* cysts than the controls.

Control   Solvent   Low   High
           control   pesticide pesticide

FIGURE 1.6 Pesticides May Weaken Tadpole Immune Systems In a laboratory experiment, some wood frog (*Rana sylvatica*) tadpoles were exposed to low or high concentrations of the pesticide esfenvalerate and then exposed to 50 *Ribeiroia* parasites per tadpole. The tadpoles were then examined for (A) numbers of eosinophils (a type of white blood cell used in the immune response) and (B) numbers of *Ribeiroia* cysts. Two types of controls against which the effects of pesticides could be compared were used: one in which only parasites were added to the tadpoles' containers ("control"), and another in which both parasites and the solvent used to dissolve the pesticide were added ("solvent control"). The intent of the second control was to check for possible effects of the solvent. (After Kiesecker 2002.)

rate of *Ribeiroia* cyst formation (**Figure 1.6**). Together, Kiesecker's laboratory and field results suggested that pesticide exposure can affect the frequency with which parasites cause deformities in amphibian populations.

Connections in nature can lead to unanticipated side effects

As we have just seen, the immediate cause of amphibian deformities is often infection by *Ribeiroia* parasites. But we also noted on p. 2 that amphibian deformities are occurring more often now than in the past. Why has the frequency of amphibian deformities increased?

One possible answer is suggested by Kiesecker's findings (2002): pesticides in ponds may decrease the ability of amphibians to ward off parasite attack, with the result that deformities are more likely in environments that contain pesticides.

The first synthetic pesticides were developed in the late 1930s, and their usage has risen dramatically since that time. Thus, it is likely that amphibian exposure to pesticides has increased considerably in recent years, which may help to explain the recent rise in the frequency of amphibian deformities. If this scenario turns out to be correct, it illustrates how any action (increased pesticide use by people) can have unanticipated side effects (more frequent deformities in amphibians).

Other environmental changes may also contribute to the observed increase in amphibian deformities. For example, scientists have reported cases in which *Ribeiroia* and frequent amphibian deformities were found in artificial ponds (used to store water for cattle or crops) that were overgrown with algae. The suspected mechanism in this case is the use of fertilizers. Rain or snowmelt can wash fertilizers from an agricultural field into an artificial pond. Fertilizer inputs often stimulate increased growth of algae, and the snails that harbor *Ribeiroia* parasites eat algae. Thus, as the algae increase, so do the snail hosts of *Ribeiroia*, which tends to increase the number of *Ribeiroia* found in the pond. Here, a chain of events that begins with increased fertilizer use by people ends with increased numbers of *Ribeiroia*, and hence increased numbers of deformed amphibians.

As illustrated by these possible causes for the rising frequency of amphibian deformities in recent years, events in the natural world are connected, one to another. One action or event can stimulate another, which in turn may cause still other events to unfold.

We live in an ecological world

The fact that events in nature are interconnected means that when people alter one aspect of the environment, we can cause other changes that we do not intend or anticipate. As we have just seen, the current evidence suggests that amphibian deformities have become more common in part because people have changed the lakes, ponds, and streams in which amphibians live. When we increased our use of pesticides and fertilizers, we did not intend to increase the frequency of deformities in frogs—nevertheless, we seem to have done just that.

The indirect effects of human actions include more than bizarre deformities in frogs. For example, some changes we are making to our local and global environment appear to have increased human health risks. The damming of rivers in Africa has created favorable habitat for snails that harbor trematode parasites that cause schistosomiasis, thereby increasing the spread of an infection that can weaken or kill people. Globally, the past few decades have seen an increase in the appearance and spread of new epidemics, such as AIDS, Lyme disease, Hantavirus pulmonary syndrome, Ebola, and West Nile fever (**Figure 1.7**). Many public health experts think that the effects of human actions on the environment have contributed to the emergence of these and other new epidemics (Weiss and McMichael 2004).

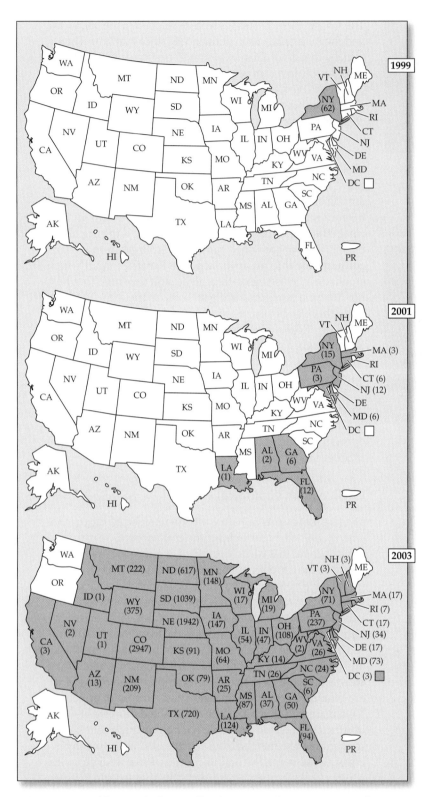

FIGURE 1.7 Rapid Spread of a Deadly Disease   Within 5 years, West Nile Fever (shaded areas) spread from its North American point of entry (New York City) to all but three of the lower 48 states of the United States (WA, OR, and ME) . Birds are a primary host for the West Nile virus, which may help to explain its rapid spread. Mosquitoes transmit the disease from birds and other animal hosts to people. Numbers in parentheses indicate the number of human cases.

If you live in a city, it can be easy to forget the extent to which everything you do depends on the natural world. Your house or apartment shelters you from the elements and keeps you warm in winter and cool in summer. Similarly, you obtain food from a grocery store, clothes from a shop or department store, water from a faucet. Ultimately, however, each of these items—and everything else you use or own—comes from or depends on the natural environment. No matter how far from the natural world our day-to-day activities take us, people, like all other organisms on Earth, are part of an interconnected web of life. Let's turn now to the study of those connections, the scientific discipline of *ecology*.

Ecology

> **CONCEPT 1.2**
> Ecology is the scientific study of interactions between organisms and their environment.

In this book, **ecology** is defined as the scientific study of interactions between organisms and their environment. This definition is meant to include the interactions of organisms with one another, because as we have just seen, organisms are an important part of one another's environment. Ecology can also be defined in a variety of other ways; for example, as the scientific study of interactions that determine the distribution (geographic location) and abundance of organisms. As will become clear as you read this book, these definitions of ecology can be related to one another, and each emphasizes different aspects of the discipline. A more important point for our purpose here, however, is that the term "ecology," as used by ecologists, refers to a scientific endeavor.

We emphasize this point because "ecology" has other meanings in its public usage. People who are not scientists may assume that an "ecologist" is an environmental activist. Some ecologists are activists, but many are not. Furthermore, as a scientific discipline, ecology is related to—yet different from—other disciplines such as environmental science. Ecology is a branch of biology, while **environmental science** is an interdisciplinary field that incorporates concepts from the natural sciences (including ecology) and the social sciences (e.g., politics, economics, ethics). Compared with ecology, environmental science is focused more specifically on how people affect the environment and how we can address environmental problems. While an ecologist might examine pollution as one of several factors that influence the reproductive success of wetland plants, an environmental scientist might try to develop a new industrial process that produces less pollution.

Public and professional ideas about ecology often differ

Surveys have shown that many members of the public think that (1) there is a "**balance of nature**," in which natural systems are stable and tend to return to an original, preferred state after a disturbance, and that (2) each species in nature has a distinct role to play in maintaining that balance. Such ideas about ecological systems can have moral or ethical implications for the people who hold them. For example, the view that each species has a distinct function can lead people to think that each species is important and irreplaceable, which in turn can cause people to feel that it is wrong to harm other species. As summarized by one interviewee in a survey on the meaning of ecology (Uddenberg et al. 1995, as quoted in Westoby 1997), "There is a certain balance in nature, and there is a place for all species. There is a reason for their existence and we are not free to exterminate them."

Public views on the balance of nature and the unique function of each species are not surprising, since these views were once held by many ecologists. Ecologists, however, now recognize that change is an integral part of natural systems and that seemingly random perturbations often play an important role in those systems. In addition—as we will see in Unit 4—the current evidence suggests that the different species in an area often respond in different ways to changing conditions, a finding at odds with the idea that each species has a distinct function within a larger, tightly knit group of species. Thus, unless they provide careful qualifications, few ecologists today speak of a balance of nature or the unique function of each species.

While early views about a balance of nature and the distinct function of each species have not stood the test of time, other ideas have. In particular, early ecologists and modern ecologists would agree that events in nature are interconnected. As a result, a change in one part of an ecological system can alter other parts of the system, including those that govern life-supporting processes such as the purification and replenishment of air, water, and soil.

The interconnections in nature form the basis for the first of seven ecological maxims that we will mention in this book; namely, "You can never do just one thing" (see **Table 1.1** for the full list). This maxim is meant to suggest that all actions have multiple effects because events in nature are interconnected. Overall, although the natural world may not be as static or as tightly woven as early ecologists may have thought, species are connected to one another. For some people, the fact that events in nature are interconnected provides an ethical imperative to protect natural systems. A person who feels an ethical obligation to protect human life, for example, may also feel an ethical obligation to protect the natural systems on which human life depends.

TABLE 1.1

Some Ecological Maxims

1. You can never do just one thing.
   Organisms interact with each other and with their physical environment. As a result, events in nature are connected and what affects one organism or place can affect others as well.

2. Everything goes somewhere.
   There is no "away" into which waste materials disappear.

3. No population can increase in size forever.
   There are limits to the growth and resource use of every population, including our own.

4. There is no free lunch.
   An organism's energy and resources are finite, and increasing inputs into one function (such as reproduction) results in a tradeoff in which there is a loss for other functions (such as growth).

5. Evolution matters.
   Organisms evolve or change over time—it is a mistake to view them as static. Evolution is an ongoing process because organisms continually face new challenges from changes in both the living and nonliving components of their environment.

6. Time matters.
   Ecosystems change over time. When we look at the world as we know it, it is easy to forget how past events may have affected our present, and how our present actions may affect the future.

7. Life would be impossible without species interactions.
   Species depend on one another for energy, nutrients, and habitat.

Ecology is broad in scope

Ecologists study interactions in nature across many levels of biological organization (**Figure 1.8**). Some ecologists are interested in how particular genes or proteins enable organisms to respond to environmental challenges. Other ecologists study how hormones influence social interactions, or how specialized tissues or organ systems allow animals to cope with extreme environments. However, even among ecologists whose research is focused on lower levels of biological organization (e.g., from molecules to organ systems), ecological studies usually emphasize one or more of the following levels: organisms, populations, communities, or ecosystems.

A **population** is a group of individuals of a single species that live in a particular area and interact with one another. Many of the central issues in ecology concern questions about how and why the locations and abundances of populations change over time. To answer such questions, it is often helpful to understand the roles played by other species. Thus, many ecologists study nature at the level of the **community**, which is an association of interacting populations of different species

Molecules are made up of atoms.

Cells are built of many types of molecules.

A **tissue** is a group of many cells with similar and coordinated functions (such as sensing odors).

Organs composed of tissues that function together form **systems**, such as the nervous system.

Molecule (DNA)

Cell (neuron)

Tissue (nervous tissue)

Organ system (nervous system)

Organism (fish)

FIGURE 1.8 Levels of Biological Organization   Life in this tropical coral reef can be organized at a number of levels, from molecules to the biosphere (which includes all of Earth's ecosystems). These levels are nested within one another, in the sense that each level is composed of groups of the entity found in the level below it.

Population (school of fish)

An **organism** is a recognizable, self-contained individual. A multicellular organism is made up of organs and organ systems.

A **population** is a group of organisms of the same species.

Communities consist of interacting populations of different species.

Community (coral reef)

Biological communities in the same geographical location form **ecosystems**. Ecosystems exchange energy and create Earth's **biosphere**.

Biosphere

that live in the same area. Communities can cover large or small areas, and they can differ greatly in terms of the numbers and types of species found within them (**Figure 1.9**).

Ecological studies at the population and community levels often examine the effects of components of the physical environment. For example, ecologists interested in factors that influence population dynamics often ask whether features of the environment, such as temperature, precipitation, nutrients, or pH, influence fertility or survivorship. Other ecologists are particularly interested in how ecosystems work. An **ecosystem** is a community of organisms plus the physical environment in which they live. An ecologist studying ecosystems might want to know how the rates at which chemicals enter the physical environment in a particular place affect the species found there. For example, in studying amphibian deformities, an ecologist might examine how the addition of fertilizers to a pond affects both algal growth and the presence of parasites that cause deformities.

Global patterns of air and water circulation (see Chapter 2) link all of the world's ecosystems into one giant ecosystem, the **biosphere**, which consists of all living organisms on Earth plus the environments in which they live. The biosphere forms the highest level of biological organization. Today, ecologists work in a variety of ways to help ensure that human actions do not impair the natural systems on which life depends. Such efforts can be directed at any level of biological organization, but as we will see in Unit 6, ecologists increasingly seek to understand the big picture: how the biosphere works.

FIGURE 1.9 A Few of Earth's Many Communities Moving counter-clockwise, these photographs show: (A) A savanna in Kenya, where a male topi antelope stands on an ant mound to attract a mate; (B) A rainforest in Ecuador; (C) Dunes in the Namib Desert, Namibia, an extremely dry desert that receives 2 to 85 mm (0.8 to 3.3 inches) of rain per year; (D) A shallow-water marine community off the west coast of Mexico.

The scale of an ecological study affects what can be learned from it

Whether they study individuals or the biosphere—or anything in between—ecologists always draw boundaries around what they observe. An ecologist interested in frog deformities might ignore the birds that migrate above the study site, while an ecologist studying bird migrations might ignore the details of what occurs in the ponds below. It is not possible or desirable to study everything at once.

When they seek to answer a particular question, ecologists must select the most appropriate dimension, or **scale**, for collecting observations. Every ecological study addresses events on some scales, but ignores events at other scales. A study on the activities of soil microorganisms, for example, might be conducted at a small spatial scale (e.g., measure-

TABLE 1.2

Key Terms for Studying Connections in Nature

Term	Definition
Adaptation	A feature of an organism that improves its ability to survive and reproduce in its environment
Natural selection	An evolutionary process in which individuals that possess particular characteristics survive or reproduce at a higher rate than other individuals because of those characteristics
Consumer	An organism that obtains its energy by eating other organisms or their remains
Producer	An organism that uses energy from an external source, such as the sun, to produce its own food without having to eat other organisms or their remains
Net primary productivity (NPP)	The amount of energy (per unit of time) that producers fix by photosynthesis or other means, minus the amount used in cellular respiration
Nutrient cycle	The cyclic movement of a nutrient between organisms and the physical environment

ments might be collected at centimeter to meter scales). For a study addressing how atmospheric pollutants affect the global climate, on the other hand, the scale of observation will be large indeed, and might include Earth's entire atmosphere. Ecological studies also differ greatly in the time scales they cover. Some studies, such as those that document how leaves respond to momentary increases in the availability of sunlight, concern events on short time scales (seconds to hours). Others, such as studies that use fossil data to show how the species present in a given area have changed over time, address events on much longer time scales (centuries to millennia or longer).

Some key terms are helpful for studying connections in nature

Whether we are discussing organisms, populations, communities, or ecosystems, all chapters of this book incorporate the principle that events in the natural world are interconnected. For example, in Unit 2, we will see how an explosion in the population size of an introduced species (the comb jelly, *Mnemiopsis*) altered the entire Black Sea ecosystem. Because we stress connections in nature in every chapter—and hence may discuss ecosystems in a chapter about organisms, or vice versa—we describe here a handful of key terms that you should learn as you begin your study of ecology. These terms are also summarized in **Table 1.2**.

A universal feature of living systems is that they change over time, or evolve. Depending on the questions or time scale of interest, **evolution** can be defined as (1) a change in the genetic characteristics of a population over time, or as (2) *descent with modification*, the process by which organisms gradually accumulate differences from their ancestors. We will discuss evolution in the context of ecology more fully in Chapter 6, but here we define two key evolutionary terms: adaptation and natural selection.

An **adaptation** is a characteristic of an organism that improves its ability to survive and reproduce within its environment. Adaptations are of critical importance to understanding how organisms function and interact with one another. As we'll see in Chapter 6, although several mechanisms can cause evolutionary change, only natural selection can produce adaptations consistently. In **natural selection**, individuals with particular characteristics tend to survive and reproduce at a higher rate than other individuals *because of those characteristics*. If the characteristics being selected for are heritable, then the offspring of individuals favored by natural selection will tend to have the same characteristics that gave their parents an advantage. As a result, the frequency of those characteristics may increase in a population over time.

Consider what happens within the body of a person taking an antibiotic. Some of the bacteria that live inside that person may possess alleles of a gene that provide resistance to the antibiotic. Because of those alleles, those bacteria will survive and reproduce at a higher rate than will nonresistant bacteria (**Figure 1.10**). In this case, the trait on which natural selection acts (antibiotic resistance) is heritable, and hence the offspring of the resistant bacteria will tend to be resistant. As a result, the proportion of resistant bacteria in the person's body will increase over time.

The remaining four key terms that we'll introduce in this section concern ecosystem processes. One way to look at how ecosystems work is to consider the movement of energy and materials through a community. Energy enters the community when an organism such as a plant or bacterium captures energy from an external source, such as the sun, and uses that energy to produce food. An organism that can produce its own food from an external energy source without having to eat other organisms or their remains is called a **producer** (such organisms are also called *primary producers* or *autotrophs*). An organism that obtains its energy by eating other organisms or their remains is called a **consumer**. The amount

1 A few bacteria (shown in blue) in this population have an allele that makes them resistant to an antibiotic.

Antibiotic

2 When the population is exposed to an antibiotic …

3 … resistant bacteria survive at a higher rate than nonresistant bacteria.

4 The resistant bacteria that survive pass their resistance on to their offspring.

5 When the population is exposed to the antibiotic again …

6 … the proportion of resistant bacteria in the population increases further.

FIGURE 1.10 Natural Selection in Action   As shown in this drawing in which a sieve represents the selective effects of an antibiotic, natural selection can cause the frequency of antibiotic resistance to increase over time.

of energy that producers capture by photosynthesis or other means, minus the amount they lose as heat in cellular respiration, is called **net primary productivity (NPP)**. NPP varies greatly from one ecosystem to another, and changes in NPP can have a large effect on ecosystem function.

FIGURE 1.11 How Ecosystems Work   Each time one organism eats another, a portion of the energy originally captured by a producer is lost as heat given off during the chemical breakdown of food by cellular respiration. As a result, energy moves through the ecosystem in a single direction and is not recycled. Nutrients such as carbon and nitrogen, on the other hand, cycle between organisms and the physical environment. This diagram is based on an ecosystem in which producers obtain energy from sunlight; in some ecosystems, producers such as bacteria obtain energy from chemicals.

Each unit of energy captured by producers is eventually lost from the ecosystem as heat (**Figure 1.11**). As a result, energy moves through ecosystems in a single direction only—it cannot be recycled. Nutrients, however, are continuously recycled from the physical environment to organisms and back again. The cyclic movement of nutrients such as nitrogen or phosphorus between organisms and the physical environment is referred to as a **nutrient cycle**. Life would cease if nutrients were not recycled because the molecules organisms need for their growth and reproduction would not be available to them.

Whether they are concerned with adaptation or NPP, populations or ecosystems, the scientists who study ecological systems have not produced a fixed body of knowledge, engraved in stone. Instead, what we know about ecology changes constantly as ideas are tested and, if necessary, revised or discarded as new information emerges. As we will see in the next section, like all branches of science, ecology is about answering questions and seeking to understand the underlying causes of natural phenomena.

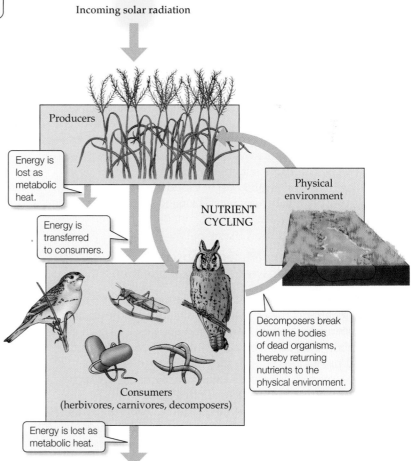

Incoming solar radiation

Producers

Energy is lost as metabolic heat.

Energy is transferred to consumers.

NUTRIENT CYCLING

Physical environment

Decomposers break down the bodies of dead organisms, thereby returning nutrients to the physical environment.

Consumers (herbivores, carnivores, decomposers)

Energy is lost as metabolic heat.

Energy is not recycled—it flows through the ecosystem

Answering Ecological Questions

CONCEPT 1.3
Ecologists evaluate competing hypotheses about natural systems with experiments, observations, and models.

The studies of amphibian deformities that we discussed earlier in this chapter illustrate several ways in which ecologists seek to answer questions about the natural world. The study by Johnson and his colleagues (1999), for example, had two key components: observational studies in the field and a controlled experiment in the laboratory. In the observational part of their work, the researchers surveyed ponds, noted the species present, and observed that deformities were found only in ponds that contained both tree frogs and a snail that harbored the *Ribeiroia* parasite. These findings suggested that *Ribeiroia* might cause deformities, so Johnson and his colleagues performed a laboratory experiment to test whether this was the case (it was).

Kiesecker's paper (2002) also described two experiments, one performed in the field, the other in the laboratory. To examine the effects of pesticides on frog deformities, Kiesecker first compared results from three ponds with detectable levels of pesticides with results from three ponds without detectable levels of pesticides. While this approach had the advantage of allowing the effects of *Ribeiroia* to be examined under different field conditions (in ponds with and without pesticides), Kiesecker could not control the conditions as precisely as he had in his laboratory experiment. The constraints of working in the field meant, for example, that he could not start out with six identical ponds, then add pesticides to three of them but not to the other three—an experiment that would test more directly whether pesticides were responsible for the results he obtained. As this example suggests, no single approach works best in all situations, so ecologists use a variety of methods when seeking to answer ecological questions.

Ecologists use experiments, observations, and models to answer ecological questions

In an ecological experiment, an investigator alters one or more features of the environment and observes the effect of that change on natural processes. When possible, such experiments include both a control group (which is not subjected to any alterations) and one or more experimental groups (see p. 4). When designing an experiment, ecologists have a range of scales to choose from (**Figure 1.12**). The scale of ecological experiments varies widely, from small-scale studies that cover a few square meters to large-scale studies in which entire ecosystems, such as lakes or forests, are manipulated.

In some cases, however, it can be difficult or impossible to perform experiments. For example, if the questions of inter-

est concern events that cover large geographic regions or occur over long periods, such as global warming, an experiment conducted in the laboratory or in small plots over one or a few years may provide little useful information. In such a situation, observational studies can be conducted in which field data are collected over large regions or long periods. Such observations can then be analyzed to determine whether a suspected underlying cause (such as an increase in temperature) shows a consistent association with particular results (such as shifts to the north in the locations of different species).

(A)

(B)

(C)

FIGURE 1.12 Ecological Experiments Experiments in ecology range from laboratory experiments (A) to small-scale field experiments conducted in natural or artificial environments (B) to large-scale experiments that alter major components of an ecosystem (C).

Let's take a closer look at the study of global warming. As we will see in Chapter 24, scientists agree that Earth's climate is warming. However, the future magnitude and effects of global warming remain uncertain. We are not sure, for example, how the geographic ranges of different species will change as a result of the projected temperature increases. There is only one Earth, so of course even if we wanted to, we could not apply different levels of global warming to replicates of the planet and then observe how species' ranges change over time in each of our experimental treatments.

Instead, we must approach this problem using a mixture of observational studies, small-scale experiments, and quantitative (mathematical or computer) models. Field observations reveal that hundreds of species have shifted their ranges north or up the sides of mountains in a manner that is consistent with the amount of global warming that has already occurred (see Chapter 24). Field observations can also be used to summarize the environmental conditions under which species are currently found, and small-scale experiments can be used to examine the performance of species under different environmental conditions. To put all of this information together, results from observational studies and small-scale experiments can be used as input data in quantitative models that predict how environmental conditions—and hence the geographic ranges of species—are likely to change depending on how much the planet actually warms in the future.

Experiments are designed and analyzed in consistent ways

When ecologists perform controlled experiments, they often take the following three additional steps to reduce the chance that variables not under the control of the experimenter will unduly influence the results of an experiment (**Figure 1.13**): (1) they **replicate** (i.e., perform more than once) each treatment, including the control; (2) they assign treatments to subjects or plots at random; and (3) they use statistical analyses to determine whether the treatments had a significant effect.

An advantage of replication is that as the number of replicates increases, it becomes less likely that the results were actually due to a variable that was not measured or controlled in the study. Imagine that Kiesecker had performed his experiment with only two ponds, one with detectable levels of pesticides, the other without. Assume that frog deformities were more common in the single pond that contained pesticides. While pesticides might have caused this result, the two ponds might have differed in many other ways, too, one or more of which might have been the real cause of the deformities. By using three ponds with pesticides and three ponds without pesticides, Kiesecker made it less likely—though not impossible—that each of the three ponds with pesticides also contained something else that increased the chance of frog deformities. In the actual experiment, Kiesecker accounted for the possible effects of some unmeasured variables: he showed, for example, that the number of snails

FIGURE 1.13 Experimental Design and Analysis   This aerial photograph shows the layout of a field study that included three steps common to many ecological experiments: replication, random assignment of treatments, and statistical analysis. This study was designed to test the effects of insect herbivores on plant growth and asexual reproduction. There were two treatments: a control, in which natural processes were left undisturbed in a set of 5 × 5 meter plots and an insect removal treatment, in which an insecticide was used to reduce the numbers of insect herbivores that fed on aboveground plant parts in another set of 5 × 5 meter plots.

and the frequency of their infection by *Ribeiroia* were similar in all six ponds, thus making it unlikely that the ponds with pesticides had many more *Ribeiroia* than the ponds without pesticides.

Ecologists also seek to limit the effects of unmeasured variables by assigning treatments at random. Suppose an experiment was designed to test whether insects that eat plants decrease the number of seeds the plants produce. One way to test this idea would be to divide an area into a series of experimental plots (see Figure 1.13). Each plot would receive one of the following two treatments: it would either be sprayed regularly with an insecticide to reduce the number of plant-eating insects, or left alone. If plant-eating insects have much of an effect on seed production, plots that are sprayed should produce more seeds. The decision as to whether a particular plot would be sprayed (or not) would be made at random at the start of the experiment. The purpose of assigning treatments at random is to make it less likely that plots that receive a particular treatment share other characteristics that might influence seed production, such as high or low levels of soil nutrients.

Finally, ecologists often use statistical analyses to determine whether their results are "significant." To understand why, let's turn again to Kiesecker's experiment. It would have been very unusual if Kiesecker had found that rates of frog deformities in ponds with pesticides were exactly equal to those in ponds without pesticides. Since the results in different experimental treatments will rarely be identical, the investigator must ask whether an observed difference is great enough to be of biological importance. Statistical methods are often used as a standardized way to help make this decision. We describe general statistical principles and one statistical method—the *t*-test—in the **Web Stats Review**. There are many different types of statistical analyses; books such as those by Zar (2006), Sokal and Rohlf (1995), or Gotelli and Ellison (2004) provide examples of what methods to use under what circumstances.

What we know about ecology is always changing

The information we describe in this book is not a static body of knowledge. Instead, like the natural world itself, our understanding of ecology is constantly changing. Like all scientists, ecologists observe nature and ask questions about how nature works. For example, when the existence of amphibian deformities became widely known in 1995, some scientists set out to answer a series of questions about those deformities. There were many things they wanted to know: How many species were afflicted by deformities? Did amphibian deformities occur in a few or many geographic regions? What caused the deformities?

The questions stimulated by the discovery of amphibian deformities illustrate the first in a series of four steps by which scientists learn about the natural world. These four steps comprise the **scientific method**, which can be summarized as follows:

1. Observe nature and ask a question about those observations.
2. Use previous knowledge or intuition to develop possible answers to that question (**hypotheses**).
3. Evaluate competing hypotheses by performing experiments, gathering carefully selected observations, or analyzing results of quantitative models.
4. Use the results of those experiments, observations, or models to modify one or more of the hypotheses, to pose new questions, or to draw conclusions about the natural world.

This four-step process is iterative and self-correcting. New observations lead to new questions, which stimulate ecologists to formulate and test new ideas about how nature works. The results from such tests can lead to new knowledge, still more questions, and the abandonment of ideas that fail to explain the results.

We've already seen some examples of how the process of scientific inquiry works: as answers to some questions about amphibian deformities were found, new questions arose, and new discoveries were made. As we'll see shortly, these discoveries have caused some early ideas to be abandoned, such as the notion that global declines in amphibian populations were caused by just one or a few factors. Overall, the research on amphibian deformities and population declines suggests that our understanding of ecological processes is—and always will be—a work in progress.

Deformity and Decline in Amphibian Populations

As we've seen in this chapter, amphibian deformities are often caused by parasites, but can also be influenced by other factors, such as exposure to pesticides or fertilizers. With respect to population decline, studies have suggested that a range of factors can cause amphibian abundances to drop; such factors include habitat loss, overexploitation, climate change, disease, parasites, pollution, introduced species, and UV light. Collectively, these studies suggest that no single factor or set of factors can explain most amphibian declines—a finding that differs from early views that a single, global factor might be responsible. Instead, the declines seem to be caused by complex factors that often act together and may vary from place to place.

For example, Audrey Hatch and Andrew Blaustein (2003) conducted a field experiment to examine the joint effects of UV light and nitrate (a fertilizer) on tadpole survival. At a high-elevation site, they found that neither exposure to UV light nor exposure to nitrate, by themselves, had any effect on the survival of Pacific tree frog tadpoles. Acting together, however, UV light and nitrate reduced the survival of the tadpoles by a significant amount (**Figure 1.14**). In contrast, when they conducted a similar experiment at a low-elevation site, UV light and nitrate acting together did not have a significant effect on tadpole survival. The simplest explanation for this result is that UV light levels at high-elevation sites were greater than those at low-elevation sites (as reported in Table 1 in Hatch and Blaustein 2003). It is also possible that the experimental frog populations may have differed in their sensitivity to UV light or nitrate; for example, frogs from low-elevation sites (where nitrate from fertilizers is common) might be less sensitive to nitrate than frogs from high-elevation sites (where nitrate is less common).

Recent discoveries about the effects of pesticides also suggest that the causes of amphibian decline are likely to be complex and variable. While pesticides may contribute to frog deformities (see pp. 5–7), many studies have failed

In the absence of nitrate, UV light did not affect tadpole survival.

In the presence of nitrate, tadpoles that were exposed to UV light survived much more poorly than those that were not.

FIGURE 1.14 Joint Effects of Nitrate and UV Light on Tadpole Survival In a field experiment set in the Cascade Mountains, ecologists found that nitrate and UV light had joint, but not individual, effects on the survivorship of Pacific tree frog (*Hyla regilla*) tadpoles. (After Hatch and Blaustein 2003.)

or survival. Rick Relyea, of the University of Pittsburgh, repeated such experiments, but with an added twist: predators. In two of six amphibian species studied, pesticide concentrations became up to 46 times more lethal if the tadpoles sensed the presence of a predator (Relyea 2003). The predators were kept separate from the tadpoles by netting, but the tadpoles could smell them.

In Relyea's experiments, the ability of some tadpole species to cope with pesticides was reduced by stress caused by the presence of a predator. The mechanism by which this works is not known. What is true of Relyea's work is true in general: we know relatively little about the mechanisms that underlie the complex responses of amphibian species to different stressors. Although we know that a broad set of factors can cause frog deformities (**Figure 1.15**), little is known about the extent to which these factors interact or how any such interactions exert their ef-

to link pesticides to decreases in the size of amphibian populations. Many of these negative findings came from laboratory studies that held other factors constant and examined the effect of pesticides alone on amphibian growth

FIGURE 1.15 Complex Causation of Amphibian Deformities As we have seen, amphibian deformities can be caused by parasites, such as *Ribeiroia*. However, other factors—many of them a result of human actions—may also be involved in amphibian deformities and declines. (After Blaustein and Johnson 2003.)

1 Rainwater washes fertilizers, livestock manure, and pesticides from agricultural operations into the water.

2 Nitrogen from fertilizers and manure provides extra nutrients for algae in the water, causing an algal bloom, thereby increasing the abundance of snails (a host of the *Ribeiroia* parasite; see Figure 1.3).

3 The presence of pesticides in the water may make amphibians less able to resist *Ribeiroia* infection.

4 Increased exposure to UV light may also make amphibians less able to resist *Ribeiroia* infection.

6 Deformities caused by *Ribeiroia* make amphibians easy targets for predators.

5 Many tadpoles infected by *Ribeiroia* die before metamorphosis.

fects. In this and many other areas of ecology, we have learned enough to solve parts of the mystery, yet more remains to be discovered.

CONNECTIONS IN NATURE

Mission Impossible?

As we stressed in the opening pages of this chapter, people have begun to realize that it is important for us to understand how nature works, if only to protect ourselves from inadvertently changing our environment in ways that cause us harm. Does the fact that the natural world is vast, complex, and interconnected mean that it is impossible to understand? Most ecologists do not think so. Our understanding of natural systems has improved greatly in the last hundred years. Ongoing efforts to understand how nature works are sure to be challenging, but such efforts are also enormously exciting and important. What we learn, and how we use that knowledge, will have a great impact on the current and future well-being of human societies. Whatever your career path, we hope this book will help you to understand the natural world in which you live, as well as how you affect—and are affected by—that world.

SUMMARY

CONCEPT 1.1 Events in the natural world are interconnected.

- Laboratory and field experiments on the effects of parasites on amphibian deformities illustrate how events in nature can be connected with one another.
- Because events in natural systems are interconnected, any action can have unanticipated side effects.
- People both depend on and affect the natural environment.

CONCEPT 1.2 Ecology is the scientific study of interactions between organisms and their environment.

- Ecology is a scientific discipline that is related to, but differs from, other disciplines such as environmental science.
- Public and professional ideas about ecology often differ.
- Ecology is broad in scope and encompasses studies at many levels of biological organization.

- All ecological studies address events on some spatial and temporal scales while ignoring events at other scales.

CONCEPT 1.3 Ecologists evaluate competing hypotheses about natural systems with experiments, observations, and models.

- In an ecological experiment, an investigator alters one or more features of the environment and observes the effect of that change on natural processes.
- Some features of the natural world are best investigated with a combination of field observations, small-scale experiments, and quantitative models.
- Experiments are designed and analyzed in consistent ways: typically, each treatment, including the control, is replicated; treatments are assigned at random; and statistical methods are used to analyze the results.
- The information in this book is not a static body of knowledge; what we know about ecology is always changing.

REVIEW QUESTIONS

1. Describe what the phrase "connections in nature" means, and explain how such connections can lead to unanticipated side effects. Illustrate your points with an example discussed in the chapter.

2. What is ecology, and what do ecologists study? If an ecologist studied the effects of a particular gene, describe how the emphasis of that researcher's work might differ from the emphasis of a geneticist or cell biologist.

3. How does the scientific method work? Include in your answer a description of a controlled experiment.

HANDS-ON PROBLEM SOLVING

| **Amphibian and Reptile Population Declines** | This Web exercise (www.sinauer.com/ecology) explores declines in population sizes of amphibians and reptiles in Costa Rica, and suggests a climatic link. You will interpret data from a recent paper showing changes in population sizes of reptile and amphibian species, as well as changes in climatic conditions. |

SUGGESTED READINGS

Blaustein, A. R. and J. M. Kiesecker. 2002. Complexity in conservation: Lessons from the global decline of amphibian populations. *Ecology Letters* 5: 597–608. *This review article argues that amphibian declines result from a broad range of factors whose effects depend on the particular environment in which a given population is found. The authors also note that we should expect different species, and even different populations of the same species, to respond differently to the same environmental stress.*

Kingsland, S. 2004. Conveying the intellectual challenge of ecology: An historical perspective. *Frontiers in Ecology and the Environment* 2: 367–374. *A brief survey of the history of ecological thought that emphasizes the importance of integrating knowledge from different fields when attempting to understand how nature works or solve difficult applied problems.*

Krebs, C. J. 1988. *The Message of Ecology.* Harper Collins, New York. *This short (195-page) book provides an excellent overview of some of the key lessons that have emerged from the scientific discipline of ecology.*

Stuart, S. N., J. S. Chanson, N. A. Cox, B. E. Young, A. S. L. Rodrigues, D. L. Fischman and R. W. Waller. 2004. Status and trends of amphibian declines and extinction worldwide. *Science* 306: 1783–1786. *A sobering paper that illustrates that while habitat loss and overexploitation often cause amphibian declines, among 435 species whose populations are declining rapidly, the cause of the decline remains unknown in 48% of the cases.*

Wilson, E. O. 2006. *Nature Revealed: Selected Writings, 1949–2006.* John Hopkins University Press, Baltimore. *This book offers an introduction to ecological thinking by providing a selection of papers written by Edward O. Wilson, a leading ecologist and evolutionary biologist. The topics, ranging from detailed studies on ants to sweeping essays about ecology and the diversity of life, make for fascinating reading.*

Worster, D. 1985. *Nature's Economy: A History of Ecological Ideas.* Cambridge University Press, Cambridge. *A well-written account of key episodes in the history of ecology. The book describes how ecological thinking has changed over time and influenced human ideas about our place in nature.*

UNIT **1**

Organisms and Their Environment

KEY CONCEPTS

CONCEPT 2.1 Climate is the most fundamental characteristic of the physical environment.

CONCEPT 2.2 Winds and ocean currents result from differences in solar radiation across the surface of Earth.

CONCEPT 2.3 Large-scale atmospheric and oceanic circulation patterns establish global patterns of temperature and precipitation.

CONCEPT 2.4 Regional climates reflect the influence of the distribution of oceans and continents, elevation, and vegetation.

CONCEPT 2.5 Seasonal and long-term climatic variation are associated with changes in Earth's position relative to the sun.

CONCEPT 2.6 Salinity, acidity, and oxygen concentrations are major determinants of the chemical environment.

CASE STUDY

Climatic Variation and Salmon

Grizzly bears of the Pacific Northwest feast seasonally on the salmon that arrive in huge numbers to spawn (reproduce) in the streams of the region (**Figure 2.1**). Salmon are *anadromous*; that is, they are born in freshwater streams, spend their adult lives in the ocean, and then return to spawn in the freshwater habitats where they were born. Grizzlies capitalize on the salmon's reproductive habits to gorge themselves on this rich food resource. These normally aggressive bears will forgo their usual territorial behavior and tolerate high densities of other bears while fishing for salmon.

It is not only bears that rely on salmon for food. Salmon have been an important part of the human economy of the Pacific Northwest for millennia. The fish were a staple of the diets of Native Americans in this region as well as a central part of their cultural and spiritual lives. Salmon are now fished commercially in the waters of the North Pacific Ocean, providing an important economic base for many coastal communities. Commercial salmon fishing is a risky venture, however. Successful reproduction for salmon is dependent on the health of the streams in which they spawn. The construction of dams, increases in stream sediments due to logging operations, water pollution, and overharvesting have all been blamed for declines in salmon populations, primarily from the California coast north to British Columbia (Walters 1995). Despite significant efforts to mitigate this environmental degradation, the recovery of salmon stocks has been marginal at best in the southern portion of the region.

Researchers, environmental advocates, and government policy experts have focused primarily on the degradation of freshwater habitat as a cause for the declines in salmon. In 1994, however, Steven Hare and Robert Francis suggested that changes in the marine environment, where salmon spend the vast majority of their adult life, could be contributing to the declines in salmon abundance. In particular, they noted that records of fish harvests covering more than a century indicated that multi-decadal periods of low or high fish production have occurred repeatedly, separated by abrupt changes in production rather than gradual transitions (**Figure 2.2**). In addition, Nathan Mantua and colleagues (Mantua et al. 1997) noted that periods of high salmon production in Alaska corresponded with periods of low salmon production at the southern end of the salmon range, particularly in Oregon and Washington. They found telling quotes from commercial fishing publications of the times

FIGURE 2.1 A Seasonal Opportunity Grizzly bears congregate in the streams and rivers of Alaska to feed on salmon migrating upstream to reproduce. The size of the salmon run each year depends in part on physical conditions in the Pacific Ocean, many miles away.

FIGURE 2.2 Changes in Salmon Harvests over Time
Records of commercial harvests of (A) sockeye salmon and (B) pink salmon in Alaska over 65 years show abrupt drops and increases in production. Red lines represent annual catch, purple lines are a statistical fit to the data. (After Hare and Francis 1994.)

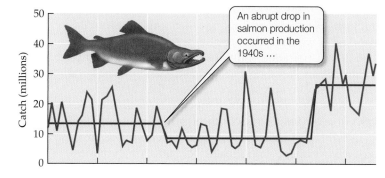

that told the same story: when the fishing was poor in Washington and Oregon, it was good in Alaska, and vice versa.

Pacific Fisherman (1915)

Never before have the Bristol Bay (Alaska) salmon packers returned to port after the season's operations so early. [i.e., a bad year, with few fish to catch]

The spring [chinook salmon] fishing season on the Columbia River [Washington and Oregon] closed at noon on August 25, and proved to be one of the best for some years.

Pacific Fisherman (1939)

The Bristol Bay Red [Alaska sockeye salmon] run was regarded as the greatest in history.

The [chinook] catch this year is one of the lowest in the history of the Columbia [Washington].

Hare and Francis hypothesized that the abrupt shifts in salmon production were associated with long-term climatic variation occurring in the North Pacific. The nature and cause(s) of these underlying climatic shifts, however, were unclear. Additional work by Mantua and colleagues found good correspondence between the decadal shifts in salmon production and changes in sea surface temperatures in the North Pacific.

How widespread is this variation in climate and its effects on salmon and the associated marine ecosystem? As we will see at the end of this chapter, the research on variation in salmon production led to the discovery of an important long-term cyclic climatic pattern that affects a large area.

The physical environment is the ultimate determinant of where organisms can live, the resources that are available to them, and the rate at which their populations can grow. Thus, an understanding of the physical environment is key to understanding all ecological phenomena, from the outcome of interactions between bacteria and fungi in the soil to the exchange of carbon dioxide between the biosphere and the atmosphere.

The physical environment includes climate, which consists of long-term trends in temperature and precipitation. The physical environment also includes radiation from the sun, which ultimately drives the climate system as well as biological energy production. The chemical environment, which includes salinity (concentrations of dissolved salts), acidity (pH), and concentrations of gases in the atmosphere and dissolved in water, is another aspect of the physical environment. Soil is an important component of the physical environment because it is a medium in which microorganisms, plants, and animals live. Soil also influences the availability

of critical resources, particularly water and nutrients. This chapter focuses on climate and the chemical environment; we will cover soils, their development, and nutrient supply in Chapter 21.

This chapter will provide a framework for characterizing the physical environment, including its variability, at a variety of spatial and temporal scales. We will begin by exploring the processes that create the climatic patterns we see at global to regional scales.

Climate

CONCEPT 2.1
Climate is the most fundamental characteristic of the physical environment.

In our daily lives, we are aware of the **weather** around us: the current temperature, humidity, precipitation, wind, and cloudiness. Weather is an important determinant of our behavior: what we wear, the activities we engage in, and our mode of transportation. **Climate** is the long-term description of weather, based on averages and variation measured over decades. Climatic variation includes the daily and seasonal cycles associated with changes in solar radiation as Earth rotates on its axis and orbits the sun. Climatic variation also includes changes over years or decades, such as large-scale cyclic weather patterns related to changes in the atmosphere and oceans (El Niño events, discussed later in this chapter, are one example). Longer-term climate change occurs as a result of changes in the intensity and distribution of solar radiation reaching Earth's surface. Earth's climate is currently changing due to changes in atmospheric concentrations of gases such as carbon dioxide that are emitted as a result of human activities. These gases reflect energy back to Earth's surface, creating a greenhouse effect (see Chapter 24).

Climate controls where and how organisms live

Where organisms live (their geographic distribution) and how they function are determined by climate. Temperature determines the rates of biochemical reactions and physiological activity for all organisms. Water supplied by precipitation is an essential resource for terrestrial organisms. Freshwater organisms are dependent on precipitation for the maintenance and quality of their habitats. Marine organisms depend on ocean currents that influence the temperature and chemistry of the ocean waters they live in.

We often characterize climate—or any aspect of the physical environment—at any location by the average conditions over time. For example, the average temperature at a location provides information on the long-term conditions organisms must face if they are to survive there. However, the distributions of organisms often reflect *extreme* conditions more than average conditions because extreme events are important determinants of mortality. Temperature and moisture extremes can impact long-lived organisms such as forest trees. For example, record high temperatures, along with a severe drought in 2000–2003, caused widespread mortality in large stands of piñon pines (*Pinus edulis*) in the southwestern United States (**Figure 2.3**). Even these long-lived plants could no longer survive in the region where they had existed for many years. Thus, the physical environment must also be characterized by its **variability**

over time, not just by average conditions, if we are to understand its ecological importance.

The *timing* of changes in the physical environment is also ecologically important. The seasonality of rainfall, for example, is an important determinant of water availability for terrestrial organisms. In regions with a "Mediterranean-type" climate, the majority of precipitation falls in winter. Although these regions receive more precipitation than most desert areas, they experience regular dry periods during the summer growing season. The lack of water during the summer reduces the growth of plants and promotes the occurrence of fires. In contrast, some grasslands have the same annual average temperature and precipitation as these Mediterranean-type ecosystems, but the precipitation is spread evenly throughout the year.

Climate also influences the rates of abiotic processes that affect organisms. The rate at which rocks and soil are broken down to supply nutrients to plants and microorganisms, for example, is determined by climate. Climate can

(A)

(B)

FIGURE 2.3 Widespread Mortality in Piñon Pines A historic drought from 2000 to 2003 killed large areas of piñon pines (*Pinus edulis*) throughout the southwestern United States. (A) Here, stands in the Jemez Mountains, New Mexico, begin to show substantial needle death due to water and temperature stress, combined with a bark beetle outbreak in October 2002. (B) By May 2004, most of the trees had died.

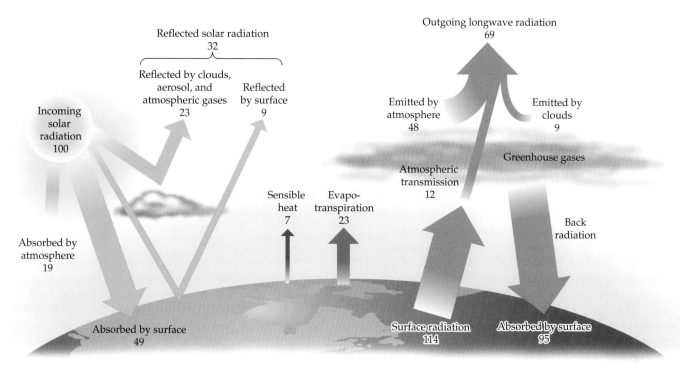

FIGURE 2.4 Earth's Radiation Balance   Average annual energy balance for Earth's surface, including gains from solar radiation, and gains and losses due to emission of infrared radiation and latent and specific heat fluxes. Gains and losses are given as percentages of the average annual incoming solar radiation at the top of Earth's atmosphere (342 W/m²). (After Kiehl and Trenberth 1997.)

also influence the rates of periodic disturbances, such as fires, rockslides, and avalanches. These events kill organisms and disrupt biological communities, but subsequently create opportunities for the establishment and growth of new organisms and communities.

Global energy balance drives the climate system

The energy that drives the global climate system is ultimately derived from solar radiation. On average, the top of Earth's atmosphere receives 342 watts (W) of solar radiation per square meter each year (**Figure 2.4**). About a third of this solar radiation is reflected back out of the atmosphere by clouds, atmospheric particles, and Earth's surface. Another fifth of the incoming solar radiation is absorbed by ozone, clouds, and water vapor in the atmosphere. The remaining half of the incoming solar radiation is absorbed by land and water at Earth's surface.

These energy gains from solar radiation must be offset by energy losses if Earth's temperature is to remain the same. Much of the solar radiation absorbed by Earth's surface is emitted to the atmosphere as infrared radiation. Earth's surface also loses energy when water at the surface evaporates. The change in phase from liquid water to vapor absorbs energy, and is a cooling process. The heat loss due to evaporation is known as **latent heat flux**. Energy is also transferred through the exchange of kinetic energy by molecules in direct contact with one another (**conduction**) or by the movement (**convection**) of currents of air (wind) and water. This energy transfer from the warm air immediately above the surface to the cooler atmosphere by convection and conduction is known as **sensible heat flux**.

Earth's surface actually releases more energy than it receives by solar radiation (see Figure 2.4) Additional energy inputs occur due to atmospheric absorption and reradiation of infrared radiation emitted by Earth. The atmosphere contains several gases, known as **radiatively active gases** or **greenhouse gases**, that absorb and reradiate the infrared radiation emitted by Earth. These gases include water vapor (H_2O), carbon dioxide (CO_2), methane (CH_4), and nitrous oxide (N_2O). Some of these greenhouse gases are produced through biological activity (e.g., CO_2, CH_4, N_2O), linking the biosphere to the climate system. Without these greenhouse gases, Earth's climate would be considerably cooler than it is (by approximately 33°C, or 59°F). As noted earlier, increases in atmospheric concentrations of greenhouse gases due to human activities are altering Earth's energy balance, changing the climate system, and causing global warming (see Chapter 24).

Our discussion of Earth's energy balance in this section has focused on average annual transfers of energy to and from Earth as a whole. But not every location on Earth receives the same amount of energy from the sun. Let's consider how these differences in solar radiation affect the movements of Earth's atmosphere and ocean waters.

Atmospheric and Oceanic Circulation

CONCEPT 2.2
Winds and ocean currents result from differences in solar radiation across the surface of Earth.

Near the equator, the sun's rays strike Earth's surface perpendicularly. Toward the North and South Poles, the angle of the sun's rays becomes steeper, so that the same amount of energy is spread over a progressively larger area of Earth's surface (**Figure 2.5**). In addition, the amount of atmosphere the rays must pass through increases toward the poles, and more radiation is reflected and absorbed before it reaches the surface. As a result, more solar energy is received per unit area in equatorial regions (between 23.5° north and south latitude) than in regions closer to the poles. This differential reception of solar radiation is the driving force for much of the dynamics of Earth's climate. In addition, the movement of Earth around the sun, in combination with the tilt of Earth's axis of rotation, results in changes in the amount of solar radiation received at any location over the course of the year. These changes are the cause of seasonal climatic variation: winter–spring–summer–fall changes at high latitudes and wet–dry shifts in tropical regions.

Atmospheric circulation cells are established in regular latitudinal patterns

When solar radiation heats Earth's surface, the surface warms and emits infrared radiation to the atmosphere, warming the air above it. As we have just seen, the heating of Earth's sur-

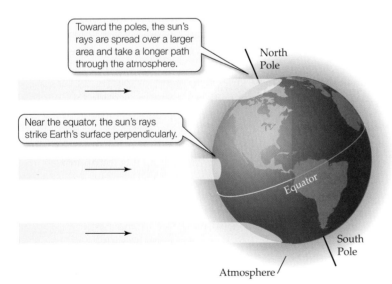

FIGURE 2.5 Latitudinal Differences in Solar Radiation at Earth's Surface   The angle of the sun's rays affects the intensity of the solar radiation that strikes Earth's surface.

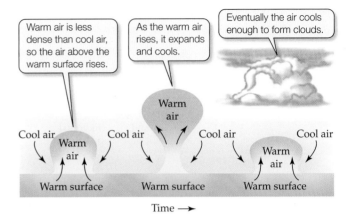

FIGURE 2.6 Surface Heating and Uplift of Air   Differential solar heating of Earth's surface leads to the uplift of pockets of air over the warmest surfaces.

face varies with latitude, and it can also vary with topography. This differential warming creates pockets of warm air surrounded by cooler air. Warm air is less dense than cool air (because it has fewer molecules per unit of volume), so as the air warms, it rises, a process called **uplift** (**Figure 2.6**). As it rises higher, the air expands because air pressure decreases with increasing elevation. The expansion of air is a cooling process, so the rising air becomes cooler. Cool air cannot hold as much water vapor as warm air, so as the air continues to rise and cool, the water vapor contained within it begins to condense into droplets and form clouds.

The condensation of water into clouds is a warming process (another form of latent heat transfer), which may act to keep the pocket of air warmer than the surrounding atmosphere and enhance its uplift, despite its cooling due to expansion. You may have observed this process on a warm summer day, as bubble-shaped cumulus clouds began to form into thunderstorms. When there is substantial heating of the surface and a progressively cooler atmosphere above the surface, the uplifted air will form thunderstorms, with wedge-shaped cloud tops. The thunderclouds reach to the boundary between the *troposphere*, the atmospheric layer above Earth's surface, and the *stratosphere*, the next atmospheric layer above the troposphere. This boundary is marked by a transition from cooler temperatures in the troposphere to warmer temperatures in the stratosphere. Thus, the air pocket ceases to rise once it reaches the warmer temperatures of the stratosphere.

The tropical regions around the equator receive the most precipitation of any area on Earth because they receive the most solar radiation, and thus experience the greatest amount of surface heating, uplift of air, and storm formation. The uplifting of air in the tropics creates bands of low atmospheric pressure relative to zones to the north and south. **Atmospheric pressure** results from the force exerted on the surface by the air molecules above it. Less dense air therefore has lower atmospheric pressure than more dense air. When air rising over the equatorial zone reaches the boundary be-

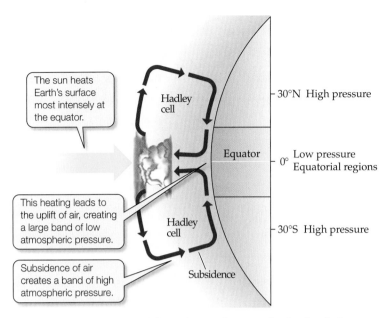

FIGURE 2.7 Equatorial Heating and Atmospheric Circulation Cells The heating of Earth's surface at the equator causes air to rise and release precipitation.

polar cells. The Ferrell cell is driven by the movement of the Hadley and polar cells and by exchange of energy between equatorial and polar air masses in a region known as the *polar front*.

These three atmospheric circulation cells establish the major climatic zones on Earth. Between 30° N and S is the **tropical zone**, the **temperate zones** are between 30 and 60° N and S, and the **polar zones** are above 60° N and S (see Figure 2.8).

Atmospheric circulation cells create surface wind patterns

The areas of high and low pressure formed by atmospheric circulation cells give rise to consistent patterns of air movement at Earth's surface, known as *prevailing winds*. The directions of those winds are influenced by the frictional forces as-

tween the troposphere and stratosphere, it flows toward the poles (**Figure 2.7**). Eventually, this poleward-moving air cools as it exchanges heat with the surrounding air and meets cooler air moving from the poles toward the equator. Once the air reaches a temperature similar to that of the surrounding atmosphere, it descends toward Earth's surface, a process known as **subsidence**. This subsidence of air creates regions of high atmospheric pressure at latitudes 30° N and S that inhibit the formation of storms. It is not surprising, therefore, that the major deserts of the world are found at these latitudes.

The equatorial uplift of air creates a large-scale, three-dimensional pattern of atmospheric circulation in each hemisphere known as a **Hadley cell**, named after George Hadley, the eighteenth-century British meteorologist and physicist who first proposed its existence. Additional atmospheric cells are formed at higher latitudes (**Figure 2.8**). The **polar cell**, as its name indicates, occurs at the North and South Poles. Cold, dense air subsides at the poles and moves toward the equator when it reaches Earth's surface. The descending air at the poles is replaced by air moving through the upper atmosphere from lower latitudes. Subsidence at the poles creates an area of high pressure, so the polar regions, despite the abundance of ice and snow on the ground, actually receive little precipitation, and are known as "polar deserts." An intermediate **Ferrell cell** (named after American meteorologist William Ferrell), exists at mid-latitudes between the Hadley and

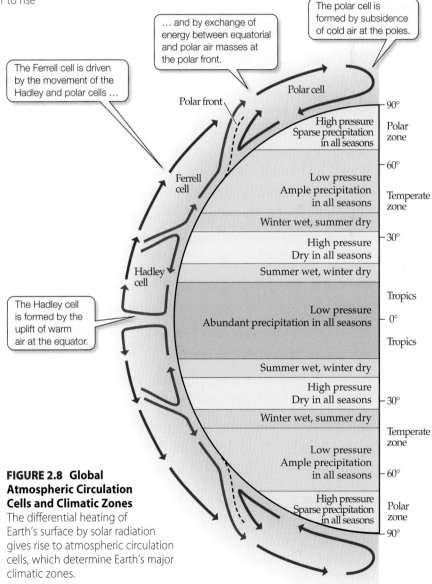

FIGURE 2.8 Global Atmospheric Circulation Cells and Climatic Zones The differential heating of Earth's surface by solar radiation gives rise to atmospheric circulation cells, which determine Earth's major climatic zones.

FIGURE 2.9 Influences on Global Wind Patterns   The Coriolis effect results from Earth's rotation. (Compare the prevailing wind directions shown by the black arrows with the atmospheric winds indicated by the red and blue arrows.)

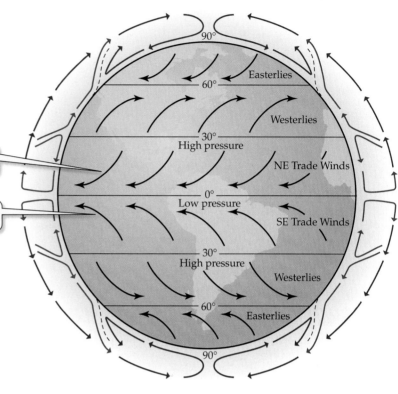

Moving air at Earth's surface is deflected to the right (clockwise) in the Northern Hemisphere…

…and to the left (counterclockwise) in the Southern Hemisphere.

sociated with the spinning of Earth on its axis. Air moving at Earth's surface is deflected to the right (clockwise) in the Northern Hemisphere and to the left (counterclockwise) in the Southern Hemisphere (**Figure 2.9**). This apparent force is known as the **Coriolis effect**. As a result of the Coriolis effect, surface winds blowing toward the equator from the high-pressure zones at 30° N and S are deflected toward the west. These winds are known as the *trade winds* because of their importance to the global transport of trade goods in sailing ships during the fifteenth through the nineteenth centuries. Winds blowing toward the poles from those zones of high pressure, called *westerlies*, are deflected toward the east.

The presence of continental land masses interspersed with oceans complicates this idealized depiction of prevailing wind patterns (**Figure 2.10**). Water has a higher **heat capacity** than land; in other words, water can absorb and store more energy without its temperature changing than land can. For this reason, ocean water tends to be cooler in summer and warmer in winter than land at the same latitude, and seasonal temperature changes over the oceans are slower and less extreme than those on land. In summer, air over the oceans is cooler and denser than that over land. As a result, semipermanent zones of high pressure form over the oceans, particularly around 30° N and S. In winter, the opposite situation exists: the air over the continents is cooler and denser than that over the oceans, so high-pressure cells develop in temperate zones over large continental areas. Because winds blow from areas of high pressure to areas of low pressure, these seasonal shifts in pressure cells influence the direction of the prevailing winds. The effect of land areas on the development of these semipermanent pressure cells is more pronounced in the Northern Hemisphere than the Southern Hemisphere because continental land masses make up a larger proportion of Earth's surface there.

Ocean currents are driven by surface winds

Wind moving across the ocean surface creates a frictional drag that moves the surface water. As a result of the Coriolis effect, the water moves at an angle to the wind, since it is deflected toward the right in the Northern Hemisphere and

to the left in the Southern Hemisphere. The pattern of ocean surface currents is similar, but not identical to the pattern of prevailing winds (**Figure 2.11**). The speed of ocean currents is usually only about 2%–3% of the wind speed. An average wind speed of 10 meters per second would therefore produce an ocean current moving at 30 centimeters per second. In the North Atlantic Ocean, current velocities may be as high as 200 centimeters per second.

Like atmospheric circulation, oceanic circulation has a vertical dimension. Generally, the surface and deep layers of ocean water do not mix because of differences in their temperature and salinity (the concentration of dissolved salts). The surface waters (those above 75–200 m) are warmer and less saline, and therefore less dense, than the deeper, cooler ocean waters. When warm tropical surface currents reach polar regions, particularly the coasts of Antarctica and Greenland, the water loses heat to the surrounding environment and becomes cooler and denser. The waters cool off enough for ice to form, which increases the salinity of the unfrozen water. This combination of cooling and greater salinity increases the density of the water, which sinks to deeper layers. These dense, downwelling currents move toward the equator, carrying cold polar water toward the warmer tropical oceans.

These deep ocean currents connect with surface currents again at zones of **upwelling**, where deep ocean waters rise to the surface. Upwelling occurs where prevailing winds blow nearly parallel to a coastline, such as off the western coasts of North and South America. The force of the wind, in combination with the Coriolis effect, causes surface waters to flow

(A) July

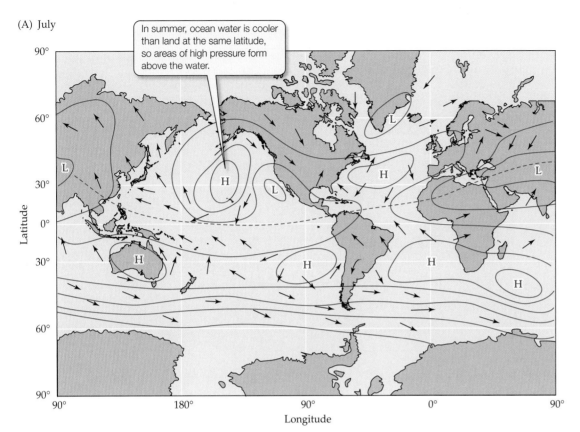

FIGURE 2.10 Prevailing Wind Patterns The difference in heat capacity between the oceans and the continents leads to seasonal changes in pressure cells that influence prevailing wind patterns. (A) July and (B) January patterns of surface pressure and prevailing winds.

(B) January

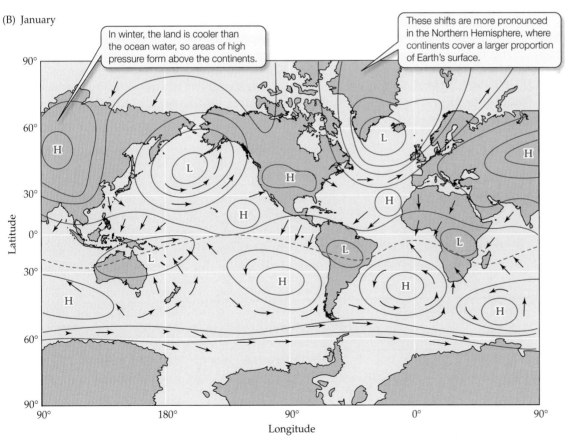

FIGURE 2.11 Global Ocean Currents   The major ocean surface currents are driven by surface winds shown in Figure 2.9, but modified by the Coriolis force.

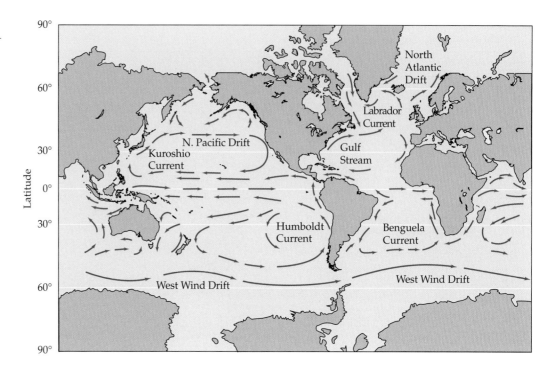

away from the coast (**Figure 2.12**), and deeper, colder ocean waters rise to replace them. Upwelling has important consequences for the local climate, creating a cooler, moister environment. Upwelling also has a strong effect on biological activity in the surface waters. When organisms in the surface waters die, their bodies, and the nutrients they contain, fall into deeper water. Thus, nutrients tend to accumulate in deep water and in sediments at the ocean bottom. Upwelling brings those nutrients back to the *photic zone*, the layer of sur-

face water where there is enough light to support photosynthesis. Upwelling zones are among the most productive open ocean ecosystems because these nutrients increase the growth of phytoplankton (small, free-floating algae and other photosynthetic organisms), which provide food for zooplankton (free-floating animals and protists), which in turn support the growth of their consumers.

Ocean currents can significantly influence the climate of the regions where they flow. For example, the Gulf Stream

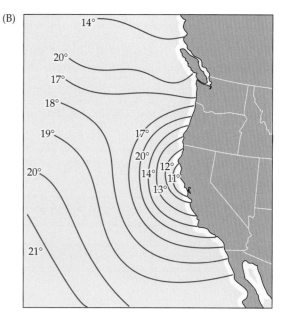

FIGURE 2.12 Upwelling of Coastal Waters   (A) Wind blowing parallel to the coast causes surface water to flow away from the coast, pulling deep water upward to replace it. (B) Upwelling influences surface water temperatures off the western coast of North America. Ocean temperatures are shown in °C.

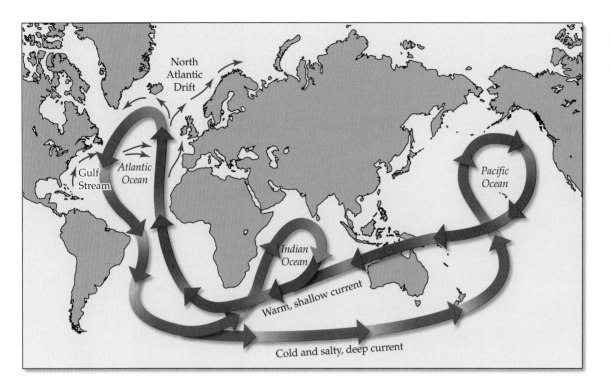

FIGURE 2.13 The Great Ocean Conveyor Belt An interconnected system of surface and deep ocean currents transfers energy between tropical and polar regions.

and North Atlantic Drift, a current system that flows from the tropical Atlantic northward to the North Atlantic (see Figure 2.11), contributes to warmer winters in Scandinavia relative to similar locations across the Atlantic in North America. In additon, winds blowing across the Atlantic pick up heat from the ocean, which also contributes to a warmer climate in Northern Europe. Winter temperatures on the west coast of Scandinavia are approximately 15°C (22°F) warmer than the coast of Labrador. This temperature difference is reflected in the vegetation: deciduous forests are more common on the Scandinavian coast, while boreal forests of spruce and pine dominate the coast of Labrador. The Gulf Current also keeps the North Atlantic ice free most of the winter, while significant sea ice forms at the same latitude off the North American coast.

Ocean currents are responsible for about 40% of the heat exchange between the tropics and the polar regions; the remaining 60% is transferred by winds. Thus, ocean currents are sometimes referred to as the "heat pumps" or "thermal conveyers" of the planet. A large system of interconnected currents that links the Pacific, Indian, and Atlantic oceans, sometimes called the "great ocean conveyor belt," is an important means of transferring heat to the polar regions (**Figure 2.13**).

Now that we have seen how the differential heating of Earth's surface generates patterns of atmospheric winds and oceanic currents, let's turn to an examination of the effects of those patterns on Earth's climates, including the patterns of temperature and precipitation.

Global Climatic Patterns

CONCEPT 2.3
Large-scale atmospheric and oceanic circulation patterns establish global patterns of temperature and precipitation.

Earth's climates reflect a variety of temperature and precipitation regimes, from the warm, wet tropics to the cold, dry Arctic and Antarctic. In this section, we examine these global patterns of temperature and precipitation and explore how both climatic averages and climatic variation are influenced by prevailing winds and oceanic currents.

Oceanic circulation and the distribution and topography of continents influence global temperatures

Given what we have learned about the global pattern of solar radiation, it is not surprising that temperatures at Earth's surface become progressively cooler from the equator toward the poles. Note, however that the *isotherms* (lines of equal temperature) in **Figure 2.14** are not exactly parallel with the lines of latitude. Two major influences alter these global patterns: ocean currents and the relative distribution of land and water. As we saw in the previous section, ocean currents contribute to a warmer climate in northern Europe relative to other locations at the same latitude. Similarly, the influence of the cold Humboldt Current on temperature is noticeable on the west coast of South America, where temperatures are cooler, and isotherms follow a more vertical pattern, than at similar latitudes elsewhere.

The difference in heat capacity between the oceans and the continents is not reflected in the annual average temperatures shown in Figure 2.14. Why is this so? Because the annual *variation* in temperatures is not depicted in that figure.

Air temperatures on land show greater seasonal variation, rising higher in summer and dropping lower in winter, than those over the oceans (**Figure 2.15**).

Elevation above sea level has an important influence on continental temperatures. Note in Figure 2.14A the large difference in temperature between the Indian subcontinent and

(A)

(B)

FIGURE 2.14 Global Average Annual Temperatures Average air temperatures over land (A) and over the oceans (B) vary across the globe.

FIGURE 2.15 Annual Seasonal Temperature Variation Seasonal temperature changes are expressed as the difference in mean monthly temperatures between the warmest and coldest months (°C).

Asia. The narrowing of the isotherms in this region is due to the influence of the Himalayas and the Tibetan Plateau. Topographic relief is extreme here, as elevation rises from about 150 m on the Ganges Plain in India to over 8,000 m in the highest peaks of the Himalayas in 200 km.

Why is it colder in mountains and highlands than in surrounding lowlands? Two factors contribute to the colder climates found at higher elevations. First, the pressure and therefore the density of air in the atmosphere decreases with increasing elevation above sea level, and as a result, the heat capacity of the air also decreases. Thus, even though the highlands may receive as much solar radiation as nearby lowlands, the heating of air by the ground surface is less effective because of the lower air density. Second, highlands exchange air more effectively with cooler air in the surrounding atmosphere. Because the atmosphere is warmed mainly by infrared radiation emitted by Earth's surface, the temperature of the atmosphere decreases with increasing distance from the ground. This decrease in temperature with increasing height above the surface is known as the **lapse rate**. In addition, wind velocity increases with increasing elevation due to less friction with the ground surface. As a result, the decrease in air temperature with increasing elevation in highlands tends to follow the lapse rate.

Patterns of atmospheric pressure and topography influence precipitation

The locations of the Hadley, Ferrell, and polar circulation cells suggest that precipitation should be highest in the tropical latitudes between 23.5° N and S and in a band at about 60° N and S, and lowest in zones around 30° N and S (see Figure 2.8). The African continent displays the pattern closest to this idealized precipitation distribution. However, there is substantial deviation from the expected latitudinal precipitation pattern in other areas, particularly in the Americas (**Figure 2.16**).

Deviations from the idealized precipitation pattern are associated with the semipermanent high-pressure and low-pressure cells discussed earlier (see Figure 2.10), as well as with large mountain chains. The pressure cells influence the movement of moist air from oceans to continents as well as storm formation. For example, high pressure over the South Pacific Ocean decreases precipitation along the central west coast of South America. One of the driest deserts in the world,

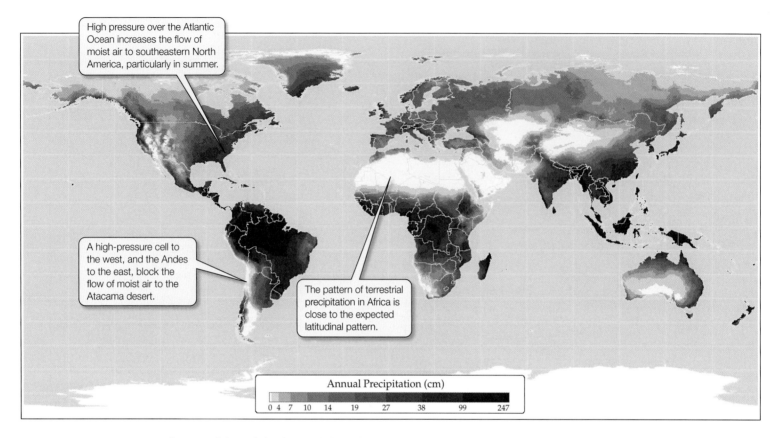

High pressure over the Atlantic Ocean increases the flow of moist air to southeastern North America, particularly in summer.

A high-pressure cell to the west, and the Andes to the east, block the flow of moist air to the Atacama desert.

The pattern of terrestrial precipitation in Africa is close to the expected latitudinal pattern.

Annual Precipitation (cm)

0   4   7   10   14   19   27   38   99   247

FIGURE 2.16 Average Annual Terrestrial Precipitation The latitudinal pattern of precipitation deviates from what would be expected based on atmospheric circulation patterns alone (see Figure 2.8).

the Atacama, located along the Pacific coast of Chile, is associated with the presence of this high-pressure cell and with the blockage of air masses moving from the east by the Andes. In contrast, high pressure over the Atlantic Ocean increases the flow of moist air to southeastern North America, particularly in summer, increasing precipitation and supporting the development of a productive deciduous forest biome.

Mountains also influence precipitation patterns by forcing air moving across them to rise, which enhances local precipitation. The effects of mountains, as well as those of oceans and vegetation, on regional climatic patterns are addressed in the next section.

Regional Climatic Influences

CONCEPT 2.4
Regional climates reflect the influence of the distribution of oceans and continents, elevation, and vegetation.

You may have noticed that as you travel from a coastal area to an inland location, the climate changes. This change in climate can be abrupt, particularly when you travel across a mountain chain. The daily variation in air temperatures increases, humidity decreases, and precipitation decreases. These climatic differences result from the effects of oceans and continents on regional energy balance and the influence of mountains on air flow and temperature. The vegetation often reflects these regional climatic differences, testifying to the effects of climate on the distributions of species and biological communities. But the vegetation has effects on the climate in turn.

Proximity to oceans influences regional climates

As we have seen, ocean water has a higher heat capacity than land. As a result, seasonal temperature changes are less over oceans than over continental areas (see Figure 2.15). Additionally, oceans provide a source of moisture for clouds and precipitation. Coastal terrestrial regions that are influenced by an adjacent ocean have a **maritime climate**. Maritime climates are characterized by little daily and seasonal variation in temperature, and they often have higher humidity than regions more distant from the coast. In the temperate zones, the influence of oceans on coastal climates tends to be accentuated on west coasts in the Northern Hemisphere and on east coasts in the Southern Hemisphere because of the prevailing wind patterns. In contrast, terrestrial areas in the middle of large continental land masses experience much greater variation in daily and seasonal temperatures, which characterizes a **continental climate**. Maritime climates occur in all climatic zones, from tropical to polar. Continental climates, on the other hand, are limited to high latitudes (primarily in the temperate zone), where large seasonal changes in solar radiation accentuate the effect of the low heat capacity of land masses.

The influence of land and water on climate can be exemplified by comparing the seasonal variation in temperature for locations at similar latitudes and elevations in Siberia (**Figure 2.17**). Sangar, a town on the Lena River in the middle of the Asian continent (65°55′ N, 127°27′ E, 92 m elevation), has a maximum average monthly temperature of 17.9°C (in July) and a minimum average monthly temperature of –39.7°C (in January), for an average seasonal variation of 57.6°C. Khatyrka, on the Pacific coast (62°03′ N, 175°16′ E, 120 m elevation), has a maximum average monthly temperature of 9.5°C (in July) and a minimum average monthly temperature of –14.2°C (in February), for an average seasonal variation of 23.7°C. The seasonal variation in temperature

in continental Sangar is therefore more than double that in coastal Khatyrka, at nearly the same latitude. In addition, both the maximum and minimum monthly temperatures are more extreme in the continental than in the maritime climate. Note too that the maximum and minimum temperatures occur slightly later in the year in the maritime climate (Khartyrka), another reflection of the high heat capacity of the ocean.

Mountains influence wind patterns and gradients in temperature and precipitation

The effects of mountains on climate are visually apparent in the elevational patterns of vegetation, particularly in arid regions. As we move up a mountain, grasslands may abruptly change to forests, and at higher elevations, forests may give way to alpine grasslands. These abrupt shifts in vegetation patterns reflect the rapid changes in climate that occur over short distances in mountains as temperatures decrease, precipitation increases, and wind speed increases with elevation. What causes these abrupt changes in climate? Mountain climates are the product of the effects of topography and elevation on air temperatures, the behavior of air masses, and their own generation of unique local wind patterns.

When air moves across Earth's surface and encounters a mountain range, the slopes of the mountain force it upward. This uplifted air cools as it rises, and water vapor condenses to form clouds and precipitation. As a result, the amount of precipitation increases with elevation. This enhancement of precipitation in mountains is particularly apparent in north–south-trending mountain ranges on the slopes that face into the prevailing wind (the *windward* slopes). In the temperate zones, the westerlies encounter the western slopes of these mountain ranges (such as the Andes and Rockies in the Americas) and drop most of their moisture there as precipitation before cresting over the summits. The loss of moisture, as well as the warming of the air as it moves downslope on the eastern slopes, dries the air mass (**Figure 2.18A**). This **rain-shadow effect** results in little precipitation and low soil moisture on the slopes facing away from the prevailing wind (the *leeward* slopes), and higher precipitation and higher soil moisture on the windward slope. The rain-shadow effect influences the types and amounts of vegetation on mountain ranges: lush, productive plant communities tend to be found on the windward slopes, and sparser, more drought-resistant vegetation on the leeward slopes (**Figure 2.18B**).

Mountains can also generate local wind and precipitation patterns. Differences in the direction that mountain slopes

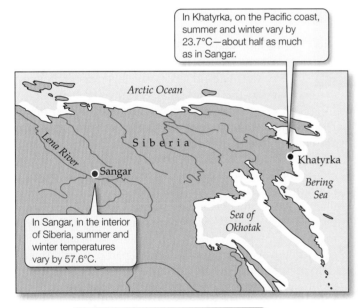

In Khatyrka, on the Pacific coast, summer and winter vary by 23.7°C—about half as much as in Sangar.

In Sangar, in the interior of Siberia, summer and winter temperatures vary by 57.6°C.

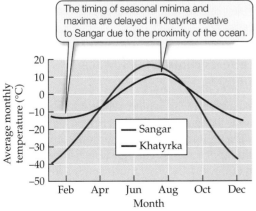

The timing of seasonal minima and maxima are delayed in Khatyrka relative to Sangar due to the proximity of the ocean.

FIGURE 2.17 Monthly Mean Temperatures in a Continental and a Maritime Climate The difference in seasonal temperature variation between two locations in Siberia at about the same latitude illustrates the effect of the high heat capacity of ocean water.

(A)

A moving air mass picks up moisture over the ocean.

On the windward slope, the air rises and cools, releasing most of its moisture as precipitation.

On the leeward slope, the dry air descends and warms, and there is little precipitation.

Altitude (m)   Temperature (°C)

3000 — −4

2250 — 2

1500 — 8

750 — 14

0 — 20

Evaporation

Moist   Arid

(B)

West-facing slope   East-facing slope

FIGURE 2.18 The Rain-Shadow Effect (A) Precipitation tends to be greater on the windward slope of a mountain range than on the leeward slope. (B) Vegetation on west-facing and east-facing slopes in the Cascade range of Washington reflects the rain shadow effect.

calized upslope winds in the mountains. Depending on the moisture content of the air and the prevailing winds at higher elevations, clouds may form on the eastern flanks of the mountains. These clouds can generate local thunderstorms that may move off the mountains and into surrounding lowlands, increasing local precipitation.

At night, the ground surface cools, and the air above it becomes denser. Nighttime cooling is more pronounced at high elevations in mountains than at lower elevations because the thinner atmosphere absorbs and reradiates less energy, and allows more heat to be lost from the ground surface. Cold, dense air behaves like water, flowing downslope and pooling in low-lying areas. As a result, valley bottoms are the coldest sites in mountainous areas during clear, calm nights. Cold air drainage influences vegetation distributions in temperate zones because of the higher frequency of subfreezing temperatures in low-lying areas. Daily upslope and nightly downslope winds are a common feature of many mountainous areas, particularly in summer when solar radiation is highest.

At continental scales, mountains influence the movement, position, and behavior of air masses, and as a result, they influence temperature patterns in surrounding lowlands. Large mountain chains, or *cordilleras*, can act to channel the movement of air masses. The Rocky Mountains, for example, steer cold Canadian air through the central part of North America to their east and inhibit its movement through the intermountain basins to their west. Storm systems passing over the leeward sides of cordilleras often strengthen before moving off to the east due to topographic effects that intensify low-pressure systems.

Vegetation affects climate via surface energy exchange

Climate, as we have seen, determines where and how organisms can live, but living organisms, in turn, have important influences on the climate system. The amount and type of vegetation influences how the ground surface interacts with solar radiation and wind and how much water it loses to the atmosphere. The capacity of a land surface to reflect solar radiation, known as its **albedo**, is influenced by the presence

face (slope exposure or *aspect*) can cause differences in the amounts of solar radiation the slopes and surrounding flatlands receive. As we saw in the case of the large-scale circulation patterns that generate Hadley cells, differences in solar heating of the ground surface can cause the uplift of air parcels that are warmer than the surrounding air. In the morning, east-facing slopes receive more solar radiation from the rising sun, and thus become warmer than the surrounding slopes and lowlands. This differential heating creates lo-

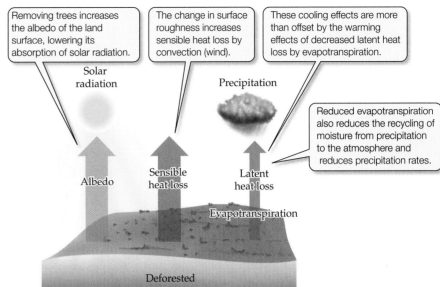

FIGURE 2.19 The Effects of Deforestation Illustrate the Influence of Vegetation on Climate The conversion of forest to pasture in the tropics results in a number of changes in energy balance. (After Foley et al. 2003.)

and type of vegetation, as well as by soil and topography. Light-colored surfaces have the highest albedo. A coniferous forest has a darker color, and thus a lower albedo, than most types of bare soil or a dormant grassland, so it absorbs more solar energy. The "roughness" of Earth's surface is also influenced by vegetation. A smooth surface, such as a continuous grassland, allows greater transfer of energy to the atmosphere by wind (sensible heat loss by convection) than a rough surface, such as mixed forest and grassland. Finally, vegetation affects the movement of water into the atmosphere. The sum of water loss through **transpiration** (evaporation of water from inside the plant) by plants and evaporation from the soil, known as **evapotranspiration**, increases with the area of leaves per unit of area of ground surface. Evapotranspiration transfers energy (as latent heat) as well as water into the atmosphere, thereby affecting air temperature and moisture.

What happens to climate when the type or amount of vegetation is altered? What are the climatic consequences, for example, of the conversion of forest to pasture? This question has particular significance because of the current high rates of deforestation in the tropics: each year, between 6 and 7 million hectares of tropical forest are cleared worldwide (Wright 2005). Loss of the trees increases the albedo of the land surface as bare soil is exposed and the trees are partially replaced with lighter-colored grasses (**Figure 2.19**). The higher albedo lowers the absorption of solar radiation, resulting in less heating of the land surface. However, the lower heat gain from solar radiation is offset by lower cooling by evapotranspiration (lower latent heat transfer) due to loss of leaf area (Foley et al. 2003). Lower evapotranspiration not only reduces surface cooling, but also leads to lower precipitation because less moisture is returned from the ground surface to the atmosphere. Thus, the outcome of tropical defor-

estation may be a warmer, drier regional climate (Foley et al. 2003). Widespread deforestation may lead to climate change that is significant enough to inhibit reforestation, and may thus lead to long-term changes in tropical ecosystems.

In Chapter 24 we will return to human effects on climate change, focusing primarily on the past two centuries. Human activities, however, are not the only cause of long-term climate change. We turn next to the natural climatic variation that has occurred throughout Earth's history.

Climatic Variation over Time

CONCEPT 2.5
Seasonal and long-term climatic variation are associated with changes in Earth's position relative to the sun.

As noted at the beginning of this chapter, understanding variation in climate is critical to understanding ecological phenomena. Climatic variation at daily to decadal time scales determines the range of environmental conditions experienced by organisms as well as the supply of the resources and habitats they need to survive. Long-term climatic variation over hundreds and thousands of years influences the evolutionary history of organisms and the development of ecosystems. As we will see, the global climate has changed substantially over the course of Earth's history. In this section, we will review climatic variation from seasonal to 100,000-year time scales.

Equinox
September 22

Winter in northern
hemisphere; summer in
southern hemisphere.

Earth's orbit

FIGURE 2.20 Seasonal Changes in Climate
As Earth revolves around the sun, its orientation relative to the sun changes due to the tilt of its axis. The resulting changes in the intensity of solar radiation create seasonal climatic variation.

Summer in northern
hemisphere; winter in
southern hemisphere.

Solstice
December 21

23.5°

North Pole

Solstice
June 21

Equinox
March 20

has a wet season from October to February and a dry season from April to August.

The temperate and polar zones experience pronounced changes in temperature associated with variation in solar radiation over the year. Summer occurs in the Northern Hemi-

Seasonality results from the annual orbit of Earth around the sun

The amount of sunlight striking any point on Earth's surface varies as Earth makes its 365-day journey around the sun. Earth rotates on an axis that is tilted at an angle of 23.5° relative to the sun's direct rays (**Figure 2.20**). Thus, the angle and intensity of the sun's rays striking any point on Earth vary as Earth orbits the sun. This influence of the tilt of Earth's axis overrides the variation associated with seasonal changes in the distance between Earth and the sun due to Earth's slightly elliptical orbit. Earth is closest to the sun in January (at a point called the perihelion, 147 million miles) and farthest away in July (at the aphelion, 152 million miles). As we will see later, however, the effect of Earth–sun distance on climate is important at much longer time scales.

Seasonal climate change in the tropics is marked primarily by changes in precipitation, rather than by changes in temperature, because the seasonal changes in solar radiation are relatively small there compared with those in the temperate and polar zones. Seasonal changes in precipitation are associated with the movement of the zone of maximum solar radiation, known as the **Intertropical Convergence Zone** or **ITCZ**, within the tropical zone as Earth orbits the sun. The ITCZ moves from 23.5° N in June to 23.5° S in January, bringing the wet season wherever it goes (**Figure 2.21**). For example, Tampico, Mexico (22°12′ N), reaches its maximum precipitation levels from July to October, and has a dry season from November to April. In contrast, Vicosa, Brazil (20°45′ S),

FIGURE 2.21 Wet and Dry Seasons and the ITCZ Seasonality of precipitation in the tropics is associated with movement of the Intertropical Convergence Zone (ITCZ) as Earth orbits the sun.

sphere in June when it is tilted towards the sun, while the Southern Hemisphere is oriented away from the sun and enters its winter. The difference in solar radiation between summer and winter increases from the tropics toward the poles. The seasonal changes in the angle of the sun affect not only the intensity of solar radiation, but also the length of the day. Above 66.5° N and S latitudes, the sun may not set for several days, weeks, or even months in summer. At these same latitudes, the sun may not rise and warm the surface of Earth during winter. Because air temperatures regularly drop below freezing during winter in the temperate and polar zones, seasonality is an important determinant of biological activity and strongly influences the distributions of organisms.

Seasonal changes in aquatic environments are associated with changes in water temperature and density

Aquatic environments in the temperate and polar regions also experience seasonal changes in temperature. Liquid water has the unique property of being most dense at 4°C. Ice has a lower density than liquid water, and therefore forms on the surfaces of liquid water bodies in winter. Because it has a higher albedo than open water, ice on the surface of a lake or polar oceans effectively prevents warming of the water below it.

Differences in water temperature with depth result in the **stratification**, or layering, of water in oceans and lakes. Stratification has important implications for aquatic organisms because it determines the movement of nutrients and oxygen. Surface waters in lakes and oceans mix freely, but are underlain by colder, denser layers that do not mix easily with the surface waters. In oceans, the surface waters mix with the subsurface layers only rarely; for example, in upwelling zones. In temperate-zone lakes, however, seasonal changes in water temperature and density result in seasonal changes in stratification (**Figure 2.22**). In summer, the surface layer, or **epilimnion**, is the warmest, and contains active populations of phytoplankton and zooplankton. The epilimnion is underlain by a zone of rapid temperature decline, called the **thermocline**. Below the thermocline is a stable layer of the densest, coldest water in the lake, known as the **hypolimnion**.

During the fall, the air above the water surface cools, and the lake loses heat to the atmosphere through emission of infrared radiation and latent (evaporation) and sensible (conduction and convection) heat transfers. As the epilimnion cools, its density increases until it is the same as that of the layers below it. Eventually, water at all depths of the lake has the same temperature and density, and winds blowing on the surface lead to a mixing of epilimnion and hypolimnion, known as lake **turnover**. This mixing is important for recycling the nutrients that are lost from the epilimnion during summer. In addition, lake turnover moves oxygen into the

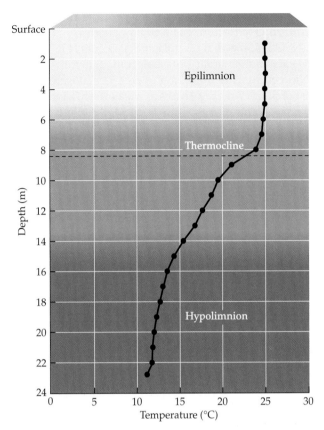

FIGURE 2.22 Lake Stratification   Lake stratification, which is most apparent in summer in temperate and polar regions, is a result of temperature effects on water density.

hypolimnion and the sediments at the lake bottom. The replenishment of oxygen, which has been used up by the respiration of aerobic bacteria during summer, increases biological activity in these lake zones. Turnover occurs in spring as well when the surface ice melts and the lake water has a uniform density once again.

Climatic variation over years and decades results from changes in atmospheric pressure cells

Peruvian fisherman and other inhabitants of the west coast of South America have long been aware of times when the normally productive ocean waters hold few fish and the weather becomes extremely wet. They named these climatic episodes "El Niño," for the Christ child, because they usually started around Christmas. El Niño events are associated with a switch (or oscillation) in the positions of high-pressure and low-pressure systems over the equatorial Pacific, leading to a weakening of the easterly trade winds that normally push warm water toward Southeast Asia. The underlying causes of this oscillation are still not well understood. Climatologists refer to this oscillation and the climate changes associated with it as the **El Niño Southern Oscillation**, or **ENSO** for short. The frequency of ENSO is somewhat irregular, but it occurs at intervals of 3–8 years and generally lasts for about 18 months. During El Niño events, the upwelling of deep ocean water off the coast of South America ceases as the easterly winds weaken, or in some events, shift to westerly winds (**Figure**

2.23A,B). ENSO also includes **La Niña** events, which are stronger-than-average phases of the normal pattern, with high pressure off the coast of South America and low pressure in the western Pacific. La Niña events usually follow El Niño events, but tend to be less frequent.

ENSO is associated (or "teleconnected") with unusual climatic conditions, even at localities distant from the tropical

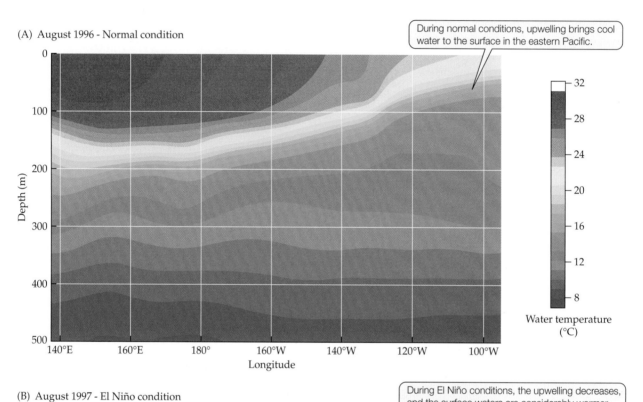

(A) August 1996 - Normal condition

During normal conditions, upwelling brings cool water to the surface in the eastern Pacific.

(B) August 1997 - El Niño condition

During El Niño conditions, the upwelling decreases, and the surface waters are considerably warmer.

FIGURE 2.23 El Niño Global Climatic Variation (A, B) Changes in ocean temperatures in equatorial waters of the Pacific under (A) normal conditions and (B) El Niño conditions. (C) El Niño events have widespread climatic effects, altering temperature and precipitation patterns at a global scale. Effects shown here are for September–March.

Pacific, through its complex interactions with broad-scale atmospheric circulation patterns (**Figure 2.23C**). El Niño events are associated with unusually dry conditions in Indonesia, other parts of Southeast Asia, and Australia. The likelihood of fires in the grasslands, shrublands, and forests of these areas increases as precipitation decreases and vegetation dries out. In contrast, in the southern United States and northern Mexico, El Niño events may increase precipitation, while the ensuing La Niña events bring drought conditions. The increase in plant growth associated with an El Niño event, followed by dry La Niña conditions, may intensify fires in the southwestern United States (Veblen et al. 2000).

Similar atmospheric pressure–ocean current oscillations occur in the North Atlantic Ocean. The **North Atlantic Oscillation** affects climatic variation in Europe, in northern Asia, and on the eastern coast of North America. A long-term oscillation in sea surface temperature and atmospheric pressure, known as the **Pacific Decadal Oscillation,** or **PDO**, has been discovered in the North Pacific. The PDO affects climate in ways similar to ENSO and can moderate or intensify the effects of ENSO. The effects of the PDO are felt primarily in northwestern North America, although southern parts of North America, Central America, Asia, and Australia may also be affected. We will return to the PDO at the end of this chapter.

Long-term climate change is associated with variation in Earth's orbital path

Antarctica today is almost completely covered with ice, and life is limited to its outer margins. Yet evidence from fossils indicates that there were once forests in interior parts of Antarctica, and that dinosaurs once roamed the continent. Based on our knowledge of the biology of trees and dinosaurs, we can conclude that there must have been a warmer climate in the Antarctic past. What could have caused the climate to be so much warmer? The latitudinal position of Antarctica has shifted over geologic time (over hundreds of millions of years) due to *continental drift*, the movement of land masses across the surface of Earth as they float on the mantle beneath the crust. Continental drift could have resulted in a warmer or cooler climate as Antarctica moved closer to or farther away from the equator. However, the age of many of the Antarctic fossils, along with a reconstruction of the movements of the continent, indicates that Antarctica was in a latitudinal position similar to today's when the forests and dinosaurs occurred there (190–65 million years ago). Thus, Earth's climate must have once been much warmer than it is now.

Many lines of evidence indicate that Earth has experienced several episodes of warmer and cooler climates over the past

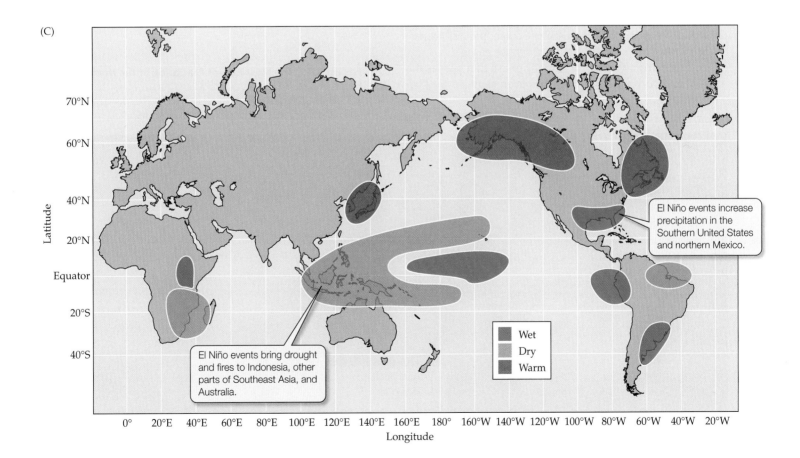

(C)

El Niño events increase precipitation in the Southern United States and northern Mexico.

El Niño events bring drought and fires to Indonesia, other parts of Southeast Asia, and Australia.

Wet
Dry
Warm

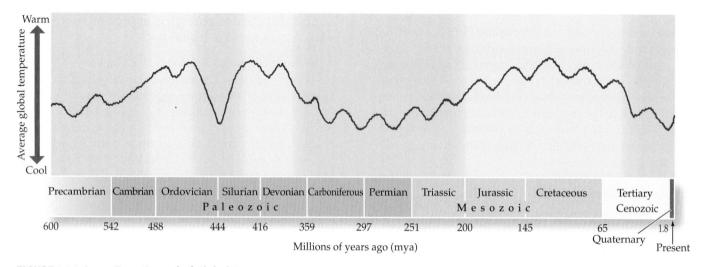

FIGURE 2.24 Long-Term Record of Global Temperature
Global climatic patterns have alternated between warm and cool
cycles for the past 500 million years.

500 million years (**Figure 2.24**). During the course of Earth's
history, the amount of radiation emitted by the sun has been
gradually increasing, so that factor does not explain these irregular cycles of long-term global climate change. Hypotheses for the causes of these climatic shifts have focused on
changes in the concentrations of greenhouse gases in the atmosphere. Warmer periods are associated with higher concentrations of greenhouse gases, while cooler periods are associated with lower concentrations of those gases.

Earth is currently in a cool climatic phase that has lasted
about 3 million years. This phase has been characterized by

regular periods of cooling, accompanied by the formation and
advance of glaciers, followed by periods of warming, accompanied by glacial melting. The peaks of glacial advance are
referred to as glacial maxima, and the periods of glacial melting and retreat are referred to interglacial periods. These glacial–interglacial cycles occur at frequencies of about 100,000
years. We are currently in an interglacial period; in the past,
these periods have lasted approximately 23,000 years. The
last glacial maximum was approximately 18,000 years ago
(**Figure 2.25**).

**FIGURE 2.25 The Most Recent Glaciation of the Northern
Hemisphere** (A) At the last glacial maximum, about 18,000 years
ago, ice sheets covered extensive areas of the Northern Hemisphere.
(B) Today's ice sheets are shown for comparison.

(A)

(B)

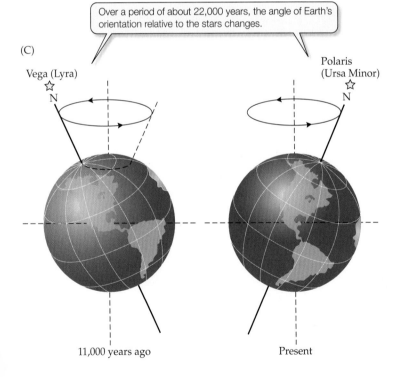

(C)

FIGURE 2.26 Milankovitch Cycles and Long-Term Climatic Variation Variations in (A) the shape of Earth's orbit around the sun, (B) the tilt of Earth's axis of rotation, and (C) the orientation of the axis to other celestial objects all affect the intensity and distribution of solar radiation striking Earth and correlate with glacial–interglacial cycles over the past 400,000 years.

at high latitudes. Currently, the shape of Earth's orbit is nearly circular, but it shifts between a circular and a more elliptical shape in regular cycles of 100,000 years (**Figure 2.26A**). When Earth has a more elliptical orbit, the distance between Earth and the sun at the aphelion is greater, and thus the intensity of solar radiation is less, accentuating seasonal variation in climate. The angle of tilt of Earth's axis also changes, in regular cycles of about 41,000 years (**Figure 2.26B**); the greater the angle of tilt, the greater the seasonal variation in solar radiation at Earth's surface. Currently, the axis is tilted at an angle of 23.5°, which is near the middle of its range of variation (24.5°–22.1°). Finally, Earth's orientation relative to other celestial bodies changes in regular cycles of about 22,000 years. Today, the North Pole is oriented toward Polaris, the North Star, but that has not always been the case; there have been other "North Stars" in the past. These changes in Earth's orientation influence the timing of the seasons by determining the hemisphere that receives more solar radiation (**Figure 2.26C**).

Long-term climatic cycles correlate well with the Milankovitch cycles (Hays et al. 1976). The dominant glacial–interglacial cycle has a period of 100,000 years, corresponding to the changes in the shape of Earth's orbit. Indications of smaller climatic fluctuations with 41,000-year and 22,000-year cycles have also been observed.

The Chemical Environment

CONCEPT 2.6
Salinity, acidity, and oxygen concentrations are major determinants of the chemical environment.

All organisms are bathed in a matrix of chemicals. Water is the primary chemical constituent of aquatic environments, along with variable amounts of dissolved salts and gases. Small differences in the concentrations of these dissolved chemicals can have important consequences for the functioning of aquatic organisms, as well as for plants and microorganisms that are dependent on the chemistry of water in the soil. Terrestrial organisms are immersed in a gaseous atmosphere that is relatively invariant, consisting primarily of nitrogen (78%), oxygen (20%), water vapor (1%), and argon (0.9%). The atmosphere also contains trace gases, including the greenhouse gases that play a critical role in Earth's energy balance, and pollutants derived from human activities,

What causes the regularity of these glacial–interglacial cycles? One hypothesis was proposed by the Serbian astrophysicist Milutin Milankovitch in the 1920s. He suggested that a combination of regular changes in the shape of Earth's orbit and the tilt of its axis, collectively known now as **Milankovitch cycles**, changed the intensity of solar radiation which can have important effects on atmospheric chemistry. We will discuss the effects of air pollutants and greenhouse gases in the final chapter of this book. In this section, we will review three common chemical factors that influence biological function in both aquatic and terrestrial ecosystems: salinity, acidity, and the availability of oxygen.

All waters contain dissolved salts

Salinity refers to the concentration of dissolved salts in water. Dissolved salts are important from a biological perspective because they influence properties of the water that affect the ability of organisms to absorb it. The salts also have direct influences on organisms as nutrients (as we will see in Chapter 21), and they may inhibit metabolic activity if their concentrations are too high or too low.

We often think about salinity in the context of ocean water because it is one of the saltiest solutions we come into contact with. Seventy percent of Earth's surface is ocean water, and the oceans contain 97% of the water on Earth. The salinity of the oceans varies between 33 to 37 parts per thousand; this variation is a result of evaporation, precipitation, and the freezing and melting of sea ice (**Figure 2.27**). The salinity of ocean surface waters is highest near the equator and lowest at high latitudes.

What are the salts that make water saline, and where do they come from? Ocean salts consist mainly of sodium, chloride, magnesium, calcium, sulfate, bicarbonate, and potassium. These salts come from gases emitted by volcanic eruptions early in Earth's history, when its crust was cooling, and from the gradual breakdown of minerals in the rocks that make up the crust. Most landlocked bodies of water become more saline over time, reflecting a balance between water inputs from precipitation, water losses due to evaporation, and inputs of salts. When these inland "seas" occur in arid areas, such as the Great Salt Lake or the Dead Sea, their salinity may exceed that of ocean water. The types of salts that contribute to salinity in these inland seas vary, reflecting the chemistry of the rock minerals that make up their basins.

FIGURE 2.27 Global Variation in Salinity at the Ocean Surface Variations in surface water salinity reflect the concentrating effect of evaporation, dilution by melting sea ice, and precipitation. (pss, practical salinity scale.)

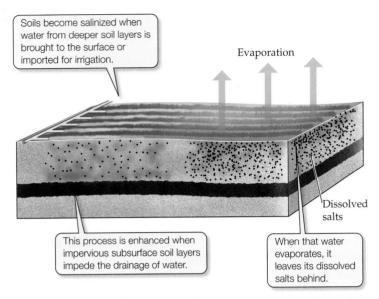

Soils become salinized when water from deeper soil layers is brought to the surface or imported for irrigation.

Evaporation

Dissolved salts

This process is enhanced when impervious subsurface soil layers impede the drainage of water.

When that water evaporates, it leaves its dissolved salts behind.

FIGURE 2.28 Salinization Salinization of soils is disrupting agricultural production in many areas, especially in arid regions.

High levels of salinity occur naturally in the water of soils adjacent to oceans, such as in salt marshes or tidal estuaries. Soils may also become more saline in arid regions as water from deeper soil layers is brought to the surface by plant roots or through pumping of groundwater for irrigation. If drainage of the water is impeded by impervious layers beneath the soil, high rates of evapotranspiration will result in a progressive buildup of salts at the soil surface. This process, known as **salinization**, occurs naturally in some desert soils, but may also occur in irrigated agricultural soils of arid regions (**Figure 2.28**). Salinization contributed to agricultural decline in ancient Mesopotamia (modern-day Iraq) and is a problem today in California's Central Valley, Australia, and other regions.

Organisms are sensitive to the acidity of their environment

Acidity, and its converse, **alkalinity**, refer to the ability of a solution to behave as an acid or a base, respectively. Acids are compounds that give up protons (H^+) to the water they are dissolved in. Bases take up protons or give up hydroxide ions (OH^-). Examples of common acids include citric, tannic, and ascorbic acids found in fruits. Examples of common bases include sodium bicarbonate (baking soda) and other carbonate minerals in rock. Acidity and alkalinity are measured in pH, which is equal to the negative of the logarithm ($-\log_{10}$) of the concentration of H^+. Thus, one pH unit represents a tenfold change in the concentration of H^+. Pure water has a neutral pH of 7.0. Solutions with pH values higher than 7.0 are alkaline, or basic, and solutions with pH values lower than 7.0 are acidic.

The pH values of water have important effects on organismal function. Changes in pH values can directly inhibit metabolic activity. The pH values of water also determine the chemistry and availability of nutrients, as we will see in Chapter 21. Animals, plants, and microorganisms have a limited range of pH values that they can tolerate. Natural levels of alkalinity tend not to be as important as acidity as a constraint on organismal function and distributions.

In the oceans, pH does not vary appreciably because the chemistry of seawater acts to buffer changes in acidity. Thus, pH tends to be most important in terrestrial and freshwater ecosystems. However, increases in atmospheric carbon dioxide (CO_2) concentrations are expected to increase the acidity of the oceans appreciably during the next century, with potentially negative effects on marine ecosystems (Orr et al. 2005).

The pH of surface waters and soils varies naturally. What causes this variation? Water can become more acidic over time through the input of acidic compounds derived from several sources, most associated with soil development. Two of the main components of soils are mineral particles from the breakdown of rocks and organic matter from the decomposition of dead plants and other organisms. Some rock types, such as granites, generate acidic salts, while other rock types generate basic salts. As soils age, they become more acidic because the basic minerals leach away more easily, and because plant organic matter adds acidic compounds to the soil. The emission of acidic pollutants into the atmosphere by the burning of fossil fuels, as well as overuse of agricultural fertilizers, can increase the acidity of soil and water. We will cover these topics in more detail in Chapter 24.

Oxygen concentrations vary with elevation, consumption, and diffusion

When life on Earth first evolved, there was no oxygen in the atmosphere, and even today, there are many organisms that are intolerant of oxygen. However, with the exception of some archaea, bacteria, and fungi, most organisms require oxygen to carry out their metabolic processes and cannot survive in **anoxic** (low-oxygen) conditions. Anoxic conditions can also promote the formation of chemicals (e.g., hydrogen sulfide) that are toxic to many organisms. In addition, oxygen levels are important for chemical reactions that determine the availability of nutrients.

Oxygen concentrations in the atmosphere have been stable at about 21% for the past 65 million years, so most terrestrial environments have invariant oxygen concentrations. However, the availability of atmospheric oxygen decreases with elevation above sea level. As we have seen, the overall density of air decreases with elevation, so there are fewer molecules of oxygen in a given volume of air at higher elevations. We will discuss the repercussions of this variation for human health in Chapter 4.

Oxygen concentrations can vary substantially in aquatic environments and in soils. The rate of diffusion of oxygen into water is slow and may not keep pace with its consumption by organisms. Waves and currents mix oxygen from the atmosphere into ocean surface waters, so its concentration is usually stable there. Oxygen concentrations are low in the deep ocean and in marine sediments, where biological uptake is greater than replenishment from surface waters. The same holds true in deep lakes, lake sediments, and flooded soils (e.g., in wetlands). Oxygen concentrations are highest in freshwater ecosystems with moving water (streams and rivers) because mixing with the atmosphere is greatest there.

CASE STUDY REVISITED

Climatic Variation and Salmon

The research of Steven Hare and Robert Francis on salmon production in the North Pacific contributed to the discovery of the Pacific Decadal Oscillation (PDO). As noted earlier, the PDO is a multi-decadal shift in sea surface temperature and atmospheric pressure. A review of existing records of sea surface temperatures during the twentieth century indicated that the PDO was associated with alter-

nating 20–30-year periods of "warm" and "cool" temperatures in the North Pacific (**Figure 2.29**). The length of the phases of the PDO differentiate it from other climatic oscillations whose phases tend to be much shorter (e.g., 18 months–2 years for ENSO). These warm and cool phases had influenced the marine ecosystems that the salmon depended on, and had thus shifted salmon production north or south, depending on the phase of the PDO.

The PDO has been linked to changes in the abundances and distributions of many marine organisms, and through its climatic effects, in the functioning of terrestrial ecosystems (Mantua and Hare 2002). Its effects have been found primarily in western North America and eastern Asia, but effects have been reported even in Australia. Thus, the influence of the PDO on climate reaches throughout the Western Hemisphere. Evidence for the existence of climatic changes associated with the PDO extends back to the 1850s, in the form of instrumental temperature records, and to the 1600s, in the form of corals and tree rings. The underlying mechanisms that cause the PDO are unclear, but its effect on climate is significant and widespread (**Table 2.1**).

FIGURE 2.29 The Pacific Decadal Oscillation and Ocean Temperatures Difference in ocean surface temperatures from average temperature during (A) the warm phase and (B) the cool phase of the Pacific Decadal Oscillation (PDO). (C) PDO phases since 1900, as estimated using a PDO index based on sea surface temperatures.

TABLE 2.1

Summary of Climatic Effects of the Pacific Decadal Oscillation (PDO)

Climatic effect	Warm phase PDO	Cool phase PDO
Ocean surface temperature in the northeastern and tropical Pacific	Above average	Below average
October–March northwestern North American air temperature	Above average	Below average
October–March southeastern U.S. air temperature	Below average	Above average
October–March southern U.S./northern Mexico precipitation	Above average	Below average
October–March northwestern North American and Great Lakes precipitation	Below average	Above average
Northwestern North American spring snowpack and water year (October–September) stream flow	Below average	Above average
Winter and spring flood risk in the Pacific Northwest	Below average	Above average

Source: Mantua 2001.

CONNECTIONS IN NATURE

Climatic Variation and Ecology

Two aspects of the PDO are particularly significant in the context of ecology. First, the realization that the PDO existed was driven initially by an attempt to understand variation in a biological phenomenon; namely, salmon abundance. This observation underscores the relationship between climate, the functioning of organisms and their growth and reproduction, and population and community processes. This relationship is one of the central themes of ecology that will form a common thread throughout this book. Ultimately, the physical environment, including climate and the myriad factors, such as the PDO, that control it, determines whether an organism can exist in a given location. Extremes in the physical environment, including

those that are driven by climatic oscillations, play a critical role in our understanding of ecological phenomena.

Second, the time scale of the climatic variation associated with the PDO is long relative to the human life span. The abrupt changes in climate and the associated ecological responses of the marine ecosystem were therefore perceived by people as unusual events. Indeed, the phases of the PDO may be longer than the life spans of most of the organisms affected by it, limiting their evolutionary responses to this cyclic climate change. As a result, from the perspective of an ecological community, the PDO represents a disturbance, an event that detrimentally affects some species' populations. However, although we don't understand what causes it, the PDO has been a part of the climate system for at least the last 400 years, and a better understanding of its effects will help us place other climatic phenomena, including global warming, in perspective.

SUMMARY

CONCEPT 2.1 Climate is the most fundamental characteristic of the physical environment.

■ Weather refers to the current conditions of temperature, precipitation, humidity, wind, and cloudiness, while climate is the long-term average weather at a given location.

■ Climate determines the geographic ranges and physiological functioning of organisms and the resources that are available to them.

■ Climate is driven by the balance of energy gains from solar radiation and reradiation from the atmosphere and energy losses due to radiation from Earth's surface, latent heat flux, and sensible heat flux.

CONCEPT 2.2 Winds and ocean currents result from differences in solar radiation across the surface of Earth.

■ Latitudinal differences in the intensity of solar radiation at Earth's surface establish atmospheric circulation cells.

■ The Coriolis effect and the differential heat capacity of ocean waters and land act on atmospheric circulation patterns to determine the direction of prevailing winds at Earth's surface.

■ Ocean currents are driven by surface winds and by differences in water temperature and salinity.

■ Winds and ocean currents transfer energy from the tropics to higher latitudes.

CONCEPT 2.3 Large-scale atmospheric and oceanic circulation patterns establish global patterns of temperature and precipitation.

- Global temperature patterns are determined by latitudinal variation in solar radiation, as well as by oceanic circulation patterns and by the distribution of continents.
- Temperature decreases as the elevation of the land surface increases.
- Global patterns of terrestrial precipitation are determined by atmospheric circulation cells, but are also influenced by semipermanent pressure cells.

CONCEPT 2.4 Regional climates reflect the influence of the distribution of oceans and continents, elevation, and vegetation.

- Seasonal variation in temperature is greater in the middle of a continent than on the coast because ocean water has a higher heat capacity than land.
- Mountains force air masses passing over them to rise and drop most of their moisture as precipitation, resulting in moister environments on windward slopes and drier environments on leeward slopes.
- Vegetation influences climate through its effects on energy exchange associated with albedo, evapotranspiration (latent heat transfer), and surface winds (sensible heat transfer).

CONCEPT 2.5 Seasonal and long-term climatic variation are associated with changes in Earth's position relative to the sun.

- The tilt of Earth's axis as it orbits the sun causes seasonal changes in precipitation in tropical regions and in temperatures in temperate and polar regions.

- Temperature-induced differences in water density result in nonmixing layers of water in oceans and lakes. In temperate-zone lakes, these layers break down in fall and spring, allowing the movement of oxygen and nutrients.
- Variations in climate over years to decades are caused by cyclic changes in atmospheric pressure systems, which have widespread effects beyond the regions where the pressure systems are located.
- Long-term climatic cycles over hundreds and thousands of years are associated with changes in the shape of Earth's orbit, the angle of tilt of its axis, and its orientation to other celestial bodies.

CONCEPT 2.6 Salinity, acidity, and oxygen concentrations are major determinants of the chemical environment.

- The salinity of Earth's waters, including water in soils, is determined by the balance between inputs of salts and gains (precipitation) and losses (evapotranspiration) of water.
- The pH of soils and surface waters is determined by inputs of salts from the breakdown of rock minerals, organic acids from plants, and pollutants.
- Oxygen concentrations are stable in most terrestrial ecosystems, but oxygen availability decreases as elevation increases. Concentrations of oxygen in aquatic ecosystems are low where consumption by organisms exceeds its slow rate of diffusion into water.

REVIEW QUESTIONS

1. Why is the variability of physical conditions potentially more important than average conditions as a determinant of ecological patterns, such as species distributions?

2. Describe the factors that determine the major latitudinal climatic zones (the tropics, temperate zones, and polar zones).

3. How do the following features affect regional terrestrial climates?
   a. Mountains
   b. Oceans

4. Why would deserts be more prone to salinization from irrigation than areas with greater precipitation?

HANDS-ON PROBLEM SOLVING

Effect of El Niño Events on Tuna Populations | This Web exercise (www.sinauer.com/ecology) shows the effect of the El Niño Southern Oscillation (ENSO) on tuna populations in the western Pacific. You will investigate whether changes in the locations of tuna populations covary with displacements of warm water that occur during ENSO events.

SUGGESTED READINGS

Ahrens, C. D. 2005. *Essentials of Meteorology: An Invitation to the Atmosphere*. 4th ed. Thomson Brooks/Cole, Belmont, CA. *An excellent introduction to the fundamentals of climate and weather.*

The following two textbooks cover the basics of the physical environment, including atmospheric, terrestrial, and aquatic systems important to ecology:

Christopherson, R. W. 2006. *Geosystems: An Introduction to Physical Geography*. 6th ed. Prentice Hall, Upper Saddle River, NJ.

Strahler, A. and A. Strahler. 2005. *Physical Geography*. 3rd ed. John Wiley and Sons, Hoboken, NJ.

Foley, J. A., M. H. Costa, C. Delire, N. Ramankutty and P. Snyder. 2003. Green surprise? How terrestrial ecosystems could affect Earth's climate. *Frontiers in Ecology and Environment* 1: 38–44. *Examines the possible feedbacks of tropical deforestation to the climate system, providing an intuitive presentation of the role of plants in surface–atmosphere energy exchange.*

Stenseth, N. C., A. Myserud, G. Ottersen, J. W. Hurrell, K. Chan and M. Lima. 2002. Ecological effects of climate fluctuations. *Science* 297: 1292–1296. *A fascinating review of the effects of climatic fluctuations associated with ENSO and the North Atlantic Oscillation on ecological factors such as population changes and predation in both terrestrial and marine ecosystems.*

CHAPTER 3 The Biosphere

CASE STUDY

The American Serengeti: Twelve Centuries of Change in the Great Plains

Today, the region covering the central part of North America, known as the Great Plains, bears little resemblance to the Serengeti Plain of Africa. Biological diversity is very low in many parts of the current landscape, which contain large stands of uniform crop plants (which are often even the same genetically) and a few species of domesticated herbivores. In the Serengeti, on the other hand, some of the largest and most diverse herds of wild animals in the world roam a picturesque savanna (**Figure 3.1**). If not for a series of important environmental changes, however, the two ecosystems might look very similar.

Temperate and high-latitude biological communities have been subjected to natural, long-term climate changes, leading to latitudinal or elevational shifts in their positions and biological composition. Eighteen thousand years ago, at the last glacial maximum of the Pleistocene epoch, ice sheets covered the northern portion of North America. Over the next 12,000 years, the climate warmed, and the ice receded. Vegetation followed the retreating ice northward and colonized the newly exposed substrate. Grasslands in the center of the continent expanded into former spruce and aspen woodlands. These grasslands contained species of grasses, sedges, and low-growing herbaceous plants similar to those found in the grasslands that exist today.

The animal inhabitants of these earlier grasslands were, however, strikingly different from today's. A diverse collection of megafauna (animals larger than 45 kg, or 100 pounds) existed in North America, rivaling the diversity found today in the Serengeti (Martin 2005) (**Figure 3.2**). Thirteen thousand years ago—a relatively short time in an evolutionary context—North American herbivores included woolly mammoths and mastodons (relatives of elephants), as well as several species of horses, camels, and giant ground sloths. Predators included saber-toothed cats with 18 cm (7-inch) incisors, cheetahs, lions, and giant short-faced bears that were larger and faster than grizzly bears.

FIGURE 3.1 The Serengeti Plain of Africa Large, diverse herds of native animals migrate across the Serengeti in search of food and water.

FIGURE 3.2 Pleistocene Animals of the Great Plains The animals of the grasslands of central North America 13,000 years ago included woolly mammoths, horses, and wolverines. Many of these large mammals went extinct within a short time between 13,000 and 10,000 years ago.

About 10,000–13,000 years ago, as the extensive grasslands of the Great Plains were developing, many of the large mammals of North America went extinct (Barnosky et al. 2004). The rapidity of the disappearance of approximately 28 genera (40–70 species) made this extinction unlike any previous extinction event during the previous 65 million years (the Cenozoic period). Another unusual aspect of this extinction was that nearly all of the animals that went extinct belonged to the same group: large mammals. The causes of the extinction are a mystery to paleontologists (scientists who investigate the development of life on Earth through the study of fossils and other ancient remains).

Several hypotheses have been proposed to account for the disappearance of the North American megafauna. Changes in the climate during the extinction period were rapid and could have led to changes in habitat or food supply that would have negatively affected the animals. Another hypothesis that has generated substantial controversy suggests that the arrival of humans in North America may have hastened the demise of the animals (Martin 1984). When this hypothesis was first proposed, it was met with widespread skepticism, and the initial supporting evidence was considered weak. Although humans first appeared in the central part of North America about 14,000 years ago, it is unclear how hunters bearing stone and wooden tools could have driven so many species of large mammals to extinction. What evidence is there to support the hypothesis that humans were involved in this extinction event?

L iving things can be found in remarkable places. Birds such as ravens, lammergeyers (Eurasian vultures), and alpine choughs (crows) fly over the highest summits of the Himalayas, over 8,000 m above sea level. Fish (e.g., *Abyssobrotula galatheae*, the "fangtooth") can be found over 8,000 m below the ocean surface. Bacteria and archaea can be found almost everywhere on Earth, in hot sulfur springs at the extreme chemical and temperature limits for life, under glaciers, and on dust particles many kilometers above Earth's surface. However, most living things occur in a range of habitats that cover a thin veneer of Earth's surface, from the tops of trees to the surface soil layers in terrestrial environments, and within 200 meters of the surface of the oceans.

The **biosphere**, the zone of life on Earth, is sandwiched between the **lithosphere**, Earth's surface crust and upper mantle that includes the tectonic plates, and the *troposphere*, the lowest layer of the atmosphere. In this chapter, we will introduce the broad categories of biota that make up the biosphere. Biological communities can be categorized at multiple scales of varying complexity. Here, we will use **biomes** as the unit to introduce the diversity of terrestrial life on Earth. The diversity of aquatic life is not as easily categorized, but we will describe several freshwater and marine biological zones, which, like terrestrial biomes, reflect the physical conditions where they are found.

Terrestrial Biomes

CONCEPT 3.1
Terrestrial biomes are characterized by the dominant growth forms of vegetation.

Biomes are large biological communities shaped by the physical environment they are found in. In particular, they reflect the climatic variation described in Chapter 2. Biomes are categorized by the most common forms of organisms, usually plants, distributed across large geographic areas. The determination of biomes does not incorporate taxonomic similarities among organisms; instead, it relies on similarities in the morphological responses of organisms to the physical environment. Biomes occur as similar biological assemblages on distant continents, indicating similar responses to climatic forces. Biomes provide a useful introduction to the diversity of life on Earth. Biomes also provide a convenient unit for modelers simulating the effects of climate change on biological communities, as well as the effects of the biota on the climate system.

Terrestrial communities vary considerably from the warm, wet tropics to the cold, dry polar regions. Tropical forests have multiple verdant layers, high growth rates, and tremendous species diversity. Lowland tropical forests in Borneo have an estimated 10,000 species of vascular plants, and most other tropical forest ecosystems have about 5,000 species. In contrast, polar deserts have a scattered cover of tiny plants clinging to the ground, reflecting a harsh climate of high winds, low temperatures, and dry soils. High-latitude Arctic communities contain about 100 species of vascular plants. Tropical rainforest vegetation may reach over 75 m in height and contain over 400,000 kg of surface biomass (the mass of all living organisms, in this case primarily plant mass), in a single hectare. Plants of polar deserts, on the other hand, rarely exceed 5 cm in height and have less than 1,000 kg of surface biomass per hectare.

Terrestrial biomes are classified by the growth form of the most abundant plants (such as trees, shrubs, or grasses) (**Figure 3.3**). In addition, characteristics of the leaves may be used, such as *deciduousness* (seasonal shedding of leaves), thickness, and *succulence* (development of fleshy water storage tissues). Why use plants rather than animals to catego-

Growth form **Environment**

Seasonally dry/moist and warm/cool

Sclerophyllous shrubs have tough, leathery leaves.

Sclerophyllous shrubs

FIGURE 3.3 Plant Growth Forms The size and morphology of a plant constitute its "growth form." The growth form is an evolutionary response to the environment, particularly climate and soil fertility.

Growth form **Environment**

Moist, seasonally warm/cool, or cool/cold on fertile soils or warm, seasonally wet/dry

Deciduous plants drop their leaves during cold or dry periods.

Deciduous trees

Growth form **Environment**

Dry, seasonally hot/cool

Succulent stems and leaves contain water storage tissues.

Cacti and shrubs; succulent stems or leaves

Growth form **Environment**

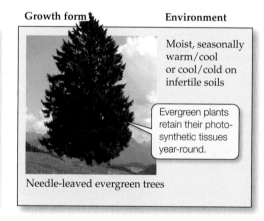

Moist, seasonally warm/cool or cool/cold on infertile soils

Evergreen plants retain their photosynthetic tissues year-round.

Needle-leaved evergreen trees

Growth form **Environment**

Moist, seasonally warm/cool, with fire

Grasses grow from the base of their leaves.

Grasses, sedges

Growth form **Environment**

Wet, warm year–round

Evergreen leaves in tropical regions carry out photosynthesis year-round.

Evergreen broad-leaved trees

Growth form **Environment**

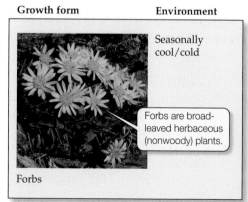

Seasonally cool/cold

Forbs are broad-leaved herbaceous (nonwoody) plants.

Forbs

rize terrestrial biomes? Plant growth forms are good indicators of the physical environment, reflecting the climatic zones discussed in Chapter 2 as well as rates of disturbance (e.g., fire frequency). In addition, plants are immobile. In order to successfully occupy a site for a long time, plants must be able to cope with its environmental extremes as well as its biological pressures, such as competition for water, nutrients, and light. Thus, plants are excellent integrators of their physical and biological environments. Animals are a less visible component of most large landscapes, and their mobility allows them to avoid exposure to adverse environmental conditions. Microorganisms (archaea, bacteria, and fungi) are important components of biomes, and their species composition reflects physical conditions in a manner similar to that of plants, but their tiny size, as well as rapid temporal and spatial changes in their composition, makes them impractical for classifying biomes.

Since their emergence from the oceans about 500 million years ago, plants have taken on a multitude of different forms in response to the selection pressures of the terrestrial environment (see Figure 3.3). These selection pressures include aridity, high and subfreezing temperatures, intense solar radiation, grazing by terrestrial animals, and crowding by neighbors. Having deciduous leaves, for example, is one solution to seasonal exposure to subfreezing temperatures or extended dry periods. Trees and shrubs invest significant energy in woody tissues in order to increase their height and ability to capture sunlight and to protect their tissues from damage by wind or large amounts of snow. Perennial grasses, unlike most other plants, can grow from the bases of their leaves and keep their vegetative and reproductive buds below the soil surface, which facilitates their tolerance of grazing, fire, high and subfreezing temperatures, and dry soils. Similar plant growth forms appear in similar climatic zones on different continents, even though the plants may not be genetically related. The evolution of similar growth forms among distantly related species in response to similar selection pressures is called **convergence**.

Terrestrial biomes reflect global patterns of precipitation and temperature

Chapter 2 described Earth's climatic zones as a consequence of atmospheric and oceanic circulation patterns resulting from the differential heating of Earth's surface by the sun. These climatic zones are major determinants of the distribution of terrestrial biomes.

The tropics (between 23.5° N and S) are characterized by high rainfall and warm, invariant temperatures. In the subtropical regions that border the tropics, rainfall becomes more seasonal, with pronounced dry and wet seasons. The major deserts of the world are associated with zones of high pressure at about 30° N and S and with the rain shadows of large mountain ranges. Subfreezing temperatures during winter are an important climatic feature of the temperate and polar zones. The amount of precipitation north and south of 40° varies depending on proximity to the ocean and the influence of mountain ranges (see Figure 2.16).

The locations of terrestrial biomes are correlated with these variations in temperature and precipitation. Temperature influences the distribution of plant growth forms directly through its effect on the physiological functioning of plants. Precipitation and temperature act in concert to influence the availability of water and its rate of loss by plants. Water availability and soil temperature are important for determining the supply of nutrients in the soil, which is also an important control on plant growth form.

The association between climatic variation and terrestrial biome distribution can be visualized using a graph of average annual precipitation and temperature (**Figure 3.4**). While these two factors predict biome distributions reasonably well, this approach fails to incorporate seasonal variation in temperature and precipitation. As we saw in Chapter 2, climatic extremes are sometimes more important in determining species distributions than mean annual conditions. For example, grasslands and shrublands have wider global distributions than indicated in Figure 3.4, occurring in regions with relatively high average annual precipitation, but with regular dry peri-

FIGURE 3.4 Biomes Vary with Mean Annual Temperature and Precipitation When plotted on a graph of temperature and precipitation, the nine major terrestrial biomes form a triangle.

ods (e.g., Mediterranean shrublands; grasslands at the margins of deciduous forests).

The potential distributions of terrestrial biomes differ from their actual distributions due to human activities

The effects of land conversion and resource extraction by humans are increasingly apparent on the land surface. These human effects are collectively described as **land use change**. Human modification of terrestrial ecosystems began at least 10,000 years ago with the use of fire as a tool to clear forests and enhance the size of game populations. The greatest changes have occurred over the last 150 years, with the onset of mechanized agriculture and logging and an exponential increase in the human population (see Figure 9.2) (Harrison and Pearce 2001). Between 50% and 60% of Earth's land surface has been altered by human activities, primarily agriculture, forestry, and livestock grazing; a smaller amount (2%–3%) has been transformed by urban development and transportation corridors (Harrison and Pearce 2001, Sander-

son et al. 2002). As a result of these human influences, the potential and the actual distributions of biomes are markedly different (**Figure 3.5**). Temperate biomes, particularly grasslands, have been transformed the most, although tropical and subtropical biomes are experiencing rapid change as well.

In the following section, we will briefly describe nine terrestrial biomes, their biological and physical characteristics, and the human activities that influence the actual amount of natural vegetation cover that remains in each biome. Each biome section begins with a map of the potential geographic distribution of the biome, an exemplory climate diagram showing the characteristic seasonal patterns of air temperature and precipitation, and a couple of photos illustrating the vegetation types that make up the biome. It is important to remember that each biome incorporates a mix of different

FIGURE 3.5 Global Biome Distributions Are Affected by Human Activities The potential distributions of biomes differ from their actual distributions because human activities have altered so much of Earth's land surface. (A) The potential global distribution of biomes. (B) The distribution of human-dominated agricultural systems. (C) The distribution of urban and suburban developed land.

(A)

Equator

Polar ice	Temperate deciduous forest	Temperate shrubland and woodland
Tundra	Temperate evergreen forest	Temperate grassland
Boreal forest	Tropical seasonal forest	Desert
Mountain zone	Tropical rainforest	

plant and animal communities. Boundaries between biomes can be complex due to variations in regional climatic influences, soil types, topography, and the occurrence of distur- bances. Thus, the boundaries of biomes portrayed here are approximate, provided to simplify the discussion of their global positions.

(B)

(C)

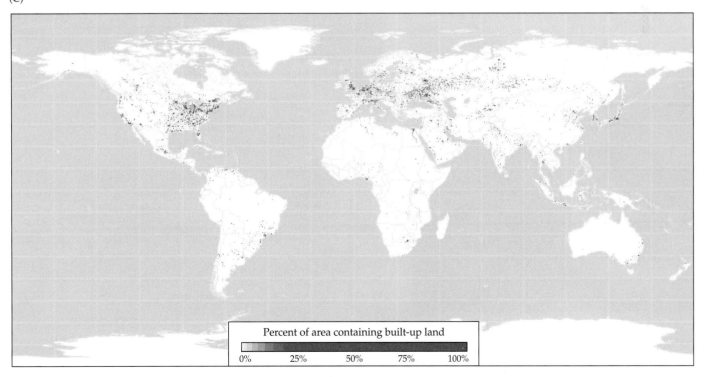

Tropical Rainforests Rainforests are aptly named, as they are found in low-latitude regions (between 10° N and S) with precipitation usually exceeding 200 cm annually and reaching over 5 m in some forests. Tropical rainforests experience warm, seasonally invariant temperatures. The abundant precipitation may be spread evenly throughout the year or occur in one or two main peaks associated with the movement of the Intertropical Convergence Zone (ITCZ) (see Figure 2.21). Seasonal climatic rhythms are generally absent from this biome, and plants grow continuously throughout the year. As indicated earlier, tropical rainforests contain a substantial amount of living plant biomass, and they include the most productive ecosystems on Earth. They contain an estimated 50% of Earth's species in only about 11% of its terrestrial vegetation cover (Dirzo and Raven 2003). Tropical rainforests occur in Central and South America, Africa, Australia, and Southeast Asia.

The tropical rainforest biome is characterized by broad-leaved evergreen and deciduous trees. Light is a key environmental factor determining the vegetation structure of this biome. Climatic conditions favoring plant growth dictate an ecological pressure to grow tall above neighboring plants or

TROPICAL RAINFORESTS

Trees, shrubs, and epiphytes crowd a rainforest below the canopy, Napo, Ecuador

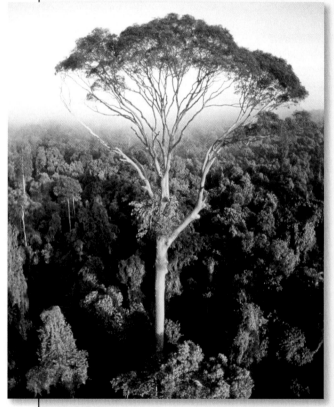

Emergent tree with epiphytes rises above the forest canopy, Borneo

Yanganbi, Zaire
0°, 487 m
Annual average temperature 24.6°C
Total annual precipitation 1,828 mm

June 19, 1975

August 1, 1986

Forest cutting proceeds outward from both sides of roads, producing a "fish skeleton" pattern.

June 22, 1992

February 7, 2001

FIGURE 3.6 Tropical Deforestation   These satellite photos show the extent of rainforest clearing in Rondônia, Brazil, over a 26-year period, 1975–2001.

to adjust physiologically to low light levels. As many as five layers of plants occur in tropical rainforests, including *emergent trees* that rise above the majority of the other trees that make up the *canopy* of the forest. This forest canopy consists primarily of the leaves of evergreen trees, which form a continuous layer approximately 30 to 40 m above the ground. Below the canopy, plants that utilize the canopy trees for support and to elevate their leaves above the ground are found draped or clinging, including *lianas* (woody vines) and *epiphytes* (plants that grow on the branches of the canopy and emergent trees). *Understory* plants grow in the shade of the canopy and emergent trees, further reducing the light that finally reaches the forest floor. *Shrubs* and *forbs* occupy the forest floor, where they must cope with the low availability of light for photosynthesis, relying on light flecks that move across the forest floor during the day.

Globally, tropical rainforests are disappearing rapidly due to logging and conversion of forests to pasture and croplands (**Figure 3.6**). Approximately half of the tropical rainforest biome has been altered. Rainforests in Africa and Southeast Asia have been altered the most, and rates of deforestation continue to be greatest in these areas (Wright 2005). In some cases, rainforests have been replaced by disturbance-maintained pastures of forage grasses, while in other cases the rainforest is regrowing. The recovery of the previous rainforest structure is uncertain, however. Rainforest soils are often nutrient-poor, and recovery of nutrient supplies may take a very long time, hindering forest regrowth.

Tropical Seasonal Forests and Savannas   Moving to the north and south of the wet tropics toward the Tropics of Capricorn (23.5° S) and Cancer (23.5° N), rainfall becomes seasonal, with pronounced wet and dry seasons associated with shifts in the ITCZ. This region is marked by a large gradient in climate primarily associated with the seasonality of rainfall. The vegetation response to this climatic gradient includes shorter stature, lower tree densities, and an increasing degree of drought deciduousness, with leaves dropping from the trees during the dry season. In addition, there is a greater abundance of grasses and shrubs with fewer trees relative to rainforests.

The tropical seasonal biome includes a complex of tree-dominated systems, including *tropical dry forests*, *thorn woodlands*, and *tropical savannas*. The frequency of fires increases with the length of the dry season, influencing the vegetation growth forms. Recurrent fires, sometimes set by humans, promote the establishment of **savannas**, communities dominated by grasses with intermixed trees and shrubs. In Africa, large herds of herbivores, such as wildebeests, zebras, elephants, and antelopes, also influence the balance between trees and grasses and act as an important force promoting the establishment of savannas. On the floodplains of the Orinoco River in South America, seasonal inundation con-

TROPICAL SEASONAL FORESTS AND SAVANNAS

Savanna in Natal, South Africa

Coxim, Brazil
18°S, 287 m
Annual average temperature 24.5°C
Total annual precipitation 1,493 mm

Semi-evergreen forest including Pijio trees (*Cavanillesia platanifolia*), during the dry season, Cerro Blanco, Ecuador

tributes to the establishment of savannas, as trees are intolerant of long periods of soil saturation. Thorn woodlands (communities dominated by widely spaced trees and shrubs) get their name from the heavy armaments of thorns on the trees, which act as a deterrent to herbivores that would consume the vegetation. Thorn woodlands typically occur in regions with climates intermediate between tropical dry forests and savannas.

Seasonal tropical forests and savannas once covered an area greater than tropical rainforests. Increasing human demand for supplies of wood and land for agriculture has resulted in rates of loss of seasonal tropical forests and savannas equal to or greater than those for tropical rainforests (Bullock et al. 1995). Less than half of this biome remains intact. Large increases in human populations in tropical dry forest regions have had a particularly large effect. Large tracts of tropical dry forest in Asia and Central and South America have been converted to cropland and pasture to meet the needs of growing human populations for food and earnings from economic exports to more developed countries.

Hot Deserts In contrast to the tropical ecosystems we have just discussed, deserts contain sparse populations of plants and animals, reflecting the constraints of high temperatures and low water availability. The subtropical positions of hot deserts generally correspond to the descending air of Hadley cells (see Figure 2.8). The descending air creates zones of high pressure around 30° N and S, which inhibit the formation of storms and their associated precipitation. The low precipitation levels, coupled with high temperatures and high rates of evapotranspiration, result in a limited supply of water for desert organisms. The major desert zones include the Sahara and Arabian deserts, the Atacama desert of Chile and Peru, and the Chihuahuan, Sonoran, and Mojave deserts of North America.

Low water availability is an important constraint on the abundance of desert plants as well as an important influence on their form and function. One of the best examples of convergence in plant form is the occurrence of stem succulence in desert plants. Stem succulence occurs in both the cacti of the Western Hemisphere and the euphorb family of the East-

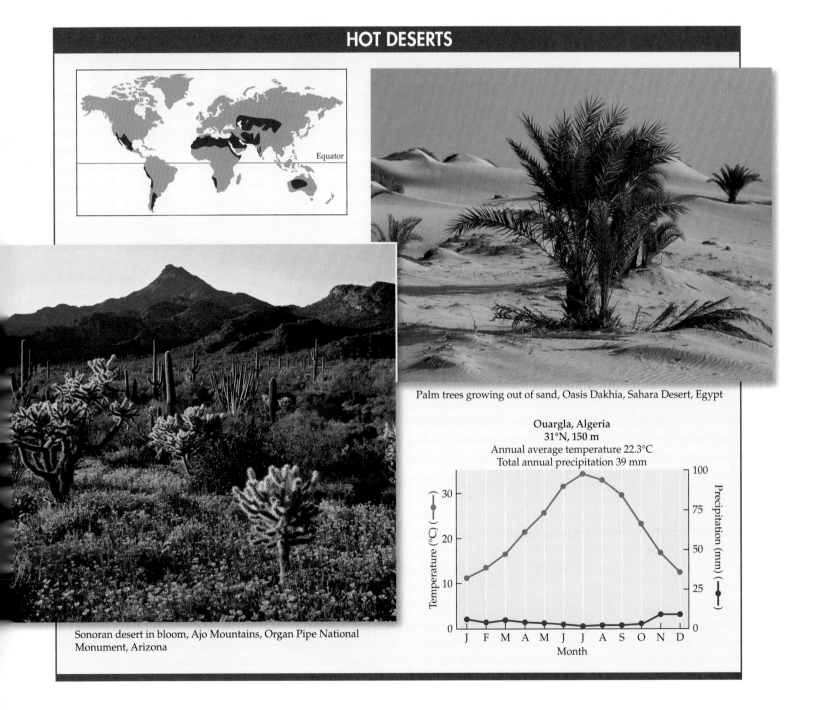

HOT DESERTS

Equator

Palm trees growing out of sand, Oasis Dakhia, Sahara Desert, Egypt

Sonoran desert in bloom, Ajo Mountains, Organ Pipe National Monument, Arizona

Ouargla, Algeria
31°N, 150 m
Annual average temperature 22.3°C
Total annual precipitation 39 mm

(A)

(B)

FIGURE 3.7 Convergence in the Forms of Desert Plants (A) The blue candle cactus (*Myrtillocactus geometrizans*) is native to the Chihuahuan Desert of Mexico. (B) The candelabra plant (*Euphorbia lactea*) is native to arid regions of Asia and India. These two plants both have succulent stems, water-conserving photosynthetic pathways, upright stems that minimize midday sun exposure, and spines that protect them from herbivores. However, the two plants are only distantly related, and these traits evolved independently in each species.

ern Hemisphere (**Figure 3.7**). Plants with succulent stems can store water in their tissues to help the plant continue to function during dry periods. Plants of the desert biome also include drought-deciduous shrubs, grasses, and short-lived annual plants that are active only after sufficient precipitation has fallen prior to the spring growing season. Although the abundance of organisms may be low, species diversity can be high in some deserts. The Sonoran Desert, for example, has over 4,500 plant species, 1,200 bee species, and 500 bird species (Nabhan and Holdsworth 1998).

Humans have used deserts for livestock grazing and agriculture for centuries. Agricultural development in desert areas is dependent on irrigation, often using water derived from distant mountains or extracted from deep underground. Unfortunately, irrigated agriculture in deserts has repeatedly failed due to salinization of soils (see p. 45). Livestock grazing in deserts is also a risky venture due to the unpredictable nature of the precipitation needed to support plant growth for herbivores. Long-term droughts in association with unsustainable grazing can result in the loss of plant cover and soil erosion, a process known as **desertification**. Desertification is a concern in populated regions at the margins of deserts, such as the Sahel to the south of the Sahara in Africa.

Temperate Grasslands Large expanses of grasslands once occurred throughout North America and Eurasia (the Great Plains and the steppes of Central Asia) at latitudes between 30° and 50° N. Southern Hemisphere grasslands (pampas) are found at similar latitudes on the east coasts of South America, New Zealand, and Africa. These vast, undulating expanses of grass-dominated landscape have often been compared to a terrestrial ocean, with wind-driven "waves" of plants bending to the gusts blowing through them.

Temperate climates have greater seasonal temperature variation than tropical climates, with an increasing period of wintertime subfreezing temperatures toward the poles. Within the temperate climatic zone, grasslands are usually associated with warm, moist summers and cold, dry winters. Precipitation in some grasslands is high enough to support forests, as at the eastern edge of the Great Plains. However, frequent fires and grazing by large herbivores such as bison prevent the establishment of trees and thus maintain the dominance of grasses in these environments. The use of fire to manage grasslands near the edges of forests was probably one of the first human activities with a widespread effect on the terrestrial biosphere.

The grasslands of the world have been a major focus for agricultural and pastoral development. In order to cope with dry conditions and acquire enough water, grasses grow more roots than stems and leaves. The rich organic matter that accumulates in the soils of many grasslands over time enhances their fertility, so these soils are particularly well suited for agricultural development. As a result, most of the fertile grasslands of central North America and Eurasia have been converted to agriculture. The diversity of the crop species grown is far less than the diversity of the grasslands they replaced. In more arid grasslands, rates of grazing by domesticated animals can exceed the capacity of the plants to produce new tissues, and grassland degradation, including desertification, has occurred. As in deserts, irrigation of some grassland soils has resulted in salinization, decreasing their fertility over time. In parts of Europe, cessation of centuries-old grazing practices has resulted in increased forest invasion into grasslands. This long legacy

Western edge of the Great Plains, Arapaho National Wildlife Refuge, Colorado

Denison, Nebraska, USA
41°N, 389 m
Annual average temperature 9.1°C
Total annual precipitation 727 mm

Grassland with camomile flowers, Altai Plateau, Russia

of grassland use for agriculture and grazing has made grasslands one of the most human-influenced biomes on Earth.

Temperate Shrublands and Woodlands   The seasonality of precipitation is an important control on the distribution of temperate biomes. Shrublands and woodlands occur in regions with a winter-dominant precipitation regime. *Mediterranean-type climates*, which occur on the west coasts of the Americas, Africa, Australia, and Europe between 30° and 40° N and S, are an example of such a climate regime. As we saw in Chapter 2, these Mediterranean-type climates are characterized by asynchrony between precipitation and the summer growing season. Precipitation falls primarily in winter, and hot, dry weather occurs throughout the late spring, sum-

mer, and fall. The vegetation of Mediterranean-type climates is characterized by evergreen shrubs and trees. Evergreen leaves allow plants to be active during cooler, wetter periods and also lower their nutrient requirements, since they do not have to develop new leaves every year. Many plants of Mediterranean-type climates have *sclerophyllous* leaves, which are tough, leathery, and stiff. The plants are well adapted to dry soils and may continue to photosynthesize and grow at reduced rates during the summer. Their sclerophyllous leaves help to deter consumption by herbivores and prevent wilting as water is lost. Sclerophyllous shrublands are found in each of the zones characterized by a Mediterranean-type climate, including the *mallee* of Australia, the *fynbos* of South Africa, the *matorral* of Chile, the

Maquis shrubland, Mediterranean coast, Majorca, Spain

Fynbos landscape with everlastings (*Helichrysum* sp.), Kogelberg Nature Reserve, South Africa

Gerona, Spain
41°N, 76 m
Annual average temperature 16.7°C
Total annual precipitation 747 mm

maquis around the Mediterranean Sea, and the *chaparral* of North America.

Fire is a common feature in Mediterranean-type shrublands, and as in some grasslands, may promote their persistence. Some of the shrubs recover after fires by resprouting from woody storage organs protected from the heat below the ground surface. Other shrubs produce seeds that germinate and grow quickly after a fire. Without regular fires at 30–40-year intervals, some shrublands may be replaced by forests of oaks, pines, junipers, or eucalypts. The regular disturbance by fire, combined with the unique climate of temperate shrublands, is thought to promote high species diversity.

Shrubland steppes and woodlands are also found in the continental interior of North America and Eurasia, associat-ed with rain shadows and seasonally cold climates. The Great Basin, for example, occupies the interior of North America between the Sierra Nevada and Cascade mountain ranges to the west and the Rocky Mountains to the east. Large expanses of sagebrush (*Artemesia tridentata*), saltbush (*Atriplex* spp.), creosote bush (*Larrea tridentata*), and piñon pine and juniper woodland occur throughout the Great Basin.

Humans have occupied temperate shrublands and woodlands for many thousands of years. While some have been converted to croplands and vineyards, their climates and nutrient-poor soils have limited the extent of agricultural and pastoral development. In the Mediterranean basin, agricultural development using irrigation was attempted and failed due to the infertile soils. Urban development has reduced the cover of shrublands in some regions (e.g., southern California). Increases in local human populations have increased the frequency of fires, which decreases the ability of shrubs to recover and may lead to their replacement by invasive annual grasses.

Temperate Deciduous Forests   Deciduous leaves are a solution to the extended periods of freezing weather in the temperate zone. Leaves are more sensitive to freezing than other plant tissues because of the high level of physiological activity associated with photosynthesis. Temperate deciduous forests occur in areas with enough rainfall to support tree growth (50–250 cm/year) and where soils are fertile enough to supply the nutrients lost when leaves are shed in the fall. Temperate deciduous forests are primarily limited to the Northern Hemisphere, as the Southern Hemisphere contains less land area and lacks extensive areas with the continental climates associated with the deciduous forest biome.

Deciduous forests occur at 30° to 50° N on the eastern and western edges of Eurasia and in eastern North America, extending inland to the continental interiors before diminishing due to lack of rainfall, and in some cases increased frequency of fires. Similar species occur on each of

TEMPERATE DECIDUOUS FORESTS

Autumn foliage prior to leaf fall, Great Smoky Mountains, Tennessee

Wellsboro, Pennsylvania, USA
41°N, 567 m
Annual average temperature 7.6°C
Total annual precipitation 848 mm

Beech forest in summer, Japan

these continents in this biome, reflecting a common biogeographic history. Oak, maple, and beech trees, for example, are all components of this forest biome on each continent. The vertical structure of the forest includes canopy tree species as well as shorter trees, shrubs, and forbs that occur below the canopy. Species diversity is lower than in tropical forests, but can include as many as 3,000 plant species (e.g., in eastern North America). Disturbances such as fire and herbivory do not play a major role in determining the development and persistence of temperate deciduous forest vegetation, although they can be important in determining its boundaries.

The temperate deciduous forest biome has been a focus for agricultural development for centuries. The fertile soils and climate are conducive to the growth of crops. Forest clearing for crop and wood production has therefore been widespread in this biome. Very little old-growth temperate deciduous forest remains on any continent. Since the early twentieth century, however, agriculture has gradually shifted away from the temperate zone toward the tropics, particularly in the Americas. Abandonment of agricultural fields has resulted in reforestation in some parts of Europe and North America. However, the species composition of the second-growth forests often differs from what was present prior

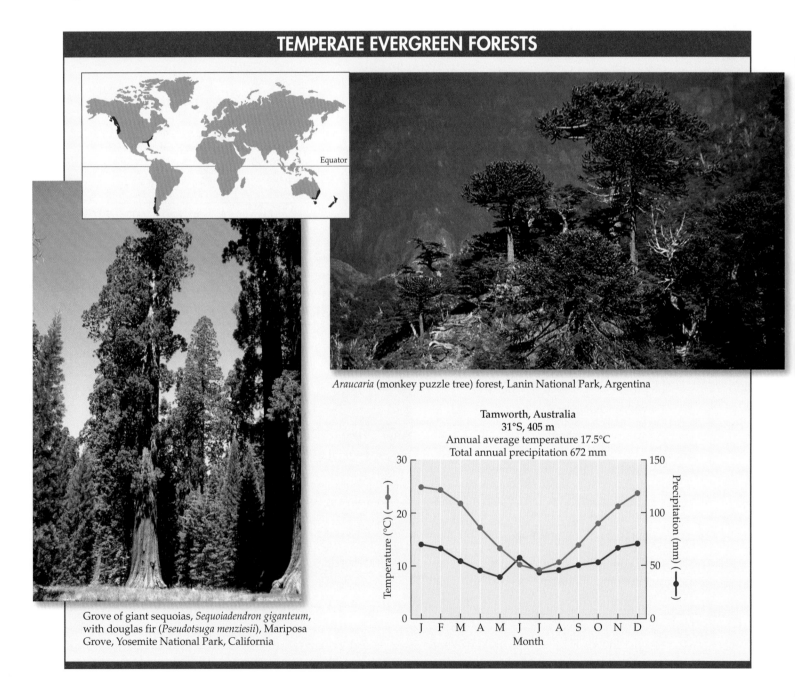

TEMPERATE EVERGREEN FORESTS

Araucaria (monkey puzzle tree) forest, Lanin National Park, Argentina

Grove of giant sequoias, *Sequoiadendron giganteum*, with douglas fir (*Pseudotsuga menziesii*), Mariposa Grove, Yosemite National Park, California

Tamworth, Australia
31°S, 405 m
Annual average temperature 17.5°C
Total annual precipitation 672 mm

FIGURE 3.8 Temperate Rainforest in Tasmania Rainforests occur in temperate zones with high (> 500 cm) precipitation and relatively mild winter temperatures. Here, understory tree ferns and canopy trees of leatherwood and myrtle crowd together near Montezuma Falls in western Tasmania, Australia.

to agricultural development. Nutrient impoverishment of soils due to long-term agricultural use is one reason for this difference. Another is the loss of some species due to introductions of invasive species. For example, chestnut blight fungus, introduced from Asia, nearly wiped out the chestnut trees (*Castanea* spp.) of North America in the early twentieth century (see Chapter 13). As a result, oak species are more widespread than they were prior to agricultural development.

Temperate Evergreen Forests Evergreen forests span a wide range of environmental conditions in the temperate zone, from warm coastal zones to cool continental and maritime climates. Precipitation also varies substantially among these forests, from 50 to 400 centimeters per year. Some temperate evergreen forests with high levels of precipitation, which are typically located on west coasts at latitudes between 45° and 50°, are referred to as "temperate rainforests" (**Figure 3.8**). Temperate evergreen forests are commonly found on nutrient-poor soils, a condition that is in-part related to the acidic nature of the leaves of the evergreen trees. Some evergreen forests are subject to regular fires at intervals of 30–60 years, which may promote their persistence.

Temperate evergreen forests are found in both the Northern and Southern hemispheres between 30° and 50° latitude. Their diversity is generally lower than that of deciduous and tropical forests. In the Northern Hemisphere, tree species include needle-leaved conifers such as pines, junipers, and Douglas fir. In the Southern Hemisphere, on the west coasts of Chile and Tasmania and in southeastern and southwestern Australia, there is greater tree diversity, including southern beeches (*Nothofagus* spp.), eucalypts, the Chilean cedar (*Austrocedrus*), and podocarps.

Evergreen trees provide a source of high-quality wood and pulp for paper production. The temperate evergreen for-

est biome has been subjected to extensive forest clearing, and little old-growth temperate evergreen forest remains. Forestry practices tend to promote sustainable use of these forests, although in some regions the planting of non-native species, such as Monterey pine in New Zealand, the uniform age and density of the trees, and losses of formerly dominant species, have created forests that are ecologically very different from their pre-logging condition. The suppression of naturally occurring fires in western North America has increased the density of forest stands, which has resulted in more intense fires when they do occur and increased the spread of insect pests (e.g., bark beetles) and pathogens. In industrialized countries, the effects of air pollution have damaged some temperate evergreen forests (see Chapter 24) and made them more susceptible to other stresses.

Boreal Forests Above 50° N, the severity of winters increases. Minimum temperatures of –50°C are common in continental locations such as Siberia, and continuous subfreezing temperatures may last up to 6 months. The extreme weather in these subarctic regions is an important determinant of the vegetation structure. Not only must the plants cope with the low air temperatures, but soils may regularly freeze, leading to the formation of **permafrost**, defined as a subsurface soil layer that remains frozen year-round for at least 3 years. Although precipitation is low, the permafrost impedes water drainage, so soils are moist to saturated.

The biome that occupies the zone between 50° and 65° N is the boreal (far northern) forest. This biome is also known as *taiga*, the Russian word for this northern forest. It is composed primarily of coniferous species, including spruces, pines, and larches (deciduous needle-leaved trees), but also includes extensive deciduous birch forests in maritime locations, particularly in Scandinavia. Conifers tend to resist dam-

Spruce trees on the margin of a bog in autumn, Denali National Park, Alaska

Fort Simpson, NWT, Canada
61°N, 169 m
Annual average temperature –4.6°C
Total annual precipitation 333 mm

Boreal forest landscape, Canadian shield, northern Quebec

age from winter freezing better than angiosperm trees, despite maintaining green leaves year-round. Although the boreal forest is found only in the Northern Hemisphere, it is one of the largest biomes in area and contains one-third of Earth's forested land.

The cold, wet condition of boreal forest soils limits the decomposition of plant material such as leaves, wood, and roots. These soils therefore contain large amounts of organic matter, as the rate of plant growth exceeds the rate of decomposition. During extensive summer droughts, conditions are conducive for forest fires, which are set off by lightning. These fires may burn both the trees and the soil (**Figure 3.9**). Soil fires may continue to burn slowly for several years, even through the cold winters. In the absence of fire, forest growth enhances permafrost formation by lowering the amount of sunlight absorbed by the soil surface. In low-lying areas, soils become saturated, killing the trees and forming extensive peat bogs.

Relative to other forest biomes, boreal forests have been little affected by human activities. Logging occurs in some regions, as does oil and gas development. These activities will pose an increasing threat to the boreal forest as demands for wood and energy increase. In addition, the large store of organic matter in the soil makes boreal forests an important component of the global carbon cycle. Climate change may result in higher rates of carbon release from boreal forest soils, enhancing atmospheric greenhouse gas concentrations and resulting in a positive feedback to global warming (see Chapter 24).

Tundra Trees cease to be the dominant vegetation beyond approximately 65° latitude. The tree line that marks the transition from boreal forest to tundra is associated with low

FIGURE 3.9 Fire in the Boreal Forest Despite its cold climate, fire is an important part of the boreal forest environment.

growing-season temperatures, although the causes of this transition are complex and can also include other climatic and soil factors. The tundra biome occurs primarily in the Arctic, but can also be found on the edges of the Antarctic Peninsula and on few islands in the Southern Ocean. The poleward decrease in temperature and precipitation across the tundra biome is associated with zones of high pressure generated by the polar atmospheric circulation cells.

The tundra biome is characterized by sedges, forbs, grasses, and low-growing shrubs such as heaths, willows, and

TUNDRA

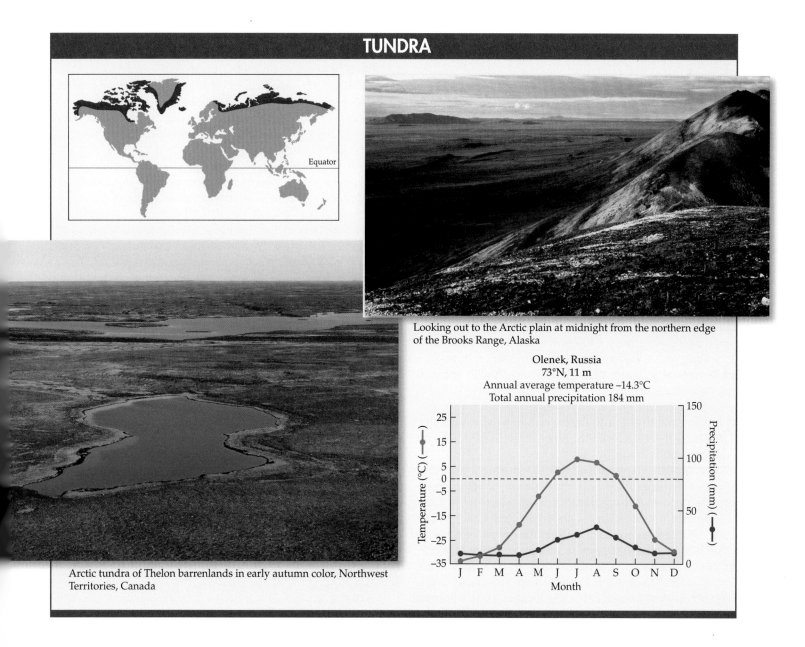

Looking out to the Arctic plain at midnight from the northern edge of the Brooks Range, Alaska

Arctic tundra of Thelon barrenlands in early autumn color, Northwest Territories, Canada

Olenek, Russia
73°N, 11 m
Annual average temperature –14.3°C
Total annual precipitation 184 mm

FIGURE 3.10 Soil Polygons and Pingo Repeated freezing and thawing of soils in cold regions causes sorting of soil particles. Over time, polygons form, with coarse materials at the periphery and the finer soils in the middle. Note also the raised hill in the background, called a pingo, caused by an intrusion of water that freezes in the subsurface permafrost zone.

birches. Lichens and mosses are also important components of this biome. Although the summer growing season is short, the days are long, with continuous periods of light for 1 to 2 months of the summer. The plants and lichens survive the long winter by going dormant, maintaining living tissues under the snow or soil, insulated from the cold air temperatures.

The tundra has several similarities to the boreal forest: temperatures are cold, precipitation is low, and permafrost is widespread. Despite the low precipitation, many tundra areas are wet, as the permafrost keeps the precipitation that does fall from percolating to deeper soil layers. Repeated freezing and thawing of surface soil layers over several decades results in sorting of soil materials according to their texture. This process forms polygons of soil at the surface, with upraised rims and depressed centers (**Figure 3.10**). Where soils are coarser or permafrost does not develop, the soils may be dry, and plants must be able to cope with low water availability. These polar deserts are most common at the highest latitudes where tundra is found.

Human inhabitants of the tundra biome are scattered in sparse settlements. As a result, it contains some of the largest pristine regions on Earth. Herds of caribou and musk oxen and predators such as wolves and brown bears, which have been extirpated throughout much of their previous range in other biomes, inhabit the tundra, and many species of migratory birds nest there. The influence of human activities on the Arctic tundra is increasing, however. Exploration and development of energy resources has accelerated. A key to limiting the effects of energy development is preventing damage to

the permafrost, which can cause long-term erosion. The Arctic has experienced significant climate change during the late twentieth and early twenty-first centuries, with warming almost double the global average. Increased losses of permafrost, catastrophic lake drainage, and a lower ability to store carbon in the soil have been linked to climate change.

Biological communities in mountains occur in elevational bands

Approximately one-fourth of Earth's land surface is mountainous. Mountains create elevational climatic gradients that change more rapidly over a given distance than those associated with changes in latitude. Temperatures in temperate continental mountain ranges, for example, decrease approximately 6.4°C for every 1,000 m increase in elevation (or 3.6 °F / 1000 feet),* a decrease equivalent to that over approximately a 13° change in latitude, or a distance of 1,400 km. As we might expect from our consideration of biomes and their close association with climate, coarse biotic assemblages similar to biomes occur in elevational bands on mountains. Finer-scale biotic distinctions are found in association with slope aspect (e.g., north-facing versus south-facing), proximity to streams, and the orientation of slopes in relation to prevailing winds (see Chapter 2).

If we sampled biological communities in a straight line (sometimes referred to as a *transect*) from the base to the summit of a temperate-zone mountain range, we would see trends similar to those we would find in moving along a latitudinal transect. An elevational transect on the eastern slope of the southern Rocky Mountains in Colorado, for example, runs from grassland to alpine vegetation across a 2,200 m increase in elevation (**Figure 3.11**). The change in climate and vegetation is very similar to the transition from grassland to Arctic tundra that occurs with a 27° increase in latitude, from Colorado to the Northwest Territories of Canada. Grasslands occur at the base of the mountains, but give way to pine savannas on the initial slopes of the mountains (the lower montane zone). Fire plays an important role in determining the vegetation structure of both grasslands and savannas. With increasing elevation, the pine savannas are replaced by denser stands of mixed pine–aspen forests (the montane zone), which resemble temperate evergreen and deciduous forest biomes. Spruces and firs make up the forests of the subalpine zone, which resemble the boreal forest biome. Mountain tree lines are similar to the transition from boreal forest to tundra, although topography can play an important role through its influence on snow distribution and avalanches. The alpine zone above the tree line includes diminutive plants such as sedges, grasses, and forbs, including some of the same species

---

*This decrease in temperature, called *lapse rate*, is dependent on the humidity of the air and local winds. A more complete discussion of temperature change in mountains can be found in Chapter 2.

FIGURE 3.11 Mountain Biological Zones An elevational transect on the eastern slope of the southern Rocky Mountains reveals climatic variation and biome-like life zones similar to those found along a latitudinal gradient between Colorado and northern Canada. (Data from Marr 1967.)

	Lower montane zone	Montane zone	Subalpine zone	Alpine zone
Median elevation (m)	1,500	2,400	3,000	3,700
Mean annual temperature (°C)	9	5.5	2.5	−3.5
Mean annual precipitation (mm)	450	600	750	1,000

that occur in the Arctic tundra. Although the alpine zone resembles the Arctic tundra, its physical environment is different, with higher wind speeds, more intense solar radiation, and lower atmospheric partial pressures[*] of CO_2.

Mountains are found on all continents and at all latitudes. As indicated in the example above, the change in climate associated with changes in elevation alters the composition of the local vegetation. However, not all of the vegetation assemblages that occur in mountains resemble major terrestrial biomes. Some mountain-influenced biological communities have no biome analogs. For example, daily temperature changes at high-elevation sites in the tropics (e.g., Mount Kilimanjaro and the tropical Andes) are greater than seasonal temperature changes (**Figure 3.12**). Subfreezing temperatures occur on most nights in the tropical alpine zone. As a result of these unique climatic conditions, tropical alpine vegetation does not resemble that of the temperate alpine zone or the Arctic tundra.

---

[*]The *partial pressure* of a gas is defined as the pressure exerted by a particular component of a mixture of gases. The concentrations of CO_2 and O_2 are the same at high elevations as they are at sea level, but their partial pressures are lower because total atmospheric pressure is lower. The exchange of a gas between an organism and the atmosphere is determined by its partial pressure rather than its concentration.

FIGURE 3.12 Tropical Alpine Plants Frailejón (*Espeletia hartwegiana*) grows in alpine grasslands, known as *páramo*, in the Columbian Andes. Its growth form, characterized by a circle of leaves (rosette), is typical of plants in the tropical alpine zones of South America and Africa. The adult leaves help to protect the developing leaves and stems at the apex of the plant from nightly frosts. Such giant rosettes are found exclusively in the tropical alpine zone and do not have analogs in the Arctic or Antarctic.

Freshwater Biological Zones

CONCEPT 3.2
Biological zones in freshwater systems are associated with the velocity, depth, temperature, clarity, and chemistry of the water.

Although they occupy a small portion of the terrestrial surface, freshwater streams, rivers, and lakes are a key component in the connection between terrestrial and marine ecosystems. Rivers and lakes process inputs of chemical elements and energy from terrestrial systems and transport them to the oceans. The biota of these freshwater systems reflect the physical characteristics of the water, including its velocity (flowing streams and rivers versus lakes and ponds), its temperature (including seasonal changes), how far light can penetrate it (clarity), and its chemistry (salinity, nutrient status, and pH).

We will briefly explore the biota and associated physical conditions found in freshwater ecosystems in the next two sections. In contrast to terrestrial biomes, for which only plants are used as indicators, the biological assemblages of freshwater ecosystems are characterized by both plants and animals, reflecting the greater proportional abundance of animals in aquatic ecosystems.

Biological communities in streams and rivers vary with stream size and location within the stream channel

Water flows downhill over the land surface in response to the force of gravity. The land surface is partly shaped by the erosional power of water, which cuts valleys as it heads toward a lake or ocean. The descending water converges into progressively larger streams and rivers, called **lotic** (flowing water) systems. The smallest streams at the highest elevations in a landscape are called *first-order streams* (**Figure 3.13**). Two first-order streams converge to form second-order streams. Large rivers such as the Nile or Mississippi are equal to or greater than sixth-order streams.

Individual streams tend to form repeated patterns of **riffles** and **pools** along their paths. Riffles are fast-moving portions of the stream flowing over coarse particles on the stream bed; pools are deeper and flow more slowly over a bed of fine sediments. Biological communities in lotic systems are associated with different physical locations within the stream and their related environments (**Figure 3.14**). Organisms that live in the flowing water of the main channel are generally swimmers, such as fish. **Benthic** (bottom-dwelling) organisms include invertebrates that consume **detritus** (dead organic matter), such as some mayfly and fly (dipteran) larvae, or hunt other organisms, such as some caddisflies and crustaceans. Benthic organisms found in the riffles of streams benefit from greater oxygenation of the turbulent water than in the pools. Some organisms, such as rotifers, copepods, and insects, are found in portions of the substratum below and adjacent to the stream, where water movement still occurs, either from the stream or from groundwater moving into the stream. This area is known as the **hyporheic zone**.

The composition of biological communities in streams and rivers changes with stream order and channel size. The *river continuum concept* was developed to describe these changes in terms of both the physical and biological characteristics of a stream (Vannote et al. 1980). This conceptual model holds that as a stream moves downslope and increases in size, the amount of detritus input from the vegetation adjacent to the stream (known as riparian vegetation) relative to the size of the stream decreases, and the particle size in the stream bed decreases, from boulders and coarse rock to fine sand. As a result, the importance of terrestrial vegetation as food sources for stream organisms decreases in the downstream direction. Coarse terrestrial detritus is most important near the stream source, while the importance of fine organic matter, algae, and rooted aquatic plants (known as **macrophytes**) increases downstream. The general feeding styles of organisms change accordingly along the stream course. *Shredders*, organisms adapted to tear up and chew leaves (e.g., some species of caddisfly larvae), are most abundant in the higher parts of the stream, while *collectors*, organisms that collect fine particles from the water (e.g., some dipteran larvae), are most abundant in the lower parts of the stream. The river continuum concept applies best to temperate river systems, but not

FIGURE 3.13 Stream Orders Stream order affects environmental conditions, biological communities, and the energy and nutrient relationships of communities within the stream.

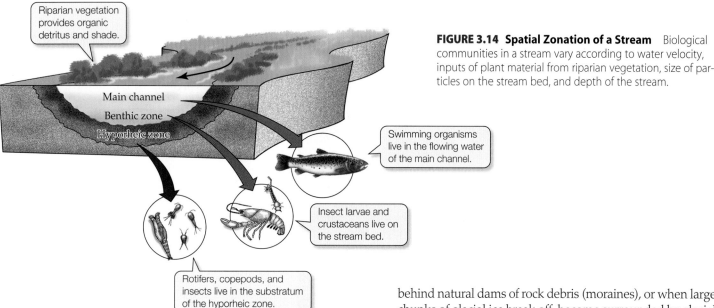

FIGURE 3.14 Spatial Zonation of a Stream Biological communities in a stream vary according to water velocity, inputs of plant material from riparian vegetation, size of particles on the stream bed, and depth of the stream.

so well to boreal/Arctic and tropical rivers or to rivers with high concentrations of dissolved organic substances (including tannic and humic acids) derived from wetlands. Nonetheless, the model provides a basis for studying biological organization in stream and river systems.

Human effects lotic ecosystems have been significant and extensive. Most fourth-order and higher-order rivers have been altered by the results of human activities, including pollution, increases in inputs of sediments, and introductions of non-native species. Streams and rivers have been used as conduits for the disposal of sewage and industrial wastes in most parts of the world inhabited by humans. These pollutants often reach levels that are toxic to many aquatic organisms. Excessive applications of fertilizers to croplands results in the leaching of nutrients into groundwater and eventually into rivers. Inputs of nitrogen and phosphorus from fertilizers alter the composition and abundances of aquatic organisms. Deforestation increases sediment inputs into nearby streams, which can reduce water clarity, alter benthic habitat, and inhibit gill function in many aquatic organisms. Introductions of non-native species, such as sport fish (e.g., bass and trout), have lowered the diversity of native species in both stream and lake ecosystems. The construction of dams on streams and rivers tremendously alters their physical and biological properties, converting them to a "stillwater" system, the topic of the next section.

Biological communities in lakes vary with depth and light penetration

Lakes and other still waters (**lentic** systems) occur where natural depressions have filled with water, or where humans have dammed rivers to form reservoirs. Lakes and ponds may be formed when glaciers gouge out depressions and leave

behind natural dams of rock debris (moraines), or when large chunks of glacial ice break off, become surrounded by glacial debris, and then melt. Most temperate and polar lakes were formed by glacial processes. Lakes may also form when meandering rivers cease to flow through a former channel, leaving a section stranded, called an *oxbow lake*. Geological phenomena, such as extinct volcanic calderas and sinkholes, leave natural depressions that may fill with water. Lakes of biological origin, in addition to reservoirs, include beaver dams and animal wallows.

Lakes vary tremendously in size, from small, ephemeral ponds to the massive Lake Baikal in Siberia, 1,600 m in depth and covering 31,000 km^2. The depth and area of a lake has important consequences for its nutrient and energy status, and therefore for the composition of its biological communities. Deep lakes with a relatively small surface area tend to be nutrient-poor compared with shallow lakes with a relatively large surface area (see Chapter 21).

Lake biological assemblages are associated with depth and degree of light penetration. The open water, or **pelagic zone**, is inhabited by **plankton**, small, often microscopic organisms that are suspended in the water (**Figure 3.15**). Photosynthetic plankton (called **phytoplankton**) are limited to the surface layers of water where there is enough light for photosynthesis, called the **photic zone**. **Zooplankton**, animal plankton, and nonphotosynthetic protista occur throughout the pelagic zone, as do other consumers such as bacteria and fungi, feeding on detritus as it falls through the water. Fish patrol the pelagic zone, scouting for food and predators that might eat them.

The nearshore zone where the photic zone reaches to the lake bottom is called the **littoral zone**. Macrophytes join with floating and benthic phytoplankton to produce energy. Fish and zooplankton also occur in the littoral zone.

In the benthic zone, detritus derived from the littoral and pelagic zones serves as an energy source for the animals, fungi, and bacteria. The benthic zone is usually the coldest part of the lake, and oxygen levels are often low.

(A) Phytoplankton

(B) Zooplankton

FIGURE 3.15 Examples of Lake Plankton (A) Aquatic green algae from a lake water sample, including spherical *Pediastrum* sp. and disk-like *Scenedesmus* sp. Note the projections at the corners of the *Scenedesmus* and at the surface of *Pediastrum*, which help to suspend the cells in the water. (B) Zooplankton include a water flea (suborder Cladocera, upper left), a seed shrimp (*Cyclocypris* sp., upper right), a copepod (*Cyclops* sp., lower right) and the nauplius larva of a copepod crustacean (lower left). Strands of algae also stretch through the picture.

Marine Biological Zones

CONCEPT 3.3
Marine biological zones are determined by ocean depth, light availability, and the stability of the bottom substratum.

Oceans cover 71% of Earth's surface and contain a rich diversity of unique biota. The vast area and volume of the oceans and their environmental uniformity make them considerably different from terrestrial ecosystems in the context of biological organization. Marine organisms are more widely dispersed, and marine biological communities are not as easily organized into broad units as terrestrial biomes are. Instead, marine biological zones tend to be coarsely categorized by physical location relative to shorelines and the ocean bottom (**Figure 3.16**). The distributions of organisms that inhabit these zones reflect differences in temperature, as we saw for terrestrial biomes, but more important factors include differences in light availability, water depth, stability of the bot-

FIGURE 3.16 Marine Biological Zones Biological zones in the ocean are categorized by water depth and by their physical locations relative to shorelines and the ocean bottom.

tom substratum, and interactions with other organisms.

In this section we will take a tour of the biological zones of the oceans, from the margins of the land to the deep, dark, cold ocean bottom. We will examine the relationships between the physical and biological factors that characterize the different zones and the major organisms found in them. As in freshwater communities, a range of plant and animal taxa are used to characterize marine biological zones.

Nearshore zones reflect the influence of tides and substratum stability

Marine biological zones adjacent to the continents are influenced by the rise and fall of ocean waters associated with tides, as well as by wave action and the influx of fresh water and terrestrial sediments from rivers. **Tides** are generated by the gravitational attraction between Earth and the moon and sun. Ocean water rises and falls in most nearshore zones twice daily. The magnitude of the tidal range is highly variable among different locations and is related to shoreline morphology and ocean bottom structure. Tidal fluctuations produce unique transition zones between terrestrial and marine environments.

Estuaries The junction of a river with the ocean is called an estuary (**Figure 3.17**). Estuaries are characterized by variations in salinity associated with the flow of fresh river water into the ocean and the influx of salt water flowing inland from the ocean as tides rise. Rivers also bring terrestrial sed-

FIGURE 3.17 Estuaries Are Found at the Junction of Rivers and Oceans The mixing of fresh and salt water gives estuaries a unique environment with varying salinity and energy and nutrients from terrestrial ecosystems.

iments containing nutrients to the ocean, and the interaction of tidal and river flows acts to trap these sediments in estuaries, enhancing the productivity of these nearshore zones. The varying salinity of the estuary is an important determinant of the organisms that occur there. Many commercially valuable fish species spend their juvenile stages in estuaries, away from fish predators that are not as tolerant of the changes in salinity. Other inhabitants of estuaries include shellfish (e.g., clams and oysters), crabs, marine worms, and

seagrasses. Estuaries are increasingly threatened by water pollution carried by rivers. Nutrients from upstream agricultural sources can cause local dead zones (see Chapter 24) and losses of biological diversity.

Salt marshes Terrestrial sediments carried to shorelines by rivers form shallow marsh zones that are dominated by emergent vascular plants (*emergent* in this case refers to plants that rise out of the water), including grasses, rushes, and broadleaved herbs (**Figure 3.18**). In these salt marshes, as in estuaries, the input of nutrients from rivers enhances biological productivity. Periodic flooding of the marsh at high tide results in a gradient of salinity in the marsh. The highest portions of the marsh are the most saline, since infrequent flooding and evaporation of water from the soil leads to a progressive buildup of salts. Salt marsh plants occur in distinct zones that reflect this salinity gradient, with the most salt-tolerant species in the highest portions of the marsh. Salt marshes provide food and protection from predators for a wide variety of animals, including fish, crabs, birds, and mammals. Organic matter trapped in salt marsh sediments may serve as a nutrient and energy source for nearby marine ecosystems.

Mangrove forests Shallow coastal estuaries and nearby mudflats of tropical and subtropical regions are inhabited by salt-tolerant evergreen trees and shrubs (**Figure 3.19**). These woody plants are collectively referred to as "mangroves," but they do not make up a single taxonomic group; mangroves include species from 16 different plant families. Mangrove roots trap mud and sediments carried by the water, which build up and modify the shoreline. Like salt marshes, man-

FIGURE 3.18 Salt Marshes are Characterized by Salt-Tolerant Vascular Plants Rooted vascular plants form salt marshes in shallow nearshore zones.

FIGURE 3.19 Salt-Tolerant Evergreen Trees and Shrubs Form Mangrove Forests The mangrove roots trap mud and sediments and provide habitat for other marine organisms.

grove forests provide nutrients to other marine ecosystems and habitat for numerous animals, both marine and terrestrial. Among the unique animals associated with mangroves are manatees, crab-eating monkeys, fishing cats, and monitor lizards. Mangrove forests are threatened by human development of coastal areas, particularly the development of shrimp farms, as well as water pollution, diversion of inland freshwater sources, and cutting of the forests for wood.

Rocky intertidal zones   Rocky shorelines provide a stable substratum to which a diverse collection of algae and animals can anchor themselves to keep from being washed away by the pounding waves (**Figure 3.20**). The physical environment of the intertidal zone alternates between marine and terrestrial with the rise and fall of the tides. Between the high-tide and low-tide marks, a host of organisms are arranged in zones associated with their tolerance for temperature changes, salinity, *desiccation* (drying out), wave action, and

FIGURE 3.20 The Rocky Intertidal Zone: Stable Substratum, Changing Conditions   Rocky shorelines provide a stable substratum to which organisms can anchor themselves, but those organisms must cope with the shift from terrestrial to marine conditions that occurs with each tide, as well as wave action. Sessile organisms must be resistant to temperature changes and desiccation. Mobile organisms often take refuge in tide pools to avoid exposure to the terrestrial environment.

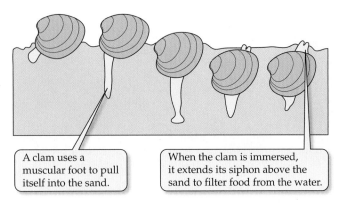

A clam uses a muscular foot to pull itself into the sand.

When the clam is immersed, it extends its siphon above the sand to filter food from the water.

FIGURE 3.21 Burrowing Clams   Clams, like most animals of sandy shorelines, live in the sandy substratum.

interactions with other organisms. *Sessile* (fixed) organisms such as barnacles, mussels, and seaweeds must cope with these stresses in order to survive. Mobile organisms, such as sea stars and sea urchins, may move to pools in order to minimize exposure to these stresses.

Sandy shores   Except for a few scurrying crabs and the occasional bit of seaweed washed ashore, sandy beaches appear devoid of life. The sandy substratum provides no stable anchoring surface as with the rocky shore, and the lack of attached seaweeds limits the supply of potential food for herbivorous animals. Wave action further limits the potential for the development of biological communities. Beneath the sand, however, invertebrates such as clams, sea worms, and mole crabs find suitable habitat to maintain their populations (**Figure 3.21**). Smaller organisms, such as polychaete worms, hydroids (small animals related to jellyfishes), and copepods live on or among the grains of sand. These organisms are protected from temperature changes and desiccation at low tide and from the turbulent water at high tide. When the sand is immersed in seawater, some of these organisms emerge to feed on detritus or other organisms, while others remain buried and filter detritus and plankton from the water.

Shallow ocean zones are diverse and productive

Near the coastline, enough light may reach the bottom of the ocean to permit the establishment of photosynthetic organisms. Like terrestrial plants, these photosynthetic organisms provide energy to support communities of animals and microorganisms as well as a physical structure that creates habitat for other organisms, including surfaces to which they can anchor and places where they can find refuge from predators. The diversity and complexity of the habitats provided by the photosynthesizers supports considerable biological diversity in these shallow ocean environments.

Coral reefs   In warm, shallow ocean waters, corals (animals related to jellyfishes), living in a close association with algal

FIGURE 3.22 A Coral Reef   Corals create habitat for a diverse assemblage of marine organisms.

FIGURE 3.23 Coral Reefs Can Be Seen from Outer Space
Long Island, in the Bahamas, was formed by coral reefs, which can be seen on the fringes of the island.

partners, form large colonies. Many corals build a skeleton-like structure by extracting calcium carbonate from seawater. Over time, these coral skeletons pile up into massive formations called reefs (**Figure 3.22**). The formation of reefs is aided by other organisms that extract other minerals from seawater, such as sponges that precipitate silica. The unique association of these reef-building organisms gives rise to a unique, structurally complex habitat that supports a rich marine community.

Coral reefs grow at rates of only a few millimeters per year, but they have shaped the face of Earth (Birkeland 1997). Over millions of years, corals have constructed thousands of kilometers of coastline and numerous islands (**Figure 3.23**). The rate of production of living biomass in coral reefs is among the highest on Earth. The accretions of coral skeletons are as much as 1,300 m thick in some places, and they currently cover a surface area of 600,000 km^2, approximately 0.2% of the ocean surface.

As many as a million species are found in coral reefs worldwide, including more than 4,000 fishes. A large number of economically important fishes rely on coral reefs for habitat, and reef fishes provide a source of food for fishes of the open ocean, such as jacks and tuna. The taxonomic and morphological diversity of animals in coral reefs is greater than in any other ecosystem on Earth (Paulay 1997). The full diversity of coral reefs however, has yet to be explored and described. The potential for development of medicines from coral reef organisms is great enough that the U.S. National Institutes of Health established a laboratory in Micronesia. Human activities threaten the health of coral reefs in a number of ways. Sediments carried by rivers can cover and kill the corals, and excess nutrients increase the growth of algae on the surface of the corals, increasing mortality. Changes in

ocean temperatures associated with climate change can result in the loss of the corals' algal partners, a condition called *bleaching*. Another threat is the increased incidence of fungal infections, possibly related to increased deposition of dust from the atmosphere associated with desertification (see Chapter 24). Increased atmospheric CO_2 poses a serious threat that could change ocean chemistry and inhibit the ability of corals to form skeletons (Orr et al. 2005).

Kelp beds   In clear, shallow (< 15 m) temperate ocean waters, large stands of seaweed, known as kelp beds or kelp "forests" (**Figure 3.24**), support a rich and dynamic community of marine life. Kelp are large brown algae of several different genera. They have specialized tissues resembling leaves (fronds), stems (stipes), and roots (holdfasts). Kelp are found where a solid substratum is available for anchoring. Residents of kelp beds include sea urchins, lobsters, mussels, abalones, numerous other seaweeds, and sea otters. Interactions among these organisms, both direct and indirect, influence the abundance of the kelp. Grazers, such as sea urchins, are capable of greatly reducing kelp abundance. Urchin abundance is tied to predation by sea otters, while sea otter abundance is in turn related to orca and human predation (see Chapter 8). In the absence of grazing, kelp beds can become so dense that light reaching the bottom of the canopy is not sufficient to support photosynthesis.

FIGURE 3.24 A Kelp Bed   Kelp are brown algae (order Laminariales) that attach themselves to the solid bottom in shallow ocean waters, providing food and habitat for many other marine organisms.

Seagrass beds   While we typically associate flowering plants with terrestrial environments, they can be important components of shallow (< 5 m) subtidal biological zones. These submerged flowering plants are called seagrasses, although they are not closely related to plants in the grass family. Morphologically, they are similar to their relatives on land, with roots, stems, and leaves as well as flowers, which are pollinated under

water. Seagrass beds are found in subtidal marine sediments composed of mud or fine sand. The plants reproduce primarily by vegetative growth, although they produce seeds as well. Marine algae and animals grow on the surfaces of the plants, and the larval stages of some organisms, such as mussels, are dependent on them for habitat. Inputs of nutrients from upstream agricultural activities can increase the density of algae in the water and on the surfaces of the seagrasses. Seagrasses are also susceptible to periodic outbreaks of fungal diseases.

Open ocean and deep benthic zones are determined by light availability and proximity to the bottom

The vastness and depth of the open ocean beyond the continental shelves, known as the pelagic zone, makes it difficult to differentiate distinct biological communities there. Light availability determines where photosynthetic organisms can occur, which in turn determines the availability of food for animals and microorganisms. Thus, the surface waters with enough light to support photosynthesis (the photic zone) contain the highest densities of organisms (see Figure 3.16). The photic zone, whose depth depends on water clarity, generally extends 200 m downward from the ocean surface. Below the photic zone, the supply of energy, mainly in the form of detritus falling from the photic zone, is much lower, and life is far less abundant.

The diversity of life in the pelagic zone varies considerably. Its **nekton** (swimming organisms capable of overcoming ocean currents) include cephalopods such as squid and octopuses, fishes, sea turtles, and mammals such as whales and porpoises. Most of the photosynthesis of the pelagic zone is carried out by phytoplankton, which include green algae,

(A) Marine phytoplankton

(B) Marine zooplankton

FIGURE 3.25 Plankton of the Pelagic Zone   (A) This sample of marine phytoplankton includes diatoms (*Odontella sinensis*, the rectangular cells with the concave ends) and dinoflagellates (*Biddulphia regia*). (B) Marine zooplankton include larval stages of different organisms, such as those shown here.

diatoms, dinoflagellates, and cyanobacteria (**Figure 3.25A**). Zooplankton include protists such as ciliates, crustaceans such as copepods and krill, and jellyfishes (**Figure 3.25B**). Many species of pelagic seabirds, including albatross, petrels, fulmars, and boobies, spend the majority of their lives flying over open ocean waters, feeding on marine prey (fish and zooplankton) and detritus found on the ocean surface.

Organisms that live in the pelagic zone are faced with the problem of overcoming the effects of gravity and water currents that could force them to progressively greater depths. Photosynthetic organisms, and those directly dependent on them as a food source, must stay in the photic zone where sunlight is sufficient to maintain photosynthesis, growth, and reproduction. Swimming is an obvious solution to this prob-

FIGURE 3.26 A Denizen of the Deep Pelagic Zone
Anglerfishes are named for their unique strategy for capturing prey. The bioluminescent organ dangling from the forehead of this female humpback anglerfish (*Melanocetus johnsoni*) is used to attract prey to a position where they are easily engulfed by her huge, toothy mouth. Some male anglerfishes attach themselves to a female once they reach maturity and spend the rest of their lives as parasites, sucking the blood of their mate.

lem, used by organisms such as fish and squid. Seaweeds such as *Sargassum* may form large, floating mats, aided by gas-filled bladders that keep them buoyant. Mats of *Sargassum* sometimes occur as floating islands that host rich and diverse biological communities. Some plankton retard their sinking by decreasing their density relative to seawater (e.g., through alteration of their chemical composition) or by adapting a shape that lowers their downward velocity (e.g., having a cell wall with projections).

Beneath the photic zone, the availability of energy decreases, and the physical environment becomes more stressful as temperatures drop and water pressure rises. As a result, organisms are few and far between. Crustaceans such as copepods graze on the rain of falling detritus from the photic zone. Crustaceans, cephalopods, and fishes are the predators of the deep sea. Some fishes take on frightening forms, appearing to be mostly mouth (**Figure 3.26**). The scientific names given to some of the sea creatures at this depth, such as "vampire squid from hell" (*Vampyroteuthis infernalis*), "stalked toad with many filaments" (*Caulophryne polynema*), and "Prince Axel's wonder fish" (*Thaumatichthys axeli*), testify to the unusual forms found there. Most deep-sea fishes have weak bone structure to reduce their weight and lack the gas bladder found in most fishes, since the high pressures would collapse it.

The ocean bottom (the benthic zone) is also very sparsely populated. Temperatures are near freezing, and pressures are great enough to crush any terrestrial organism. Conversely, if deep-sea creatures adapted to these high pressures are brought to the surface, their bodies may expand and burst. The sediments of the benthic zone, which are rich in organic matter, contain bacteria and protists as well as sea worms. Sea stars and sea cucumbers graze the ocean floor, consuming the organic matter or organisms in the sediments, or filtering food from the water. Bioluminescence is also used by benthic predators to lure prey. Because of the logistic difficulties involved, the deep-sea benthic zone is one of the least explored and poorly understood marine biological zones.

CASE STUDY REVISITED

The American Serengeti: Twelve Centuries of Change in the Great Plains

---

Humans have been implicated in several major biological changes in the grasslands of the world. One of the earliest was the disappearance of large mammals from North America during the late Pleistocene. Paul Martin, an early proponent of this hypothesis, noted the strong correspondence between extinction events on several continents and the arrival of humans on those continents, principally Europe, North and South America, and Australia (Martin 1984, 2005). Martin suggested that the rapidity of the extinctions and the greater proportion of large animals that disappeared reflected the hunting efficiency of the early humans. Larger animals have lower reproductive potential than smaller animals and they cannot recover from increases in predation as quickly. The hypothesis therefore took on the unfortunate label of "overkill hypothesis."

Since it was first proposed, the overkill hypothesis has received increasing support. Archeological research has uncovered numerous butchering sites containing remains of extinct animals, as well as spearheads and scrape marks on bones made by tools found at the sites. Other strong evidence indicates that human arrival on small, isolated oceanic islands led to large numbers of extinctions, due to increased predation by humans and by other animals they introduced (e.g., rats and snakes). While most scientists accept that humans had a role in some of the continental extinctions in the late Pleistocene, other mechanisms have been proposed as well. These include the spread of diseases carried by humans and possibly the domesticated dogs that accompanied them (MacPhee and Marx 1997). Another hypothesis suggests that the loss of some animals on which other species depended, such as mastodons, led to more widespread extinctions (Owen-Smith 1987). No one hypothesis explains the extinctions of all of the megafauna on all of the continents, and a combination of climate change and the arrival of humans probably contributed to their demise (Barnosky et al. 2004).

Although the diversity of large mammals on the Great Plains was greatly diminished following the Pleistocene, large mammals were still abundant. Bison may have numbered as many as 30 million, and numerous elk (wapiti), pronghorn, and deer roamed the plains. These animals continued to be hunted by humans, who also began to use fire on the eastern edge of the Great Plains as a tool for managing the habitat of their prey, as well as for small-scale agriculture (Delcourt et al. 1998). The writings of travelers to the Great Plains in the early 1800s indicate that the western edge of the eastern deciduous forest was significantly farther east than it is today, probably due to the influence of human-set fires.

Between 1700 and 1900, ecological changes occurred in the Great Plains that profoundly transformed both the plants and the animals. The reintroduction of horses into North America by Spanish explorers facilitated the development of a Native American culture centered on the hunting of bison. The arrival of Euro-Americans, and their subsequent conflicts with Native Americans, led to the near-extinction of bison and other large Plains animals by the late 1800s (**Figure 3.27**). With the arrival of domesticated cattle and mechanized agriculture after 1850, the Great Plains became a domesticated landscape. The moister, eastern tallgrass prairie was converted into monocultures

FIGURE 3.27 Buffalo Hunting   The arrival of large numbers of Euro-Americans in the Great Plains in the nineteenth century led to mass slaughter of bison, which was facilitated by the construction of railroad lines and the use of high-powered rifles.

of corn, wheat, soybeans, and other crops; today, only 1% of that grassland remains. A larger proportion of the mixed-grass and short-grass prairies to the west remained intact, but overgrazing and unsustainable agricultural practices led to serious degradation some of these areas during the "Dust Bowl" of the 1930s, when drought and massive windstorms resulted in substantial losses of fertile topsoil.

CONNECTIONS IN NATURE

Long-Term Ecological Research

Grasslands are one of the most heavily impacted terrestrial biomes, but most terrestrial biomes and marine biological zones across the globe are experiencing changes due to human activities. Even remote, seemingly pristine areas are subject to the effects of climate change and air pollution. Recognizing the effects of human activities on these systems, as well as our incomplete understanding of those effects, the U.S. National Science Foundation initiated a network of long-term ecological research (LTER) sites in 1980. Initially consisting of 5 sites, the network has grown to 26 sites representing a diversity of terrestrial ecosystems, from tropical to polar, as well as marine systems, croplands, and urban centers (http://www.lternet.edu/) (**Figure 3.28**). The mission of the LTER program is "to provide the scientific

FIGURE 3.28 Long-Term Ecological Research Programs
Twenty-six research sites constitute the U.S. Long-Term Ecological Research network, encompassing deserts, grasslands, forests, mountains, lakes, estuaries, agricultural systems, and cities. Researchers measure long-term changes in ecosystems, and perform experiments, to better understand ecological dynamics over decades to centuries.

community, policy makers, and society with the knowledge and predictive understanding necessary to conserve, protect, and manage the nation's ecosystems, their biodiversity, and the services they provide." The formation of the U.S. LTER program has spurred the formation of an international network of LTER sites, facilitating international collaborative research to better understand Earth's ecological systems.

The Konza Prairie LTER site is located in the Flint Hills of northeastern Kansas in a remnant tallgrass prairie. Research at the Konza site has focused on understanding the interactive roles of fire, grazing, and climate in the tallgrass prairie ecosystem. This research has included experiments varying the frequencies of fire and grazing in large landscape units (**Figure 3.29**). Researchers have examined the potential effects of climate change by varying how much and when precipitation falls. Results from this research have provided important insights into how climate change might affect the diversity and ecosystem function of grasslands, indicating that extremes in rainfall are important controls on these properties (Knapp et al. 2002). These and other experiments will help us better understand and protect native grassland biodiversity in the face of changing human land use patterns and climate change.

(A)

(B)

(C)

FIGURE 3.29 Research at the Konza Prairie LTER Site   Long-term research and experiments are investigating the effects of the frequencies of (A) grazing, (B) fire, and (C) precipitation on the diversity and functioning of the tallgrass prairie ecosystem.

SUMMARY

CONCEPT 3.1 Terrestrial biomes are characterized by the dominant growth forms of vegetation.

- Terrestrial biomes and their associated plant growth forms reflect global patterns of precipitation and temperature.

- The potential and actual distributions of terrestrial biomes differ due to human influences, particularly conversion of land for agriculture and grazing.

- There are nine major terrestrial biomes: rainforests, seasonal forests and savannas, and deserts in tropical and subtropical climates; grasslands, shrublands and woodlands, deciduous forests, and evergreen forests in the temperate zone, and boreal forests and tundra in polar regions.

- Biological communities in mountains occur in elevational bands associated with gradients in temperature and precipitation.

CONCEPT 3.2 Biological zones in freshwater systems are associated with the velocity, depth, and chemistry of the water.

- Biological communities in streams and rivers vary with stream order and location within the stream channel.

- Biological communities in lakes vary with depth and the amount of light penetration.

CONCEPT 3.3 Marine biological zones are determined by ocean depth, light availability, and the stability of the bottom substratum.

- Estuaries, salt marshes, and mangrove forests occur in shallow zones at the margins between terrestrial and marine systems, and are influenced by the import of fresh water and sediments by nearby rivers.

- Biological communities at the shoreline reflect the influence of tides and the stability of the substratum (sandy versus rocky).

- Coral reefs and kelp and seagrass beds are productive communities with high diversity associated with habitat complexity provided by their photosynthesizers.

- Biological communities of the open ocean and deep benthic zones contain sparse populations of organisms, whose distributions are determined by light availability and proximity to the bottom.

REVIEW QUESTIONS

1. Why are terrestrial biomes characterized using the growth forms of the dominant plants that occupy them?

2. The potential and actual distributions of terrestrial biomes differ considerably due to human transformation of the land surface. What biomes have been affected the most, and why?

3. As rivers flow from their source to the oceans, what physical changes occur that affect the distribution of their biological communities?

4. Why do ocean depth and the stability of the substratum play important roles in the determination of marine biological communities?

HANDS-ON PROBLEM SOLVING

Losses of Coral Reefs | This Web exercise (www.sinauer.com/ecology) quantifies losses in coral reefs, a marine biome that is especially threatened. You will explore the rate, timing, and geographical extent of losses of coral cover in reefs in the Indo-Pacific region.

SUGGESTED READINGS

Harrison, P. and F. Pearce. 2001. *AAAS Atlas of Population and the Environment*. University of California Press, Berkeley. *This monograph, commissioned by the American Association for the Advancement of Science, reviews the widespread effects of resource extraction and pollution on biomes .*

Levinton, J. S. 2001. *Marine Biology; Function, Biodiversity, Ecology.* 2nd ed. Oxford University Press, New York. *An excellent textbook introduction to marine biology, including information on the organisms making up marine communities and the underlying ecological factors determining variation in their distribution.*

Martin, P. S. 2005. *Twilight of the Mammoths: Ice Age Extinctions and the Rewilding of America.* University of California Press, Berkeley. *An engaging book examining the controversial role of humans in the rapid extinctions of large animals over the past 50,000 years.*

Prentice, I. C., W. Cramer, S. P. Harrison, R. Leemans, R. A. Monserud and A. M. Solomon. 1992. A global biome model based on plant physiology and dominance, soil properties and climate. *Journal of Biogeography* 19: 117–134. *This paper describes a modeling approach to classifying biomes based on the morphological and physiological characteristics of plants, climatic data, and soil properties. This type of approach is useful in determining the potential distribution of biome types, since the actual biome distributions may be greatly altered by human activities.*

The following two textbooks on freshwater ecology provide an overview of lotic and lentic communities, their component organisms, and the abiotic and biotic factors determining where they are found.

Wetzel, R. G. 2001. *Limnology: Lake and River Ecosystems.* Academic Press, San Diego, CA.

Dodson, S. I. 2004. *Introduction to Limnology.* McGraw-Hill, New York.

Whittaker, R. H. 1975. *Communities and Ecosystems.* Macmillan, London. *Robert Whittaker was one of the first ecologists to quantitatively address the distribution of biomes and communities in relation to the physical environment. This book is notable for being one of the first ecological texts including quantitative (i.e., mathematical) treatments of many topics in ecology. The book contains one of the first presentations of the relationship between the physical environment and biomes.*

CHAPTER 4

Coping with Environmental Variation: Temperature and Water

KEY CONCEPTS

CONCEPT 4.1 Each species has a range of environmental tolerances that determines its potential geographic distribution.

CONCEPT 4.2 The temperature of organisms is determined by exchanges of energy with the external environment.

CONCEPT 4.3 The water balance of organisms is determined by exchanges of water and solutes with the external environment.

FIGURE 4.1 A Frozen Frog Wood frogs (*Rana sylvatica*) spend winter in a partially frozen state without breathing and with no circulation or heartbeat.

CASE STUDY

Frozen Frogs

In the movie *2001: A Space Odyssey*, astronauts traveling to Jupiter are placed in a state of suspended animation while being transported across vast distances of space to prevent them from aging. The idea of suspended animation—life being put on hold temporarily—has captured the imagination and hopes of people waiting for medical science to develop ways to cure untreatable diseases or reverse the ravages of aging. *Cryonics* is the preservation of the bodies of deceased people at subfreezing temperatures with the goal of eventually bringing them back to life and restoring them to good health. Proponents of cryonics exist throughout the world, some more visible than others. In Nederland, Colorado, each year, there is a "Frozen Dead Guy Days" festival, considered to be the "Mardi Gras of cryonics." This festival commemorates the efforts of a former resident who had his grandfather frozen immediately after his death from heart failure, hoping that one day his grandfather could be brought back to life and given a heart transplant (as also documented in the movie *Grandpa's Still in the Tuff Shed*).

To some, cryonics seems far-fetched, a thing of science fiction. Bringing life to a halt and then restarting it after a long period of quiescence doesn't seem plausible. Yet, strange tales from nature provide examples of life apparently springing out of death. While seeking the existence of the Northwest Passage in the boreal and Arctic zones of Canada in 1769–1772, the English explorer Samuel Hearne found frogs under shallow layers of leaves and moss during the winter "frozen as hard as ice, in which state the legs are as easily broken off as a pipe-stem" (Hearne 1911) (**Figure 4.1**). Hearne wrapped the frogs in animal skins and placed them next to his campfire. Within hours, the rock-hard amphibians came to life and began hopping around. The American naturalist John Burroughs found frozen frogs under a shallow cover of dead leaves in a New York forest during winter. Return visits to the same locations over a period of months indicated that the frogs hadn't moved, yet by spring they had disappeared. Could a complex organism like a frog, with a sophisticated circulatory and nervous system, have achieved cryonic preservation as a result of evolutionary forces responding to a harsh winter climate?

Organisms of temperate and polar zones face tremendous challenges imposed by a seasonal climate that includes subfreezing temperatures in winter. Amphibians are unlikely candidates to have solved this challenge by allowing their bodies to partially freeze.

FIGURE 4.2 Northern Exposure Wood frogs (*Rana sylvatica*) and boreal chorus frogs (*Pseudacris maculata*) have geographic ranges that extend into the boreal forest and tundra biomes. (After Pinder et al. 1992.)

Legend:
- *Rana sylvatica*
- *Pseudacris maculata*

Aside from their aforementioned complex organ–tissue systems, amphibians are "cold-blooded" (generate little heat internally) and, as a group, first evolved in tropical and subtropical biomes. Yet two frog species, wood frogs (*Rana sylvatica*) and boreal chorus frogs (*Pseudacris maculata*), live in the Arctic tundra biome (**Figure 4.2**) (Pinder et al. 1992). These frogs survive extended periods of subfreezing air temperatures in shallow burrows in a semi-frozen state, with no heartbeat, no blood circulation, and no breathing. Among the vertebrates, only a few species of amphibians (four frogs and one salamander) and one turtle species can survive a long winter in a semi-frozen state. Freezing in most organisms results in substantial damage to tissues as ice crystals perforate cell membranes and organelles. How do these vertebrates survive being frozen without turning to mush once they thaw out, then reinitiate their blood circulation and breathing once spring arrives?

Spruce trees (*Picea obovata*) of the boreal forest of Siberia experience the extreme range of seasonal temperatures characteristic of a continental climate. In winter, air temperatures regularly drop below –50°C (–58°F), and in summer, they reach 30°C (86°F). As an immobile tree, the spruce lacks the option that humans have to get a coat to keep warm in winter or a fan to cool off in summer. The spruce must *tolerate* these temperature extremes, surviving the 80°C (144°F) seasonal change in its tissue temperatures, rather than *avoiding* them through some behavior or physiological activity. These two options for coping with environmental change, **tolerance** and **avoidance**, provide a useful framework for thinking about how organisms cope with the environmental extremes they face.

The range of physical environmental conditions described in Chapter 2 establishes the variation in biomes and marine biological zones described in Chapter 3. In this and the next chapter, we will examine more thoroughly the interactions between organisms and the physical environment that influence their survival and persistence, and therefore their geographic distributions. Our focus in this chapter will be on temperature and water, two key physical environmental drivers of organismal function and distribution.

Response to Environmental Variation

CONCEPT 4.1
Each species has a range of environmental tolerances that determines its potential geographic distribution.

A fundamental principle in ecology and biogeography is that the geographic ranges of species are related to constraints imposed by the physical and biological environments. In this section, we will discuss the general concepts related to organismal responses to the physical environment before delving into more specific responses to variation in temperature and water availability in the following sections.

Species distributions reflect environmental influences on energy acquisition and physiological tolerances

The potential geographic range of an organism is ultimately determined by the physical environment, which influences an organism's success in two important ways. First, the organism's ability to obtain energy and resources required to

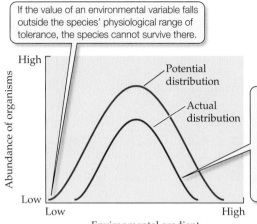

If the value of an environmental variable falls outside the species' physiological range of tolerance, the species cannot survive there.

The actual distribution of an organism is different from its potential distribution due to interactions with other organisms, such as competition, that may constrain the geographic range.

FIGURE 4.3 Abundance Varies across Environmental Gradients The abundance of an organism reaches a maximum value at some theoretical optimal value across an environmental gradient, and drops off at either end at values that constrain the potential geographic distribution of the organism.

tribution of a species is also related to other factors, such as disturbance and interactions with other organisms such as competition (**Figure 4.3**).

As indicated in the introduction to this chapter and in the discussion of biomes in Chapter 3, the immobility of plants makes them good indicators of the physical environment. Farmers are acutely aware of the effects of extreme events on

FIGURE 4.4 Climate and Aspen Distribution Aspen's (*Populus tremuloides*) geographic distribution is associated with climate. (A) Predicted distribution, based on climatic effects on survival and reproduction observed in natural populations, mapped with the actual distribution. (B) Factors limiting the distribution of aspen, estimated based on climatic effects observed in natural populations. (After Morin et al. 2007.)

maintain its metabolic functions, and therefore to grow and reproduce, is influenced by its physical environment. Rates of photosynthesis and abundances of prey, for example, are controlled by environmental conditions. A species' ability to acquire energy, grow, and reproduce, and therefore to maintain a viable population, is constrained at the limits of its geographic range. Second, a more immediate effect on an organism's success is the influence of extreme environmental conditions on survival. If temperature, water supply, chemical concentrations, or some other physical environmental condition is outside the physiological range that an organism can tolerate, the organism will die. These two factors—energy acquisition and physical environmental tolerance limits—are not mutually exclusive, as energy supply influences an organism's ability to tolerate environmental extremes. It is important to keep in mind that the *actual* geographic dis-the survival of crop plants, which are often grown outside the geographic ranges where they evolved. Freezing events or extreme droughts can result in catastrophic crop losses. Aspen (*Populus tremuloides*) provides a good example of a native species with a geographic range related to climatic tolerance. Aspen occurs in boreal forests and mountain highlands throughout North America. Its geographic distribution can be accurately predicted based on the effects of climate on its survival and reproduction (Morin et al. 2007) (**Figure 4.4A**). These climatic factors include the effects of low temperatures on its reproductive success and the effects of drought and low temperatures on its survival (**Figure 4.4B**). The range of climatic conditions under which a species occurs—its *climate envelope*—provides a useful tool for predicting its response to climate change (see Chapter 24).

Individuals respond to environmental variation through acclimatization

Physiological processes such as growth or photosynthesis have a set of optimal environmental conditions most conducive to their functioning. Deviations from the optimum will cause a decrease in the rate of the process (**Figure 4.5**). **Stress** is the condition in which an environmental change results in a decrease in the rate of an important physiological process, thereby lowering the potential for an organism's survival, growth, or reproduction. For example, when you travel to high elevations, typically above 2,400 m (8,000 feet), the lower partial pressure of oxygen in the atmosphere (see p. 69) results in the delivery of less oxygen to your tissues by your circulatory system. This condition, known as *hypoxia*, results when the amount of oxygen picked up by hemoglobin molecules in your blood decreases. Hypoxia causes "altitude sickness," a type of physiological stress, decreasing your ability to exercise and think clearly and making you feel nauseated.

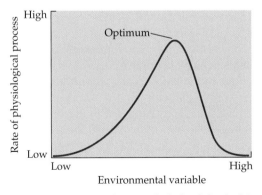

FIGURE 4.5 Environmental Control of Physiological Processes
The rates of physiological processes, such as growth or photosynthesis, are greatest under a set of optimal environmental conditions (e.g., optimal temperature, optimal water availability). Changes from the optimal values cause a decrease in the physiological process.

Many organisms have the ability to adjust their physiology, morphology, or behavior to lessen the effect of an environmental change and minimize the associated stress. This kind of adjustment, known as **acclimatization**,[*] is usually a short-term, reversible process. Your body acclimatizes to high elevations if you remain there for several weeks (but only below 5,500 m or 18,000 feet). Acclimatization to high elevations involves higher breathing rates, greater production of red blood cells and associated hemoglobin, and higher pressure in the pulmonary arteries to circulate blood into areas of the lung that are not used at lower elevations (Hochochka and Somero 2002). The outcome of these physiological changes is the delivery of more oxygen to your tissues. The acclimatization process reverses when you return to lower elevations.

Populations respond to environmental variation through adaptation

Throughout a species' geographic range, individual populations often occur in unique environments (e.g., cool climates, saline soils) that may have initially been stressful to the organisms when they first occupied them. Genetic variation among the individuals within the population in physiological, morphological, or behavioral traits that influenced their survival and functioning in the new environment would have led to natural selection favoring those individuals whose traits made them best able to cope with the new conditions. The underlying genetic basis for these traits would have resulted in a long-term shift in the genetic makeup of the population as the abundance of individuals with the favored traits increased (see Chapter 6). Such traits are known as **adaptations**. Through many generations, these unique, genetically based solutions to environmental stress would have become more frequent in the population.

Adaptation is similar to acclimatization in that both involve a change that minimizes the influence of a stress on the functioning of an organism, and the ability to acclimatize represents a type of adaptation. However, adaptation differs from acclimatization in being a long-term, genetic response of a population to environmental stress that increases its survival and reproductive success under the stressful conditions (**Figure 4.6**). Populations with adaptations to unique environments are called **ecotypes**. Ecotypes may be responses to both abiotic (e.g., temperature, water availability, soil type, salinity) and biotic environmental factors (e.g., competition, predation). Ecotypes can eventually become separate species as the physiology and morphology of individuals in different populations diverge and the populations eventually become reproductively isolated.

---

[*]Animal physiologists use the term "acclimatization" to refer to the short-term response of an animal to changes in the physical environment under field conditions and "acclimation" to refer to the short-term response under controlled laboratory conditions.

FIGURE 4.6 Organismal Responses to Stress Organisms respond to stress over different time scales. (After Lambers et al. 1998.)

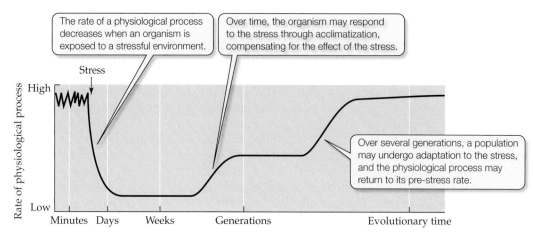

Returning to our previous example, human populations have lived continuously in the Andean highlands for at least 10,000 years. When the Spanish first settled there alongside the indigenous people in the sixteenth and seventeenth centuries, their birth rates were low for 2–3 generations, probably due to poor oxygen supply to developing fetuses (Ward et al. 1995). The same held true for the domesticated animals they brought with them. This observation provides anecdotal evidence for the adaptive response of the indigenous Andean populations to the low-oxygen conditions at high elevations. Research in the twentieth century showed that their adaptations to high elevations include higher red blood cell production and greater lung capacity (Ward et al. 1995).

Adaptations to environmental stress can vary among populations. In other words, the solution to the environmental problem may not be the same for each population. This is demonstrated in a comparison of native human populations from the Andean, Tibetan, and Ethiopian highlands. The adaptations to high elevations in Andean populations (high red blood cell concentration and large lung volume) are not the same as those found in Tibetan or Ethiopian populations (Beall 2006). Tibetan populations have red blood cell concentrations similar to populations at sea level, lower blood oxygen concentrations relative to populations at sea level, but have a higher breathing rate, enhancing the exchange of oxygen with the blood system. Populations living at high elevation in Ethiopia also do not show elevated level of red blood cells, yet remarkably their blood oxygen concentrations are similar to populations from sea level, higher than those found in the Andean and Tibetan populations (Beall 2006). The underlying mechanism for the maintenance of high blood oxygen concentrations in the high elevation Ethiopian populations has not been elucidated. Thus, there are three different ways that human populations have adapted to hypoxic stress imposed by living at high elevations.

Acclimatization and adaptation are not "free;" they require an investment of energy and resources by the organism (as in the maxim "there is no free lunch" presented in Chapter 1). They represent possible trade-offs with other functions of the organism that may also affect its reproduction and survival. Acclimatization and adaptation must therefore increase the survival and reproductive success of the organism under the specific environmental conditions in order to be favored over other patterns of energy and resource investment. (Trade-offs in energy and resource allocation are discussed in Chapter 7.)

In the remaining two sections, we will examine the factors that determine organisms' temperatures, water content, and water uptake. We will then consider examples of acclimatization and adaptation that allow organisms to function in the face of varying temperatures and water availability.

Variation in Temperature

CONCEPT 4.2
The temperature of organisms is determined by exchanges of energy with the external environment.

Environmental temperatures vary greatly throughout the biosphere, as we saw in Chapter 2. The boreal forest described earlier in this chapter represents one extreme of seasonal variation, with as much as an 80°C (144°F) swing from summer to winter. Tropical forests, on the other hand, experience far less seasonal variation in temperature, about 15°C (22°F). Soil environments, which are home to many species of microorganisms, plant roots, and animals, are buffered from aboveground environmental temperature extremes, although soil surface temperatures may change as much as or more than air temperatures. Aquatic environments also experience temperature changes over seasonal and daily time scales. Pelagic marine environments tend to have very little temporal variation in temperature due to the ocean's massive volume and heat capacity. In contrast, tide pools experience large variations in water temperature as the tides rise and fall, with as much as a 20°C (36°F) change over a 5-hour period.

The survival and functioning of organisms is strongly tied to their internal temperatures (**Figure 4.7**). The extreme upper

FIGURE 4.7 Temperature Ranges for Life on Earth

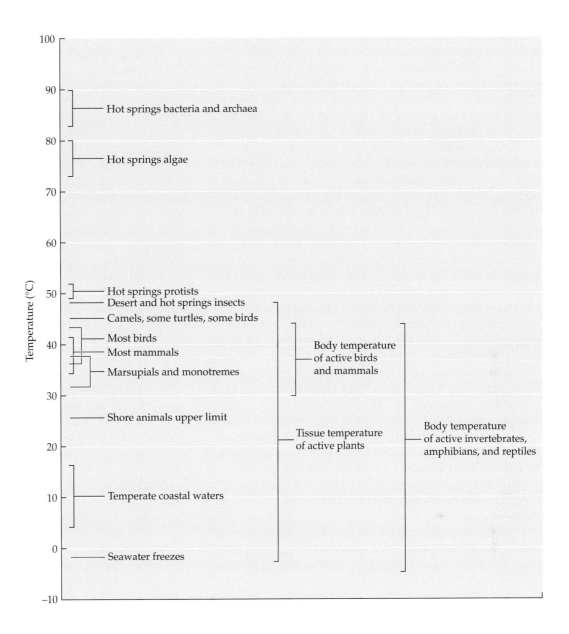

limit for metabolically active multicellular plants and animals is about 50°C (122°F), although some single-celled algae can function at temperatures as high as 70°C (158°F), and some archaea and bacteria that live in hot springs can function at 90°C (194°F) (Willmer et al. 2005). The extreme lower limit for organismal function is tied to the temperature at which water in cells freezes, typically between –2°C and –5°C (28°F– 23°F). Some organisms can survive periods of extreme heat or cold by entering a state of *dormancy*, in which little or no metabolic activity occurs.

The temperature of an organism is related to the balance between gains of energy from and losses of energy to the external environment. Thus, organisms must either tolerate changes in their internal temperature as the temperature of the external environment changes or modify their internal temperature by some physiological, morphological, or behav-

ioral means. Environmental temperatures—particularly their extremes—are therefore important determinants of the distributions of organisms, as demonstrated by the relationship between biomes and global climatic patterns discussed in the previous two chapters.

Temperature controls physiological activity

The metabolic generation of chemical energy and the manufacture of components for growth and reproduction depend on biochemical reactions that are temperature-sensitive. Each reaction has an optimal temperature that is related to the activity of *enzymes*, protein-based molecules that catalyze biochemical reactions. Enzymes are structurally stable under a limited range of temperatures. At high temperatures, the constituent proteins lose their structural integrity, or become *de-*

natured, as their bonds break. Most enzymes become denatured at temperatures ranging between 40°C and 70°C (104°F–158°F), but enzymes in bacteria inhabiting hot springs can remain stable at temperatures up to 100°C (212°F). The upper lethal temperature for most organisms is lower than the temperature at which their enzymes become denatured, probably because metabolic coordination among biochemical pathways is lost at these temperatures. The extreme lower limit for enzyme activity is about –5°C (23°F) (Willmer et al. 2005). Antarctic fish and crustaceans may reach body temperatures of –2°C because the salt concentration of the seawater in which they live lowers its freezing point, and soil microbes are active at temperatures as low as –5°C.

Temperature also determines the rates of physiological processes by influencing the properties of membranes, particularly at low temperatures. Cell and organelle membranes are composed of two layers of lipid molecules. At low temperatures, these membranes can solidify, and proteins and enzymes embedded in them can lose their function, affecting processes such as mitochondrial respiration and photosynthesis. Membranes also lose their function as filters when they solidify, leaking cellular metabolites. Tropical plants may suffer loss of function associated with membrane disruption at temperatures as high as 10°C (50°F), while alpine plants can function at temperatures close to freezing. The sensitivity of membrane function to low temperatures appears to be related to the chemical composition of the membrane lipid molecules. Plants of cooler climates have a higher proportion of unsaturated membrane lipids (with greater numbers of double bonds between carbon molecules) than plants of warmer climates.

Finally, temperature influences physiological processes in terrestrial organisms by affecting water availability. As we saw in Chapter 2, the warmer the air, the more water vapor it can hold. As a result, the rate at which terrestrial organisms lose water from their bodies is related to air temperature. We will return to this point later when we discuss how organisms cope with variations in water availability.

Organisms influence their temperature by modifying energy balance

On a hot day, jumping into a swimming pool and then sitting in the shade with a light wind brings relief from the oppressive heat. Elephants follow a similar routine, swimming and spraying water on their backs with their trunks. This kind of behavior facilitates heat loss in several ways. The direct contact of your warm skin with the cool water causes energy to be lost through the process of *conduction*, the transfer of energy from warmer, more rapidly moving molecules to cooler, more slowly moving molecules. When water and air move across the surface of your body, heat energy is carried away via *convection*. The change in the state of water from liquid to vapor as it evaporates on the surface of your skin ab-

sorbs heat (*latent heat transfer*). Less obvious is the heat lost when you move into the shade and radiate heat to cooler surrounding objects.

The balance between inputs and outputs of energy determines whether the temperature of any object, living or not, will increase or decrease. By adjusting their exchange of energy with the environment, organisms can influence their temperature and therefore their physiological activity. Avoidance of adverse temperatures through behavioral and morphological modifications of energy balance is a strategy used by both plants and animals. Plants and animals also exhibit varying degrees of tolerance of changes in body temperature. Archaea, bacteria, protists, and algae cannot avoid changes in their temperature when the environmental temperature changes. They must tolerate variations in temperature through biochemical modifications. When temperatures occur outside of their range of tolerance, microorganisms often survive as resistant spores.

Let's examine some examples of modification of energy balance in plants and animals to better understand the adaptations that help them cope with temperature variation.

Plant modification of energy balance   Among plants, temperature stress is experienced mainly by those that live in terrestrial environments. Marine and aquatic plants usually experience temperatures within the range that is conducive to their physiological functioning. The factors involved in the energy balance of terrestrial plants are shown in **Figure 4.8**. Energy inputs that warm the plant include sunlight and longwave (infrared) radiation from surrounding objects. If the ground and air are warmer than the plant, energy inputs also include conduction and convection. Losses of energy include the emission of infrared radiation to the surrounding environment. Heat loss also occurs through evaporation and transpiration (evaporation of water from inside the plant), collectively referred to as *evapotranspiration*. If the ground and air are cooler than the plant, energy losses also include conduction and convection.

We can put these inputs and outputs together to determine whether the temperature of the plant is changing:

$$\Delta H_{plant} = SR + IR_{in} - IR_{out} \pm H_{conv} \pm H_{cond} - H_{et} \quad \textbf{(4.1)}$$

where ΔH_{plant} is the heat energy change of the plant, SR is solar radiation, IR_{in} is the input of infrared radiation, IR_{out} is the output of infrared radiation, H_{conv} is convective heat transfer, H_{cond} is conductive heat transfer, and H_{et} is heat transfer through evapotranspiration. A negligible loss of energy occurs through the plant's use of solar radiation for photosynthesis. If the plant is warmer than the surrounding air, then H_{conv} and H_{cond} are negative. If the sum of the energy inputs exceeds the sum of the outputs, then ΔH_{plant} is positive, and the plant's temperature is increasing. Conversely, if more heat is being lost than gained, then ΔH_{plant} is negative, and the plant's temperature is decreasing.

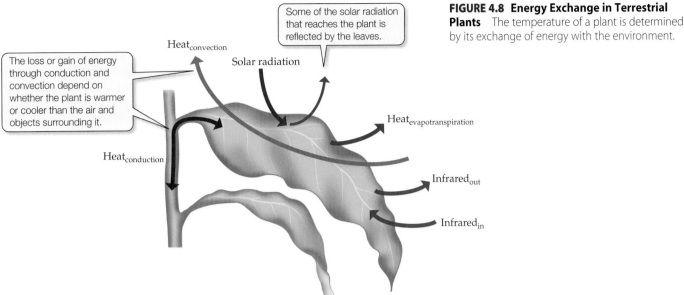

FIGURE 4.8 Energy Exchange in Terrestrial Plants   The temperature of a plant is determined by its exchange of energy with the environment.

Plants can modify their energy balance to adjust their temperature by adjusting these energy inputs and outputs. The most important and common adjustments include changes in the rate of transpirational water loss that cools the leaf, changes in surface reflective properties that alter the amount of solar radiation absorbed by the plant, and changes in convective heat transfer by changing surface roughness. Leaves are most often associated with these adjustments because they are the primary photosynthetic organs of the plant and typically are the most temperature-sensitive tissue.

Transpiration is an effective evaporative cooling mechanism for leaves. This is especially evident in the canopies of tropical forests, which are subject to warm air temperatures and high levels of solar radiation. Without transpirational cooling, the leaves of tropical canopy plants could reach lethal temperatures over 45°C (> 113°F). The rate of transpiration is controlled by specialized *guard cells* surrounding a pore,

called a **stomate**. Stomates are the gateway for both transpirational water loss and the uptake of carbon dioxide from the atmosphere (we will return to the latter function in the next chapter). Variation in the degree of stomatal opening, as well as the number of stomates, which can vary between shade and sun leaves, control the rate of transpiration and therefore exert an important control on leaf temperature (**Figure 4.9**).

Transpiration requires a steady supply of water. Where the supply of water in the soil is limited—as it is over a substantial part of Earth's land surface—transpiration is not a reliable

FIGURE 4.9 Stomates Control Leaf Temperature by Controlling Transpiration   (A) Specialized guard cells surrounding the stomates control their degree of opening. Open stomates allow CO_2 to diffuse in for photosynthesis and allow water to transpire out, cooling the leaves. (B) Leaf temperatures vary according to the degree of stomatal opening. The plant on the right has open stomates and is transpiring freely, while the plant on the left, kept under identical conditions, has closed stomates, a lower transpiration rate, and a temperature 1°C – 2°C (2°F – 4°F) higher, as indicated by thermal infrared imaging.

cooling mechanism. As we saw in Chapter 3, some plants shed their leaves during dry seasons, thereby avoiding both temperature stress and water stress. However, the high resource requirement for producing leaves may favor protecting existing leaves rather than shedding them. Plants that maintain their leaves during long dry periods require mechanisms other than transpiration to dissipate heat energy. One option is to alter the reflective properties of the leaf with hairs on the surface, a condition known as **pubescence**. The light color associated with white hairs lowers the amount of solar radiation absorbed by the leaf surface. Note, however, that pubescence can also lower the effectiveness of convective heat loss and thus represents a trade-off between two opposing heat exchange mechanisms.

One of the best studies addressing the adaptive significance of leaf pubescence for temperature regulation has focused on shrubs of the genus *Encelia*. Jim Ehleringer and his colleagues have demonstrated significant differences among species of *Encelia* (in the composite or daisy family) in leaf pubescence in relation to the temperature and water regimes of their native habitats. *Encelia farinosa*, a native of the Sonoran and Mojave deserts, maintains a high amount of leaf pubescence relative to *Encelia* shrubs from moister, cooler environments. Ehleringer and Craig Cook evaluated the relative roles of leaf pubescence and transpiration in the cooling of leaves of *E. farinosa* and two other species whose leaves lack pubescence: *E. frutescens*, which occurs in desert washes, and

E. californica, native to the cooler, moister coastal sage community of California and Baja California (Ehleringer and Cook 1990). To control for environmental variation that could influence the morphology and physiology of the plants, they grew plants of each species from seed in "common gardens" in the Sonoran Desert and on the California coast. Half of their experimental plants were watered, and the other half were left under natural conditions. They measured leaf temperatures, along with the degree of stomatal opening and the amount of sunlight absorbed.

The three *Encelia* species showed few differences in leaf temperature and stomatal opening when grown in the cooler, moister California coastal garden. In the desert garden, however, *E. californica* and *E. frutescens* shed their leaves during the hot summer months under natural conditions, while *E. farinosa* did not. *Encelia frutescens* did not shed its leaves when the shrubs were watered, and its leaves maintained sublethal temperatures through high transpiration rates. *Encelia farinosa* leaves reflected about twice as much solar radiation as the other species (**Figure 4.10A**), which facilitated the shrub's ability to maintain leaf temperatures lower than the air temperature.

Ehleringer and Cook's common garden experiment provides correlative evidence of the adaptive value of leaf pubescence to *Encelia farinosa* under hot desert conditions. Additional work by Darren Sandquist and Ehleringer has supported its adaptive value, indicating that natural selection

(A)

FIGURE 4.10 Sunlight, Seasonal Changes, and Leaf Pubescence (A) Solar heating of leaves varies according to the amount of pubescence on those leaves. The leaves of *Encelia farinosa* absorb less solar radiation than the leaves of two nonpubescent species: *E. californica*, native to the coastal sage community of California, and *E. frutescens*, an inhabitant of moister desert wash communities. *Encelia farinosa* is therefore less dependent on transpiration for leaf cooling than the other two species. (B) *Encelia farinosa* produces greater amounts of pubescence on its leaves in summer than in winter, representing acclimatization to hot summer temperatures. Photos are scanning electron micrographs of leaf cross-sections. (A after Ehleringer and Cook 1990.)

has acted on variation in pubescence among populations (ecotypes) of *E. farinosa*. Populations in drier environments have more leaf pubescence, and absorb less solar radiation, than populations from moister environments (Sandquist and Ehleringer 2003).

In addition to varying among species and populations, leaf pubescence can also vary seasonally in *E. farinosa*, exemplifying acclimatization to environmental conditions. *Encelia farinosa* shrubs produce smaller, more pubescent leaves in summer and larger, less pubescent leaves in winter (**Figure 4.10B**). There are costs to being pubescent, associated with the construction of the hairs and the loss of solar radiation that can be used for photosynthesis. Thus, when cooler temperatures occur or when reliable soil water is present, *E. farinosa* plants construct leaves with fewer hairs.

Heat can be lost from a leaf by convection when the air temperature is lower than the temperature of the leaf. The effectiveness of convective heat loss is related to the speed of the air moving across a surface. As the moving air encounters more resistance closer to the surface of an object, the flow becomes more turbulent, forming eddies (**Figure 4.11**). This zone of turbulent flow, called the **boundary layer**, lowers convective heat loss. The thickness of the boundary layer on a leaf is related to its size and its surface roughness. Small, smooth leaves have thin boundary layers and lose heat more effectively than large or rough leaves. This relationship between the boundary layer and convective heat loss contributes to the rarity of large leaves in desert ecosystems.

Excessive heat loss by convection can be a problem for plants (and animals) in cold, windy environments such as alpine ecosystems. Convection is the most important source of heat loss from the land surface in temperate alpine environments, and high winds can shred leaves in exposed sites. Most alpine plants hug the ground surface to avoid the high wind velocities. Some alpine plants produce a layer of insulating hair on their surface to lower convective heat loss. The

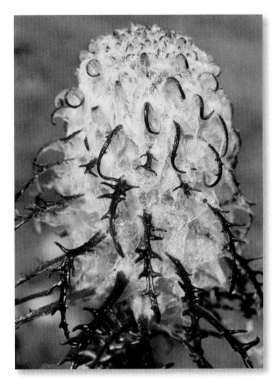

FIGURE 4.12 A Woolly Plant of the Himalayas The snow lotus (*Saussurea medusa*) has dense pubescence surrounding its emergent flowering stems, providing them with thermal insulation.

snow lotus of the Himalayas (*Saussurea medusa*) produces a series of very densely pubescent leaves that surround the flowers of the plant (**Figure 4.12**). Although they project above the ground surface and are exposed to more wind than ground-hugging plants, the flowers of *S. medusa* remain as much as 20°C (36°F) warmer than the air by absorbing and retaining solar radiation (Tsukaya et al. 2002). The plant not only maintains warmer photosynthetic tissues, but also provides a warm environment for potential pollinators, which are in short supply in cold, windy alpine environments.

Animal modification of energy balance Animals are subject to the same heat exchange factors described for plants in Equation 4.1, with a key difference: some animals—in particular, birds and mammals—have the ability to generate heat internally. As a result, another term is needed in the energy balance equation representing this metabolic heat generation:

$$\Delta H_{animal} = SR + IR_{in} - IR_{out} \pm H_{conv} \pm H_{cond} - H_{evap} + H_{met} \quad (4.2)$$

where ΔH_{animal} is the heat energy change of the animal, SR is solar radiation, IR_{in} is the input of infrared radiation, IR_{out} is the output of infrared radiation, H_{conv} is convective heat transfer, H_{cond} is conductive heat transfer, H_{evap} is heat transfer through evaporation, and H_{met} is metabolic heat generation. In contrast to plants, evaporative heat loss is not wide-

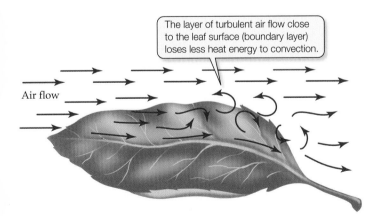

The layer of turbulent air flow close to the leaf surface (boundary layer) loses less heat energy to convection.

Air flow

FIGURE 4.11 A Leaf Boundary Layer Air flowing across the surface of a leaf is subject to friction, causing the flow to become turbulent, lowering convective heat loss from the leaf to the surrounding air.

FIGURE 4.13 Internal Heat Generation as a Defense Bees can generate heat by contracting their flight muscles. Japanese honeybees (*Apis cerana*) use internal heat generation as a defense against hornets (*Vespa mandarina*) that attack bee colonies. (A) When a hornet enters a nest, honeybees swarm the larger invader. (B) The defensive ball of bees surrounding an invading hornet generates enough heat that temperatures in the center exceed the upper lethal temperature for the hornet (about 47°C), resulting in the invader's death.

(A)

(B)

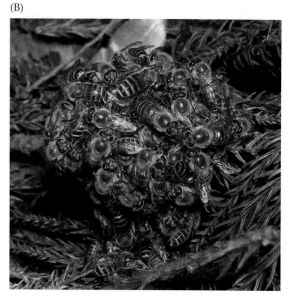

spread among animals. Notable examples of evaporative cooling in animals include sweating in humans, panting by dogs and other animals, and licking of the body by some marsupials under conditions of extreme heat.

The internal generation of heat by some animals represents a major ecological advance. Animals capable of metabolic heat generation can maintain relatively constant internal temperatures near the optimum for physiological functioning under a wide range of external temperatures, and as a result, can expand their geographic ranges. Varying degrees of reliance on internal heat generation exist throughout the animal kingdom. Animals that primarily regulate their body temperature through energy exchange with the external environment, which includes the majority of animal species, are called **ectotherms**. Animals that rely primarily on internal heat generation, which are called **endotherms**, include, but are not limited to, birds and mammals. Internal heat generation is also found in fish (e.g., tuna), insects (e.g., bees, which generate heat for metabolic function as well as for defense; **Figure 4.13**), and even a few plant species (e.g., skunk cabbage, *Symplocarpus foetidus*, warms its flowers using metabolically generated heat during the spring).

Temperature regulation and tolerance in ectotherms Generally, ectotherms have a greater tolerance for variation in their body temperature than endotherms, presumably because they cannot change their body temperatures as readily as endotherms. The exchange of heat with the environment, whether for cooling or heating, depends on the amount of surface area relative to the volume of the animal. A larger surface area relative to volume allows greater heat exchange, but makes it harder to maintain a constant internal temperature in the face of variable external temperatures.

A smaller surface area relative to volume decreases the animal's ability to gain or lose heat. This relationship between surface area and volume imposes a constraint on the body size and shape of ectothermic animals. For a semi-spherical body shape, as body size increases, the surface area-to-volume ratio decreases, and the ability to exchange heat with the environment decreases. As a result, large ectothermic animals are considered improbable, and that conclusion has led to speculation that large dinosaurs may have had some degree of endothermy.

Small aquatic ectotherms generally remain at the same temperature as the surrounding water. Some larger aquatic animals, however, demonstrate an ability to maintain a body temperature significantly warmer than that of the surrounding water. For example, skipjack tuna (*Katsuwonis pelamis*) use muscle activity, in conjunction with heat exchange between blood vessels, to maintain a body temperature as much as 14°C (25°F) warmer than the surrounding seawater (**Figure 4.14**). Other large oceanic fishes use similar circulatory heat exchange mechanisms to keep their muscles warm. This is particularly important for predatory species that depend on rapid acceleration to capture prey, which is aided by having warmer muscles.

The mobility of many terrestrial ectotherms allows them to move to places that are warmer or cooler than they are to adjust their body temperature. Basking in the sun or moving into the shade allows these animals to adjust their energy gains and losses via solar radiation, conduction, and infrared radiation. For example, reptiles and insects emerging from hiding places after a cool night will bask in the sun to warm their bodies prior to initiating their daily activities (**Figure 4.15**). This basking behavior, however, increases their risk of being found by predators. Many of these animals rely on

(A)

(B)

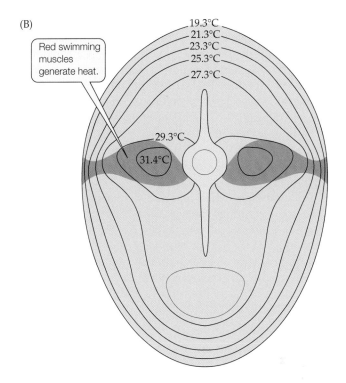

Red swimming muscles generate heat.

19.3°C
21.3°C
23.3°C
25.3°C
27.3°C
29.3°C
31.4°C

Red swimming muscle

Vertebra

FIGURE 4.14 Internal Heat Generation by Tuna (A) Heat generated in the red swimming muscles of the skipjack tuna, used for cruising through the water, warms blood flowing through them, which is carried toward the body surface in veins. Those veins run parallel to arteries carrying cool oxygenated blood from the gills, warming that blood before it reaches the swimming muscles. (B) A cross-section of the tuna shows that its core temperatures remain warmer than that of the surrounding water.

camouflage (also called *crypsis*) to escape detection while basking.

In temperate and polar regions, temperatures drop below freezing for extended periods. Ectotherms inhabiting these regions must either avoid or tolerate exposure to subfreezing conditions. Avoidance may take the form of seasonal migration (e.g., moving to a lower latitude) or movement to local microsites where temperatures stay at or above freezing (e.g., burrowing into the soil). Tolerance of subfreezing temperatures involves minimizing the damage associated with ice formation in cells and tissues. If ice forms as crystals, it will puncture cell membranes and disrupt metabolic functioning. Some insects inhabiting cold climates contain high concentrations of glycerol, a chemical compound that minimizes the formation of ice crystals and lowers the freezing

point of body fluids. These insects spend winter in a semi-frozen state, emerging in spring when temperatures are more conducive to physiological activity. Vertebrate ectotherms generally do not tolerate freezing to the degree that invertebrate ectotherms do because of their larger size and greater physiological complexity. A very few amphibians, however, can survive being partially frozen, as described in the Case Study at the opening of this chapter.

Temperature regulation and tolerance in endotherms As noted earlier, despite their tolerating a narrower range of body temperatures than ectotherms (30°C–45°C, 86°F–113°F), the ability of endotherms to generate heat internally has allowed

FIGURE 4.15 Mobile Animals Can Use Behavior to Adjust Their Body Temperature This ornate tree lizard (*Urosaurus ornatus*) has moved to a sunny location where it can absorb solar radiation to raise its body temperature to a range suitable for undertaking its daily activities. Basking behavior can expose animals to the risk of predation, but camouflage (crypsis) helps minimize this risk.

them to greatly expand their geographic ranges. Endotherms can remain active at subfreezing temperatures, something that most ectotherms cannot do. The cost of being endothermic is a high demand for energy (food) to support metabolic heat production. The rate of metabolic activity in endotherms is associated with the external temperature and the rate of heat loss. The rate of heat loss, in turn, is related to body size due to its influence on surface area-to-volume ratio. Small endotherms have higher metabolic rates, and require more energy and higher feeding rates, than large endotherms.

Endothermic animals maintain a constant resting metabolic rate over a range of environmental temperatures known as the **thermoneutral zone**. When the environmental temperature drops to a point at which heat loss is greater than metabolic heat production, the body temperature begins to drop, triggering an increase in metabolic heat generation. This

point is called the **lower critical temperature** (**Figure 4.16A**). The thermoneutral zone and the lower critical temperature differ among mammal species (**Figure 4.16B**). As one would expect, animals from the Arctic have lower critical temperatures below those of animals from tropical regions. Note also that the rate of metabolic activity increases more rapidly below the lower critical temperature in tropical than in Arctic mammals.

What is the reason for these differences between animals of different biomes? For endothermy to work efficiently, animals must be able to retain body heat. Thus, the evolution of endothermy in birds and mammals required insulation: feathers, fur, and fat. These insulating layers provide a barrier limiting conductive (and in some cases, convective) heat loss. Fur and feathers insulate primarily by providing a layer of still air, similar to a boundary layer, adjacent to the skin. Differences in insulation help explain the differences in lower critical temperatures and the speed of metabolic rate increases among the endotherms in Figure 4.16B. Arctic mammals generally maintain thicker fur than mammals of warmer climates. In warm climates, the ability to cool off through conduction and convection is inhibited by insulation, and thick fur can be an impediment to maintaining an optimal body temperature. Some endotherms show seasonal acclimatization to temperature by growing thicker fur in winter, which is shed when temperatures get warmer (a fact that most pet owners know well). Our human ancestors lost much of their hairy insulating layer about 2 million years ago, when they inhabited the hot tropical regions of Africa (Jablonski 2006).

Cold climates are tough on small endotherms. Small mammals, by necessity, have thin fur, since thick fur would inhibit their mobility. The high demand for metabolic energy below

(A)

When the environmental temperature drops below the lower critical temperature, the metabolic rate begins to increase in order to generate more heat.

Thermoneutral zone

The resting metabolic rate remains the same as long as the environmental temperature is within the thermoneutral zone.

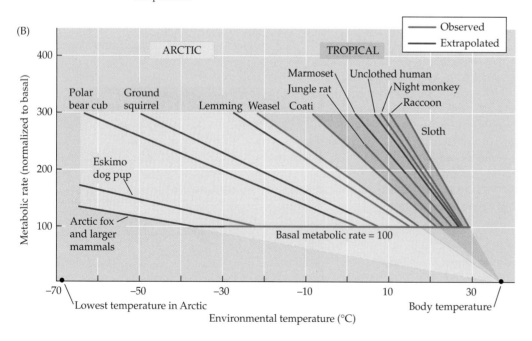

FIGURE 4.16 Metabolic Rates in Endotherms Vary with Environmental Temperatures (A) An endotherm's resting metabolic rate stays constant throughout a range of temperatures known as the *thermoneutral zone*. When environmental temperatures reach a lower limit, known as the *lower critical temperature*, its metabolic rate increases to generate additional heat. (B) The thermoneutral zones and lower critical temperatures of endotherms vary with their habitats. The lower critical temperatures of Arctic endotherms are lower than those of tropical endotherms, and their metabolic rates increase more slowly below those lower critical temperatures, as shown by the shallower slopes of the curves. (After Scholander et al. 1950.)

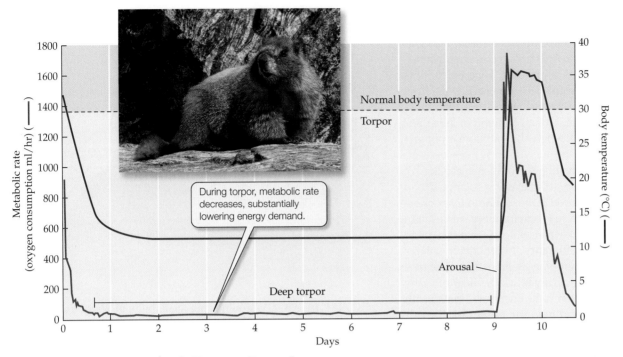

During torpor, metabolic rate decreases, substantially lowering energy demand.

FIGURE 4.17 Long-Term Torpor in Marmots   Torpor allows marmots (*Marmota flaviventris*) to conserve energy during winter, when food is scarce and the demand for metabolic energy to keep warm is high.

the lower critical temperature, the low insulation values of their fur, and their low capacity to store energy make small mammals improbable residents of polar, alpine, and temperate habitats. However, the faunas of many of these cold climates contain many small endotherms, sometimes in high abundances. What explains this apparent discrepancy? Small endotherms, such as rodents and hummingbirds, are able to alter their lower critical temperature during cold periods by entering a state of dormancy known as **torpor**. The body temperatures of animals in torpor may drop as much as 20°C (36°F) below their normal temperatures. The metabolic rate of an animal in torpor is 50%–90% lower than its normal metabolic rate, providing substantial energy savings (Schmidt-Nielsen 1997). However, energy is still needed to arouse the animal from torpor and bring its body temperature back up to its usual set point. Thus, the length of time an animal can remain in torpor is limited by its reserves of energy. Small endotherms may regularly undergo a daily torpor to minimize energy needed during cold nights. Torpor lasting several weeks during the winter is possible only for animals that have access to enough food and can store enough energy reserves, such as marmots (**Figure 4.17**). Torpor is somewhat rare in polar climates, due to the lack of food to provide enough energy storage in fat to get through the winter without eating. Some large animals, such as bears, enter into a long-term winter sleep, during which the body temperature decreases only slightly, rather than going into torpor.

Variation in Water Availability

CONCEPT 4.3
The water balance of organisms is determined by exchanges of water and solutes with the external environment.

Water is essential for life. Water is the medium in which all biochemical reactions necessary for physiological functioning occur. Water has unique properties that make it a universal solvent for biologically important solutes. The range of organismal water content conducive to physiological functioning is relatively narrow, between 60% and 90% of body mass. Maintaining an optimal water content is a challenge primarily to organisms of freshwater and terrestrial environments. Marine organisms seldom gain or lose too much water because they exist in a medium that is conducive to maintaining water balance: the oceans in which life first evolved. Terrestrial organisms face the problem of losing water to a dry atmosphere, while freshwater organisms may lose solutes to and gain water from their environment. The evolution of freshwater and terrestrial organisms is very much a story of dealing with the need to maintain water balance. In this section, we will review some basic concepts related to water and solute balance and provide some examples of how freshwater and terrestrial organisms maintain a water balance that is conducive to physiological functioning.

Water flows along energy gradients

Water flows along gradients of energy, from high energy to low energy. What is an energy gradient in the context of water? Gravity represents one example that is intuitively obvious—liquid water flows downhill, following a gradient of potential energy. Another type of energy influencing water movement is pressure. When elephants spray water out of their trunks, the water is flowing from a condition of higher energy, inside the trunk with muscles exerting pressure on it, to a condition of lower energy, outside of the trunk where that pressure is not acting on it.

Other, less obvious factors that influence the flow of water are important to organismal water balance. When solutes are dissolved in water, the solution loses energy. Thus, if the water in a cell contains more solutes than the water surrounding it, water will flow into the cell to equilibrate the energy difference. Alternatively, solutes may be lost to the surrounding medium, but most biological membranes selectively block the flow of many solutes. In biological systems, the energy associated with dissolved solutes is called **osmotic potential**. The energy associated with gravity is called **gravitational potential**, but in a biological context it is important in water movement only in very tall trees. The energy associated with the exertion of pressure is called **pressure** (or *turgor*) **potential**. Finally, the energy associated with attractive forces on the surfaces of large molecules inside cells or on the surfaces of soil particles is called **matric potential**.

The sum of these energy components within an aqueous system determines its overall water energy status, or **water potential**. The water potential of a system can be defined mathematically as

$$\Psi = \psi_o + \psi_p + \psi_m \qquad (4.3)$$

where Ψ is the total water potential of the system (in units of pressure; usually megapascals, MPa), ψ_o is the osmotic potential (a negative value, because it lowers the energy status of the water), ψ_p is the pressure potential (a positive value if pressure is exerted on the system; a negative value if the system is under tension), and ψ_m is the matric potential (a negative value). Water will always move from a system of higher Ψ to a system of lower Ψ, following the energy gradient. This terminology is most often used in plant, microbial, and soil systems, but it works in animal systems as well.

The atmosphere has a water potential that is related to humidity. From a biological perspective, air with a humidity of less than 98% represents a system with very low water potential, so the gradient in water potential between most terrestrial organisms and the atmosphere is very high. Without some barrier to water movement, terrestrial organisms would lose water rapidly to the atmosphere. Any force that impedes the movement of water (or other substances such as CO_2) moving along an energy gradient is called **resistance**.* Barriers that increase organisms' resistance to water loss include the waxy cuticles of plants and insects and the skin of amphibians, reptiles, birds, and mammals.

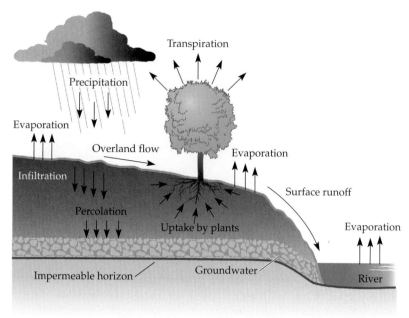

FIGURE 4.18 What Determines the Availability of Water from the Soil? The water content of soil is determined by the balance between water inputs (precipitation and overland flow of water) and outputs (percolation to deeper layers, evapotranspiration) and by the capacity of the soil to hold water. Soil water storage capacity and the rate of percolation are dependent on soil texture.

Terrestrial plants and microorganisms rely on water uptake from soils to replace the water they lose to the atmosphere. Soils are important reservoirs of water that support a multitude of ecological functions, as we will see throughout this book. The water potential of most soils is determined by its matric potential, which is related to the attractive forces of the soil particles. The osmotic potential of some soils can be important, particularly where dissolved salts are found, such as near marine environments or where salinization (see p. 45) has occurred. The amount of water that can be stored in the soil is related to a balance between inputs and outputs, soil texture, and topography (**Figure 4.18**). Inputs include precipitation and overland flow of water. Outputs include percolation to deeper layers below the plant rooting zone and evapotranspiration. Sandy soils store less water than fine-textured soils, but fine soil particles also have a higher matric potential, and thus hold onto water more tightly. Soils with mixed coarse and fine particles are generally most efficient in storing water and supplying it to plants and soil organisms. When the volume of water in the soil drops below a certain point (25% of total soil mass in fine-textured soils and 5% in sandy soils), the matric forces are strong enough that most of the remaining water is unavailable to most plants and soil microorganisms.

---

*Many physiologists prefer using "conductance" rather than "resistance" to express the influence of a barrier on the movement of water and gases between an organism and its environment. Mathematically, conductance is the reciprocal of resistance.

Controls on water balance in microorganisms   Single-celled microorganisms, including archaea, bacteria, algae, and protists, are active primarily in aqueous environments. Their water balance is dependent on the water potential of the surrounding environment, which is determined mainly by its osmotic potential. In most marine and freshwater ecosystems, the osmotic potential of the environment changes little over time, and organisms in these environments are not challenged by changes in water potential. Some environments, however, such as estuaries, tide pools, saline lakes, and soils, experience frequent changes in osmotic potential due to evaporation or variable influxes of fresh and salt water. Microorganisms in these environments must respond to these changes by altering their cellular osmotic potential if they are to maintain a water balance suitable for physiological functioning. They accomplish this through **osmotic adjustment**, an acclimatization response that involves changing the solute concentration, and thus the osmotic potential, of the cell. Some organisms synthesize organic solutes to adjust their osmotic potential, which also help to stabilize enzymes. Some organisms use inorganic salts from the surrounding medium for osmotic adjustment. The ability to adjust osmotic potential in response to external water potential varies substantially among microorganisms: some completely lack this ability, while others (such as *Halobacterium* spp.) can adjust to even the extremely saline conditions in saline lakes of dry climates.

As noted above, terrestrial environments are too dry for any organism that does not have a barrier to prevent cellular water loss to the atmosphere. Many microorganisms avoid exposure to dry conditions by forming resistant spores, encasing themselves in a protective coating that prevents water loss to the dry environment. Some microorganisms with filamentous forms, such as fungi and yeasts, are very tolerant of low water potentials and can grow in dry environments. Most terrestrial microorganisms, however, are found in soils, which have a higher water content and humidity than the air above them.

Controls on water balance in plants   One of the distinguishing characteristics of plants is a rigid cell wall composed of cellulose. Bacteria and fungi also have cell walls, composed of materials such as chitin (found in fungi) or peptidoglycans and lipopolysaccharides (found in bacteria). Cell walls are important to water balance because they result in the development of a positive **turgor pressure**. When water moves into a plant cell following a gradient in water potential, it causes the cell to expand and press against the cell wall, which resists the pressure due to its rigidity (**Figure 4.19**). Turgor pressure has been described as the hydrostatic skeleton of the terrestrial plant world, and it is an important force for growth in plants, promoting cell division. When non-woody plants lose turgor pressure due to dehydration, they wilt. Wilting is generally a sign that a terrestrial plant is experiencing water stress.

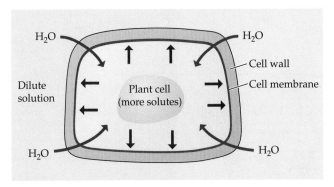

FIGURE 4.19 Turgor Pressure in Plant Cells   When a plant cell is surrounded by water with a solute concentration lower than its own, water moves into the cell following the gradient in water potential, while solutes in the cell are prevented from moving out by the cell membrane. The increasing amount of water in the cell causes the cell to expand, pressing against the cell wall. The pressure generated by this expansion is turgor pressure, an important structural component of plants.

Plants take up water from sources with a water potential higher than their own. For aquatic plants, the source is the surrounding aqueous medium. In freshwater environments, the presence of solutes in the plant's cells creates a water potential gradient from the surrounding water to the plant. In marine environments, plants must lower their water potential below that of seawater to take up water. Marine plants, as well as terrestrial plants of salt marshes and saline soils, adjust their osmotic potential in a manner similar to microorganisms by synthesizing solutes and taking up inorganic salts from their environment. Inorganic salts must be taken up selectively, however, since some (e.g., Na^+, Cl^-) can be toxic at high concentrations. The cell membranes of plants act as a solute filter, determining the amounts and types of solutes that move into and out of the plant.

Terrestrial plants acquire water from the soil through their roots, as well as from beneficial fungi that grow into their roots from the soil, called *mycorrhizae* (see Chapter 14). The earliest land plants, which had not yet evolved roots, used mycorrhizae to take up water and nutrients from the soil. The majority of land plants today use a combination of roots and mycorrhizae to take up water. Only the finest roots can take up water from the soil, since older, thicker roots develop a water-resistant waxy coating that limits their ability to absorb water as well as to lose water to the soil. Mycorrhizae provide a greater surface area for absorption of water and nutrients for the plant, and allow greater exploration of the soil for these resources. Mycorrhizal fungi obtain energy from the plant.

Plants lose water by transpiration when their stomates open to allow CO_2 from the atmosphere to diffuse into their leaves. Water moves out through the stomates following the

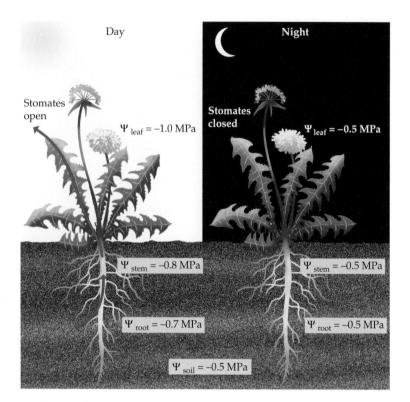

Day

Night

Stomates open

$\Psi_{leaf} = -1.0$ MPa

Stomates closed

$\Psi_{leaf} = -0.5$ MPa

$\Psi_{stem} = -0.8$ MPa

$\Psi_{stem} = -0.5$ MPa

$\Psi_{root} = -0.7$ MPa

$\Psi_{root} = -0.5$ MPa

$\Psi_{soil} = -0.5$ MPa

FIGURE 4.20 The Daily Cycle of Dehydration and Rehydration During the day, when the stomates open to take up CO_2 for photosynthesis, transpiration results in water potential gradients from leaf to stem, stem to roots, and roots to soil. At night, when the stomates are closed, these water potential gradients are relaxed as the plant rehydrates.

To avoid reaching detrimentally or even lethally low water content levels, plants must restrict their transpirational water losses. If leaf cells become so dehydrated that turgor is lost, the stomates close. This level of water stress can be detrimental to the plant, causing impairment of physiological functions such as photosynthesis. Some plants of seasonally dry environments shed their leaves during long dry periods to eliminate transpirational water loss (see p. 58). Others have a signaling system that helps prevent the onset of water stress. As the soil dries out, the roots send a hormonal signal (abscisic acid) to the stomates that triggers a decrease in the degree of opening, lowering the rate of water loss. Plants of dry environments,

water potential gradient from the inside of the leaf (100% relative humidity) to the air. As we saw in the previous section, transpiration is an important cooling mechanism for leaves. The plant must replace the water lost by transpiration, however, if it is to avoid water stress. As a leaf loses water, the water potential of its cells decreases, creating a water potential gradient between the leaf and the xylem in the stem to which it is attached, so water moves through the xylem into the leaf. In this way, while the plant is transpiring, it creates a gradient of decreasing water potential from the soil through the roots and stems to the leaves (**Figure 4.20**). Water flows from moist soils, which have the highest water potential, into the roots, the xylem, and eventually the leaves, from which it is lost to the atmosphere via transpiration.

Because there is greater resistance to the movement of water into the roots and through the xylem than out through the stomates, the water supply from the soil cannot keep up with water loss by transpiration. As a result, the water content of the plant decreases during the day. At night, however, the stomates close, and the water supply from the soil rehydrates the plant until it reaches near-equilibrium with the soil water potential. This daily cycle of daytime dehydration and nighttime rehydration can go on indefinitely if the supply of water in the soil is adequate. When precipitation is not sufficient to recharge the soil water, however, the supply may become depleted through evapotranspiration. The water content of the plant will then decrease as its osmotic potential increases, and its turgor pressure will decrease as its cells become dehydrated (**Figure 4.21**).

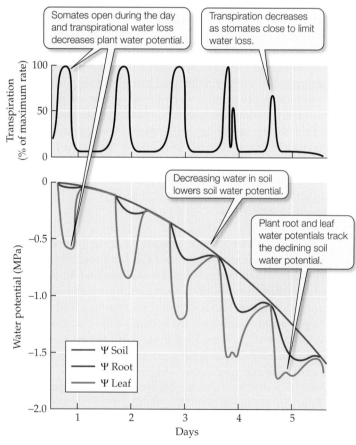

Somates open during the day and transpirational water loss decreases plant water potential.

Transpiration decreases as stomates close to limit water loss.

Transpiration (% of maximum rate)

100

50

0

Decreasing water in soil lowers soil water potential.

Plant root and leaf water potentials track the declining soil water potential.

Water potential (MPa)

0

−0.5

−1.0

−1.5

−2.0

Ψ Soil
Ψ Root
Ψ Leaf

1   2   3   4   5

Days

FIGURE 4.21 Depletion of Soil Water Without recharge of soil water, transpiration will deplete pools of soil water, leading to progressive drying of the soil and a decrease in soil water potential.

such as deserts, grasslands, and Mediterranean-type ecosystems, generally have better control of stomatal opening than plants growing in wetter climates. These plants also have thicker waxy coatings (cuticles) on their leaves to prevent water loss through the nonporous regions of the leaves. Additionally, plants of dry environments maintain a higher ratio of root biomass relative to the biomass of stems and leaves than plants of moister environments, enhancing the rate of water supply to transpiring tissues (Mokany et al. 2006) (**Figure 4.22**). Some plants are capable of acclimatization by altering the growth of their roots to match the availability of soil moisture and nutrients.

When conditions are extremely dry, plant water potentials can by very low. The xylem can be under high tension (negative ψ_p), which can pull air into the water column, causing a disruption in the plant's plumbing, termed **cavitation**. Cavitation can also occur in woody plants of cold climates. During winter, water in the xylem may freeze. Because air is not as soluble in ice as it is in liquid water, bubbles form when the water in the xylem freezes. In spring, when the water thaws, the bubbles can coalesce and form a larger bubble the width of the xylem tube, breaking the water column. A cavitated xylem element is nonfunctional, since the water column must be intact for water to flow through the xylem. There are multiple xylem columns in a plant, so cavitation in a few does not result in a catastrophic loss of water supply. If enough xylem columns are lost to cavitation, however, some tissues will die. Refilling of cavitated xylem can occur in some species once the water supply increases.

Can plants have too much water? Technically, no, but wet soils inhibit the diffusion of oxygen, and can cause anoxic conditions in saturated soils. Thus, waterlogged soils inhibit aerobic respiration in roots. Very moist soils also enhance the growth of harmful fungal species that can damage roots. Ironically, the combination of these factors can lead to root death, and plants in waterlogged soils can wilt due to a loss of their water supply. Adaptations to low oxygen concentrations in wet soils include air channels in root tissues (called *aerenchy-*

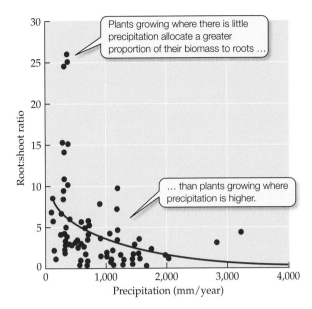

FIGURE 4.22 Allocation of Growth to Roots versus Shoots Is Associated with Precipitation Levels   The ratio of root biomass to leaf and stem (shoot) biomass increases with decreasing precipitation in shrubland and grassland biomes. Allocation of more biomass to roots in dry soils provides more water uptake capacity to support leaf function. (After Mokany et al. 2006.)

ma) as well as specialized roots (e.g., *pneumatophores*) in plants such as mangroves (see p. 73) that grow vertically above the water or waterlogged soil the plant is growing in.

Controls on water balance in animals   Multicellular animals face the same challenges plants and microorganisms do in maintaining water balance. Water losses and gains in animals, however, are governed by a more diverse set of exchanges than in plants and microorganisms (**Figure 4.23**). Many animals have the added complexity of specialized organs for gas exchange, digestion, excretion, and circulation, which create

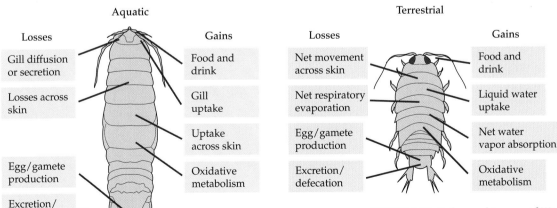

FIGURE 4.23 Gains and Losses of Water and Solutes in Aquatic and Terrestrial Animals

areas of localized water and solute exchange as well as gradients of water and solutes within the animal's body. Most animals are mobile and can seek out environments conducive to maintaining a favorable water and solute balance, an option not available to most microorganisms and plants.

Aquatic environments may be more saline (*hyperosmotic* to the animal; *hyper*, "greater"), of similar salinity (*isoosmotic*; *iso*, "same"), or less saline (*hypoosmotic*; *hypo*, "less") than an animal's cells or blood, so salt balance is intimately tied to water balance. Marine invertebrates rarely face problems with water balance because they tend to be isoosmotic, except in environments where salinity can change, such as estuaries and tide pools. Under conditions of varying salinity, animals must be able to adjust their gains and losses of water and solutes to maintain favorable water and solute balances. A marine animal that lacks this ability will die if transferred to brackish or fresh water.

Although most marine invertebrates are isoosmotic to seawater, the specific types of solutes in their bodies can vary. Many invertebrates that are capable of osmotic adjustment to changes in the solute concentration of their environment do so by exchanging solutes with the surrounding medium.

Like plants, these animals selectively control the exchange of specific solutes with their environment, because some external solutes are toxic at the concentrations at which they are found in seawater, and because some internal solutes are needed for biochemical reactions. Jellyfish, squid, and crabs, for example, have sodium (Na^+) and chloride (Cl^-) concentrations similar to those of seawater, but their sulfate (SO_4^{2-}) concentrations may be one-half to one-fourth of those found in seawater.

Marine vertebrates include animals that are isoosmotic and hypoosmotic to seawater. The cartilaginous fishes, including the sharks and rays, have blood solute concentrations similar to those of seawater, although, as in invertebrates, their concentrations of specific solutes differ from those in seawater. In contrast, marine bony (teleost) fish and mammals evolved in fresh water and later moved into marine environments. Their blood is hypoosmotic to seawater. Fish exchange water and salts with their environment across the gills, which are also the organs of O_2 and CO_2 exchange, and through drinking and eating (**Figure 4.24A**). Salts that diffuse into or are ingested by marine teleost fishes must be continuously excreted in urine and through the gills against an

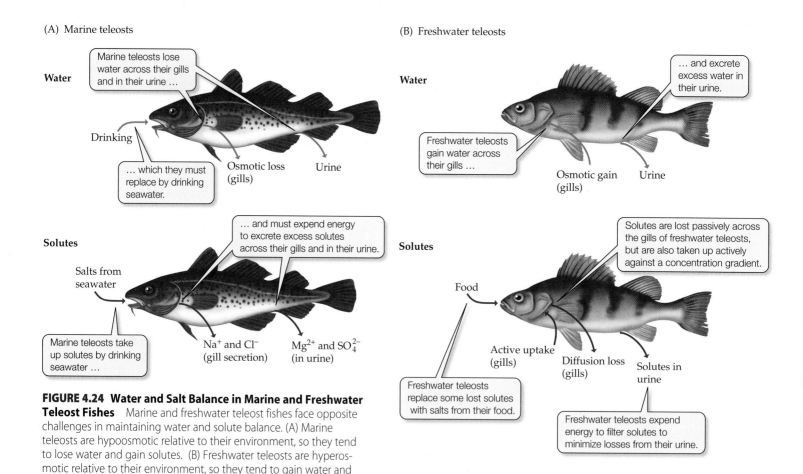

FIGURE 4.24 Water and Salt Balance in Marine and Freshwater Teleost Fishes Marine and freshwater teleost fishes face opposite challenges in maintaining water and solute balance. (A) Marine teleosts are hypoosmotic relative to their environment, so they tend to lose water and gain solutes. (B) Freshwater teleosts are hyperosmotic relative to their environment, so they tend to gain water and lose solutes.

osmotic gradient, which requires an expenditure of energy. Water lost across the gills must be replaced by drinking. Marine mammals, such as whales and porpoises, produce urine that is hyperosmotic to seawater and avoid drinking seawater to minimize salt uptake.

Freshwater animals are hyperosmotic to their environment; therefore, they tend to gain water and lose salts. Most salt exchange occurs at the respiratory exchange surfaces, including the skin of some invertebrates (e.g., freshwater worms) and the gills of many vertebrates and invertebrates. These animals must compensate for salt losses by taking up solutes in their food, and some groups, such as teleost fishes, must take up solutes actively through the gills against an osmotic gradient (**Figure 4.24B**). Excess water is excreted as dilute urine, from which the excretory system actively removes solutes to minimize their loss.

Terrestrial animals face the challenge of exchanging gases (respiratory O_2 and CO_2) in a dry environment with a very low water potential. Their key approaches to avoiding evaporative water loss and exposure to water stress are minimizing water loss by increasing skin resistance or living in environments where they can compensate for high water losses with high water intake. Risks and compromises are faced with both approaches. A high resistance to water loss may compromise the ability to exchange gases with the atmosphere. Reliance on a steady water supply puts the animal at risk if the source of water fails (e.g., during a severe drought). Tolerance for water loss varies substantially among groups of terrestrial animals. Generally, invertebrates have a higher tolerance for water loss than vertebrates. Within the vertebrates, amphibians have a higher tolerance for, but lower resistance to, to water loss than mammals and birds (**Table 4.1**).

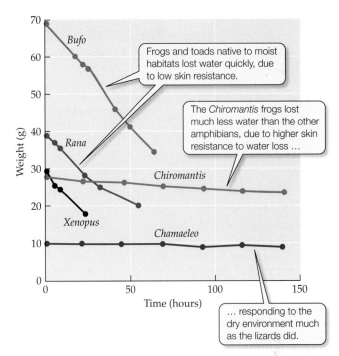

FIGURE 4.25 Resistance to Water Loss Varies among Frogs and Toads Amphibians were kept under uniform dry environmental conditions (25°C, 20%–30% relative humidity) to examine the rate of water loss, measured as weight loss. A lizard (*Chamaeleo*) was also tested for comparative purposes.

TABLE 4.1

Tolerances for Water Loss in Selected Animal Groups

Group	Weight loss (%)
INVERTEBRATES	
Mollusks	35–80
Crabs	15–18
Insects	25–75
VERTEBRATES	
Frogs	28–48
Small birds	4–8
Rodents	12–15
Human	10–12
Camel	30

Source: Willmer et al. 2005.

Note: Values are maximum percentages of body weight lost as water that can be tolerated, from a range of exemplary species in each group.

Amphibians, including frogs, toads, and salamanders, rely primarily on stable water supplies to meet their water balance. They can be found in a wide variety of biomes, from tropical rainforests to deserts, as long as there is a reliable source of water, such as regular rains or ponds. Amphibians depend on gas exchange through the skin to a greater degree than other terrestrial vertebrates. Therefore, amphibian skin is often thin, with a low resistance to water loss. However, some amphibian species have responded to dry environments by developing specialized skin with higher resistance to water loss. For example, the southern foam-nest frog (*Chiromantis xerampelina*), which occurs throughout Africa, has skin that resists water loss in a manner similar to that of lizards (**Figure 4.25**). To compensate for reduced gas exchange through the skin, *Chiromantis* frogs have higher breathing rates (Stinner and Shoemaker 1987). As a group, tree frogs have higher skin resistance to water loss than ground frogs, reflecting their drier environment. Some ground frogs of seasonally dry environments, such as the northern snapping frog (*Cyclorana australis*) of Australia, form a "cocoon" of mucous secretions consisting of proteins and fats that lower their rates of water loss.

Reptiles have been extremely successful at inhabiting dry environments. The thick skin of desert snakes and lizards provides protection for the internal organs as well as an ef-

TABLE 4.2

Resistance of External Cover (Skin, Cuticle) on Animals to Water Loss

Group	Resistance (seconds/centimeter)
Crabs (marine)	6–14
Fish	2–35
Frogs	3–100
Earthworms	9
Birds	50–158
Desert tortoises	120
Desert lizards	1,400
Desert scorpions, spiders	1,300–4,000

Source: Willmer et al. 2002.

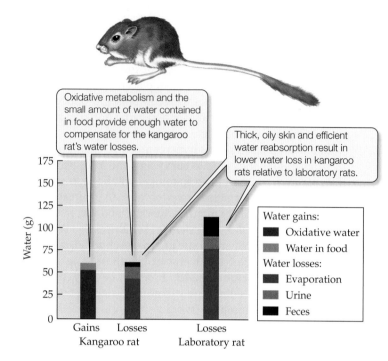

FIGURE 4.26 Water Balance in a Kangaroo Rat Under dry laboratory conditions (25°C, 25% relative humidity), kangaroo rats, native to deserts of the southwestern U.S., do not require liquid water to survive. (After Tracy and Walsberg 2002.)

fective barrier to water loss. The outer skin has multiple layers of dead cells with a fatty coating and is overlain by plates or scales. These layers give reptilian skin a very high resistance to water loss. Mammals and birds have skin anatomy similar to that of reptiles, but have hair or feathers covering the skin rather than scales. The presence of sweat glands in mammals represents a trade-off between resistance to water loss and evaporative cooling. The highest resistances to water loss among terrestrial animals are found in the arthropods (e.g., insects and spiders), which are characterized by an outer exoskeleton made of hard chitin and coated with waxy hydrocarbons that make it impervious to water movement (**Table 4.2**).

An instructive example of how animals use a variety of integrated adaptations to cope with arid environments involves the kangaroo rats (*Dipodomys* spp.) found throughout the deserts of North America. A combination of efficient water use and low rates of water loss greatly diminishes these rodents' water requirements (Schmidt-Nielsen and Schmidt-Nielsen 1951) (**Figure 4.26**). Kangaroo rats rarely drink water. A large proportion of their water requirement is met by eating dry seeds and generating water oxidatively; that is, by metabolically converting carbohydrates and fats into water and CO_2 (Schmidt-Nielsen 1964). The animals also consume more water-rich foods, such as insects or succulent vegetation, if they are available. Kangaroo rats minimize water loss through several physiological and behavioral adaptations. During the hottest periods of the year, they are active only at night, when air temperatures are lowest and humidities highest. During the day, they stay in their underground burrows, which are cooler and more humid than the desert surface. In some parts of their geographic range, however, kangaroo rats experience high enough temperatures even in their burrows to expose them to significant evaporative water losses (Tracy and Walsberg 2002). To

increase their resistance to these losses, kangaroo rats have thicker, oilier skin, with fewer sweat glands, than related rodents of moister environments. They minimize water losses in their urine and feces through effective removal of water by their kidneys and intestines. Kangaroo rats produce some of the most concentrated urine of any animal. The combination of these characteristics allows kangaroo rats to inhabit very arid environments without exposure to water stress, even without access to water for drinking.

CASE STUDY REVISITED

Frozen Frogs

The existence of amphibians above the Arctic Circle seems improbable, given their reliance on a steady supply of liquid water to maintain their water balance and the high potential for damage to complex organisms associated with freezing. Several problems must be overcome in order for organisms to survive freezing. First, when water freezes, it forms needle-like crystals that can penetrate and damage or destroy cell membranes and organelles. Second, the supply of oxygen to tissues is severely restricted due to the lack of circulation and breathing. Finally, as ice forms, pure water is pulled from cells, resulting in cell shrinkage and an increase in solute concentration. Any one of these factors, or all of them working in combination, will kill tissues and organisms in subfreezing temperatures. Yet the frogs described in the Case

Study, as well as many species of invertebrates, can tolerate the freezing of a substantial amount of their body water.

Wood frogs and other freeze-tolerant amphibians spend winter in shallow depressions under leaves, moss, or logs, which don't protect them from subfreezing temperatures. Several adaptations facilitate the survival of these amphibians through the winter and allow them to emerge from their frozen state in spring unharmed. Freezing of water in these animals is limited to the spaces outside the cells. A substantial proportion of their body water, from 35% to 65% in "fully frozen" frogs, freezes (Pinder et al. 1992). If more than 65% of their body water is frozen, most animals will die due to excessive cell shrinkage. The formation of ice outside the cells is enhanced by the existence of ice-nucleating proteins that serve as the site of slow, controlled ice formation (Storey 1990). Solute concentrations in the unfrozen cells increase as the cells lose water to extracellular ice formation. In addition, freeze-tolerant animals synthesize additional solutes, including glucose and glycerol derived from the breakdown of liver glycogen. The resulting increase in solute concentrations lowers the freezing point inside the cells, allowing the intracellular solution to remain liquid at subfreezing temperatures. The concentrated solutes also stabilize the cell volume and the structures of organelles, proteins, and enzymes. As freezing proceeds, the frog's heart stops, and its lungs cease to pump air. Once it reaches this semi-stable state of partial freezing, the frog can remain frozen for several weeks, as long as the temperature does not drop below about –5°C (23°F). Although their winter "quarters" are not far below the surface of the ground, the insulating cover of leaves and snow keeps the frogs above this temperature.

The freezing process is initiated in wood frogs within minutes of ice formation within the animal, although the full process occurs over several days to weeks (Layne and Lee 1995). Thawing, on the other hand, may be rapid, with normal body functioning returning within 10 hours (a video of a thawing frog can be seen at http://www.units.muohio.edu/cryolab/projects/WoodFrogFreezingSurvivalhigh.htm). This amazing amphibian feat of spending winter in a semi-frozen state and emerging unharmed in spring has provided information to medical science that has facilitated the preservation of human tissues and organs at low temperatures (Costanzo et al. 1995), as well as optimism to proponents of whole-body cryonics, who hope that someday Grandpa can finally leave the Tuff Shed.

Desiccation Tolerance, Body Size, and Rarity

While subfreezing temperatures are an important constraint on the distribution and functioning of organisms in high-latitude and high-elevation biomes, low water availability is a more widespread challenge. Arid conditions can potentially occur in most terrestrial biomes, and they regularly occur over more than 60% of the land surface. As we have seen, the majority of organisms, particularly animals, avoid exposure to dry conditions and rely on minimizing water losses to the environment. However, some organisms can tolerate arid conditions, much in the same way the frozen frogs tolerate subfreezing winter conditions, by entering a state of suspended animation while allowing themselves to dry out. This adaptive approach is common in microorganisms, including bacteria, fungi, and protists, but is also found in some multicellular animals and some plants, including mosses, liverworts, and a few flowering plants (Alpert 2006).

Desiccation-tolerant organisms can survive extreme dehydration, losing 80%–90% of their water as they equilibrate with the humidity of the air, then regain metabolic function shortly after they are rehydrated (**Figure 4.27**). As the cells dry out, the organisms synthesize sugars, which are the key

(A)

(B)

(C)

FIGURE 4.27 Desiccation-Tolerant Organisms
(A) The leaves of this club moss (*Selaginella lepidophylla*) reach a very low moisture content during prolonged periods without rain. (B) Within 6 hours of receiving water, the leaves are functional and carrying out photosynthesis. (C) Water bears (tardigrades) are small invertebrates (less than 1 mm in length) found in aquatic environments, including lakes and ponds, soils, the water films on vegetation, and the ocean. Water bears curl up and cease metabolism when they and their environment dry up.

to protecting the cell and organelle structure (Alpert 2006). Once dehydration proceeds beyond a particular threshold, metabolism ceases, and the sugars and the small amount of remaining water form a glassy coating over the cellular constituents. As with recovery from freezing, the recovery from dehydration is rapid, occurring in hours to days.

The prevalence of dry conditions in terrestrial environments suggests that desiccation tolerance should be more common than it is. Why aren't more plants and animals desiccation-tolerant? A clue to this puzzle may be the small size of the organisms that are desiccation-tolerant (Alpert 2006). Small organisms (less than 5 mm in animals) do not require structural reinforcement, such as a skeletal system, which would restrict the necessary shrinking of the organism as it dehydrates. In addition, water loss during dehydration must be slow enough to allow the adaptive response of sugar synthesis to occur, but not so slow that the organism spends a long time with a low water content while metabolism is still occurring, which can cause physiological stress. Small organisms have surface area-to-volume ratios and thicknesses favorable for the water loss rates required.

These arguments explain why small size is favored in desiccation-tolerant organisms, but not why they are rare. The two characteristics—small size and rarity—are intimately linked. As we will see later in this book, small size is often associated with slow growth rates and poor competitive ability under conditions of low resource availability. Thus, natural selection for desiccation tolerance may involve trade-offs with other ecological characteristics, such as competitive ability, that might prevent these organisms from being successful in competitive environments.

SUMMARY

CONCEPT 4.1 Each species has a range of environmental tolerances that determines its potential geographic distribution.

- The physical environment affects a species' ability to obtain energy and resources, thereby determining its growth and reproduction and, more immediately, its ability to survive the extremes of that environment. The physical environment is therefore the ultimate constraint on a species' geographic distribution.

- Individual organisms can respond to environmental change through acclimatization, a short-term response that enhances the organism's ability to survive, grow, and reproduce under stressful environmental conditions.

- A population may respond to unique environmental conditions through natural selection for physiological, morphological, and behavioral traits, known as adaptations, that enhance its survival, growth, and reproduction under those conditions.

CONCEPT 4.2 The temperature of organisms is determined by exchanges of energy with the external environment.

- Temperature exerts important controls on physiological processes through its effects on enzymes and membranes.

- Gains of energy from and losses of energy to the external environment determine an organism's temperature. Modifying this exchange of energy with the environment allows an organism to control its temperature.

- Terrestrial plants modify their energy balance by controlling transpiration, increasing or decreasing absorption of solar radiation, and adjusting the effectiveness of convective heat loss.

- Animals modify their energy balance mainly through behavior, adjusting the effectiveness of convective heat loss, and for endothermic animals, generation of internal heat.

CONCEPT 4.3 The water balance of organisms is determined by exchanges of water and solutes with the external environment.

- Water flows along energy gradients determined by solute concentrations (osmotic potential), pressure or tension (pressure potential), and the attractive force of surfaces (matric potential).

- Plants and microorganisms can influence water potential by adjusting the solute concentration in their cells (osmotic adjustment).

- Terrestrial organisms can alter their gains or losses of water by adjusting their resistance to water movement, as by the opening or closing of stomates in plants or variations in the thickness of skin in animals.

- Aquatic animals that are hypoosmotic to the surrounding water must expend energy to excrete salts against an osmotic gradient. On the other hand, aquatic animals that are hyperosmotic to their environment must take up solutes from the environment to compensate for solute losses to the surrounding water.

REVIEW QUESTIONS

1. Organisms subjected to stressful conditions exhibit different degrees of tolerance for environmental stresses. How does tolerance for variation in tissue temperatures vary among plants, ectothermic animals, and endothermic animals? What factors influence the differences in tolerance among these groups? Can plants exhibit avoidance of temperature extremes?

2. Arctic foxes exhibit large seasonal changes in the thickness of their fur, while African bat-eared foxes lack this response. Use this example to describe the concepts of acclimatization and adaptation. Assume that the bat-eared fox evolved earlier than the Arctic fox.

3. Organismal adaptations to environmental conditions often affect multiple ecological functions, leading to associated trade-offs. The following are two different trade-offs to consider. a. Plants transpire water through their stomates. What effects does this have on temperature regulation in leaves? What is the trade-off of transpirational temperature regulation in terms of leaf physiological function? b. Animals can more effectively warm their bodies by absorbing solar radiation if they are a dark color. Many animals, however, are not dark-colored, but instead have a coloration close to that of their habitat (camouflage, as in the case of the basking lizard in Figure 4.14). What trade-off exists between animal coloration and heat exchange?

4. List several ways in which plants and animals in terrestrial environments influence their resistance to water loss to the atmosphere.

HANDS-ON PROBLEM SOLVING

Thermal Adaptations in Urban Ants | This Web exercise (www.sinauer.com/ecology) demonstrates thermal adaptations in ants that live in cities. You will investigate whether adaptations to heat in urban leaf-cutter ants reduces tolerance to cold temperatures.

SUGGESTED READINGS

Alpert, P. 2006. Constraints of tolerance: Why are desiccation-tolerant organisms so small or rare? *Journal of Experimental Biology* 209: 1575–1584. *This fascinating paper explores the reasons why tolerance for severe desiccation is rare among organisms. It provides a compelling argument that a trade-off with growth, common in many adaptations to extreme environments, limits the frequency of desiccation tolerance.*

Gould, S. J. and R. C. Lewontin. 1979. The spandrels of San Marco and the Panglossian paradigm: A critique of the adaptationist programme. *Proceedings of the Royal Society of London B*, 205: 581–598. *Following several decades of descriptive research that assumed traits found in organisms were optimal adaptations to their environment, Stephen Jay Gould and Richard Lewontin wrote this amusing essay criticizing such an approach. Using analogies from architecture and anthropology, they suggested that trade-offs, both evolutionary and ecological, explain the existence of many nonadaptive traits in nature.*

Lambers, H., F. S. Chapin III and T. L. Pons. 1998. *Plant Physiological Ecology*. Springer, New York. *Provides a useful introduction to plant responses to environmental variation, including both physical and biotic components of the plant's environment.*

Morin, X., C. Augspurger and I. Chuine. 2007. Process-based modeling of species' distributions: What limits temperate tree species' range boundaries? *Ecology* 88: 2280–2291. *The geographic distributions of species have been modeled using a number of approaches linking climate and several ecological aspects of the species. This paper provides a good example of a model linking the survival and reproductive success of a species, and its subsequent geographic distribution, with climate.*

The following two texts provide excellent coverage of the challenges animals face in a variable physical environment and the adaptations that have evolved to help them cope.

Schmidt-Nielsen, K. 1997. *Animal Physiology, Adaptation and Environment*. Cambridge University Press, Cambridge.

Willmer, P., G. Stone and I. Johnston. 2005. *Environmental Physiology of Animals*. Blackwell Publishing, Malden, MA.

Storey, K. B. 1990. Life in a frozen state: Adaptive strategies for natural freeze tolerance in amphibians and reptiles. *American Journal of Physiology* 258: R559–R568. *An interesting review of freezing tolerance in amphibians and reptiles, this article outlines the challenges associated with tissue freezing and thawing and their solutions.*

5 Coping with Environmental Variation: Energy

KEY CONCEPTS

CONCEPT 5.1 Organisms obtain energy from sunlight, from inorganic chemical compounds, or through the consumption of organic compounds.

CONCEPT 5.2 Radiant and chemical energy captured by autotrophs is converted into stored energy in carbon–carbon bonds.

CONCEPT 5.3 Environmental constraints resulted in the evolution of biochemical pathways that improve the efficiency of photosynthesis.

CONCEPT 5.4 Heterotrophs have evolved mechanisms to acquire and assimilate energy efficiently from a variety of organic sources.

CASE STUDY

Toolmaking Crows

Humans employ a multitude of tools to enhance our ability to gather food to meet our energy needs. We use a highly mechanized system of planting, fertilizing, and harvesting crops to feed ourselves or the livestock that we consume. For thousands of years, we have used specialized tools for hunting prey, including spears, bows and arrows, and rifles. We view our toolmaking capacity as something that differentiates us from other animals.

However, humans are not alone in using tools to enhance their food acquisition ability. In the 1920s, Wolfgang Köhler, a psychologist studying the behavior of chimpanzees, observed that chimps in captivity made tools to retrieve bananas stashed in areas that were difficult to reach (Köhler 1927). Jane Goodall, a prominent primatologist, reported observing chimpanzees in the wild using grass blades and plant stems to "fish" for termites in holes in the ground and in decaying wood (**Figure 5.1**). Although these reports challenged the commonly held belief that modern humans were the only toolmakers, it was perhaps comforting to those clinging to this notion that the observations were associated with one of our closest extant relatives. No one would ever have suspected similar behavior in birds, touted as one of the least intelligent vertebrates, as evidenced by the dubious insult "bird brain" exchanged between humans.

Among the families of birds, the corvids, which includes crows, ravens, magpies, jays, and jackdaws, have entered our cultural heritage with a reputation for being clever. However, the discovery that crows use food-collecting tools manufactured from plants was still quite unexpected. Gavin Hunt reported in 1996 that the crows (*Corvus monoeduloides*) of New Caledonia, an island in the South Pacific, used tools to snag insects, spiders, and other arthropods and pull them from the wood of living and decomposing trees (Hunt 1996) (**Figure 5.2A**). Hunt found that two types of tools were being used by

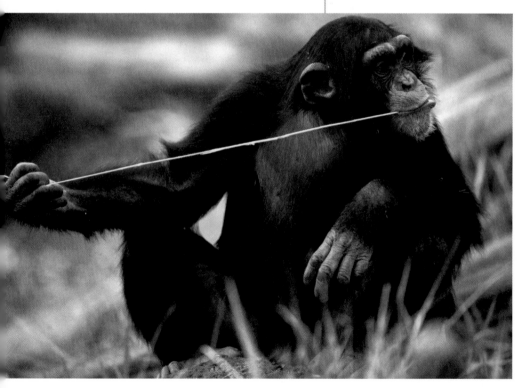

FIGURE 5.1 Nonhuman Tool Use
A chimpanzee uses a plant stem as a tool to forage for insects. Chimpanzees were the first non-human animals observed using tools to forage for food.

(A)

(B)

FIGURE 5.2 Tools Manufactured by New Caledonian Crows
(A) Crows use the tools they make to probe for food in the cavities and crevices of trees. (B) Hooked twig tools, made from shoots of trees. The crows use their bills to form the hook while holding the stick with their feet. (B after Hunt 1996.)

the crows. The first was a hooked twig, fashioned from a shoot stripped of its leaves and bark (**Figure 5.2B**). The second was a "stepped-cut" serrated edge clipped from a *Pandanus* tree. The tools were therefore manufactured, rather than just collected from materials lying on the ground.

Hunt described a unique hunting style used by the New Caledonian crows. Birds probed tree cavities or areas of dense foliage using their tools as extensions of their bills. The birds used the tools repeatedly, carrying them from tree to tree while they foraged for food. The presence of hooks on both types of tools demonstrated an innovative element that increased the birds' efficiency in extracting prey from their refuges in the trees. The

tools also appeared to be uniform in their construction; Hunt examined 55 tools manufactured by different birds and found that they differed little. When New Caledonian crows were captured and brought into the laboratory, they made hooked tools from wire, and experiments showed that the tools increased their food retrieval efficiency (Weir et al. 2002).

Toolmaking at a skill level equivalent to that shown by the crows appeared in humans only in the late Stone Age, approximately 450,000 years ago (Mellars 1989). How could birds have achieved a similar level of sophistication in their tool construction? The high numbers of New Caledonian crows using tools, and the consistency in the construction of the tools, indicate a cultural phenomenon, that is a skill learned socially within a population of animals, never before observed in birds. Was toolmaking a learned behavior, passed on from crow to crow? Hunt's discovery posed intriguing questions bridging psychology, anthropology, and behavioral ecology.

Energy is the most basic requirement for all organisms. Growth, physiological maintenance, and reproduction all depend on energy acquisition. Because organisms are complex systems, cessation of energy input leads to loss of biological functioning. Enzyme systems fail without replacement proteins, cell membranes degrade, and organelles cease to operate without energy to maintain and repair them. Growth and maintenance also require a steady supply of nutrients to meet physiological requirements, as we will see in Chapter 21. In this chapter, we will review the different ways that organisms acquire energy to meet the demands of cellular maintenance, growth, reproduction, and survival. We'll focus on the major mechanisms that allow organisms to obtain energy from their environment, including the capture of sunlight and chemical energy, and the acquisition and use of organic compounds synthesized by other organisms.

Sources of Energy

CONCEPT 5.1
Organisms obtain energy from sunlight, from inorganic chemical compounds, or through the consumption of organic compounds.

We sense energy in our environment in a variety of forms. Light from the sun, a form of *radiant energy*, illuminates our world and warms our bodies. Objects that are cold or warm to our touch have different amounts of *kinetic energy*, associated with the motion of the molecules that make up the objects. A grasshopper eating a leaf and a coyote eating a meadow vole both represent the transfer of *chemical energy*, which is stored in the food that is being consumed. Radiant and chemical energy are the forms organisms use to meet the demands of growth and maintenance, while kinetic energy is important for determining the rate of activity and metabolic energy demand of organisms. A cold endotherm needs chemical energy to warm its body to the optimal temperature for physiological functioning. It does this by "burning" chemical energy from its food during cellular respiration. Ultimately, this food was derived from the radiant energy of sunlight, converted to chemical energy by plants. Even the energy used to support industrial development, fuel our cars, and heat our homes originated ultimately with photosynthesis, which produced the plants that became the oil we pump out of the ground.

Autotrophs are organisms that assimilate energy from sunlight (*photosynthetic* organisms) or from inorganic chemical compounds[*] in their environment (*chemosynthetic* archaea and bacteria). Autotrophs convert the energy of sunlight or inorganic compounds into chemical energy stored in the carbon–carbon bonds of organic compounds, typically carbohydrates. **Heterotrophs** are organisms that obtain their energy by consuming energy-rich organic compounds from other organisms—energy that ultimately originated with organic compounds synthesized by autotrophs. Heterotrophs include organisms that consume nonliving organic matter, such as earthworms and fungi feeding on soil detritus derived from plant litter or bacteria in a lake consuming dissolved organic compounds. Heterotrophs also include organisms that consume live hosts, but do not necessarily kill them (parasites and herbivores), as well as consumers (predators) that capture and kill their food source (prey).

On the surface, the distinction between autotrophs and heterotrophs would seem to be clear-cut: all plants are autotrophs, all animals and fungi are heterotrophs, and archaea and bacteria can be both autotrophs and heterotrophs. The situation is not always this simple, however. Some plants have lost their photosynthetic function and obtain their energy by parasitizing other plants. Such plants, known as *holoparasites* (*holo*, "entire, whole"), have no photosynthetic pigments, and are heterotrophs. Dodder (genus *Cuscuta*, with approximately 150 different species), for example, is a common plant parasite found throughout the world (**Figure 5.3A,B**) and is considered a major pest of agricultural species. Dodder attaches to its host plant by growing in spirals around the stem and penetrates the phloem of the host, using modified roots called haustoria, to take up carbohydrates. One group of holoparasitic plants, which includes some orchid species, parasitizes fungi that use soil detritus for energy. Other plants, known as *hemiparasites*, are photosynthetic, but obtain some of their energy, as well as nutrients and water, from host plants (**Figure 5.3C**). Parasitism is widespread among plants and is found in approximately 15 families.

---

[*]Organic compounds have carbon–hydrogen bonds and are usually biologically synthesized; all other compounds are considered inorganic compounds.

(A)

(B)

(C)

FIGURE 5.3 Plant Parasites (A) Dodder (*Cuscuta* sp.) is a holoparasite that lacks chlorophyll, shown here wrapped around the stem of a goldenrod plant. (B) Increasing amounts of dodder (*Cuscuta europaea*) biomass result in decreasing growth of its host plant, stinging nettle (*Urtica dioica*). (C) Mistletoe is a hemiparasite. This juniper mistletoe (*Phoradendron juniperinum*), growing on the branches of a Utah juniper (*Juniperus osteosperma*), is using its host to supply it with water and nutrients as well as some of its energy, despite having green photosynthetic tissues. (B after Koskela et al. 2002.)

FIGURE 5.4 Green Sea Slug A lettuce sea slug (*Elysia crispata*), showing the green color associated with the chloroplasts in its digestive system. The chloroplasts can supply enough energy to the sea slug to maintain it for several months without food.

Animals can act as autotrophs, although this phenomenon is relatively rare. Their photosynthetic capacity is acquired by consuming photosynthetic organisms or by living with them in a close relationship known as a *symbiosis* (see Chapter 14). Some sea slugs, for example, have fully functional chloroplasts that supply them with carbohydrates through photosynthesis. These animals, in the order Ascoglossa, take up intact chloroplasts from the algae they feed on into their digestive cells (**Figure 5.4**). The chloroplasts are maintained intact for up to several months, providing energy as well as camouflage to the sea slug.

In the next two sections, we'll take a more detailed look at the mechanisms autotrophs use to capture energy and at some of the adaptations that make that process more efficient. We'll do the same for heterotrophs in the final section of this chapter.

Autotrophy

> **CONCEPT 5.2**
> Radiant and chemical energy captured by autotrophs is converted into stored energy in carbon–carbon bonds.

The vast majority of the autotrophic production of chemical energy on Earth occurs through **photosynthesis**, a process that uses sunlight to provide the energy needed to take up CO_2 and synthesize carbon compounds, principally carbohydrates or fats. Although its contribution to the overall energy picture is lower, **chemosynthesis** (also known as *chemolithotrophy*), a process that uses energy from inorganic compounds to produce carbohydrates, is important to some key bacteria involved in nutrient cycling (see Chapter 21) and in some unique ecosystems, such as ocean vent communities (see Chapter 19). Because the energy derived from photosynthesis and chemosynthesis is stored in the carbon–carbon bonds

of the organic compounds produced by these processes, carbon is often used as an energy currency.

Chemosynthesis fixes energy from inorganic compounds

The earliest autotrophs on Earth were probably chemosynthetic bacteria or archaea that evolved when the composition of the atmosphere was markedly different than it is today: low in oxygen, but rich in hydrogen, with significant amounts of CO_2 and methane (CH_4). A diverse group of archaea and bacteria still use energy from inorganic elements and compounds in their environment to take up CO_2 and synthesize carbohydrates. Chemosynthetic bacteria are often named according to the inorganic substrate they use for energy (**Table 5.1**).

During chemosynthesis, the organisms obtain electrons from the inorganic substrate, or in other words, they *oxidize*[*] the inorganic substrate. The electrons are used to generate two high-energy compounds, adenosine triphosphate (ATP) and nicotinamide adenine dinucleotide phosphate (NADPH). Energy from ATP and NADPH is subsequently used to take up, or "fix," CO_2. **Fixation** of carbon from the CO_2 is used to synthesize carbohydrates or other carbon-containing molecules, which are then used for energy storage or biosynthesis (manufacture of chemical compounds, membranes, organelles, and tissues). Alternatively, some bacteria can use the electrons from the inorganic substrate directly to fix CO_2. The biochemical pathway used most commonly to fix CO_2 is the **Calvin cycle**, named for Melvin Calvin, the biochemist who first described it. The Calvin cycle is catalyzed by several enzymes, and occurs in both chemosynthetic and photosynthetic organisms.

---

[*]Oxidation–reduction reactions involve the exchange of electrons between chemical compounds. The compound giving up electrons is oxidized, while the compound accepting electrons is reduced.

TABLE 5.1

Inorganic Substrates Used by Chemosynthetic Bacteria as Electron Donors for CO_2 Fixation

Substrate (chemical formula)	Type of bacteria
Ammonium (NH_4^+)	Nitrifying bacteria
Nitrite (NO_2^-)	Nitrifying bacteria
Hydrogen sulfide (H_2S/HS^-)	Sulfur bacteria (purple and green)
Sulfur (S)	Sulfur bacteria (purple and green)
Ferrous iron (Fe^{2+})	Iron bacteria
Hydrogen (H_2)	Hydrogen bacteria
Phosphite (HPO_3^{2-})	Phosphite bacteria

Source: After Madigan and Martinko 2005.

One of the most widespread and ecologically important groups of chemosynthetic organisms is the nitrifying bacteria (e.g., *Nitrosomonas*, *Nitrobacter*). These bacteria are found in both aquatic and terrestrial ecosystems. In a two-step process, nitrifying bacteria convert ammonium (NH_4^+) to nitrite (NO_2^-), then oxidize it to nitrate (NO_3^-). These chemical conversions of nitrogen compounds are an important component of nitrogen cycling and plant nutrition, and we will discuss them in more detail in Chapter 21. Another important chemosynthetic group is the sulfur bacteria, associated with volcanic deposits, sulfur hot springs, and acid mine wastes. Sulfur bacteria initially use the higher-energy forms of sulfur, H_2S and HS^- (hydrogen sulfide), producing elemental sulfur, which is insoluble and highly visible (**Figure 5.5**). Once the H_2S and HS^- are exhausted, the bacteria use elemental S as an electron source, producing SO_4^{2-} (sulfate).

Photosynthesis is the powerhouse for life on Earth

As noted earlier, the vast majority of biologically available energy on Earth is derived from the conversion of sunlight into energy-rich carbon compounds by photosynthesis. Photosynthetic organisms include some archaea, bacteria, and protists and most algae and plants. In plants, photosynthesis is usually associated with leaves, but may also occur in stem and reproductive tissues. Here, we will briefly review the major steps of plant photosynthesis and consider some ecologically relevant constraints on photosynthetic rates. In the next subsection, we will examine some variations in plant photosynthetic pathways.

Light and dark reactions Photosynthesis has two major steps. The first is the harvesting of light energy, which is used to split water to provide electrons for generating ATP and NADPH. This step is often referred to as the "light reaction" of photosynthesis. The second step is the fixation of CO_2 using the Calvin Cycle and the synthesis of carbohydrates. This step is often referred to as the "dark reaction" of photosynthesis, despite its daytime occurrence in most plants.

Light harvesting is accomplished by chlorophyll and accessory pigments. Chlorophyll gives photosynthetic organisms their green appearance because it absorbs red and blue light and reflects green wavelengths (**Figure 5.6**). Plants and photosynthetic bacteria have similar chlorophyll pigments, but they absorb light at slightly different wavelengths. Additional pigments associated with photosynthesis, called accessory pigments, include the carotenoids, which are characteristically red, yellow, or orange in appearance. In addition to helping to harvest light energy, accessory pigments may help prevent damage to photosynthetic organisms from intense solar radiation (see Box 5.1). These photosynthetic pigments are embedded in a membrane, along with other molecules involved in the light reaction. In plants, this membrane is associated with specialized organelles called chloroplasts, while in photosynthetic bacteria the pigments are embedded in the cytosolic (cell) membrane. The pigment molecules are arrayed like antennas, with each array containing between 50 and 300 molecules. The pigments absorb energy from discrete units of light, called photons. The energy from sunlight is used to split water and provide electrons. These electrons are passed on to molecular complexes on the membranes, where they are used to synthesize ATP and NADPH.

FIGURE 5.5 Sulfur Deposits from Chemosynthetic Bacteria Sulfur bacteria thrive in hot springs with water temperatures as high as 110°C. They use the hydrogen sulfide–rich waters to generate chemical energy, leaving behind elemental sulfur.

Chlorophylls absorb blue and red wavelengths and reflect green wavelengths, giving photosynthetic tissues their characteristic color.

Carotenoids absorb blue and green wavelengths and reflect red, yellow, and orange wavelengths.

FIGURE 5.6 Absorption Spectra of Plant Photosynthetic Pigments Plants typically contain several light-absorbing pigments, each of which absorbs light of different wavelengths.

The splitting of water to provide electrons for the light reaction generates oxygen (O_2). The evolution of photosynthesis, and the accompanying release of O_2 into the atmosphere, was a critical step in the development of the chemistry of the modern atmosphere and lithosphere as well as the evolution of life on Earth. Atmospheric oxygen led to the creation of an ozone layer high in the atmosphere that shields organisms from high-energy ultraviolet radiation. The evolution of aerobic respiration, in which O_2 is used as an electron acceptor, facilitated significant evolutionary advances for life on Earth.

The energy captured by the light reaction provides the high-energy compounds needed to fix CO_2 using the Calvin cycle. CO_2 is taken up from the atmosphere through the stomates of vascular plants, or diffuses across the cell membranes in nonvascular plants, algae, and photosynthetic bacteria and archaea. A key enzyme associated with the Calvin cycle is ribulose 1,5 bisphosphate carboxylase/oxygenase,

thankfully usually referred to by its abbreviation, rubisco. Rubsico, the most abundant enzyme on Earth, catalyzes the uptake of CO_2 and the synthesis of a three-carbon compound (phosphoglyceraldehyde, or PGA). PGA is eventually converted into a six-carbon sugar (sucrose in most plants). The net reaction of photosynthesis is

$$6\ CO_2 + 6\ H_2O \rightarrow C_6H_{12}O_6 + 6\ O_2 \qquad \textbf{(5.1)}$$

Environmental constraints and solutions   The rate of photosynthesis determines the supply of energy and substrates for biosynthesis by photosynthetic organisms, which in turn influences their growth and reproduction, often equated with their ecological success. Thus, environmental controls on the rate of photosynthesis are a key topic in physiological ecology. It should be noted, however, that net energy (carbon) gain and its relation to growth is also influenced by CO_2 losses associated with cellular respiration.

Light is clearly an important factor in rates of photosynthesis in terrestrial and aquatic habitats. The relationship between light levels and a plant's photosynthetic rate can be portrayed by a *light response curve* (**Figure 5.7A**). When there is enough light that the plant's photosynthetic CO_2 uptake

FIGURE 5.7 Plant Responses to Variations in Light Levels   (A) The light response curve. Photosynthetic uptake of CO_2 increases as light increases until a light saturation point is reached. Integrated over the course of the day, the average light level the plant experiences is near the transition between light limitation and light saturation. (B) Spearscale (*Atriplex triangularis*) plants grown in different light environments in growth chambers exhibited acclimation to the light levels. Their light response curves indicate that adjustments in the light saturation point have occurred. Small, but ecologically significant, changes in the light compensation point also occur in many species, facilitating CO_2 uptake at low light levels. (C, D) Golden banner (*Thermopsis montana*) leaves adjust morphologically to changes in light levels. Leaves grown at high light levels (C) are thicker, with greater numbers of chloroplasts, than leaves grown in low light (D). (B after Bjorkman 1981.)

is balanced by its CO_2 loss by respiration, the plant is said to have reached the *light compensation point*. As light levels increase, photosynthetic rates also increase; in other words, photosynthesis is *limited* by the availability of light. Photosynthetic rates level off at a *light saturation point*, which is typically reached at a level well below full sunlight, which falls on a leaf surface only a small fraction of the time.

How do plants cope with light variation in their environment? How would an understory forest plant, for example, respond to shading by canopy trees? Could that plant acclimatize to more light if the canopy tree fell and allowed full sunlight to reach the ground? In a series of classic studies using controlled growth conditions, Olle Bjorkman demonstrated that acclimatization to different light levels involves

shifts in light response curves (Bjorkman 1981) (**Figure 5.7B**). Morphological changes associated with this acclimatization include altering the thickness of leaves and varying the number of chloroplasts available to harvest light (**Figure 5.7C**). Photosynthetic organisms may also alter the density of their light-harvesting pigments—a strategy analogous to changing the size of the antenna on a radio—and the amounts of photosynthetic enzymes available for the dark reaction. Bacteria are especially well adapted to photosynthesis at low light levels, which allows them to thrive in dimly lit environments such as relatively deep ocean water (up to about 20 m).

Water availability is an important control on the supply of CO_2 for photosynthesis in terrestrial plants. As we saw in Chapter 4, low water availability results in closure of the stomates, restricting the entry of CO_2 into the leaf. Stomatal control represents an important trade-off for the plant: water conservation versus energy gain through photosynthesis as well as cooling of the leaf. Keeping stomates open while tissues lose water can permanently impair physiological processes in the leaf. Closing stomates, however, not only limits photosynthetic CO_2 uptake, but also increases the chances of light damaging the photosynthetic membranes. When the Calvin cycle is not operating, energy continues to accumulate in the light-harvesting arrays. If enough energy builds up, it can damage the photosynthetic membranes. Plants have evolved a number of ways of dissipating this energy safely, including the use of carotenoids to release it as heat (**Box 5.1**).

Acclimatization and adaptation to temperature variation are associated with the properties of the Calvin cycle enzymes and the chemistry of the photosynthetic membranes. Photosynthetic organisms have different forms of the same photosynthetic enzymes that operate best under the environmental temperatures where they occur, resulting in markedly different temperature ranges for photosynthesis in organisms from different climates (**Figure 5.8A**). Lichens and plants of Arctic and alpine environments can photosynthesize at temperatures close to freezing, while desert

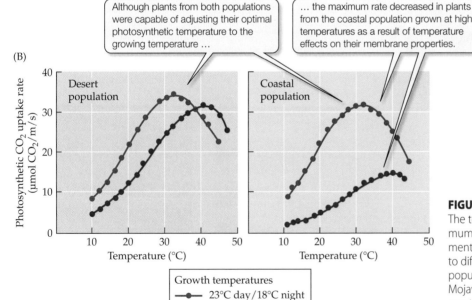

FIGURE 5.8 Photosynthetic Responses to Temperature (A) The temperatures at which a plants and lichens reach their maximum photosynthetic rates correspond to the range of environmental temperatures in the plant's native habitat. (B) Acclimation to different growth temperature regimes by plants from different populations of *Atriplex lentiformis*, a shrub that occurs in the Mojave Desert and in coastal zones of California. The growth temperatures are representive of the habitats native to the species. (A after Lambers et al. 1998; B after Pearcy 1977.)

BOX 5.1

How Do Plants Cope with Too Much Light?

Light is obviously a key ingredient in photosynthetic energy capture. The amount of light a plant captures, however, can exceed the plant's ability to use it, particularly when other stresses limit photosynthesis, as when water stress lowers the uptake of CO_2 through the stomates or low temperatures limit enzyme activity. Excess light energy can generate toxic oxygen molecules that damage the membranes of the plant, a condition known as *photoinhibition*.

Plants have evolved several mechanisms to avoid photoinhibition. Over periods of hours to days, some plants can move their leaf blades away from the sun, or curl their leaves, lowering their exposure to intense solar radiation. In addition, the chloroplasts of some plants can migrate within the cell to increase self-shading (**Figure A**). A widely observed mechanism for dissipating light energy is the xanthophyll cycle (Demmig-Adams and Adams 1996), in which carotenoid pigments are converted from one form to another. Some forms (zeaxanthin and antheraxanthin) are more efficient at dissipating light energy as heat than other forms (violaxanthin) (**Figure B**). Under stressful environmental conditions, greater conversion of the heat-dissipating forms of the pigments will occur, preventing the onset of photoinhibition.

(I) Darkness   (II) Weak blue light   (III) Strong blue light

FIGURE A Chloroplast Movements in Response to Light   Chloroplasts within mesophyll cells of duckweed (*Lemna*) can shift their position within the cell to absorb more (I and II) or less (III) light. This mobility of chloroplasts helps to enhance photosynthesis under low light levels and avoid damage under intense light.

FIGURE B The Xanthophyll Cycle and Dissipation of Light Energy   When leaves are irradiated with intense light, the carotenoid pigment violaxanthin is converted into two other carotenoid pigments, antheraxanthin and zeaxanthin. Zeaxanthin and antheraxanthin absorb light energy and dissipate it as heat. This chemical conversion, known as the xanthophyll cycle, helps prevent the buildup of excess light energy and subsequent damage to the chloroplast membranes. (After Demmig-Adams and Adams 1996.)

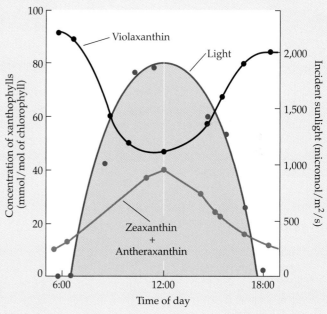

plants may have their highest photosynthetic rates at temperatures that are hot enough to denature most other plants' enzymes (40°C–50°C, 104°F–122°F). Plants that acclimatize to changes in temperature synthesize different forms of photosynthetic enzymes with different temperature optima (**Figure 5.8B**). Temperature also influences the fluidity of the cell and organelle membranes (see Chapter 4). Cold sensitivity in plants of tropical and subtropical biomes is associated with loss of membrane fluidity, which inhibits the functioning of

the light-harvesting molecules embedded in the chloroplast membranes. High temperatures, particularly in combination with intense sunlight, can damage photosynthetic membranes.

The nutrient concentrations of leaves reflect their photosynthetic potential because most of the nitrogen in plants is associated with rubisco and other photosynthetic enzymes. Thus, higher amounts of nitrogen in the leaf are correlated with higher photosynthetic rates. Why, then, don't all plants allocate more nitrogen to their leaves to increase their photosynthetic capacity? There are two main reasons. First, the supply of nitrogen is low relative to the demand for growth and metabolic functions such as photosynthesis. (see Chapter 21). Second, increasing the nitrogen concentration of a leaf increases the risk that herbivores will consume the leaf, as plant-eating animals are also nitrogen-starved (see Chapter 12). Plants must balance the competing demands of photosynthesis, growth, and protection from herbivores.

Photosynthetic Pathways

CONCEPT 5.3
Environmental constraints resulted in the evolution of biochemical pathways that improve the efficiency of photosynthesis.

As we have just seen, rates of photosynthesis are subject to environmental constraints, particularly water availability in the case of terrestrial plants. In addition, an apparent biochemical inefficiency in the initial step of the Calvin cycle limits energy gain by photosynthetic organisms. In this section, we will examine some evolutionary responses to these environmental constraints on photosynthesis. We will describe two specialized photosynthetic pathways, the Hatch–Slack–Kortschack (C_4) pathway and crassulacean acid metabolism (CAM), that make photosynthesis more efficient. Plants that lack these specialized pathways use the **C_3 photosynthetic pathway**. The C_3 and C_4 photosynthetic pathways take their names from the number of carbon atoms in their first stable chemical products. First, we'll examine photorespiration, a process that operates in opposition to the Calvin cycle and lowers its efficiency.

Photorespiration lowers the efficiency of photosynthesis

Earlier, we mentioned that a key enzyme in the Calvin cycle is rubisco and noted that the "o" in the abbreviation stands for "oxygenase." Rubisco can catalyze two competing reactions. One is a carboxylase reaction, in which CO_2 is taken up, leading to the synthesis of sugars and the release of O_2 (i.e., photosynthesis; see Equation 5.1). The other is an oxygenase reaction, in which O_2 is taken up, leading to the breakdown of carbon compounds and the release of CO_2. This oxygenase reaction is part of a process called **photorespiration**, which results in a net loss of energy.

The balance between photosynthesis and photorespiration is related to two main factors: (1) the ratio of O_2 to CO_2 in the atmosphere and (2) temperature. As the atmospheric concentration of CO_2 decreases relative to that of O_2, the rate of photorespiration increases relative to the rate of photosynthesis. Since the evolution of C_3 photosynthesis over 3 billion years

ago, atmospheric CO_2 concentrations have changed repeatedly over periods of hundreds of thousands of years in response to major global geological and climatic events (see Chapter 24). These atmospheric shifts in CO_2 concentrations would have influenced the balance between photosynthesis and photorespiration. Furthermore, as temperatures increase, the rate of O_2 uptake catalyzed by rubisco increases relative to the rate of CO_2 uptake, so photorespiration increases more rapidly at high temperatures than photosynthesis does. Thus, energy loss due to photorespiration is particularly acute at high temperatures and low atmospheric CO_2 concentrations.

If photorespiration is detrimental to the functioning of photosynthetic organisms, why hasn't a new form of rubisco evolved that minimizes uptake of O_2? Is it possible that photorespiration provides some benefit to the plant? A possible clue comes from experiments with *Arabidopsis thaliana*: plants with a genetic mutation that knocks out photorespiration die under normal light and CO_2 conditions (Ogren 1984). One hypothesis for a potential benefit of photorespiration is that it protects the plant from damage to the photosynthetic machinery at high light levels. This hypothesis is supported by the results of a study that used genetically altered tobacco plants. Akiko Kozaki and Go Takeba used

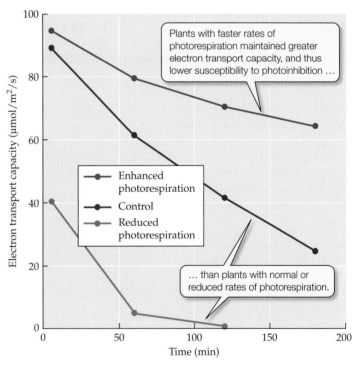

FIGURE 5.9 Does Photorespiration Protect Plants from Damage by Intense Light? The ability of plants to process light energy for photosynthesis (electron transport capacity) under conditions that promote damage to photosynthetic membranes (high light, low CO_2 concentrations) is greater in transgenically altered plants with high rates of photorespiration than in control plants, or plants transgenically altered with low rates of photorespiration. (After Kozaki and Takeba 1996.)

plants that had been altered to elevate or lower rates of photorespiration (Kozaki and Takeba 1996). They subjected these experimental plants to high-intensity light and recorded the damage to their photosynthetic machinery. Plants with higher rates of photorespiration showed less damage than control plants with normal rates of photorespiration (**Figure 5.9**) or plants with depressed rates of photorespiration.

Despite this possibility that photorespiration plays a role in protecting plants from damage at high light levels, there are hypothetical conditions in which the decrease in photosynthetic CO_2 uptake it causes may be a serious problem for the plant. If atmospheric CO_2 concentrations are low and temperatures high, photosynthetic energy gain may not keep pace with photorespiratory energy loss. Such conditions existed 7 million years ago, at about the time when plants with a unique biochemical pathway, C_4 photosynthesis, first appeared (Cerling et al. 1997).

Corn (*Zea mays*), an important crop species

Globe amaranth (*Gomphrena globosa*), a C_4 forb
native to Central and South America

C_4 photosynthesis lowers photorespiratory energy loss

The **C_4 photosynthetic pathway**, sometimes called the Hatch–Slack–Kortschack pathway to commemorate the plant physiologists who discovered it, reduces photorespiration. C_4 photosynthesis evolved independently several times in different plant species. It is found in 18 different plant families but is most closely associated with the grass family (**Figure 5.10**). Well-known examples of crop plants with the C_4 pathway include corn, sugarcane, and sorghum.

C_4 photosynthesis involves both biochemical and morphological specialization. The biochemical specialization can be thought of as a pump that provides high concentrations of CO_2 to the Calvin cycle. This greater supply of CO_2 lowers the rate of O_2 uptake by rubisco, substantially reducing photorespiration. The morphological specialization involves spatial separation of the regions in the leaf where CO_2 is taken up (mesophyll) and where the Calvin cycle operates (bundle sheath), in order to increase the concentration of CO_2 where rubisco is found.

In C_4 plants, CO_2 is initially taken up by an enzyme called phosphoenol pyruvate carboxylase, or PEPcase, that has a greater capacity to take up CO_2 than rubisco and lacks oxy-

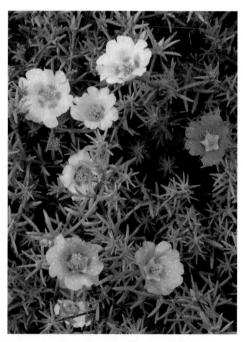

Rose moss (*Portulaca grandiflora*), a C_4 forb
native to Uruguay

FIGURE 5.10 Examples of Plants with the C_4 Photosynthetic Pathway The C_4 photosynthetic pathway has evolved multiple times, and is found in plants in 18 different families encompassing a variety of growth forms.

C$_3$ leaf

- Upper epidermis
- Mesophyll cell
- Xylem and phloem
- Bundle sheath cell
- Spongy mesophyll cell
- Lower epidermis
- Stomate

C$_4$ leaf

- Upper epidermis
- Mesophyll cell
- Xylem and phloem
- Bundle sheath cell
- Spongy mesophyll cell
- Lower epidermis
- Stomate

genase activity. PEPcase fixes CO_2 in the mesophyll tissue of the plant (**Figure 5.11**). Once the CO_2 is taken up, a four-carbon compound is synthesized and transported to a group of cells surrounding the vascular tissues (xylem and phloem), known as the bundle sheath, where the Calvin cycle occurs. The four-carbon compound is broken down in the bundle sheath cells, releasing CO_2 to the Calvin cycle, and a three-carbon compound is transported back to the mesophyll to continue the C$_4$ cycle. The bundle sheath is surrounded by a waxy coating that keeps CO_2 from diffusing out. As a result, CO_2 concentrations inside the bundle sheath may reach a high of 5,000 parts per million while external CO_2 concentrations are only 380 parts per million. Additional energy in the form of ATP must be expended to operate the C$_4$ photosynthetic pathway, but the increased efficiency of CO_2 fixation compensates for the higher energy requirement.

As is apparent from the discussion above, plants with the C$_4$ photosynthetic pathway can photosynthesize at higher rates than C$_3$ plants under environmental conditions that elevate rates of photorespiration, such as high temperatures. In addition, most C$_4$ plants have lower rates of transpiration at a given photosynthetic rate (known as *water use efficiency*) than C$_3$ plants. This difference is due to the ability of PEPcase to take up CO_2 under the lower CO_2 concentrations that exist when stomates are not fully open.

If we assumed that photosynthetic rates determine ecological success (abundance and geographic range), we could use climatic patterns to predict where C$_4$ plants should pre-

The high concentration of CO_2 in the bundle sheath limits the amount of O_2 taken up by rubisco.

The waxy coating surrounding the bundle sheath lowers the outward diffusion of CO_2.

Calvin cycle

2 ... and transported to the bundle sheath as a four-carbon compound, where it is released as CO_2.

CO_2

Sugars

C$_4$ PEP case

1 CO_2 is taken up in the mesophyll by PEPcase ...

C$_3$

CO_2

Stomate

CO_2

3 A three-carbon compound is transported back to the mesophyll to continue the C$_4$ cycle.

- Xylem and phloem
- Bundle sheath cell
- Waxy coating
- Spongy mesophyll cell
- Lower epidermis

FIGURE 5.11 Morphological Specialization in C$_4$ Plants Unlike C$_3$ plants, C$_4$ plants have well-differentiated mesophyll and bundle sheath tissues in their leaves. This spatial separation of CO_2 uptake and the Calvin cycle effectively eliminates photorespiration.

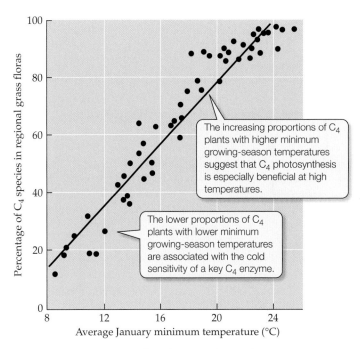

FIGURE 5.12 C$_4$ Plant Abundance and Growing-Season Temperature The proportions of C$_4$ plants in Australian grass- and sedge-dominated communities changed along latitudinal and elevational gradients. (After Henderson et al. 1995.)

dominate over C$_3$ plants. Such an analysis would be over simplistic, however, as multiple factors other than temperature influence the biogeography of C$_3$ and C$_4$ plants, including abiotic factors such as light levels and biotic factors such as competitive ability and the pool of available species. However, analyses of similar communities across latitudinal or elevational gradients provide support for the benefit of C$_4$ photosynthesis at high temperatures and for its role in C$_4$ plant distribution (Ehleringer et al. 1997). In particular, studies of grass- and sedge-dominated communities in Australia suggest a close correlation between growing-season temperature and the proportion of C$_3$ and C$_4$ species in the community (**Figure 5.12**). As atmospheric CO$_2$ concentrations continue to increase, photorespiration rates are likely to decrease, and the advantages of C$_4$ over C$_3$ photosynthesis may be diminished, leading to changes in the proportions of C$_3$ and C$_4$ plants.

CAM photosynthesis enhances water conservation

When plants first "invaded" the terrestrial environment, they evolved adaptations to restrict water losses to a dry atmosphere. Among these adaptations is a unique photosynthetic pathway, called **crassulacean acid metabolism (CAM)**, which occurs in over 10,000 plant species belonging to 33 families. While C$_4$ photosynthesis separates CO$_2$ uptake and

the Calvin cycle spatially in the mesophyll and bundle sheath, CAM separates these two steps temporally (**Figure 5.13**). CAM plants open their stomates at night, when C$_3$ and C$_4$ plants have their stomates closed. Because air temperatures are cooler at night, humidities are higher. The higher humidity at night results in a lower water potential gradient between the leaf and the air (Chapter 4), so the plant loses less water by transpiration than it would during the day. CAM plants close their stomates during the day, when the potential for water loss is highest.

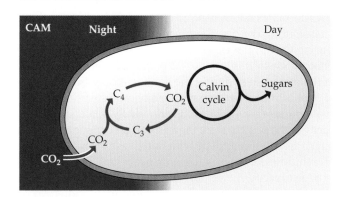

FIGURE 5.13 C$_3$ versus C$_4$ versus CAM Photosynthesis All three photosynthetic pathways fix carbon and produce sugar, but C$_4$ photosynthesis separates these steps spatially, while CAM separates them temporally.

Night: Stomata opened   **Day: Stomata closed**

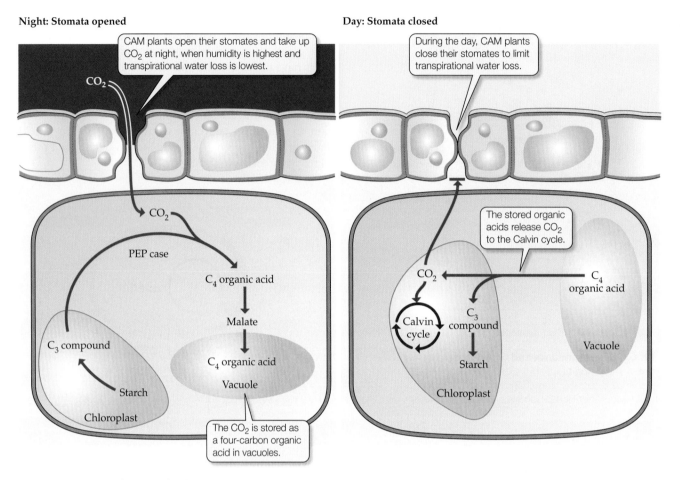

FIGURE 5.14 CAM Photosynthesis   CAM plants take up CO_2 at night and run the Calvin cycle during the day.

During the night, CAM plants take up CO_2 using PEPcase and incorporate it into a four-carbon organic acid, which is stored in vacuoles (**Figure 5.14**). The resulting increase in acidity in the plant tissues during the night is characteristic of CAM plants and is used to estimate their photosynthetic rates. During the day, the organic acid is broken down, releasing CO_2 to the Calvin cycle. CO_2 concentrations in the photosynthetic tissues of CAM plants are higher than those in the atmosphere during the day, increasing the efficiency of photosynthesis as they suppress photorespiration. Photosynthetic rates in CAM plants are usually related to the capacity of the plant to store the four-carbon organic acid. CAM plants are often succulent, with thick, fleshy leaves or stems. The thickness of the tissues enhances their nighttime acid storage capacity.

CAM plants are typically associated with arid environments (**Figure 5.15**). While some CAM plants are clearly linked to dry climates, including deserts and Mediterranean-type ecosystems, many others are found in the more humid tropics. Tropical CAM plants are typically epiphytes, growing

on the branches of trees, without access to the abundant water stored in the soil. These epiphytes rely on rainfall for their water and may be subjected to long periods without access to water.

CAM is also found in some aquatic plants such as quillwort (*Isoetes*), which is closely related to the club mosses. This observation suggests that water conservation was probably not the only driving force for the evolution of CAM, which evolved independently in several different plant groups in response to different selective forces. The rate of CO_2 diffusion into water is low, and CAM has been hypothesized to facilitate the uptake of CO_2 at the low concentrations found in the aquatic environment.

A unique property of some CAM plant species is the ability to switch between C_3 and CAM photosynthesis, known as **facultative CAM**. When conditions are favorable for daytime gas exchange (i.e., abundant water is available), these plants utilize the C_3 photosynthetic pathway, which allows greater carbon gain than CAM. As conditions become more arid or more saline, and therefore inhibitory to C_3 photosynthesis, the plants switch over to CAM. The reversibility of the transition from C_3 to CAM varies among species. For example, the common ice plant (*Mesembryanthemum crystallinum*), which has been intensively studied as a facultative CAM

Crassula tabularis, a stem succulent plant native to the desert of Namibia

FIGURE 5.15 Examples of Plants with CAM Photosynthesis

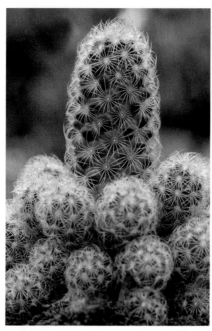

Golden star cactus (*Mammillaria elongata*), native to Mexico

Pineapple (*Anana comosus*), a bromeliad native to Central and South American tropical forests

model system, undergoes an irreversible transition from C_3 to CAM when salinity increases or the soil dries out (Osmond et al. 1982). In contrast, some species in the genus *Clusia* can switch relatively rapidly between C_3 and CAM (Borland et al. 1992). These plants start out as epiphytes in canopy trees, but eventually grow toward the base of their host tree, eventually strangling it and taking on a tree growth form. This capacity to switch between C_3 and CAM facilitates the change from an epiphyte to a tree form, and it supports continued photosynthesis during the transition from wet season to dry season characteristic of some tropical locations.

Now that we have reviewed the ways in which autotrophs acquire energy, let's turn our attention to how that energy is acquired by heterotrophs.

Heterotrophy

CONCEPT 5.4
Heterotrophs have evolved mechanisms to acquire and assimilate energy efficiently from a variety of organic sources.

The first organisms on Earth were probably heterotrophs that consumed amino acids and sugars that formed spontaneously in the early atmosphere and rained down on the surface. The diversity of heterotrophic strategies for obtaining energy has expanded tremendously since that time. The organic matter that provides energy for heterotrophs includes living and freshly killed organisms and organic material derived from dead organisms in various stages of decomposition (see Chapter 19). In this section, we will examine the ways in which heterotrophs obtain energy. There is a wide range of variation in the complexity of heterotrophic energy acquisition and assimilation processes associated with variation in heterotroph body size and physiology. In Chapters 12 and 13, we will take a more in-depth look at the various types of consumers (predators, herbivores, and parasites) and how the food they consume affects their growth and reproduction as well as the distributions and abundances of the consumers themselves and their food resources (prey and hosts).

Food sources differ in their chemistry and availability

Heterotrophs consume energy-rich organic compounds (food) from their environment and convert them into usable chemical energy, primarily ATP, during glycolysis, a metabolic process that breaks down sugars and fats. The heterotroph's energy gain from food depends on the chemistry of the food, which determines its digestibility and amount of energy per unit mass it contains. The effort invested in finding and obtaining the food also influences how much benefit the heterotroph gets from consuming it. For example, microorganisms that consume detritus in the soil invest little energy in obtaining food. However, the energy content of this decomposing plant matter is low compared with the energy content of live organisms. Living food items are rarer than detri-

FIGURE 5.16 Variations on a Theme: Insect Mouthparts
Differences in the morphology of insect mouth parts reflect different strategies for effectively capturing and consuming the food types they prefer.

The "sponging" mouthparts of blowflies are used to soak up partially digested food.

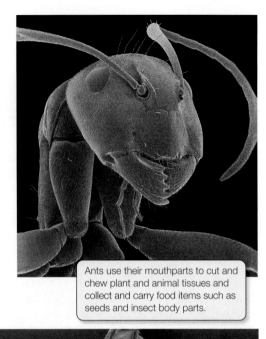

Ants use their mouthparts to cut and chew plant and animal tissues and collect and carry food items such as seeds and insect body parts.

Female mosquitoes use piercing mouthparts to penetrate the skin of animals and extract blood.

tus, and they may have defense mechanisms that consumers must expend energy to overcome. Thus, a cheetah hunting a gazelle invests substantial energy in finding, chasing, capturing, and killing its prey, but it obtains a substantial, energy-rich meal if the hunt is successful.

The chemistry of food is associated with the cell types of the organisms from which it is derived. Animal cells are generally more energy-rich than plant, fungal, or bacterial cells, which tend to have higher concentrations of structural components, such as cell walls, that are not easily digested. Most food consumed by heterotrophs consists of complex chemical compounds that must be chemically transformed into simpler compounds before they can be used as energy sources. Proteins, carbohydrates, and fats are broken down into their component amino acids, simple sugars, and fatty acids before being utilized as energy sources. Each of these compounds has a different yield of energy. Fats are richer in energy than carbohydrates per unit mass, which in turn provide more energy than amino acids. However, amino acids also provide nitrogen, a nutrient often in high demand. In addition, some metabolic functions require specific energy-containing compounds. Flight in insects, for example, has a high energy demand, and some insects have fat storage bodies to supply the lipids required for initiation of flight. Humans require carbohydrates to fuel brain activity, which explains why low blood sugar can lead to poor cognitive ability.

Heterotrophs obtain food using diverse strategies

Heterotrophs vary in size from archaea and bacteria (as small as 0.5 µm) to blue whales (up to 25 m long). The ratio of their body size to the food they ingest varies widely, but generally increases as body size increases. Bacteria may be bathed in their food, while food for larger heterotrophs is usually more diffuse and smaller relative to the consumer. Feeding methods and the complexity of food absorption are accordingly very diverse among heterotrophs.

Prokaryotic heterotrophs typically absorb food directly through their cell membranes. Archaea, bacteria, and fungi excrete enzymes into the environment to break down organic matter, acting in effect to digest their food outside their bodies. Heterotrophic bacteria have adapted to a wide variety of organic energy sources and produce a large number of enzymes capable of breaking down organic compounds. This capacity of microorganisms to use different energy sources has been exploited in environmental waste management as an approach to cleaning up toxic chemical wastes, a process known as *bioremediation*. Spills of fuels, pesticides, sewage, and other toxins have been effectively contained by using microorganisms to break down these harmful compounds.

The evolution of feeding in protists and animals is a story of increasing complexity in the ingestion and digestion of food. Small protozoans, such as amoebas and ciliates, ingest food particles into their cells, where the food is digested internally. With the advent of multicellular animals, specialized tissues for absorption, digestion, transport, and excretion evolved, and the efficiency of energy assimilation increased. Animals display tremendous diversity in their specialized morphological and physiological feeding adaptations, which reflect the diversity of the foods they consume. Several examples serve to demonstrate this diversification.

Insects display tremendous diversity in facial appearance. All insects have the same basic set of mouthparts, consisting of several paired appendages that are used to seize, handle, and consume their food (**Figure 5.16**). Morphological variation in these mouthparts reflects the feeding specializations that have evolved within different insect groups. Common houseflies have "sponging" mouthparts that release saliva onto their food, then soak up and ingest the partially digested solution. Female mosquitoes and aphids have piercing and sucking mouthparts for extracting fluids from different living targets—blood from animals and sap from plants, respectively. Biting flies have razor-sharp appendages for cutting through skin to draw blood for drinking, similar to the cutting mouthparts of insects that consume leaves.

Birds display similar, although somewhat less complex, variation in their mouthparts (bills) relative to insects. Birds use their bills in a multitude of ways to capture, manipulate, and consume their food (**Figure 5.17**). The morphology of a bird's bill is closely associated with the taxonomic group to which the bird belongs. In other words, the flat bills of ducks and the hooked bills of raptors vary little within the taxonomic group. However, subtle differences among the bills of closely related species reflect slight differences in food acquisition and handling. This variation in bill morphology reflects adaptations that help to optimize food acquisition and minimize competition among groups of birds.

Craig Benkman studied the relationship between differences in bill morphology among crossbills and differences in

Black-capped chickadees have short, thin bills suitable for capturing small insects.

Pileated woodpeckers have pointed stout bills for hitting and probing in wood.

American avocet bills are long and thin for probing mud and sand for small subsurface animals.

Bald eagle bills are hooked for tearing the flesh of their prey.

Malachite kingfishers have long, sturdy bills suitable for plucking small fish from the water.

FIGURE 5.17 Variations on a Theme: Bird Bills   Bird bill morphology is associated with feeding behavior of the species, enhancing the capture of preferred food resources.

(A)

(B)

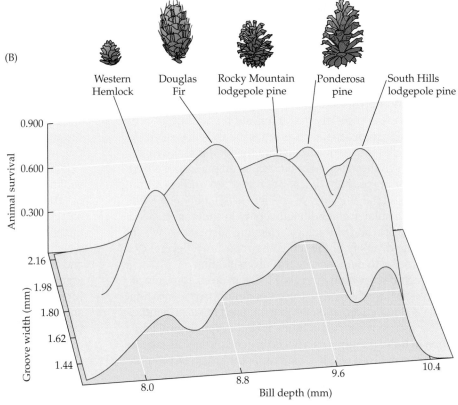

FIGURE 5.18 Crossbill Morphology, Food Preference, and Survival (A) Red crossbill (*Loxia curvirostra*). (B) A three-dimensional plot of Craig Benkman's data shows the relationship between bill morphology (groove widths and bill depth) and annual survival of five crossbill species. Each species shows "adaptive peaks" in association with the conifer species they preferentially feed on; that is, higher survival occurred with the species that the bill morphology was best suited to exploit. The cones shown are drawn to relative scale. (After Benkman 2003.)

the conifer seeds they use as food (Benkman 1993, 2003). As their name indicates, crossbills have unique asymmetrical bills with crossing tips (**Figure 5.18A**). Crossbills are adept at using their bills to open the cones of coniferous trees and pull out seeds for consumption. Differences in their bill morphologies had been hypothesized to relate to the morphologies of the cones they obtain seeds from.

Benkman tested this hypothesis experimentally using captive and wild birds from five species of the red crossbill species complex (*Loxia curvirostra*). He showed in a series of studies that a bird's speed of seed extraction from a given conifer's cone was associated with its bill depth, and that its speed of seed husking was associated with the width of the groove in the bill where the seed is held (Benkman 1993, 2003). Each crossbill species extracted and husked an individual conifer species' seeds more efficiently than other tree species' seeds. The study showed a positive correlation between a species' bill depth and the depth in the cone of its preferred conifer species at which the seeds are held. Benkman found that the annual survival rate for each species was related to its feeding efficiency, which varied according to the different conifer species. When he put these results together, Benkman found a series of five "adaptive peaks," showing that each species' bill morphology was associated with the conifer that it fed most efficiently on, providing the highest survival rate (**Figure 5.18B**). Benkman (2003) concluded that red crossbills are currently undergoing evolutionary divergence (speciation) as a result of se-

lection associated with differences in available food resources and the effects of those differences on bill morphology.

Optimal foraging theory predicts that foraging animals will maximize their energy gain

The availability of food varies significantly over space and time. Food availability may be greater in some locations than in others. For example, plant or animal habitat preferences result in some areas of a landscape having a higher density of individuals than surrounding areas, due to higher water or nutrient availability associated with differences in microclimate. If energy is in short supply, then animals moving through a heterogeneous landscape should invest the majority of their time in acquiring the highest-quality food resources where they are most abundant, and which are the shortest distance away. Such behavior would increase the amount of energy obtained and reduce the risks involved, such as becoming food for another animal. These ideas are the essence of the theory of **optimal foraging**, which proposes that animals will maximize the amount of energy acquired per unit time, energy, and risk involved in finding food. Optimal foraging theory relies on the assumption that evolution acts on the behavior of animals to maximize their energy gain.

We can envision the consequences of foraging behavior with a simple conceptual model (**Figure 5.19**) that describes the benefit an animal gets from its food based on the cumulative amount of effort invested. The benefit is a function of both the quality of the food and the amount of energy invested in obtaining it. An animal's success in acquiring food in-

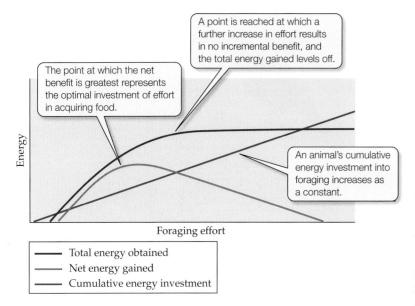

The point at which the net benefit is greatest represents the optimal investment of effort in acquiring food.

A point is reached at which a further increase in effort results in no incremental benefit, and the total energy gained levels off.

An animal's cumulative energy investment into foraging increases as a constant.

Energy

Foraging effort

——— Total energy obtained
——— Net energy gained
——— Cumulative energy investment

FIGURE 5.19 Conceptual Model of Optimal Foraging The benefit (net energy gain) obtained from the effort invested into foraging is related to the energy gain from the food acquired minus the cumulative energy invested in acquiring that food (effort). This simple model can be used to test whether animals forage in a manner that results in the maximum benefit, using estimates for effort and benefit.

creases with the effort it invests—time spent searching for and extracting or disarming prey. At some point, however, a further increase in effort results in no incremental benefit, and the net energy obtained begins to decrease. Several factors may be responsible for this decrease, including a limitation on how much food can be carried or ingested.

While this model is rather simple, it provides a basis for making quantitative predictions about animal foraging behavior. Similar models have been used to derive hypotheses that can be tested under field or laboratory conditions. We could change the independent variable (the x-axis) to the size of the food item and the dependent variable (y-axis) to growth, then test whether there is an optimum food size that provides the greatest benefit per unit energy invested in obtaining and consuming it. If optimal foraging is an adaptation to limited food supplies, then we must be able to relate our dependent variable to the survival and reproduction of the animal.

Tests of optimal foraging theory Several studies support some aspects of optimal foraging theory. Many of these studies focus on the selection of food items of different sizes by foragers and the energetic costs of searching for and handling prey. For example, research with the Eurasian oystercatcher (*Haematopus ostralegus*), a shorebird that eats bivalves (e.g., clams and mussels), has demonstrated that it selects prey of the size that provides the most energy for the effort, despite the relatively low abundance of this size class (Meire and Ervynck 1986, **Figure 5.20**). Oystercatchers must find a bivalve buried in the sand, lift it out, and open it before they can eat it. For bivalves below a certain size, the net energy benefit from this effort is marginal, setting a lower limit on the bivalve size selected by the oystercatchers. Larger bivalves have thicker shells and require more effort to open, setting an upper limit on the bivalve size selected by the oystercatchers.

Crows also exhibit feeding behavior that appears to support optimal foraging theory. As we saw in the Case Study, crows are clever birds, and they have developed a trick to circumvent the problem oystercatchers had with opening large bivalves. Northwestern crows (*Corvus caurinus*) that forage for shellfish in intertidal and sandy shore zones use gravity as a means to open their shells. Crows pick up shellfish (including whelks and bivalves) in their bills, fly into the air, and drop them on rocky surfaces to crack them open. They consume the shellfish once its protective shell has been shattered.

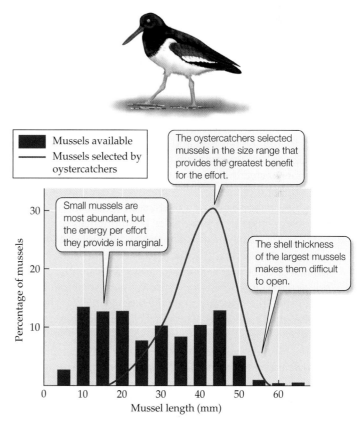

■ Mussels available
—— Mussels selected by oystercatchers

The oystercatchers selected mussels in the size range that provides the greatest benefit for the effort.

Small mussels are most abundant, but the energy per effort they provide is marginal.

The shell thickness of the largest mussels makes them difficult to open.

Percentage of mussels

Mussel length (mm)

FIGURE 5.20 Food Size Selection in Oystercatchers Researchers determined the optimal food size range (25–60 cm) for Eurasian oystercatchers (*Haemotopus astralegus*) feeding on mussels (*Mytilus edulis*) using estimates of foraging effort (handling time, effort required to open the shell, and success rate) and benefit (energy gain) for mussels of each size class. They found that oystercatchers in the field selected mussels in the optimal size range, despite the greater abundance of smaller mussels. (After Meire and Ervynck 1986.)

FIGURE 5.21 Optimal Food Selection by Clam-Dropping Crows
(A) Richardson and Verbeek estimated energy gains for Northwestern crows (*Corvus caurinus*) foraging on Japanese littleneck clams (*Venerupis japonica*) as the energy gained by clams of different sizes minus the energy expended to find a clam, dig it up, carry it into the air and drop it to crack it open, and extract it from its shell. (B) The percentage of clams found by the crows that they dropped and ate as a function of clam size. (After Richardson and Verbeek 1986.)

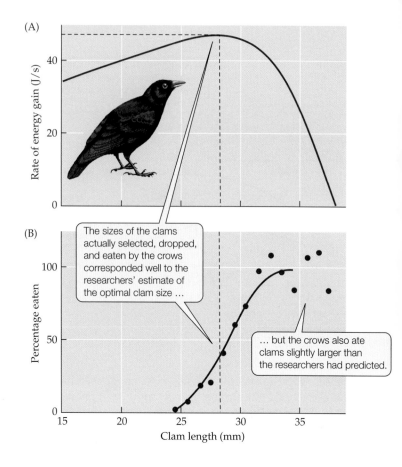

What determines which shellfish the crows will select to drop from on high? Different sizes of bivalves crack open equally well, but the net energy benefit of larger bivalves is greater, although the time spent handling them is also greater (Richardson and Verbeek 1986). Howard Richardson and Nicolaas Verbeek estimated the energy costs associated with finding Japanese littleneck clams (*Venerupis japonica*), digging them up, carrying them into the air and dropping them, and extracting the body from the shell. They also estimated the amount of energy gained from consumption of clams of different sizes and the abundances of clams of different sizes in the crows' environment. These measures allowed them to estimate the clam size that provided the maximum net energy gain for the crows (**Figure 5.21A**). Their estimate of the optimal clam size was very close to the size of the average clam selected, dropped, and eaten by crows in the wild (**Figure 5.21B**). Selecting clams smaller than the optimal size would result in too much energy investment for marginal energy gain, while larger clams were too rare to rely on and took more effort to find. Richardson and Verbeek found that the crows sometimes dug up clams smaller than the optimal size, but rejected them without investing the energy in dropping them.

Additional research demonstrated that shellfish size selection by crows overrides the influences of food quality. When provided access to different size classes of both clams (higher energy content) and whelks (lower energy content), crows consistently selected the larger shellfish, regardless of their identity, despite the greater potential energy gain of selecting small clams over large whelks (O'Brien et al. 2005).

Food density and the marginal value theorem Another aspect of optimal foraging theory considers the habitat in which an animal forages as a heterogeneous landscape containing patches of different amounts of food. To optimize its energy gain, an animal should forage in the most productive patches—that is, those with the highest food density—until those patches decline to the same food density as the area surrounding them. We can also consider the benefit obtained by a foraging animal from the perspective of time spent in a patch (**Figure 5.22**).

Once it finds a productive patch, its rate of energy gain is initially high, but decreases and eventually becomes marginal as the forager depletes the food supply. A foraging animal should stay in a patch until the time when the rate of energy gain in that patch has declined to the average rate for the

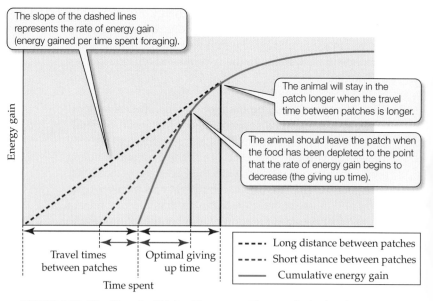

FIGURE 5.22 The Marginal Value Theorem The marginal value theorem assumes that a foraging animal will encounter food patches of varying density. The time the animal spends in a patch should optimize its rate of energy gain per amount of time spent foraging.

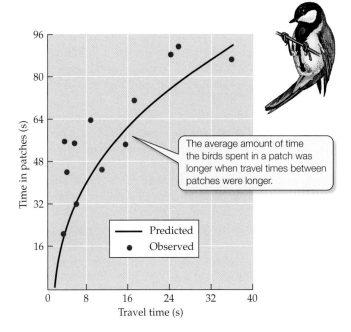

The average amount of time the birds spent in a patch was longer when travel times between patches were longer.

FIGURE 5.23 Effect of Travel Time between Patches on Giving Up Time Cowie predicted the effect of travel time between patches on the average amount of time great tits (*Parus major*) foraging in a laboratory setup would spend in a patch, based on the marginal value theorem. The solid line is the predicted foraging time based on the marginal value theorem and the energy expended traveling between patches. The birds' behavior matched his predictions fairly well. (After Cowie 1977.)

habitat (known as the *giving up time*), then depart for another patch. This conceptual model, called the **marginal value theorem**, was initially developed by Eric Charnov (Charnov 1976). It can be used to evaluate the influences of distance between patches, the quality of the food in the patch, and the energy extraction efficiency of the animal on the giving up time.

One of the predictions of the marginal value theorem is that the longer the travel time between food patches, the longer the animal should spend in a patch (see Figure 5.22).

This prediction was tested soon after Charnov published his model by Richard Cowie (1977). Cowie used a laboratory setup with great tits (*Parus major*) in a "forest" composed of wooden dowels. The food "patches" consisted of sawdust-filled plastic cups containing mealworms. The "travel time" among patches was manipulated by placing cardboard covers on top of the food cups and adjusting the ease with which they could be removed. Cowie used the marginal value theorem to predict the amount of time the birds should spend in the patches based on the travel time between them. His results matched his predictions very well (**Figure 5.23**). Similar results have been obtained from other laboratory experiments that have tested the predictions of the marginal value theorem using bird and insect predators.

Despite their success in testing the predictions of optimal foraging theory and the marginal value theorem, laboratory experiments have been criticized for lacking any connection to "real-world" foraging animals. A test of the marginal value theorem in a natural setting was performed by James Munger, who studied the behavior of horned lizards (*Phrynosoma* spp.) foraging for ants in the Chihuahuan Desert (Munger 1984) (**Figure 5.24**). Ant densities vary substantially across the desert landscape, depending on proximity to ant nests and the foraging trails used by worker ants. Munger observed the rate of ant consumption by horned lizards in various patches and compared that overall rate with the ant consumption rate at the time the lizards abandoned the patches. Based on the marginal value theorem, he predicted

FIGURE 5.24 Giving Up Time in a Natural Setting Horned lizards encountering patches of ants will remain in those patches until their rate of ant consumption approaches that in the surrounding habitat, as predicted by the marginal value theorem. These data have been corrected by removing observations of patch abandonment by lizards for reasons other than food depletion (e.g., thermoregulation). (After Munger 1984.)

that the consumption rate at the giving up time should be equal to the overall consumption rate for the habitat, and that it should be higher at times when the overall food (ant) density in the habitat was higher. Munger's predictions were largely borne out (see Figure 5.24).

While ample evidence supports some aspects of optimal foraging theory, significant criticisms have been expressed. Optimal foraging best describes the foraging behavior of animals that feed on immobile prey and applies less well to animals feeding on mobile prey (Sih and Christensen 2001). In addition, the assumptions that energy is always in short supply, and that a shortage of energy dictates foraging behavior, may not always be correct. Carnivores, in particular, may not lack for food resources to the degree assumed in optimal foraging models (Jeschke 2007). Furthermore, resources other than energy may be involved in the selection of food items, particularly nutrients such as nitrogen and sodium. Additional considerations for foragers include the risks of exposure to their own predators. Finally, the defensive behavior of prey influences the costs and benefits of prey to foragers. We will discuss predator–prey interactions in more detail in Chapter 12.

CASE STUDY REVISITED
Toolmaking Crows

We've seen that foraging animals often display behavioral as well as morphological specializations that increase their efficiency in harvesting food. The specialized bill of crossbills demonstrates an adaptive response that improves their feeding efficiency. The use of flight and gravity by crows to crack open shells displays a greater reliance on behavior to increase their feeding efficiency. Significant uncertainty exists about whether an underlying genetic basis or learning is responsible for behaviors that increase foraging efficiency.

The unexpected discovery of toolmaking in crows adds to the debate about the adaptive nature of animal behaviors. Are the New Caledonian crows inheriting their knowledge of toolmaking, are they predisposed to learn how to make tools, or does the ability to learn toolmaking occur widely among other bird species? To address this question, Ben Kenward and colleagues reared New Caledonian crows in captivity, without exposure to adult birds. Some of the birds received "tutoring" in toolmaking and tool use by human foster parents, while another control group did not (Kenward et al. 2005). To evaluate the birds' toolmaking abilities, the researchers placed supplemental food in tight crevices in the birds' aviaries, where it was not accessible to the birds without the assistance of tools. Twigs and leaves were also left in their aviaries. The captive crows developed the ability to make and use tools to retrieve the food in the crevices, whether they had been tutored or not (**Figure 5.25**). Kenward and colleagues concluded that the ability of the New Caledonian crows to manufacture tools was partially inherited, rather than an acquired skill learned from adult birds in the wild. Very similar results were reported for experiments with captive woodpecker finches, birds endemic to the Galápagos archipelago that use twigs and cactus spines to forage for arthropods (Tebbich et al. 2001).

An additional twist to the crow toolmaking story is the apparent variation in tool styles among different groups of birds on New Caledonia. In other words, there appears to be the potential for technological evolution in the styles of tools manufactured among different crow populations. Gavin Hunt and Russell Gray conducted a survey of 21 sites on New Caledonia and examined 5,550 different stepped-cut tools constructed by crows from *Pandanus* tree leaves (Hunt and Gray 2003). They found three distinct widths of tools: wide, narrow, and stepped. The tools found at a given site were very similar, and the geographic ranges of the tool types showed little overlap. There were no apparent correlations between where a tool type was found and local ecological factors such as forest structure

FIGURE 5.25 Untutored Tool Use in Captive Crows A captive New Caledonian crow (*Corvus monoeduloides*) uses a stick tool to retrieve food from artificial crevices, despite never having been exposed to tool use, either by humans or by other birds.

or climate. Hunt and Gray suggested that the three tool designs were derived from a single original tool (of the wide type) subjected to additional modifications, including additional stripping of leaf material. Their study suggests ongoing innovation in toolmaking by the New Caledonian crows. This crow engineering challenges our traditional view of technological advancement in nonhuman animals.

CONNECTIONS IN NATURE
Tool Use and Innovative Foraging

How widespread is tool use among birds and other non-primate animals? Many anecdotes of toolmaking and other innovative foraging techniques have been reported, but few have been examined thoroughly. The orange-winged sitella of Australia uses sticks to forage for insect larvae, much like the crows on New Caledonia. Egyptian vultures crack open ostrich eggs using rocks. There are additional reports of tool use by insects, mammals, and other bird species (Beck 1980). The multitude of reports involving a wide range of animal species thoroughly dispels the notion of human monopoly on tool use.

A unique report of a green heron (*Butorides virescens*) that used bait to lure fish came from a Florida lake in 1957. Several observers noted that the heron would collect bread fed to it and put it back in the water rather than eating it. When fish came to feed on the bread, the heron speared

them with its bill. When a flock of American coots showed up to feed on the bread, the heron chased them away, grabbed the bread, and waited for the coots to leave before continuing to fish (Lovell 1958). Why didn't the heron just eat the bread, and not go through all the trouble to attract the fish? The energy and nutrient content of fish is far greater than that of bread. Although the energy investment in catching fish is higher, the net benefit of the fish is greater. It is unclear how the heron could have achieved the ability to fish, or to make decisions about which food item to consume. However, similar reports of bait use to attract fish have been documented for other heron species and for great egrets.

Learned behavior is important for toolmaking in some species. A notable example comes from studies of bottlenose dolphins in Shark Bay, Australia. Researchers observed that some dolphins swim with sponges plucked from the ocean floor on their noses (technically, rostra) (**Figure 5.26**). The sponges appear to aid the dolphins when they probe the seafloor for fish to eat, protecting their sensitive rostra from stinging animals, such as stonefish, and sharp objects. The group of dolphins displaying this innovation is part of a larger group under study to examine dolphin genetics and population dynamics. The researchers' knowledge of genetics and family structure of the dolphins allowed them to address the question of whether this unique behavior was learned or inherited (Krützen et al. 2005). Michael Krützen and colleagues found that the majority of "sponging" dolphins were female. They reasoned that a single sex-linked gene (the kind of genetic basis one might expect for a trait occurring in only one sex) was a highly unlikely cause for a complex trait like sponging. A comparison of the genetic fingerprints of individuals that sponged with those of nonsponging dolphins indicated that most of the sponging occurred within a single family line. The combination of these results led Krützen and colleagues to conclude that sponging was a learned behavior passed from mother to daughter. This finding supports the idea of a cultural phenomenon in animals that influences the efficiency of their feeding behavior and challenges the notion that cultural learning is unique to humans.

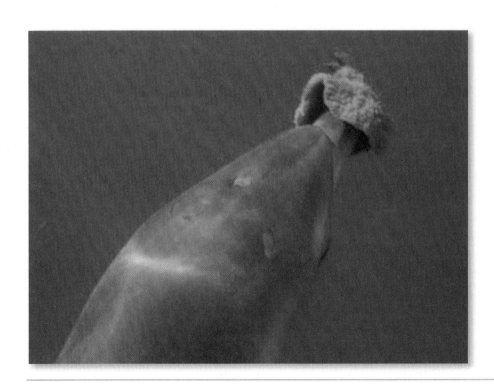

FIGURE 5.26 Dolphin Nose Gear at Shark Bay, Australia A bottlenose dolphin wears a sponge on its rostrum to protect it while foraging on the seafloor.

SUMMARY

CONCEPT 5.1 Organisms obtain energy from sunlight, from inorganic chemical compounds, or through the consumption of organic compounds.

- Autotrophs convert energy from sunlight (photosynthesis) or inorganic chemicals (chemosynthesis) into energy stored in the carbon–carbon bonds of carbohydrates and fats.

- Heterotrophs acquire energy by consuming organic compounds from other organisms, living or dead.

CONCEPT 5.2 Radiant and chemical energy captured by autotrophs is converted into stored energy in carbon–carbon bonds.

- During chemosynthesis, bacteria and archaea oxidize inorganic substrates to obtain energy, which they use to fix CO_2 and synthesize compounds for energy storage.

- Photosynthesis has two main steps, the absorption of sunlight by pigments to produce energy in the form of ATP and NADPH (the light reaction), and the use of that energy in fixation of CO_2 by the Calvin cycle, and subsequent synthesis of carbohydrates (the dark reactions).

- Photosynthetic responses to variation in light levels, water availability, and nutrient availability include both short-term acclimatization and long-term adaptation.

CONCEPT 5.3 Environmental constraints resulted in the evolution of biochemical pathways that improve the efficiency of photosynthesis.

- Photorespiration operates in opposition to photosynthesis, lowering the rate of energy gain, particularly at high temperatures and low atmospheric CO_2 concentrations.

- The C_4 photosynthetic pathway concentrates CO_2 at the site of the Calvin cycle, minimizing the amount of photorespiration.

- CAM plants reduce transpirational water loss by opening the stomates at night to take up CO_2 and releasing it to the Calvin cycle during the day, when the stomates are closed.

CONCEPT 5.4 Heterotrophs have evolved mechanisms to acquire and assimilate energy efficiently from a variety of organic sources.

- Variations in the chemistry and availability of food determine how much energy heterotrophs gain from different food sources.

- Heterotrophs display tremendous diversity in morphological and physiological adaptations that enhance their efficiency of energy acquisition and assimilation.

- Optimal foraging theory predicts that animals will obtain the maximum amount of energy per unit of time and per unit of energy invested in seeking, capturing, and extracting food resources.

- The marginal value theorem suggests that an animal foraging in a heterogeneous environment should remain in a food patch until it depletes the food to the level found in the habitat as a whole.

REVIEW QUESTIONS

1. Define autotrophy and heterotrophy and provide a few examples of each that illustrate the diversity of the ways in which organisms obtain energy.

2. How do plants acclimatize to a changing light environment? Can plants have too much light?

3. Describe the biochemical and morphological specialization of the C_4 photosynthetic pathway. What are the ecological advantages of C_4 photosynthesis over C_3 photosynthesis?

4. How does the CAM photosynthetic pathway influence water loss from plants?

5. What are the trade-offs associated with heterotrophic consumption of live animals versus dead plant material?

6. Two species of birds forage for insects that live in shrubs that have a clumped, patchy distribution throughout their habitat. The birds have the same ability to locate, capture, and consume the insects. However, one species (species A) uses less energy to fly from patch to patch than the other species (species B). Which bird would the marginal value theorem predict would spend more time in each patch, and why?

Optimal Foraging | This Web exercise (www.sinauer.com/ecology) illustrates optimal foraging. You will manipulate the foraging decision-rules of a predator to explore how different foraging strategies vary in the benefits they return.

SUGGESTED READINGS

Benkman, C. W. 2003. Divergent selection drives the adaptive radiation of crossbills. *Evolution* 57: 1176–1181. *A fascinating example of how adaptation associated with feeding behavior can eventually lead to speciation in birds.*

Ehleringer, J. R. and R. K. Monson. 1993. Evolutionary and ecological aspects of photosynthetic pathway variation. *Annual Review of Ecology and Systematics* 24: 411–439. *A thorough consideration of the evolution of different plant photosynthetic pathways and the ecological selective forces that favored their appearance and determine their habitat distributions.*

Kamil, A. C., J. R. Krebs and H. R. Pulliam, eds. 1987. *Foraging Behavior: Ecological, Ethological and Psychological Approaches.* Garland STPM Press, New York. *A collection of papers addressing numerous aspects of foraging behavior in animals. Many of the papers address optimal foraging, both its theoretical aspects and empirical tests in the laboratory and field.*

The following two textbooks are very useful references on the biology of bacteria and fungi, including the multitude of mechanisms they use to obtain energy:

Staley, J. T., R. P. Gunsalus, S. Lory and J. J. Perry. 2007. *Microbial Life.* 2nd ed. Sinauer Associates, Sunderland, MA.

Madigan, M. T. and J. M. Martinko. 2006. 11th ed. *Brock Biology of Microorganisms.* Prentice Hall, Upper Saddle River, NJ.

Sage, R. F. and R. K. Monson, eds. 1999. C_4 *Plant Biology.* Academic Press, San Diego, CA. *An excellent account of all aspects of C_4 plants, including their evolution, physiology, ecology, and biogeography.*

CASE STUDY

Trophy Hunting and Inadvertent Evolution

Bighorn sheep (*Ovis canadensis*) are magnificent animals, beautifully suited for life in the rugged mountains in which they are found. Despite their substantial size (males can weigh up to 127 kg, or 280 pounds), these sheep can balance on ledges a mere 5 cm (2 inches) wide and can leap 6 meters (20 feet) from one ledge to another. Bighorn sheep are also noted for the male's large curl of horns, which are used in combat over females (**Figure 6.1**). Rams run at speeds of up to 20 mph and crash their heads into each other, battling over the right to mate with a female.

Ram horns have been collected as trophies for many centuries. Over the last 200 years, however, human actions such as habitat encroachment, hunting, and the introduction of domesticated cattle have reduced populations of bighorn sheep by 90%. As a result, the hunting of bighorn sheep has been restricted throughout North America. These restrictions make a world-class trophy ram (one with a large, full curl of horns) extremely valuable: permits to shoot one of these rams, which are sold at auction, frequently cost over $100,000.

Although funds raised by the auction of hunting permits are used to preserve bighorn sheep habitat, scientists have expressed concern that trophy hunting has negative effects on bighorn sheep populations. Trophy hunting removes the largest and strongest males—the very males that would tend to sire large numbers of healthy offspring, thus helping sheep populations to recover. For example, Coltman et al. (2003) found that over a 30-year period, in a population from which about 10% of the males were removed by hunting each year, both the average size of males and the average size of their horns decreased (**Figure 6.2**).

Hunting and fishing appear to be having similar effects on other species. Wildlife scientists have suggested that the observed decrease in the proportion of African elephants that have tusks has been a response to the preference of poachers for elephants with large tusks. Similarly, by targeting older and larger fish, commercial fishing for cod has led to a reduction in the age and size at which these fish mature. This happens because cod that mature at a younger age and smaller size are more likely to reproduce

FIGURE 6.1 Fighting over the Right to Mate Large males with a full curl of horns—the same males prized by trophy hunters—are the most successful in contests over the right to mate with females.

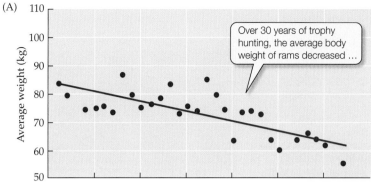

FIGURE 6.2 Trophy Hunting Decreases Ram Body and Horn Size Coltman and colleagues tracked the body weights (A) and horn lengths (B) of rams in a bighorn sheep population on Ram Mountain (Alberta, Canada) that was subjected to trophy hunting over a 30-year period. (After Coltman et al. 2003.)

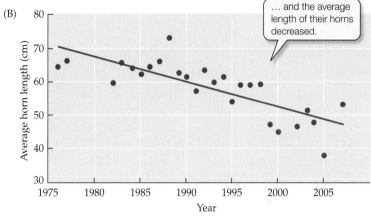

before they are caught than are fish that mature when they are older and larger.

Harvesting has even affected sex determination in some species. Consider the cold-water rock shrimp of the northern Atlantic Ocean. Individuals of this species are born as males; later in life, when they are large enough to carry a clutch of eggs effectively, these males change into females. Beginning in the late 1950s, cold-water rock shrimp were harvested intensively. As in most shrimp fisheries, individuals with long tails were preferred. Thus, because females are larger than males, almost all of the shrimp harvested initially were females. As a result, females started to become rare. The shortage of females provided an advantage to those few individuals already present in the population that changed their sex from male to female at a smaller size. The resulting small females (and the genes that encoded the "early" switch) spread in the population, causing the abundance of females to bounce back but decreasing the number of offspring per female since small females carry eggs less efficiently than larger females.

The unintended effects of human harvesting on bighorn sheep, elephants, cod, and shrimp illustrate how populations can change, or *evolve*, over time. What biological mechanisms cause these evolutionary changes? Do human actions other than harvesting produce evolutionary change?

A s news reports often emphasize, humans have a large effect on the environment. We change the global climate, pollute the water and air, convert large tracts of natural habitat into farmland and urban areas, drain wetlands, and reduce the population sizes of species that we hunt for food (e.g., fishes) or use as resources (e.g., trees). Although we have taken steps to limit some of the damage we cause to biological communities, human actions have a pervasive consequence that we have barely begun to recognize, much less address: we cause evolutionary change.

In this chapter, we'll examine what evolution is, and we'll see how it both affects ecological interactions and is affected by them. At the close of the chapter, we'll focus specifically on how humans cause evolutionary change. Our goal in this chapter is not to provide comprehensive coverage of evolution—for that, see textbooks on evolution such as Freeman and Herron (2007) or Futuyma (2005). Instead, our aim is to show that ecology and evolution are interconnected, a theme to which we will return in later chapters of this book. We'll begin by considering two ways in which evolution can be defined.

What Is Evolution?

CONCEPT 6.1
Evolution can be viewed as genetic change over time or as a process of descent with modification.

In the most general sense, biological evolution is change in organisms over time. Evolution includes the relatively small fluctuations that occur continually within populations, as when the genetic makeup of a population changes from one year to the next. But evolution can also refer to the larger changes that occur as species gradually become increasingly different from their ancestors. Let's explore these two ways of looking at evolution in more detail, focusing first on genetic fluctuations (*allele frequency change*) and then on how organisms accumulate differences from their ancestors (*descent with modification*).

Evolution is allele frequency change

The researchers who studied the effect of trophy hunting on the body size and horn size of male bighorn sheep have

(A)

UNIQUE FOSSILS
These 10-million-year-old fossils of the threespine stickleback (*Gasterosteus doryssus*) were collected from a lake bed in Nevada (USA). The fossils could be dated to the nearest 250 years because the rocks in which they were found showed exceptionally clear annual layers of sediments.

DESCENT
Fossil evidence shows that *G. doryssus* colonized the Nevada lake 10 million years ago. Over the next 16,000 years, many of the bones in these fishes did not change in size, shape, or position. The overall bone structure of these fossil fishes is similar because of common descent—the fishes descended from the original colonists and hence shared many characteristics with them.

MODIFICATION
The fossil sticklebacks also show how organisms become modified from their ancestors over time. For example, in less than 5,000 years the pelvic bone—originally the largest single bone in the body of these fishes—became greatly reduced in size. This reduction occurred at multiple sites and probably resulted from natural selection.

This individual had a complete pelvic bone (pelvic score = 3).

Diagram of a complete pelvic bone

The pelvic bone of this fish originally had a shape similar to that shown in the inset diagram—however, the pelvic bone became flattened and twisted into the rock as the fish fossilized (pelvic score = 2.5).

The pelvic bone of this individual is greatly reduced (pelvic score = 1).

This fish had no pelvic bone (pelvic score = 0). Such individuals first appear in the fossil record 11,000 years after the lake was colonized.

(B)

After 3,000 years, the average pelvic score began to drop substantially.

After 15,000 years, most members of the *G. doryssus* population had a highly reduced pelvis, causing the average pelvic score to hover near 1.

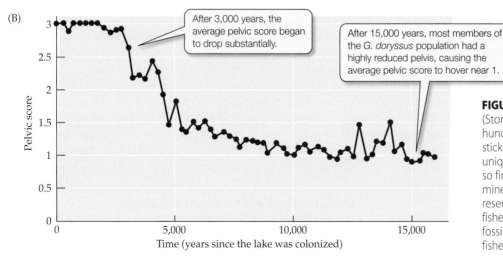

Pelvic score (y-axis): 0, 0.5, 1, 1.5, 2, 2.5, 3
Time (years since the lake was colonized) (x-axis): 0, 5,000, 10,000, 15,000

FIGURE 6.3 Descent with Modification Michael Bell (Stony Brook University) and colleagues have analyzed hundreds of 10-million-year-old fossils of the threespine stickleback fish, *Gasterosteus doryssus*. Their specimens are unique in that the fossil bed in which they were found is so finely layered that the ages of the fossils can be determined to the nearest 250-year interval of time. (A) Representative fossils showing how the pelvic bone of these fishes became reduced over time; the scale bar for each fossil is 1 cm. (B) The average pelvic score of the fossil fishes at different points in time. (After Bell et al. 2006.)

shown that horn size is a heritable trait. This means that rams with large horns tend to have offspring that have large horns, and that rams with small horns tend to have offspring that have small horns. Because trophy hunting selectively eliminates rams with large horns, it favors rams whose genetic characteristics lead to the production of small horns. Hence, it seems likely that trophy hunting is causing the genetic characteristics of the bighorn sheep population to change, or evolve, over time.

As suggested by the trophy hunting example, biologists often define evolution in terms of genetic change. To make such a definition more precise (and to introduce some concepts and terms that will be used throughout this chapter), let's review some principles from your introductory biology class:

■ Genes are composed of DNA, and they specify how to build (encode) proteins.

■ A given gene can have two or more forms (known as **alleles**) that result in the production of different versions of the protein that the gene encodes.

■ We can designate the **genotype** (genetic makeup) of an individual with letters that represent the individual's two copies of each gene (one inherited from its mother, the other from its father). If a gene has two alleles, designated *A* and *a*, the individual could be of genotype *AA*, *Aa*, or *aa*.

With these principles as background, we can define **evolution** as change over time in the *frequencies* (proportions) of different alleles in a population. To illustrate how this definition is applied, consider a gene with two alleles (*A* and *a*) and a population of 1,000 individuals, in which there are 360 individuals of genotype *AA*, 480 of genotype *Aa*, and 160 of genotype *aa*. The frequency of the *a* allele in this population is 0.4, or 40%;[*] hence, since there are only two alleles in the population (*A* and *a*), the frequency of the *A* allele must be 0.6, or 60%. If the frequency of the *a* allele were to change over time—say, from 40% to 71%—then the population would have evolved at that gene.

Evolution is descent with modification

In many parts of this book, when we refer to evolution, we will be referring to allele frequency change over time. But evolution can also be defined more broadly as *descent with modification*. At the heart of this definition is the observation that populations accumulate differences over time, and hence, when a new species forms, it differs from its ancestors. The new species, however, may differ from its ancestors in a rel-

atively small number of ways. As a result, a new species not only differs from its ancestors, it also resembles its ancestors because it descended from them and continues to share many characteristics with them. Hence, when evolution occurs, both *descent* (shared ancestry, resulting in shared characteristics) and *modification* (the accumulation of differences) can be observed, as seen in the fossil fishes in **Figure 6.3**.

Charles Darwin (1859) used the phrase "descent with modification" to summarize the evolutionary process in his book *The Origin of Species*. Darwin proposed that populations accumulate differences over time primarily by **natural selection**, the process by which individuals with certain heritable characteristics survive and reproduce more successfully than individuals with other heritable characteristics. In the Case Study, we've already seen several examples of natural selection at work. In bighorn sheep, trophy hunting has favored rams with small horns, while in the cod fishery, harvesting practices have selected for individuals that matured at a younger age and smaller size.

How can natural selection explain the accumulation of differences between populations? Darwin argued that if two populations experience different environmental conditions, individuals with one set of characteristics may be favored by natural selection in one population, while individuals with a different set of characteristics may be favored in the other population (**Figure 6.4**). By favoring individuals with different heritable characteristics in different populations, natural selection can cause populations to diverge genetically from one another over time; that is, to accumulate more and more genetic differences. Thus, natural selection can be responsible for the "modification" part of "descent with modification."

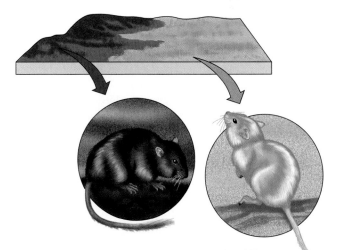

FIGURE 6.4 Natural Selection Can Result in Differences between Populations   Populations of rock pocket mice (*Chaetodipus intermedius*) that live on dark lava formations in Arizona and New Mexico have dark coats, while nearby populations that live on light-colored rocks have light coats. In each population, natural selection has favored individuals whose coat colors match their surroundings, making them less visible to predators.

---

[*]There are 1,000 individuals in the population, and each individual carries two alleles. Thus, there are a total of 2,000 alleles in the population. Each individual of genotype *AA* has zero *a* alleles, each individual of genotype *Aa* has one *a* allele, and each individual of genotype *aa* has two *a* alleles. Thus, the frequency of the *a* allele is (0 × 360 + 1 × 480 + 2 × 160)/2,000 = 0.4

Populations evolve, individuals do not

Natural selection acts as a sorting process, favoring individuals with some heritable traits (e.g., those with small horns) over others (e.g., those with large horns). Individuals with the favored traits tend to leave more offspring than individuals with other traits. As a result, from one generation to the next, a greater proportion of the individuals in the population will have the traits favored by natural selection. This process ensures that the allele frequencies of the population will change over time, and hence, that the population will evolve. But the *individuals* in the population do not evolve—they either have the traits favored by selection or they don't.

Mechanisms of Evolution

> **CONCEPT 6.2**
> Natural selection, genetic drift, and gene flow can cause allele frequencies in a population to change over time.

Although natural selection is often the cause of evolutionary change, it is not the only one. In this section, we'll examine four key processes that influence evolution: mutation, natural selection, genetic drift, and gene flow. As we'll see, mutation is the source of the new alleles on which all of evolution depends, while natural selection, genetic drift, and gene flow are the main mechanisms that cause allele frequencies to change over time.

Mutation generates the raw material for evolution

Individuals in populations differ from one another in their observable characteristics, or **phenotype** (**Figure 6.5**). Many aspects of an organism's phenotype, including its physical features, metabolism, growth rate, susceptibility to disease, and behavior, are influenced by its genes. As a result, individuals differ from one another in part because they have different alleles for genes that influence their phenotype. These different alleles arise by **mutation**, change in the DNA of a gene that can result from copying errors during cell division, mechanical damage when molecules and cell structures collide with DNA, exposure to certain chemicals (mutagens), or exposure to high-energy forms of radiation such as ultraviolet light and X rays. As we'll see in Chapter 7, the environment can also affect an organism's phenotype; for example, a plant growing in nutrient-rich soil may tend to be larger than another individual of its species growing in nutrient-poor soil, even if both have the same alleles for genes that influence size. In this chapter, however, we will focus on phenotypic differences that result from genetic, not environmental, factors.

The formation of new alleles by mutation is critical to evolution. In a hypothetical species in which there was no mutation, each gene would have only one allele, and all members of a population would be genetically identical. If this

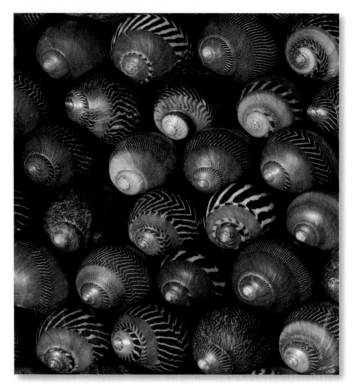

FIGURE 6.5 Individuals in Populations Differ from One Another   These *Neritina communis* snails, native to mangrove swamps of the western Pacific, show considerable variation in the color and banding pattern of their shells. Individuals in a population of these snails would also be likely to differ in other morphological traits, as well as many biochemical, behavioral, and physiological traits.

were true, evolution could not occur—allele frequencies cannot possibly change over time unless the individuals in a population differ genetically. You may recall from your introductory biology class that the individuals in a population may differ genetically not only because of mutation, but also because of *recombination*, the production of offspring that have combinations of alleles that differ from those in either of their parents. We can think of mutation as providing the raw material (new alleles) on which evolution is based, and recombination as rearranging that raw material into unique new combinations. Together, these processes provide the genetic variation on which evolution depends.

Despite its importance to evolution, mutation occurs too rarely to be the direct cause of significant allele frequency change over short periods of time. Mutations typically occur at rates of 10^{-4} to 10^{-6} new mutations per gene per generation (Hartl and Clark 1997). In other words, in each generation, we can expect one mutation to occur in every 10,000 to 1,000,000 copies of a gene. At these rates, in one generation, mutation causes virtually no change in the allele frequencies of a population. Eventually, mutation can cause appreciable allele frequency change, but it takes thousands of generations for it to do so. Overall, in terms of its direct effects, mutation is a weak agent of allele frequency change. But because it pro-

(A) Directional selection

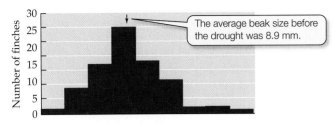

Finches hatched in 1976, the year before the drought

The average beak size before the drought was 8.9 mm.

Finches hatched in 1978, the year after the drought

In just 2 years, the average beak size increased to 9.7 mm.

Beak size (depth in mm)

(B) Stabilizing selection

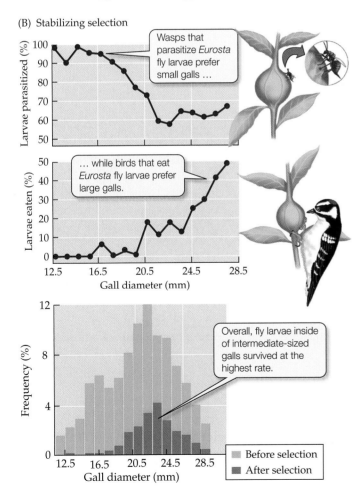

Wasps that parasitize *Eurosta* fly larvae prefer small galls …

… while birds that eat *Eurosta* fly larvae prefer large galls.

Gall diameter (mm)

Overall, fly larvae inside of intermediate-sized galls survived at the highest rate.

Gall diameter (mm)

Before selection
After selection

(C) Disruptive selection

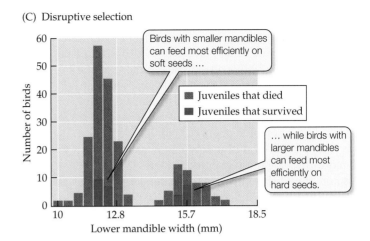

Birds with smaller mandibles can feed most efficiently on soft seeds …

Juveniles that died
Juveniles that survived

… while birds with larger mandibles can feed most efficiently on hard seeds.

Lower mandible width (mm)

FIGURE 6.6 Three Types of Natural Selection (A) Directional selection favors individuals at one phenotypic extreme. A prolonged drought in the Galápagos archipelago resulted in directional selection on the beak size of the seed-eating medium ground finch (*Geospiza fortis*). As a result of the drought, most of the available seeds were large and hard to crack, so birds with large beaks, which could more easily crack those seeds, had an advantage over birds with smaller beaks. (B) Stabilizing selection favors individuals with an intermediate phenotype. *Eurosta* flies parasitize goldenrod plants, causing the plant to produce a gall in which the fly larva matures as it feeds on the plant. The preferences of *Eurosta*'s own predators and parasites result in stabilizing selection on gall size. Wasps that parasitize and kill the fly larvae prefer small galls, while birds that eat the fly larvae prefer large galls, so larvae in galls of intermediate size have an advantage. (C) Disruptive selection favors individuals at both extremes. African seedcrackers (*Pyrenestes ostrinus*) depend on two major food plants in their environment. Birds with smaller mandible sizes can feed on one plant's soft seeds most efficiently, while birds with larger mandibles can feed on the other plant's hard seeds most efficiently. Thus, individuals with mandible sizes that are either relatively small or relatively large have an advantage. (A after Grant and Grant 2003; B after Weis and Abrahamson 1986; C after Smith et al.1993.)

vides the new alleles on which natural selection and other mechanisms of evolution can act, mutation is central to the evolutionary process.

Natural selection increases the frequencies of advantageous alleles

Natural selection occurs when individuals with particular heritable traits tend to leave more offspring than individuals with other heritable traits. As mentioned above, Darwin believed that natural selection was the most important agent of evolutionary change, and his view is well supported by genetic (e.g., Rieseberg et al. 2002) and ecological studies (e.g., Grant 1986; Endler 1995).

Natural selection typically acts on the whole organism: individuals with certain phenotypic traits survive and reproduce more successfully than other individuals with different traits. Depending on what traits are favored, we can categorize natural selection into three types (**Figure 6.6**). **Directional selection** occurs when individuals with one extreme of a heritable phenotypic trait (for example, large size) are favored over other individuals (small and medium-sized individuals). In **stabilizing selection**, individuals with an intermediate phenotype (for example, medium-sized individuals) are favored, while in **disruptive selection**, individuals with a phenotype at either extreme are favored (for example, small and large individuals have an advantage over medium-sized individuals). In all three types of natural selection, the fundamental process is the same: some individuals have heritable phenotypic traits that give them an advantage in survival or reproduction, causing them to leave more offspring than other individuals.

It is important to bear in mind that natural selection operates on aspects of the phenotype that have a genetic basis. In other words, it favors individuals with alleles that result in phenotypic traits that make them more likely to survive and reproduce than are individuals with other alleles. In some cases, the end result of this process is that most or all of the individuals in a population have the allele favored by natural selection. A well-studied example occurs in the bar-headed goose (*Anser indicus*), which migrates over the Himalayas. These birds have evolved a version of the oxygen transport protein hemoglobin that has an unusually high affinity for oxygen and hence provides an advantage in this low-oxygen, high-altitude environment (Weber 2007). The allele that encodes this version of the hemoglobin protein has evolved independently in other species of geese that fly over the Andes in South America.

At the population level, natural selection can cause the frequency of an allele that confers an advantage to increase over time—as has occurred in the bar-headed goose. We'll consider the consequences of such increases in the frequencies of advantageous alleles later in the chapter. But first, we'll look at two other mechanisms that can cause allele frequencies to change, *genetic drift* and *gene flow*.

Genetic drift is the effect of chance events

Allele frequencies in populations can be influenced by chance events. Imagine a population of ten wildflowers in which three individuals have genotype *AA*, four have genotype *Aa*, and three have genotype *aa*. Thus, the initial frequency of the *A* allele is 50%, as is the frequency of the *a* allele. Assume that the *A* and *a* alleles encode two different versions of a protein that function equally well. Although neither allele is more advantageous than the other (and hence natural selection does not affect this gene), chance events could alter their frequencies. For example, imagine that a moose walking through the woods happened to step on four of the wildflowers, two of genotype *AA* and two of genotype *Aa*, killing them—but not harming any of the three wildflowers of genotype *aa*. In one generation, the frequency of the *a* allele in the population would increase from 50% to 67% *by chance alone*. Similar genetic accidents can occur when wind moves pollen from plant to plant, influencing which individuals produce offspring and thereby affecting the allele frequencies in the next generation.

When chance events determine which alleles are passed from one generation to the next, **genetic drift** is said to occur. Although chance events occur in populations of all sizes, genetic drift alters allele frequencies significantly only in small populations. To see why, imagine that our wildflower population had 10,000 individuals—3,000 of genotype *AA*, 4,000 of genotype *Aa*, and 3,000 of genotype *aa*. If moose stepped on a random sample of 40% of the individuals in this larger population, there is virtually no possibility that all of the 3,000 individuals of genotype *aa* would be spared. Instead, it is likely that many individuals of each genotype would be killed—and hence that the frequencies of the *A* and *a* alleles would change little, if at all.

Genetic drift has four related effects on evolution in small populations:

1. Because it acts by chance alone, genetic drift causes allele frequencies to fluctuate randomly over time (**Figure 6.7**). When this occurs, eventually some alleles disappear from the population, while others reach **fixation** (a frequency of 100%).
2. By causing alleles to be lost from a population, genetic drift reduces the genetic variation of the population, making the individuals within the population more similar genetically to one another.

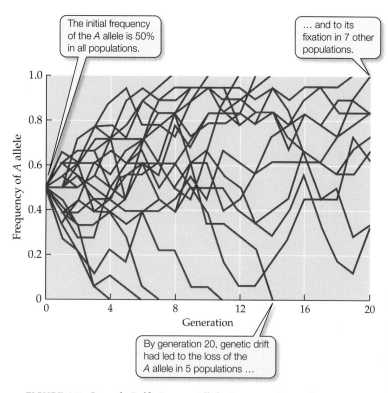

The initial frequency of the *A* allele is 50% in all populations.

... and to its fixation in 7 other populations.

By generation 20, genetic drift had led to the loss of the *A* allele in 5 populations ...

FIGURE 6.7 Genetic Drift Causes Allele Frequencies to Fluctuate at Random Results of a computer simulation of genetic drift in 20 populations, each with 9 diploid individuals (18 alleles). In small populations such as these, genetic drift has rapid effects.

(A)

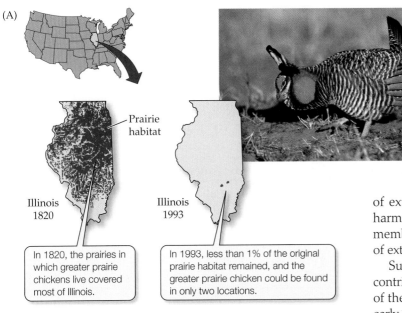

Prairie
habitat

Illinois
1820

Illinois
1993

In 1820, the prairies in which greater prairie chickens live covered most of Illinois.

In 1993, less than 1% of the original prairie habitat remained, and the greater prairie chicken could be found in only two locations.

By 1993, the greater prairie chicken population in Illinois had dropped to fewer than 50 birds and was experiencing the negative effects of genetic drift on small populations.

(B)

| | Illinois | | | | |
	1933	1993	Kansas	Nebraska	Minnesota
Population size	25,000	<50	750,000	75,000–200,000	4,000
No. of alleles at six genes	31	22	35	35	32
Percentage of eggs that hatch	93	<50	99	96	85

FIGURE 6.8 Harmful Effects of Genetic Drift
(A) The Illinois population of greater prairie chickens dropped from millions of birds in the 1800s to 25,000 in 1933, and, finally, to fewer than 50 birds in 1993. Genetic drift in the small 1993 population led to a loss of alleles at six genes and to a rise in the frequencies of harmful alleles, reducing egg hatching rates to less than 50%. (B) The table compares the 1993 Illinois populations with historical populations in Illinois and with populations in Kansas, Nebraska, and Minnesota, none of which experienced as severe a drop in population size. (B after Bouzat et al. 1998.)

of extinction. Likewise, an increase in the frequency of harmful alleles in a population can hinder the ability of its members to survive or reproduce, again increasing the risk of extinction.

Such negative effects of genetic drift are thought to have contributed to the near-extinction of the Illinois populations of the greater prairie chicken (*Tympanuchus cupido*). In the early 1800s, there were millions of these birds in Illinois. Over time, their numbers plummeted as more than 99% of the prairie habitat on which they depend was converted to farmland and other uses. By 1993, fewer than 50 greater prairie chickens remained in Illinois. By comparing the DNA of birds in the 1993 Illinois population with that of birds that lived in Illinois in the 1930s (obtained from museum specimens), Juan Bouzat and colleagues (1998) showed that the drop in the population size had reduced the genetic variation of the population (**Figure 6.8**). In addition, more than 50% of the eggs laid by birds in the 1993 Illinois population failed to hatch, suggesting that genetic drift had led to the fixation of harmful alleles. This interpretation was strengthened by the results of experiments begun in 1992: when greater prairie chickens from other populations were brought to Illinois, new alleles entered the Illinois population, and egg hatching rates increased from less than 50% to more than 90% in just 5 years (Westemeier et al. 1998).

Gene flow is the exchange of alleles between populations

Gene flow occurs when alleles are transferred from one population to another via the movement of individuals or gametes (e.g., plant pollen). Gene flow has two important effects. First, by transferring alleles between populations, it tends to make populations more similar to one another genetically. This homogenizing effect of gene flow is one reason that individuals in different populations of the same species resemble one another—alleles are exchanged often enough that relatively few differences accumulate between the populations.

Second, gene flow can introduce new alleles to a population. When this occurs, gene flow acts in a manner similar to mutation (although mutation remains the ultimate source

3. Genetic drift can increase the frequency of a harmful allele. This may seem counterintuitive, because in general, genetic drift acts on alleles that neither harm nor benefit the organism, and we would expect natural selection to reduce the frequency of a harmful allele. However, if the population size is very small and the allele has only slightly deleterious effects, genetic drift can "overrule" the effects of natural selection, causing the harmful allele to increase or decrease in frequency by chance alone.

4. Genetic drift can increase genetic differences between populations because chance events may cause an allele to reach fixation in one population, yet be lost from another population.

The second and third effects can have dire consequences for small populations. A loss of genetic variation can reduce the capacity of a population to evolve in response to changing environmental conditions—potentially placing it at risk

FIGURE 6.9 Gene Flow: Setting the Stage for Selection for Insecticide Resistance In this idealized diagram, an allele that causes resistance to organophosphate insecticides arises by mutation in one population and then spreads (by gene flow) to two other populations. When mosquitoes in the two other populations are exposed to the insecticide, natural selection causes the frequency of the resistance allele to increase rapidly.

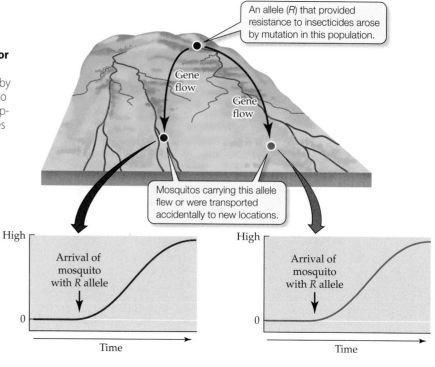

of *all* new alleles). This effect of gene flow has considerable applied importance. For example, before the 1960s, the mosquito *Culex pipiens* was not resistant to organophosphate insecticides. This mosquito transmits malaria and West Nile virus, so insecticides were often used to destroy its populations. In the late 1960s, however, new alleles that provided resistance to organophosphate insecticides were produced by mutation in a few *C. pipiens* populations, probably in Africa or Asia (Raymond et al. 1998). Mosquitoes carrying these alleles were blown by storms or transported accidentally by humans to new locations, where they bred with mosquitoes from the local populations. In populations of mosquitoes exposed to insecticides, once these alleles had entered the population by gene flow, their frequency increased rapidly because insecticide resistance was favored by natural selection (**Figure 6.9**). The global spread of these alleles by gene flow has allowed billions of mosquitoes to survive the application of insecticides that otherwise would have killed them.

Evolutionary change that results in a closer match between organisms and the conditions of their environment, such as the increase in the frequency of insecticide resistance in a *C. pipiens* population exposed to insecticides, is an example of adaptive evolution, the topic we'll consider next.

Adaptive Evolution

CONCEPT 6.3
Natural selection is the only evolutionary mechanism that consistently causes adaptive evolution.

The natural world is filled with striking examples of organisms that are well suited for life in their environments (**Figure 6.10**). This match between organisms and their environments highlights their **adaptations**, which are features of organisms that improve their ability to survive and reproduce in their environment (see Chapter 4). Examples of adaptations include remarkable features like those shown in Figure 6.10, but also include less visually striking characteristics—such as a photosynthetic enzyme in a desert plant that can function at high temperatures that would denature most enzymes. There are literally millions of other examples of adaptations. How do these adaptations arise?

Adaptations result from natural selection

Unlike genetic drift, natural selection is not a random process. Instead, when natural selection operates, individuals with certain alleles consistently leave more offspring than do individuals with other alleles. By consistently favoring individuals with some alleles over individuals with other alleles, natural selection causes **adaptive evolution**, a process of change in which traits that confer advantages tend to increase in frequency over time. This process tends to increase the effectiveness of an adaptation that selection acts on, causing the match between organisms and their environments to improve over time. Although gene flow and genetic drift *can* improve the effectiveness of an adaptation (by increasing the frequency of an advantageous allele), they can also do the reverse (by increasing the frequency of a disadvantageous allele). Thus, natural selection is the only evolutionary mechanism that consistently results in adaptive evolution.

An example of adaptive evolution is provided by changes in populations of the soapberry bug (*Jadera haematoloma*) (Carroll and Boyd 1992; Carroll et al. 1997). This insect uses its needle-like beak to feed on seeds located within the fruits of several different plant species. In southern Florida, soapberry bug populations feed on the seeds of the insect's native host, the balloon vine (*Cardiospermum corundum*). Balloon vines, however, are rare in central Florida. Thus, in that region, soapberry bugs do not feed on balloon vines, but instead feed on the seeds of a species introduced from eastern Asia, the golden rain tree (*Koelreuteria elegans*). A few specimens of the golden rain tree were brought to Florida in 1926, but it was

FIGURE 6.10 A Gallery of Adaptations   (A) This small fish (2 cm), the pygmy seahorse (*Hippocampus bargibanti*), is only known to occur on corals of the genus *Muricella* in the western Pacific. Due to its effective camouflage, this species was discovered only after the coral on which the original specimen was living was collected and placed in an aquarium. (B) A Parson's chameleon (*Chamaeleo parsonii*) extending its tongue to capture an insect. The tongue fires rapidly from the mouth, hitting the prey item in about 30 thousandths of a second. The tongue, which has a sticky tip against which prey become immobilized, is then drawn quickly back into the mouth, where the captured prey is crushed by powerful jaws. (C) This Morgan's sphinx moth (*Xanthopan morgani*) has a long "tongue" (technically, a proboscis) that is specifically adapted for feeding on nectar found at the bottom of a Comet orchid's (*Angraecum sesquipedale*) 10-inch-long floral tube. Both species are found only in Madagascar. (D) Temperatures of 64°C – 87°C (147°F – 188°F) are common at Grand Prismatic Spring, Yellowstone National Park, USA. Only a few algae and prokaryotes (bacteria and archaea), collectively known as "hyperthermophiles," have the structural and biochemical adaptations necessary for life in water this hot. Extensive orange and brown microbial mats radiate from the hot spring.

not commonly planted until the 1950s. The oldest golden rain trees in the central Florida populations studied by Carroll and colleagues were 35 years old, suggesting that the soapberry bugs there have fed on this species for 35 years or less.

Soapberry bugs feed most efficiently when the length of the beak matches the depth to which it must pierce a fruit to reach the seeds. Since golden rain tree fruits are smaller than balloon vine fruits, the introduction of the golden rain tree 35 years ago can be viewed as an unplanned natural experiment on the effect of natural selection on the insect's beak length. Carroll and Boyd predicted that as a result of natural selection, beak lengths would evolve to be *shorter* in soapberry bug populations that fed on golden rain tree fruits than in populations that fed on the native host, balloon vine. Carroll and Boyd also studied soapberry bugs in Oklahoma and Louisiana, where the insect had begun to feed on several other new host plants that had been introduced within the past 100 years. However, in Oklahoma and Louisiana, the fruits of the introduced hosts were larger than those of the native hosts, leading to the prediction that in those two states, the beak lengths of insects that ate the introduced species would be *longer* than those of insects that ate the native species.

In all three locations, Carroll and Boyd found that soapberry bug beak lengths evolved in the direction predicted by fruit size, decreasing in central Florida (**Figure 6.11**) and increasing in both Oklahoma and Louisiana. The changes in beak length were substantial: average beak lengths dropped by 26% in central Florida and increased by 8% (on one introduced host species) and 17% (on another introduced host species) in Oklahoma and Louisiana. In addition, Carroll et al. (1997) showed that beak length is a heritable characteristic, so the observed changes in beak length must have been due at least in part to changes in the frequencies of alleles that affect beak length. Thus, we can conclude that in a relatively short time (35–100 years), natural selection in soapberry bug populations caused adaptive evolution in which a characteristic of the organism (beak length) evolved to match an aspect of its environment (fruit size) more closely.

Adaptive evolution can occur rapidly

Soapberry bugs are not unique: studies on populations of a wide range of other organisms show that natural selection can lead to rapid increases in the frequency of advantageous traits. Examples include the evolution of increased antibiotic resistance in bacteria (in days to months); increased insecticide resistance in insects (in months to years); drabber coloration in guppies, which makes them harder for visually hunting predators to find (several years); and increased beak size in *Geospiza* finches (several years; see Figure 6.6A). These and dozens of other examples are described in more detail by Endler (1986) and Thompson (1998).

Rapid adaptive evolution has also been documented on a continental scale. A number of studies have focused on **clines**: gradual changes in a characteristic of an organism over a geographic region. Many species of terrestrial vertebrates show a cline in which the average body size of individuals increases in a regular manner as latitude increases (Mayr 1956). Similarly, in the fruit fly *Drosophila melanogaster*, the alcohol dehydrogenase (*Adh*) gene exhibits a cline in which the Adh^S allele decreases in frequency as latitude increases (**Figure 6.12A**); this pattern has been found in both the Northern and Southern hemispheres. Previous studies have shown that this cline results from natural selection on the Adh^S allele, which is less effective in the colder temperatures at higher latitudes and hence is less common there. Over the past 20 years in coastal Australia, the *Adh* cline has shifted about 4° in latitude toward the South Pole (Umina et al. 2005), a movement of roughly 400 km (**Figure 6.12B**). During the same period, mean temperatures in the region

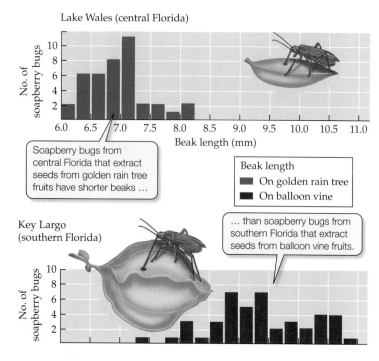

FIGURE 6.11 Rapid Adaptive Evolution in Soapberry Bugs
Soapberry bug populations in southern Florida feed on the seeds of their native host, the balloon vine, while soapberry bug populations in central Florida feed on the seeds of the golden rain tree, a plant introduced in the 1950s, which has smaller fruits than the balloon vine. The beak lengths of insects feeding on the golden rain tree decreased by 26% in 35 years, providing a better match to the smaller fruits of the introduced plant. (After Carroll and Boyd 1992.)

have increased by 0.5°C. Thus, the 4° shift in latitude appears to be an adaptive increase in the frequency of the Adh^S allele in response to climate change.

Gene flow can limit local adaptation

Although many populations are strikingly well matched to their environments, others are not. Gene flow is one of the factors that can limit the extent to which a population is adapted to its local environment. For example, some plant species have tolerant genotypes that can grow on soils at former mine sites containing high concentrations of heavy metals; such soils are toxic to intolerant genotypes. On normal soils, the tolerant genotypes grow poorly compared with intolerant genotypes. Thus, we would expect the frequencies of tolerant genotypes to approach 100% on mine soils (where they are advantageous) and 0% on normal soils (where they are disadvantageous). Researchers found that a population of the bentgrass *Agrostis tenuis* growing on mine soils was dominated by tolerant genotypes, as expected. However, a population growing on normal soils downwind from the mine site contained more tolerant genotypes than expected

FIGURE 6.12 Rapid Adaptive Evolution on a Continental Scale
The *Adh* gene encodes a metabolically important enzyme, alcohol dehydrogenase. Previous field and laboratory studies indicate that the *Adh^S* allele of this gene is selected against in cooler environments. (A) Frequencies of the *Adh^S* allele in coastal Australian *Drosophila melanogaster* populations in 1979–1982 and in 2002–2004. (B) Regression lines calculated from the data in part A show that between 1979–1982 and 2002–2004, the cline of the *Adh^S* allele shifted 4° toward the South Pole as the region's average temperatures increased by 0.5°C. (After Umina et al. 2005.)

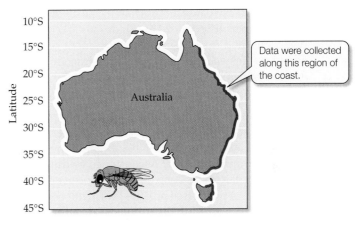

(McNeilly 1968). Bentgrass is wind-pollinated, and each year, pollen from the plants growing on mine soils carried alleles for heavy metal tolerance into the population growing on normal soils, preventing that population from becoming fully adapted to its local conditions. The population growing on mine soils also received pollen from plants growing on normal soils. In this case, however, gene flow had relatively little effect on allele frequencies because selection against intolerant genotypes was so strong (they survived poorly on mine soils). In general, whenever alleles are transferred between populations that live in different environments, the extent to which adaptive evolution occurs in each population depends on whether natural selection is strong enough to overcome the effects of ongoing gene flow.

Adaptations are not perfect

As we have just seen, gene flow can limit the extent to which a population is adapted to its local environment. But even when gene flow does not have this effect, natural selection does not result in a perfect match between organisms and their environments. In part, this occurs because an organism's environment is not static—it is a moving target, because the abiotic and biotic components of the environment change continually over time. In addition, all organisms face a number of constraints on their evolution:

- *Lack of genetic variation.* If none of the individuals in a population has a beneficial allele for a particular gene that influences survival and reproduction, adaptive evolution cannot occur at that gene. For example, the mosquito *Culex pipiens* initially lacked alleles that provided resistance to organophosphate insecticides (see p. 138. For decades, this lack of genetic variation prevented adaptive evolution in response to insecticides, allowing humans to destroy mosquito populations with impunity—at least up until the time when insecticide resistance alleles arose by mutation and spread by gene flow. Note that in this and in all other cases, advantageous alleles arise by chance—they are not produced as needed or "on demand."

- *Evolutionary history.* Natural selection does not craft the adaptations of an organism from scratch. Instead, if the

necessary genetic variation is present, selection works by modifying the traits already present in an organism. In addition, organisms have certain characteristics and lack others because of their ancestry. It would be advantageous, for example, for an aquatic mammal such as a dolphin to be able to breathe under water. Dolphins lack this capacity, however, in part because of constraints imposed by their evolutionary history: they evolved from terrestrial vertebrates that had lungs and breathed air. Natural selection can bring about great changes—witness the mode of life and streamlined body form of the dolphin—but it does so by modifying traits that are already present in the organism, not by creating advantageous traits de novo.

- *Ecological trade-offs.* To survive and reproduce, organisms must perform many essential functions, such as acquir-

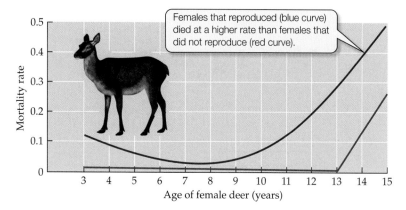

FIGURE 6.13 **A Trade-off between Reproduction and Survival** Female red deer that reproduced had a lower chance of surviving to the next year than did females that did not reproduce. (After Clutton-Brock et al. 1983.)

ing food, escaping predators, warding off disease, and finding mates. Energy and resources are required for each of these essential functions. Hence, as suggested by the maxim "There is no free lunch" (see Table 1.1), organisms face trade-offs in which the ability to perform one function can cause a reduction in the ability to perform another (**Figure 6.13**). Trade-offs occur in all organisms, and they ensure that adaptations will never be perfect. Instead, adaptations represent compromises in the abilities of organisms to perform many different and sometimes conflicting functions.

Despite these pervasive constraints, adaptive evolution is a key component of the evolutionary process. What does the importance of adaptive evolution tell us about the link between ecology and evolution? As illustrated by the changes in the beak lengths of soapberry bugs that accompanied their shifts to new host plants (see pp. 138–140), natural selection, and the adaptive evolution that results, is driven by the interactions of organisms with one another and with their environment. Any such interaction is an ecological interaction, and hence ecology serves as a basis for understanding natural selection. Next, we'll consider how ecological interactions influence broader evolutionary changes, such as the formation of new species and the great changes that have occurred during the long history of life on Earth.

The Evolutionary History of Life

CONCEPT 6.4
Long-term patterns of evolution are shaped by large-scale processes such as speciation, mass extinction, and adaptive radiation.

Earth is home to roughly 2 million **species** (groups of organisms whose members have similar characteristics and can in-

terbreed) that have been named by taxonomists—and to millions more that have yet to be discovered or named. This tremendous diversity of species serves as a foundation for all of ecology, which, as we saw in Chapter 1, is the study of how species interact with one another and with their environment. But the causation runs both ways: while it is true that ecological interactions are influenced by the diversity of species, it is also true that ecological interactions have influenced the number of species alive today. To see why, let's examine the origin of species and some of the other processes that have affected the history of life on Earth.

The diversity of life results from speciation

Each of the millions of species alive today originated by **speciation**, the process by which one species splits into two or more species. Speciation most commonly occurs when a barrier prevents gene flow between two or more populations of a species. The barrier may be geographic, as when a new population becomes established far from the parental population, or it may be ecological, as when some members of an insect population begin to feed on a new host plant. When a barrier to gene flow is established between populations, they diverge genetically over time (**Figure 6.14**). New species can also form in several other ways, such as when members of two different species mate to produce fertile hybrid offspring (see p. 146 for an example in sunflowers). Whether it is produced by divergence, hybridization, or other means, the key step in the formation of a new species is the evolution of barriers that prevent its members from breeding with members of the parental species. Such reproductive barriers arise when a population accumulates so many genetic differences that its members cannot produce viable, fertile offspring if they mate with members of the parental species.

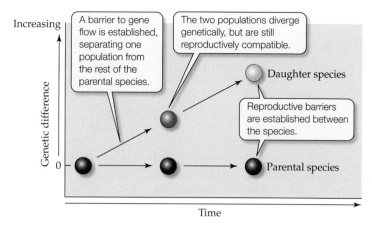

FIGURE 6.14 **Speciation by Genetic Divergence** Once genetic divergence begins, the time required for speciation varies tremendously, from a single generation (perhaps a single year), to a few thousand years, to millions of years in most cases.

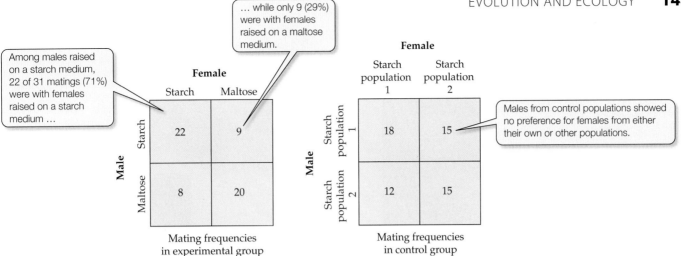

FIGURE 6.15 Reproductive Barriers Can Be a By-product of Selection   After 1 year (about 40 generations) in which experimental populations of *Drosophila pseudoobscura* fruit flies were selected for growth on different sources of food, most matings occurred between flies selected to feed on the same food source. No such mating preference was observed in control populations that were not subjected to selection, regardless of whether the control populations were reared on starch (shown here) or maltose (not shown). (After Dodd 1989.)

The accumulation of genetic differences that lead to the formation of a new species can be an accidental by-product of selection. For example, an experiment with fruit flies demonstrated the beginnings of reproductive barriers between populations selected for growth on different sources of food, but no such barriers were observed between control populations that had not been subjected to selection (**Figure 6.15**). Natural selection has produced similar changes, both in plant populations growing on soils with differing concentrations of heavy metals (Macnair and Christie 1983) and in frog populations living in different environments with different temperatures (Moore 1957). In some cases, the trait favored by selection is the same trait that drives speciation. For example, the mosquitofish (*Gambusia hubbsi*) lives in small pools in the Bahamas. Some of these pools contain fish species that eat *Gambusia*, while others lack such predators. In pools with predatory fishes, the mosquitofish have evolved a body shape that allows high-speed escape swimming (Langerhans et al. 2007). Female mosquitofish prefer to mate with males whose body shape is similar to their own. Thus, natural selection favors different body shapes in mosquitofish depending on the presence or absence of predators, and these different body shapes drive the early stages of speciation through their effects on mate choice.

In each of these cases, reproductive barriers arose as a by-product of selection in response to a feature of the environment, such as food source, heavy metal concentration, temperature, or presence of predators. As we've seen, genetic drift can also promote the accumulation of genetic differences between populations (see pp. 136–137). Thus, like natural selection, genetic drift can ultimately lead to the evolution of reproductive barriers, and hence to the formation of new species. In contrast to selection and drift, gene flow *always* acts to slow down or prevent the formation of new species because populations that exchange many alleles tend to remain genetically similar to one another, making it less likely that reproductive barriers will evolve.

Mass extinctions and adaptive radiations have shaped long-term patterns of evolution

Thus far in this chapter, much of our focus has been on the *process* of evolution—the mechanisms by which evolutionary change occurs. But evolution can also be viewed as an observed *pattern* of change. Evolutionary patterns are revealed by observations of the natural world, such as data on the changing allele frequencies of a population over time. If we wish to broaden our scope and look at patterns of evolution over long time scales, we can turn to the fossil record, which shows that life on Earth has changed greatly over time (**Figure 6.16**).

The earliest known fossils are those of 3.5 billion-year-old bacteria, and the most ancient fossils of multicellular organisms are 1.2 billion years old. Complex animals with bilateral symmetry (in which the body has two equal but opposite halves, as in most living animals) did not appear until about 600 million years ago (Chen et al. 2004). These and many other great changes in the history of life resulted from descent with modification as new species arose that differed from their ancestors. Over millions of years, these differences gradually accumulated, leading eventually to the formation of major new groups of organisms, such as terrestrial plants, amphibians, and reptiles. For example, the fossil record contains dozens of fossil species that show how one such group, the mammals, arose over a 120 million-year period (300 to 180 million years ago) from an earlier group of terrestrial vertebrates, the synapsids (Allin and Hopson 1992; Sidor 2003). The fossil record also shows cases in which a rise to prominence of one group of organisms was associated with a decline in another group. For

(A)

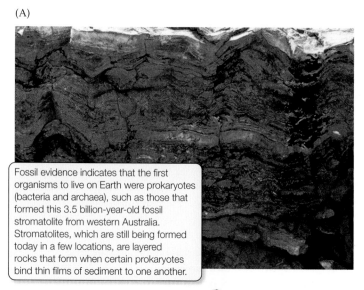

Fossil evidence indicates that the first organisms to live on Earth were prokaryotes (bacteria and archaea), such as those that formed this 3.5 billion-year-old fossil stromatolite from western Australia. Stromatolites, which are still being formed today in a few locations, are layered rocks that form when certain prokaryotes bind thin films of sediment to one another.

(B)

The oldest fossils of eukaryotes are 2.1 billion years old. Other early evidence of eukaryotes include this 1.5-billion-year-old fossil of the unicellular alga *Tappania* (left). Over time, some microscopic eukaryotes evolved anti-predator defenses, as seen below in this 625-million-year-old fossil from the Doushantuo Formation in South China.

(F)

In 2006, researchers reported the discovery of this 380 million-year-old fossil of *Tiktaalik roseae*, one of dozens of fossil species that document the origin of tetrapods from lobe-finned fishes. The tetrapods are four-legged terrestrial \vertebrates whose living members include amphibians, reptiles, and mammals.

(C)

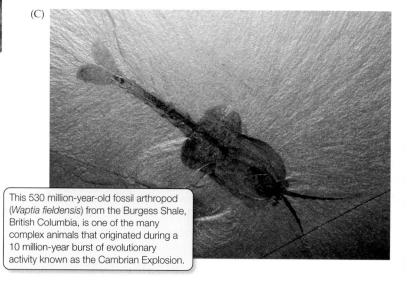

This 530 million-year-old fossil arthropod (*Waptia fieldensis*) from the Burgess Shale, British Columbia, is one of the many complex animals that originated during a 10 million-year burst of evolutionary activity known as the Cambrian Explosion.

(E)

This 300 million-year-old fossil horsetail (*Equisetum* spp.) closely resembles living members of the genus. Fossil evidence indicates that plants began the colonization of land roughly 450 million years ago, with the oldest fossil ancestors of today's vascular plants dating to 420 million years ago.

(D)

A lineage of armored, jawed vertebrates known as *placoderms* first appeared in the fossil record 470 million years ago. Most placoderms were less than a meter in length, but the group also included fearsome predators, such as the 4-ton, 10-m-long *Dunkleosteus* shown here.

FIGURE 6.16 Life Has Evolved Greatly over Time

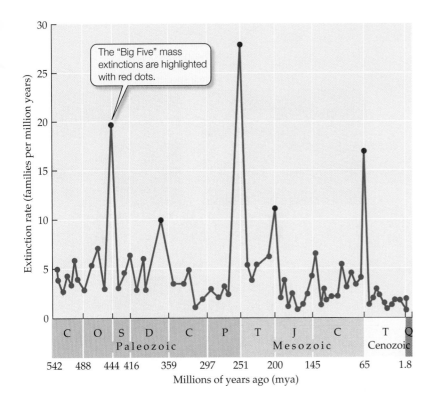

The "Big Five" mass extinctions are highlighted with red dots.

FIGURE 6.17 The "Big Five" Mass Extinctions Five peaks in extinction rates are revealed by a graph of extinction rates over time in families of marine invertebrates.

million years ago and may have been caused by a large asteroid that struck Earth, setting in motion cataclysmic environmental changes that led to the demise of dinosaurs and many other organisms.

Each of the five mass extinctions was followed by great increases in the diversity of some of the surviving groups of organisms; mammals, for example, increased greatly in diversity after the extinction of dinosaurs. Mass extinctions can promote such increases in diversity by removing competitor groups, thus allowing the survivors to expand into new habitats or new ways of life. Great increases in diversity can also occur when a group of organisms evolves major new adaptations, such as the stems and waxy cuticles on leaves that provided early terrestrial plants with support against gravity and protection from desiccation. Whether stimulated by a mass extinction or by the evolution of new adaptations, an event in which a group of organisms gives rise to many new species that expand into new habitats or new ecological roles in a relatively short time is referred to as an **adaptive radiation**.

What can we learn about ecology and evolution from mass extinctions, adaptive radiations, and other great changes in the history of life? First, biological communities are devastated by mass extinction events (**Figure 6.18**). Second, after a mass extinction occurs, it takes millions of years for adaptive radiations to increase the diversity of life to the levels seen prior to the mass extinction (Jablonski 1995). This observation has sobering implications for the recovery of life on Earth if current trends continue and human activities cause a sixth mass extinction (see Chapter 22).

example, 265 million years ago, reptiles replaced amphibians as the dominant group of terrestrial vertebrates, and then, 65 million years ago, the reptiles were replaced in turn by the mammals.

The rise and fall of different groups of organisms over time—as illustrated by the amphibians, reptiles, and mammals—has been heavily influenced by mass extinctions and adaptive radiations. The fossil record documents five **mass extinction** events in which large proportions of Earth's species were driven to extinction worldwide in a relatively short time—a few million years or less, sometimes much less (**Figure 6.17**). The most recent mass extinction occurred 65

FIGURE 6.18 Devastating Effects of a Mass Extinction An ancient seabed (A) before and (B) after the end-Permian mass extinction 251 million years ago. Over 90% of marine species were driven to extinction, radically altering this biological community and many others worldwide.

Third, many of the great changes in the history of life appear to have been caused by ecological factors. Fossils from the first 60 million years after the origin of complex animals indicate that all of them were small or soft-bodied, or both, and that all of the larger animals were herbivores, filter feeders, or scavengers. Ten million years later, this safe, soft-bodied world disappeared forever with the appearance of large, well-armed, mobile predators and large, well-defended prey. This major step in the history of life appears to have resulted from an "arms race" between predators and prey. Early predators equipped with claws and other adaptations for capturing large prey provided powerful selection pressure that favored heavily armored prey species—and that armor, in turn, promoted further increases in the effectiveness of the predators.

Since that time, ecological interactions have continued to shape the history of life. There are many examples in which the rise to prominence of one new group of organisms has led to increases in the diversity of other organisms—such as those that can escape from, eat, or compete effectively with the new group (Benton and Emerson 2007). A specific example can be seen in flowering plants. The petals of a flower can be symmetrical in only one direction (bilaterally symmetrical, as in peas and orchids) or they can be symmetrical in all directions (radially symmetrical, as in magnolias and daisies). Sargent (2004) found that plant species with bilaterally symmetrical flowers gave rise to new species more rapidly than did closely related species with radially symmetrical flowers. Sargent attributed this observation to the greater specificity of the plant–pollinator interaction in plants with bilaterally symmetrical flowers. Typically, an insect pollinator can harvest nectar from a bilaterally symmetrical flower only if it enters the flower from a single direction. As the insect collects nectar from flowers of the same species, the fact that there is only one approach to the nectar increases the chance that pollen will be deposited on a part of the insect's body that will come into contact with the female reproductive parts of the next flower the insect visits. As a result, gene flow may be reduced more rapidly in diverging populations with bilaterally symmetrical flowers than in diverging populations with radially symmetrical flowers, and this relatively rapid drop in gene flow could lead to the higher speciation rates observed in plants with radially symmetrical flowers.

Joint Effects of Ecology and Evolution

CONCEPT 6.5
Ecological interactions and evolution exert a profound influence on one another.

We turn now to a more detailed look at an idea that we have already encountered several times in this chapter: while ecological interactions influence evolution, evolution also in-

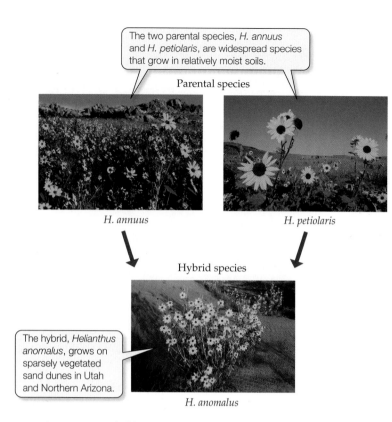

The two parental species, *H. annuus* and *H. petiolaris*, are widespread species that grow in relatively moist soils.

Parental species

H. annuus   *H. petiolaris*

Hybrid species

The hybrid, *Helianthus anomalus*, grows on sparsely vegetated sand dunes in Utah and Northern Arizona.

H. anomalus

FIGURE 6.19 Hybrids That Live in New Environments   Two sunflower species, *Helianthus annuus* and *H. petiolaris*, gave rise to a new hybrid species, *H. anomalus*. This species grows in a different, and more extreme, environment than either of the two parental species.

fluences ecological interactions. Ecological and evolutionary interactions can, in fact, be so closely related as to be entangled. Consider the sunflower species *Helianthus anomalus*. This species originated from a speciation event in which two other sunflowers, *H. annuus* and *H. petiolaris*, produced hybrid offspring. As Loren Rieseberg and colleagues have shown in a series of experiments and genetic analyses (Rieseberg et al. 2003), the new gene combinations generated by hybridization appear to have facilitated a major ecological shift in *H. anomalus*. This hybrid species grows in a different and more extreme environment than its parental species do (**Figure 6.19**). This ecological shift illustrates how evolution influences ecology. Simultaneously, however, the speciation event that gave rise to *H. anomalus* was driven by natural selection for life under different ecological conditions, showing how ecology influences evolution. Such joint ecological and evolutionary effects are common—as we should expect, since both evolution and ecology depend on how organisms interact with one another and with their environment.

Ecological interactions can cause evolutionary change

Much of the drama of the natural world stems from the efforts of organisms to do three things: to eat, to avoid being

(A)

(B)

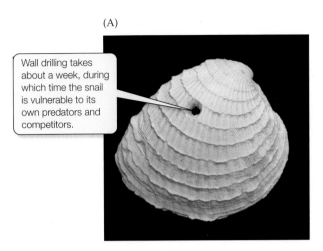

Wall drilling takes about a week, during which time the snail is vulnerable to its own predators and competitors.

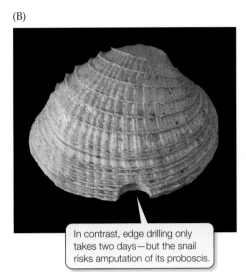

In contrast, edge drilling only takes two days—but the snail risks amputation of its proboscis.

FIGURE 6.20 Evidence of Ancient Predatory Behavior Predatory snails can use either of two strategies when drilling into clam shells, each of which has certain advantages and disadvantages. These fossils show the outcome when a predatory snail drills through (A) the wall or (B) the edge of a clam shell.

eaten, and to reproduce. As organisms interact with one another in this drama, a web of ecological interactions results. These interactions can drive evolutionary change. We've already seen how predator–prey interactions (via natural selection) caused large-scale, reciprocal evolution in which predators became more efficient at capturing prey and prey became more adept at escaping their predators. Evolution can also result from a broad range of other ecological interactions, including competition, herbivory, parasitism, and mutualism (see Unit 3).

Studies of speciation have led to a similar conclusion: it is common for speciation to be caused by ecological factors (Schluter 1998; Funk et al. 2006). The effect of ecology on evolution is also clear from studies of relatively small evolutionary changes in populations. Examples discussed earlier in this chapter include the effects of directional selection in bighorn sheep, *Geospiza* finches, and soapberry bugs (see Figures 6.2, 6.6A, and 6.11) and the effects of genetic drift in greater prairie chickens (see Figure 6.8); in this last example, genetic drift resulted from an ecological change (loss of suitable habitat) that caused a drop in population size.

Evolution can alter ecological interactions

Whenever a group of organisms evolves a new, highly effective adaptation, the outcome of ecological interactions may change, and that change may have a ripple effect that alters the entire community. For example, if a predator evolves a new way of capturing prey, some prey species may be driven to extinction, while others may decrease in abundance, migrate to other areas, or evolve new ways to cope with the more efficient predator. Similar changes can occur among species that compete for resources; we will discuss one such example in Chapter 11, in which evolutionary changes in experimental populations of one fly species reversed the outcome of its competitive interactions with another fly species.

Processes that drive the patterns of evolution over long time scales can also have a profound effect on ecological interactions. Mass extinctions, for example, are major events in which large numbers of species are lost and all aspects of biological communities are changed (see Figure 6.18). Here, we'll take a closer look at the effects of a regional extinction event that occurred 2 million years ago, removing 70% of species from marine communities in the western Atlantic Ocean.

Dietl et al. (2004) studied the effect of this extinction event on predatory snails that drill through the shells of clams and other bivalves. Experiments with modern relatives of these snails that survived the extinction showed that it takes these snails roughly a week to drill through the wall of a clam shell (**Figure 6.20A**). Once a hole is drilled, the snail eats the clam's soft body parts. As it drills through a shell wall, a predatory snail faces two risks: it may lose its food to other snails, and it may lose its life by being eaten by a fish or a crab. Predatory snails sometimes drill through the edge of a shell, rather than the wall (**Figure 6.20B**). Drilling through the edge of a shell decreases the time it takes to penetrate the shell by a factor of three. But edge drilling is not risk-free: as a snail drills through the edge of a shell, the bivalve may close its shell on the snail, amputating its feeding organ (proboscis).

In a species-rich community, edge drilling can be advantageous because it decreases the time that the snail is vulnerable to its competitors and predators. Reasoning that edge drilling would provide less of an advantage after an extinction event (because the snails would face fewer competitor and predator species), Dietl and colleagues predicted that edge drilling would be uncommon after the regional extinction in the western Atlantic 2 million years ago. This prediction was confirmed in a survey of fossils from before and after the extinction: edge drilling was found in the species-rich communities that lived 5 million to 2 million years ago, but it stopped altogether after the extinction event. Thus, the ex-

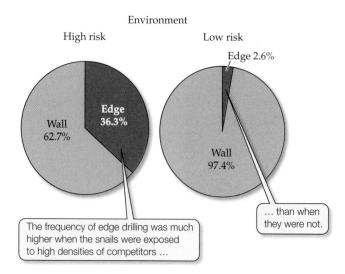

FIGURE 6.21 Edge Drilling Increases in High-Risk Environments   In laboratory experiments, researchers manipulated the densities of modern predatory snails and observed the responses of their predatory behavior to high-risk environments (with high densities of competitors) and low-risk environments (with low densities of competitors). (After Dietl et al. 2004.)

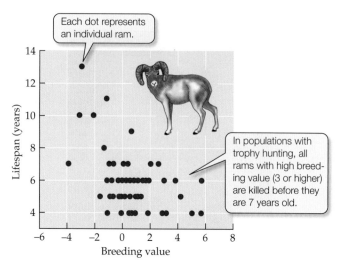

FIGURE 6.22 Longevity and Breeding Value in Bighorn Sheep In populations with no hunting, rams with large horns typically begin to sire large numbers of offspring after they are six years old. However, in populations where trophy hunting occurs, rams with large horns and high "breeding value"—those whose genetic characteristics are predicted to yield offspring with large horns—rarely live beyond their sixth or seventh year. (After Coltman et al. 2003.)

tinction appears to have altered ecological interactions: it reduced the competition and predation experienced by shell-drilling snails, leading to a reduction in the frequency of edge drilling. This interpretation was strengthened by the results of experiments with modern snails shown in **Figure 6.21**.

CASE STUDY REVISITED

Trophy Hunting and Inadvertent Evolution

Trophy hunters of bighorn sheep prefer to kill large males that carry a full curl of horns. The majority of these males are killed when they are between 4 and 6 years old (**Figure 6.22**), often before they have sired many offspring. As a result, hunting decreases the chance that alleles carried by males with a full curl of horns will be passed on to the next generation—and that, in turn, has caused horn sizes to decrease over a 30-year period (see Figure 6.2). Overall, then, trophy hunting has inadvertently caused directional selection in bighorn sheep, favoring small males with small horns and changing allele frequencies in the population over time.

Humans have caused unintended evolutionary changes in a wide variety of other populations. Consider our use of antibiotics against bacteria that cause diseases and lethal infections. When antibiotics were first discovered (ca. 1940), they were hailed as miracle drugs. At first, bacterial populations were highly vulnerable to antibiotics, and their use saved many human lives. But the use of antibiotics

provided a strong source of directional selection, leading to the evolution of antibiotic resistance in bacterial populations (see Figure 1.10). For example, in 1941, a dose of 40,000 units of penicillin per day could cure a person of pneumonia, while today, a patient infected with a resistant strain of pneumonia-causing bacteria might receive a dose of 24 *million* units of penicillin per day and still die. Antibiotic resistance also has enormous financial costs. The total magnitude of these costs is not known, but in the United States alone, efforts to cure patients infected with antibiotic-resistant strains of just one bacterial species (*Staphylococcus aureus*) result in an estimated $24–$31 billion in medical expenses *each year* (Palumbi 2001).

Although antibiotics still save many lives, one or more species of bacteria are resistant to every antibiotic now in use—making it vital that we develop new antibiotics and slow the evolution of resistance to antibiotics now in use. One way to achieve the latter goal would be to use antibiotics less frequently, thus reducing the extent to which bacterial populations are exposed to antibiotics and hence selected for resistance. We could, for example, limit the prescribing of antibiotics for people with colds and other respiratory infections, since these conditions are usually caused by viruses, and viruses are not affected by antibiotics.

We have seen throughout this chapter that human actions such as trophy hunting and antibiotic use serve as a source of selection and hence may cause evolutionary change. But does our influence on evolution extend beyond cases in which we selectively kill other organisms?

CONNECTIONS IN NATURE

The Human Impact on Evolution

Many human actions alter the environment and hence have the potential to change the course of evolution. As we've seen, actions such as trophy hunting and antibiotic use are themselves a powerful source of selection. Other human actions, such as emissions of pollutants or introductions of invasive species, change aspects of the abiotic or biotic environment. By changing the environment, we alter the selection pressures that organisms face, and hence again may cause evolutionary change. Still other human actions, such as *habitat fragmentation* (in which portions of a species' habitat are destroyed, leaving spatially isolated fragments of the original habitat), can have a series of evolutionary effects (**Figure 6.23**).

Human actions that affect the environment can alter each of the three main mechanisms of evolution: natural selection, genetic drift, and gene flow. Because we know with certainty that our actions are causing great changes to environments worldwide, we can infer that they are also causing evolutionary changes in populations worldwide. We also know that our actions have the potential to alter patterns of evolution over long time scales. As we will see in Chapter 22, human actions such as habitat fragmentation, overharvesting, and introductions of invasive species are among the main reasons why Earth is undergoing a biodiversity crisis. The extinction rate of species today is 100 to 1,000 times higher than the usual or "background" extinction rate seen in the fossil record. Extinction is forever. Thus, when human actions drive a species to extinction, the future course of evolution is altered in a way that cannot be reversed. Many scientists think that if current trends continue, humans will cause a sixth mass extinction in the next few centuries or millennia. If that happens, our actions will greatly and irreversibly change the evolutionary history of life on Earth.

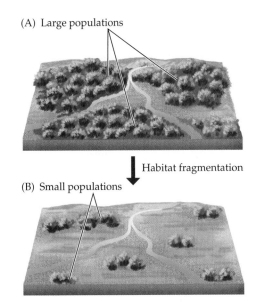

	Unfragmented habitat	Fragmented habitat
Population size	Large	Small
Distance between populations	Short	Long
Genetic drift	Low impact	High impact
Genetic variation within populations	High	Low
Gene flow	High	Low

FIGURE 6.23 Evolutionary Effects of Habitat Fragmentation on a Hypothetical Species (A) Prior to habitat fragmentation, there are many individuals in the species' populations, and the distances between populations are short. (B) When human activities remove large portions of the species' habitat, its population sizes shrink, and the distances between populations increase, causing evolutionary changes that increase the species' risk of extinction and decrease its potential for adaptive evolution.

SUMMARY

CONCEPT 6.1 Evolution can be viewed as genetic change over time or as a process of descent with modification.

- Biologists often define evolution in a relatively narrow sense as change over time in the frequencies of alleles in a population.

- Evolution can also be viewed as descent with modification, a process in which populations accumulate differences over time and hence differ from their ancestors.

- Natural selection modifies populations by favoring individuals with some heritable traits over others.

- Although individuals are acted on by natural selection, an individual does not evolve—it either has a favored trait or it does not. Only populations evolve.

CONCEPT 6.2 Natural selection, genetic drift, and gene flow can cause allele frequencies in a population to change over time.

- Mutation and recombination are the sources of new alleles and new combinations of alleles, thereby providing the genetic variation on which evolution depends.

- Natural selection occurs when individuals with certain heritable phenotypic traits survive and reproduce more successfully than individuals with other traits.

- Genetic drift, which occurs when chance events determine which alleles are passed from one generation to the next, can have negative effects on small populations.
- Gene flow, the transfer of alleles between populations, makes populations more similar to one another genetically and can introduce new alleles into populations.

CONCEPT 6.3 Natural selection is the only evolutionary mechanism that consistently causes adaptive evolution.

- By consistently favoring individuals that have advantageous alleles over individuals that have other alleles, natural selection can cause adaptive evolution, in which the frequency of an advantageous trait in a population increases over time.
- Natural selection can increase the frequency of advantageous traits rapidly—in days to years, depending on the generation time of the organism under selection.
- Gene flow can limit the extent to which a population is adapted to its local environment.
- Constraints on adaptive evolution result from factors such as lack of genetic variation, evolutionary history, and ecological trade-offs.

CONCEPT 6.4 Long-term patterns of evolution are shaped by large-scale processes such as speciation, mass extinction, and adaptive radiation.

- The diversity of life has resulted from speciation, the process by which one species splits into two or more species.
- Biological communities are devastated by mass extinctions, global events in which large proportions of Earth's species are driven to extinction in a relatively short time.
- An adaptive radiation occurs when a group of organisms gives rise to many new species that expand into new habitat or fill new ecological roles.
- Adaptive radiations can be stimulated by the removal of competitor groups by a mass extinction or by the evolution of a major new adaptation.

CONCEPT 6.5 Ecological interactions and evolution exert a profound influence on one another.

- Ecological interactions among organisms and between organisms and their environment can cause evolutionary change, ranging from allele frequency change in populations to the formation of new species.
- Similarly, evolutionary change can alter the outcome of ecological interactions, thus having a large influence on biological communities.

REVIEW QUESTIONS

1. Natural selection acts on individuals, yet one of the points made in this chapter is that *populations evolve, but individuals do not*. Explain how natural selection works and why the italicized statement is true.

2. What causes adaptive evolution? Explain in your answer why each of the three primary mechanisms of allele frequency change in populations causes or does not cause adaptive evolution.

3. What large-scale processes determine patterns of evolution observed over long time scales? Explain how each process that you describe has this effect.

4. Explain why ecological interactions and evolutionary change have joint effects, each affecting the other.

HANDS-ON PROBLEM SOLVING

Effects of Natural Selection and Genetic Drift | This Web exercise (www.sinauer.com/ecology) demonstrates how natural selection and genetic drift can alter the frequencies of alleles in populations. You will investigate the effects of manipulating population sizes (and thus the strength of genetic drift) and strengths of selection.

SUGGESTED READINGS

Cowen, R. 2005. *History of Life*. Blackwell, Oxford. *A lively tour through the major events in the history of life, containing clear descriptions of both what happened and how we know what we know. Hallmarks of this highly readable text include penetrating analyses of controversial ideas and cogent summaries of differing views on topics that are still under active debate.*

Coyne, J. A and H. A. Orr. 2004. *Speciation*. Sinauer Associates, Sunderland, MA. *This clear and authoritative text on speciation was written by two of the leading researchers in the field.*

Darwin, C. 1859. *On the Origin of Species by Means of Natural Selection*. John Murray, London. *One of the most influential books ever published, Darwin's book still a must-read for biologists today.*

The following two textbooks on evolutionary biology provide more information on many of the topics covered here.

Freeman, S. and J. C. Herron. 2007. *Evolutionary Analysis*. 4th ed. Pearson Prentice Hall, Upper Saddle River, NJ.

Futuyma, D. J. 2005. *Evolution*. Sinauer Associates, Sunderland, MA.

Heath, D. D, J. W. Heath, C. A. Bryden, R. M. Johnson and C. W. Fox. 2003. Rapid evolution of egg size in captive salmon. *Science* 299: 1738–1740. *A cautionary tale about how inadvertent evolution that results from captive breeding and release programs can harm the very populations those programs were designed to protect—in this case, struggling populations of wild salmon.*

Orr, M. R. and T. B. Smith. 1998. Ecology and speciation. *Trends in Ecology and Evolution* 13: 502–506. *A review article that summarizes field and laboratory evidence indicating that ecological factors can cause rapid adaptive evolution leading to the formation of new species.*

Temeles, E. J. and W. J. Kress. 2003. Adaptation in a plant–hummingbird association. *Science* 300: 630–633. *One in a series of excellent studies by Temeles and colleagues on natural selection and coevolution (reciprocal evolutionary change in interacting species) in the hummingbirds of the Caribbean islands.*

Wilson, D. S. 2007. *Evolution for Everyone: How Darwin's Theory Can Change the Way We Think about Our Lives*. Delacorte Press, New York. *This thought-provoking book presents a clear description of the evolutionary process and argues that an understanding of evolution can lead to new ways of thinking about topics that range from infanticide to infidelity to altruism.*

UNIT **2** Populations

CHAPTER 7 Life History Analyses

FIGURE 7.1 Offspring Vary in Size and Number Organisms produce a tremendous range of offspring numbers and sizes. A rhinoceros produces a single calf that weighs 40–65 kilograms (90–140 pounds). On the other end of the spectrum, many plants produce hundreds to thousands of millimeter-long seeds.

CASE STUDY

Nemo Grows Up

Birds do it, bees do it, even educated fleas do it—they all produce offspring that perpetuate their species. But beyond that basic fact of life, the offspring produced by different organisms vary in every way imaginable. A grass plant produces seeds a few millimeters long that can wait buried in the soil for years until conditions are right for germination. A sea star spews hundreds of thousands of microscopic eggs that develop adrift in the ocean. A rhinoceros produces one calf that develops in her womb for 16–18 months, can walk well several days after birth, but requires more than a year of care before it becomes fully independent (**Figure 7.1**).

Even this broad range of possibilities barely begins to describe the different ways in which organisms reproduce. In popular media, we humans often depict other animals as having family lives similar to ours. For example, in Pixar's animated film *Finding Nemo*, clownfish live in families with a mother, a father, and several young offspring. When Nemo the clownfish loses his mother to a predator, his father takes over the duties of raising him. But in a more realistic version of this story, after losing his mate Nemo's father would have done something less predictable: he would have changed sex and become a female.

Actually, the correspondence between the movie and true biology breaks down long before Nemo loses his mother, so let's go back to the beginning. Clownfish spend their entire adult lives within a single sea anemone (**Figure 7.2**). Anemones can be thought of as modified upside-down jellyfish with a central mouth ringed by stinging tentacles. In what appears to be a mutually beneficial relationship, the anemone protects the clownfish by stinging their predators, but the clownfish themselves are not stung. The clownfish, in turn, may help the anemone by eating its parasites or driving away predators that eat anemones. Two to six clownfish typically inhabit a single anemone, but they are far from a traditional human family—in fact, they are usually not related to one another.

The clownfish that live in an anemone interact according to a strict pecking order that is based on size. The largest fish in the anemone is a female. The next fish in the hierarchy, the second largest, is the breeding male. The remaining fish in each anemone are nonbreeding males. If the female dies, as in Nemo's story, the male undergoes a growth spurt and changes sex to become a female, and the largest nonbreeder becomes the new breeding male.

FIGURE 7.2 Life in a Sea Anemone Clownfish (*Amphiprion percula*) form hierarchical groups of unrelated individuals that live and reproduce among the tentacles of their anemone host (*Heteractis magnifica*).

The breeding male clownfish mates with the female and cares for the fertilized eggs until they hatch. The hatchling fish leave the anemone to live in the open ocean, away from the predator-infested reef. The young fish eventually return to the reef and develop into juveniles. Then they must find an anemone to inhabit. When juvenile fish enter an anemone, the resident fish allow them to stay there only if there is room. If there is no room, the young fish is expelled and returns to the dangers of an exposed existence on the reef.

This life cycle, with its expulsions, hierarchies, and sex changes, is certainly as colorful as the fish that live it. But why do clownfish engage in these complicated machinations just to produce more clownfish? Organisms have arrived at a vast array of solutions to the basic problem of reproduction. As we will see, these solutions are often well-suited for meeting the challenges and constraints of the environment where a species lives.

Human history is a record of past events. Your personal history might consist of a series of details about the course of your life: your birth weight, when you started walking and talking, your adult height, and other relevant information about your development. Similarly, an individual organism's **life history** is a record of events and landmarks relating to its growth, development, reproduction, and survival.

In this chapter, we'll discuss characteristics that define the life history of a organism including age and size at sexual maturity, amount and timing of reproduction, and survival and mortality rates. As we'll see, the timing and nature of these events, and therefore the life history itself, are products of adaptation to the environment in which the organism lives. We'll also consider how biologists analyze life history patterns in order to understand the trade-offs, constraints, and selection pressures imposed on different stages of an organism's life cycle.

Life History Diversity

CONCEPT 7.1
Life history patterns vary within and among species.

The study of life histories is concerned with categorizing variation in life history traits and analyzing the causes of that variation. In order to understand such analyses, we must first examine some of the broad life history patterns found within and among species.

Individuals within species differ in their life histories

Individual members of a species grow and mature at different rates. These differences may result from genetic variation or from differences in environmental conditions. Practically speaking, however, most variation results from a combination of genetic and environmental factors.

Individual variation Individual differences in life history traits are ubiquitous. Think about your own life experiences and those of your family and friends. Some members of your social group reached developmental milestones such as puberty significantly earlier or later than others. Different families in your neighborhood had different numbers of children with different age gaps between them. Despite this variation, it is possible to make some generalizations about life histories in *Homo sapiens*: for example, women typically have one baby at a time, reproduction usually occurs between the ages of 15 and 45, and so on. Similar generalizations can be made for other species. The **life history strategy** of a species is the overall pattern in the timing and nature of life history events averaged across all the individuals in the species (**Figure 7.3**). The life history strategy is shaped by the way the organism divides its time and energy between growth, reproduction, and survival.

Some life history variation within species is determined genetically. Genetically influenced traits often can be recognized as those that are more similar within families than between them. Again, these kinds of traits are familiar in humans: for example, siblings often are similar in appearance and reach

FIGURE 7.3 Life History Strategy The timing and nature of life history events shapes the overall life cycle of an organism. Although life history "decisions" are presented here as questions, natural selection, rather than the individual organism, chooses from among the available options to determine the life history strategy.

similar adult heights and weights. The same is true in other organisms. For example, in bluegrass (*Poa annua*), life history traits such as age at first reproduction, growth rate, and number of flowers produced are similar among sibling plants (Law et al. 1977). As with any other trait, heritable variation in life history traits is the raw material on which natural selection acts. Selection favors individuals whose life history traits result in their having a better chance of surviving and reproducing than do individuals with other life history traits.

Much of life history analysis is concerned with explaining how and why life history patterns have evolved to their present states. Ecologists sometimes describe life histories as **optimal**—that is, resulting in a maximization of fitness. However, as in other areas of ecology, it is important to remember that optimal life histories are theoretical ideals. Although real life histories often serve organisms well in the environments in which they have evolved, they are not necessarily perfectly adapted to maximize fitness, particularly when environmental conditions change.

Phenotypic plasticity A single genotype may produce different phenotypes for a particular trait under different environmental conditions. This phenomenon, known as **phenotypic plasticity**, is pervasive—almost every trait shows some degree of plasticity, and life history traits are no exception. For example, most plants and animals grow at different rates depending on temperature. This occurs because development typically speeds up as the temperature warms, then slows down again due to heat stress as the temperature approaches the organism's upper physiological limit (see Chapter 4 for more discussion of the effects of temperature).

Changes in life history traits often translate into changes in adult morphology. Thus, plasticity in life histories can be a source of plasticity in other traits. Slower growth under cooler conditions, for example, may lead to a smaller adult size or to differences in adult shape. Callaway and colleagues (1994) showed that ponderosa pine (*Pinus ponderosa*) trees grown in cool, moist climates allocate more biomass to leaf growth relative to sapwood production than do those in warmer desert climates. *Allocation* describes the relative amounts of energy or resources that an organism devotes to different tasks. The result of allocation differences in ponderosa pines is that trees grown in different environments differ in their adult shape and size. Desert trees are shorter and squatter, with fewer branches. They also have lower photosynthetic rates and consume less CO_2 because they have fewer leaves (**Figure 7.4**).

Phenotypic plasticity that responds to temperature variation often produces a continuous range of growth rates. Other types of phenotypic plasticity produce discrete types, or **morphs**, with few or no intermediate forms. This particular type of phenotypic plasticity, in which a single genotype produces multiple, discrete phenotypes, is called **polyphenism**. An example of polyphenism occurs in the spadefoot toad (*Spea multiplicata*). Populations of spadefoot toad tadpoles in Arizona ponds contain both omnivore morphs, which feed on detritus and algae, and larger carnivore morphs, which feed on fairy shrimp and on other tadpoles (**Figure 7.5**). Pfennig (1992) showed that omnivore tadpoles can turn into carnivores when fed on shrimp and tadpoles. Carnivore tadpoles grow faster and are more likely to metamorphose before the ponds where they live dry

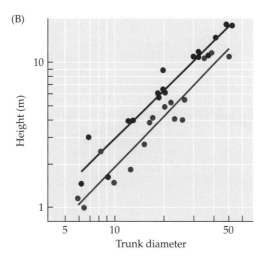

FIGURE 7.4 Plasticity of Growth Form in Ponderosa Pines (A) Ponderosa pine trees in cool, moist climates allocate more resources to leaf production than do trees in desert climates. (B) Desert trees are shorter than those grown in cooler climates, but for a given height they have thicker trunks. (After Callaway et al. 1994.)

up. Thus, the rapidly growing carnivores are favored in ephemeral ponds, while the more slowly growing omnivores are favored in ponds that persist longer, because they metamorphose in better condition and thus have better chances of survival as juveniles. The differing body shapes of omnivores and carnivores result from differences in the relative growth rates of different body parts: carnivores have bigger mouths and stronger jaw muscles because of accelerated growth in those areas.

The differences in morphology between separate morphs of spadefoot toads and ponderosa pines both result from vari-ation in the relative growth rates of different body parts. In the pines, the relative growth of leaves versus sapwood determines body shape, while in the toads, the relative growth of the jaw versus the rest of the body determines whether the tadpole is a carnivore or an omnivore. These patterns are examples of **allometry**, or differential growth of body parts that results in a change in shape or proportion with size. Allometry is a very common mechanism of variation within and among species.

When thinking about examples such as spadefoot toad polyphenism, it is tempting to assume that phenotypic plasticity is adaptive—that the ability to produce different phenotypes in response to changing environmental conditions helps individuals survive and reproduce. This may be true in many instances, but adaptation must be demonstrated rather than assumed. For example, it may be adaptive for ponderosa pines to be stockier and have fewer leaves in hot, dry climates because these characteristics could help reduce water loss. However, adaptation would have to be demonstrated by measuring and comparing the survival and reproductive rates of stockier and taller trees in the desert environment. In some instances, phenotypic plasticity may be a simple physiological response, not an adaptive response shaped by natural selection. For example, changes in growth rate due to temperature variation may occur because chemical reactions are slower at lower temperatures, and thus metabolism and growth are necessarily slower.

(A) Omnivore morph

(B) Carnivore morph

Omnivores feed on the pond bottom on detritus

Carnivores feed in the water column on fairy shrimp

FIGURE 7.5 Polyphenism in Spadefoot Toad Tadpoles Spadefoot toad (*Spea multiplicata*) tadpoles can develop into small-headed omnivores (A) or large-headed carnivores (B), depending on the food they consume early in development. Later in development, omnivores and carnivores feed on different food sources that are located in different portions of their habitat.

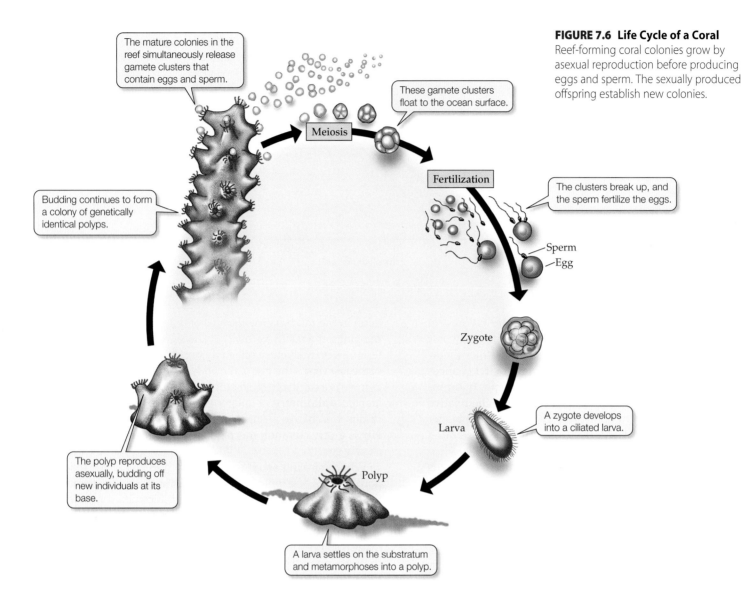

The mature colonies in the reef simultaneously release gamete clusters that contain eggs and sperm.

Budding continues to form a colony of genetically identical polyps.

The polyp reproduces asexually, budding off new individuals at its base.

Meiosis

These gamete clusters float to the ocean surface.

Fertilization

The clusters break up, and the sperm fertilize the eggs.

Sperm

Egg

Zygote

A zygote develops into a ciliated larva.

Larva

Polyp

A larva settles on the substratum and metamorphoses into a polyp.

FIGURE 7.6 Life Cycle of a Coral Reef-forming coral colonies grow by asexual reproduction before producing eggs and sperm. The sexually produced offspring establish new colonies.

Mode of reproduction is a basic life history trait

At the most basic level, evolutionary success is determined by successful reproduction. Despite this universal reality, organisms have evolved vastly different mechanisms for reproducing—from simple asexual splitting to complex mating rituals and intricate pollination systems.

Asexual reproduction The first organisms to evolve on Earth probably reproduced asexually by cell division. Meiosis, recombination, and sexual reproduction arose sometime later. Today, all prokaryotes and many protists reproduce by cell division. Sexual reproduction is the norm in multicellular organisms, but many are also capable of reproducing asexually. For example, after they are initiated by a (sexually produced) founding polyp, coral colonies grow by asexual reproduction. Each individual polyp in a colony is produced when a multicellular bud splits off from a parent polyp to

form a new polyp; as a result, each polyp is an identical clone of the founding polyp. Once the colony has grown to a certain size and conditions are right, the polyps reproduce sexually, producing offspring that develop into polyps that start their own new colonies of clones (**Figure 7.6**).

Sexual reproduction and anisogamy Sex has some clear benefits, including recombination, which promotes genetic variation and may provide protection against disease. Sex also has some disadvantages. Because meiosis produces haploid gametes that contain half the genetic content of the parent, a sexually reproducing organism can transmit only half of its genetic material to each offspring, whereas asexual reproduction allows transmission of the entire genome. In addition, the growth rate of sexually reproducing populations is only half that of asexually reproducing ones, all else being equal (**Figure 7.7**).

(A)

(B)

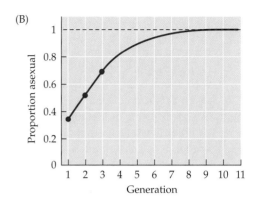

FIGURE 7.7 The Cost of Sex One cost of sex is referred to as the *cost of males*. Imagine a population in which there are sexual and asexual individuals. Each sexual or asexual female can produce four offspring per generation, but half of the offspring produced by the sexual females are male and must pair with a female to produce offspring. Under these conditions, the asexual individuals will reproduce more rapidly (A), and (B) dominate the population in less than ten generations.

Sexual reproduction first evolved in single-celled protists. Protists such as the green alga *Chlamydomonas reinhardii* (**Figure 7.8A**) have two different *mating types*, analogous to males and females, but they both produce gametes that are the same size. The production of equal-sized gametes is called

FIGURE 7.8 Isogamy and Anisogamy (A) An isogamous species: Two gametes of the single-celled alga *Chlamydomonas* fusing. (B) An anisogamous species: Fertilization of a human egg, showing the difference in size between egg and sperm.

isogamy. In most multicellular organisms, the two types of gametes are different sizes—a condition called **anisogamy**. Typically, the eggs are much larger than the sperm and contain more cellular and nutritional provisions for the developing embryo. The sperm are small and may be motile (**Figure 7.8B**). As we will see later in this chapter, these differences between the sexes in gamete size influence their other reproductive characteristics, such as size at first reproduction and the timing of sex changes.

Life cycles are often complex

The small, early stages of animal life cycles often look and behave completely differently from adult stages. They frequently eat different foods and prefer different habitats. For example, coral reef fishes such as the damselfish *Chromis atripectoralis* start life as larvae only a few millimeters long. The larvae live and grow in the open ocean, feeding on planktonic algae. When they have grown to about a centimeter in length, they return to the reef and begin to eat larger food items. This life cycle may have evolved in response to high levels of predation on larval fish that stay on the reef; larvae that spend more time growing in the open ocean may have better chances of survival.

As corals and coral reef fishes both demonstrate, life cycles can involve stages that have different body forms and live in different habitats. A **complex life cycle** is a life cycle in which there are at least two distinct stages that differ in their habitat, physiology, or morphology. In many cases, the transitions between stages in complex life cycles are abrupt. For example, many organisms undergo **metamorphosis**, an abrupt transition in form from the larval to the juvenile stage that is sometimes accompanied by a change in habitat.

Because most vertebrates have simple life cycles that lack an abrupt transition between habitats or forms, we tend to think of metamorphosis as an exotic and strange process—the stuff of Kafka and science fiction. However, complex life cycles and metamorphosis are the norm rather than the exception among animals (Jagersten 1972). The majority of marine invertebrates produce microscopic larvae that swim in the open ocean before settling to the bottom at metamorphosis.

Many insects also undergo metamorphosis—from caterpillars to moths, grubs to beetles, maggots to flies, and aquatic larvae to dragonflies, mayflies, and caddisflies. Vertebrates with complex life cycles include some fishes and most amphibians.

One possible reason for the prevalence of complex life cycles is that offspring, because of their small size, often experience the environment very differently from their parents. Even simple physical conditions affect organisms of different sizes differently. A seed is easily carried away by wind that has little effect on its parent tree, and a tadpole is more strongly affected by surface tension and viscosity than an adult frog. As we will see in Section 7.4, complex life cycles and metamorphosis often result when offspring and parents are subjected to very different selection pressures.

Werner (1988) calculated that of the 33 phyla of animals recognized at that time, 25 contained at least some subgroups that have complex life cycles. He also noted that about 80% of all animal species undergo metamorphosis at some time in their life cycle (**Figure 7.9**). Complex life cycles have been lost in some species within groups in which such cycles are considered the norm. The resulting simple life cycles are sometimes referred to as **direct development** because they go directly from fertilized egg to juvenile without passing through a larval stage. For example, the members of one group of salamanders, the plethodontids, have lost the gilled aquatic larval stage that is typical of salamanders. Instead, they lay their eggs on land, where they hatch directly into small, lunged juveniles.

Many parasites have evolved intricate and complex life cycles with one or more specialized stages for each host that they inhabit. For example, the worm parasite *Ribeiroia* has 3 specialized stages (see Figure 1.3), while the Chinese liver fluke passes through six stages in the course of a life cycle that moves between three hosts: snails, fish, and humans.

Complex life cycles also occur in many types of algae, plants, and protists, reaching some of their most elaborate forms in these groups. Most algae and all plants have complex life cycles in which a multicellular diploid **sporophyte** alternates with a multicellular haploid **gametophyte**. The sporophyte produces haploid spores that disperse and grow into gametophytes, and the gametophyte produces haploid gametes that combine in fertilization to form zygotes that grow into sporophytes (**Figure 7.10**). This life cycle, called **alternation of generations**, has been elaborated on in different plant and algal groups. In mosses, the gametophyte is larger, but in algae and most plants, the sporophyte is the dominant stage of the life cycle.

FIGURE 7.9 The Pervasiveness of Complex Life Cycles   Most groups of animals include members that undergo metamorphosis. Familiar examples are insects such as (A) the mealworm beetle (*Tenebrio molitor*), which develops from a grub larva that lives in soil. Most marine invertebrates have free-swimming larval stages, including echinoderms such as (B) the common sea star (*Asterias rubens*).

Life History Continua

> **CONCEPT 7.2**
> Reproductive patterns can be categorized along several continua.

Categorizing the vast diversity of reproductive patterns is one of the challenges facing biologists who study life histories. Several classification schemes have been proposed for organizing these diverse patterns into a coherent structure. Most of these schemes make broad generalizations about reproductive patterns and attempt to place them along continua that are anchored by extremes at each end. In this section, we examine some of the most prominent of these schemes and discuss how they relate to one another.

Some organisms reproduce only once while others reproduce multiple times

One way of classifying the reproductive diversity of organisms is by examining the number of reproductive bouts in an individual's lifetime. **Semelparous** species reproduce only once in a lifetime, while **iteroparous** species have the capacity to reproduce multiple times.

Many annual plant species are semelparous—after they germinate from a seed they reproduce once and die. A more

FIGURE 7.10 Alternation of Generations in a Fern
The fern has a complex life cycle that is typical of plants, with an alternation of generations between a diploid sporophyte stage and haploid gametophyte stage.

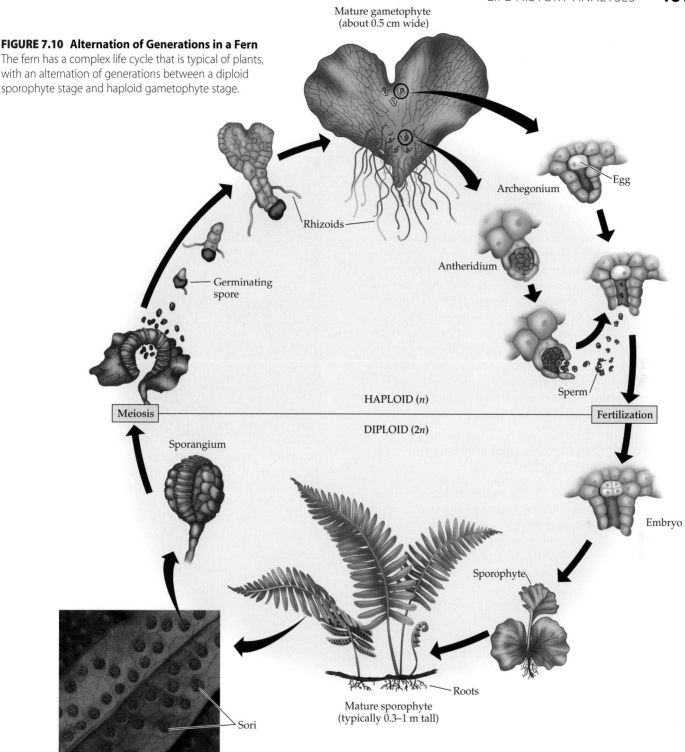

Mature gametophyte
(about 0.5 cm wide)

Egg

Archegonium

Rhizoids

Antheridium

Germinating spore

Sperm

HAPLOID (*n*)

Meiosis

Fertilization

DIPLOID (2*n*)

Sporangium

Embryo

Sporophyte

Sori

Roots

Mature sporophyte
(typically 0.3–1 m tall)

Microsporum sp.

complex example of a semelparous plant is the century plant (a common name applied to several species in the genus *Agave*) from the North American deserts. This plant has a prolonged stage of vegetative growth that lasts up to 25 years before it undergoes a single intensive bout of sexual repro-

duction. When it is ready to reproduce, the century plant produces a single stalk of flowers that may be up to 20 feet tall and towers over the rest of the plant. After pollination, the flowers produce clumps of seeds that drop off and take root around the parent plant. The portion of the plant that produced the tall stalk of flowers dies after this event and hence it is semelparous. However, at the genetic level, a century plant individual does not die when it flowers because the plant also reproduces asexually, producing genetically identical clones that surround the original plant (**Figure 7.11**). In this sense, century plants are not semelparous after all—the clones survive after the flowering event and will eventually flower themselves.

An interesting example of a semelparous animal is the giant Pacific octopus (*Enteroctopus dofleini*), which can reach 25 feet in length and weigh nearly 400 pounds. The female of this marine invertebrate species lays a single clutch containing tens of thousands of fertilized eggs. She then broods the eggs for up to 6 months. During this time, the female does not feed at all, and she is a constant presence over her offspring, cleaning and ventilating them. The female dies shortly after the eggs hatch, having exhausted herself in this intense period of parental investment.

Most organisms do not invest so heavily in single reproductive events. Iteroparous organisms have multiple bouts of reproduction over the course of a lifetime. Examples of iteroparous plants are long-lived trees such as pines and spruces. Among animals, most large mammals are iteroparous. Of course, iteroparity can take a variety of forms, from annual plants that flower twice in a season and then die to trees that reproduce every year for centuries. We turn now to the first of several proposals that attempt to explain why organisms show such wide variation in how they reproduce.

Live fast and die young or slow and steady wins the race?

The most famous scheme for classifying reproductive diversity was also one of the first proposed. In 1967, Robert MacArthur and Edward O. Wilson coined the terms r-selection and K-selection to describe two ends of a continuum of reproductive patterns. The r in the term r-selection refers to the *intrinsic rate of increase* of a population and is a measure of how rapidly a population can grow. The term **r-selection** refers to selection for high population growth rates. This type of selection occurs in environments where population density is low—for example, in recently disturbed habitats that are being recolonized. In this type of habitat, any genotype that can grow and reproduce rapidly (however inefficiently) will be favored. In contrast, **K-selection** refers to selection for slower rates of increase in populations that are at or approaching K, the carrying capacity or population size limit for the environment in which they live (see Chapter 9 for a more in-depth discussion of both r and K). K-selection occurs under crowded conditions, where genotypes that can efficiently convert food into offspring are favored. By definition, K-selected populations do not have high population growth rates because they are already near the carrying capacity for their environment and competition for resources is fierce.

One way to think of the r–K continuum is as a spectrum of population growth rates, from fast to slow. Organisms at the r-selected end of the continuum tend to have short life spans, rapid development, early maturation, low parental in-

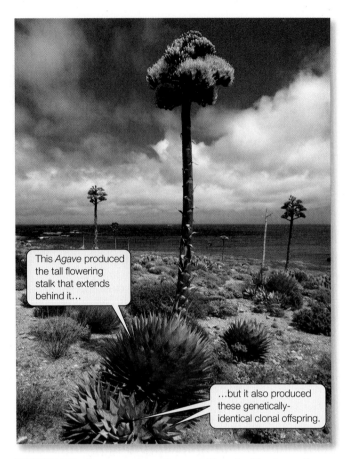

This *Agave* produced the tall flowering stalk that extends behind it…

…but it also produced these genetically-identical clonal offspring.

FIGURE 7.11 *Agave*—**a Semelparous Plant?** The *Agave* that produced the tall flowering stalk dies shortly after it flowers—as such it can be viewed as semelparous. But the *Agave* that flowered also produced genetically-identical clonal offspring—thus, the genetic individual lives on after it flowers and in that sense is not semelparous after all.

vestment, and high rates of reproduction. Examples of this fast end of the continuum include most insects, small short-lived vertebrates such as mice, and weedy plant species. In contrast, K-selected species tend to be long-lived, develop slowly, delay maturation, invest heavily in each offspring, and have low rates of reproduction. Examples of this slow end of the continuum include large mammals such as elephants and whales, reptiles such as tortoises and crocodiles, and long-lived plant species such as oak and maple trees.

Like most classification schemes, the r–K continuum tends to emphasize the extremes. However, most life histories are intermediate between these extremes, and the distinction between r-selection and K-selection is perhaps most useful in comparing life histories in closely related species or species living in similar environments. For example, Braby (2002) compared three species of Australian butterflies in the genus *Mycalesis*. The species that occurs in drier, less predictable habitats shows more r-selected characteristics, including more rapid development, earlier reproduction, production of many small eggs, and more rapid population growth. In contrast, the two species found in more predictable, wet forest habitats are more *K*-selected.

Plant life histories can be classified based on environmental characteristics

In the late 1970s, Peter Grime (1977) developed a classification system specifically for plant life histories. The success of a plant species in a given habitat, he argued, is limited by two factors: stress and disturbance. Grime chose to define **stress** very broadly as any external abiotic factor that limits growth. Under this definition, examples of stress include extreme temperatures, shading, low nutrient levels, water shortages, and any other characteristics of the abiotic environment that reduce vegetative growth. He defined **disturbance** as any process that destroys plant biomass. Sources of disturbance include both biotic sources such as herbivory and abiotic sources such as fire.

If we consider that in a given habitat stress and disturbance may each be either high or low, then there are four possible habitat types: high stress and high disturbance, low stress and high disturbance, low stress and low disturbance, and high stress and low disturbance. If we further consider that habitats with high stress *and* high disturbance often will not be suitable for plant species, then there are three main habitat types to which plants may adapt. In 1977, Grime developed a model for understanding the life history patterns that correspond to these three habitat types: competitive plants (low stress, low disturbance), stress-tolerant plants (high stress, low disturbance), and ruderal plants (low stress, high disturbance) (**Figure 7.12**).

Grime defined competition between plants in a very specific manner as "the tendency of neighboring plants to utilize the same quantum of light, ion of a mineral nutrient, molecule of water, or volume of space." Under conditions of low stress and low disturbance, **competitive plants** that are superior in their ability to acquire light, minerals, water, and space should have a selective advantage.

Under conditions in which stress (in any form) is high and disturbance is low, stress-tolerant plants become ecologically dominant. Although stressful conditions may vary widely across habitats, Grime identified several features of **stress-tolerant plants**, including but not limited to slow growth rates, evergreen foliage, absence of phenotypic plasticity, slow rates of water and nutrient use, and low palatability to herbivores. Environments favoring stress-tolerant plants might include places where water or nutrients are scarce or temperature conditions are extreme.

Grime classified plants that are adapted to environments with high levels of disturbance and low levels of stress as **ruderals**. The ruderal strategy generally includes short life spans, rapid growth rates, heavy investment in seed production, and seeds that can survive in the ground for long periods until conditions are right for rapid germination and growth. Ruderal species are often called "weedy" species and are adapted for brief periods of intense exploitation of favorable habitats after disturbance has removed competitors.

To summarize, the triangular model put forward by Grime posits that natural selection has resulted in three distinct yet very broad categories of life history strategies in plants. Although Grime focused on describing these three extreme strategies, he also recognized that intermediate strategies are commonly found. The combination of any two of these strategies yields what Grime termed secondary strategies, such as competitive ruderals or stress-tolerant competitors (see Figure 7.12). Many plants fall into one of these secondary strategies and may still be described in the context of Grime's model.

How can Grime's model be reconciled with the *r*–*K* continuum? Grime presented an admittedly oversimplified way of placing the competitive, stress-tolerant, and ruderal strategies within the framework of *r*- and *K*-selection. In this scheme, ruderal plants occupy the same space as *r*-selected species, while stress-tolerant plants correspond to *K*-selected species. Competitive plants then occupy the middle region of the *r*–*K* continuum. For example, weedy species such as dandelions can be categorized as both ruderal and *r*-selected, while some long-lived tree species are competitive

FIGURE 7.12 Grime's Triangular Model Grime categorized plant life histories within a triangle whose axes measure the degree of competition, disturbance, and stress in the habitat type to which they are adapted. C = competitive; R = ruderal; S = stress-tolerant. (After Grime 1977.)

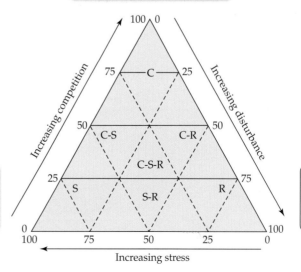

Competitive plants, such as birch trees, are adapted to low-stress, low-disturbance habitats.

Stress-tolerant plants, such as ivies, are adapted to high-stress, low disturbance habitats.

Ruderal plants, such as dandelions, are adapted to high-disturbance, low-stress habitats.

and others are stress-tolerant and *K*-selected. Grime argued that the linear arrangement of the *r*–*K* continuum can be misleading and is less realistic than his scheme of three primary strategies and several secondary strategies.

Life histories can be classified independent of size and time

Unlike the classification schemes discussed above, a recent approach by Eric Charnov (2002) organizes life histories in a manner that removes the influence of size and time. As we saw in our discussion of the *r*–*K* continuum, size and time play a critical role in traditional classifications of life histories. For example, *r*-selected species are characterized as smaller and more short-lived than *K*-selected species. But if we could control for the effects of body size and life span, then we could ask whether closely related taxa experience similar selection pressures independent of those factors.

Using this approach, the life histories of all organisms can be placed in a "life history cube." Their locations within the cube are determined by three dimensionless axes (**Figure 7.13**): (1) the size of offspring relative to adults, (2) the reproductive life span divided by the time to reach maturity, and (3) the adult reproductive effort per unit of adult mortality. The first two axes correct for the effects of adult size and time to reach maturity on offspring size and reproductive life span. The third axis is a measure of **reproductive effort**, or the quantity of energy and resources devoted to reproduction, corrected to take into account the costs of reproduction. Costs of reproduction could include slower growth rates, in-

creased vulnerability to predation, or increased physiological stress due to reproduction.

All three axes are dimensionless ratios in which the units in the numerator of the ratio (e.g., offspring size in *grams*) are identical to and hence cancel those in the denominator (e.g., adult size, also in *grams*). This innovation allowed Charnov to show that within a group such as the mammals, when size and time are removed from the analysis, the life histories of very small species such as rodents occupy the same portion of the cube as those of large species such as elephants. This result suggests that similar selection pressures are operating on mammals over a very wide range of sizes. In addition, Charnov's approach shows that differences between more distantly related taxa are often very striking, even when sizes, and life spans, are similar. For example, birds fall at the upper extremes of all three axes, while fish fall at the opposite ends, suggesting that birds are investing more energy in larger offspring over a longer portion of their lives than are fish.

It is probably inappropriate to think of this dimensionless approach as better or worse than schemes that incorporate time and size. Indeed, Charnov's approach has been questioned by Nee et al. (2005), who argue that the similarities in dimensionless life history parameters within taxonomic groups may be an artifact of Charnov's mathematical methods. In any case, there are many ways to organize the vast diversity of life history patterns. The classification scheme that is most useful in any given case will depend on the organisms and questions of interest. For example, the *r*–*K* continuum has a long history of use in relating life history characteristics to population growth characteristics, whereas Grime's scheme may be most appropriate for comparisons between plant taxa. Alternatively, Charnov's dimensionless axes may be most helpful when comparing life histories across broad ranges of taxonomy or size.

Trade-Offs

CONCEPT 7.3
There are trade-offs between life history traits.

As indicated by the maxim, "There is no free lunch" (see Table 1.1), no organism can invest unlimited amounts of energy in growth and reproduction. **Trade-offs** occur when organisms allocate their limited energy or other resources to one structure or function at the expense of another. As we'll see, trade-offs shape and constrain life history evolution.

There is a trade-off between size and number of offspring

Simply stated, the larger an organism's investment in each individual offspring, the fewer offspring it can produce, assuming that it has finite resources available for reproduction.

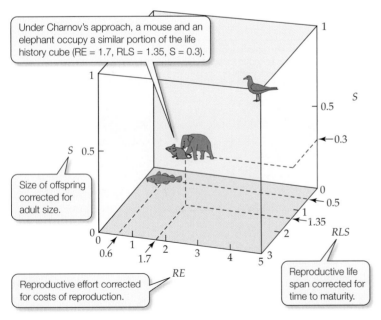

Under Charnov's approach, a mouse and an elephant occupy a similar portion of the life history cube (RE = 1.7, RLS = 1.35, S = 0.3).

Size of offspring corrected for adult size.

Reproductive effort corrected for costs of reproduction.

Reproductive life span corrected for time to maturity.

FIGURE 7.13 Charnov's Life History Cube   Charnov proposed categorizing life histories within a dimensionless cube. (After Charnov 2002.)

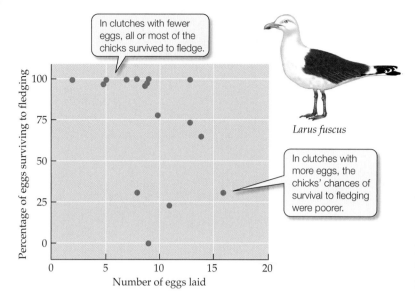

FIGURE 7.14 Clutch Size and Survival Nager and colleagues showed that when lesser black-backed gulls are manipulated experimentally to produce larger clutches of eggs, their offspring have reduced chances of survival to fledging. (After Nager et al. 2000.)

tion and removal of eggs in order to test Lack's hypothesis. For example, Nager and colleagues (2000) artificially increased the number of eggs laid in each clutch by the lesser black-backed gull (*Larus fuscus*). They did this by removing eggs from nests, which stimulated the females to lay more eggs. They found that the increased clutch size resulted in a lower lipid content in later-produced eggs. The eggs were then placed in surrogate nests to hatch and fledge. Chicks from larger clutches had reduced survivorship to fledging (**Figure 7.14**).

Size–number trade-offs in organisms without parental care
Across the animal kingdom, parental care like that provided by birds and some other vertebrates is relatively rare. In organisms that do not provide parental care, resources invested in *propagules* (eggs or seeds) are the main measure of reproductive investment. In this case, the size of the propagule is the primary measure of parental investment and is traded off against the number of propagules produced in a reproductive bout. Two examples of groups that generally lack parental care are terrestrial plants and marine invertebrates. In plants, the size of the seeds that a species produces is negatively correlated with the number of seeds it produces (**Figure 7.15**).

Investment in offspring includes energy, resources, time, and the loss of chances to engage in alternative activities such as foraging.

We can think of the trade-off between the size and number of offspring in the context of a continuum like those presented in the previous section. At one end of the continuum are organisms that produce very large numbers of small offspring, and at the other end are those that produce very small numbers of large offspring. As before, these extremes represent only two ends of a potentially continuous distribution of reproductive patterns.

Lack clutch size The classic example of the trade-off between investment per offspring and number of offspring occurs in birds and was first described by David Lack in 1947. Lack asserted that clutch size (the number of eggs per reproductive bout) in birds is limited by the maximum number of young that the parents can raise at one time. If the parents rear fewer than this maximum number, then they will reduce their genetic representation in future generations. If they attempt to rear more than this maximum number, then they risk starvation and death for the entire clutch, in which case they will leave no representation in future generations.

Lack made careful observations of the breeding biology of bird species from the poles to the tropics. What struck him was that clutch size varied with latitude: at higher latitudes, birds could rear greater numbers of offspring. He hypothesized that the reason for larger clutches at higher latitudes was that such latitudes had longer periods of daylight during the breeding season than did lower latitudes. These longer days allowed parents more time for foraging, and they could therefore feed greater numbers of offspring.

The term "Lack clutch size" refers to the maximum number of offspring that a parent can successfully raise to maturity. Lack hypothesized that this most productive clutch size should be the clutch size that is most commonly observed in nature. Clutch size can be manipulated in birds by the addi-

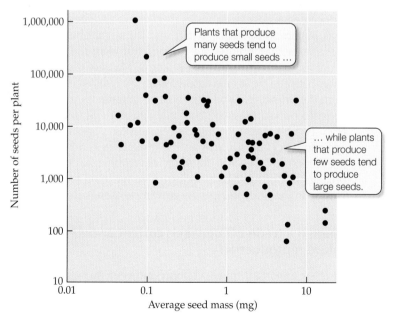

FIGURE 7.15 Seed Size–Seed Number Trade-Offs in Plants (After Stevens 1932.)

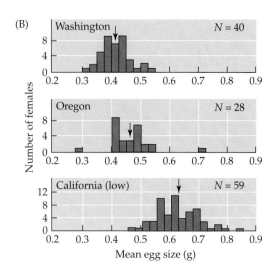

FIGURE 7.16 Egg Size–Egg Number Trade-Off in Fence Lizards Western fence lizards (*Sceloporus occidentalis*) in northern populations produced larger clutches (A) and smaller eggs (B) than those in southern populations. The arrow points to the average for each population. (After Sinervo 1990.)

The size–number trade-off also applies to variation within species. The western fence lizard (*Sceloporus occidentalis*), which is common throughout the coastal mountains of the western United States, does not provide parental care. Barry Sinervo (1990) found that lizard populations farther to the north laid more eggs per clutch (Washington: 12 eggs/clutch vs. California: 7 eggs/clutch), but laid smaller eggs (Washington: 0.40 g vs. California: 0.65 g) (**Figure 7.16**).

In order to determine the consequences of egg size for offspring performance, Sinervo raised fence lizard eggs in the laboratory. He artificially reduced the size of some of the eggs by using a syringe to remove some yolk from them. To control for any possible effects of this method on egg development, he inserted a syringe into some other eggs, but did not remove any yolk. These eggs that had been poked, but not reduced, developed at the same rate as unmanipulated eggs, suggesting that insertion of the syringe was not the cause of differences between unmanipulated and reduced eggs. The reduced eggs developed faster than unmanipulated, larger eggs, but produced smaller hatchlings. These small hatchlings grew faster than their larger siblings, but were not able to sprint as fast to escape from predators. Many of the differences between the lizards hatched from the reduced eggs and from the unmanipulated eggs echoed differences observed between populations with naturally differing egg sizes. Sinervo speculated that the differences between populations in egg and hatchling size may be the result of selection favoring faster sprint speeds in the south, where there may be more predators, or of selection favoring earlier hatching and faster growth in the north, where the growing season is shorter.

There is a trade-off between current and future reproduction

Another important trade-off occurs between an organism's current reproductive investment and the likelihood of future successful reproduction. This trade-off highlights the fact that maximizing current reproductive success does not always result in a maximization of lifetime reproductive success.

Age at sexual maturity The earlier an iteroparous organism reproduces, the more times it can reproduce over the course of its lifetime. However, not all reproductive events are equally successful, and often the number of offspring produced in a reproductive event increases with both the size and age of the organism.

A real-life example of an increase in reproductive output with age occurs in the Atlantic cod (*Gadus morhua*) (**Figure 7.17**). Normally, female cod become sexually mature at a length of about 60 cm. By the time they reach 80 cm, their fecundity is approximately 2 million eggs per year. However, by the time they reach 120 cm, their annual reproductive output is about 15 million eggs! This pattern has had important consequences for the cod fisheries of the Atlantic, in which overexploitation has actually resulted in evolutionary change in the life history of cod. Fishing has selectively removed older and larger fish because those fish are more likely to be retained in the mesh of fishing nets. This selection pressure has led to significant reductions in growth rates and in age and size at maturity in the remaining population. Removing the larger and older fish, which also have the highest reproductive potential, has resulted in a reduction in the total quality and quantity of egg production. This evolutionary change may persist even if overfishing ends and may delay or prevent the recovery of cod populations (Marteinsdottir and Begg 2002).

In situations in which reproductive bouts become more productive with age, it may be advantageous to delay sexual maturation and focus energetic investment on growth and survival instead. For example, imagine a fish with a 5-year life span whose reproductive output increases with age such that in year 1 it produces 10 offspring, in year 2 it produces 20 offspring, in year 3, 30 offspring, and so on. But now consider what would happen if the same fish could delay maturation by one year and devote the energy it would have invested in reproduction to growth. Now, by year 2, it may have grown to a large enough size that its reproductive output is 30 offspring. The size gain it made by delaying reproduction also carries over into subsequent years: in year 3 it produces 40 offspring, in year 4, 50 offspring, and in year 5, 60. In this

FIGURE 7.17 Evolving to Reproduce at an Earlier Age and Smaller Size Fishing has dramatically reduced population sizes of the Atlantic cod (*Gadus morhua*). In addition, fishing preferentially harvests large cod fishes (because smaller individuals can escape through the mesh of fishing nets). As a result, cod now reach sexual maturity at an earlier age and smaller size—an evolutionary change that may hinder the recovery of cod populations because smaller fish produce far fewer offspring than larger fish.

scenario, if it reproduces in year 1, the fish produces a total of 150 offspring over its 5-year life span, while by delaying first reproduction for a year, it can produce a total of 180 offspring, despite having one fewer reproductive bout. In this example, we assume that our hypothetical fish has a good chance of surviving for 5 years. However, if the fish has poor odds of surviving to reach larger sizes—because of overfishing or for any other reason—it is unlikely to realize the benefits of delaying reproduction. Under those circumstances, early reproduction would be favored.

Under what conditions should an organism allocate its energy to growth rather than to reproduction? A combination of a long life span, high adult survival rates, and increasing fecundity with body size means that early investment in growth should be paid back by the success of future reproductive bouts. In contrast, if rates of adult survival from one time period to the next are low, then the benefits of future reproductive bouts may never be realized. In this case, investing in current reproduction rather than growth should lead to higher lifetime reproductive success. Of course, most organisms cannot make informed decisions about their likelihood of future reproduction. But natural selection can and does shape the life history of a species—as we've seen in the way fishing pressures have selected for cod that mature earlier and at smaller sizes and that grow more slowly.

Senescence A decline in the fitness of an organism with age as a result of physiological deterioration is called **senescence**. The onset of senescence can set an upper age limit for reproductive events. Most semelparous organisms undergo very rapid senescence and death following a reproductive bout. In some iteroparous organisms, such as humans, senescence is a much longer stage. Many humans live well beyond their ca-

pacity for reproduction and then experience a gradual decline in physiological function with age. Because postreproductive individuals have already made their contributions to the next generation, there is no way for selection to remove deleterious mutations that are not expressed until a late age. However, selection for delayed senescence could occur if postreproductive individuals contribute significantly to the survival of their offspring and their offspring's offspring through parental and grandparental care or by contributing to the success of the social group in other ways. Some large social mammals, such as African elephants, do make such contributions.

Senescence may occur earlier in populations that experience high mortality rates due to disease or predation. A possible explanation of this phenomenon is provided by the mutation accumulation hypothesis of Medawar (1952). When few individuals survive long enough for selection to act against deleterious mutations that are expressed late in life, these mutations should accumulate in the population. Medawar's hypothesis is sometimes invoked to explain the increased prevalence of cancer in humans during the last 100 years. Before the advent of modern medicine and antibiotics, the argument goes, few people lived long enough to develop cancer, and so cancers that occur late in life were not selected against.

David Reznick and his colleagues (2004) have recently shown that wild populations of guppies that live in low-mortality environments exhibit delayed senescence relative to populations from high-mortality environments. They suggested that in populations where mortality is high due to predation or starvation, guppies may be investing less energy in immune system development and maintenance, resulting in higher rates of senescence due to disease. This interpretation is consistent with the mutation accumulation hypothesis: immune system mutations tend to accumulate when mortality is high due to other causes.

Life Cycle Evolution

CONCEPT 7.4
Organisms face different selection pressures at different life cycle stages.

In Section 7.1, we saw that size may vary greatly over the course of the life cycle. This variation leads to differences in habitat, preferred food, and vulnerability to predation. These differences mean that different morphologies and behaviors are adaptive at different life cycle stages. Differences in selection pressures over the course of the life cycle are responsible for some of the most distinctive patterns in the life histories of organisms.

Small size has benefits and drawbacks

Small, early life cycle stages are generally more vulnerable to predation because there are more predators that are big

Egg   Head

FIGURE 7.18 Parental Investment in the Kiwi This X-ray photograph shows the size of a kiwi egg in proportion to the female's body size.

enough to consume them (although for some predators, small prey may be more difficult to detect). They are also more vulnerable to competition for food and environmental perturbations that diminish food supply because they have little storage capacity for nutrients to help them withstand starvation. These vulnerabilities are typically counteracted by behavioral, morphological, and physiological adaptations. On the other hand, in some groups of organisms, small, mobile early stages are able to perform essential functions that are not possible for large adult stages. Here we'll examine how organisms protect small life history stages and the important benefits those stages can provide.

Parental investment   One common mechanism for protecting small, vulnerable offspring is parental care. Birds and mammals are the most familiar examples of parental care because they invest large amounts of time and energy in protecting and feeding their relatively helpless offspring. Female kiwis, for example, produce one very yolky egg at a time; the egg is so large that it makes up 15%–20% of the bird's body size (**Figure 7.18**). During the month that it takes her to make the egg, the female kiwi eats about three times as much as when she is not producing an egg. Some fishes, reptiles, amphibians, and invertebrates also guard or brood their embryos and hatchlings, protecting them until they are big enough to be less vulnerable.

In organisms that do not provide parental care, the parents' main investment in their offspring is the provisioning of the eggs or embryos. Animals add yolk to their eggs, which helps their offspring survive and grow through the small, vulnerable stages of life. In many invertebrate groups, species with yolkier eggs develop more rapidly and require less food during development than those with less yolky eggs. Another pattern common among invertebrates is investment in energetically expensive egg capsules that protect the offspring during development. Plants provision the fertilized embryos

in their seeds by adding **endosperm**, nutrient-rich material that sustains the developing embryo and often the young seedling. The starchy, white part of corn kernels and the milk and meat of coconuts are examples of endosperm.

Dispersal and diapause   Although small offspring are vulnerable to many hazards, they are also well suited to serve several very important purposes, including dispersal and diapause. Dispersal is a key feature in the life history of all organisms, providing a number of potential advantages. For example, dispersal can reduce the amount of competition among close relatives, as well as allow organisms to reach new areas in which to grow and reproduce. In some circumstances, dispersal can also increase the chance of escaping regions of high mortality, as when pathogens and other natural enemies are common at the location from which dispersal occurs.

Plants, fungi, and many marine invertebrates that are sessile or move very little as adults typically disperse as they reproduce. The small pollen, seeds, spores, or larvae of these organisms can be carried long distances by water or wind or, in the case of pollen and seeds, by pollinating or dispersing animals. In general, smaller propagules disperse more readily and can travel farther in a given amount of time.

The ability of an organism to disperse can have important evolutionary consequences. For example, Hansen (1978) compared the fossil records of gastropod species with typical swimming larvae with those of species that had lost their swimming larval stages and developed directly into crawling juveniles. He found that the species without swimming larvae tended to have smaller geographic distributions and were more prone to extinction (**Figure 7.19**). Hansen attributed these differences to differences in dispersal ability. Species with swimming larvae were able to move greater distances and hence had broadly distributed populations that were less vulnerable to random events that could lead to extinction.

Small size also makes eggs and embryos well suited to **diapause**—a state of suspended animation or dormancy in which an organism can survive unfavorable conditions. Many seeds are capable of long periods of dormancy before germination. Many protists and some animals can also undergo diapause. Amoeboid protists enter diapause by forming a hard cyst or protective casing around themselves, which allows them to survive in a desiccated state. The brine shrimp eggs that children purchase as "sea monkeys" are also in a state of diapause that allows them to survive out of water, often for years. Small seeds, eggs, and embryos are better suited to diapause than large multicellular organisms because they do not have to expend as much metabolic energy to stay alive.

Complex life cycles may result from stage-specific selection pressures

Organisms with complex life cycles have multiple stages, each adapted to its habitat and habits. This flexibility may be one

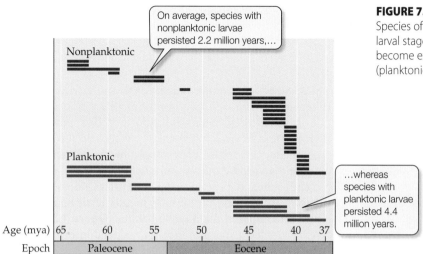

On average, species with nonplanktonic larvae persisted 2.2 million years,...

...whereas species with planktonic larvae persisted 4.4 million years.

FIGURE 7.19 Developmental Mode and Species Longevity Species of marine snails that develop directly without a swimming larval stage (nonplanktonic) disperse less well (not shown) and become extinct more rapidly than those with swimming larvae (planktonic). (After Hansen 1978.)

of the reasons that complex life cycles are so common in so many groups of organisms. Because different life history stages can evolve independently in response to size- and habitat-specific selection pressures, complex life cycles can minimize the drawbacks of small, vulnerable early stages.

Larval function and adaptation As we have seen, embryonic and larval stages in some organisms are adapted to serve specialized functions, such as dispersal and diapause. Functional specialization of particular stages in the life cycle is a common feature of complex life cycles. Having multiple stages with largely independent morphological features can result in a pairing of particular functions with particular stages. Such a pairing can reduce some of the trade-offs that come along with simultaneously optimizing multiple functions.

A classic example of this type of specialization occurs in many insects with complex life cycles. Such insects spend their entire larval stage in a very small area—sometimes on a single plant. Insect larvae such as caterpillars and grubs are specialized eating and growing machines. They spend almost all of their time taking in food and turning it into body mass, without forming many complex morphological structures other than jaws. Once they have accumulated sufficient mass, these larvae metamorphose into adult butterflies, moths, and beetles, whose main function is often to disperse, find a mate, and reproduce. In extreme cases, such as in mayflies, the adults are incapable of feeding and live only the few hours or days it takes them to reproduce.

Marine invertebrate larvae are also specialized for feeding, although they serve this function while dispersing on ocean currents. For example, the larvae of many mollusks, such as snails and clams, and echinoderms, such as sea urchins and sea stars, have intricate feeding structures that cover most of the larval body. These structures, called ciliated bands, are ridges covered in cilia that beat in coordinated patterns to catch tiny food particles and move them, like a conveyer belt, toward the mouth. The ciliated bands wind and fold their way around the larval body, which often has extra lobes or arms that support and elongate the ciliated band. In sea urchins, the longer the larval arms, and the longer the ciliated band, the more efficiently the larvae are able to feed (Hart and Strathmann 1994).

Other specialized larval structures prevent small life cycle stages from becoming food for other organisms. For example, the skeletal structures of sea urchin larvae, which support the larval arms, the head spine of crab larvae (**Figure 7.20**), and the setae or bristles of polychaete worm larvae deter predators by making the larva a large and uncomfortable mouthful.

Timing of life cycle shifts Most organisms with complex life cycles use different habitats and food resources at different stages. However, size-dependent differences in habitat and food preferences are not confined to organisms in which

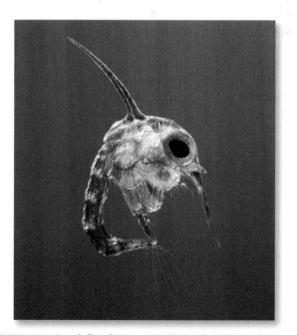

FIGURE 7.20 Specialized Structures in Marine Invertebrate Larvae Crab larvae have defensive head spines that make them a more difficult mouthful for fish.

(A)

(B)

there are abrupt transitions between stages. Even when growth and morphological change are gradual, different sized and aged individuals of the same species may have very different ecological roles. Werner and Gilliam (1984) coined the term **ontogenetic niche** for a size- or stage-specific ecological role.

In organisms with complex life cycles, an abrupt metamorphosis occurs at the transition between stages, thus minimizing the time spent in vulnerable stages that are intermediate between larva and adult. In theory, there should be an optimal time at which to undergo metamorphosis, or any change in ontogenetic niche, in order to maximize survival over the course of the life cycle. Werner suggested that metamorphosis should occur when the organism reaches a size at which conditions are more favorable for its survival or growth in the adult habitat than in the larval habitat.

Dahlgren and Eggleston (2000) applied Werner's theory to an ontogenetic niche shift that occurs in the Nassau grouper (*Epinephelus striatus*), an endangered coral reef fish that spends its juvenile stages in and around large clumps of algae. Smaller juveniles spend their time hiding within the algae, whereas larger ones spend their time in rocky habitats near algal clumps. By tethering and enclosing juvenile fish of different sizes in the two habitats, Dahlgren and Eggleston were able to measure mortality and growth rates in each habitat. They found that smaller juveniles were very vulnerable to predation in the rocky habitats, while the larger juveniles were less vulnerable and were able to grow faster in the rocky habitats. Thus, the niche shift in this species ap-

pears to be timed to maximize growth and survival, as Werner predicted.

In some cases, the larval habitat may be so favorable for growth and survival that metamorphosis is delayed—or even eliminated altogether. For example, most salamanders have aquatic larvae that metamorphose into terrestrial adults, but some salamanders, such as the mole salamander *Ambystoma talpoideum*, can become sexually mature while retaining gills and remaining in the aquatic habitat (**Figure 7.21**). These aquatic, gilled adults are referred to as **paedomorphic**, which means that they result from a delay of some developmental events (loss of gills, development of lungs) relative to sexual maturation. In the mole salamander, both aquatic paedomorphic adults and terrestrial metamorphic adults can exist in the same population of salamanders. The frequency of paedomorphosis in these mixed populations seems to depend on factors such as predation, food availability, and competition—all of which influence survival and growth in the aquatic habitat.

CASE STUDY REVISITED

Nemo Grows Up

Why does a male clownfish that has lost his mate become a female rather than simply finding a new partner? One reason is that, as in cod, the number of eggs a clownfish can produce is proportional to its body size. Thus, larger females produce more eggs and presumably have a better chance of having some of their offspring survive. Smaller individuals are more easily able to make sperm cells, which take fewer resources to produce. For these reasons, in clownfish and many other animals, females are larger than males.

Changes in sex during the course of the life cycle, called **sequential hermaphroditism**, are found in 18 fish families and in many invertebrate groups. Researchers have hypothesized that these sex changes should be timed to take advantage of the high reproductive potential of different sexes at different sizes, and in some cases they appear to do so (**Figure 7.22**). This hypothesis helps to explain sex changes in clownfish and the timing of those changes relative to size, but it leaves unanswered the question of how and why growth is regulated to maintain a hierarchy of clownfish within each anemone.

As a graduate student at Cornell University, Peter Buston (2003b) conducted experiments on a clownfish species

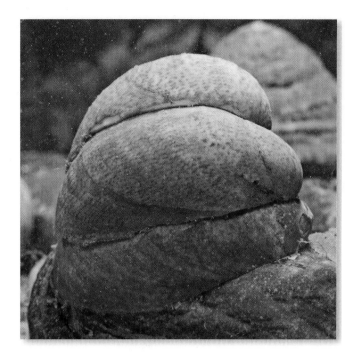

FIGURE 7.22 Sequential Hermaphroditism The slipper limpet (*Crepidula fornicata*) is a marine snail that exhibits sequential hermaphroditism. Slipper limpets live in stacks attached to rocks or shells. The largest individuals, at the bottom of the stack, are females, while the smaller upper individuals are males. As the males grow larger, they eventually reach a size (around 3 to 4 cm) at which they can produce more surviving offspring as females than as males; this is the size at which the sex change occurs. The limpets have an extraordinary ability to cling to one another and the substratum by suction to avoid being removed by predators.

FIGURE 7.23 Clownfish Size Hierarchies Clownfish within an anemone regulate their growth to maintain a hierarchy in which each rank of fish belongs to a distinct size class. Anemones may be home to between one and six fish, and the size of each fish is determined by that fish's rank and the size of the group in which it lives. (After Buston 2003b.)

(*A. percula*) that lives on reefs in Papua New Guinea. He found that each clownfish strictly maintains the size hierarchy by remaining smaller than the fish ahead of it in line and bigger than the one behind it (**Figure 7.23**). If a fish grows to be too close in size to one of its anemone-mates, a fight results, which usually ends in the smaller fish being killed or expelled from the anemone. Buston suggests that the clownfish regulate their own growth to prevent such conflicts.

Buston also manipulated clownfish groups by removing the breeding males from anemones and measuring the growth of the remaining individuals. He found that the largest nonbreeding fish grew only enough to take the place of the breeding male, and avoided growing too big and threatening the female's dominance. Similarly, the next largest nonbreeder grew only enough to take the place of the fish that had become the breeding male, and so on. Thus, clownfish avoid conflict within their social groups by exerting remarkable control over their growth rates and reproductive status.

CONNECTIONS IN NATURE

Territoriality, Competition, and Life History

The physiology of clownfish growth regulation is not entirely clear, but a more pressing ecological and evolutionary question is why the size hierarchy is maintained. What makes small clownfish bide their time as nonbreeding adults under the dominance of a single breeding female and male? The answer seems to lie in the clownfish's complete dependence on the protection of anemones for survival.

Clownfish are brightly colored, and they are poor swimmers. Outside the anemone's stinging tentacles, they are easy prey for larger fish on the reef. Thus, expulsion from the anemone is a death sentence. So the stakes are very high in conflicts between fish within an anemone: the loser will probably die, probably without reproducing. This situation exerts strong selection pressure on the fish to avoid conflicts by regulating their growth. In evolutionary terms, growth regulation mechanisms have evolved because individuals that avoid growing to a size that necessitates conflict with other fish have higher survival and reproductive rates. Buston (2003a, 2004) showed that the same selection pressures favor remaining in an anemone as a nonbreeder rather than trying to leave the anemone and find a new one. Anemones are a limited resource for the clownfish, and those that bide their time once they find an anemone experience higher lifetime reproductive success.

The scarcity of anemones also results in competition among clownfish at a key stage in their life history. After the eggs hatch within an anemone, the young disperse and spend their early life stages away from the predator-rich reef. When they return to the reef, their survival depends on their choice of an anemone. The number of fish in an anemone is generally correlated with the anemone's size. However, Buston found that at any given time some anemones are undersaturated, meaning that they have room for more fish. If a juvenile fish is lucky enough to enter such an anemone, it is allowed to stay, and it enters the line of succession toward becoming a breeder. If the juvenile enters a saturated anemone, however, it is expelled, and it usually dies before it can find another anemone. Similar settlement lotteries play out in many organisms that live in crowded habitats and compete for space. For example, in environments such as tropical rainforests, where many long-lived tree species compete for limited space and sunlight, ecologists have hypothesized that the success of any one seed or seedling depends on chance events, such as the death of a nearby large tree creating a gap in the canopy (Denslow 1987). Specialized life histories like that of the clownfish appear to be one way to maximize reproductive success in such highly competitive environments.

SUMMARY

CONCEPT 7.1 Life history patterns vary within and among species.

- Life histories are diverse, varying between individuals of the same species as well as between species. The source of this variation may be genetic or environmental.

- Organisms may reproduce sexually or asexually, and sometimes the same organism can do both.

- Most organisms have complex life cycles with multiple stages that differ in size, morphology, and habitat.

CONCEPT 7.2 Reproductive patterns can be categorized along several continua.

- Semelparous species reproduce only once in a lifetime, while iteroparous species reproduce multiple times.

- The terms r-selection and K-selection describe two ends of a continuum of reproductive patterns based on population growth rates.

- Grime's triangular model categorizes plant life histories by the degree of competition, disturbance, and stress in the habitat type to which they are adapted.

- Charnov's life history cube attempts to remove the effects of size and time in order to compare life histories across a broader diversity of organisms.

CONCEPT 7.3 There are trade-offs between life history traits.

- There is a trade-off between offspring size and number, such that organisms tend to produce larger numbers of relatively small offspring or smaller numbers of relatively large offspring.

- There is a trade-off between investment in current reproduction and investment in future reproductive bouts.

- As organisms age, they experience a decline in their physiological capabilities, termed senescence, the onset of which can set an upper age limit for reproductive events.

CONCEPT 7.4 Organisms face different selection pressures at different life cycle stages.

- The small sizes of early life cycle stages make them vulnerable to predation and food shortages.

- Small life cycle stages are well suited to serve some important functions, such as dispersal and diapause.

- Complex life cycles allow life histories to evolve in response to differences in selection pressures on different sizes and stages.

REVIEW QUESTIONS

1. Many closely related animal species produce eggs of vastly different sizes. As discussed in Section 7.3, one trade-off of producing larger eggs is that fewer eggs can be produced. Despite the apparent simplicity of this trade-off, it is still unclear why both strategies (many small eggs and few large eggs) are maintained in closely related species. What are some other life history traits besides offspring number that might also be correlated with egg size, and under what environmental conditions might those traits be advantageous? Can you think of any reasons why species that live in the same habitats continue to exhibit reproductive patterns that vary so widely?

2. Some animals exhibit both sexual and asexual reproduction, depending on the environmental conditions they experience. Rotifers are a classic example of this phenomenon. Females can produce diploid eggs by mitosis that hatch soon after release. In this manner, rotifer populations can double within hours. These same females, under other conditions, can produce haploid eggs that form males if unfertilized and form females if fertilized. What might be the reason for the maintenance in rotifers of both sexual and asexual reproduction?

3. The Nassau grouper (*Epinephelus striatus*) is popular in the live-fish trade in Asia. Restaurant-goers in Hong Kong and other Asian cities can pick the grouper they want steamed for dinner from a selection of live fish swimming in tanks. Adult groupers can grow to extremely large sizes (up to 3 feet long and 55 pounds), but those favored by restaurants are plate-sized juvenile and young adult fish. How would you expect the removal of these younger, smaller fish to affect the life history evolution of the remaining population? How might parameters such as age and size at reproduction and investment in growth versus reproduction evolve in response to fishing pressures?

HANDS-ON PROBLEM SOLVING

The Growth/ Reproduction Tradeoff | This Web exercise (www.sinauer.com/ecology) explores the tradeoff that species must make between growth and reproduction. You will investigate the effects of manipulating the set point at which fish start allocating resources into reproduction rather than further growth, under different levels of predation.

SUGGESTED READINGS

Agrawal, A. A., C. Laforsch and R. Tollrian. 1999. Transgenerational induction of defenses in animals and plants. *Nature* 401: 60–63. *In this article, Agrawal and colleagues show how the exposure of one generation to predators can influence both the phenotype of the exposed organism (animal or plant) as well as the phenotype of the offspring that organism produces.*

Grosberg, R. K and R. R. Strathmann. 1998. One cell, two cell, red cell, blue cell: The persistence of a unicellular stage in multicellular life histories. *Trends in Ecology and Evolution* 13: 112–116. *This article reviews the theories for the retention of a unicellular stage (i.e., the egg) in the life histories of multicellular organisms.*

Lindquist, N. and M. E. Hay. 1995. Can small rare prey be chemically defended? The case for marine larvae. *Ecology* 76: 1347–1358. *The use of chemical deterrents to predation in the early-stage embryos and larvae of marine invertebrates is examined in this elegant set of experiments by Lindquist and Hay. In particular, they show convincingly for the first time that the chemicals present in the larvae of marine invertebrates can significantly decrease the fitness of the organisms that consume them.*

Parker, G. A., R. R. Baker and V. G. F. Smith. 1972. The origin and evolution of gamete dimorphism and the male–female phenomenon. *Journal of Theoretical Biology* 36: 529–553. *In this paper, the authors explore the theoretical foundations for the evolution of anisogamy and sexes.*

Pechenik, J. A., D. E. Wendt and J. N. Jarrett. 1998. Metamorphosis is not a new beginning: Larval experience influences juvenile performance. *BioScience* 48: 901–911. *In this review article, the authors present evidence that the experience of larval stages during development can strongly influence postmetamorphic stages. This observation provides convincing evidence for a fitness linkage between disparate life cycle stages.*

The following two books, published nearly simultaneously, go into greater detail on many of the topics covered in Chapter 7. In particular, these books bring together both theoretical models and empirical data to provide broad overviews of the field of life history analysis.

Roff, D. A. 1992. *The Evolution of Life Histories: Theory and Analysis.* Chapman and Hall, London. (See next entry.)

Stearns, S. C. 1992. *The Evolution of Life Histories.* Oxford University Press, Oxford.

Via, S., R. Gomulkiewicz, G. De Jong, S. M. Scheiner, C. D. Schlichting and P. H. Van Tienderen. 1995. Adaptive phenotypic plasticity: Consensus and controversy. *Trends in Ecology and Evolution* 10: 212–217. *A useful review paper on the conceptual framework for phenotypic plasticity with an emphasis on cases of plasticity that are adaptive in nature.*

8 Population Distribution and Abundance

KEY CONCEPTS

CONCEPT 8.1 Populations are dynamic entities that vary in size over time and space.

CONCEPT 8.2 The distributions and abundances of organisms are limited by habitat suitability, historical factors, and dispersal.

CONCEPT 8.3 Many species have a patchy distribution of populations across their geographic range.

CONCEPT 8.4 The dispersion of individuals within a population depends on the location of essential resources, dispersal, and behavioral interactions.

CONCEPT 8.5 Population abundances and distributions can be estimated with area-based counts, mark–recapture methods, and niche modeling.

FIGURE 8.1 Key Players in the Forests of the Deep   The kelp *Nereocystis* is one of several species that make up the kelp forests found off the coasts of some Aleutian islands. Research shows that the presence or absence of kelp forests at these islands is influenced by the sea otter, *Enhydra lutris.*

CASE STUDY

From Kelp Forest to Urchin Barren

Stretching over 1,600 kilometers across the Pacific Ocean to the west of Alaska, the mountainous Aleutian Islands are often shrouded in fog and battered by violent storms. The islands have few large trees, and except on the eastern islands that once were connected to the mainland, they have none of the terrestrial mammals that are found on the mainland, such as brown bears, caribou, or lemmings. There is abundant marine wildlife in the surrounding waters, however, including seabirds, sea otters, whales, and a variety of fish and invertebrates (**Figure 8.1**).

Although there are few trees on land, the nearshore waters of some Aleutian islands harbor fascinating marine communities known as "kelp forests," made up of brown algae such as *Laminaria* and *Nereocystis*. Dense clusters of kelp rise from attachments on the sea bottom toward the surface, producing what feels like an underwater forest. Other, nearby islands do not have kelp forests. Instead, the bottoms of their nearshore waters are carpeted with sea urchins and support few kelp or other large algae. Areas with large numbers of urchins are called "urchin barrens," since they contain far fewer species than do kelp forests. Why are some islands surrounded by kelp forests and others by urchin barrens?

One possibility is that islands with kelp forests differ from islands without kelp forests in terms of climate, ocean currents, tidal patterns, or physical features such as underwater rock surfaces. But no such differences have been found, leaving us to look for other reasons why some islands have kelp forests while others do not. Because urchins feed on algae and can eat vast quantities of it, investigators suspected that grazing by urchins might prevent the formation of kelp forests.

This idea has been tested in two ways. First, studies in the Aleutian Islands and elsewhere along the Alaskan coast consistently showed that kelp forests were not found in regions where there were many large urchins. Although such correlations do not prove that urchins suppress kelp forests, the fact that a number of studies found the same result suggested that urchins might determine where kelp forests are located. Second, the importance of urchins was tested in an experiment that measured kelp densities in 50 m^2 plots containing urchins and in similar, nearby 50 m^2 plots from which urchins were removed (Duggins 1980). There were no kelp in any of the plots at the start of the experiment. Kelp densities remained at zero in the plots with urchins. In the plots from which urchins had been removed, however, the density of *Laminaria* rose to 21 individuals per square meter in the first year, and reached 105 individuals in the second year (**Figure 8.2**). *Laminaria* is a dominant member of kelp forests, so these results suggested that kelp forests would return in the absence of urchins.

These and other results indicate that the presence or absence of urchins is one of the answers to the question of why some Aleutian islands have kelp forests and others do not. But this answer just shifts the question from what determines the locations of kelp forests to what determines the locations of urchins. As we'll see, a more complete answer to our question about why kelp forests are found in some areas but not others turns out to depend on the voracious feeding habits of sea otters, which themselves may have become a meal of last resort for killer whales.

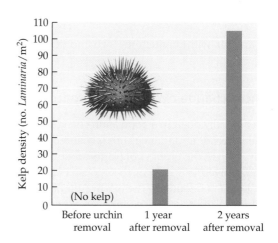

FIGURE 8.2 Do Sea Urchins Limit the Distribution of Kelp Forests? Mean densities of the kelp *Laminaria* increased dramatically after urchins were removed. (After Duggins 1980.)

I n the Case Study, we discussed why kelp forests are found in some places and urchin barrens in others. That discussion focused on a fundamental ecological question: what determines a species' **distribution**, the geographic area where individuals of a species are present? In this chapter and throughout Unit 2, we will also be concerned with the related issue of what determines **abundance**, the number of individuals in a species that are found in a given area. These two issues are related, since the distribution of a species can be viewed as a map of all areas where the abundance of the species is greater than zero.

Ecologists often seek to understand the factors that determine the distributions and abundances of organisms. One reason for this is the desire to know how nature works. The abundances and distributions of populations can change greatly over time and space; in other words, populations are *dynamic*. Thus, our ability to predict such changes provides a "measuring stick" of how well we understand events in nature. Knowledge of the factors that influence population distribution and abundance also has considerable applied importance. As we have seen, populations of many species are decreasing due to human actions, such as habitat destruction and pollution (these declines will be addressed in more detail in Chapter 22).

We can best prevent the decline of populations we want to protect if we have a clear understanding of what determines their distribution and abundance. Thus, when a government agency develops a plan for the recovery of an endangered species, it must address questions such as, Where are populations of the endangered species currently found? Are there other areas of suitable but unoccupied habitat for the species? What steps could be taken to increase the abundance of the population? We'll begin our exploration of factors that influence population distribution and abundance by describing populations in more detail.

Populations

CONCEPT 8.1
Populations are dynamic entities that vary in size over time and space.

A **population** is a group of individuals of the same species that live within a particular area and interact with one another. To explore this definition further, what exactly do we mean by "interact?" In species with sexual reproduction, we may be most interested in the group of individuals that interact by interbreeding. In species that reproduce asexually, such as dandelions or the fish *Poecilia formosa*, we may be primarily interested in other kinds of interactions, such as competition for common sources of food. Furthermore, our definition of a population also focuses on the area over which members of a species interact. If that area is known, as in a population of lizards that live on and move throughout a small island, we can report population abundance either as **population size** (the number of individuals in the population) or as **population density** (the number of individuals per unit of area). For example, if there were 2,500 lizards on a 20-hectare island (roughly 50 acres), the population size would be 2,500 lizards, and the population density would be 125 lizards per hectare.

In some cases, the total area occupied by a population is not known. For example, when little is known about how far a sexually reproducing organism or its gametes (e.g., plant pollen) can travel, it is difficult to estimate the area over which individuals interbreed frequently and hence represent a single population. For asexual species, similar problems are encountered when we try to estimate the area over which interactions other than interbreeding occur. When the area occupied by a population is not fully known, ecologists use the best available information about the biology of the species to delimit an area within which the size and density of the population can be estimated.

Abundances change over time and space

The number of individuals in a population changes over time. This is true whether abundances are measured on a small spatial scale, such as the number of plants found in a restricted area along a riverbank, or a much larger spatial scale, such as the number of cod found in the North Atlantic Ocean. At any given time, population abundances also differ from one place to another. Some populations differ little in abundance over time and space; others differ considerably.

For example, Richard Root and Naomi Cappuccino studied abundances of 23 species of herbivorous insects that fed on tall goldenrod (*Solidago altissima*). They studied these insects for at least 6 consecutive years at each of 22 sites located in the Finger Lakes region of New York State (**Figure 8.3**). These sites were no more than 75 km (47 miles) apart; hence, in any given year, all the sites experienced roughly the same climatic conditions. Nevertheless, insect abundances

varied considerably from one site to another and from one year to the next. For some species, such as the ball gall fly (*Eurosta solidaginis*), abundances varied relatively little. The maximum abundance reached by *Eurosta* over a 6-year period varied sixfold across the 22 sites, from 0.05 insects per stem at the site with the fewest individuals to 0.3 insects per stem the site with the most individuals. Maximum abundances of other species, however, such as the beetle *Trirhabda virgata*, varied much more (by a factor of 336), ranging from 0.03 to 10.1 insects per stem. Overall, *Trirhabda* populations varied considerably in abundance, both from one site to another and over time (see Figure 8.3). We'll return to this subject in Chapters 9 and 10, where we'll delve into factors that cause populations to fluctuate in abundance.

As we'll see in the next section, populations are dynamic in another sense: individuals move from one population to another, sometimes traveling great distances.

FIGURE 8.3 Abundances are Dynamic Abundances over time of the beetle *Trirhabda virgata* feeding on tall goldenrod (*Solidago altissima*) at Montezuma, Maple Island, and Hector, three of the 22 sites studied by Richard Root and Naomi Cappuccino. Study sites are indicated by data points on the map; five study sites were located close to one another at the position marked with an asterisk. (After Root and Cappucino 1992.)

Dispersal links populations

Organisms differ greatly in their capacity for movement. In plants, for example, dispersal occurs when seeds, which contain the embryo of an individual of the next generation, move away from the parent plant. Although events such as storms can transport seeds long distances (hundreds of meters to kilometers), dispersal distances in plants are usually small (one to a few tens of meters). In some cases, typical seed dispersal distances are so small that they hardly count as movement. The forest plant *Viola odorata*, for example, has a maximum observed dispersal distance of 0.02 m. When typical dispersal distances are small, we would expect populations of interacting individuals to be found in small, not large, areas. At the other end of the spectrum, some whale species travel tens of thousands of kilometers in a single year. Over-

all, the spatial extent of populations varies tremendously, from very small in organisms that disperse little to very large in species that travel great distances.

Finally, bear in mind that a population may consist of a series of patches or fragments that are spatially isolated from one another but linked by dispersal. This patchy structure can result from features of the physical environment, as we'll see later in this chapter. It can also result from human actions that subdivide once continuous populations into fragments. For example, heaths in England once covered large, continuous areas, but over the past 200 years the development of farms and urban areas has greatly reduced the extent of these plants (**Figure 8.4**). In some cases, the fragmentation results in patches that are so isolated that little dispersal can occur among them, thus breaking a single large population into a series of much smaller populations. In Chapter 10, we will explore the occurrence and consequences of patchy population structures ("metapopulations") in more detail.

What are individuals?

As we've seen, a population may cover a single area whose extent depends on the capacity of the species for dispersal, or it may cover a series of spatially isolated patches linked by dispersal. For the many species whose dispersal capabilities are poorly understood, it can be challenging to determine the spatial extent of a population. In addition, for many organisms, it can even be hard to determine what constitutes the individuals of the population.

How can there be confusion over what is an individual? Consider the aspen trees in **Figure 8.5**. Like many plant

FIGURE 8.4 Fragmentation of Dorset Heathlands   Heathlands in Dorset, England, reached their maximum extent in Roman times, 2,000 years before the present. Over the course of the last 250 years, their decline accelerated: the total area of heathlands shrank from 300 km² to less than 60 km² and the number of patches increased greatly. The distribution of heathlands in 1759 is shown in yellow and the distribution in 1978 is shown in red. (After Webb and Haskins 1980.)

FIGURE 8.5 Aspen Groves—One Tree or Many?   These quaking aspen (*Populus tremuloides*) growing in Yellowstone National Park could represent over 20 different genetic individuals, each established from a seed. However, it is also possible that each of these aspens is actually part of one "tree," having been produced asexually from the root buds of a single genetic individual.

species, an aspen tree can produce genetically identical copies of itself, or **clones**. Aspens produce clones by forming new plants from root buds, while species such as clover and strawberries do so by forming new plants from buds located on horizontal stems, or "runners" (**Figure 8.6**). Among animals, many corals, sea anemones, and hydroids can form clones of genetically identical individuals, as can some frogs, fishes, lizards, and many insects (see Figure 8.6). Some plant clones can grow to enormous sizes (e.g., 81 ha or 200 acres in aspen clones) or live for extremely long periods (e.g., 43,000 years in the king's holly, a Tasmanian shrub).

To cope with the complications that result from the formation of clones, biologists who study such organisms define individuals in several different ways. For example, an individual can be defined as the product of a single fertilization event. Under this definition, a grove of genetically identical aspen trees is actually a single genetic individual, or **genet**. However, members of a genet are often physiologically independent of one another, and may in fact compete for resources. Such actually or potentially independent members of a genet are called **ramets**. In strawberries, for example, a rooted plant is considered a ramet because it can persist even if it is not connected to the rest of its genet (see Figure 8.6). Whether we view a patch of strawberries or a grove of aspen trees as one individual or many depends on what we are interested in. If we are interested in evolutionary change over time, the genet level may be most appropriate. In contrast, if we are interested in how independent physiological units compete, the ramet level may be most appropriate.

Now that we've defined populations and considered some of the factors that complicate their study, let's turn to the factors that influence where they are found and how many individuals they contain.

FIGURE 8.6 Plants and Animals that Form Clones Many plants and animals reproduce asexually thereby forming clones of genetically identical individuals. Examples include *budding* (in which clonal offspring detach from the parent), *apomixis* (in which clonal offspring are produced from unfertilized eggs; also known as parthenogenesis), and *horizontal spread* (in which clonal offspring are produced as the organism grows).

Distribution and Abundance

> **CONCEPT 8.2**
> The distributions and abundances of organisms are limited by habitat suitability, historical factors, and dispersal.

Many different factors can influence the distributions and abundances of organisms. We'll survey these factors by grouping them into three categories: habitat suitability, historical factors (such as evolutionary history and continental drift), and dispersal.

Habitat suitability limits distribution and abundance

Good and poor places to live exist for all species. A desert species is not likely to perform well in the Arctic, or vice versa. Thus, the distribution and abundance of a species should be influenced strongly by the presence of suitable habitat. What factors make habitat suitable?

Abiotic features of the environment As we discussed in Unit 1, aspects of the abiotic (nonliving) environment, such as moisture, temperature, sunlight, soil pH, salt concentration, and available nutrients, set limits on whether a habitat will be suitable for a particular species. Some species can tolerate a broad range of physical conditions, while others have more narrow requirements.

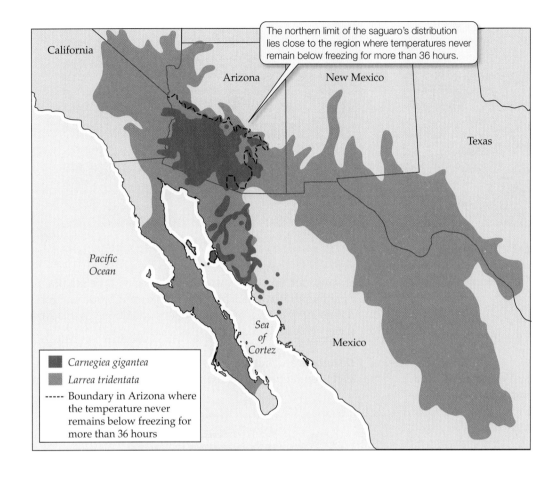

The northern limit of the saguaro's distribution lies close to the region where temperatures never remain below freezing for more than 36 hours.

FIGURE 8.7 The Distributions of Two Drought-Tolerant Plants The geographic distribution of the creosote bush (*Larrea tridentata*) is much larger than that of the saguaro cactus (*Carnegiea gigantea*). (After Hunziker et al. 1972.)

Legend:
- *Carnegiea gigantea*
- *Larrea tridentata*
- Boundary in Arizona where the temperature never remains below freezing for more than 36 hours

The creosote bush (*Larrea tridentata*), for example, has a broad distribution in North American deserts, ranging across much of the southwestern United States and northwestern and central Mexico (**Figure 8.7**). Creosote bush is very tolerant of arid conditions: it uses water rapidly when it is available, then shuts down its metabolic processes during periods of extended drought. Creosote bush also tolerates cold well, so its populations thrive in high-desert regions where temperatures can remain below freezing for several days.

The saguaro cactus (*Carnegiea gigantea*), on the other hand, has a more limited distribution. Like creosote bush, saguaro flourishes under arid conditions, but it achieves its drought tolerance in different ways. Although saguaro does not have typical leaves, its thorns are actually modified leaves whose low surface area reduces water loss. Furthermore, during wet periods, saguaro stores water in its massive trunk and arms, saving it for use during times of drought. However, saguaro cannot tolerate cold: it is killed when temperatures remain below freezing for 36 hours or more. The importance of saguaro's sensitivity to cold is revealed by its geographic distribution: the northern limit of its distribution corresponds closely to a boundary beyond which temperatures occasionally remain below freezing for at least 36 hours (see Figure 8.7).

Biotic features of the environment The biotic environment has an equally important effect on population distribution and abundance. Obviously, species that depend completely on one or a few other species for their growth, reproduction, or survival cannot live where the species on which they depend are absent. Organisms can also be excluded from an area by herbivores, predators, competitors, parasites, or pathogens, any of which can greatly reduce the survival or reproduction of members of a population.

A dramatic example of such exclusion is provided by the biological control of *Opuntia stricta*, an introduced cactus that spread rapidly to cover large areas in Queensland and New South Wales, Australia. The cactus was first imported from the southern United States in 1839 and planted as hedge. Within 40 years, *O. stricta* had become a pest species, and by 1925 it had spread to cover 243,000 km^2. The cactus can grow up to 2 m high, and in many areas it covered the ground with dense, thorny thickets, making the rangelands it occupied useless (**Figure 8.8A**). In the hope of controlling the cactus, an Argentinean moth (*Cactoblastis cactorum*) known to feed

(A)

(B)

FIGURE 8.8 Herbivores Can Limit Plant Distributions In Australia, the moth *Cactoblastis cactorum* was used to control populations of an introduced cactus, *Opuntia stricta*. (A) A dense thicket of *Opuntia* 2 months before the release of the moth. (B) The same stand 3 years later, after the moth had killed the cacti by feeding on their growing tips. (From Department of Natural Resources, Queensland, Australia.)

on *O. stricta* was released in 1926. By 1931, billions of cacti had been destroyed by the moth (**Figure 8.8B**). Since 1940, the cactus has persisted in small numbers, but its distribution and abundance have both been greatly reduced by the moth. Although the introduction of *Cactoblastis* as a means of biological pest control appears to have been a great success, such introductions must be undertaken cautiously because they can lead to unintended consequences, such as damage to native species (Louda et al. 1997).

Interactions between abiotic and biotic features In many cases, abiotic and biotic features of the environment act together to determine the distribution and abundance of a population. For example, the barnacle *Semibalanus balanoides* cannot survive when summer air temperatures are above 25°C, and it cannot reproduce if winter air temperatures do not remain below 10°C for 20 days or more. On the Pacific coast of North America, temperatures are such that *S. balanoides* could be found 1,600 kilometers farther south than it currently is. But this barnacle is absent from the region shown in purple in **Figure 8.9,** presumably because competition from other species of barnacles prevents it from living in what would otherwise be suitable habitat. To the north, as temperatures become increasingly colder, a point is reached where *S. bal-*

anoides outcompetes the other barnacles and maintains healthy populations. Thus, the abiotic and biotic environments interact to determine where populations of this barnacle are found.

Disturbance Finally, the distribution of some organisms depends on regular forms of disturbance. A **disturbance** is an

Semibalanus balanoides

FIGURE 8.9 Joint Effects of Temperature and Competition on Barnacle Distribution Although temperatures are suitable for the barnacle *Semibalanus balanoides* as far south as the region shaded purple on the map, it is excluded from this region by its competitors. *S. balanoides* is found in the region shaded red, where temperatures are colder and it is a superior competitor.

event that kills or damages some individuals and thereby creates opportunities for other individuals to grow and reproduce. Many plant species, for example, persist in an area only if there are periodic fires. If humans prevent fires, such species are replaced by species that are not as tolerant of fire but that are superior competitors in the absence of fire. Floods, windstorms, and droughts are other forms of disturbance that can exclude some species but give others an advantage.

History and dispersal limit distribution and abundance

Populations persist only in regions of suitable habitat, but they are not found in all such regions. History and dispersal also play an important role in their distributions and abundances.

Evolution and continental drift Events in the evolutionary and geological history of Earth can have a profound effect on where organisms live today. Why, for example, are polar bears found in the Arctic but not in Antarctica? Polar bears hunt on ice packs and eat seals, both of which abound in Antarctica. In part, the answer to our question lies in an accident of history. Fossil and genetic evidence indicate that polar bears evolved from brown bears in the Arctic; hence they are in the Arctic because they originated there. As for their absence from Antarctica, polar bears can travel over 1,000 km in a year, but it appears they cannot or will not cross the tropical regions that separate the Arctic from Antarctica. Thus, the distribution of polar bear populations is influenced by evolutionary history and dispersal, as well as by the presence of suitable habitat.

History also plays a key role in many groups of organisms whose curious distributions puzzled biologists for nearly a hundred years. Consider Alfred Russel Wallace's observation that the animals of a region can differ considerably across relatively short geographic distances (Wallace 1860). The mammals of the Philippines, for example, are more similar to those in Africa (88% overlap at the family level) than they are to those in New Guinea (64% overlap), despite the fact that Africa is 5,500 km away and New Guinea is only 750 km away. No explanation for this and other similar observations could be found until the discovery of *continental drift*, the gradual movement of continents over time (see Figure 17.10). This discovery led to the realization that the Philippines and New Guinea are on different tectonic plates and have been in close geographic contact for a relatively short time (**Figure 8.10**). We'll return to the role of evolution and continental drift in Chapter 17, where we'll discuss biogeographic and historical approaches to community ecology.

Dispersal On both large and small spatial scales, **dispersal limitation** can prevent species from reaching areas of suitable habitat, as we have just seen in the case of the polar bear. To give another example, the Hawaiian Islands have only one native terrestrial mammal, the Hawaiian hoary bat (*Lasiurus cinereus*), which was able to fly to the islands. No other land mammals were able to disperse to Hawaii on their own, although cats, pigs, wild dogs, rats, goats, mongooses, and other mammals now thrive in Hawaii following their introduction to the islands by people.

Dispersal limitation can also prevent populations from expanding to nearby areas of apparently suitable habitat. Richard Primack and ShiLi Miao (1992) performed an experiment to test whether local dispersal limitation was a

30 million years before present

The region that would later give rise to the Philippines has long been connected by land to regions close to Africa,...

Land mass
Continental shelf

...but until recently it was separated by a broad expanse of ocean from the Australian tectonic plate (which contains New Guinea).

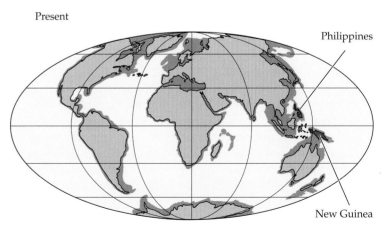

Present

Philippines

New Guinea

FIGURE 8.10 Continental Drift Affects the Distribution of Organisms Due to continental drift, the Philippine region has only recently been located close to the New Guinea region, but has long been connected by land to Africa. As a result, the mammals of the Philippines are more similar to those of Africa than those in New Guinea.

FIGURE 8.11 Populations Can Expand after Experimental Dispersal One of three study populations of *Impatiens capensis* that expanded over time after seeds were placed in unoccupied sites. (A) In 1987, a hundred seeds from an established population located 75 m away were scattered within 1 m of the black dot. The map shows the distribution of the population in the years 1988–1990; the distribution for 1991 was not mapped. (B) The graph shows the number of adults in the population for the years 1988–1991. Note that although the abundance dropped in 1990, some individuals had spread far enough that the population covered the larger area shown in (A). (After Primack and Miao 1992.)

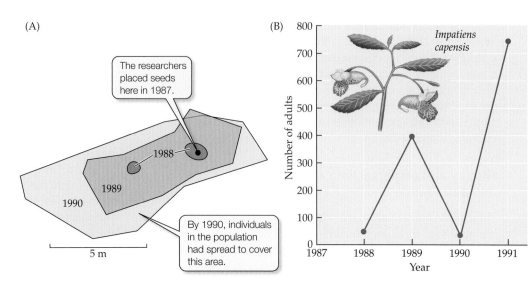

(A)

The researchers placed seeds here in 1987.

1988

1989

1990

By 1990, individuals in the population had spread to cover this area.

5 m

(B)

Impatiens capensis

factor in the distributions of *Impatiens capensis* and three other annual plant species. They placed seeds of each species in 34 unoccupied but apparently suitable sites that were 40 to 600 meters from existing populations. At 3 of the 34 test sites, new populations became established, persisted for four generations, and slowly expanded to cover a larger area (**Figure 8.11**).

In addition to its effect on distribution, dispersal is also related to population abundance. When individuals disperse from one population to another, the density of the population they leave decreases and the density of the population they join increases. While dispersal clearly affects density in this way, dispersal can also be affected by density. For example, many species of aphids produce winged forms (which are capable of dispersing) in response to crowding (Harrison 1980). For some aphids, the frequency of winged offspring in a population depends on the degree of crowding experienced by their mothers, indicating that dispersal can increase in response to conditions existing before birth. In other aphids, the percentage of larvae that develop wings increases as larval density increases; for these species, dispersal increases in response to conditions experienced after birth.

The relationship between dispersal and population abundance can also be seen in an experiment on a desert pupfish, *Cyprinodon macularius* (**Figure 8.12**). These fish live in small desert pools that often lack competitor species and predators. Wildlife biologists Thomas McMahon and Jerry Tash compared the abundance, reproduction, and mortality of desert pupfish in two sets of artificial pools, some from which dispersal was not possible ("closed" pools) and some from

which dispersal was possible ("open" pools). The study began in May of 1982, when five individuals were placed in each of the experimental pools.

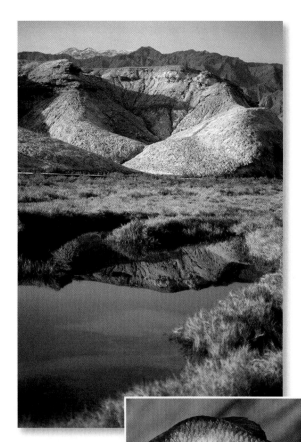

FIGURE 8.12 Desert Pupfish Habitat Desert pupfishes are excellent study organisms because the relatively simple environmental conditions of their natural habitat can be mimicked in experimental pools. The behavior, reproduction, and abundances of the fish in such pools cannot be distinguished from fish in natural pools.

Initially, all of the pupfish populations increased rapidly in size, peaking after 4 months at a mean of 276 fish in the closed pools and 195 fish in the open pools. In both closed and open pools, abundances then dropped over the first winter. In the closed pools, the drop in population size resulted from mortality (an average of 133 individuals per pool died). In the open pools, mortality (an average of 22 individuals per pool died) and dispersal (84% of the young dispersed) both contributed to the winter population decline. In the second year of the study, females in the closed pools produced an average of 1 offspring, whereas females in the open pools produced an average of 7 offspring (**Table 8.1**). Mortality remained high in the closed pools throughout the study; the death rate was nearly six times higher in the closed pools than in the open pools.

Overall, there was less overcrowding in the open pools, and the fish that remained in those pools had relatively low death rates and relatively high reproductive rates. The fish that dispersed were in poorer condition than the fish that remained, suggesting that they may have had less access to resources. In addition, many of the fish that dispersed were small males, which in desert pupfish are known to compete poorly for mates. Dispersal may play a similar role in natural populations of desert pupfish. Following heavy rains, the small pools in which these fish live are connected to one another by temporary streams. Dispersal is a risky enterprise for fish that live in desert pools, but it does not mean certain death. The results of McMahon and Tash's experiment (1988) suggest that although it is risky, dispersal may provide a fish with a greater chance for survival and reproduction than it would have in a crowded pool with limited food.

Geographic Range

CONCEPT 8.3
Many species have a patchy distribution of populations across their geographic range.

As we saw in the previous section, a number of factors—including the presence of suitable habitat, geological and evolutionary history, and dispersal—can limit the distribution of a species. Indeed, we expect that in many cases, several or perhaps all of these factors will operate at the same time to influence where a species is found. For some species, the net effect of these interacting factors will limit the species to a small geographic region, while for other species, this is not true. But whether a species inhabits a small or a large geographic area, no species can live everywhere, since much of Earth will not provide it with suitable habitat.

Geographic ranges vary in size among species

Although no species is found everywhere, there is considerable variation in the size of species' geographic ranges. A species' **geographic range** is the entire geographic region over which that species is found. Examples of species with a small geographic range include the Devil's Hole pupfish (*Cyprinodon diabolis*), which lives in a single desert pool (7 × 3 m across and 15 m deep). Many tropical plants also have a small geographic range. This latter point was illustrated dramatically in 1978, when 90 new tropical plant species were discovered on a single mountain ridge in Ecuador, each

TABLE 8.1

Effects of Dispersal on Reproduction and Mortality of Desert Pupfish

	Open Pools		Closed Pools	
	Mean	(SE)	Mean	(SE)
Offspring production per pool				
1982	283	(6.7)	269	(23.4)
1983	117	(60.7)	19	(6.5)
Combined years	350	(56.3)	287	(29.8)
Number mature females per pool in 1983	18	(3.6)	19	(2.6)
Number offspring per female in 1983	7		1	
Total no. dead offspring per pool (1982 and 1983 combined)	36	(11.8)	164	(29.3)
Percentage of offspring that died	10		57	

Source: Data from McMahon and Tash 1988.

Note: Values were measured in artificial pools that were open to dispersal versus artificial pools that were closed to dispersal (there were four pools of each type). Mean values and the associated standard error (SE) of the mean are given. The SE of the mean provides a measure of how much the data varied from one pool to the next (see **Web Stats Review**).

Fir branches covered with monarchs at a Mexican overwintering site.

Spring range
Summer range
Northern limit of milkweed plants

FIGURE 8.13 Monarch Migrations Monarch butterflies that overwinter in central Mexico are found only in oyamel fir (*Abies religiosa*) forests. These forests are located in mountainous regions that make up less than 0.5% of Mexico's land area. In the spring, monarchs migrate north from Mexico to breed throughout much of eastern North America. Because monarch caterpillars depend on milkweed plants (*Asclepias*) for food, the northern limit of the monarch's summer breeding range (shown in yellow) closely matches the northern limit of milkweeds. This diagram shows only the migration patterns for eastern North American populations of monarchs; the smaller populations of monarchs found west of the Rockies overwinter along the Pacific coast of California. (After Brower 1996.)

with a geographic range that was restricted to that ridge. Other species, such as coyotes, live over most of one continent (North America), while still others, such as gray wolves, live on small portions of several continents (North America and Eurasia). Relatively few species are found on all or most of the world's continents. Notable exceptions include humans, Norway rats, and the bacteria *Escherichia coli*, which lives in the intestinal tracts of reptiles, birds, and mammals (including humans) and thus is found wherever its host organisms are found.

The distribution of a species includes the areas it occupies during all its life stages. It is particularly important to keep this in mind for species that migrate and for species whose biology is poorly understood. For example, if we wish to protect monarch butterfly populations, we must ensure that conditions are favorable for them in both their summer breeding grounds and their overwintering sites (**Figure 8.13**). In some cases, we understand an organism's distribution poorly because it has life stages that are hard to find or study; this is true for many fungi, plants, and insects. We may know, for example, under what conditions the adult organism lives, yet have no idea where or how other life stages live. That was, in fact, long the case for the monarch butterfly. Biologists knew that these butterflies arrived each spring in eastern North America from the south, but it took almost 120 years of effort (from 1857 to 1975) before the location of the overwintering sites was discovered in mountains west of Mexico City, Mexico.

Populations have patchy distributions

Even within a species' geographic range, much of the habitat is not suitable for the species. As a result, populations tend to have a patchy structure, an observation that holds at both large and small spatial scales. At the largest spatial scales, climate constrains where populations of a species are located, while at smaller spatial scales factors such as topography, soil type, and the presence or absence of other species prevent populations from being spread evenly across the landscape.

A vivid example of patchiness at different spatial scales is provided by work Ralph Erickson performed on the shrub *Clematis fremontii* (**Figure 8.14**). This species is found in portions of Kansas, Nebraska, and Missouri. Erickson studied a variety of *Clematis fremontii* named *riehlii*, which is found only in a small region of eastern Missouri. *Clematis fremontii* var. *riehlii* is restricted to areas of dry, rocky soil that support few trees in otherwise wooded areas; such areas are called barrens or glades. The glades on which *Clematis fremontii* var. *riehlii* is found occur on outcrops of limestone located on south- or west-facing slopes. As shown in Figure 8.14, these glades are clustered, or grouped in clumps, when viewed across the entire geographic range of the shrub. The distribution of glades remains clustered at progressively smaller scales as well. Individual plants are also found in clusters, both as groups of individuals within a glade and as aggregates of individuals within one of these groups.

Clematis fremontii var. *riehlii* requires a particular habitat that is found only in portions of its geographic range; hence

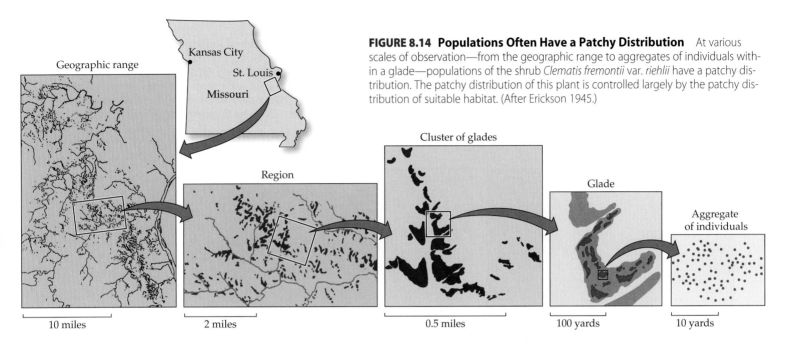

FIGURE 8.14 Populations Often Have a Patchy Distribution At various scales of observation—from the geographic range to aggregates of individuals within a glade—populations of the shrub *Clematis fremontii* var. *riehlii* have a patchy distribution. The patchy distribution of this plant is controlled largely by the patchy distribution of suitable habitat. (After Erickson 1945.)

FIGURE 8.15 Abundance Varies Throughout a Species' Geographic Range The map shows abundances of the red kangaroo (*Macropus rufus*) throughout its range in Australia. These data were based on aerial surveys conducted from 1980 to 1982. (After Caughley et al. 1987.)

its populations have a very patchy distribution. Other species tolerate a broader range of habitats, but their abundances still vary throughout their geographic range. The red kangaroo (*Macropus rufus*) of Australia illustrates this point. The density of red kangaroos varies throughout their geographic range, which includes several regions of high density and several areas where red kangaroos are not found (**Figure 8.15**). In some cases, population density tends to be greatest in the center of the range, declining gradually toward the boundaries. This result has been found for many North American species, including plants, intertidal invertebrates, and terrestrial vertebrates. The mean density of the indigo bunting (*Passerina cyanea*), for example, declined smoothly when measured at various points along four lines that extended from the center of the range of this bird species. At smaller spatial scales, similar patterns have been found for the abundances of trees at different elevations along the sides of mountains and for mean densities of invertebrates found in the intertidal zone of the Gulf of California.

The analysis of abundances at small spatial scales brings us to our next topic, the location of individuals within populations.

Dispersion within Populations

CONCEPT 8.4

The dispersion of individuals within a population depends on the location of essential resources, dispersal, and behavioral interactions.

The aggregates of *Clematis fremontii* var. *riehlii* individuals found in glades provide an example of the **dispersion**, or

Nearly regular   Random   Clumped

(A)   (B)

FIGURE 8.16 Dispersion of Individuals within Populations   Some organisms, such as creosote bush (*Larrea tridentata*) in the Mojave Desert of North America (A), have a nearly regular dispersion. This plant probably has such a dispersion because individuals compete for limited water supplies. Other organisms have a random dispersion, and the majority of organisms, like these conifers (B), have a clumped dispersion.

spatial arrangement, of individuals within a population. We can recognize three basic patterns in which individuals of a population are positioned with respect to one another (**Figure 8.16**). In some cases, members of a population have a **regular dispersion**, in which individuals are relatively evenly spaced throughout their habitat. In other cases, individuals show a **random dispersion**, similar to what would occur if individuals were positioned at locations selected at random. Finally, as in *Clematis fremontii* var. *riehlii*, individuals may be grouped together to form a **clumped dispersion**. In the field, clumped dispersions are more common than either regular or random dispersions.

Resources and dispersal affect dispersion within populations

A variety of processes can cause individuals to have a regular, random, or clumped dispersion. Consider an aquatic plant that grows poorly unless a particular set of environmental conditions are met (e.g., the right combination of light, salinity, temperature, and soil nutrients). In such a case, the spatial arrangement of individuals within a population would be likely to match the spatial arrangement of conditions suitable for growth. Since environmental conditions often vary at random or are clumped in space, our hypothetical plant might be expected to have a random or clumped dispersion. Random or clumped dispersions can also occur as a result of dispersal (e.g., short dispersal distances can cause individuals to clump together).

Individual behavior affects dispersion within populations

Interactions among individual organisms also influence dispersion patterns. Individuals may repel one another (to produce nearly regular dispersions) or attract one another (to produce clumped dispersions). Both of these tendencies can be seen in the Seychelles warbler (*Acrocephalus sechellensis*), an endangered songbird. In the 1950s, this bird nearly went extinct: its total world population was reduced to just 26 individuals located on Cousin Island in the Seychelles, a group of islands off the east coast of Africa. After the Seychelles warbler was legally protected in 1968, the Cousin Island population increased to about 300 birds, and the species was introduced successfully to two other islands.

Seychelles warblers are territorial. A breeding pair defends its territory against other birds of the species, thereby causing the dispersion of individuals in the population to be somewhat regular (**Figure 8.17A**). But not all territories are equal: some are of higher quality than others because they provide more food (e.g., insects). Birds that live in a high-quality territory live longer and produce more young. In addition, a breeding pair that lives in a high-quality territory often receives help from some of its offspring born in previous years. In a behavioral pattern known as **cooperative breeding**, these younger birds postpone breeding and instead help their parents raise offspring by performing such activities as nest building, feeding the young, defending the territory, and mobbing predators. Thus, high-quality sites serve to attract additional birds (offspring from previous

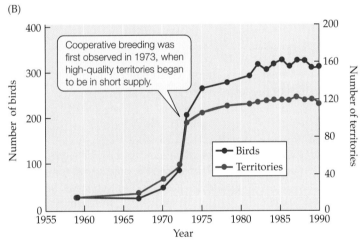

FIGURE 8.17 Territorial Behavior Affects Dispersion within Populations (A) The mean quality across years of Seychelles warbler territories on Cousin Island from 1986 to 1990. Territory quality was calculated from data on territory size, plant cover, and insect abundance and grouped into three categories (high, medium, and low). High-quality territories were clustered inland; these territories had high vegetation cover, little wind, and abundant insects. Coastal areas had lower-quality territories because salt spray led to defoliation, which lowered insect abundance. (B) Increases in the number of Seychelles warblers and occupied territories on Cousin Island from 1959 to 1990. (After Komdeur 1992.)

this prediction: as the population of Seychelles warblers increased from its low of 26, young birds did not start engaging in cooperative breeding until 1973, when high-quality territories began to be in short supply (**Figure 8.17B**). Similar results were found in an experiment in which warblers were transferred from Cousin Island to a nearby island with apparently suitable habitat, but no Seychelles warblers. When the birds were first transferred, no cooperative breeding was observed. However, after 2 years, all of the high-quality territories were occupied, and, as predicted, cooperative breeding began at that time.

Estimating Abundances and Distributions

CONCEPT 8.5

Population abundances and distributions can be estimated with area-based counts, mark–recapture methods, and niche modeling.

The most direct way to determine how many individuals live in an area is to count all of them. This is possible in some cases, as for the Seychelles warblers on Cousin Island or the *Clematis fremontii* var. *riehlii* in a Missouri glade, and other organisms that are confined to a small area, are easy to see, or do not move. But complete counts of organisms are often difficult or impossible. Consider the chinch bug (*Blissus leucopterus*), which attacks cereal crops such as corn and wheat. This species can cover large areas and reach densities that exceed 5,000 individuals per square meter, making it impractical to count all the individuals in a population. In such cases, ecologists often rely on one of two methods to estimate the actual abundance, or **absolute population size**: area-based counts and mark–recapture methods.

Ecologists estimate abundance with area-based counts and mark–recapture methods

Area-based counts are widely used to estimate the population sizes of plants and other relatively immobile organisms. These methods may make use of a **quadrat**, which is a sampling area (or volume) of any size or shape, such as a 1 m²

years). Because the high-quality sites attract more birds and are aggregated toward one end of the island, differences in territory quality make the dispersion of individuals in the population more clumped than it otherwise would be.

Why do these young birds help their parents? The answer seems to hinge on the availability of high-quality territories. When high-quality territories are scarce, a young bird born in a high-quality territory produces more offspring over its lifetime if it remains with its parents and helps them for several years than if it leaves the nest right away and breeds in a territory of lower quality (Komdeur 1992). From this observation, we might predict that young birds should not help their parents if there is no shortage of high-quality breeding sites. Observations on Cousin Island are consistent with

(A)

(B)

FIGURE 8.18 Estimating Absolute Population Size
(A) For stationary organisms such as plants, area-based methods are often used to estimate absolute population size. (B) Mark–recapture methods are commonly used to estimate the abundances of mobile organisms, such as the horseshoe crabs being tagged in this photograph.

circular plot used to count small plants, a 0.1 ha plot used to count trees, or a soil core of a certain diameter and depth used to count soil organisms (**Figure 8.18A**). Abundances are counted in a series of quadrats, and the resulting numbers are used to estimate the total population size. For example, if we counted 40, 10, 70, 80, and 50 chinch bugs in five 10 cm × 10 cm quadrats (i.e., five 0.01 m² quadrats), we would estimate that there were an average of

$$\frac{\left(40+10+70+80+50\right)/5}{0.01} = 5000 \qquad (8.1)$$

chinch bugs per square meter. Thus, for a large agricultural field, such as a 400 ha (ca. 1,000 acre) field of corn, there would be an estimated 20 billion chinch bugs in the population.

Area-based methods work well if individuals can be counted accurately within the quadrats, and if the quadrats provide a good representation of the entire area covered by the population. To help ensure that the latter condition is met, ecologists use as many quadrats as is feasible, and they often place these quadrats at locations selected at random from the entire area covered by the population. Quadrats can also be placed in a variety of other ways, such as at evenly spaced locations along a rectangular grid.

Mark–recapture methods are often used to estimate the abundance of mobile organisms (**Figure 8.18B**). To use this approach, a subset of the individuals in a population are captured, marked (as with a tag or dot of paint) so that they can be recognized at a later time, and released. After the marked individuals have been given enough time to recover and move throughout the population, individuals from the population are captured a second time, and the number of marked individuals found in the second capture is used to estimate the total population size.

Imagine, for example, that we capture 23 butterflies from a meadow, which we then mark and release. Several days later, we sample the meadow again, this time catching 15 butterflies, of which 4 are marked. In our first sample, we caught and marked $M = 23$ butterflies from a total population of unknown size (N); thus, we initially caught a proportion M/N of the butterflies in the field. The second time butterflies were sampled, we caught $C = 15$ butterflies, of which 4 were marked and hence were recaptured ($R = 4$). Assuming that no butterfly births, deaths, or movements into or out of the meadow have occurred since our first sample, the proportion of marked individuals captured in our second sample (R/C) should equal the original proportion, M/N. Thus, we have the equation

$$\frac{M}{N} = \frac{R}{C} \qquad (8.2)$$

We can rearrange that equation to estimate the total number of butterflies in the meadow as

$$N = (M \times C)/R \qquad (8.3)$$

which in this case would equal $(23 \times 15)/4 = 86$.

In some cases, the available data may not provide a way to estimate the absolute population size. We may know, for example, how many cougar tracks are found in two areas, how many fish are caught per unit of effort (e.g., the number of hooks trawled each day), or how many birds have been observed by the roadside while driving a standard distance. In each of these examples, our data provide an index of population size that is related in an unknown way to the actual population size. Such data can be compared from one time period or place to another, thereby allowing an estimate of **relative population size**. While useful, information on relative population sizes must be interpreted carefully. Data such as the number of cougar tracks observed depend not only on population density but also on animal activity: more tracks will be found in areas where animals are more active. Thus, if twice as many tracks were found in area A as in area B, it would not be correct to conclude that area A had twice as many animals—there could be more or fewer than that, depending on whether animals were more active in area A or area B.

We'll close this section with two examples of how abundance and distribution data are collected and used, the first of which highlights how long-term ecological data sets can contribute to efforts to solve important applied problems.

Ecologists used abundance data to track down a mysterious disease

In the spring of 1993, a healthy young man had difficulty breathing and collapsed on the way to his fiancée's funeral. He was rushed to a hospital in Gallup, New Mexico, but died soon thereafter. While reviewing his case, medical personnel discovered that his fiancée had died in a similar way. Initially, she had flulike symptoms, but after a few days she began to have trouble breathing, and within hours, she was dead.

Puzzled by the symptoms and concerned by the fact that two people who knew each other well had died within a few days of each other, the New Mexico Office of Medical Investigations searched for similar cases throughout the Four Corners region of the American Southwest. Within hours, three additional cases had been found. No one had ever seen this disease before. There was no cure or successful treatment, and the mortality rate was 70%. An outbreak of a lethal, previously unknown disease appeared to be in progress, leading New Mexico officials to contact the U.S. Centers for Disease Control (CDC).

Within a month, the CDC had identified the disease agent as a new strain of hantavirus, and learned that the virus was carried by the deer mouse, *Peromyscus maniculatus*. Seeking more information about the new disease, now known as Hantavirus Pulmonary Syndrome or HPS, the CDC turned to ecologists who had been studying mouse populations in the Southwest. Deer mouse specimens collected between 1979 and 1992 revealed that the virus had been present in the area for more than 10 years prior to the outbreak. Why, then, did the outbreak occur in 1993 and not before?

Researchers were able to address this question using data on the abundances of *Peromyscus* species collected at the nearby Sevilleta National Wildlife Refuge since 1989. Their data indicated that the densities of several species of *Peromyscus* had increased 3–20-fold between 1992 and 1993. Next, a series of satellite images of the vegetation taken over that same period was used to develop an index of how much plant matter was available to *Peromyscus* species for food. When that index was compared with precipitation data, the results suggested that unusually high rainfall from September 1991 through May 1992 had led to enhanced plant growth in spring 1992 (**Figure 8.19**). In turn, the high levels of plant growth produced abundant food for rodents (seeds, nuts, berries, green plant matter, arthropods), which allowed mouse populations to increase in size by 1993—the year of the outbreak.

As a result of these and other supporting data, ecologists and CDC workers hypothesized that unusually high levels of precipitation in 1992 led to enhanced plant growth, and thus high mouse densities, in 1993. Because mouse densities were higher than usual, the chance that people would come into contact with mice was probably increased. Rodents shed hantavirus in their urine, feces, and saliva; hence high mouse numbers leading to increased mouse–human contact was thought to be the cause of the 1993 disease outbreak.

FIGURE 8.19 Causing the Outbreak? From Rain to Plants to Mice The outbreak of Hantavirus Pulmonary Syndrome in the American Southwest in 1993 may have been caused by a series of interconnected events. (After Yates et al. 2002.)

We now know that the hypothesis outlined in the previous paragraph is only partly correct. For *Peromyscus* to spread HPS to people, the mice must be infected with hantavirus, which is not always the case during years of high mouse density. Disease outbreaks are most likely when both mouse densities and mouse infection rates are high, but the factors that cause that outcome remain only partially understood. It is known that the density of infected deer mice is positively correlated with the total deer mouse density in the previous year. However, it is also clear that the actual risk to people varies greatly with location and depends on such factors as habitat type (which can influence mouse movements), microclimate (e.g., in arid regions, nearby areas often experience very different amounts of rainfall), and local food abundance. Overall, we now know enough about this new disease to predict periods of heightened risk to human populations, but more remains to be learned before we can predict the specific locations most at risk.

How can we predict exactly where infected deer mice will be found? We turn next to a technique that has proved helpful in predicting the locations of other organisms: niche models.

Niche models can be used to predict where species can be found

To determine the geographic distribution of a species, scientists record all locations where the species is found. Most of our examples thus far in this chapter have involved species whose distributions are well understood. However, there are many species whose geographic ranges are not yet known. How do we know where to look for them? Furthermore, ecologists often want to predict *future* distributions of species—for example, when a pest species begins to spread after it has been introduced to a new geographic region. Scientists and policymakers face similar challenges when they seek to predict how distributions of species will change in response to global warming.

How can we predict the current or future distributions of species? One way is to characterize a species' **ecological niche**, the physical and biological conditions that the species needs to grow, survive, and reproduce. Ecologists can use data collected from areas where the species is known to occur to characterize its ecological niche. These data can then be used to construct a **niche model**, a predictive tool that models the environmental conditions occupied by a species based on the conditions at localities it is known to occupy.

Investigators from the United States and Mexico used such an approach to predict the distributions of chameleons in Madagascar whose habitat is threatened by human development (Raxworthy et al. 2003). In this study, investigators obtained information about plant cover (based on satellite images), temperature, precipitation, topography (elevation, slope, aspect), and hydrology (water flow, tendency to pool) from government and commercial sources. Values for these environmental variables were recorded for each of a series of 1×1 km^2 areas (referred to as "grid cells") that covered all of Madagascar. Next, for each of 11 chameleon species, the investigators developed rules that described the environmental conditions where each species was most likely to be found; we'll refer to these rules as "habitat rules."

To develop the habitat rules for each species, the investigators used a computer program and entered the values for environmental variables in grid cells where that species was known to occur; the program then compared those values with the environmental values for grid cells selected at random from a map of Madagascar (**Figure 8.20**). The program searched for accurate habitat rules using a flexible approach known as GARP (Genetic Algorithm for Rule-set Prediction). This search mechanism works by changing habitat rules in a way that mimics the occurrence of genetic mutations (random changes in the DNA sequence of an organism) and natural selection. For example, initially a habitat rule might state that a species should be found in regions where the temperature ranges from 15°C to 25°C and the elevation ranges from 300 to 550 m. Using GARP, this rule might "mutate" at random to have a temperature range of 15°–30°C and an elevation range of 300–500 m. Alternatively, a habitat rule might experience a "deletion"—for instance, the elevation requirement might be dropped. Habitat rules might also exchange components, again at random; thus, a 300–500 m elevation requirement might be exchanged with a 125–175 cm per year precipitation requirement found in a different rule. As the rules change over time in the program, new rules that improve the ability of the program to predict where species are found are retained, and other rules are discarded.

For the Madagascar chameleons, the accuracy of the habitat rules developed using GARP was tested with chameleon location data that had not been entered into the program. The rules performed well, correctly predicting where chameleons would be found 75%–85% of the time. Next, the habitat rules were used to predict the geographic ranges of each of the 11 chameleon species, information that will be useful to land managers seeking to protect chameleon habitat. Finally, the researchers investigated an interesting "error" in the predic-

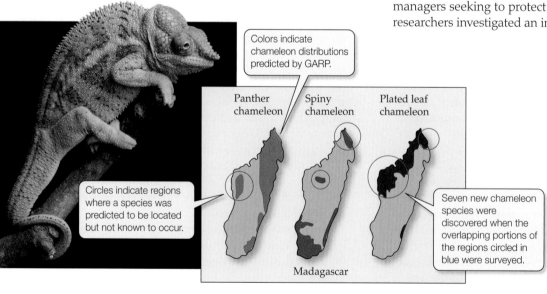

Colors indicate chameleon distributions predicted by GARP.

Circles indicate regions where a species was predicted to be located but not known to occur.

Panther chameleon

Spiny chameleon

Plated leaf chameleon

Seven new chameleon species were discovered when the overlapping portions of the regions circled in blue were surveyed.

Madagascar

FIGURE 8.20 Predicted Distributions of Madagascar Chameleons The distribution predicted by GARP is shown for the Panther chameleon (*Furcifer pardalis*), the Spiny chameleon (*F. verrucosus*), and the Plated leaf chameleon (*Brookesia stumpffi*), three of the eleven Madagascar chameleon species studied by Raxworthy et al. All 11 of the predicted distributions proved accurate with independent test data (see text). (After Raxworthy et al. 2003.)

tions: there were several overlapping areas in which the program predicted that two or more of the 11 species would be found, but in which none of the species were known to occur. When researchers surveyed two of these overlapping areas (see Figure 8.20), seven previously unknown chameleon species were discovered. In contrast, more intensive surveys conducted during the same period but at sites outside these overlapping areas yielded only two new species. Thus, the scientists were able to predict both the distributions of known chameleon species and the locations of habitats that were good for other chameleons, and the latter prediction led to the discovery of seven new chameleon species.

From Kelp Forest to Urchin Barren

When sea urchins graze kelp so heavily that kelp forests disappear and are replaced by urchin barrens, what happens next? We might expect that the urchins would starve because they have destroyed their food source, kelp. However, field studies show that urchin barrens can persist for years on end because urchins can use food sources other than kelp, including benthic diatoms, less preferred algae (including hard, encrusting forms that cover rock surfaces), and detritus. In addition, when food is extremely scarce, urchins can reduce their metabolic rate, reabsorb their sex organs (forgoing reproduction but increasing their chances of survival), and absorb dissolved nutrients directly from seawater.

As tough and resilient as urchins are, they are vulnerable to predation by sea otters, which function as impressive urchin-eating machines. Otters need to eat large quantities of food each day because they have a high metabolic rate and they store little energy as fat. In captivity, where conditions are relatively benign, otters eat 20%–23% of their body weight in food each day. Urchins are a favorite food of otters, and since there are 20–30 otters per square kilometer around some Aleutian islands, the potential exists for otters to consume enormous quantities of urchins. These facts, coupled with the observation that urchins often are common only when otters are absent, led investigators to suspect that otters might control the locations of urchins, and hence the locations of kelp forests.

To test this hypothesis, Estes and Duggins (1995) compared sites with and without otters, both in the Aleutian Islands and along the coast of southern Alaska. Confirming the results of previous studies, they found that sites where otters had been present for a long time usually had many kelp and few urchins, whereas sites without otters usually had many urchins and few kelp. Estes and Duggins also collected data from sites colonized by otters during the course of their study. At sites in southern Alaska, the arrival

of otters had a rapid and dramatic effect: within 2 years, urchins virtually disappeared, and kelp densities increased dramatically (**Figure 8.21**). At Aleutian Islands sites, however, kelp recovered more slowly after the return of otters. At these sites, otters ate most of the large urchins (which they prefer over small urchins), reducing urchin biomass by an average of 50% and allowing kelp to begin to recover at some sites. However, in a twist that did not occur at the southern Alaska sites, the arrival of new urchin larvae

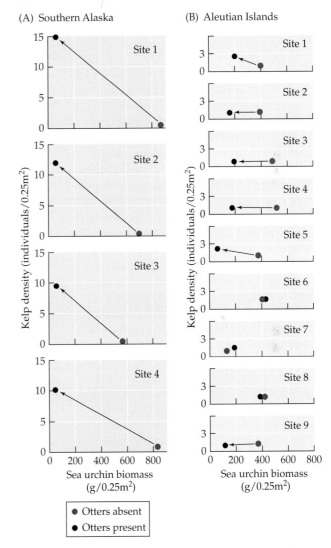

FIGURE 8.21 The Effect of Otters on Urchins and Kelp
Plots of kelp density versus sea urchin biomass measured before and 2 years after the return of otters. (A) Two years after otters colonized four sites in southern Alaska, urchins had declined considerably and kelp had increased substantially at all four sites. (B) Two years after otters colonized nine sites in the Aleutian Islands, sea urchin biomass had declined at six of the sites, but kelp showed clear signs of recovery only at sites 1 and 5. Arrows indicate a decline in urchin biomass and (only in some sites) an increase in kelp density in the presence of otters. (After Estes and Duggins 1995.)

FIGURE 8.22 Orca Predation on Otters May Have Led to Kelp Decline Declines in otter abundance (% maximum number) over time (A) are associated with (B) a rise in urchin biomass (g/0.25 m²), (C) an increase in urchin grazing intensity on kelp (% loss per day), and (D) a decrease in kelp density (individuals/0.25 m²). The proposed mechanisms for these changes are illustrated to the right of the graphs. Prior to otter decline (pre-1990), otters had a strong effect on urchins (indicated by the thick arrow), and urchins had a weak effect on kelp (indicated by the thin arrow). According to the hypothesis shown here, once orcas began to eat otters in the 1990s, they had a strong effect on otters (late 1990s), weakening the effect of otters on urchins, strengthening the effect of urchins on kelp, and leading to the decline of kelp. (After Estes et al. 1998.)

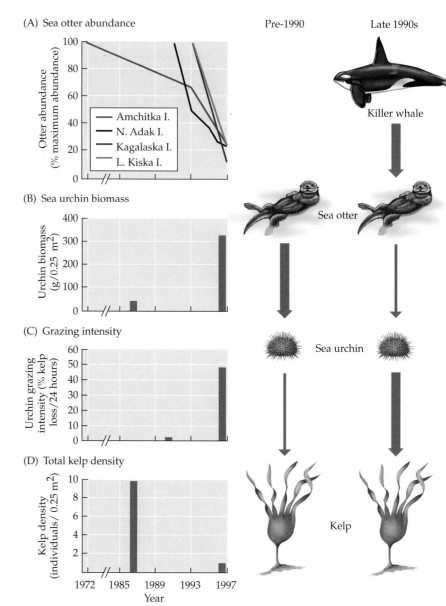

(most likely via ocean currents) provided a steady supply of small urchins. These small urchins slowed the rate at which kelp forests replaced urchin barrens.

Historically, sea otters were abundant throughout the North Pacific, but by 1900 they had been hunted (for fur) to near extinction. By 1911, when international treaties protected the sea otter, only about a thousand otters remained, less than 1% of their original numbers. Scattered colonies of otters survived around some Aleutian islands, causing the observed pattern of kelp forests around some islands, urchin barrens around others. By the 1970s, otter populations had recovered to between 110,000 and 160,000 individuals. In the 1990s, however, there was a sudden and unexpected decline in otter populations. Urchins made a comeback, and kelp densities were reduced (**Figure 8.22**). The question now became, what caused the decline of sea otter populations?

James Estes and his colleagues have suggested that otters declined because of increased predation from *Orcinus orca*, or killer whales. It is not known why killer whales began to eat more otters. Some researchers have argued that this change was part of a chain of events that began when commercial whaling drove populations of large whales to low numbers (Springer et al. 2003). Their idea, while still speculative, is that once their preferred prey (large whales) became rare, killer whales began to hunt a series of other species (first harbor seals, then fur seals, then sea lions), which then also declined in numbers. Other researchers dispute the connection between commercial whaling and the decline of seals and sea lions, suggesting that seal and sea lion populations declined for other reasons, such as a lack of food due to reduced fish populations in the open ocean (DeMaster et al. 2006). Whatever the cause, it was in the 1990s, when populations

of harbor seals, fur seals, and sea lions had all declined to low levels, that killer whales were first seen attacking otters. Otters and killer whales had been observed in close proximity for decades, but within 10 years of the first known attack, otter populations crashed.

CONNECTIONS IN NATURE

From Urchins to Ecosystems

Urchins, otters, and perhaps killer whales and people play an important role in determining the distribution of kelp forests. What are the ramifications of their distribution? Do kelp forests have a strong effect on nearshore ecosystems?

Indeed they do. Kelp forests are among most productive ecosystems in the world, rivaling tropical forests in the amount of new biomass they produce each year (up to 2,000 grams of carbon per square meter per year). Kelp strands grow from their base, and their tips are constantly "eroded" by wave action and other physical forces. Thus, much of their biomass ends up as floating bits of detritus, which provides food for suspension feeders such as barnacles and mussels that filter food from water. As a result, barnacles and mussels grow more rapidly and are more abundant in a kelp forest than in an urchin barren. Carbon-13 labeling studies have shown that the sugars kelp produce by photosynthesis provide a food source for a wide range of species (Duggins et al. 1989). In addition, kelp forests serve as nurseries for the young of many species and as havens from predators for the adults of still more species.

Finally, kelp forests located close to shore protect the shore from waves, thereby reducing water turbulence in intertidal zones. Because the water is less turbulent, sedimentation rates increase, which harms filter feeders (including barnacles and mussels) located in intertidal zones. For example, the density of mussels in the intertidal zone increases from 6 individuals per square meter when kelp is present to 1,400 individuals per square meter when kelp is absent. Although barnacles and mussels are more abundant in deeper waters when kelp is present, kelp has the opposite effect on these species in the intertidal zone.

Overall, kelp has a large effect on nearshore ecosystems. Hence, the effects of otters on urchins matter: otters (and perhaps killer whales and people) set into motion a chain of events that alters fundamental aspects of the marine ecosystem. The possible connection to killer whales also illustrates the potential for ecosystems to be connected across huge spatial and temporal scales. As we saw, the possible reasons for the decline of seals and sea lions include commercial whaling and decreases in ocean fish populations. Such changes in the open ocean may, in turn, have caused killer whales to begin to eat otters, thereby having a large effect on nearshore island ecosystems.

SUMMARY

CONCEPT 8.1 Populations are dynamic entities that vary in size over time and space.

- Populations are groups of individuals of a species that live in the same area and interact with one another.

- The number of individuals in a population changes over time and from one place to another.

- Dispersal can link the populations of a species to one another.

- In species that can reproduce asexually, the members of a population can be defined in terms of genetic individuals (genets) or physiological individuals (ramets).

CONCEPT 8.2 The distributions and abundances of organisms are limited by habitat suitability, historical factors, and dispersal.

- The presence of suitable habitat limits the distribution and abundance of organisms.

- The suitability of habitat depends on abiotic and biotic features of the environment, the interaction between abiotic and biotic factors, and disturbance.

- Dispersal and events in the evolutionary and geological history of Earth also influence the distribution and abundance of organisms.

CONCEPT 8.3 Many species have a patchy distribution of populations across their geographic range.

- No species is found everywhere, since much of Earth is not suitable habitat for its populations.

- Geographic ranges vary considerably in size from one species to another.

- Populations often have a patchy structure at both small and large spatial scales.

CONCEPT 8.4 The dispersion of individuals within a population depends on the location of essential resources, dispersal, and behavioral interactions.

- The dispersion of individuals within a population may be regular, random, or clumped. In the field, clumped dispersions are most common.

- A random or clumped dispersion may match the spatial arrangement of important resources. Clumping may also result from short dispersal distances.

- Behavioral interactions in which individuals repel or attract one another can affect the dispersion of individuals within a population.

CONCEPT 8.5 Population abundances and distributions can be estimated with area-based counts, mark–recapture methods, and niche modeling.

- The most direct way to determine the number of individuals in a population is to count all of them. When this is not possible or practical, area-based counts or mark–recapture methods can be used to estimate the number of individuals in a population.

- The geographic distribution of an organism can be analyzed in terms of its ecological niche, the physical and biological conditions of the environment that the organism needs to grow, survive, and reproduce.

- Niche models can be used to estimate the distribution of an organism when we have insufficient data on its geographic range, or when we want to predict the future locations of its populations.

REVIEW QUESTIONS

1. Describe some of the complicating factors that can be encountered in studying a population.

2. No species is found everywhere on Earth. Why? Your answer should include an explanation of why organisms often are not found in all places where you might expect them to thrive.

3. What are niche models? Describe how such a model could be used to predict the future distribution of an organism that is spreading into a new geographic region.

4. Sea otters can eat 20%–23% of their body weight in food each day. An average sea otter weighs 23 kg (roughly 50 pounds), and there are 20–30 otters per square kilometer where they are present. An average sea urchin weighs 0.55 kg. Assuming the otters eat only sea urchins, use these data to calculate a conservative estimate of the number of sea urchins per square kilometer that an otter population would be expected to eat each year.

HANDS-ON PROBLEM SOLVING

Abiotic and Biotic Factors | This Web exercise (www.sinauer.com/ecology) explores how the interaction of abiotic and biotic factors affects the abundance and distribution of lizards. You will interpret data to determine how well lizard populations are able to survive a hurricane on islands with predators and on islands without predators.

SUGGESTED READINGS

Lomolino, M. V., B. R. Riddle and J. H. Brown. 2006. *Biogeography*. 3rd ed. Sinauer Associates, Sunderland, MA.

Brower, L. P and S. B. Malcolm. 1991. Animal migrations: Endangered phenomena. *American Zoologist* 31: 265–276. *This paper describes how the spectacular annual migrations of monarch butterflies and other species are threatened by human actions that reduce population abundance.*

Estes, J. A. and D. O. Duggins. 1995. Sea otters and kelp forests in Alaska: Generality and variation in a community ecology paradigm. *Ecological Monographs* 65: 75–100. *An authoritative study by two leading researchers that provides further details about the kelp forest system described in the Case Study.*

Jackson, J. B. C., L. W. Buss and R. E. Cook, eds. 1983. *Population Biology and Evolution of Clonal Organisms.* Yale University Press, New Haven. *This outstanding volume of contributed chapters delves into the fascinating biological consequences of clonality in plants and animals that reproduce asexually to form clones.*

Peterson, A. T. and D. A. Vieglais. 2001. Predicting species invasions using ecological niche modeling. *BioScience* 51: 363–371. *Focusing on the problem of species invasions, this paper provides* an excellent introduction to how niche models can be used to predict the future distribution of organisms.

Ridley, H. N. 1930. *The Dispersal of Plants throughout the World*. Reeve, Ashford, U.K. *An unparalleled description of the many unique ways by which plants regularly (albeit somewhat rarely) disperse long distances and hence can colonize distant locations.*

Root, R. B. 1967. The niche exploitation pattern of the blue-grey gnat-catcher. *Ecological Monographs* 37: 317–350. *This classic paper is often cited for its introduction of a new ecological concept ("guilds," groups of species that obtain food or use other environmental resources in similar ways). While the guild concept is important, the paper is even more valuable for its thoughtful and superb description of the ecological niche of a particular species, the blue-grey gnat-catcher.*

Yates, T. L., H. N. Mills, C. A. Parmenter, T. G. Ksiazek, R. R. Parmenter, J. R. Vande Castle, C. H. Calisher, S. T. Nichol, K. D. Abbott, J. C. Young et al. 2002. The ecology and evolutionary history of an emergent disease: Hantavirus pulmonary syndrome. *BioScience* 52: 989–998. *An accessible and engaging summary of what we know—and how we learned it—about a new lethal disease, hantavirus pulmonary syndrome.*

9 Population Growth and Regulation

KEY CONCEPTS

CONCEPT 9.1 Life tables show how survival and reproductive rates vary with age, size, or life cycle stage.

CONCEPT 9.2 Life table data can be used to project the future age structure, size, and growth rate of a population.

CONCEPT 9.3 Populations can grow exponentially when conditions are favorable, but exponential growth cannot continue indefinitely.

CONCEPT 9.4 Population size can be determined by density-dependent and density-independent factors.

CONCEPT 9.5 The logistic equation incorporates limits to growth and shows how a population may stabilize at a maximum size, the carrying capacity.

FIGURE 9.1 Transforming the Planet
Irrigation systems that pivot around a central point create red circles of healthy vegetation in this satellite image of croplands near Garden City, Kansas.

CASE STUDY

Human Population Growth

Viewed from space, Earth appears as a beautiful ball of blue and white in a vast sea of black. If we narrow our focus by using satellite images to explore the surface of this beautiful ball in more detail, we find clear signs of human impacts across the globe. These signs range from changes in the distribution of the world's forests, to rivers that once meandered but now flow straight in channels dug by people, to surrealistic patterns formed by farm fields (**Figure 9.1**).

People have a large effect on the global environment for two underlying reasons: our population has grown explosively, and so has our use of energy and resources. The human population crossed the 6.6 billion mark in 2007, more than double the 3 billion people alive in 1960 (**Figure 9.2**). Our use of energy and resources has grown even more rapidly. From 1860 to 1991, for example, the human population quadrupled in size, but our energy consumption increased 93-fold.

The addition of more than 3 billion people since 1960 is nothing short of remarkable. For thousands of years our population grew relatively slowly, reaching 1 billion for the first time in 1825 (Cohen 1995). The time we took to reach the 1 billion mark puts the current growth of our population in perspective: it took roughly 200,000 years (from the origin of our species to 1825) for the human population to reach its first billion, but now we are adding 1 billion people every 13 years. When did we switch from relatively slow to explosive population growth?

No one knows for sure, since it is difficult to estimate population sizes from long ago. According to the best information we have, by 1550 there were roughly 500 million people alive, and the population was doubling every 275 years. By the time we reached our first billion in 1825, the human population was growing very rapidly: it doubled from 1 to 2 billion by 1930, in just 105 years. Forty-five years later, it had doubled again, reaching 4 billion in 1975, at which time it was growing at an annual rate of nearly 2%. To appreciate what that means, a population with a 2% annual growth rate doubles in size every 35 years. If that rate of growth could be sustained, we would reach 8 billion in 2010, 16 billion in 2045, and 32 billion in 2080. Should there be 32 billion people in 2080, a long-lived person born in 1975 and living to 2080 would have seen the human population increase eightfold in their lifetime.

What do you think the world would be like with 32 billion people? Already, with fewer than 7 billion people, we have transformed the planet. It is unlikely, however, that there will be 32 billion people on Earth in 2080. Over the last 40 years, the annual rate of human population growth has slowed considerably, from a high of 2% per

FIGURE 9.2 Explosive Growth of the Human Population Human population growth was relatively slow until 1825, but our population has grown extremely rapidly since that time.

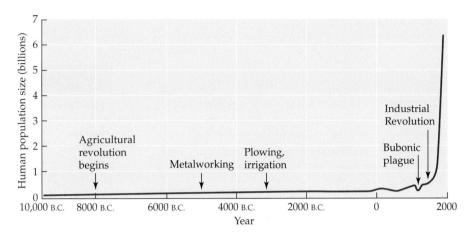

year to the present growth rate of 1.21% annually. Currently, the human population increases by about 80 million people per year, more than 9,100 people each hour. Six countries account for half of the annual growth: India (21%), China (12%), Pakistan (5%), and Bangladesh, Nigeria, and the United States (4% each). If the current annual growth rate of 1.21% were maintained from now until 2080, there would be roughly 16 billion people on Earth in 2080.

Could Earth support 16 billion people? Will there be that many people in 2080? Or will annual growth rates continue to fall? We'll return to these questions at the close of this chapter.

One of the ecological maxims introduced in Table 1.1 reads: "No population can increase in size forever." Logically, this maxim has to be true, since Earth is a finite planet, and hence it cannot support ever-increasing numbers of any species. The limits imposed by a finite planet restrict what otherwise appears to be a universal feature of all species: a capacity for rapid population growth. As we saw in the Case Study, the human population is increasing rapidly. Other species, such as giant puffballs, have an even more impressive capacity to increase their numbers. These fungi each produce so many offspring (7 *trillion*)

that if all of them reached adulthood, the descendants of two individuals would weigh more than the entire planet in just two generations. But Earth is not covered with giant puffballs, or even people. The challenge for ecologists—whether they are studying giant puffballs, people, or any other organism— is to understand what factors promote population growth and what factors limit population growth.

What we learn can surprise us. We may find, for example, that current methods of protecting an endangered species work poorly. Such was the case for loggerhead sea turtles, a rare species whose young often die as they crawl to the sea after hatching from nests dug in the sand (**Figure 9.3**). Efforts to protect loggerheads initially focused on protecting newborns. However, researchers found that even if newborn survival could be increased to 100%, loggerhead populations would continue to decline. Fortunately, the researchers were able to use the methods described in this chapter to identify more effective ways to protect loggerheads (see Box 9.1).

How do scientists reach such conclusions? What data are needed? As we seek to understand population growth and its limits, some of the most powerful tools at our disposal are based on life tables, the topic we explore next.

Life Tables

CONCEPT 9.1
Life tables show how survival and reproductive rates vary with age, size, or life cycle stage.

Information about patterns of births and deaths in a population is essential if we want to understand current population trends or predict future population sizes. To obtain such information for an annual plant, for example, we could mark (e.g., with a numbered tag) a large number of individual

FIGURE 9.3 Dash to the Sea In the nesting environment, the eggs and hatchlings of loggerhead sea turtles face threats from a variety of factors, including predators, beach development, and artificial lighting (which can disorient the hatchlings, preventing them from reaching the sea). Loggerheads also face threats in the marine environment from predators, commercial fisheries (turtles can be caught accidentally in nets and tarps), boat collisions, and chemical and solid-waste pollutants.

seedlings shortly after they germinated. We could then follow the fate of each seedling over the course of the growing season. By recording whether each seedling was alive or dead at various points in time, we could estimate how the chance of surviving from one time period to the next varied with plant age. Similarly, by recording how many seeds each plant produced at different times, we could estimate how reproduction varied with age.

Life tables can be based on age, size, or life cycle stage

Scientists have collected data on patterns of birth and death for plants, people, sea turtles, and many other organisms, and they have used these data to construct life tables. A **life table** provides a summary of how survival and reproductive rates vary with the age of the organisms. For example, **Table 9.1** shows data collected for the grass *Poa annua*. These data were collected by marking 843 naturally germinating seedlings and then following their fates over time. The second column on the left, labeled N_x, shows the number of individuals alive at age x, where x was measured in 3-month periods. As individuals die over time, N_x decreases steadily from the original 843 individuals, reaching 0 at $x = 8$ (24 months).

The next two columns in Table 9.1, S_x and l_x, are calculated from the N_x data. S_x is the age-specific **survival rate**, which is the chance that an individual of age x will survive to be age $x + 1$. S_2, for example, equals 0.6, which indicates that, on average, an individual of age $x = 2$ (6 months) has a 60% chance of surviving to reach age $x = 3$ (9 months). The next column, l_x, represents **survivorship**, which is the proportion of individuals that survive from birth (age 0) to age x. For example, l_3 equals 0.375, indicating that 37.5% of newborns survive to reach age $x = 3$. The final column, F_x, represents **fecundity**, which is the average number of offspring produced by a female while she is of age x.

Table 9.1 is an example of a **cohort life table**, in which the fate of a group of individuals born during the same time period (a "cohort") is followed from birth to death. Cohort life tables are often used for plants or other sessile organisms because individuals can be marked and followed over time relatively easily. For organisms that are highly mobile or have long life spans (e.g., trees that live much longer than people), however, it is hard to observe the fate of individuals from birth to death. In some of these cases, a **static life table** can be used, in which the survival and reproduction of individuals of different ages during a single time period are recorded. To use this method, one must be able to estimate the age of the organisms under observation. This is very difficult in some species, but for others, reliable indicators of age are known, such as annual growth rings in tree wood and fish scales and tooth wear in deer. Once age has been estimated, age-specific birth rates can be determined by counting how many offspring the individuals of different ages produce. Age-spe-

TABLE 9.1

Life Table for the Grass *Poa annua*

Age (in 3-month periods)[a] x	Number alive[b] N_x	Survival rate[c] S_x	Survivorship[d] l_x	Fecundity[e] F_x
0	843	0.856	1.000	0
1	722	0.730	0.856	300
2	527	0.600	0.625	620
3	316	0.456	0.375	430
4	144	0.375	0.171	210
5	54	0.278	0.064	60
6	15	0.200	0.018	30
7	3	0.000	0.004	10
8	0		0.000	

Source: Data from Table 1.1 in Begon et al. 1996.

[a]Age (x) is measured in 3-month periods, so an individual of age $x = 5$, for example, is 15 months old.

[b]N_x = number of individuals alive at age x.

[c]S_x = proportion of individuals of age x that survive to age $x + 1$; $S_x = N_{x+1}/N_x$.

[d]l_x = proportion of individuals that survive from birth (age 0) to age x; $l_x = N_x/N_0$.

[e]F_x = average number of offspring born to a female while she is of age x.

cific death rates can also be determined from a static life table (see Review Question 1), but only if we assume that survival rates have remained constant during the entire time that the individuals in the population have been alive—an assumption that may not be correct.

In discussing life tables, we have emphasized the importance of age because in many species birth and death rates differ greatly between individuals of different ages. For other kinds of organisms, age is less important. In many plant species, for example, if conditions are favorable, a seedling may grow to full size relatively rapidly and reproduce at a young age. If conditions are not favorable, however, the plant may remain small for years and reproduce little or not at all; if conditions become favorable at a later time, the plant may then grow to full size and reproduce. For such species, whether an individual reproduces or not is more closely related to size than to age. When birth and death rates correlate poorly with age—or when age is difficult to measure—life tables based on the size or the life cycle stage (e.g., newborn, juvenile, adult) of individuals in the population can be constructed.

Extensive life table data exist for people

Many economic, sociological, and medical applications rely on human life table data. Life insurance companies, for ex-

TABLE 9.2

Survivorship, Fecundity, and Life Expectancy by Age for U.S. Females in 2001

Age (in years) x	Survivorship l_x	Fecundity F_x	Life expectancy (at age x)
0–4	0.993	0.0	77.3
5–9	0.992	0.0	72.4
10–14	0.992	0.002	67.4
15–19	0.990	0.113	62.5
20–24	0.988	0.265	57.7
25–29	0.985	0.283	52.8
30–34	0.982	0.230	48.0
35–39	0.977	0.102	43.2
40–44	0.970	0.020	38.5
45–49	0.959	0.001	33.9
50–54	0.944	0.0	29.4
55–59	0.921	0.0	25.1
60–64	0.885	0.0	21.0
65–69	0.831	0.0	17.2
70–74	0.754	0.0	13.7
75–79	0.647	0.0	10.5
80–84	0.502	0.0	7.8
85–89	0.323	0.0	5.7
90–94	0.157	0.0	4.2
95–99	0.052	0.0	3.1
≥100	0.029	0.0	2.8

Source: Martin et al. (2003) and Arias (2004).

ample, use census data to construct static life tables that provide a snapshot of current survival rates; they use these data to determine the premiums they charge customers of different ages. Let's consider two examples of human life tables, one from the United States, the other from Gambia.

The Centers for Disease Control and Prevention periodically release reports that provide life table data for people in the United States. Reports released in 2003 and 2004 provide information on the survivorship (l_x), fecundity (F_x), and life expectancy (expected number of years of life remaining) of U.S. girls and women of different ages (**Table 9.2**). To make their interpretation easier, such results can be graphed, as in **Figure 9.4**, which plots l_x data for U.S. females in 2001. This curve shows what careful examination of Table 9.2 also reveals: survival probabilities for U.S. females remain high for many years and do not begin to drop sharply until around age 70.

The data from the United States are in stark contrast to data from Gambia, a country located on the west coast of Africa. Moore et al. (1997) analyzed birth and death records for 3,102 people born in three Gambian villages between 1949 and 1994. They found that the season of birth had a lasting impact on adult survivorship: children born during the annual "hungry season" (July–October, when food stored from the previous year is in low supply) had lower survivorship than children born at other times of the year (see Figure 9.4). Their data also reveal large differences between the survivorship of people in Gambia and in the United States. For example, only 47%–62% (depending on season of birth) of Gambians survived to reach age 45, whereas more than 96% of U.S. females survived to that age. Viewed another way, the proportion of Gambians born in the hungry season that live to age 45 (47%) is roughly the same as the proportion of U.S. females that live to age 84 (50%).

There are three types of survivorship curves

Data from life tables, such as those for U.S. females and Gambian villagers, can be graphed as a **survivorship curve**. In such a curve, survivorship (l_x) data are used to plot the number of individuals from a hypothetical cohort (typically, of 1,000 individuals) that will survive to reach different ages. Results from studies on a variety of species suggest that survivorship curves can be classified into three general types (**Figure 9.5**).

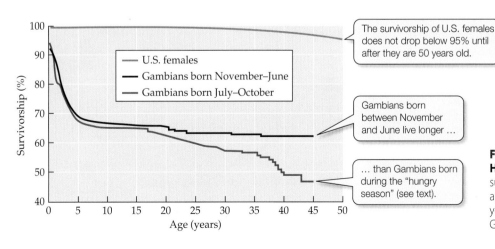

The survivorship of U.S. females does not drop below 95% until after they are 50 years old.

Gambians born between November and June live longer ...

... than Gambians born during the "hungry season" (see text).

FIGURE 9.4 Survivorship Varies among Human Populations In the United States, survivorship does not drop greatly until old age. In Gambia, many people die at much younger ages. (USA data from Arias 2004; Gambia data from S. E. Moore et al. 1997.)

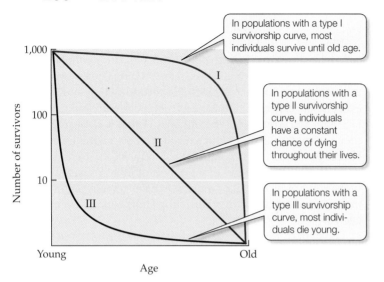

In populations with a type I survivorship curve, most individuals survive until old age.

In populations with a type II survivorship curve, individuals have a constant chance of dying throughout their lives.

In populations with a type III survivorship curve, most individuals die young.

FIGURE 9.5 Three Types of Survivorship Curves The *y* axis of a survivorship curve is usually plotted on a logarithmic scale, as has been done here.

ly, in populations with a **type III survivorship curve**, individuals die at very high rates when they are young, but those that reach adulthood survive well later in life. Type III survivorship curves—the most common type observed in the field—are often found in species that produce large numbers of young. Examples include oysters, marine corals, most known insects, and many plants, including the desert shrub *Cleome droserifolia* (**Figure 9.6C**), in which a population of 2,000 adults produces some 20 million seeds each year (roughly 10,000 seeds per adult), but only about 800 of those seeds survive to become juvenile plants.

In populations with a **type I survivorship curve**, newborns, juveniles, and young adults all have high survival rates; death rates do not begin to increase greatly until old age. Examples of populations with type I survivorship curves include U.S. females (see Figure 9.4) and Dall mountain sheep (**Figure 9.6A**). In populations with a **type II survivorship curve**, individuals experience a constant chance of surviving from one age to the next throughout their lives. Some bird species have a type II survivorship curve (**Figure 9.6B**), as do mud turtles (after their second year), some fish, and some plant species. Final-

FIGURE 9.6 Species with Type I, II, and III Survivorship Curves Survivorship curves in (A) Dall mountain sheep (*Ovis dalli*), (B) the song thrush, (*Turdus ericetorum*), and (C) the desert shrub *Cleome droserifolia*. (A, B after Deevey 1947; C after Hegazy 1990.)

(B) Type II

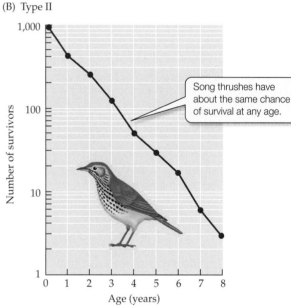

Song thrushes have about the same chance of survival at any age.

(A) Type I

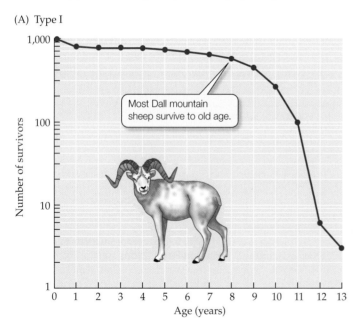

Most Dall mountain sheep survive to old age.

(C) Type III

Out of a million *Cleome droserifolia* seeds, only 39 survive to produce 1-year-old plants.

So far, we have discussed survivorship curves as if they were constant for each species, but that is not necessarily the case. Survivorship curves can vary among populations of a species, between males and females in a population, and among cohorts of a population that experience different environmental conditions (see Figure 9.4). In fact, by comparing birth and death rates in groups of individuals that experience different conditions, we can assess the effects of those conditions on populations. As we'll see in the next section, we can also use birth and death rates to predict how the size and composition of a population will change over time.

Age Structure

> **CONCEPT 9.2**
> Life table data can be used to project the future age structure, size, and growth rate of a population.

Members of a population whose ages fall within a specified range are said to be part of the same *age class*. Age class 1, for instance, might include all individuals who are at least 1 year old, but who are not yet 2 years old. Once individuals have been categorized in this way, a population can be described by its **age structure**, the proportions of the population in each age class. Imagine a population of a hypothetical organism in which all members die before they reach 3 years of age. In this population, every individual will be 0 ("newborns," which includes all individuals less than 1 year old), 1, or 2 years old. If there are 100 individuals in the population, and 20 are newborns, 30 are 1-year-olds, and 50 are 2-year-olds, then the age structure would be 0.2 in age class 0, 0.3 in age class 1, and 0.5 in age class 2. Age structure is a key feature of a population, in part because it influences whether a population increases or decreases in size.

Age structure influences how rapidly populations grow

Imagine two human populations of the same size and with the same survival and fecundity rates, but with different age structures. If one of the populations had many people older than 55, while the other had many people between ages 15 and 30, we would expect the second population to grow more rapidly than the first because it contains more individuals of reproductive age. Conversely, human populations that are growing rapidly typically have a greater percentage of people in younger age classes than do populations that are growing slowly or are in decline (**Figure 9.7**). In general, age structure influences how rapidly any population grows—at least initially, as we'll see next by examining a hypothetical life table in some detail.

Age structure and population size can be predicted from life table data

Table 9.3 shows survival and fecundity rates for a population of a hypothetical organism that reproduces in the spring and dies before it reaches 3 years of age. How will the age structure and size of this population change over time?

To answer these questions, we'll represent the number of individuals in age class 0 (the newborns) by n_0, the number in age class 1 (the 1-year-olds) by n_1, and the number in age class 2 (the 2-year-olds) by n_2. Assume that our population begins with a total of 100 individuals, of which 20 are in age class 0 ($n_0 = 20$), 30 are in age class 1 ($n_1 = 30$), and 50 are in age class 2 ($n_2 = 50$). Assume further that all mortality occurs over the winter, before the next spring breeding season, and that individuals are counted immediately after the breeding season. We can now use the information in Table 9.3 to pre-

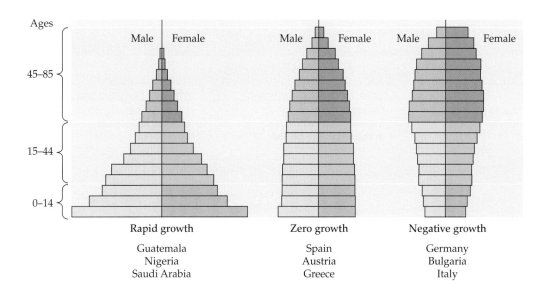

FIGURE 9.7 Age Structure Influences Growth Rate in Human Populations These diagrams show the general shapes of age structures that are typical of human populations with rapid, zero or nearly zero, and negative annual growth rates. Colors indicate reproductive status, with the main reproductive ages (15–44) shown in green. (After Miller 2007.)

TABLE 9.3

Life Table for a Hypothetical Organism*

Age x	Survival rate S_x	Survivorship l_x	Fecundity F_x
0	0.3	1.00	0
1	0.8	0.30	2
2	0.0	0.24	4
3	—	0.0	—

*The organism reproduces in the spring and dies before it is 3 years old.

dict how many individuals our population will have in the following year.

To do this, we must calculate two things: (1) the number of individuals that will survive to the next time period (in this case, to the next year's breeding season), and (2) the number of newborns those survivors will produce in the next time period. To calculate the number of individuals that will survive to the next time period, we multiply the number of individuals in each age class by the survival rate for that age class (**Table 9.4**). Thus, to determine the number of 0-year-olds that will survive to be 1-year-olds in the following year, we multiply $n_0 = 20$ by $S_0 = 0.3$ (see Table 9.3) to get 6 1-year-olds. Similarly, to determine the number of 1-year-olds that will survive to be 2-year-olds in the following year, we multiply $n_1 = 30$ by $S_1 = 0.8$ to get 24 2-year-olds. Finally, to determine the number of newborns in the following year, we note that, on average, each 1-year-old has 2 offspring and each 2-year-old has 4 offspring (see Table 9.3). Since there will be 6 1-year-olds and 24 2-year-olds in the next year, the number of offspring produced at that time will be (6 × 2 + 24

× 4) = 108. Thus, in the following year, the total population size (N) will be N = 138 (108 newborns + 6 1-year-olds + 24 2-year-olds).

Now that we have predicted how the age structure and size of our hypothetical population will change in one year, we can extend those calculations to future years. Just as we used the data in Table 9.3 to predict that the population would increase from 100 in the first year (at time $t = 0$) to 138 individuals in the next year (at time $t = 1$), we can perform calculations similar to those in Table 9.4 to determine how many individuals will be in each age class at time $t = 2$, then time $t = 3$, then time $t = 4$, and so on. We show the results of such calculations in **Figure 9.8**.

In Figure 9.8A, the numbers of individuals in each age class and in the population as a whole are plotted over time. Examining the curve for total population size (N), we see that with one exception ($t = 2$), the population rises steadily from its initial value of $N = 100$, reaching 1,361 individuals at time $t = 10$. By time $t = 4$, the numbers of individuals in each age class also rise steadily. In the first few years of population growth, however, the numbers of individuals in the different age classes vary considerably. For example, from its initial value of 50, n_2 drops to 24 individuals at time $t = 1$ and just 5 individuals at time $t = 2$; n_2 then rises substantially, reaching 136 individuals by time $t = 10$.

By time $t = 8$, the four curves in Figure 9.8A are roughly parallel to one another, indicating that all three age classes—and the total population size, N—are increasing at nearly the same rate from one year to the next. To examine this yearly rate of increase further, we can calculate the ratio of the population size in year $t + 1$, denoted N_{t+1}, to the population size in year t, denoted N_t. This ratio, which provides a measure of the yearly population growth rate, is designated by the Greek symbol λ (pronounced "lambda"):

$$\lambda = \frac{N_{t+1}}{N_t}$$

TABLE 9.4

A Two-Step Method for Projecting the Size of the Hypothetical Population in Table 9.3

(A)

(B)

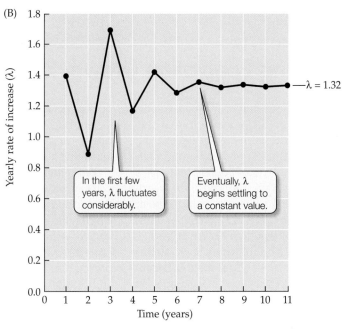

FIGURE 9.8 Growth of a Hypothetical Population The procedure described in Table 9.4 was used to calculate the growth of a hypothetical population that has the life table data shown in Table 9.3. These graphs plot (A) the number of individuals in each of the three age classes (n_0, n_1, n_2), as well as the total population size (N), at different times, and (B) the yearly rate of increase (λ) for the total population. Note that the number of individuals in (A) is plotted on a log scale.

Thus, to determine, for example, how the population will change in size from time $t = 0$ to time $t = 1$, we take the ratio of $N_1 = 138$ to $N_0 = 100$, which gives a yearly population growth rate of $\lambda = 1.38$.

As shown in Figure 9.8B, the values for the growth rate λ fluctuate over the first few years, ranging from a low of $\lambda = 0.88$ at $t = 2$ to a high of $\lambda = 1.69$ at $t = 3$. Eventually, λ settles to a constant value of $\lambda = 1.32$; if we were to calculate population sizes for additional time periods, we would find that λ remained equal to 1.32 from time $t = 10$ forward. In addition, if we were to calculate λ for any of the age classes (e.g., calculate the ratio of the size of age class 1 (n_1) at time $t + 1$ to its size at time t), we would find that by time $t = 10$, λ would equal 1.32 for each age class.

Populations grow at fixed rates when age-specific birth and death rates do not change over time

In our calculations in the previous section, we assumed that the age-specific survival and fecundity rates given in Table 9.3 were constant over time. Whenever this is true, the popula-

tion ultimately grows at a fixed rate from one year to the next. Our hypothetical population eventually grew at the fixed rate of $\lambda = 1.32$, as did each of its age classes. Because the population and each of its age classes increase by a constant multiplier every year, the proportion of individuals in each age class remains constant as long as the multiplier (λ) remains constant. When the age structure of a population does not change from one year to the next, the population is said to have a **stable age distribution**. In our example, the stable age distribution was 0.73 in age class 0, 0.17 in age class 1, and 0.10 in age class 2.

It is important to realize that if the survival and fecundity rates were different from those in Table 9.3, we would obtain different values for both the population growth rate (λ) and the stable age distribution. To illustrate this point, we'll change the average number of offspring produced by 1-year-olds (F_1) from 2.0 to 5.07, but keep all other S_x and F_x values equal to those in Table 9.3. With this single change, calculations like those in Table 9.4 show that λ changes from 1.32 to 2.0. This change also alters the stable age distribution from 0.73 in age class 0, 0.17 in age class 1, and 0.10 in age class 2 to 0.83 in age class 0, 0.12 in age class 1, and 0.05 in age class 2.

As our change to F_1 illustrates, any factor that alters the survival or fecundity of individuals can change the population growth rate. In applied settings, ecologists and managers often seek to identify the age-specific birth and death rates that most strongly influence the population growth rate. Identification of such rates can help in the development of management practices that decrease the size of a pest population or increase the size of an endangered population (**Box 9.1**).

BOX 9.1

Estimating Population Growth Rates in a Threatened Species

Loggerhead sea turtles (*Caretta caretta*; **Figure A**) are large marine turtles that lay eggs in nests that adult females dig into sandy beaches. Newly hatched baby turtles weigh just 20 g (0.04 pounds) and have a shell length of 4.5 cm (1.8 inches). They reach adulthood after 20–30 years, at which point they can weigh up to 227 kg (500 pounds) and have a shell length of 122 cm (4 feet).

Loggerhead sea turtles have been listed as a threatened species under the U.S. Endangered Species Act since 1978. Many species eat loggerhead eggs or hatchlings, and the juveniles and adults are eaten by large marine predators such as tiger sharks and orcas. Loggerheads also face threats from people, including the destruction of nesting sites by development as well as commercial fisheries (in whose nets sea turtles can become trapped and drown).

Early efforts to protect loggerhead sea turtles focused on the egg and hatchling stages, which suffer extensive mortality and are relatively easy to protect. To evaluate this approach, Crouse et al. (1987) and Crowder et al. (1994) used life table data to determine how population growth rates would change if new management practices improved the survival and fecundity rates of turtles of various ages. Their findings suggested that even if hatchling survival were increased to 100%, loggerhead populations would continue to decline. Instead, they found that the population growth rate was most responsive to decreasing the mortality of older juveniles and adults.

The results obtained by Crouse, Crowder and colleagues prompted the enactment of laws requiring Turtle Excluder Devices (TEDs) to be installed in shrimp nets (**Figure B**). TEDs function as a hatch through which juvenile and adult sea turtles can escape when caught in a net. Shrimp nets were singled out because the data suggested that shrimping accounted for more loggerhead deaths (from 5,000 to 50,000 deaths/year) than all other human activities combined.

Loggerheads are most easily counted when they nest, yet it takes 20–30 years for turtles to become sexually mature. As a result, it will be decades before we know whether TED regulations help turtle populations to increase in size. But early results are encouraging: the number of turtles killed in nets declined by about 44% after the TED regulations were implemented.

FIGURE B Turtle Excluder Device (TED)

FIGURE A Loggerhead Sea Turtle

Exponential Growth

CONCEPT 9.3
Populations can grow exponentially when conditions are favorable, but exponential growth cannot continue indefinitely.

Some species of bacteria, such as the mammalian gut mutualist *E. coli*, can double in number every 20–30 minutes when provided with abundant food. Although bacteria reproduce with phenomenal rapidity, many other organisms also can produce large numbers of offspring, as illustrated by giant puffballs and the desert shrub *Cleome droserifolia*. In such cases, if even a small fraction of those offspring survive to reproduce, the population can increase in size very quickly. Even populations of people and other organisms that produce relatively few young can grow rapidly. In general, populations can grow rapidly whenever individuals leave an average of more than one offspring over substantial periods of time. In this section, we describe geometric growth and exponential growth, two related patterns of population growth that can lead to rapid increases in population size.

Populations grow geometrically when reproduction occurs at regular time intervals

Some species, such as annual plants (and our hypothetical species in the previous section), reproduce in synchrony at regular time intervals. These regular time intervals are referred to as *discrete time periods*. If a population of such a species changes in size by a constant proportion from one discrete time period to the next, **geometric growth** is said to occur. The fact that the population changes in size by a constant proportion means that the number of individuals added to the population can become larger with each time period. As a result, the population can grow in size by ever-increasing amounts. When plotted on a graph, this growth pattern forms a J-shaped set of points (**Figure 9.9A**).

Mathematically, we can describe geometric growth as

$$N_{t+1} = \lambda N_t \qquad (9.1)$$

where N_t is the population size after t generations or, equivalently, after t discrete time periods (e.g., t years if there is one generation per year), and λ is any number greater than zero. In Equation 9.1, λ serves as a multiplier that allows us to predict the size of the population in the next time period. We'll refer to λ as the **geometric population growth rate**; λ is also known as the (per capita) **finite rate of increase**. We use this terminology by convention, but it can be confusing: we can see from Equation 9.1 that when the population "growth" rate λ is between 0 and 1, the population does not grow, but rather decreases in size over time.

Geometric growth can also be represented by a second equation,

$$N_t = \lambda^t N_0 \qquad (9.2)$$

where N_0 is the initial population size.

The two equations for geometric growth (Equations 9.1 and 9.2) are equivalent in that each can be derived from the other (see **Web Extension 9.1**). Which one you use depends on what you are interested in. If you want to predict the population size in the next time period, and you know λ and the current population size, you can use Equation 9.1. If you know the population size in both the current and previous time periods, you can also rearrange Equation 9.1 to get an estimate for λ ($\lambda = N_{t+1}/N_t$). Finally, you can use Equation 9.2 to predict the size of the population after any number of discrete time periods. If $\lambda = 2$, for example, then after 12 time periods, a population that begins with $N_0 = 10$ individuals will have $N_{12} = 2^{12}N_0$ individuals, which (as you can determine by using a calculator with a y^x function) equals $4{,}096 \times 10$, or $40{,}960$.

Populations grow exponentially when reproduction occurs continuously

In contrast to the pattern described in the previous section, individuals in many species—including people—do not reproduce in synchrony at discrete time periods. When individuals within a species reproduce at varying points in time, generations can overlap, and the species is said to reproduce

(A)

(B)

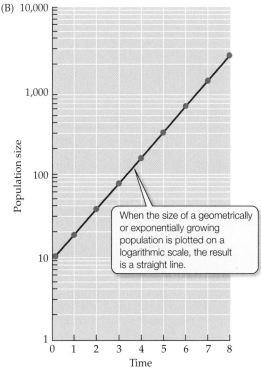

FIGURE 9.9 Geometric and Exponential Growth (A) The solid circles plot the size of a geometrically growing population that begins with 10 individuals and doubles in each discrete time period (i.e., $N_0 = 10$ and $\lambda = 2$). The solid curve plots exponential growth in a comparable population that reproduces continuously, also beginning with 10 individuals and having a growth rate of $r = \ln(2) = 0.69$. (B) When the population sizes represented by the solid circles and solid curve in A are plotted on a logarithmic scale, the result is a straight line.

continuously. When a population of a species with continuous reproduction changes in size by a constant proportion at each instant in time, we refer to the growth that results as **exponential growth**. Mathematically, exponential growth can be described by the following two equations:

$$\frac{dN}{dt} = rN \qquad (9.3)$$

and

$$N(t) = N(0)e^{rt} \qquad (9.4)$$

where $N(t)$ is the population size at each moment in time, t.

In Equation 9.3, dN/dt represents the rate of change in population size at each instant in time; we see from the equation that dN/dt equals a constant rate (r) multiplied by the current population size, N. Thus, the multiplier r provides a measure of how rapidly a population can grow; r is called the **exponential population growth rate** or the (per capita) **intrinsic rate of increase**.

As we did for Equation 9.2, we can use Equation 9.4 to predict the size of an exponentially growing population at any time t, provided we have an estimate for r and know $N(0)$, the initial population size. The "e" in Equation 9.4 is a constant, approximately equal to 2.718 ["e" is the base of the natural logarithm, $\ln(x)$]. We can calculate e^{rt} using the function e^x, which can be found on many calculators.

When plotted on a graph, the exponential growth pattern, like the geometric growth pattern, forms a J-shaped curve. Exponential growth and geometric growth are similar in that we can draw an exponential growth curve through the discrete points of a population that grows geometrically (see Figure 9.9A). Because exponential and geometric growth curves overlap, both types of growth are sometimes lumped together for simplicity and referred to as "exponential growth."

Geometric and exponential growth curves overlap because Equations 9.2 and 9.4 are similar in form, except that λ in Equation 9.2 is replaced by e^r in Equation 9.4. Thus, if we want to compare the results of discrete and continuous time growth models, we can calculate λ from r, or vice versa:

$$\lambda = e^r$$

$$r = \ln(\lambda)$$

where $\ln(\lambda)$ is the natural logarithm of λ, or $\log_e(\lambda)$. For example, if $\lambda = 2$ (as in Figure 9.9A), an equivalent value for r would be $r = \ln(2)$, which is approximately 0.69. **Figure 9.9B** illustrates a simple way to determine whether a population really is growing geometrically (or exponentially): plot the natural logarithm of population size versus time, and if the result is a straight line, the population is increasing by geometric or exponential growth.

Finally, look at Equations 9.1 and 9.3 for a moment. In Equation 9.1, which value of λ will ensure that the population does not change in size from one time period to the next? Similarly, in Equation 9.3, which value of r causes the population to remain fixed in size? The answers are $\lambda = 1$ (because then $N_{t+1} = N_t$) and $r = 0$ (because then the rate at which the population size changes is 0). When $\lambda < 1$ (or $r < 0$), the population will decline to extinction, whereas when $\lambda > 1$ (or $r > 0$), the population will increase exponentially (or geometrically) to form a J-shaped curve (**Figure 9.10**).

How can we estimate a population's growth rate (r or λ)? There are a variety of methods for doing this (see Caswell 2001), one of which we discussed on pp. 202–203: use life table data to predict future population sizes, graph those predicted population sizes versus time, and estimate the growth rate (λ) from the graph. Ecologists often wish to estimate λ (or r) from life table data, since they can then determine how fast a population is growing. Life table data can also be used to calculate two other measures of population growth, the doubling time and the net reproductive rate.

Doubling times and net reproductive rates provide useful measures of population growth

The **doubling time** (t_d) of a population refers to the number of years it will take the population to double in size. As interested readers can confirm [by solving Equation 9.4 for the time it takes a population to increase from its initial size, $N(0)$, to $2N(0)$], doubling times can be estimated as

$$t_d = \ln(2)/r \qquad (9.5)$$

where r is the exponential growth rate.

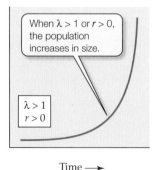

When $\lambda < 1$ or $r < 0$, the population decreases in size.

$\lambda < 1$
$r < 0$

When $\lambda = 1$ or $r = 0$, the population size does not change.

$\lambda = 1$
$r = 0$

When $\lambda > 1$ or $r > 0$, the population increases in size.

$\lambda > 1$
$r > 0$

Population size →   Time →   Time →   Time →

FIGURE 9.10 How Population Growth Rates Affect Population Size Depending on the value of λ or r, a population will decrease in size, remain the same size, or increase in size.

As we've seen, r can be estimated from life table data. We can also use life table data (broken down by age class) to calculate the **net reproductive rate** (R_0), the mean number of offspring produced by an individual during its lifetime. R_0 is calculated as

$$R_0 = \sum_{x_{\text{first}}}^{x_{\text{last}}} l_x F_x \qquad (9.6)$$

where x is age, x_{first} is the age of first reproduction, x_{last} is the age of last reproduction, and l_x and F_x are survivorship and fecundity, respectively, as defined in Table 9.1. Note that to estimate R_0, we multiply by l_x because the likelihood of surviving to each reproductive age is just as important as the number of offspring produced at that age (F_x). To check your understanding of Equation 9.6, use Table 9.1 to calculate an estimate of R_0 for the grass *Poa annua*; your calculations should yield $R_0 = 845.9$.

Whenever R_0 is greater than 1, measured from one generation to the next, λ will be greater than 1 (and $r > 0$). Under these conditions, populations have the potential to increase greatly in size, as we'll see in the next section.

Populations can grow rapidly because they increase by multiplication

Equations 9.1 and 9.3 show that populations increase by multiplication, not addition: at each point in time, the population changes in size according to the multiplier λ or r. As a result, populations have the potential to add large numbers of individuals rapidly whenever $\lambda > 1$ or $r > 0$. The principle at work here is the same one that applies to interest on a savings account. Even when the interest rate is low, you can earn a lot of money each year if you have a large amount deposited in the bank, because savings, like populations, grow by multiplication. Similarly, the fact that populations grow by multiplication means that even a growth rate that appears to be small can cause the size of a population to increase rapidly.

Consider our own population. On p. 197, we stated that the current annual growth rate of the human population was 1.21%. Such a growth rate implies that $r = 0.0121$, a value that seems close to 0. If we set the year 2007 as time $t = 0$, we have $N(0) = 6.6$ billion, the size of the human population in year 2007. Plugging these values of r and $N(0)$ into Equation 9.4, we calculate that the population size 1 year later should be $N(1) = 6.6 \times e^{0.0121}$, which rounds to 6.68 billion people. Thus, in 2007, the human population was increasing by about 80 million people per year (6.68 billion − 6.6 billion = 0.08 billion = 80 million). Since populations grow by multiplication, if r remained constant at 0.0121 for an extended period of time, the yearly increments to the human population would become astronomical. For example, after 210 years, there would be over 80 billion people, and our population would be increasing in size by more than a billion people *each year*.

Turning from humans to other species, what do field data reveal about the growth rates of their populations? Some species, such as the woodland herb *Asarum canadense* (wild ginger), have maximum observed values of λ that are close to 1 ($\lambda = 1.01$ in young forests, $\lambda = 1.1$ in mature forests) (**Figure 9.11**). Similar values were observed for a population of 25 reindeer introduced to Saint Paul Island off the coast of Alaska in 1911. After 27 years, the population had increased from 25 to 2,046 individuals, which (when we solve for λ in Equation 9.2) yields $\lambda = 1.18$.

Considerably higher annual population growth rates have been observed in many species, including western grey kangaroos ($\lambda = 1.9$), field voles ($\lambda = 24$), and rice weevils ($\lambda = 10^{17}$), insect pests of rice and other grains. Some bacteria, such as *E. coli*, can double in number every 30 minutes, resulting in the unimaginably high annual population growth rate of $\lambda = 10^{5,274}$.

Recall that when $\lambda > 1$ (or $r > 0$) for an extended period of time, populations will increase exponentially in size, forming a J-shaped curve like that in Figure 9.9A. In natural populations, $\lambda > 1$ (or $r > 0$) when the temperature, precipitation, available nutrients, and other key resources are favorable for growth, survival, and reproduction. But can such favorable conditions last for long?

There are limits to the growth of populations

An argument from basic principles suggests that the answer to the question we just posed is "No." Physicists estimate that the known universe contains a total of 10^{80} atoms. Yet if

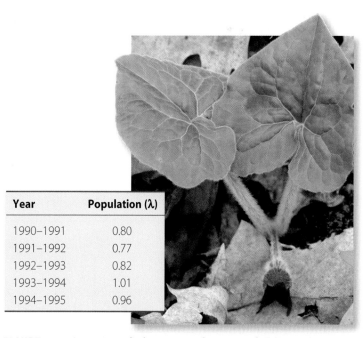

Year	Population (λ)
1990–1991	0.80
1991–1992	0.77
1992–1993	0.82
1993–1994	1.01
1994–1995	0.96

FIGURE 9.11 Some Populations Have Slow Growth Rates The growth rates of a population of wild ginger (*Asarum canadense*) in a young forest vary from year to year. Growth rates are often less than 1.0, suggesting that the population will decline in size unless conditions improve. (Data from Damman and Cain 1998.)

favorable conditions persisted for long periods of time, allowing λ to remain greater than 1, even populations of relatively slowly growing species would eventually increase to more than 10^{80} individuals. For example, based on *Asarum*'s population growth rate of λ = 1.01 in young forests, a population that began with 2 plants would have more than 10^{82} plants after 19,000 years. For an extremely rapidly growing species such as *E. coli*, the numbers are even more absurd: it would take only 6 days for a population that began with a single bacterium to exceed 10^{80} individuals.

No population could ever come close to having 10^{80} individuals because there would be no atoms with which to construct their bodies. Thus, exponential growth cannot continue indefinitely. There are limits to population growth, which cause it to slow and eventually stop. We'll look at some of those limits in the following section.

Effects of Density

CONCEPT 9.4
Population size can be determined by density-dependent and density-independent factors.

Although λ > 1 for all populations under ideal conditions (leading to exponential growth), conditions in nature are rarely ideal for long. For example, Damman and Cain (1998) calculated the geometric growth rate (λ) in each of 5 years for a population of the woodland herb *Asarum canadense* located in a young forest. As mentioned above, the maximum growth rate was λ = 1.01. During the other 4 years, however, values for λ ranged from 0.77 to 0.96 (see Figure 9.11). Thus, far from threatening to overrun the planet with its offspring, we would expect this population to decline in the long run, unless conditions changed for the better.

What factors cause λ to fluctuate over time? We can explore answers to this question by asking whether the population growth rate changes independently of density or as a function of density.

Density-independent factors can determine population size

In many species, year-to-year variation in weather leads to dramatic changes in abundance, and hence in population growth rates. For example, Davidson and Andrewartha (1948) studied how weather affected the insect *Thrips imaginis*. By correlating weather conditions with population sizes over a 14-year period, Davidson and Andrewartha showed that yearly fluctuations in population size could be predicted accurately by an equation that used temperature and rainfall data (**Figure 9.12**).

Factors such as temperature and precipitation, as well as catastrophic events such as floods or hurricanes, are often referred to as **density-independent** factors, meaning that their effects on birth and death rates are independent of the number of individuals in the population. In terms of population growth rates, per capita growth rates (λ or *r*) are density-independent when they are not a function of population density (**Figure 9.13**).

As the *Thrips imaginis* data suggest, density-independent factors can have major effects on population size from one year to the next. In principle, such factors could account entirely for year-to-year fluctuations in the size of a population. But density-independent factors do not tend to increase the size of populations when they are small and decrease the size of populations when they are large. A factor that did consistently lead to such changes would cause the per capita growth rate to change as a function of density—that is, to be density-*dependent*, not density-independent.

Density-dependent factors regulate population size

In many cases, the factors that influence population size are **density-dependent**, which means that they cause birth rates, death rates, and dispersal rates to change as the density of the population changes (see Figure 9.13). As densities increase, it is common for birth rates to decrease, death rates to increase, and dispersal from the population (emigration) to increase—all of which tend to decrease population size. When densities decrease, the opposite occurs: birth rates tend to increase and death and emigration rates decrease.

When one or more density-dependent factors cause population to increase when numbers are low and decrease when num-

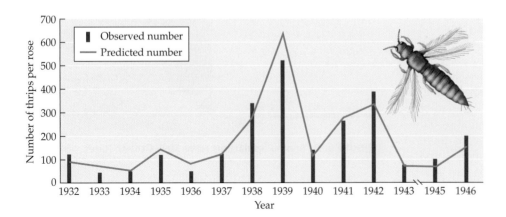

FIGURE 9.12 Weather Can Influence Population Size The observed mean number of thrips per rose was predicted accurately by an equation based on four weather-related variables (data collected in Adelaide Australia). (After Davidson and Andrewartha 1948.)

FIGURE 9.13 Comparing Density Dependence and Density Independence In the example of density dependence shown here, population growth rates decrease as the density increases; however, it is also possible for population growth rates to increase as the density increases.

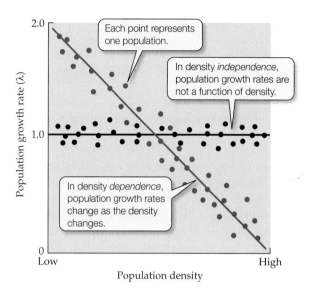

bers are high, **population regulation** is said to occur. Ultimately, when the density of any species becomes high enough, density-dependent factors decrease population size because food, space, or other essential resources are in short supply. Note that "regulation" has a particular meaning here, referring to the effects of factors that tend to increase λ or r when the population size is small and decrease λ or r when the population size is large. Density-independent factors can have large effects on population size, but they do not *regulate* population size because they do not consistently increase population size when it is small and decrease population size when it is large. Thus, by definition, only density-dependent factors can regulate population size.

Density dependence has been observed in many populations

Density dependence can often be detected in natural populations. For example, in a study that combined field observations with controlled experiments, Peter Arcese and James Smith (1988) examined the effect of population density on reproduction in the song sparrow (*Melospiza melodia*) on Mandarte Island, British Columbia. They found that the number of eggs laid per female decreased with density, as did the number of young that survived long enough to become independent of their parents (**Figure 9.14A**). Because

(A)

(B)

FIGURE 9.14 Examples of Density Dependence in Natural Populations (A) Numbers of young song sparrows reared to independence on Mandarte Island; the number next to each point indicates the year. (B) Death rates in soybeans at 93 days after planting at various densities. (C) Death rates in flour beetles at various egg densities. (A after Arcese and Smith 1988; B after Yoda et al. 1963; C after Bellows 1981.)

(C)

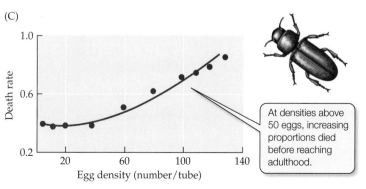

Mandarte Island is small and the birds were likely to suffer food shortages at high densities, Arcese and Smith predicted that if they provided food to a subset of nesting pairs when densities were high, the birds that were fed should be able to rear more young to independence. That is exactly what happened: nesting pairs that were fed produced nearly four times as many independent young as did control birds that were not fed.

In addition to the effect of density on reproduction, density-dependent mortality has been observed in many populations. For example, when Yoda et al. (1963) planted soybeans (*Glycine soja*) at various densities, they found that at the highest initial planting densities, many of the seedlings had died by 93 days of age (**Figure 9.14B**). Similarly, in an experiment where eggs of the flour beetle *Tribolium confusum* were placed in glass tubes (each with 0.5 g of food), death rates increased as the density of eggs increased—again revealing density dependence (**Figure 9.14C**).

In some cases, density dependence can be detected even in populations whose abundance is largely controlled by factors such as temperature or precipitation. For example, Frederick Smith (1961) used Davidson and Andrewartha's data (1948) to test whether changes in thrips abundance were density-dependent. To do this, Smith plotted the change in population size from one time period to the next versus the size of the population at the start of the time period. He predicted that if density-dependent factors were important, he would find a negative trend (that is, the increase in the thrips population would be large at low densities and small at high densities). In contrast, if density-de-

pendent factors were not important, there would be no relationship between the change in population size and the starting population size. Smith performed this analysis for a variety of time periods, typically obtaining a negative trend, which indicated that density-dependent factors were indeed important (**Figure 9.15**). Exactly what those factors are is unknown, but they may have included predators, parasites, or even weather. Density-dependent mortality can result from weather if, for example, there are a limited number of locations where individuals can take shelter during cold winters.

When birth, death or dispersal rates show strong density dependence, population growth rates (λ or r) may decline as densities increase. Such an effect was observed in populations of the grass *Poa annua* and the water flea *Daphnia pulex* (**Figure 9.16**). Eventually, if densities become high enough to cause λ to equal 1 (or r to equal 0), the popula-

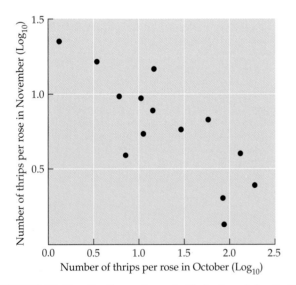

FIGURE 9.15 Density Dependence in *Thrips imaginis* Smith's reexamination of Davidson and Andrewartha's thrips data (1948) showed that the number of individuals added to the population decreased as the density of the population increased, as expected if density-dependent factors were important. (After Smith 1961.)

FIGURE 9.16 Population Growth Rates May Decline at High Densities (A) The geometric population growth rate (λ) of the grass *Poa annua* is density-dependent, as is (B) the exponential population growth rate (r) of the water flea *Daphnia pulex*. (A after Law 1975; B after Frank et al. 1957.)

tion stops growing entirely; if λ becomes less than 1 (or *r* < 0), the population declines. As we'll see in the next section, such density-dependent changes in the per capita growth rate can cause a population to reach a stable, maximum population size.

Logistic Growth

CONCEPT 9.5
The logistic equation incorporates limits to growth and shows how a population may stabilize at a maximum size, the carrying capacity.

In some cases, the growth of a population can be represented by an S-shaped curve (**Figure 9.17**). Such a population exhibits **logistic growth** in which its abundance increases rapidly at first, then stabilizes at a population size known as the **carrying capacity** (the maximum population size that can be supported indefinitely by the environment). The growth rate of the population decreases as the population size nears the carrying capacity because resources such as food, water, or space begin to be in short supply. At the carrying capacity, the growth rate is zero, so the population size does not change.

The logistic equation models density-dependent population growth

To see how the idea of a carrying capacity can be represented in a mathematical model of population growth, let's reconsider Figure 9.16. The data in that figure show that per capita growth rates (*r* or λ) decreased approximately as a straight line as population densities increased. In contrast, *r* is assumed to be constant in the exponential growth equation, $dN/dt = rN$. As we've seen, a constant value of $r > 0$ allows for unlimited growth in population size. To modify the exponential growth equation to make it more realistic, we replace the assumption that *r* is constant with the assumption that *r* declines in a straight line as density (*N*) increases. When we do this (see **Web Extension 9.2**), we obtain the *logistic equation*:

$$\frac{dN}{dt} = rN\left(1 - \frac{N}{K}\right) \qquad (9.7)$$

where *N* is population density, *r* is the (per capita) population growth rate under ideal conditions, and *K* is the density at which the population stops increasing in size. *K* can be interpreted as the carrying capacity of the population, while the term $(1 - N/K)$ can be viewed as representing the effect of factors that reduce the population growth rate from the constant rate (*r*) seen in exponential growth.

Logistic growth is similar to, but slightly slower than, exponential growth when densities are low (**Figure 9.18**). When *N* is small, the term $(1 - N/K)$ is close to 1, and hence

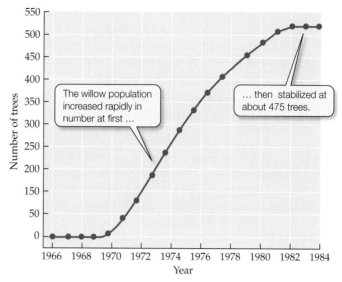

FIGURE 9.17 An S-shaped Growth Curve in a Natural Population At a site in Australia, heavy grazing by rabbits had prevented willows from colonizing the area. The rabbits were removed in 1954. When willows colonized the area in 1966, ecologists tracked the growth of their population. (After Alliende and Harper 1989.)

a population that grows logistically grows at a (per capita) rate close to *r*. As the population density increases, however, logistic and exponential growth differ greatly. In logistic growth, the rate at which the population changes in size (*dN/dt*) approaches zero as the population size nears the carrying capacity, *K*. As a result, over time, the population size approaches *K* gradually, eventually remaining constant in size with *K* individuals in the population.

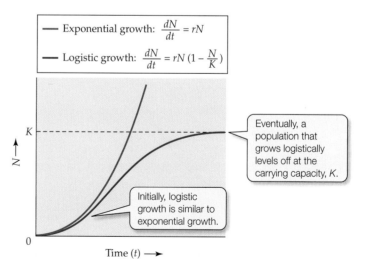

FIGURE 9.18 Logistic and Exponential Growth Compared Over time, logistic growth differs greatly from the unlimited growth of a population that increases exponentially.

In Chapter 10 we'll discuss the extent to which the growth of natural populations can be described by the S-shaped curve that results from the logistic equation; here, we examine efforts to fit the logistic equation to U.S. census data.

Can logistic growth predict the carrying capacity of the U.S. population?

In a groundbreaking paper published in 1920, Pearl and Reed examined the fit of several different mathematical models to U.S. census data for the period 1790–1910. Several of the approaches they tested did a good job of matching the historical data, but none included limits to the eventual size of the U.S. population. To address this shortcoming, they derived the logistic equation, which, unknown to them, had been first described in 1838 by the Belgian P. F. Verhulst. Pearl and Reed argued that the logistic equation provided a sensible way to represent population growth because it included limits to growth. When they fit the census data to the logistic curve, they obtained an excellent match, from which they estimated that the U.S. population had a carrying capacity of $K = 197,274,000$ people (**Figure 9.19**).

The logistic curve estimated by Pearl and Reed provides a good fit to U.S. population data through 1950. After that

time, however, the actual population size differed considerably from Pearl and Reed's projections. By 1967, the carrying capacity (197 million) they had predicted had been surpassed. Pearl and Reed intended their estimate of the carrying capacity to represent the number of people that could be supported in the United States in a self-sufficient manner. They recognized that if conditions changed—for example, if agricultural productivity increased or if more resources were imported from other countries—the population could increase beyond 197 million. These and other changes have occurred, leading some ecologists and demographers to shift their focus from the human carrying capacity to the total area (both within and outside a nation) required to support a human population (the "ecological footprint," discussed on p. 214). As we will see, the footprint approach turns the carrying capacity concept on its head: instead of asking how many people a given area can support, it asks how much area is required to support a given population.

Human Population Growth

Media reports often state that the human population is growing exponentially. As we saw in this chapter, a simple way to determine whether a population is growing exponentially is to plot the natural logarithm of population size versus time. If a straight line results, the population is growing exponentially. When we plot the natural logarithm of human population size versus time for the last 2,000 years, we see that our population sizes deviate considerably from the straight line expected in exponential growth (**Figure 9.20**). In fact, as fast as exponential growth is, the human population has increased even more rapidly than that.

The faster-than-exponential nature of human population growth is also evident from an examination of doubling times for the human population, as we saw in the Case Study. Recall that in a population that grows exponentially, the doubling time remains constant. The fact that our population's doubling time dropped from roughly 1,400 years in 5000 B.C. to a mere 45 years in 1930 indicates that our population increased more rapidly than expected in exponential growth.

Projecting into the future, we can predict how long it will take our current population to double in size. To do this, the doubling time is estimated from the relation $t_d = \ln(2)/r$ (see Equation 9.5), where r is the current growth rate of the human population. Such estimates have shown that the human population was doubling most rapidly in 1960, with a doubling time of 35 years. Since then, the doubling time has increased (because r has decreased), reaching 57 years in 2007. So, returning to the question we asked on p. 197 (whether there would be 16 billion people in 2080), the an-

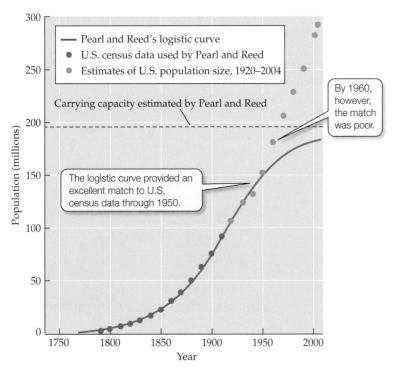

FIGURE 9.19 Fitting a Logistic Curve to the U.S. Population Size In 1920, Pearl and Reed fitted a logistic curve to U.S. census data from 1790 to 1910; the estimated carrying capacity (K) was 197 million people. (Data through 1910 from Pearl and Reed 1920; other data from Statistical Abstracts, U.S. Census Bureau.)

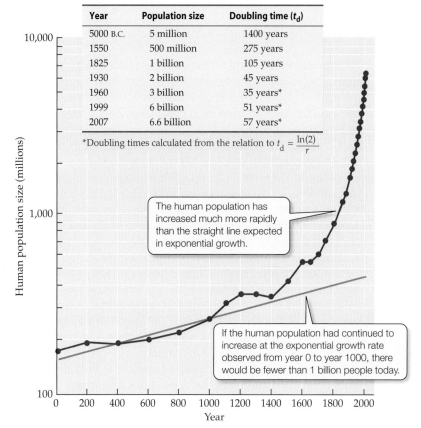

Year	Population size	Doubling time (t_d)
5000 B.C.	5 million	1400 years
1550	500 million	275 years
1825	1 billion	105 years
1930	2 billion	45 years
1960	3 billion	35 years*
1999	6 billion	51 years*
2007	6.6 billion	57 years*

*Doubling times calculated from the relation to $t_d = \dfrac{\ln(2)}{r}$

> The human population has increased much more rapidly than the straight line expected in exponential growth.

> If the human population had continued to increase at the exponential growth rate observed from year 0 to year 1000, there would be fewer than 1 billion people today.

FIGURE 9.20 Faster than Exponential   A plot of the logarithm of the human population size over the last 2,000 years differs dramatically from the straight line expected in exponential growth.

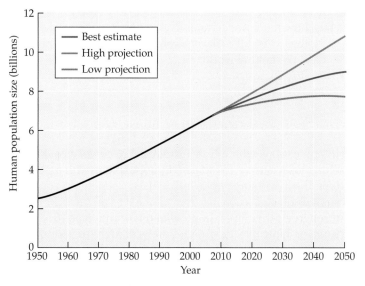

FIGURE 9.21 United Nations Projections of Human Population Size   The human population is expected to increase to 9 .2 billion by 2050; low and high projections range from 7.8 billion to 10.8 billion. (After Population Division of the United Nations 2007.)

swer is "Probably not." United Nations projections indicate that population growth rates are likely to continue to fall over the next 50 years, leading to a predicted population size of 9.2 billion in 2050 (**Figure 9.21**). Extending that curve out to 2080 suggests that there will be roughly 10 billion people in that year. If these projections turn out to be correct, or nearly so, what will the future be like with that many people? Is 10 billion above the carrying capacity of the human population?

To answer these questions, we must determine the carrying capacity of the human population, but that is trickier than it may first appear. Many researchers have estimated the human carrying capacity (see Cohen 1995). Their estimates range from fewer than 1 billion to more than 1,000 billion (**Figure 9.22**). This large variation is due in part to the fact that many different methods—from logistic models to calculations based on crop production and human energy requirements—have been used to estimate human carrying capacity. In addition, different researchers have made different assumptions about how people would live and how technology would influence our future, assumptions that have a large effect on the estimated carrying capacity.

For example, using the "ecological footprint" approach described below, it has been estimated that Earth could support about 5 billion people indefinitely based on the amount of resources used by an average person in 1999 (Wackernagel et al. 2002). However, if everyone used the amount of resources used by people in the United States in 1999, the world could support only 1.2 billion people. Toward the other end of the spectrum, if everyone used the amount of resources used by people in India in 1999, the world could support over 14 billion people. Thus, as we suggested in the introductory pages

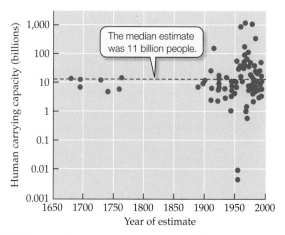

> The median estimate was 11 billion people.

FIGURE 9.22 The Human Carrying Capacity   Estimates of the carrying capacity of the human population, grouped by the dates when those estimates were made. (After Cohen 1995.)

of this chapter, issues of human population size and resource use are linked inextricably: more people means that more resources will be used, but the degree to which our growing population affects the environment depends on the amount of resources used by each person.

CONNECTIONS IN NATURE

Your Ecological Footprint

When you turn on a light, purchase an appliance, drive a car, or eat fruit imported from another country, you may not think about the effects your actions have on the natural world. How, for example, does driving to the store to get groceries affect forests or coral reefs?

To answer this question, we must account for the resources required to support our actions. The grains we eat require farmland; the wood products we use require natural forests or plantations; the fish we eat require productive fishing grounds; the machines and appliances we purchase require raw materials and energy to build, as well as energy for their operation. Ultimately, every aspect of our economy depends on the land and waters of Earth. Recognizing this, William Rees proposed that we measure the environmental impact of a population as its **ecological footprint**, which is the total area of productive ecosystems required to support that population (Rees 1992).

Ecological footprints are calculated from national statistics on agricultural productivity, production of goods, resource use, population size, and pollution. The area required to support these activities is then estimated, often in a direct way. For example, the land required to support wheat consumption in 1993 by people in Italy was estimated by dividing the amount of wheat consumed (26,087,912 tons) by the amount of wheat produced per unit of land (2.744 tons/ha), resulting in 9,507,257 hectares, or 0.167 hectares per person (Wackernagel et al. 1999).

Methods of calculating ecological footprints are still being refined, but early results are sobering. In 1999, there were 11.5 billion hectares of productive land available, and the ecological footprint of an average person was 2.3 hectares per person (Wackernagel et al. 2002). This result suggests that Earth could have supported 5 billion people (11.5 billion ha/2.3 ha per person) for a long time. In fact, the human population in 1999 was 6 billion, a 20% overshoot of its carrying capacity. An overshoot of this magnitude indicates that in 1999, environmental resources were being used more rapidly than they could be regenerated, a pattern of use that cannot be sustained. Similar calculations can be made for individual nations. In the United States, for example, the average ecological footprint was 9.7 hectares per person in 1999. Since there were 1,800 million hectares of productive land available in the nation, this calculation suggests that the carrying capacity of the United States in 1999 was 186 million people (1,800 million ha/9.7 ha per person); the actual population was 279 million, a 50% overshoot.

Human use of resources changes from year to year, depending on population size, per capita rates of resource use, and technology (i.e., the efficiency of production). In addition, the total area of productive ecosystems available to support our activities changes over time due to factors such as gain or loss of farmland, destruction of natural habitat, pollution, and extinction of species. As a result, our ecological footprint changes over time. People have now begun to use our changing footprint as a way to assess whether our current population size and resource use can be sustained for a long time. This approach highlights the fact that all of our actions—what we eat, how big a house we buy, how much we drive or fly, the goods we purchase (e.g., clothes, cars, MP3 players, cell phones)—depend on and affect the natural world.

SUMMARY

CONCEPT 9.1   Life tables show how survival and reproductive rates vary with age, size, or life cycle stage.

- Cohort life tables can be constructed from data on the fates of individuals born during the same time period and used to calculate age-specific survival, survivorship, and fecundity.

- In highly mobile or long-lived organisms, a static life table may be constructed from data on the survival and fecundity of individuals of different ages during a single time period.

- In species for which age correlates poorly with survival and fecundity, life tables based on size or life cycle stage may be constructed.

- In populations with a type I survivorship curve, most individuals survive to old age. In populations with a type II survivorship curve, individuals experience a constant chance of surviving from one age to the next throughout their lives. In populations with a type III survivorship curve, death rates are very high for young individuals, but adults survive well later in life. Of the three types, type III is the most common.

CONCEPT 9.2   Life table data can be used to project the future age structure, size, and growth rate of a population.

- The age structure of a population influences the growth rate of that population over time.

- Populations eventually grow at fixed rates when age-specific survival rates and fecundities do not change over time.
- Any factor that changes age-specific survival rates or fecundities may alter the population growth rate.

CONCEPT 9.3 Populations can grow exponentially when conditions are favorable, but exponential growth cannot continue indefinitely.

- Geometric growth occurs when a population of individuals that reproduce in synchrony at discrete time periods changes in size by a constant proportion from one discrete time period to the next.
- Exponential growth occurs when a population with continuous reproduction changes in size by a constant proportion at each instant in time.
- Populations have the potential to increase rapidly in size because they grow by multiplication, not by addition.
- All populations experience limits to growth, which ensure that exponential growth cannot continue indefinitely.

CONCEPT 9.4 Population size can be determined by density-dependent and density-independent factors.

- In many species, density-independent factors, such as temperature or precipitation, play a major role in determining year-to-year changes in population size.
- When the density of any species becomes high enough, a lack of food, space, or other resources causes birth rates to decrease, death rates to increase, or dispersal to increase.
- Population regulation occurs when one or more density-dependent factors tend to increase population size

when densities are low and decrease population size when densities are high.

CONCEPT 9.5 The logistic equation incorporates limits to growth and shows how a population may stabilize at a maximum size, the carrying capacity.

- In some species, changes in population size over time can be described by an S-shaped curve in which the population increases rapidly at first, then stabilizes at a maximum level, the carrying capacity.
- The logistic equation can be used to represent density-dependent population growth.
- Logistic population growth provides a close fit to the size of the U.S. population up to 1950; after that time, the growth rate of the U.S. population was considerably greater than expected in logistic growth.

CASE STUDY/CASE STUDY REVISITED
Human Population Growth

- Over the past 2,000 years, the human population has increased in size even more rapidly than would occur by exponential growth.
- Estimates of the carrying capacity of the human population vary widely, from fewer than 1 billion people to more than 1,000 billion people.
- The carrying capacity concept applies poorly to human populations that import resources from outside the area in which the population is found.
- Ecological footprint analyses based on available productive land area and current patterns of resource use suggest that the global human population is 20% greater than the maximum number that could be sustained for a long time.

REVIEW QUESTIONS

1. For a field ecology project, you count the number of individuals of different ages found in the population during a single time period. There are 100 newborns, 40 1-year-olds, 15 2-year-olds, 5 3-year-olds, and 0 4-year-olds.

   a. Use these data to fill in the N_x, S_x, and l_x columns of a static life table.

   b. Explain the difference between a static life table and a cohort life table.

2. Age class (x)

Age class (x)	l_x	S_x	F_x
0	1.0	1/3	0
1	0.333	1/2	3
2	0.167	0	2
3	0		0

   a. A population that grows according to the life table given in this problem, initially has 50 0-year-olds, 50 1-year-olds, and 50 2-year-olds (i.e., at time 0, N_0 = 50, N_1 = 50, and N_2 = 50). Calculate how the age distribution changes from time t = 0 to time t = 6. What is your best estimate or the growth rate (λ) and stable age distribution of this population?

   b. A second population also grows according to the life table given in this problem, but has a different initial age distribution: N_0 = 80, N_1 = 50, N_2 = 20. Calculate how the age distribution changes from t = 0 to t = 6, and estimate the growth rate (λ) and stable age distribution of this population.

   c. Compare your answers to parts (a) and (b). Without performing further calculations, estimate the growth rate (λ) and stable age distribution of a third population that also grows according to the life table in this problem, but has an initial age distribution of $N_0 = 10$, $N_1 = 50$, $N_2 = 90$.

3. A population of insects triples every year. Initially, there were 40 insects.

   a. How many insects would there be after 4 years?

   b. How many insects would there be after 27 years? (Write your answer to this question as an equation.)

c. The habitat of the insect is degraded such that the population growth rate (λ) changes from 3.0 to 0.75. If there were 100 insects in the population when its habitat became degraded, how many insects would there be after 3 years?

4. What is the distinction between factors that regulate population size and factors that determine population size?

5. Calculate your ecological footprint at http://www.myfootprint.org/.

HANDS-ON PROBLEM SOLVING

Density-dependent and -independent Factors

This Web exercise (www.sinauer.com/ecology) explores the effects of density-dependent and density-independent factors on populations. You will analyze data from populations of arctic ground squirrels to determine these effects on weaning success, over-wintering survival, and population growth rate.

SUGGESTED READINGS

Cappuccino, N. and P. W. Price, eds. 1995. *Population Dynamics: New Approaches and Synthesis*. Academic Press, New York. *The collection of papers in this edited volume provides an excellent, in-depth overview of many of the topics covered in this chapter.*

Chambers, N., C. Simmons and M. Wackernagel. 2001. *Sharing Nature's Interest: Ecological Footprints as an Indicator of Sustainability*. Earthscan Publications Ltd, London. *A comprehensive description of how to determine the ecological footprint of a population, and what we can learn from making such calculations.*

Cohen, J. E. 1995. *How Many People Can The Earth Support?* W. W. Norton, New York. *A definitive work that addresses two of the key issues of our time: the growth and the carrying capacity of the human population.*

Deevey, E. S. 1947. Life tables for natural populations. *Quarterly Review of Biology* 22: 283–314. *An important early paper that illustrated for ecologists how field data could be used in demographic analyses (including life table analyses), thereby enriching our understanding of studies on the abundance of natural populations.*

Lack, D. 1954. *The Natural Regulation of Animal Numbers*. Oxford University Press, London. *This seminal book addresses a central theme of both this chapter and much of population ecology: what regulates the numbers of individuals found in populations?*

Wilson, E. O. and W. H. Bossert. 1971. *A Primer of Population Biology*. Sinauer Associates, Sunderland, MA. *This short, classic text uses concrete examples and step-by-step solutions to problems to develop a clear understanding of life table analyses and models of population growth.*

CHAPTER 10 Population Dynamics

CASE STUDY

A Sea in Trouble

In the 1980s, the comb jelly *Mnemiopsis leidyi* (**Figure 10.1**) was introduced accidentally into the Black Sea, most likely by the discharge of ballast water from cargo ships. The timing of this invasion could hardly have been worse. At that time, the Black Sea ecosystem was already in decline due to increased nutrient inputs from the Danube, Dnieper, and Dniester rivers. The increased supply of nutrients had devastating effects across the northern part of the Black Sea, where the waters are shallow (less than 200 m) and prone to problems that stem from *eutrophication* (the addition of nutrients to water). As nutrient concentrations increased in these shallow waters, phytoplankton abundance increased, water clarity decreased, oxygen concentrations dropped, and fish populations experienced massive die-offs. Nutrient concentrations in deeper portions of the Black Sea also rose, causing increased phytoplankton abundance, but not fish die-offs.

Such was the situation when *Mnemiopsis* arrived. This marine invertebrate species is a voracious predator of zooplankton, fish eggs, and young fish. Furthermore, *Mnemiopsis* continues to feed even when it is completely full, causing it to regurgitate large quantities of prey stuck in balls of mucus. Small prey encased in mucus survive poorly. As a result, the negative effect of *Mnemiopsis* on its prey outstrips even its considerable ability to digest food.

When well fed, an individual *Mnemiopsis* can produce up to 8,000 offspring just 13 days after its own birth. Following its arrival in the Black Sea in the early 1980s, *Mnemiopsis* gradually increased in numbers. Then, in 1989, *Mnemiopsis* populations exploded in size (**Figure 10.2**), reaching astonishing biomass levels (1.5–2.0 kg/m^2) throughout the entire Black Sea. The total biomass of *Mnemiopsis* in the Black Sea was estimated at 800 million tons (live weight) in 1989, far greater than the world's entire annual commercial fish catch, which has never exceeded 95 million tons.

The enormous numbers of *Mnemiopsis* present in 1989, and again in 1990, compounded the effects of the Black Sea's ongoing nutrient enrichment problem. *Mnemiopsis* ate huge quantities of zooplankton, causing their populations to crash. Zooplankton eat phytoplankton, so *Mnemiopsis* indirectly caused phytoplankton populations to increase even more than they already had due to nutrient enrichment. Upon their deaths, the large numbers of phytoplankton and *Mnemiopsis* provided food for bacterial decomposers. The bacteria used oxygen to decompose their bodies, so as bacterial activity increased, oxygen levels decreased, harming some fish populations. In addition, by devouring the food supplies (zooplankton), eggs, and young of important commercial fishes such as anchovies, the explosion in *Mnemiopsis* numbers led to a rapid decline in fish catches

FIGURE 10.1 A Potent Invader The comb jelly *Mnemiopsis leidyi* spread from ocean waters along the east coast of North America to the Black Sea, wreaking havoc to the new ecosystem upon its arrival.

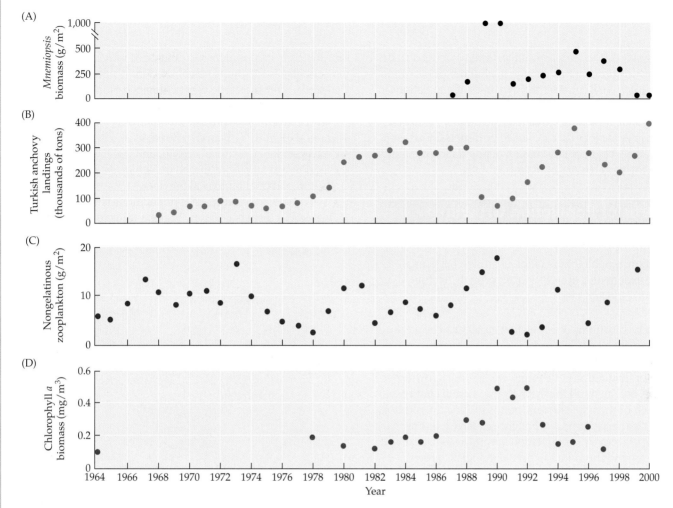

FIGURE 10.2 Changes in the Black Sea Ecosystem The graphs track long-term changes in four components of the Black Sea ecosystem: (A) mean biomass of the invasive species *Mnemiopsis leidyi* (first measured in 1987), (B) Turkish anchovy landings (Turkish fishermen have garnered most of the Black Sea anchovy catch since 1980), (C) mean biomass of nongelatinous zooplankton, and (D) mean biomass of chlorophyll *a* (an indicator of phytoplankton abundance). (After Kideys 2002.)

(see Figure 10.2B), causing enormous damage to the Turkish fishing industry.

The combined negative effects of nutrient enrichment and invasion by *Mnemiopsis* posed a serious threat to the Black Sea ecosystem. Although it covers a large surface area (over 423,000 km²), the Black Sea is nearly land-locked and exchanges little of its water each year with other ocean waters. In addition, the Black Sea is unusual in that only the top 150–200 m of its waters (ca. 10% of its average depth) contain oxygen, effectively making the entire sea "shallow" for species that require oxygen. Its limited water exchange and anoxic deep waters make the Black Sea particularly vulnerable to the negative effects of nutrient enrichment.

Native predators and parasites had failed to regulate *Mnemiopsis* populations. Thus, in the early 1990s, the future of the Black Sea ecosystem looked bleak. Fortunately, the story is different today: *Mnemiopsis* and phytoplankton populations have fallen, paving the way for the recovery of the Black Sea. How did this happen?

As we saw in the previous two chapters, populations can change in size as a result of four processes: birth, death, immigration, and emigration. We can summarize the effects of these four processes on population size with the following equation:

$$N_{t+1} = N_t + B - D + I - E$$

where N_t is the population size at time t, B is the number of births, D is the number of deaths, I is the number of immigrants, and E is the number of emigrants between time t and time $t + 1$. As implied by this equation, populations are open and dynamic entities. Individuals can move from one population to another, and the number of individuals in a population can change from one time period to the next.

Ecologists use the term *population dynamics* to refer to the ways in which populations change in abundance over time. In this chapter, we'll consider the dynamics of populations in more detail, placing special emphasis on two kinds of populations: small populations that face the threat of extinction, and sets of populations that are linked by dispersal (metapopulations). We'll begin our discussion of population dynamics by surveying patterns of population growth.

Patterns of Population Growth

CONCEPT 10.1
Populations exhibit a wide range of growth patterns, including exponential growth, logistic growth, fluctuations, and regular cycles.

Most observed patterns of population growth can be grouped into four major types: exponential growth, logistic growth, population fluctuations, and regular population cycles (a special type of fluctuation). Although we list four growth patterns here, bear in mind that a single population could experience each of them at different points in time. In addition, these four patterns are not mutually exclusive. For example, as we will see shortly, a population may grow logistically yet fluctuate around the values expected in logistic growth.

Exponential growth can occur when conditions are favorable

In the first pattern, exponential growth, a population increases (or decreases) by a constant proportion at each point in time. As we saw in Chapter 9, exponential growth can be represented by either of two related equations, $N_{t+1} = \lambda N_t$ or $dN/dt = rN$, the choice of which depends on whether reproduction occurs at discrete time periods (the first equation) or is continuous (the second equation).

Exponential growth cannot continue indefinitely (see Chapter 9), but when conditions are favorable, a population can increase exponentially for a limited time. Such periods of exponential growth can occur within the established range of a species, as when good weather occurs for several years running. They can also occur when a species reaches a new geographic area, either by dispersing on its own or with the aid of people. If conditions are favorable in the new environment, the population can increase rapidly until density-dependent factors act to regulate its numbers.

An example of how dispersal can lead to exponential growth is provided by the cattle egret subspecies *Bubulcus ibis ibis*. These birds originally lived in the Mediterranean region and in parts of central and southern Africa. Since the late 1800s and early 1900s, however, they have colonized new regions on their own, including South Africa, South America, and North America. Typically, after the subspecies reached a

new area, its population in that area would increase exponentially as it became established in its new habitat. For example, after the cattle egret colonized Florida in the 1950s, its populations there grew exponentially for several decades, with an estimated intrinsic rate of increase of $r = 0.11$ for the period 1956–1971 (van den Bosch et al. 1992). By the 1980s, exponential growth had ceased, and cattle egret numbers in Florida had stabilized (based on survey data reported by the USGS National Biological Service).

The spread of cattle egret populations illustrates common aspects of dispersal to new regions (**Figure 10.3**). As in this example, species typically colonize new geographic regions by long-distance or **jump dispersal** events, then local pop-

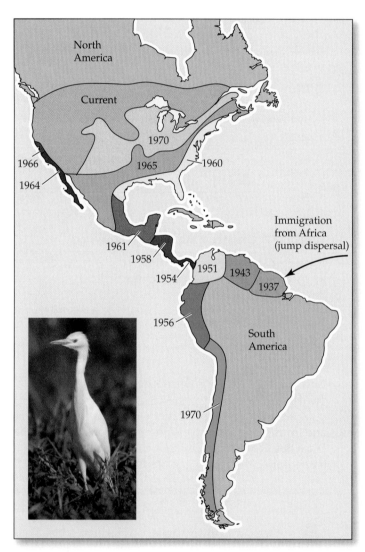

FIGURE 10.3 Colonizing the New World The cattle egret *Bubulcus ibis ibis* dispersed from Africa to South America in the late 1800s. Once it established colonies in the northeastern region of South America, it then spread rapidly to other parts of South and North America. The contour lines and dates show the edges of the cattle egret's range at different times. (After Smith 1974.)

ulations in the new region increase in size and expand (by relatively short-distance dispersal events) to occupy new areas of suitable habitat. In some cases, such populations grow more slowly as they approach a stable, maximum size.

In logistic growth, the population approaches an equilibrium

Some populations appear to reach a relatively stable, or equilibrium, population size that changes little over time. For example, in a letter written in 1778, the pastor and naturalist Gilbert White observed how many swifts (*Micropus apus*) were found each year in his village of Selborne (in southern England): "The number that I constantly find are eight pairs; about half of which reside in the church, and the rest in some of the lowest and meanest thatched cottages" (White 1789). In 1983, over 200 years later, several ecologists returned to Selborne and found 12 pairs of swifts, a number similar to that in White's time.

The 1778 and 1983 observations suggest that the swifts of Selborne may have reached, and then varied little from, a carrying capacity of roughly 10 nesting pairs. We cannot be sure

that is correct, because the number of swifts in Selborne may have fluctuated greatly during the two centuries between White's and the more recent observations. There are, however, species for which more complete data sets are available. Those data show that the number of individuals in some populations first increases in size, then fluctuates by a small amount around what appears to be the carrying capacity of the population (**Figure 10.4**). Such populations exhibit the second pattern of population growth, logistic growth.

With few exceptions (see Figure 9.17), population growth does not match the predictions of logistic growth precisely. For example, the graph of sheep abundance in Tasmania over time (see Figure 10.4) is only roughly similar to the characteristic S shape of a logistic growth curve (see Figure 9.18). As this example suggests, when applied to natural populations, the term "logistic growth" is used broadly to indicate any population whose numbers increase initially, but then level off at a maximum population size, the carrying capacity.

This lack of a perfect match is not surprising, given that in the logistic equation,

$$\frac{dN}{dt} = rN\left(1 - \frac{N}{K}\right)$$

the carrying capacity, K, is assumed to be a constant. For K to be a constant, the birth and death rates that occur at any particular density must not change from one year to the next (**Figure 10.5A**), a condition that rarely holds in nature. A more realistic scenario is shown in **Figure 10.5B**. Here, at any

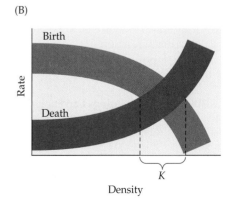

FIGURE 10.5 Why We Expect Carrying Capacity to Fluctuate The carrying capacity (K) of a population is the population size at which the birth rate equals the death rate. We can find K graphically by plotting the density at which the birth rate curve intersects the death rate curve. (A) Here we assume that the carrying capacity is constant over time; that is, at any given density, birth rates and death rates do not change over time, hence the two curves intersect at a single point, K. In (B), both the birth rate and death rate vary over time, as indicated by the broad bands. As a result, the birth and death rate curves can intersect at a broad range of values, causing K to take on a range of values (shown in red).

FIGURE 10.4 Population Growth Can Resemble a Logistic Curve Population growth in some species shows an exact match to a logistic curve (see Figure 9.17). More often, a species shows a pattern of growth (a rise in abundance, followed by a roughly stable population size) in which the match to a logistic curve is very rough, as seen here for sheep introduced to the island of Tasmania. (After Davidson 1938.)

> Phytoplankton abundance sometimes increased or decreased precipitously in just a few days.

FIGURE 10.6 Population Fluctuations Variation in phytoplankton abundance in water samples taken from Lake Erie during 1962, showing fluctuations above and below the overall mean abundance of 2250 cells/cm³ (indicated by the dashed line). (After Davis 1964.)

ample, census data for the bordered white moth (*Bupalus piniarius*) collected from 1882 to 1940 in a German pine forest showed that the densities reached during outbreaks were up to 30,000 times as great as the lowest density observed. Rapid variations in population sizes over time have also been found in many aquatic systems. The biomass of the comb jelly *Mnemiopsis*, for example, increased more than a thousandfold during a 2-year outbreak in the Black Sea (1989–1990), whereas zooplankton abundance changed much less drastically over a 35-year period (see Figure 10.2A,C). In species that reproduce very rapidly, abundances can change greatly over the course of days.

Many different factors can cause the size of a population to fluctuate. The increase in zooplankton populations in the Black Sea in the early 1980s probably occurred because their prey (phytoplankton) had increased in abundance (see Figure 10.2). Then, in 1991, zooplankton numbers plummeted, probably because of the spectacular abundance of their predator (*Mnemiopsis*) during the previous two years. The rapid changes in phytoplankton abundance shown in Figure 10.6 could reflect changes in a wide range of environmental factors, including nutrient supplies, temperature, and predator abundance. Overall, population fluctuations can be caused both by density-independent factors (e.g., weather) and by density-dependent factors (e.g., food limitation, natural enemies). In some cases, such factors produce a striking type of fluctuation: population cycles.

particular density, either the birth rate or the death rate may take on a range of values, reflecting the fact that environmental conditions change over time. As a result, the carrying capacity (*K*) also changes over time. When such a situation occurs, we would expect the population size to fluctuate around an average value of *K*, as observed for sheep in Tasmania (see Figure 10.4).

All populations fluctuate in size

Another characteristic of the sheep population in Tasmania is seen in all populations: their size rises and falls over time, illustrating the third and most common pattern of population growth, **population fluctuations**. In some populations, fluctuations occur as increases or decreases in abundance from an overall mean value (**Figure 10.6**). In other populations, fluctuations occur as deviations from a population growth pattern, such as exponential or logistic growth. If, for example, the growth of a population *exactly* matched a logistic growth pattern, the population would not be said to fluctuate. But if population abundances rose above and fell below those expected in logistic growth (as in the Tasmanian sheep), the population would be said to fluctuate.

In some cases, population fluctuations are relatively small, as seen in Figure 10.4. In other cases, the number of individuals in a population can explode at certain times, causing a **population outbreak** (**Figure 10.7**). In one such ex-

FIGURE 10.7 Populations Can Explode in Numbers All species have the potential for exponential growth. Hence, when conditions are favorable, a population outbreak can occur in which numbers increase very rapidly. The cockroaches covering the kitchen in this exhibit from the National Museum of Natural History could have been produced from a single pregnant female in a few generations. Female cockroaches can lay up to 80 eggs every few months.

Some species exhibit population cycles

In the fourth pattern of population growth, some species show **population cycles** in which alternating periods of high and low abundance occur after nearly constant intervals of time. Such cycles have been observed in populations of small rodents such as lemmings and voles, whose abundances typically reach a peak every 3–5 years (**Figure 10.8**).

Population cycles are among the most intriguing patterns observed in nature. After all, what factors can cause numbers to fluctuate greatly over time, yet maintain a high degree of regularity? Possible answers to this question include both internal factors, such as hormonal or behavioral changes in response to crowding, and external factors, such as weather, food supplies, or predators. Gilg et al. (2003) used a combination of field observations and mathematical models to argue that the 4-year cycle of collared lemmings (*Dicrostonyx groenlandicus*) in Greenland is driven by predators, one of which, the stoat, specializes on lemmings (see Figure 10.8). Other investigators have suggested that cycles of the Norwegian lemming (*Lemmus lemmus*) are caused by interactions between lemmings and their food plants. Similarly, a number of studies (e.g., Korpimäki and Norrdahl 1998) have implicated predators as the driving force behind cycles of field voles in Scandinavia, but Graham and Lambin (2002), in a large-scale field experiment, showed that predator removal had no effect on field vole cycles in England.

As the conflicting results described in the previous paragraph suggest, a universal cause of population cycles in small rodents has not emerged. As in the case of the factors that influence amphibian declines (see Chapter 1), it may be that the ecological mechanisms that drive population cycles dif-

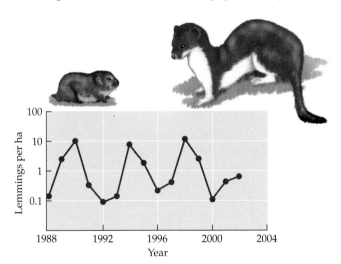

FIGURE 10.8 A Population Cycle In northern Greenland, collared lemming (*Dicrostonyx groenlandicus*, above left) abundance tends to rise and fall every 4 years. In this location, the population cycle appears to be driven by predators, the most important of which is the stoat (*Mustela erminea*, above right). In other regions, lemming cycles may be driven by food supply. (After Gilg et al. 2003.)

fer from place to place and from one species to another. We'll return to an exploration of the factors that influence population cycles when we consider one of the most famous cycles of all, the hare–lynx cycle (see Chapter 12).

Delayed Density Dependence

CONCEPT 10.2
Delayed density dependence can cause populations to fluctuate in size.

Although relatively few populations exhibit regular population cycles, all populations fluctuate in size to some degree. As we've seen, such fluctuations can result from a variety of factors, including changes in food supply, temperature, or predator abundance. Population fluctuations can also be caused by delayed density dependence, the effects of which we examine here.

The effect of population density is often delayed in time

Delays, or time lags, are an important feature of interactions in nature. For example, when a predator or parasite feeds, it does not produce offspring immediately; thus, there is a built-in delay in the effect of food supply on birth rates. As a result, it is common for the number of individuals born in a given time period to be influenced by the population densities or other conditions that were present several time periods ago.

Beginning in the 1920s and 1930s, ecologists examined such time lags with mathematical models that incorporated **delayed density dependence** (delays in the effect that density has on population size). Results from these models indicated that delayed density dependence can contribute to population fluctuations.

Intuitively, this finding makes sense. Consider a population of predators that reproduce more slowly than their prey. If there are few predators initially, the prey population may increase rapidly in size. As a result, the predator population may also increase, reaching a point at which there are a large number of adult predators that survive well and produce many offspring. However, if the large population of predators eats so many prey that the prey population decreases sharply in size, there may be few prey available for the next generation of predators. In such a case, a mismatch in predator and prey numbers (high predator numbers, low prey numbers) occurs because there is a time lag in the response of predator numbers to prey numbers. When such a mismatch takes place, the predators may survive or reproduce poorly, and their numbers may drop. If prey numbers then increase (perhaps because there are now fewer predators), predator numbers may first rebound, then fall again due to the built-in time lag. Thus, in principle at least, it seems reasonable that a delay in the response of predators to prey density could cause predator numbers to fluctuate over time.

To examine further how delayed density dependence affects population fluctuations, consider the following version of the logistic equation, modified to include a time lag, τ:

$$\frac{dN}{dt} = rN\left[1 - \frac{N_{(t-\tau)}}{K}\right] \qquad (10.1)$$

In this equation, dN/dt is the growth rate of the population at time t, r is the (per capita) population growth rate under ideal conditions, N is the population size at time t, and K is the carrying capacity. Equation 10.1 is identical to the logistic equation (see Equation 9.7) except that within the term $(1 - N/K)$, N has been replaced by $N_{(t-\tau)}$. Recall from Chapter 9 that the term $(1 - N/K)$ represents the effect of factors that reduce population growth from the exponential growth curve. Incorporating $N_{(t-\tau)}$ within this term indicates that the population growth rate (dN/dt) is affected by the population size at time $t - \tau$ in the past.

When we incorporate delayed density dependence into the logistic equation in this way, population fluctuations can re-

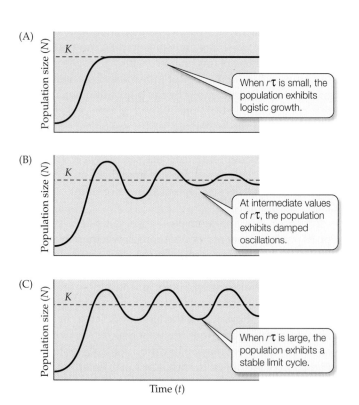

FIGURE 10.9 Logistic Growth Curves with Delayed Density Dependence Depending on the values of the population growth rate (r) and the time lag (τ), adding delayed density dependence to the logistic equation can result in (A) an S-shaped logistic growth curve; (B) fluctuations about the carrying capacity, K, that become smaller over time; or (C) a regular cycle of ongoing fluctuations about the carrying capacity.

sult (**Figure 10.9**). Robert May (1976) studied the behavior of Equation 10.1 and showed that the occurrence of such fluctuations depends on the values of the parameters r and τ. Specifically, he found that when the product of those parameters is "small" ($0 < r\tau < 0.368$), the population increases smoothly to the carrying capacity, K (Figure 10.9A). At intermediate values of $r\tau$ ($0.368 < r\tau < 1.57$), delayed density dependence causes the population to fluctuate in size (Figure 10.9B). The result is a pattern of **damped oscillations**, in which the deviations from the carrying capacity gradually get smaller over time. Finally, when $r\tau$ is "large" ($r\tau > 1.57$), the population exhibits a cycle in which it fluctuates indefinitely about the carrying capacity, K (Figure 10.9C); such a pattern is called a **stable limit cycle**.

The results in Figure 10.9 show that when delayed density dependence is introduced into the logistic equation (via the time lag, τ), population fluctuations can result. We turn now to laboratory experiments to see how delayed density dependence works in real populations.

Delayed density dependence affects blowfly populations

In the 1950s, A. J. Nicholson performed a series of pioneering laboratory experiments on density dependence in blowflies. These insects are both decomposers and parasites in that they feed on dead animals but also attack living hosts, including mammals and birds. Nicholson studied *Lucilia cuprina*—the sheep blowfly—so named because it is an important agricultural pest of sheep. Before they can lay eggs, the females of this species need a protein meal (which they usually get from animal dung or carcasses). Once they have fed, the females attack living sheep by laying their eggs near the tail or near open wounds or sores. Small white maggots hatch from those eggs and feed on dung attached to the skin or on exposed flesh. As they feed, the maggots grow larger and more voracious. At a certain point, the maggots burrow inside the sheep, where they feed on its internal tissues, causing severe lesions and sometimes death. Death can be caused directly by the maggots or by infections that spread through the lesions. The sheep blowfly's full life cycle (from egg to egg) can be completed in as little as 7 days.

In several of his experiments, Nicholson examined the effect of delayed density dependence on blowfly population dynamics. In the first of the two experiments that we will consider here, Nicholson provided adult blowflies with unlimited food, but restricted maggots to 50 g of liver per day. Because adults had abundant food, each female was able to lay many eggs. Thus, when there were many adults, enormous numbers of eggs were produced. When those eggs hatched, however, lack of food caused most or all of the maggots to die before they reached adulthood (**Figure 10.10A**). As a result, few adults were produced, and the adult population invariably declined after reaching a peak. Eventually, the number of adults in the population reached such low levels that the few eggs

FIGURE 10.10 Nicholson's Blowflies Variation over time in the abundance of sheep blowflies in laboratory experiments conducted by A. J. Nicholson. (A) Adult blowflies were supplied with unlimited food, maggots with limited food. As a result, few or no adults were produced from the many eggs laid during periods of maximum adult abundance because the many maggots that hatched from those eggs had little or no food to eat. (B) Experimental conditions were as in part A until roughly halfway through the experiment, when the food supply for adults was limited. (After Nicholson 1957.)

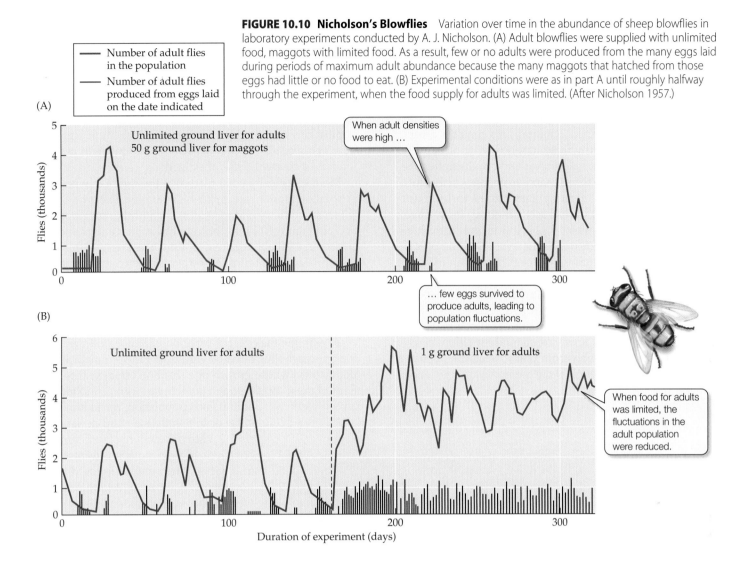

they produced were able to give rise to a new generation of adults. Once this happened, the number of adults would begin to rise again, then crash, repeating the cycle just described.

Nicholson argued that delayed density dependence caused the number of adult blowflies to rise and fall repeatedly in this experiment. His reasoning was that because adults had unlimited food, the negative effects of high adult densities were not felt until a later time, that is, when the maggots hatched and began to feed. To test this idea, Nicholson performed a second experiment in which he removed some of the effects of delayed density dependence by providing both adults and maggots with a limited amount of food. When he did this, the adult population size no longer repeatedly rose to high levels, then crashed to near zero. Instead, the number of adults increased and then fluctuated around an average of about 4,000 flies (**Figure 10.10B**). Taken together, the results shown in Figure 10.10 suggest that delayed density dependence can play a role in causing the pronounced fluctuations seen in some populations.

Delayed density dependence and other factors can cause a population to fluctuate in size because they can cause the growth, survival, or reproduction of individuals to vary over time, and that, in turn, can cause the population growth rate (λ) to vary significantly from one time period to the next. Next, we'll explore how such fluctuations in λ affect the risk that a population will become extinct.

Population Extinction

CONCEPT 10.3
The risk of extinction increases greatly in small populations.

Populations can be driven to extinction by many different factors, including predictable, or *deterministic*, changes in the environment. Consider a population of fish that colonizes a temporary pond (one that forms during the rainy season but then dries out completely at other times of the year). The fish may thrive for a while, but as the water level drops, they are doomed. While such deterministic extinctions are both common and important, they are not our focus here. In this section, we'll look at how fluctuation in the population growth rate, population size, and chance events in the environment affect a population's risk of extinction.

To set the stage for this discussion, let's consider a version of the geometric growth equation (see Equation 9.1: $N_{t+1} = \lambda N_t$) that includes random variation in the finite rate of increase, λ. From year to year, random changes in features of the environment (e.g., weather) may cause λ to fluctuate. There will be good years, in which λ is above its average value, and bad years, in which λ is below its average value. Imagine a population for which the average value of λ is greater than 1. If λ fluctuates little over time, then in most years λ will be greater than 1, and hence the population will usually increase in size. Under these circumstances, the population will deviate only slightly from a geometric growth pattern, and it will face little or no risk of extinction. However, if random variation in environmental conditions causes λ to change considerably from year to year, the population will fluctuate in size. What are the implications of such fluctuations?

In a model, fluctuation in the population growth rate increases the risk of extinction

To show what happens when λ fluctuates, we performed computer simulations of geometric growth for three populations in which λ was allowed to fluctuate at random. If we examine the results in **Figure 10.11**, we see that two of the populations recovered from low numbers, but one went extinct. These results support what common sense tells us: fluctuations increase the risk of extinction.

More specifically, what matters is the extent to which the population growth rate fluctuates over time. The growth rate (λ) changed from year to year in each of the three populations in Figure 10.11, as indicated by the increases and decreases in their abundance over time. The extent of this variation in λ was determined by a variable controlled in the simulations, the standard deviation (σ) of the growth rate. In each of the three simulated populations shown in Figure 10.11, σ was set equal to 0.4. (See the **Web Stats Review** for more information about the standard deviation.)

To examine the effect of variation in λ more fully, we used the approach illustrated by Figure 10.11 to simulate 10,000 populations whose growth rate (λ) varied from year to year. Each of these 10,000 populations began with 10 individuals, and in each population, λ had a median value of 1.05 and a standard deviation of $\sigma = 0.2$. With this amount of variation in λ, only 0.3% of the populations went extinct in 70 years. When σ was increased to 0.4, however, 17% of the 10,000 populations went extinct in 70 years, and when it was increased still further to $\sigma = 0.8$, 53% of the populations went extinct.

The take-home message provided by these simulations is that when variable environmental conditions cause a population's growth rate to fluctuate considerably over time, its risk of extinction increases. This effect, however, is dependent

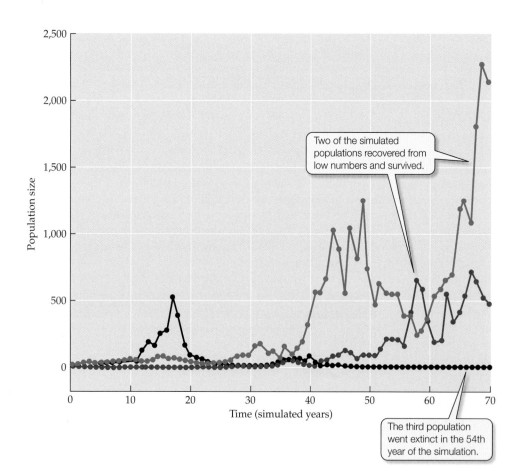

Two of the simulated populations recovered from low numbers and survived.

The third population went extinct in the 54th year of the simulation.

FIGURE 10.11 Fluctuations Can Drive Small Populations Extinct   Simulated growth of three populations whose abundance changed according to a geometric growth equation ($N_{t+1} = \lambda N_t$) in which the value of λ varied at random from year to year. This variation in λ over time was included in the model to simulate random variation in environmental conditions. Each simulated population began with 10 individuals. For each of the three simulated populations shown here, the median value of λ was 1.05 and the standard deviation of λ was 0.4. Over time, roughly 70% of the simulated values of λ were between $\lambda = 0.70$ and $\lambda = 1.57$.

on the size of the population, with small populations being particularly at risk.

Small populations are at much greater risk of extinction than large populations

The size of a population has a strong effect on its risk of extinction. As we saw in the previous section, when we simulated the fates of 10,000 populations in which the standard deviation of λ was 0.8, 53% of populations that begin with 10 individuals become extinct in 70 years. However, if we increase the initial population size to 100 and perform the simulation again, the chance of extinction drops from 53% to 29%. For initial population sizes of 1,000 or 10,000 individuals, the chance of extinction continues to fall (to 14% and 6%, respectively).

The simulation results we have just described suggest that small populations are much more prone to extinction than large populations. To see why, imagine that poor weather caused the growth rate (λ) of a population to be between 0.2 and 0.5 for 3 years in a row (thus causing the population to shrink considerably in size for each of these three years). In such a situation, a population with 10 individuals could easily be driven to extinction, whereas a population with 10,000 individuals would have survivors when conditions improved.

Similar patterns have been observed in real populations. For example, Jones and Diamond (1976) studied extinction in bird populations on the Channel Islands, located off the coast of California. By combining data from published articles (from 1868 on), museum records, unpublished field observations, and their own fieldwork, they showed that population size had a strong effect on the chance of extinction (**Figure 10.12**). They found that 39% of populations with fewer than 10 breeding pairs went extinct, whereas they observed no extinctions in populations with over 1,000 breeding pairs. Similar work by Pimm et al. (1988) showed that small populations can go extinct very rapidly: on islands off the coast of Britain, bird populations with 2 or fewer nesting pairs had a mean time to extinction of 1.6 years, while populations with 5–12 nesting pairs had a mean time to extinction of 7.5 years.

These findings for birds have been confirmed in other groups of organisms, including mammals, lizards, and insects. Overall, field data indicate that the risk of extinction increases greatly when the population size is small. But what are the factors that place small populations at risk?

Chance events can drive small populations to extinction

If we could predict in advance how factors that influence the survival and reproduction of individuals would vary from one year to the next, we could also predict how population growth rates would change over time. But we cannot do this

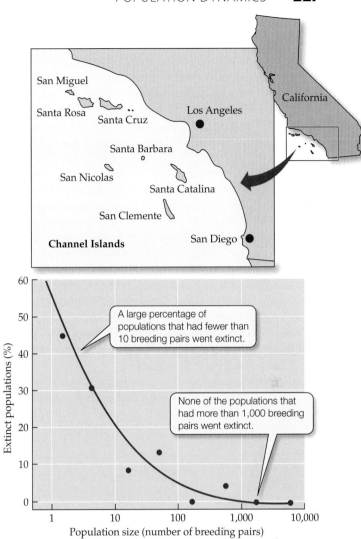

FIGURE 10.12 Extinction in Small Populations Among bird populations on the Channel Islands, the percentage of populations that went extinct declined rapidly as the number of breeding pairs in the population increased. (After Jones and Diamond 1976.)

in real populations. There are too many factors that we cannot predict accurately, including variation in environmental conditions (e.g., temperature and rainfall) and variation in the fates of individuals (e.g., a pollinator such as a bee might land on plant A or plant B, thereby giving one individual, but not the other, the chance to reproduce). For simplicity, in the discussion that follows we'll refer to such unpredictable events as "chance events," even though they may ultimately have a deterministic cause. In particular, we'll consider the role of chance genetic, demographic, and environmental events in making small populations vulnerable to extinction.

Genetic factors can threaten small populations Small populations can encounter problems associated with genetic drift and inbreeding. Recall from Chapter 6 that **genetic drift** is the process by which chance events influence which alleles are passed on to the next generation. Genetic drift can occur in many ways, including chance events associated with whether individuals reproduce or die. Imagine, for example, that an elephant walks through a population of 10 small plants, 50% of which have white flowers (genotype *aa*) and 50% of which have red flowers (*AA*). If the elephant happens to crush more red-flowered than white-flowered plants, then by chance alone, there will be more copies of the *a* allele than of the *A* allele in the next generation. This scenario is just one of many possible examples of how genetic drift can cause allele frequencies to change at random from one generation to the next.

Genetic drift has little effect on large populations, but in small populations it can cause losses of genetic variation over time. For example, if genetic drift causes the frequency of two alleles (e.g., *A* and *a*) to change at random in each generation, one allele may eventually increase to a frequency of 100%, while the other is lost (see Figure 6.7). Drift can reduce the genetic variability of small populations rapidly: for example, after 10 generations, roughly 40% of the original genetic variation is lost in a population of 10 individuals, while 95% is lost in a population of 2 individuals.

Small populations are vulnerable to the effects of genetic drift for three reasons. First, when drift leads to a loss of genetic variability, the ability of a population to respond (via natural selection) to future environmental change is limited. Second, genetic drift can cause harmful alleles to occur at high frequencies, often causing individuals to suffer poor reproductive success (as in the case of the greater prairie chicken on p. 137). Third, small populations show a high frequency of **inbreeding** (mating between related individuals). Inbreeding is common in small populations because after several generations at a small population size, most of the individuals in the population will be closely related to one another (to see why, solve Review Question 3). Inbreeding tends to increase the frequency of homozygotes, including those that have two copies of a harmful allele. Thus, like genetic drift, inbreeding can lead to reduced reproductive success, causing birth rates, and hence population growth rates, to drop.

The combined negative effects of genetic drift and inbreeding appear to have reduced the fertility of male lions that live on the floor of the Ngorongoro Crater, Tanzania (**Figure 10.13**). From 1957 to 1961, there were 60–75 lions living in the crater, but in 1962 an extraordinary outbreak of biting flies caused all but 9 females and 1 male to die. Seven males immigrated into the crater in 1964–1965, but no further immigration has occurred since that time. The population has increased in size since the 1962 crash. From 1975 to 1990, for example, the population fluctuated between 75 and 125 individuals. However, genetic analyses indicated that all of these individuals were descendants of just 15 lions (Packer et al. 1991). In a population of 15 individuals, genetic drift and inbreeding have powerful effects. Those effects appear to be the reason why the crater population has less genetic variation and more frequent sperm abnormalities than the large population of lions found nearby on the Serengeti Plain (see Figure 10.13).

Demographic factors can threaten small populations

For an individual, reproduction and survival and reproduction are all-or-nothing events: an individual either survives or it does not, and it either reproduces or it does not. At the population level, we can transform such all or nothing events into a probability that survival or reproduction will occur. For ex-

FIGURE 10.13 A Plague of Flies In 1962, the population of lions in the 260 km² (100 square miles) Ngorongoro Crater of Tanzania was nearly driven to extinction by a catastrophic outbreak of biting flies similar to those on the face of this male. Lions became covered with infected sores and eventually could not hunt, causing many to die. In the population that descended from the few survivors, genetic drift and inbreeding have led to frequent sperm abnormalities, such as this "two-headed" sperm.

ample, if 70 out of 100 individuals in a population survive from one year to the next, then (on average) each individual in the population has a 70% chance of survival.

In a small population, however, chance events related to the survival and reproduction of individuals (**demographic stochasticity**) can result in outcomes that differ from what such averages would lead us to expect. Consider a population of 10 individuals for which previous data indicate that on average, each individual has a 70% probability of surviving from one year to the next. Many chance events can cause the percentage of individuals that actually do survive to be higher or lower than this. For example, if a storm happened to kill 6 members of the population, the observed survival rate (40%) would be considerably lower than the expected 70%. By affecting survival and reproductive rates in this way, demographic stochasticity can cause the size of a small population to fluctuate over time. In one year the population may grow, while the next it may decrease in size—perhaps so drastically that extinction results.

In contrast, when the population size is large, there is little risk of extinction from demographic stochasticity. The fundamental reason for this has to do with laws of probability. You are, for example, much more likely to receive zero heads if you toss a fair coin 3 times than if you toss the same coin 300 times. Similarly, when considering the demographic fate of individuals, chance events are much more likely to cause reproductive failure or poor survival in small populations than in large populations. If each individual in a population has a 1/3 chance of producing zero offspring, then if there are 2 individuals in the population, there is an 11% chance ($0.33 \times 0.33 = 0.33^2 = 0.11$) that no offspring will be produced—driving the population to extinction in one generation. Although demographic stochasticity could cause a population of 30 individuals to fluctuate in size (perhaps leading to eventual extinction), there is essentially no chance (0.33^{30}) that it could cause the population to go extinct in a single generation.

Demographic stochasticity is one of several factors that can cause small populations to experience Allee effects. **Allee effects** occur when the population growth rate (r or λ) *decreases* as the population density decreases, perhaps because individuals have difficulty finding mates at low population densities (**Figure 10.14**). This phenomenon reverses the usual assumption that r and λ tend to *increase* as population density decreases (see Figure 9.16). Allee effects can be disastrous for small populations. If demographic stochasticity or any other factor decreases the population size, Allee effects can cause the population growth rate to drop, which causes the population size to decrease even further in a downward spiral toward extinction. We'll return to this issue in Chapter 22.

Environmental variation can threaten small populations As we have stressed repeatedly in this book, environmental conditions vary from year to year. Such variation can affect birth and death rates, leading to variations in population size. Our focus here will be on how two types of chance events—environmental stochasticity and natural catastrophes—can cause extinction in small populations.

Environmental stochasticity refers to erratic or unpredictable changes in the environment. In the simulations described above (see Figure 10.11 and related text on pp. 226–227), we've already seen that (1) variation in environmental conditions that causes fluctuations in population growth rates can

(A)

(B)

(C)

(D)

FIGURE 10.14 Allee Effects Can Threaten Small Populations As shown in laboratory experiments with the flour beetle *Tribolium*, population growth rates can decrease as the population density decreases (A). Known as the Allee effect, this phenomenon can cause the size of a small population to drop, leading to further drops in reproductive rates and a downward spiral toward extinction. Allee effects can be important in animals such as (B) bluefin tuna (*Thunnus thynnus*) that form schools or herds whose protective or early warning systems function poorly at small population sizes, and in (C) fig trees (*Ficus* sp.), (D) fruit flies (*Anastrepha ludens*), and many other species in which individuals have difficulty finding mates at low population densities. (After Courchamp et al. 1999).

lead to population size fluctuations and thus an increased risk of extinction, and (2) such environmental variation is more likely to cause extinction when the population size is small. Many species face such risks from environmental stochasticity. For example, census data on female grizzly bears (*Ursus arctos horribilis*) in Yellowstone National Park showed that the average population growth rate (*r*) was approximately 0.02, but that it varied from year to year. Despite the fact that the population tends to grow in size (because *r* > 0), researchers using a mathematical model found that random variation in environmental conditions could place the Yellowstone grizzly bear population at high risk of extinction, especially if the population size were to drop to 30 or 40 females from its 1997 level of 99 females (**Figure 10.15**).

Environmental and demographic stochasticity differ in a fundamental way. Environmental stochasticity refers to changes in the average birth or death rates of a population that occur from year to year because of random changes in environmental conditions: sometimes there are good years and sometimes there are bad years. In contrast, in demographic stochasticity, the average (population-level) birth and death rates may be constant across years, but the actual fates of individuals differ due to the random nature of whether each individual reproduces or not, and survives or not.

Finally, populations also face risks from extreme environmental events such as floods, fires, severe windstorms, or out-

breaks of disease or natural enemies. Even though they occur rarely, such **natural catastrophes** can eliminate or drastically reduce the size of populations that otherwise would seem large enough to be at little risk. For example, disease outbreaks have resulted in mass mortality in populations of sea urchins (up to 98% of the individuals in some populations) and Baikal seals (killing about 2,500 of a population of 3,000 seals).

Natural catastrophes also played a key role in the extinction of the heath hen, *Tympanuchus cupido cupido.* This bird was once abundant from Virginia to New England. By 1908, hunting and habitat destruction had reduced its population to 50 birds, all on the island of Martha's Vineyard, where a 1,600-acre reserve was established for its protection. Initially, the population thrived, increasing in size to 2,000 birds by 1915. A population of 2,000 seems large enough to be nearly "bulletproof" against problems that threaten small populations, including genetic drift and inbreeding, demographic stochasticity, and environmental stochasticity. However, a series of disasters struck between 1916 and 1920, including a fire that destroyed many nests, unusually cold weather, a disease outbreak, and a boom in the population of goshawks, a predator of heath hens. Due to the combined effect of these events, the heath hen population dropped to 50 birds by 1920 and never recovered. In 1932, the last heath hen died.

With the benefit of hindsight, we can see that heath hens were vulnerable in 1908 because they all lived in a single population. More typically, members of a species are found in multiple populations, which are often isolated from one another by regions of unsuitable habitat.

Metapopulations

CONCEPT 10.4
Many species have a metapopulation structure in which sets of spatially isolated populations are linked by dispersal.

The checkered landscapes visible from the air vividly demonstrate that the world is a patchy place (see Figure 9.1). The patchy nature of the landscape ensures that for many species, areas of suitable habitat do not cover large continuous regions, but rather exist as a series of favorable sites that are spatially isolated from one another. As a result, the populations of a species are often scattered across the landscape, each in an area of favorable habitat but separated from one another by hundreds of meters or more.

Sometimes these spatially isolated populations are not linked by dispersal and hence do not affect one another's population dynamics. In many cases, however, seemingly isolated populations do affect one another's dynamics because individuals (or gametes) occasionally disperse from one population to another. Such a group of interacting populations is called a **metapopulation**. Literally, the term "metapopulation"

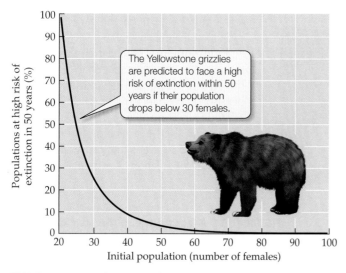

The Yellowstone grizzlies are predicted to face a high risk of extinction within 50 years if their population drops below 30 females.

FIGURE 10.15 Environmental Stochasticity and Population Size
This graph plots the risk that the Yellowstone grizzly bear population will be close to extinction in 50 years versus the initial population size (number of females). By studying 39 consecutive years of census data, researchers found that the average population growth rate of Yellowstone grizzlies was *r* = 0.02—a rate that would lead to explosive growth if it remained constant from year to year. The risk of extinction was calculated from a mathematical model that examined the effect of environmental stochasticity by incorporating the variation in *r* observed over the 39 years of data. (After Morris and Doak 2002.)

(A)

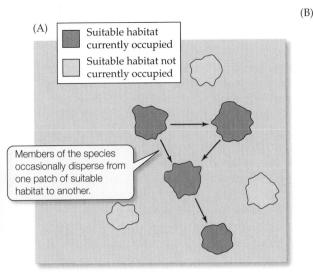

Suitable habitat currently occupied

Suitable habitat not currently occupied

Members of the species occasionally disperse from one patch of suitable habitat to another.

(B)

FIGURE 10.16 The Metapopulation Concept A metapopulation is a set of spatially isolated populations linked by dispersal. In (A), seven patches of suitable habitat for a species are diagrammed, four of which are currently occupied and three of which are not. The area outside of these seven patches represents unsuitable habitat. In (B), a satellite image shows a group of lakes in northern Siberia that can be connected to one another by temporary streams that form after the snow melts or after periods of heavy rainfall.

refers to a population of populations, but it is usually defined in a more particular sense as a set of spatially isolated populations linked to one another by dispersal (**Figure 10.16**).

Metapopulations are characterized by repeated extinctions and colonizations

As ecologists have long recognized, populations of some species are prone to extinction for two reasons: (1) the landscapes they live in are patchy (making dispersal between populations difficult), and (2) environmental conditions often change in a rapid and unpredictable manner. Metaphorically, we can think of such populations as a set of "blinking lights" that wink on and off, seemingly at random, as patches of suitable habitat are colonized and their populations then go extinct. Although the individual populations may be doomed to extinction, the species as a whole persists because the collection of populations—the metapopulation—includes populations that are going extinct and new populations established by colonization.

Building on this idea of random extinctions and colonizations, Richard Levins (1969, 1970) represented metapopulation dynamics in terms of the extinction and colonization of habitat patches:

$$\frac{dp}{dt} = cp(1-p) - ep \tag{10.2}$$

where p represents the proportion of habitat patches that are occupied at time t, while c and e are the patch colonization and patch extinction rates, respectively.

In deriving Equation 10.2, Levins made a number of assumptions: (1) that there is a very large (infinite) number of identical habitat patches, (2) that all patches have an equal chance of receiving colonists (hence the spatial arrangement of the patches does not matter), (3) that all patches have an equal chance of extinction, and (4) that once a patch is colonized, its population increases to its carrying capacity much

more rapidly than the rates at which extinction and colonization occur (this assumption allows population dynamics within patches to be ignored).

Although some of the assumptions of Equation 10.2 are not realistic, it leads to a simple but fundamental insight: for a metapopulation to persist for a long time, the ratio e/c must be less than 1 (see **Web Extension 10.1** for a description of how this result was obtained). In words, this means that some patches will be occupied as long as the colonization rate is greater than the extinction rate; otherwise, the metapopulation will collapse and all populations in it will become extinct. Levins's groundbreaking approach focused attention on a number of key issues, such as how to estimate factors that influence patch colonization and extinction, the importance of the spatial arrangement of suitable patches, the extent to which the landscape between habitat patches affects dispersal, and the vexing problem of how to determine whether empty patches are suitable habitat or not. Levins's rule for persistence also has applied importance, as we will see shortly.

A metapopulation can go extinct even when suitable habitat remains

Human actions (such as land development) often convert large tracts of habitat into a set of spatially isolated habitat fragments (see Figure 8.4). Such **habitat fragmentation** can cause a species to have a metapopulation structure where it did not have one before. If land development continues and the habitat becomes still more fragmented, the metapopulation's colonization rate (c) may decrease because patches become more isolated and hence harder to reach by dispersal. Further habitat fragmentation also causes the patches that remain to become smaller (as in Figure 8.4); as a result, the extinction rate (e) may increase because small patches have small populations which have a higher risk of extinction. Both of these trends (an increase in e and a decrease in c) cause the ratio e/c to increase. Thus, if too much habitat is removed, the

FIGURE 10.17 The Northern Spotted Owl The northern spotted owl (*Strix occidentalis caurina*) thrives in old-growth forests of the Pacific northwest; such forests include those that have never been cut, or have not been cut for 200 years or more.

ratio *e/c* may shift suddenly from less than 1 to greater than 1, thereby dooming all populations, and the metapopulation, to eventual extinction, even though some habitat remains.

The idea that all populations in a metapopulation might go extinct while suitable habitat remains was developed further in work on the northern spotted owl (**Figure 10.17**). The northern spotted owl (*Strix occidentalis caurina*) is found in the Pacific Northwest region of North America. It lives in old-growth forest, where nesting pairs establish large territories that range in size from 12 to 30 km^2 (territories are larger in poor-quality habitat). Lande (1988) modified the Levins model to include a description of how owls might search for vacant "patches," which were interpreted as sites suitable for individual territories. Lande estimated that the entire metapopulation would collapse if the logging of old-growth forest were to reduce the fraction of habitat patches that were suitable for the owl to less than 20%. This result had a powerful impact: it illustrated how a species might go extinct if its habitat dropped below a critical threshold (in this case, 20% suitable habitat), and it contributed to the 1990 listing of the northern spotted owl as a threatened species in the United States.

Extinction and colonization rates often vary among patches

As Lande's work on the northern spotted owl shows, the metapopulation approach has become increasingly important in applied ecology. But real metapopulations often violate the assumptions of the Levins model, leading ecologists to use more complex models (see Hanski 1999) when addressing applied questions. For example, patches often differ considerably in population size and in the ease with which

they can be reached by dispersal. As a result, extinction and colonization rates may vary greatly among patches. It is also common for extinction and colonization to be influenced not only by chance events, but also by nonrandom components of the environment. Here we'll consider how such complications affect butterfly and frog metapopulations in the field.

A terrestrial metapopulation: Butterflies in grasslands In the early 1900s, the skipper butterfly, *Hesperia comma*, was found on grazed calcareous grasslands (i.e., alkaline grasslands found on limestone or chalk outcrops) throughout a broad range of the United Kingdom. Starting in the 1950s, however, calcareous grasslands became overgrown because cattle stocks were reduced and because the myxomatosis virus killed off another important grazer, the introduced European rabbit (*Oryctolagus cuniculus*). As a result, *H. comma* populations began to decline. By the mid-1970s, the butterfly was found in only ten restricted regions, a very small fraction of its original range.

Things began to pick up for the butterfly in the early 1980s. By this time, habitat conditions had improved because livestock had been reintroduced and rabbit populations had recovered. In 1982, Chris Thomas and Terésa Jones observed many habitat patches that appeared suitable but did not support *H. comma* populations. To follow the recolonization of these patches over time, Thomas and Jones documented the locations of all patches containing *H. comma* populations and of all patches that appeared suitable for, but were not occupied by, *H. comma*. In 1991, they surveyed the patches again and noted which ones were occupied at that time. Their results highlight two important features of many metapopulations: isolation by distance and the effect of patch area (or population size—small patches tend to have small population sizes).

Isolation by distance occurs when patches located far away from occupied patches are less likely to be colonized than are nearby patches. In *H. comma*, distance from occupied patches had a strong effect on whether patches vacant in 1982 were colonized by 1991: few patches separated by more than 2 km from an occupied patch were colonized during that period (**Figure 10.18**). Patch area also affected the chance of colonization: the majority of colonized patches were at least 0.1 ha in size. Patch area could have affected colonization rates directly because small patches may be harder to find than large patches. In addition, *H. comma* could have colonized small patches, but then suffered extinction in those patches by 1991 due to the problems associated with small population size that we described in the previous section; such patches would appear never to have been colonized because the sites were not sampled between 1982 and 1991.

Thomas and Jones also found that the chance of extinction was highest in small patches (most likely because their populations were small) and in patches that were far from an occupied patch. Isolation by distance can affect the chance of

FIGURE 10.18 Colonization in a Butterfly Metapopulation
Colonization of suitable habitat from 1982 to 1991 by the skipper butterfly *Hesperia comma* was influenced by patch area and patch isolation (distance to the nearest occupied patch). Each red or green circle represents a patch of suitable habitat that was not occupied by *H. comma* in 1982. The lines show the combinations of patch area and patch isolation for which there was a 90%, 50%, and 10% chance of colonization (as calculated from a statistical analysis of the data). (After Thomas and Jones 1993.)

extinction because a patch that is near an occupied patch may receive immigrants repeatedly, thereby increasing the local population size and making extinction less likely. This tendency for high rates of immigration to protect a population from extinction (by reducing the problems associated with small population size) is known as the **rescue effect** (Brown and Kodric-Brown 1977).

An aquatic metapopulation: Frogs in ponds In Sweden, the pool frog (*Rana lessonae*) is found in about 60 ponds located along the Baltic coast. Researchers reported in the 1960s that many ponds in the general area where *R. lessonae* occurs were not occupied by the frog. To understand why pool frogs are present in some ponds but not others, Per Sjögren-Gulve measured several environmental characteristics of the ponds, including lime content, alkalinity, distance to the nearest occupied pond, pond area, water temperature, and the presence or absence of pike (a predatory fish). He used those characteristics to analyze survey data that showed which ponds were occupied at different times (in 1962, 1983, and 1987) (**Figure 10.19**).

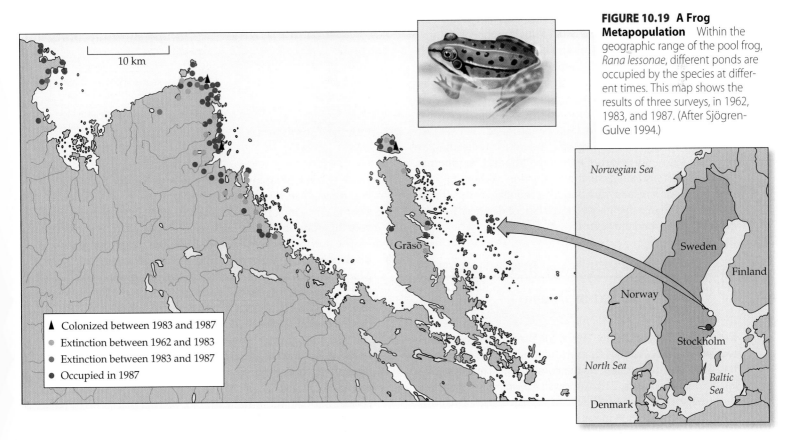

FIGURE 10.19 A Frog Metapopulation Within the geographic range of the pool frog, *Rana lessonae*, different ponds are occupied by the species at different times. This map shows the results of three surveys, in 1962, 1983, and 1987. (After Sjögren-Gulve 1994.)

FIGURE 10.20 Uplifting Shapes the Pool Frog Metapopulation
Diagram 1 illustrates that over time, the uplifting of Earth's crust causes the shoreline of Sweden to extend farther into the Baltic Sea. This process influences the spatial pattern of colonization and extinction in the pool frog (*Rana lessonae*) metapopulation. Diagrams 2–4 show the changes that might take place over 200–400 years. (After Sjögren-Gulve 1994.)

The elevation of Sweden's Baltic coast is rising at a rate of about 60–80 cm per century.

As new areas of land emerge from the sea, ponds form in low-lying areas and are colonized by pool frogs.

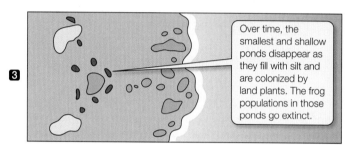

Over time, the smallest and shallow ponds disappear as they fill with silt and are colonized by land plants. The frog populations in those ponds go extinct.

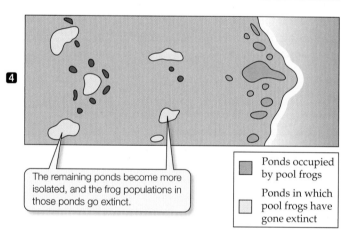

The remaining ponds become more isolated, and the frog populations in those ponds go extinct.

■ Ponds occupied by pool frogs

□ Ponds in which pool frogs have gone extinct

Sjögren-Gulve's results (1994) provided clear evidence of repeated extinctions and colonizations, as would be expected for a species with a metapopulation structure. However, his results also suggested that features of the environment may have influenced the probabilities of colonization and extinction in particular ponds, an important difference from the assumptions of the Levins model. For example, ponds that were far away from occupied ponds experienced low colonization rates and high extinction rates. Pond temperature also played an important role: relatively warm ponds were more likely to be colonized successfully because breeding success was greater in such ponds. Other environmental variables (e.g., the presence of pike and water alkalinity) had little direct effect.

In addition, the spatial pattern of pond extinctions led Sjögren-Gulve to speculate that long-term environmental changes may have influenced extinction rates in the pool frog metapopulation. Roughly 10,000 years ago, as the last glaciers of the Ice Age retreated, an enormous weight was removed from the land surface of the region in Sweden where pool frogs are found. The removal of this weight caused uplifting (i.e., an increase in altitude) of Sweden's Baltic coast, which continues today at a rate of 60–80 cm per century. As a result of this uplifting, new areas of land are gradually emerging from the sea, thereby isolating portions of what once were bays to form ponds in which pool frogs can thrive (**Figure 10.20**). Over time, these ponds are filled in by silt and overtaken by vegetation, shrinking their size and leading to the local extinction of pool frog populations. Overall, this plausible scenario suggests that extinction in the pool frog metapopulation results from both deterministic events (pond formation and disappearance) and a lack of dispersal among older, spatially isolated ponds.

CASE STUDY REVISITED

A Sea in Trouble

In the late 1980s and early 1990s, the Black Sea ecosystem was under severe duress from the combined effects of eutrophication and invasion by the comb jelly, *Mnemiopsis leidyi*, as described on p. 218. Although *Mnemiopsis* numbers declined sharply in 1991, they rose steadily again from 1992 to 1995, and then maintained high levels for several years—about 250 g/m², which translates to over 105 million tons of *Mnemiopsis* throughout the Black Sea. The situation did not look promising. But by 1999, matters were very different: the Black Sea was on the way to recovery.

The events that set the stage for the recovery of the Black Sea actually began prior to the first onslaught of *Mnemiopsis*. In the mid- to late 1980s, the amounts of nutrients added to the Black Sea began to level off. From 1991 to 1997, nutrient inputs actually declined, probably due to hard economic times in former Soviet Union countries, coupled with national and international efforts to reduce nutrient inputs. The reduction had rapid effects: after 1992, phosphate concentrations in the Black Sea declined, phytoplankton biomass began to fall, water clarity increased, and zooplankton abundance increased. *Mnemiopsis* still posed a threat, however, as evidenced by its high biomass and by

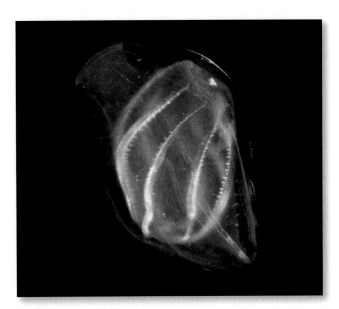

FIGURE 10.21 Invader versus Invader Another invasive comb jelly species, the predator *Beroe*, brought *Mnemiopsis* under control, thus contributing to the recovery of the Black Sea ecosystem.

falling anchovy catches from 1995 to 1998 (see Figure 10.2). Scientists and government officials were gearing up to combat the threat from *Mnemiopsis* when the problem was inadvertently solved by the arrival of another comb jelly, the predator *Beroe* (**Figure 10.21**).

Beroe arrived in 1997. Like *Mnemiopsis*, *Beroe* was probably brought to the Black Sea in the ballast water of ships from the Atlantic. *Beroe* feeds almost exclusively on *Mnemiopsis*. It is such an effective predator that within 2 years of its arrival, *Mnemiopsis* numbers plummeted (see Figure 10.2A). Following the sharp decline in *Mnemiopsis*, the Black Sea population of *Beroe* also crashed, presumably because it depended on *Mnemiopsis* for food. The fall of *Mnemiopsis* led to a rebound in zooplankton abundance (which had dropped again from 1994 to 1996) and to increases in the population sizes of several native jellyfish species. In addition, after the *Mnemiopsis* population crashed, there was an increase in the anchovy catch and in field counts of anchovy egg densities. Overall, the decline of *Mnemiopsis* helped speed the recovery of the Black Sea ecosystem, including the fisheries on which people depend for food and income.

CONNECTIONS IN NATURE

From Bottom to Top, and Back Again

The decrease in nutrient inputs and the control of *Mnemiopsis* by *Beroe* had rapid, beneficial effects on the entire Black Sea ecosystem. The speed and magnitude of the Black Sea ecosystem's recovery provide a source of hope, suggesting that it may be possible to solve large problems in other aquatic communities. Note, however, that ecologists rarely attempt to solve such problems by deliberately introducing new predators, such as *Beroe*, because such introductions often have unanticipated negative effects.

The details of the fall and rise of the Black Sea ecosystem also illustrate two important types of causation in ecological communities, bottom-up and top-down controls. The fall of the Black Sea ecosystem began when increased nutrient inputs caused problems associated with eutrophication: increased phytoplankton abundance, increased bacterial abundance, decreased oxygen concentrations, and fish die-offs (see p. 218). The effect of adding nutrients to the Black Sea illustrates **bottom-up control**, which occurs when the abundance of a population is limited by nutrient supply or by the availability of food organisms—in this case, prior to nutrient enrichment, phytoplankton abundance was limited by the supply of nutrients.

Ecosystems are also affected by **top-down control**, which occurs when the abundance of a population is limited by predators. Top-down control appears to have played a major role in both the fall and rise of the Black Sea ecosystem. The arrival of the voracious predator *Mnemiopsis* altered many key features of the ecosystem (e.g., zooplankton abundance, phytoplankton abundance, fish abundance), illustrating how top-down control contributed to the deterioration of conditions in the Black Sea. Top-down control also had a major effect on ecosystem recovery: it took another predator, *Beroe*, to rein in *Mnemiopsis*. In many cases, as in the Black Sea, bottom-up and top-down controls interact to shape how ecosystems work. We'll return to bottom-up and top-down controls in Units 4 and 5, where we consider these important topics in more detail.

SUMMARY

CONCEPT 10.1 Populations exhibit a wide range of growth patterns, including exponential growth, logistic growth, fluctuations, and regular cycles.

- Of the four major patterns of population growth, the first, exponential growth, can occur for a limited time when conditions are favorable.

- The second pattern, logistic growth, is found in populations that increase initially and then level off at a maximum population size, the carrying capacity.

- The third pattern, population fluctuations, is found in all populations. Some populations fluctuate greatly over time; others fluctuate relatively little.

■ The fourth pattern, regular population cycles, is a special type of fluctuation in which alternating periods of high and low abundance occur after nearly constant intervals of time.

CONCEPT 10.2 Delayed density dependence can cause populations to fluctuate in size.

■ There is often a time lag between a change in population density and the effect that change has on future population densities.

■ A version of the logistic equation that includes a time lag suggests that delayed density dependence can produce several types of population fluctuations, including damped oscillations and stable limit cycles.

■ A series of pioneering experiments by A. J. Nicholson indicated that delayed density dependence was a cause of fluctuations in laboratory blowfly populations.

CONCEPT 10.3 The risk of extinction increases greatly in small populations.

■ The risk of extinction increases in populations whose growth rate (λ) varies considerably from one year to the next.

■ Small populations are at much greater risk of extinction than large populations.

■ Small populations can be driven to extinction by chance events associated with genetic drift and inbreeding, demographic stochasticity, environmental stochasticity, and natural catastrophes.

CONCEPT 10.4 Many species have a metapopulation structure in which sets of spatially isolated populations are linked by dispersal.

■ Metapopulations are characterized by repeated extinctions and colonizations.

■ A metapopulation can be doomed to extinction even when suitable habitat remains.

■ Extinction and colonization rates often vary among a metapopulation's patches.

REVIEW QUESTIONS

1. Describe a factor that can cause time lags in the responses of natural populations to changes in population density. How do such time lags affect changes in abundance over time?

2. Summarize how chance events can threaten small populations.

3. A population consists of four unrelated individuals, two females (F1 and F2) and two males (M1 and M2). Individuals live only one year, and they mate only once, producing two offspring (one female, one male) from each mating. Individuals avoid mating with relatives if possible.

   a. Starting with individuals F1, F2, M1, and M2 as the parent generation, can the first two generations of offspring be born to parents that are not related to one another? You may find it helpful to construct a

diagram to illustrate the two generations of parents and their offspring.

   b. If the second generation of offspring become parents, how many of the matings in this third generation of parents can occur between unrelated individuals? Generalizing from your results, is inbreeding likely to be common or uncommon in small populations?

4. a. Explain how a metapopulation can become extinct while suitable habitat remains.

   b. Imagine that human actions create a metapopulation from what was once continuous habitat. If many small and two large habitat patches remain, what arrangement of those patches would make it most likely that the metapopulation would *not* persist?

HANDS-ON PROBLEM SOLVING

Population Dynamics | This Web exercise (www.sinauer.com/ecology) explores how changes in population growth rates and the extent of delayed density dependence affect population dynamics. You will manipulate the values of these two factors, and run a simulation to determine their effects on a population.

SUGGESTED READINGS

Cappuccino, N. and P. W. Price, eds. 1995. *Population Dynamics: New Approaches and Synthesis*. Academic Press, New York. *The collection of papers in this edited volume provides an excellent, in-depth overview of many of the topics covered in this chapter.*

Courchamp, F., T. Clutton-Brock and B. Grenfell. 1999. Inverse density dependence and the Allee effect. *Trends in Ecology and Evolution* 14: 405–410. *This article suggests that Allee effects may be more common than traditionally thought—a finding of considerable relevance for efforts to protect small populations from extinction.*

Hanski, I. 1999. *Metapopulation Ecology*. Oxford University Press, New York. *This book provides an overview of metapopulation ecology, giving readers a thorough grounding in the fundamental principles, current research, and key applications of this increasingly important field.*

Logan, J. A., J. Régnière and J. A. Powell. 2002. Assessing the impacts of global warming on forest pest dynamics. *Frontiers in Ecology and the Environment* 1: 130–137. *As pointed out in this article, insect pests currently have nearly five times the economic impact of fire on North American forests. The article also surveys existing data and results from global warming models to produce a sobering prediction: the severity of insect pest outbreaks will increase as the global climate warms.*

Myers, J. H. 1993. Population outbreaks in forest Lepidoptera. *American Scientist* 81: 240–251. *A survey of population fluctuations in forest butterflies and moths, focusing on factors that cause the remarkable "boom-and-bust" cycles found in some of these species.*

Primack, R. B. 2006. *Essentials of Conservation Biology*. 4th ed. Sinauer Associates, Sunderland, MA. *This well-written, up-to-date text includes excellent chapters on the demographic and genetic problems faced by small populations.*

Shaffer, M. L. 1981. Minimum population sizes for species conservation. *BioScience* 31: 131–134. *This accessible and highly influential paper begins by defining four types of uncertainty that threaten small populations (those due to genetic factors, demographic stochasticity, environmental stochasticity, and natural catastrophes), then goes on to describe methods that are now widely used to estimate how extinction risk varies with population size.*

Turchin, P. 2003. *Complex Population Dynamics: A Theoretical/Empirical Synthesis*. Princeton University Press, Princeton, NJ. *An advanced text on population dynamics, written by a leading researcher in the field.*

UNIT **3**

Interactions among Organisms

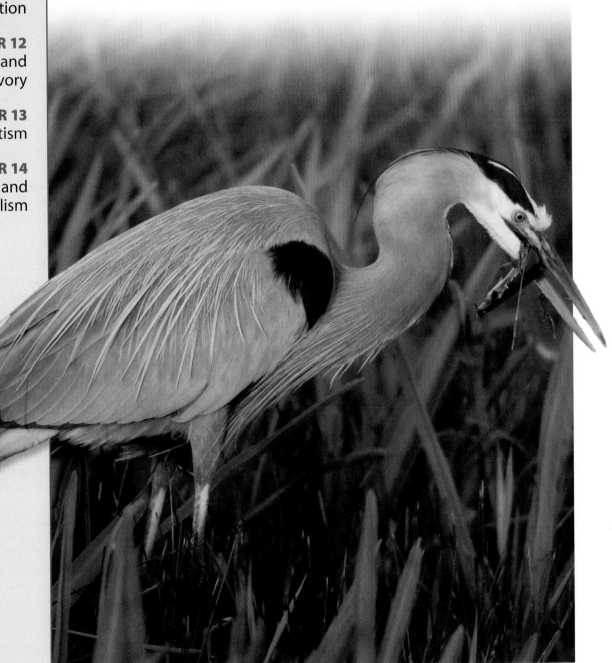

CHAPTER 11 Competition

CASE STUDY

Competition in Plants that Eat Animals

Despite repeated reports that plants could eat animals, early scientists were skeptical of those claims. Charles Darwin (1875) laid their doubts to rest by providing clear experimental evidence of carnivory by plants. Today, more than 600 species of plants that eat animals have been identified, including bladderworts, sundews, pitcher plants, and the well-known Venus flytrap.

Plants use a variety of mechanisms to eat animals. The Venus flytrap has modified leaves that look like fanged jaws, yet attract insects (and occasionally, frogs) with a sweet-smelling nectar (**Figure 11.1**). The inner surface of the leaf has touch-sensitive hairs; if an insect trips those hairs, the leaf snaps shut in half a second. Once the insect is captured, the trap tightens further, forming an airtight seal around its victim, which is digested over the course of 5–12 days.

Other plants lack moving parts, yet still can eat animals. Consider pitcher plants, which can use nectar or visual cues to lure insects into a pitcher-shaped trap. The inside of the pitcher often has downward-facing hairs, which make it easy for the insect to crawl in, but hard to crawl out. What's more, in many pitcher plants, once they are about halfway down, the insects encounter a layer of flaky wax. An insect that steps onto this wax is doomed: the wax sticks to its feet, causing it to lose its grip and tumble into a vat that contains either water (in which it drowns) or deadly digestive juices.

Why do some plants eat animals? The answer may relate to the subject of this chapter: competition. Plants are rooted in the ground and cannot move in search of food. As a result, competition among plants can be intense where soil nutrients are scarce. Many carnivorous plants are found only in environments with nutrient-poor soils. Furthermore, evolutionary relationships among plants reveal that in nutrient-poor environments, the ability to eat animals has evolved multiple times, in a variety of independent plant lineages. Overall, these observations suggest that carnivory in plants may be an adaptation for life in nutrient-poor environments—perhaps providing a way to avoid competing with other plants for soil nutrients.

Turning to field evidence, does competition have a large effect on plants that eat animals? Typically, the root systems of carnivorous plants are less extensive than the root systems of noncarnivorous plants that live in the same area.

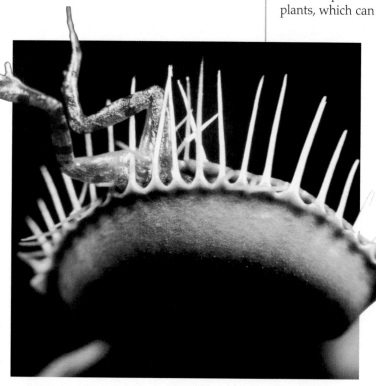

FIGURE 11.1 A Plant that Eats Animals Although the Venus flytrap typically captures insects, it can also feed on other animals—such as this hapless frog.

Hence, competition for soil resources could severely affect carnivorous plants, especially if animal prey were not available or were in short supply.

To test this idea, Stephen Brewer measured how the growth of the pitcher plant *Sarracenia alata* was affected when he cut off its access to prey and when he reduced noncarnivorous competitor species by weeding and clipping. His results show that growth rates in *S. alata* increased dramatically when competitors were reduced (**Figure 11.2**), suggesting that competition had an important effect. But further examination of Figure 11.2 reveals that matters are not as simple as they may first appear. Although the effect of competition for soil nutrients might be expected to increase when the plants were

FIGURE 11.2 Competition Decreases Growth in a Carnivorous Plant To test the effects of competition on the carnivorous pitcher plant, *Sarracenia alata*, an experiment was performed in which (1) the noncarnivorous plants growing next to some *S. alata* plants were weeded and clipped ("neighbors reduced"), and (2) the pitchers of some *S. alata* plants were covered to deprive these plants of animal prey. Relative growth rates were calculated as the logarithm of the mass of a plant at 56 weeks divided by its initial mass. (After Brewer 2003.)

deprived of prey, that did not happen. Instead, it seems that pitcher plants can sidestep some of the negative effects of competition with noncarnivorous plants. How do they do it?

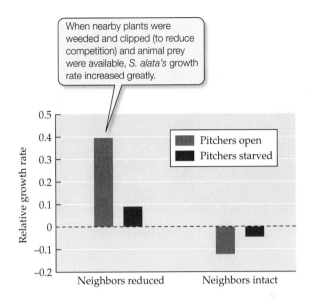

When nearby plants were weeded and clipped (to reduce competition) and animal prey were available, *S. alata*'s growth rate increased greatly.

In 1917, A. G. Tansley reported results from a series of experiments designed to explain the distribution in Britain of two species of bedstraw plants, *Galium hercynicum* and *G. pumilum* (then known as *G. sylvestre*). *G. hercynicum* was restricted to acidic soils, *G. pumilum* to calcareous soils (soils high in calcium, such as those derived from limestone). Even in places where the two species grew within inches of each other, each remained confined to its characteristic soil type. In his experiments, Tansley found that when grown alone, each species could survive on both acidic and calcareous soils. However, when he grew the species together on acidic soils, only *G. hercynicum* survived, while if he grew them together on calcareous soils, only *G. pumilum* survived. Tansley concluded that the two species competed with one another, and that when grown on its native soil type, each species drove the other to extinction. He also inferred that the restriction in nature of *G. hercynicum* and *G. pumilum* to a particular soil type resulted from competition between these species.

Tansley's work on bedstraw is one of the first experiments ever performed on **competition**, an interaction between two species in which each is harmed when they both use a resource that limits their ability to grow and reproduce. Competition can also occur between individuals of a single species, in which case it is referred to as **intraspecific competition**. As a result of intraspecific competition, the resources available to members of a high-density population can be reduced to such an extent that growth, survival, or fertility decreases or migration increases. Thus, intraspecific competition can cause density-dependent reductions in population size, a topic we addressed in Chapter 9. This chapter

focuses on competition between members of different species, or **interspecific competition**. We begin by describing how competition occurs.

Competition for Resources

CONCEPT 11.1
Competition occurs between species that share the use of a resource that limits the growth, survival, or reproduction of each species.

Organisms compete for **resources**, which are features of the environment that are required for growth, survival, or reproduction and which can be consumed or otherwise used to the point of depletion. What differentiates a resource from other components of the environment that organisms need to survive?

Organisms compete for resources such as food, water, light, and space

Food is the most obvious example of a resource. Whether we consider the soil nutrients used by bacteria, the lichens eaten in winter by reindeer, or the meat consumed by Nicholson's blowflies (see p. 224), food is essential for life. When food is scarce, population growth rates plummet.

In terrestrial ecosystems—especially arid ones—water is also a resource. Food and water are intuitive examples of resources because organisms literally consume them: they are

FIGURE 11.3 Space Can Be a Limiting Resource Competition for space can be intense when space is in short supply, as seen for the corals in this coral reef off the coast of Zanzibar.

taken into the body and used to support metabolic activities. But an organism does not need to eat or drink a substance for it to be a resource. Plants "consume" light in the sense that they use it to produce food and that they can deplete (by shading) the supply available to other plants. Space can also be viewed as a resource. Plants, algae, and sessile animals (e.g., barnacles and corals) require space to grow, and competition for space can be intense (**Figure 11.3**). Mobile animals also compete for space as they seek access to good areas in which to hunt or attract mates, or to places that provide refuge from heat, cold, or predators. Although space is not consumed in the sense of being eaten, organisms can fill all the available space—thus depleting it—and when they do, population growth rates decrease. Thus, we can think of space as a resource that is consumed, analogous to food, water, or light.

Species are also strongly influenced by features of the environment that are not consumed, including temperature, salinity, and pH. Organisms may require particular temperatures for growth, survival, or reproduction, but they do not consume or deplete temperatures; hence, temperature is not a resource. The same is true for salinity and pH. Ecologists refer to features of the environment that affect population growth rates but are not consumed or depleted as **physical factors**; physical factors are covered in detail in Chapters 2–5. The same substance can be a physical factor for some organisms and a resource for others. For example, terrestrial mammals and insects consume oxygen but usually do not deplete it; for them, oxygen is a physical factor, not a resource. However, in some aquatic and soil environments, organisms may consume oxygen more rapidly than it is replenished; for such organisms, oxygen is a resource, not a physical factor.

Competing organisms reduce the availability of resources

As implied by our discussion so far, organisms can consume resources to such an extent that resource availability drops and population growth rates fall. In a laboratory experiment, Tilman et al. (1981) examined competition for silica (SiO_2) in freshwater diatoms (single-celled algae). Diatoms use silica to construct their cell walls. Tilman and colleagues grew two diatom species, *Synedra ulna* and *Asterionella formosa*, alone and in competition with each other and measured how diatom population densities and silica concentrations in the water changed over time. When grown alone, each species reduced the resource (silica) to a low and approximately constant concentration; each species also reached a stable population size (**Figure 11.4**). *Synedra* had a lower stable population size than *Asterionella*, and also reduced silica to lower levels than did *Asterionella*. When the two species competed with each other, *Synedra* drove *Asterionella* to extinction, apparently because it reduced silica to such low levels that *Asterionella* could not survive.

Competition can increase in intensity when resources are scarce

Plants can compete for aboveground resources, such as light, as well as for belowground resources, such as soil nutrients. Researchers have suggested that the relative importance of aboveground and belowground competition in plants might change depending on whether aboveground or belowground resources are scarcer: belowground competition, for example, might be expected to increase in importance when the competing plants are growing in nutrient-poor soils. Scott Wilson and David Tilman tested this idea by performing transplant experiments with *Schizachyrium scoparium*, a perennial grass species native to their study site in Minnesota.

Wilson and Tilman selected a series of 5 × 5 m plots of natural vegetation growing in low-nitrogen, sandy soils. For 3 years, they treated half of the plots with high-nitrogen fertilizer each year. This 3-year period gave the plant communities in the fertilized plots time to adjust to the experimentally imposed change in soil nitrogen levels. At the end of the 3-year period, they planted *S. scoparium* individuals in the plots.

Once they were added to the high-nitrogen (fertilized) and low-nitrogen (unfertilized) plots, *S. scoparium* were grown under three treatments: (1) with neighbors left intact, (2) with neighbor roots left intact but neighbor shoots tied back, or (3) with neighbor roots and shoots both removed. In treatment 1 there was both belowground and aboveground competition, while in treatment 3 there was no competition. In treatment 2, the tied-back neighbor shoots did not shade *S. scoparium*, but the act of tying did not appear to affect neighbor roots, so that treatment was interpreted as belowground competition only. Aboveground competition

FIGURE 11.4 Competing Organisms Can Deplete Resources
David Tilman and his colleagues demonstrated competition between two diatom species for silica by growing them alone and in competition with each other. *Synedra* reduced silica concentrations to lower levels than did *Asterionella* (compare the green curve in part B to that in part A). This result may explain why *Synedra* outcompeted *Asterionella* when the two species were grown together (C). (After Tilman et al. 1981.)

was estimated as the amount of competition in treatment 1 minus the amount in treatment 2.

Wilson and Tilman found that while the total competition (the sum of belowground and aboveground competition) did not differ between the low- and high-nitrogen plots, belowground competition was most intense in the low-nitrogen plots (**Figure 11.5A**). They also found that aboveground competition for light increased when light levels were low (**Figure 11.5B**). Thus, their work demonstrates that the intensity of competition can increase when the resource being competed for becomes scarce.

Competition for resources is common in natural communities

The two experiments we have just described provide specific examples of competition for resources such as nutrients and light. More generally, how important is competition in ecological communities? To answer this question, results from many studies must be compiled and analyzed. The findings of three such analyses indicate that competition is common. For example, Schoener (1983) examined the results of 164 published studies on competition and found that of 390 species studied, 76% showed effects of competition under

FIGURE 11.5 Resource Availability Affects the Intensity of Competition (A) Transplant experiments with the grass *Schizachyrium scoparium* showed that belowground competition between plant species for nutrients increases in intensity when soil nutrients are scarce. (B) Similarly, aboveground competition for light increases as light levels decrease. (After Wilson and Tilman 1993.)

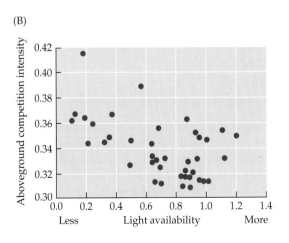

some circumstances, and 57% showed effects of competition under all circumstances tested. Connell (1983) examined the results of 72 studies and found that competition was important for 50% of 215 species. Gurevitch et al. (1992) took a different approach: they did not report the percentage of species for which competition was important, but rather analyzed the magnitude of competitive effects found for 93 species in 46 studies published between 1980 and 1989. They showed that competition had significant (though variable) effects on a wide range of organisms, including carnivores, herbivores, and producers such as plants.

Surveys such as those by Schoener, Connell, and Gurevitch et al. face potential sources of bias, including those that result from a failure to publish studies that show no significant effects, and from a tendency for investigators to study "interesting" species (i.e., those they suspect will show competition). Despite such potential sources of bias, the fact that hundreds of studies have documented effects of competition makes it clear that competition is common—though not ubiquitous—in nature.

General Features of Competition

CONCEPT 11.2
Competition, whether direct or indirect, can limit the distributions and abundances of competing species.

From the earliest beginnings of ecology as a field of science, ecologists thought that competition between species was important in natural communities. For example, although he often focused on intraspecific competition, Charles Darwin (1859) also argued that competition between species could influence both evolutionary processes and the species found in different geographic regions. Tansley took Darwin's reasoning one step further by conducting experiments designed to demonstrate that competition could limit the distributions of species. In this section, we examine recent studies that have followed Tansley's lead in documenting how competition affects the distributions and abundances of organisms. But first we'll describe direct and indirect forms of competition, along with several other general features of competitive interactions.

Species may compete directly or indirectly

In some cases, individuals perform antagonistic actions that directly interfere with the ability of their competitors to use a limiting resource. More typically, however, **exploitation competition** occurs in which species compete indirectly through their mutual effects on the availability of a shared resource. In this case, competition occurs simply because individuals reduce the availability of a resource as they use it. We have already discussed several examples of exploitation

competition, including Brewer's work on pitcher plants (see p. 241) and Tilman et al.'s work on diatoms (see Figure 11.4). Hence, this section focuses on **interference competition**, which occurs when species compete directly for access to a resource that both require, such as food or space.

The antagonistic actions that characterize interference competition are perhaps most familiar in mobile animals, as when two predators fight over prey that one of them has caught. Similarly, herbivores such as voles may aggressively exclude other voles from preferred habitat, and members of warring ant colonies may kidnap and slaughter one another. Interference competition can also occur among sessile animals. For example, as it grows, the acorn barnacle *Semibalanus balanoides* often crushes or smothers nearby individuals of another barnacle species, *Chthamalus stellatus*. As a result, *Semibalanus* directly prevents *Chthamalus* from living in most portions of a rocky intertidal zone (on p. 246 we'll provide more information about competition in these barnacles).

Interference competition may also take the form of **allelopathy**, in which individuals of one species release toxins that harm individuals of other species. A possible example of allelopathy involves the release of toxins by spotted knapweed (*Centaurea maculosa*). This invasive plant species, which is native to Eurasia, was introduced accidentally into North America in the late 1800s as a contaminant in crop seeds. Since then, it has spread rapidly through the United States and Canada, becoming common in western and northern regions.

Spotted knapweed causes substantial economic damage each year, crowding out native plants and reducing the value of rangeland. In Montana alone, *C. maculosa* has infested more than 2 million acres and causes an estimated $42 million in annual losses. One reason for spotted knapweed's success is that cattle avoid eating it, giving it an edge over plants that they like to eat, such as native grasses. In addition, *C. maculosa* appears to engage in chemical warfare by releasing a toxin called catechin into surrounding soils. In the laboratory, Bais et al. (2003) have shown that catechin reduces the germination and growth of native grasses (**Figure 11.6**). Bais and colleagues also examined the mechanisms by which catechin affected native plants. By combining an experimental approach with genetic analyses, they discovered that exposure to catechin activates hundreds of genes in the native grasses that compete with spotted knapweed. Some of those genes cause the root cells of the grasses to commit "suicide," hence killing the roots and providing spotted knapweed with a potential competitive advantage.

Competition is often asymmetrical

When two species compete for a resource that is in short supply, each obtains less of the resource than it could have obtained if the competitor were not present. Because competition reduces the resources available for growth, survival, and reproduction for both species, each species is harmed

FIGURE 11.6 Chemical Warfare in Plants
(A) The invasive spotted knapweed (*Centaurea maculosa*) has outcompeted native plants and spread to cover large regions in western North America. A possible reason for spotted knapweed's success is that it secretes a toxin called catechin into the soil. In laboratory studies, catechin reduced (B) the germination rates and (C) the growth rates of two of *C. maculosa's* competitors, the native grasses *Festuca* and *Koeleria*. (B,C after Bais et al. 2003.)

tinction. Before it goes extinct, the inferior competitor uses some resources, and hence reduces the resources available to the superior competitor. If we were to estimate how strongly each competitor affected the other, however, the effect of the superior competitor would be much greater than the effect of the inferior competitor.

Competition can occur between closely or distantly related species

We've seen that competition can occur between pairs of closely related species, such as the bedstraw species studied by Tansley. Brown and Davidson (1977) performed experiments with rodents and ants to test whether competition also occurs between groups of more distantly related species. They suspected that rodents and ants might compete because both eat the seeds of desert plants, and the sizes of the seeds they eat overlap considerably (**Figure 11.7**).

Brown and Davidson established experimental plots (each about 1,000 m² in area) in a desert region near Tucson, Arizona. Their experiment lasted 3 years and used four treatments: (1) plots in which a ¼-inch wire mesh fence excluded seed-eating rodents, and where rodents within the fence were removed by trapping; (2) plots in which seed-eating ants were excluded by applying insecticides; (3) plots in which both rodents and ants were excluded by fencing, trapping, and applying insecticides; and (4) plots in which both rodents and ants were left undisturbed (the control).

to some extent. In many cases, however, the effects of competition are unequal, or asymmetrical: one species is harmed more than the other. This asymmetry is especially clear in situations in which one competitor drives the other to ex-

FIGURE 11.7 Ants and Rodents Compete for Seeds There is extensive overlap in the sizes of seeds eaten by ants and by rodents. Removal experiments showed that these two unrelated groups compete for this food source. (After Brown and Davidson 1977.)

Their results indicated that rodents and ants compete for food. Relative to the control plots, the number of ant colonies increased by 71% in the plots from which rodents were excluded, and rodents increased by 18% in number and 24% in biomass in the plots from which ants were excluded. When both rodents and ants were excluded (treatment 3), the availability of seeds increased by 450% compared with all other plots. Treatments 1 (no rodents), 2 (no ants) and 4 (the control, with both rodents and ants present) all resulted in similar densities of seeds. These results suggested that when either rodents or ants were removed, the group that remained ate roughly as many seeds as rodents and ants combined ate in the control plots. Thus, under natural conditions, each group would be expected to eat fewer seeds in the presence of the other group than it could eat when alone.

It is not surprising that species as different as ants and rodents compete. After all, people differ greatly from bacteria, fungi, and insects, yet we compete with these organisms for food in farm fields, in grain storage bins—even in our refrigerators. Overall, irrespective of whether they are closely or distantly related, organisms can compete if they share the use of a resource whose supply is limited.

Competition can determine the local and geographic distributions of species

As we saw earlier in this chapter, Schoener (1983), Connell (1983), and Gurevitch et al. (1992) analyzed the results of hundreds of field studies and found that competition had important effects on many, but not all, species. Typically, in species for which it was important, competition reduced the growth, survival, or reproduction of one or both competitors, thus limiting their abundance. In some cases, competition also affected where the species lived. Such results indicate that competition can limit the distributions and abundances of species. As we'll see, when competition influences the distributions of species, it may restrict species to certain portions of a local region, or it may determine the geographic distributions of species (or both).

Competition and physical factors limit the local distributions of intertidal barnacles In a series of classic experiments, Joseph Connell (1961a,b) examined factors that influenced the distribution, survival, and reproduction of two barnacle species, *Chthamalus stellatus* and *Semibalanus balanoides* (formerly known as *Balanus balanoides*). The larvae of these and other barnacle species drift through ocean waters, then settle on rocks or other surfaces (such as boat hulls), where they mature into adults, forming a hard outer shell.

At Connell's study site along the coast of Scotland, the distributions of *Chthamalus* and *Semibalanus* larvae overlapped considerably: the larvae of both species were found on rocks located throughout the upper and middle intertidal zones. However, adult *Chthamalus* usually were found only near the top of the intertidal zone, whereas adult *Semibalanus* were not found there, but were found throughout the rest of the intertidal zone (**Figure 11.8**). What caused adult *Chthamalus* to be excluded from all but the top of the intertidal zone? Similarly, what excluded adult *Semibalanus* from the top of the intertidal zone?

To answer these questions, Connell examined the effects of competition and of physical factors such as desiccation (drying out due to exposure to air, which is greatest in the upper intertidal zone). To test the importance of competition, he chose some individual young barnacles of each species that had settled in each zone and removed all nearby members of the other species; for other focal individuals, he left nearby members of the other species in place. He found that competition with *Semibalanus* excluded *Chthamalus* from all but the top of the intertidal zone. As they grew, *Semibalanus* smothered (by growing on top of), removed (by growing underneath, hence prying off the rocks), and crushed the *Chthamalus*. Averaging across all regions of the intertidal zone, only 14% of *Chthamalus* survived their first year when faced with competition from *Semibalanus*, whereas 72% survived where Connell had removed *Semibalanus*. *Chthamalus* that survived a year of competition with *Semibalanus* were small and reproduced poorly.

FIGURE 11.8 Squeezed Out by Competition
Removal experiments at a field site in Scotland showed that competition determines the local distribution of the barnacle *Chthamalus stellatus*.

Semibalanus, in contrast, was not affected strongly by competition with *Chthamalus*. However, whether *Chthamalus* was removed or not, *Semibalanus* dried out and survived poorly near the top of the intertidal zone. Thus, *Semibalanus* appears to have been excluded from that zone not by competition, but by its sensitivity to desiccation.

Competition can determine the geographic distributions of species   As observed for Tansley's bedstraw plants and Connell's barnacles, competition can restrict the local distribution of a species to a particular set of environmental conditions—the bedstraws, for example, could be growing inches away from each other, but each species was restricted to a particular soil type. Competition has also been shown to prevent a wide range of species, including mammals, marine invertebrates, birds, and plants, from occupying geographic regions in which they otherwise would thrive.

In some cases, a "natural experiment" provides evidence that competition affects geographic distributions. The term *natural experiment* refers to a situation in nature that is similar in effect to a controlled removal experiment like the one Connell performed on barnacles. Consider a hypothetical pair of species that compete with each other, species A and species B. Assume that, simply by chance, some regions of suitable habitat are occupied only by species A, while other similar regions are occupied only by species B, and still other regions are occupied by both species. If the absence (i.e., "removal") of species A consistently expands the areas in which species B is found (or vice versa), that would suggest that competition affects the geographic distribution of these species.

Such a result was found for chipmunks in the genus *Neotamias* (previously known as *Eutamias*). These chipmunks live in forests on mountains in the southwestern United States, where mountain ranges are separated from one another by desert flatlands. Patterson (1980, 1981) studied the distribution of *Neotamias* chipmunks and found that when a species lived alone on a mountain range, it consistently occupied a broader range of habitats and elevations than when it lived with a competitor species (**Figure 11.9**). As in Connell's removal experiments, this result suggests that competition may have prevented some *Neotamias* chipmunk species from living in areas of otherwise suitable habitat.

As illustrated by *Neotamias* chipmunks and Connell's barnacles, a species may limit the distribution of a competitor without driving the competitor to extinction. On the other hand, Tilman's experiments with diatoms suggest that competition can lead to the complete removal of one of the competing species. Why do competitors coexist in some situations, but not others?

Competitive Exclusion

CONCEPT 11.3
Competing species are more likely to coexist when they use resources in different ways.

Connell's work on barnacles provides a starting point for understanding the different outcomes that can occur when species compete for resources. In habitats where both species could thrive, *Semibalanus* typically drove *Chthamalus* to extinction, yet *Chthamalus* could persist because it could grow and reproduce in habitats that were too dry for *Semibalanus*. The picture that emerges from Connell's results is this: if the overall ecological requirements of a species—its *ecological niche* (see Chapter 8)—are very similar to those of a superior competitor, that competitor may drive it to extinction. We turn now to experiments, field observations, and mathematical models that examine this idea. In particular, we will see how the similarity of the resources the competitors use affects the likelihood that one will drive the other to extinction.

Complete competitors cannot coexist

In the 1930s, the Russian ecologist G. F. Gause performed laboratory experiments on competition using three species

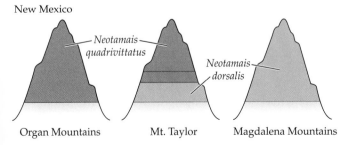

FIGURE 11.9 A "Natural Experiment" on Competition between Chipmunks Observations of the distributions of *Neotamias* chipmunks in the southwestern United States suggest that competition may restrict the habitats in which they live.

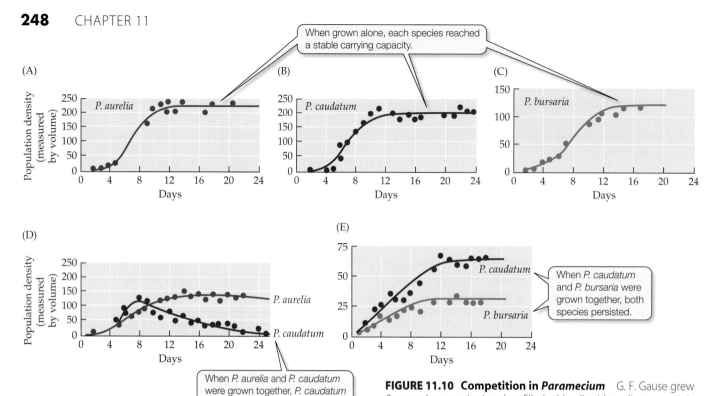

FIGURE 11.10 Competition in *Paramecium* G. F. Gause grew *Paramecium* species in tubes filled with a liquid medium containing bacteria and yeast cells. *P. aurelia* and *P. caudatum* both fed mainly on floating bacteria, while *P. bursaria* fed mainly on yeast cells. (After Gause 1934b.)

of the single-celled protist *Paramecium*. He constructed miniature aquatic ecosystems by growing *Paramecium* in tubes filled with a liquid medium that contained bacteria and yeast cells as a food supply. He found that populations of each of the three *Paramecium* species reached a stable carrying capacity when grown alone (**Figure 11.10A–C**). But when pairs of these species competed with each other, several different outcomes were obtained.

For example, when *P. aurelia* was grown in competition with *P. caudatum*, *P. aurelia* drove *P. caudatum* to extinction (**Figure 11.10D**). These species may have been unable to coexist because they both fed primarily on bacteria floating in the medium, which led to considerable overlap in their food requirements. In contrast, when *P. caudatum* was grown with *P. bursaria*, neither species drove the other to extinction (**Figure 11.10E**). Although *P. caudatum* and *P. bursaria* coexisted, it was clear that they competed for resources, because the carrying capacity of each was lowered by the presence of the other. Gause suggested that *P. caudatum* and *P. bursaria* coexisted because they tended to use different resources: *P. caudatum* usually ate bacteria floating in the medium, while *P. bursaria* usually fed on yeast cells that settled to the bottom of the tubes.

Experiments with a wide range of other species (e.g., algae, flour beetles, plants, and flies) have yielded similar results: one species drives the other to extinction unless the two species use the available resources in different ways. Such results led to the formulation of the **competitive exclusion principle**, which states that two species that use a limiting resource in the same way cannot coexist indefinitely. As we'll see next, field observations are consistent with this explanation of why competitive exclusion occurs in some cases, but not others.

Competitors may coexist when they use resources differently

In natural communities, many species use the same limiting resources, yet manage to coexist with one another. This observation does not violate the competitive exclusion principle because a key feature of that principle is that species must use limiting resources in the same way. Field studies, however, often reveal differences in how species use limiting resources. Such differences are referred to as **resource partitioning**.

Thomas Schoener studied resource partitioning in four lowland *Anolis* lizards that live on the West Indian island of Jamaica: *A. opalinus*, *A. grahami*, *A. lineatopus*, and *A. valencienni*. Although these species all live together in trees and shrubs and eat similar foods, Schoener found differences in the height and thickness of their perches (**Figure 11.11**), and in the time they spent in sun or shade (not shown).

Specifically, Schoener found that within each species, individuals of different sizes used different portions of their habitat: larger individuals were found on thicker perches. When different species were compared, however, the reverse trend was found: the larger the species, the thinner the perch-

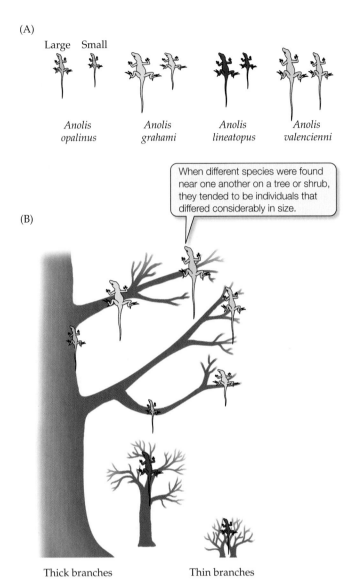

(A)

Large Small

Anolis opalinus   *Anolis grahami*   *Anolis lineatopus*   *Anolis valencienni*

(B)

When different species were found near one another on a tree or shrub, they tended to be individuals that differed considerably in size.

Thick branches   Thin branches

FIGURE 11.11 Resource Partitioning in Lizards In Jamaica, Thomas Schoener studied the use of space by males of the four species of *Anolis* lizards shown in (A). Within each species, larger individuals tend to use thicker perches, but among species, smaller lizards use thicker perches. (B) A diagram illustrating some of the common positions in which these lizard species might be found in trees and shrubs. The lizards are scaled according to their actual sizes, but the colors used here are fanciful. (After Schoener 1974.)

es it used. This result may sound odd, but the lizards managed it because the large lizard species had evolved long bodies with short legs, which enabled them to distribute their weight well on thin perches. Overall, the differences within and among *Anolis* species meant that when different species were found near one another on a tree or shrub, they tended to be species that differed considerably in size (see Figure 11.11). As a result, they were likely to consume different foods

and hence to compete less intensely than they otherwise would.

Resource partitioning has been found in many other species, including protists, birds, fishes, crustaceans, and plants. Overall, studies of resource partitioning suggest that species can coexist if they use resources in different ways—a general inference that is also supported by results from mathematical models of competition.

Competition can be modeled by modifying the logistic equation

Working independently of each other, A. J. Lotka (1932) and Vito Volterra (1926) both modeled competition by modifying the logistic equation; we describe their approach in **Web Extension 11.1**. Their equation, now known as the **Lotka–Volterra competition model**, can be written as

$$\frac{dN_1}{dt} = r_1 N_1 \left(1 - \frac{\left(N_1 + \alpha N_2 \right)}{K_1} \right)$$

$$\frac{dN_2}{dt} = r_2 N_2 \left(1 - \frac{\left(N_2 + \beta N_1 \right)}{K_2} \right)$$

(11.1)

In these equations, N_1 is the population density of species 1, r_1 is the intrinsic rate of increase of species 1, and K_1 is the carrying capacity of species 1; N_2, r_2, and K_2 are similarly defined for species 2. The **competition coefficients** (α and β) are constants that describe the effect of one species on the other: α is the effect of species 2 on species 1, and β is the effect of species 1 on species 2. We explore the meaning of these competition coefficients in more detail in **Box 11.1**.

Analyses of Equation 11.1 support the idea that *competitive exclusion is likely when competing species require very similar resources*. In the remainder of this section, we'll describe the two-step process by which this support is obtained. First, we'll see how Equation 11.1 can be employed to predict changes in the densities of species 1 and 2 over time; then we will relate those changes to the ways in which the two species use resources.

To complete the first step of this process, a computer could be programmed to solve Equation 11.1, thereby predicting the population densities of species 1 and 2 at different times. Here, however, we use a graphical approach to examine the conditions under which each species is expected to increase or decrease in abundance.

We begin by determining when the population of each species stops changing in size (**Box 11.2**). Equations 1 and 2 in Box 11.2 are straight lines, written with N_2 as a function of N_1. These lines are called **zero population growth isoclines**, so named because the population does not increase or decrease in size for any combination of N_1 and N_2 that lies on these lines. We can use these zero growth isoclines to determine the conditions under which each species will increase

BOX 11.1

What Do the Competition Coefficients α and β Represent?

Viewed in a general way, the competition coefficient α represents the overall competitive effect of species 2 on species 1, while β represents the overall effect of species 1 on species 2. When competition occurs, α and β are each greater than zero.

More specifically, α measures the extent to which the use of resources by an individual of species 2 decreases the per capita growth rate of species 1. When α = 1, individuals of the two species are identical in their effects—adding an individual of either species to the environment decreases the growth rate of species 1 by *exactly* the same amount. In contrast, when

α < 1, an individual of species 2 decreases the growth of species 1 by a *smaller* amount than does an individual of species 1. Finally, when α > 1, an individual of species 2 decreases the growth of species 1 by a *larger* amount than does an individual of species 1.

Similar reasoning can be applied to β, only in that case we would be considering whether an individual of species 1 has an equal, smaller, or larger effect on the per capita growth rate of species 2 than does an individual of species 2.

We can also think of α and β as "translation terms," each of which converts the number of

individuals of one species into the number of individuals of the other species that has an equivalent effect on population growth rates. For example, if α = 3, each individual of species 2 decreases the growth of species 1 by the same amount as would three individuals of species 1. Thus, if there are 100 individuals of species 2, it would take 300 individuals of species 1 to decrease the growth rate of species 1 by the same amount as do the 100 individuals of species 2 (that is, α = 3 and N_2 = 100, so it takes αN_2 = 3 × 100 = 300 individuals of species 1 to have an equivalent effect).

---

or decrease in abundance. For example, in **Figure 11.12A**, at point A to the right of the N_1 isocline, there are N_1^A individuals of species 1. Because N_1^A is greater than the number of individuals that would produce zero population growth (N_1^*), the density of species 1 will decrease at point A on the graph. This is true for the entire shaded region: the density of species 1 decreases for all points to the right of the N_1 isocline. In contrast, when the density of species 1 is to the left of the

N_1 isocline, the density of species 1 increases. Similar reasoning applies to species 2, only in this case—because N_2 is plotted on the *y* axis—the density of species 2 decreases in regions above the N_2 isocline and increases in regions below the N_2 isocline (**Figure 11.12B**).

The graphical approach we have just described can be used to predict the end result of competition between species. To do this, we plot the N_1 and N_2 isoclines together. Because there

BOX 11.2

When Do Competing Populations Stop Changing in Size?

As described in Chapter 9, the derivative dN/dt represents the rate at which a population grows over time. When $dN/dt = 0$, the population stops growing and hence remains constant in size unless perturbed in some way (e.g., if many individuals die in a catastrophic fire or storm). Hence, according to Equation 11.1, the population density of species 1 does not change over time when

$$\frac{dN_1}{dt} = r_1 N_1 \left(1 - \frac{(N_1 + \alpha N_2)}{K_1} \right) = 0$$

In this equation, $dN_1/dt = 0$ when $N_1 = 0$ or when

$$\left(1 - \frac{(N_1 + \alpha N_2)}{K_1} \right) = 0$$

The first case ($N_1 = 0$) just states the obvious (a population with no members does not grow), so we rearrange the equation for the second case to find

$$N_2 = \frac{K_1}{\alpha} - \frac{1}{\alpha} N_1 \qquad (1)$$

If we use a similar approach for species 2, we find that $dN_2/dt = 0$ when

$$N_2 = K_2 - \beta N_1 \qquad (2)$$

Equations 1 and 2 represent straight lines when values of N_2 are plotted on the *y* axis and values of N_1 are plotted on the *x* axis of a graph (as we do in Figure 11.12 on the facing page). The *y* intercept of Equation 1 is K_1/α, and the slope of Equation 1 is $-1/\alpha$, indicating that the

line falls from left to right. For Equation 2, the *y* intercept and slope are K_2 and $-\beta$, respectively.

Equation 1 specifies all the combinations of the densities of species 1 and 2 for which the population density of species 1 does not change over time—this is true because we obtained Equation 1 by *assuming* that the population density of species 1 does not change (i.e., that $dN_1/dt = 0$). Similarly, Equation 2 specifies all the combinations of the densities of species 1 and 2 for which the population density of species 2 does not change. Curves such as Equations 1 and 2 that are obtained by setting $dN/dt = 0$ are known as *zero population growth isoclines*.

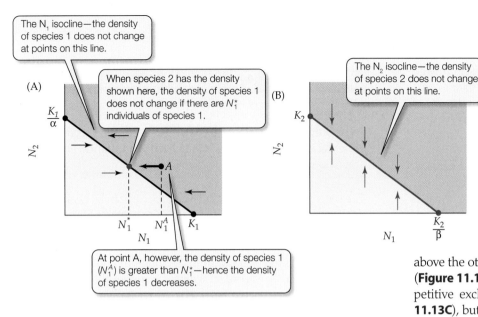

(A) The N₁ isocline—the density of species 1 does not change at points on this line.

When species 2 has the density shown here, the density of species 1 does not change if there are N_1^* individuals of species 1.

At point A, however, the density of species 1 (N_1^A) is greater than N_1^*—hence the density of species 1 decreases.

(B) The N₂ isocline—the density of species 2 does not change at points on this line.

FIGURE 11.12 Graphical Analyses of Competition The zero population growth isoclines from the Lotka–Volterra competition model can be used to predict population growth rates in competing species. (A) The N_1 isocline. The density of species 1 increases at all points in the lightly shaded region to the left of the isocline and decreases at all points in the shaded region to the right of the isocline. (B) The N_2 isocline. The density of species 2 increases at all points in the lightly shaded region and decreases at all points in the shaded region.

are four possible ways that the N_1 and N_2 isoclines can be arranged relative to each other, we must make four different graphs. In two of these graphs, the isoclines do not cross, and competitive exclusion results: depending on which isocline is

above the other, either species 1 (**Figure 11.13A**) or species 2 (**Figure 11.13B**) always drives the other to extinction. Competitive exclusion also occurs in the third graph (**Figure 11.13C**), but which species "wins" depends on whether the changing population densities of the two species first enter the region shown in yellow (in which case species 2 drives species 1 to extinction) or the region shown in blue (in which case species 1 drives species 2 to extinction). Finally, **Figure 11.13D**

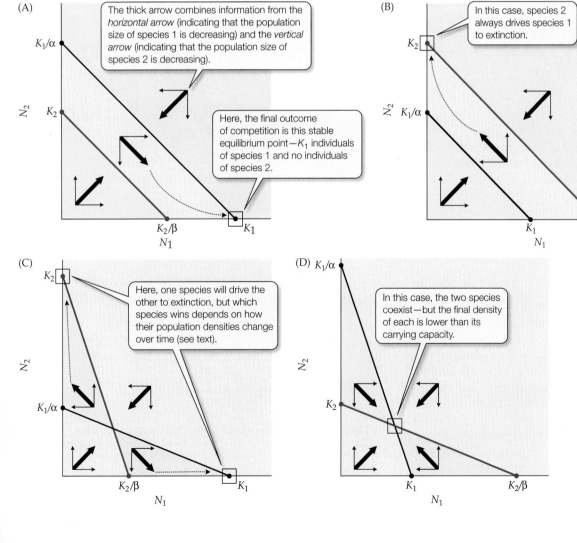

(A) The thick arrow combines information from the *horizontal* arrow (indicating that the population size of species 1 is decreasing) and the *vertical* arrow (indicating that the population size of species 2 is decreasing).

Here, the final outcome of competition is this stable equilibrium point—K_1 individuals of species 1 and no individuals of species 2.

(B) In this case, species 2 always drives species 1 to extinction.

(C) Here, one species will drive the other to extinction, but which species wins depends on how their population densities change over time (see text).

(D) In this case, the two species coexist—but the final density of each is lower than its carrying capacity.

FIGURE 11.13 Outcome of Competition in the Lotka–Volterra Competition Model The outcome of competition differs depending on the arrangement of the N_1 and N_2 isoclines. (A) Competitive exclusion of species 2 by species 1; species 1 always wins. (B) Competitive exclusion of species 1 by species 2; species 2 always wins. (C) The two species cannot coexist; either species 1 or species 2 wins. (D) Species 1 and species 2 coexist. Each box indicates a *stable equilibrium point*—a combination of densities of the two species that once obtained, does not change over time.

shows the only case in which the two species coexist. Although in this case neither species drives the other to extinction, competition still has an effect: the final or equilibrium density of each species (indicated by the box in the figure) is lower than its carrying capacity, as in Gause's experiments with *Paramecium* (compare Figures 11.10B,C, and E).

Now that we've seen the four possible outcomes predicted by the Lotka–Volterra competition model, let's revisit our goal of using the model to explain why competitive exclusion occurs in some situations, but not others. To do this, we'll focus on the single case in which competitive exclusion does *not* occur. As described in **Web Extension 11.2**, we can use Figure 11.13D to show that coexistence occurs when the values of α, β, K_1, and K_2 are such that the following inequality holds:

$$\alpha < \frac{K_1}{K_2} < \frac{1}{\beta} \qquad \textbf{(11.2)}$$

To see what we can learn from this inequality, consider a situation in which the competing species are equally strong competitors, suggesting that $\alpha = \beta$. If the two species are also very similar in how they use resources, an individual of species 1 will have nearly the same effect on the growth rate of species 2 as would an individual of species 2 (and vice-versa). Thus, when the two species use resources in very similar ways, α and β should be close to 1 (see Box 11.1).

Suppose, for example, that $\alpha = \beta = 0.95$. If we substitute these values for α and β into Equation 11.2, we obtain

$$0.95 < \frac{K_1}{K_2} < 1.053$$

This result suggests that when competing species are very similar in how they use resources, coexistence is predicted only within a narrow range of values for the carrying capacities, K_1 and K_2.

In contrast, if the competing species differ greatly in how they use resources, α and β will differ considerably from 1. To illustrate this case, suppose that $\alpha = \beta = 0.1$. In this situation, coexistence is predicted within a much broader range of carrying capacities, namely,

$$0.1 < \frac{K_1}{K_2} < 10$$

As you can demonstrate on your own, other values for the competition coefficients α and β yield similar results. Taken together, such analyses of the Lotka–Volterra competition model suggest the following refinement of the competitive exclusion principle: competing species are more likely to coexist (and hence competitive exclusion is *less* likely) when they use resources in very different ways.

A variety of factors can influence how species divide their use of resources, thereby preventing one competitor from driving the other to extinction. As we'll see in the next section, some of these factors can alter the outcome of competition entirely, turning the inferior competitor into the superior one.

Altering the Outcome of Competition

CONCEPT 11.4
The outcome of competition can be altered by environmental conditions, species interactions, disturbance, and evolution.

The outcome of competition between species can be changed by a broad suite of factors, including physical features of the environment, interactions with other species, disturbance, and evolution. For example, a difference in the conditions of the environment—as might occur from one place to another—can cause a *competitive reversal*, in which the species that was the inferior competitor in one habitat becomes the superior competitor in another. Cases in which the outcome of competition is different under different conditions include Tansley's bedstraws (see p. 241) and the results shown in Figure 8.8.

Interactions with other species can have the same effect. The presence of herbivores has been shown to reverse the outcome of competition between species of encrusting marine algae (Steneck et al. 1991) and between ragwort (*Senecio jacobaea*) and other plant species (**Figure 11.14**). What is true

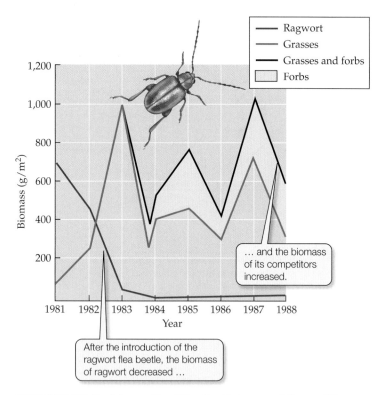

FIGURE 11.14 Herbivores Can Alter the Outcome of Competition Ragwort flea beetles are herbivores that feed on ragwort (*Senecio jacobaea*), an invasive plant species. In the absence of the flea beetle, ragwort is a superior competitor. The graph tracks the biomasses of ragwort, grasses, and forbs (broad-leaved herbaceous plants) at a site in western Oregon after the beetle was introduced there in 1980. (After McEvoy et al. 1991.)

of herbivores is also true of predators, pathogens, and mutualists: an increase or decrease in the abundance of such species can change the outcome of competition among the species with which they interact.

In later chapters, we'll explore many examples in which species interactions alter competitive outcomes—sometimes preventing a superior competitor from driving other species to extinction. Here, we focus on the effects of disturbance and evolution.

Disturbance can prevent competition from running its course

As we saw in Chapter 8, a *disturbance* such as a fire or major storm may kill or damage some individuals while creating opportunities for others. Some species can persist in an area only if such disturbances occur regularly. Forests, for example, contain some herbaceous plant species that require abundant sunlight and are therefore found only in areas where wind or fire has created an opening in the tree canopy. Over time, a population of such plants is doomed: as trees recolonize the area, shade increases to a point at which the species cannot persist. Such species are called **fugitive species** because they must disperse from one place to another as environmental conditions change.

Robert Paine has described how periodic disturbance allows a fugitive algal species, the sea palm *Postelsia palmae-formis*, to coexist with a competitively dominant species, the

mussel *Mytilus californianus*. Sea palm is a brown alga that lives in the intertidal zone and must attach itself to rocks to grow. It competes for attachment space with mussels. Although a sea palm can outcompete an individual mussel (by growing on top of it), the sea palm is eventually displaced by other mussels that grow in from the side.

Competition with mussels causes sea palm populations to decline over time (**Figure 11.15**). Hence, if competition ran its course, mussels would drive sea palm populations to extinction. That is exactly what happens on low-disturbance shorelines (with a mean rate of 1.7 disturbances per year), where waves only occasionally tear patches of mussels from the rocks. However, sea palms can persist in shoreline areas where high-energy waves remove mussels more frequently (with a mean rate of 7.7 disturbances per year) thereby creating temporary openings for sea palm individuals. Similarly, in ecosystems such as grasslands or savannas, in which plants or animals depend on periodic fires to lower the abundance of a superior competitor, disturbance prevents competition from running its course and thus promotes species coexistence.

Evolution by natural selection can alter the outcome of competition

When two species compete for resources, natural selection may favor individuals whose phenotype either (1) allows them to be superior competitors or (2) allows them to partition the available resources, thus decreasing the intensity of

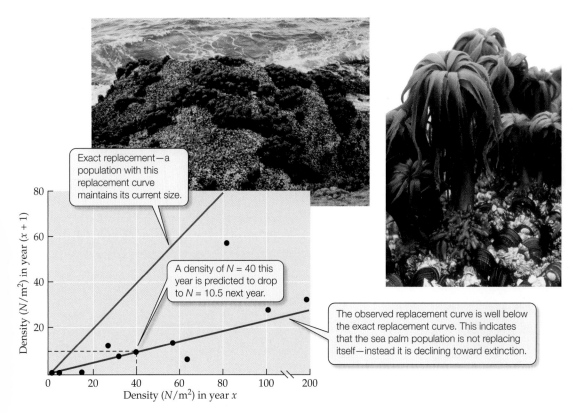

Exact replacement—a population with this replacement curve maintains its current size.

A density of N = 40 this year is predicted to drop to N = 10.5 next year.

The observed replacement curve is well below the exact replacement curve. This indicates that the sea palm population is not replacing itself—instead it is declining toward extinction.

FIGURE 11.15 Population Decline in an Inferior Competitor In this graph, each point represents an observed change in density (*N*, the number of individuals per m²) from one year (year *x*) to the next (year *x* + 1) at sites where sea palms are growing in competition with mussels. These points can be used to estimate a *replacement curve* that shows the extent to which sea palm individuals replace themselves over time. In exact replacement, the density at a given site does not change from one year to the next. (After Paine 1979.)

competition. In principle, then, competition has the potential to cause evolutionary change, and evolution has the potential to alter the outcome of competition. The interplay between competition and evolution can be observed in both experimental and observational studies.

Evolutionary change can reverse the outcome of competition

Pimentel and colleagues studied the effect of natural selection on two competing fly species, the housefly *Musca domestica* and the green blowfly *Lucilia sericata* (referred to as *Phaenicia sericata* in Pimentel et al. 1965). The larvae of these species compete for foods such as decaying animal carcasses, garbage, and the feces of chickens, humans, and carnivores. Houseflies and blowflies have similar life cycles (taking about 14 days to complete) and can be reared in laboratory cages, which makes them excellent study organisms.

Pimentel and colleagues reared houseflies and blowflies in laboratory cages to study competition between them. Larvae were fed a gel (consisting of nonfat dried milk, brewer's yeast, and agar) topped with a gram of liver; adults were fed a lump of sugar. The researchers released 100 male and 100 female houseflies into one end of an experimental arena that consisted of 16 cages linked by tubes through which the flies could crawl; the same number of blowflies were released into the opposite end of the arena (**Figure 11.16A**). The housefly initially appeared to be the superior competitor, increasing rapidly in density while the blowfly was restricted to a few cages along one side of the arena. Over time, however, the situation changed dramatically: by week 45, the blowfly population began to increase in size; by week 55, there were more blowflies than houseflies; and by week 65, the houseflies had been driven to extinction (**Figure 11.16B**).

Shortly before the blowfly population began to increase in size, Pimentel and colleagues removed individuals of both species from the 16-cell cage to test whether evolutionary change had occurred. To perform this test, they established 5 cages for each of the following four treatments: (1) "wild" (newly caught) houseflies competing against wild blowflies; (2) wild houseflies competing against "experimental" blowflies (drawn from the 16-cell cage); (3) experimental houseflies competing against wild blowflies; and (4) experimental houseflies competing against experimental blowflies.

In this experiment, treatments 3 and 4 used houseflies that had had a chance to evolve in competition with blowflies; similarly, treatments 2 and 4 used blowflies that had had a chance to evolve in competition with houseflies. The results showed that wild and experimental housefly populations usually outcompeted wild blowfly populations (7 out of 10 times). The situation was reversed, however, when experimental blowflies were used: the experimental blowfly populations always outcompeted housefly populations (10 out of 10 times). These results suggest that the blowflies raised in competition with houseflies had evolved to become better competitors, although the underlying mechanisms of competition and the associated genetic changes are not known.

Character displacement can be an important force in natural communities

When species compete for resources, natural selection may favor not only individuals that are superior competitors, but also individuals whose pattern of re-

FIGURE 11.16 A Competitive Reversal (A) The 16-cell cage used by Pimentel and colleagues in their competition experiments on houseflies and blowflies. (B) After a period of initial decline, blowflies outcompeted houseflies in the 16-cell cage. Results from additional single-cage experiments suggest that the blowfly evolved to become a better competitor during the first 38 weeks of the 16-cage experiment. (After Pimentel et al. 1965.)

(A)

(B)

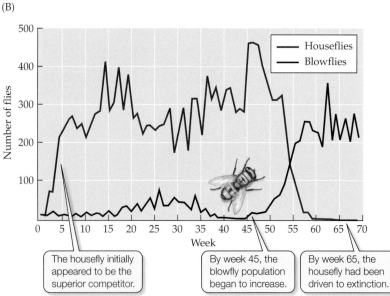

The housefly initially appeared to be the superior competitor.

By week 45, the blowfly population began to increase.

By week 65, the housefly had been driven to extinction.

source use differs from that of the competitor species. For example, when two fish species live apart from each other (each in its own lake), the two species may catch prey of similar size. If some factor (such as dispersal) were to cause the two species to live in the same lake, their use of resources would overlap considerably (**Figure 11.17A**). In such a situation, selection might favor individuals of species 1 whose morphology was such that they ate smaller prey, hence reducing competition with species 2; similarly, selection might favor individuals of species 2 that ate larger prey, hence reducing competition with species 1 (**Figure 11.17B**). Over time, such selection pressures could cause species 1 and species 2 to evolve to become more different when they live together than when they live apart. Such a process illustrates **character displacement**, which occurs when competition causes the forms of competing species to evolve to become more different over time.

Character displacement appears to have occurred in two species of finches on the Galápagos archipelago. Specifically, the beak sizes of the two species, and hence the sizes of the seeds the birds eat, are more different on islands where both species live than on islands that have only one of the two species (**Figure 11.18**). Field observations suggest that these two finch species probably differ when they live together because of competition, not because of other factors such as differences in food supplies (Schluter et al. 1985).

Data suggestive of character displacement have also been observed in plants, frogs, fishes, lizards, birds, and crabs: in each of these groups, there are pairs of species that consistently differ more where they live together than where they live apart. Additional evidence is needed, however, if we are to make a strong argument that such differences result from

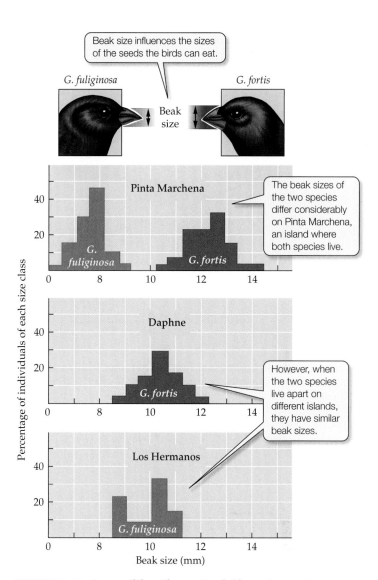

FIGURE 11.18 Competition Shapes Beak Size Competition between two species of Galápagos finches, *Geospiza fuliginosa* and *G. fortis*, may have had a selective effect on the sizes of their beaks.

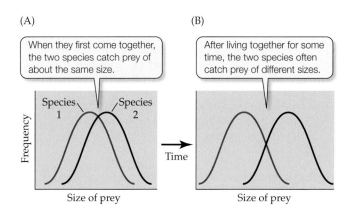

FIGURE 11.17 Character Displacement Competition for resources can cause competing species to become more different over time. Imagine that two fish species that once lived apart and tended to catch prey of about the same size, are brought together in a single lake. (A) When the two species first come together, there is considerable overlap in the resources they use. (B) As the two species evolve over time, their form might change such that they tend to catch prey of different sizes.

competition (as opposed to other factors). Strong support for the role of character displacement can come from experiments designed to test whether competition occurs and has a selective effect on morphology. Such experiments were conducted on sticklebacks of the genus *Gasterosteus*, a group of fishes whose morphology varies most when different species live in the same lake (Schluter 1994). The results indicated that individuals whose morphology differs considerably from the morphology of their competitors had a selective advantage: they grew more rapidly than did individuals whose morphology was similar to that of their competitors (**Figure 11.19**). Thus, in these fishes, we can make a strong case for competition as the cause of the observed morphological differences—that is, that character displacement has occurred.

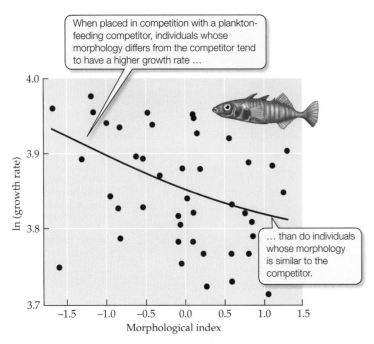

FIGURE 11.19 An Experimental Test of Character Displacement
The growth rates of individuals of a solitary species of stickleback (one that lived alone in a lake) were measured after they were placed in competition with plankton-feeding sticklebacks. The morphological index gauges whether individuals of the solitary species are more similar to sticklebacks that feed on the lake bottom (morphological index values near −1.5) or more similar to plankton-feeding sticklebacks (morphological index values near +1.5). (After Schluter 1994.)

CASE STUDY REVISITED

Competition in Plants that Eat Animals

In plants, belowground competition can be especially intense in nutrient-poor soils (see Figure 11.5). Carnivorous plants live in such soils, and their root systems are usually less well developed than the root systems of their noncarnivorous neighbors. These observations suggest that carnivorous plants may be poor competitors for soil nutrients and hence may rely on eating animals to obtain the nutrients they need for growth. They also suggest that carnivorous plants might be especially hard-hit by belowground competition if they were denied access to their unique, alternative source of nutrients (animal prey).

When Stephen Brewer studied the effect of competition on the pitcher plant *Sarracenia alata* (see p. 241), he performed an experiment with four treatments: (1) nearby, noncarnivorous plants ("neighbors") were weeded and clipped (to reduce competition), and pitchers were covered (to deprive or "starve" *S. alata* of animal prey); (2) neighbors were weeded and clipped, and pitchers were left open; (3) neighbors were left intact, and pitchers were

starved; and (4) neighbors were left intact, and pitchers were open (the control).

Contrary to what would be expected if competition for nutrients were important, *S. alata* was not especially hard-hit when neighbors were left intact and pitchers were starved. In fact, when neighbors were intact, *S. alata* had about the same growth rate regardless of whether they had access to prey or not. Moreover, *S. alata* did not increase root or pitcher production when neighbors were intact and pitchers were starved. Collectively, these results suggested that there is relatively little competition between *S. alata* and noncarnivorous plants for soil nutrients.

Competition for light appeared to be more important. Brewer found that neighbors reduced the availability of light to *S. alata* by a factor of 10. When shaded by neighbors, *S. alata* showed a shade-avoidance growth response: pitcher height increased at the expense of pitcher volume. In addition, recall that the reduction of neighbors greatly increased growth rates in *S. alata*, but only when the pitcher plants could capture animal prey (see Figure 11.2). Thus, prey deprivation reduced the ability of the pitcher plants to grow in response to increased light following competitor reduction. Finally, Brewer observed that where neighbors were left intact, natural differences in the light available to pitcher plants had no effect on their growth rates when the prey were excluded (**Figure 11.20**). However, in plants that

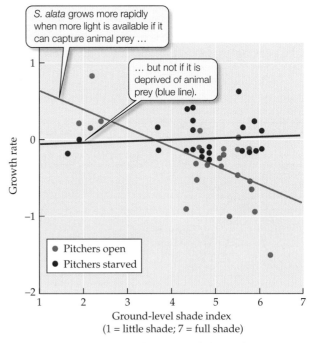

FIGURE 11.20 Interaction between Light and Prey Availability The pitcher plant *Sarracenia alata* grows more rapidly when more light is available, but only when pitchers are open, allowing it to eat animals and thereby obtain the nutrients it needs to support rapid growth. (After Brewer 2003.)

had access to prey, growth rates increased as the amount of light increased. Hence, just as they did when neighbors were reduced, *S. alata* responded to high light levels when neighbors were intact by growing more rapidly—but only if prey were available to supply the extra nutrients they needed for such growth.

Overall, it appears that *S. alata* competes with its neighbors for light, but avoids competition for soil nutrients by eating animals and by using changes in light levels as a cue for growth. When light levels are low—as would be the case when it is shaded by competitors—*S. alata* grows little, and hence requires few nutrients. In such a situation, prey deprivation has little effect because the plant does not need extra nutrients. When light levels are high, however—as would occur after a fire or whenever few competitors are present—*S. alata* is stimulated to grow. Under these circumstances, prey deprivation has a major effect because animal prey supply most of the nutrients that *S. alata* uses for growth.

CONNECTIONS IN NATURE

The Paradox of Diversity

As we've seen, some field data show that superior competitors can drive inferior competitors extinct—which is exactly what the competitive exclusion principle states *should* happen whenever two or more species use the same set of limiting resources. Natural communities, however, contain many species that share the use of scarce resources without driving one another to extinction. In the context of Brewer's experiments, let's reconsider why superior competitors do not always drive inferior competitors to extinction.

The concept of resource partitioning suggests that coexistence in nutrient-poor environments could occur if species avoided competition for scarce resources by acquiring nutrients in different ways. This idea helped to motivate Brewer's study: he wanted to know whether differences in their means of nutrient acquisition could explain the coexistence of carnivorous and noncarnivorous plants. To find out, Brewer deprived carnivorous plants of their unique source of nutrients (animal prey), thus increasing the overlap between the ways in which carnivorous and noncarnivorous plants acquired nutrients. If competition for nutrients was important, pitcher plants that were deprived of prey should have experienced more severe competitive effects, or they should have compensated for reduced nutrient intake by increasing their production of roots or pitchers. Neither of these outcomes occurred, so Brewer sought other explanations of species coexistence.

Environmental variation provides a second mechanism for the coexistence of species: if environmental conditions fluctuate over space or time (or both), species may coexist if different species are superior competitors under different environmental conditions. The bedstraw example (discussed on p. 241) illustrates how differences in soils can alter the outcome of competition, thus promoting coexistence in environments that vary over space. With respect to variation over time, an inferior competitor could persist whenever competition failed to run its course. Consider a species such as the sea palm (see p. 253) that competes poorly, but tolerates disturbance well. Such a species can persist if a disturbance periodically "resets the clock" by decreasing the abundance of a superior competitor before that species drives the inferior competitor to extinction. Such a scenario may also apply to the pitcher plant *S. alata*: it lives in a fire-prone habitat, tolerates fire well, and uses changes in light levels as a cue for growth. As a result, *S. alata* grows primarily when its competitors are absent or reduced (e.g., after a fire). This growth strategy may allow *S. alata* to persist with noncarnivorous plants that can outcompete it for both light and scarce soil nutrients.

SUMMARY

CONCEPT 11.1 Competition occurs between species that share the use of a resource that limits the growth, survival, or reproduction of each species.

- Organisms compete for resources such as food, water, light and space, all of which can be consumed or otherwise used to the point of depletion.

- When organisms compete, they reduce the availability of resources.

- If resource levels become sufficiently low, the intensity of competition can increase.

- Competition for resources is common in natural communities.

CONCEPT 11.2 Competition, whether direct or indirect, can limit the distributions and abundances of competing species.

- The most common form of competition is exploitation competition, which occurs when species compete indirectly as they share the use of a limiting resource. Species also compete directly for access to resources in interference competition.

- Competition is often asymmetrical, affecting one competitor more strongly than the other.

- Competition can occur between closely or distantly related species.

- Competition can determine the abundance of competing species.

- Competition can determine both where a species lives within a local habitat and the extent of its geographic distribution.

CONCEPT 11.3   Competing species are more likely to coexist when they use resources in different ways.

- The competitive exclusion principle states that if competing species use the same limiting resource in the same way, they cannot coexist.

- Field studies have revealed examples of resource partitioning, in which competing species use one or more shared resources in different ways.

- In the Lotka–Volterra competition model, the effects of competition are modeled by modifying the logistic equation.

- Graphical analyses of the Lotka–Volterra competition model suggest a refinement of the competitive exclu-

sion principle: the more similar two competing species are in their use of resources, the more likely it is that one will drive the other to extinction.

CONCEPT 11.4   The outcome of competition can be altered by environmental conditions, species interactions, disturbance, and evolution.

- The outcome of competition can be altered by changes in the physical conditions of the environment, or by changes in the competitors' interactions with other species.

- Periodic disturbances that remove a superior competitor can allow an inferior competitor to persist.

- Evolutionary change within a population of an inferior competitor that causes it to become a superior competitor can reverse the outcome of competition.

- In character displacement, competition causes the forms of competing species to evolve to become more different from each other over time, thereby reducing the intensity of competition.

REVIEW QUESTIONS

1. a. Explain the difference between a resource and a physical factor.

   b. Plant species require nitrogen to grow and reproduce. Assume that a single application of high-nitrogen fertilizer is added to a low-nitrogen, sandy soil in which two plant species are found. Predict how the intensity of competition for soil nitrogen will change over time (and explain your prediction).

2. List four general features of competition described in Section 11.2 and provide an example of each general feature.

3. a. As described in Section 11.3, laboratory experiments, field observations, and mathematical models have been used to help explain why competing species coexist in some situations, but not others. Describe a result from each approach that helps to explain when competing species are likely to coexist. Taken together, do these three approaches to studying competition yield similar or different explana-

tions for why coexistence occurs in some situations, but not others?

   b. If $\alpha = 0.8$, $\beta = 1.6$, $N_1 = 140$, and $N_2 = 230$, are individuals of species 1 or species 2 having a greater effect on the per capita growth rate of species 2?

   c. Based on graphical analyses of the Lotka–Volterra competition model, evaluate the following statement: If $\alpha < \beta$, species 1 will always drive species 2 to extinction. Explain your answer.

4. Suppose that each of 20 meadows contains a population of plant species 1, a population of plant species 2, or a population of both plant species. Species 1 and 2 are known to compete with each other, and each meadow is separated from the others by areas in which neither species 1 or species 2 can grow or survive.

   a. List three possible reasons why the meadows contain different combinations of the two plant species.

   b. Describe an experiment that would help to evaluate one or more of the reasons you have listed.

HANDS-ON PROBLEM SOLVING

Character Displacement | This Web exercise (www.sinauer.com/ecology) explores character displacement and phenotypic plasticity for a polyphenism in a spadefoot toad species. You will interpret data collected on different populations of toads and the results of common garden experiments to assess genetic effects.

SUGGESTED READINGS

Gause, G. F. 1934b. *The Struggle for Existence*. Williams & Wilkins, Baltimore, MD. *In a mere 163 pages, this classic book altered how many ecologists think about mathematical models and opened new areas of ecological research by illustrating how laboratory experiments could be used to test key ideas about competition between species that share limiting resources.*

Hairston, G., F. B. Smith and L. B. Slobodkin. 1960. Community structure, population control, and competition. *American Naturalist* 91: 421–425. *This groundbreaking paper offers a concise explanation for a variety of fundamental observations about nature—including why the world is green. A must read for all ecologists.*

Hardin, G. 1960. The competitive exclusion principle. *Science* 131: 1292–1297. *This influential paper clarified the meaning of the competitive exclusion principle, helping to shape how ecologists think of competition in natural communities.*

Holt, R. D. and J. H. Lawton. 1994. The ecological consequences of shared natural enemies. *Annual Review of Ecology and Systematics* 25: 495–520. *This article reviews an important phenomenon known as* apparent competition. *Apparent competition can occur when two species do not compete directly with one another—but each does interact directly with a third species, such as a predator that preys on each of the two species. As described in this interesting synthetic article, such indirect interactions can lead to population dynamics that look very much like the results of direct competition (e.g., increasing numbers of one species are associated with decreasing numbers of the other species).*

MacArthur, R. H. 1958. Population ecology of some warblers of northeastern coniferous forests. *Ecology* 39: 599–619. *This classic paper provides a careful and well-reasoned case study on how a group of seemingly very similar warblers are able to coexist in the coniferous forests in which they live.*

Seabloom, E. W., W. S. Harpole, O. J. Reichman and D. Tilman. 2003. Invasion, competitive dominance, and resource use by exotic and native California grassland species. *Proceedings of the National Academy of Sciences USA* 100: 13384–13389. *Invasive plant species pose a worldwide threat to native plants. This paper evaluates whether the success of invasive plants stems from their being superior competitors and/or better able to respond to disturbance. The results suggest that invasive species are not always superior competitors and indicate that relatively simple steps could be taken to restore some native plant communities.*

12 Predation and Herbivory

CASE STUDY

Snowshoe Hare Cycles

In 1899, a fur trader in northern Ontario reported to the Hudson's Bay Company that "Indians are bringing poor hunts. They have been starving all spring. Rabbits being scarce" (Winterhalder 1980). The "hunts" referred to were pelts of fur-bearing animals trapped by members of the Ojibwa tribe, while the "rabbits" were actually snowshoe hares (**Figure 12.1**). Collectively, 200 years of such reports showed that hare populations increased and decreased regularly, peaking about every 10 years and then falling to very low levels. When hares were abundant, the Ojibwa had enough food to spend time trapping for pelts, which they then traded to the Hudson's Bay Company. But when hares were scarce, tribal members concentrated on gathering food, rather than trapping animals that provided pelts, but little meat.

Beginning in the early 1900s, wildlife biologists used the Hudson's Bay Company's careful records to graph the abundances of both snowshoe hares and their lynx predators (**Figure 12.2A**). The regular cycles they found were striking, stimulating over 80 years of subsequent research. Snowshoe hares constitute a major portion of the lynx diet, so it was not surprising that numbers of lynx should rise and fall with numbers of hares. But what drove the cyclic fluctuations in the hare population? Adding to the mystery, hare population sizes rise and fall in synchrony across broad regions of the Canadian forest, so explanations of hare cycles should account for that synchrony as well.

One approach to finding the cause of hare population cycles is to document the changes in birth, death, and dispersal rates that are associated with increasing or declining numbers of hares. Dispersal plays a relatively small role: it may alter local population sizes, but hares do not move far enough to account for the simultaneous changes in their abundance seen across broad geographic regions. In contrast, consistent patterns of birth and death rates have been found across different regions of Canada. Snowshoe hares can raise up to three or four litters over the summer, with an average of five young per litter. Hare reproductive rates reach their highest levels (ca. 18 young/female) several years before hare density reaches a maximum. They then begin to fall, reaching their lowest levels 2–3 years after hare density peaks (**Figure 12.2B**). Hare survival rates show a similar pattern: they are highest several years before hare density peaks, then they fall and do not rise again until several years after hare density peaks.

Together, the changes over time in hare birth and survival rates drive the hare population cycle. But what causes these rates to change? Several hypotheses have been pro-

FIGURE 12.1 Predater and Prey A snowshoe hare (*Lepus americanus*) and its specialist predator, the Canadian lynx (*Lynx canadensus*).

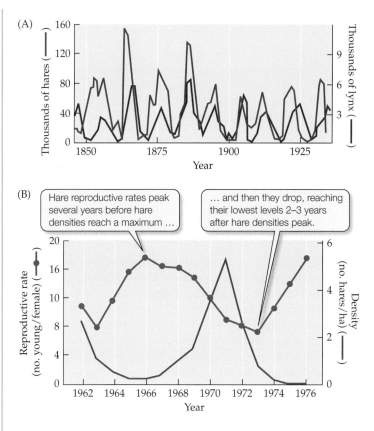

(A)

(B)

Hare reproductive rates peak several years before hare densities reach a maximum ...

... and then they drop, reaching their lowest levels 2–3 years after hare densities peak.

FIGURE 12.2 Hare Population Cycles and Reproductive Rates (A) Historical trapping data from the Hudson's Bay Company indicate that numbers of hares fluctuate in a 10-year cycle, and that numbers of lynx closely track the hare cycle. (B) The highest hare reproductive rates do not coincide with the highest hare densities. (B after Cary and Keith 1979.)

tation, and studies have shown that food can be limiting at peak hare densities (up to 2,300 hares/km^2). Two observations, however, indicate that food alone does not drive hare cycle: first, some declining hare populations do not lack food, and second, the experimental addition of high-quality food does not prevent hare populations from declining.

A second hypothesis focuses on predation. Many hares (up to 95% of those that die) are killed by predators such as lynx, coyotes, and birds of prey. In addition, lynx and coyotes kill more hares per day during the peak and decline phases of the hare cycle than during the increase phase. But questions remain. The killing of hares by predators explains the drop in survival rates as hare numbers decline, but by itself does not explain (1) why hare birth rates drop during the decline phase of the cycle, (2) why hare numbers sometimes rebound slowly after predator numbers plummet, or (3) why the physical condition of hares worsens as hares decrease in numbers. What other factors are at work?

posed, one of which focuses on food supplies. Large numbers of hares consume prodigious amounts of vege-

O ver half of the species on Earth sustain themselves by feeding on other organisms. Some kill other organisms, then eat them, while others "graze" on living organisms by eating their tissues or internal fluids. As we will see, the millions of species that feed on other organisms interact with their food organisms in a rich variety of ways. But all of these interactions share a common feature: **exploitation**, a relationship in which one organism benefits by feeding on, and thus directly harming, another.

Over the course of this and the next chapter, we will consider three broad categories of organisms that harm the organisms they feed on: herbivores, predators, and parasites (**Figure 12.3**). An **herbivore** eats the tissues or internal fluids of living plants or algae, often feeding on many different individuals. A **predator** kills and eats other organisms, re-

(A)

(B)

(C)

FIGURE 12.3 Three Ways to Eat Other Organisms (A) Herbivores such as the giraffe (*Giraffa camelopardalis*) eat leaves or other plant parts. (B) This predatory insect, a green lacewing (*Chrysopa oculata*) larva, can kill and consume large numbers of its aphid prey; the aphids also face a threat from the plant's glandular trichomes (visible here in red), hairlike structures that protect against insect herbivores by secreting a sticky and often toxic substance. (C) This isopod parasite (*Anilocra haemuli*) attaches to and feeds upon the tissues of its host, the Creole fish (*Paranthias furcifer*).

ferred to as its *prey*. A **parasite** typically lives in or on another organism (its *host*), feeding on parts of the host, such as its tissues or body fluids. Parasites harm, but usually do not kill, the organisms they feed on; some parasites (called *pathogens*) cause disease. Individual parasites usually feed on only one or a few host individuals.

These definitions sound simple, and it is easy to think of examples: an insect that eats a plant leaf, a lion that kills and eats a zebra, a tapeworm that robs a dog of the nutrients in its digested food. But the natural world defies such simple categorization. Is an insect that spends its entire life eating the leaves of one tree an herbivore, a parasite, or both? Or consider those prototypical herbivores, sheep: they get most of their food from plants, but they have also been known to eat the helpless young of ground-nesting birds. Conversely, predators can act like herbivores: wolves, for example, will eat berries, nuts, and leaves. And some organisms do not fit neatly into any category. **Parasitoids** are insects that typically lay one or a few eggs on or in another insect (the host) (**Figure 12.4**). After they hatch from their eggs, the parasitoid larvae remain with the host, which they eat and usually kill. Parasitoids can be considered unusual parasites (because they consume most or all of their host, almost always killing it) or unusual predators (because over the course of their lives they eat only one host individual, killing it slowly).

Despite these and other complications, we will approach the rich variety of exploitative interactions by organizing Chapters 12 and 13 around three basic types: predation, herbivory, and parasitism. This chapter is focused on predation and herbivory, the next on parasitism and disease. We begin with a section that explores some aspects of the natural history of predators and herbivores.

Predation and Herbivory

CONCEPT 12.1
Most predators have broad diets, whereas a majority of herbivores have relatively narrow diets.

This chapter covers predators and herbivores, but not parasites, for several reasons. Parasites often spend their entire lives living with and eating a single host individual, whereas predators and herbivores more typically eat several to many different individuals. In many cases, multiple generations of parasites live within an individual host, which makes them more likely to show special adaptations to hosts than are most predators or herbivores. Finally, some parasites are pathogens, the dynamics of which are different enough to merit special consideration.

Although they share some similarities, predators and herbivores also differ greatly from each other. The most obvious difference is that predators invariably kill their prey, while herbivores usually do not kill their food plants, at least not immediately. Here too there are exceptions, however. Herbivores that eat seeds can be considered predators: each seed contains a unique genetic individual (the product of a sperm cell fusing with an egg cell) that is killed when the seed is eaten. And, as we will see later in this chapter, some herbivores can strip entire regions bare of vegetation, causing many plants to die.

We begin with an overview of how predators obtain food and what they eat, and then follow with a similar discussion of herbivores.

Some predators move in search of prey, others sit and wait

Many predators forage throughout their habitat, moving about in search of prey. Examples of such species include wolves, sharks, and hawks. Other predators, called **sit-and-wait predators**, remain in one place and attack prey that move within striking distance (as do moray eels and sessile filter feeders, such as barnacles) or enter a trap (such as a spider's web or the modified leaf of a carnivorous plant).

In either case, predators tend to concentrate their efforts in areas that yield abundant prey. Birds that eat insects forsake patches with few prey to forage in patches where more food is available. Wolf packs follow the seasonal migrations of elk herds. Sit-and-wait predators, including spiders, relocate more often from areas where prey are scarce to areas where prey are abundant than vice versa. As a result, areas with more prey also have more webs.

> When fully developed, the wasp emerges from this exit hole in the (now dead) aphid.

FIGURE 12.4 Are Parasitoids Predators or Parasites? Parasitoids such as the wasp *Aphidius testaceipes*, shown here depositing an egg into an aphid, can be considered unusual predators because during their lifetime they eat and slowly kill only one prey individual. Parasitoids can also be viewed as unusual parasites that eat all or most of their host, thereby killing it.

Many predators have broad diets

Most predators eat prey in relation to their availability across a broad range of prey species, without showing a preference for any particular species. A predator shows a preference for a particular prey species if it eats that species more often than would be expected based on that prey's availability. Some predators do show a strong preference for certain prey species, and in that sense they can be considered specialist predators. Even when snowshoe hares are not common, for example, lynx and coyotes eat more hares than would be expected based on prey availability; for example, when hares constitute only 20% of the available food, they constitute 60%–80% of the diet of lynx and coyotes.

Some predators concentrate their foraging on whatever prey is most plentiful. When researchers provided guppies with two kinds of prey, the guppies ate disproportionate amounts of whichever prey was most abundant: fruit flies (floating on the water surface) or tubificids (aquatic worms found on the bottom) (**Figure 12.5**). Predators like these guppies that focus on abundant prey tend to "switch" from one prey species to another. Such switching may occur because the predator forms a search image of the most common prey type and hence tends to orient toward that prey, or because learning enables it to become increasingly efficient at capturing the most common prey type. As we saw in Chapter 5, in some cases predators switch from one prey species to another in a manner consistent with the predictions of optimal foraging theory.

Many herbivores specialize on particular plant parts

Plants have a simple body plan, with three main organs (roots, stems, leaves) linked to one another by a series of vessels (xylem and phloem) that transport water and nutrients. Seed production typically occurs in modified leaves, such as the flowers of flowering plants. With this simple body plan as their target food source, herbivores can be grouped according to what part of the plant they eat. Some herbivores that are large relative to their food plants eat all aboveground parts. Most herbivores, however, specialize on particular plant parts, such as leaves, roots, stems, seeds, or internal fluids (e.g., nutrient-containing sap).

More herbivores eat leaves than any other plant part. Leaves are abundant and are often available year-round; it is also common for leaves to be more nutritious than other plant parts (**Figure 12.6**). Herbivores that eat leaves range from large browsers such as deer or giraffes, to caterpillars and herbivo-

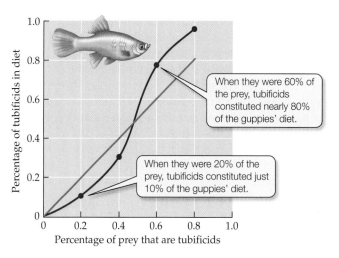

FIGURE 12.5 A Predator That Switches to the Most Abundant Prey Guppy predators focused their foraging efforts on whichever prey species was most common in their habitat: tubificids (aquatic worms) or fruit flies. The solid green line indicates the results that would have been expected if the guppies had captured tubificids according to their availability (instead of switching to whichever prey species was more abundant). (After Murdoch et al. 1975.)

When they were 60% of the prey, tubificids constituted nearly 80% of the guppies' diet.

When they were 20% of the prey, tubificids constituted just 10% of the guppies' diet.

rous fishes, to tiny "leaf miners" that enter a leaf and eat it from the inside.

By removing photosynthetic material, leaf-eating herbivores can reduce the growth, survival, or reproduction of their food plants. Belowground herbivory can also have major effects on plants, as illustrated by the 40% reduction in growth observed in bush lupine plants after 3 months of herbivory by root-killing ghost moth (*Hepialus californicus*) caterpillars.

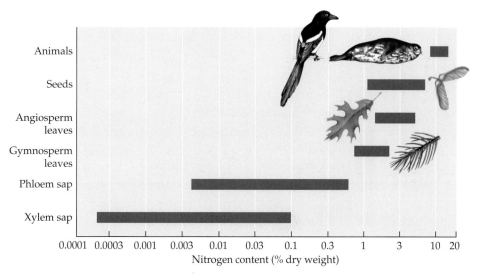

FIGURE 12.6 The Nitrogen Content of Plant Parts Varies Considerably Nitrogen is an essential component of any animal's diet. Leaves tend to have the highest nitrogen concentrations of any plant parts, other than seeds. Compared with those in animal bodies, however, nitrogen concentrations in plant parts are low. (After Mattson 1980.)

Similarly, herbivores that eat seeds can have large effects on plant reproductive success, sometimes reducing it to zero. The effects of herbivores that feed on internal fluids are not always obvious (because visible plant parts are not removed), but they can be considerable. For example, Dixon (1971) showed that although the lime aphid *Eucallipterus tiliae* did not reduce aboveground growth in lime trees during the year of infestation, the roots of trees infested with aphids did not grow that year, and a year later, their leaf production dropped by 40%.

Most herbivores have relatively narrow diets

While most predators eat a broad range of prey species, the majority of herbivores feed on a comparatively restricted set of plant species. This statement is true largely because of the insects: there are an enormous number of herbivorous insect species, and most of them live on and eat only one (or a few) plant species. For example, most species of agromyzid flies, whose larvae are leaf miners, feed on only one or a few plant species (**Figure 12.7**). Similar results have been found for leaf-feeding beetles in the genus *Blepharida*: among 37 species of these beetles, 25 feed on a single plant species, 10 feed on 2–4 plant species, and 2 feed on a relatively broad suite of plants (12–14 species) (Becerra 2007).

However, there are also numerous examples of herbivores that eat many plant species. Grasshoppers, for example, graze on a broad range of plant species, and even among the leaf

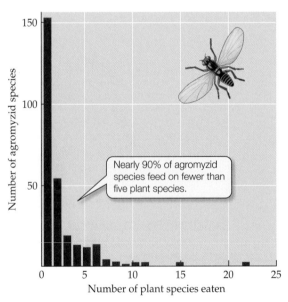

FIGURE 12.7 Most Agromyzid Flies Have Narrow Diets
The larvae of agromyzid flies are leaf miners—they live inside leaves and feed on leaf tissue. The overwhelming majority of these flies have narrow diets, feeding on only one or a few plant species. (After Spencer 1972.)

miners shown in Figure 12.7, several species eat more than 10 different plants. Large browsers, such as deer, often switch from one tree or shrub species to another; they also eat all or most of the aboveground parts of many herbaceous plant species. The golden apple snail (*Pomacea canaliculata*) is a voracious generalist, capable of removing all the large plants from wetlands; the snail then survives by eating algae and detritus.

Now that we have discussed several general features of predation and herbivory, let's explore some of those features in more detail, focusing on adaptations found in predators, herbivores, and their food organisms.

Adaptations

CONCEPT 12.2
Organisms have evolved a wide range of adaptations that help them capture food and avoid being eaten.

Among the other challenges it faces, an organism must obtain food while striving not to become food for another organism. This ongoing drama has resulted in the evolution of a dazzling array of defensive mechanisms in prey animals and food plants—as well as novel ways of overcoming those defenses in predators and herbivores.

Predators and herbivores exert strong selection on their food organisms

Life on Earth changed radically with the appearance of the first macroscopic predators roughly 530 million years ago. Before that time, the seas were dominated by soft-bodied organisms. Within a few million years after the evolution of these first predators, however, many herbivores had evolved formidable defenses, such as body armor and spines. The safe, soft-bodied world was gone forever.

This explosive increase in prey defenses occurred because predators exert strong selection pressure on their prey: if prey are not well defended, they die. While the effects of herbivory are usually not as dramatic (because death often does not immediately ensue), herbivores exert similar selection pressure on plants. We turn now to the rich variety of defensive mechanisms that have evolved in response to predation and herbivory.

Escaping predators: Physical defenses, toxins, mimicry, and behavioral responses Many prey species have physical features that reduce their chances of being killed by predators (**Figure 12.8A**). Such physical defenses include large size (e.g., elephants), a body plan designed for rapid or agile movement (e.g., the legs of a gazelle), and body armor (e.g., a snail or ant-eater).

Other species use poisons to defend themselves against predation. Species that contain powerful toxins are often

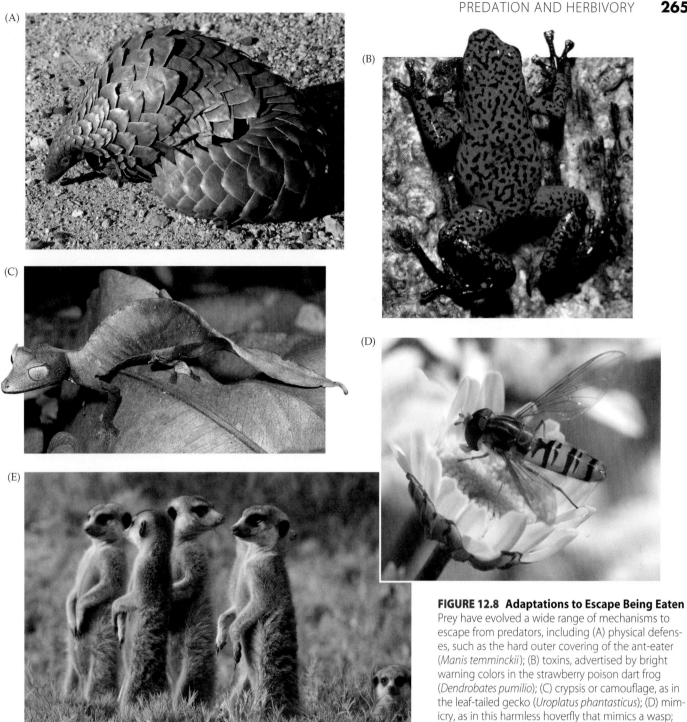

(A)

(B)

(C)

(D)

(E)

FIGURE 12.8 Adaptations to Escape Being Eaten
Prey have evolved a wide range of mechanisms to escape from predators, including (A) physical defenses, such as the hard outer covering of the ant-eater (*Manis temminckii*); (B) toxins, advertised by bright warning colors in the strawberry poison dart frog (*Dendrobates pumilio*); (C) crypsis or camouflage, as in the leaf-tailed gecko (*Uroplatus phantasticus*); (D) mimicry, as in this harmless hoverfly that mimics a wasp; and (E) behavioral defenses, as in these meerkats (*Suricata suricatta*) on the lookout in Botswana.

brightly colored (**Figure 12.8B**). Such **warning (aposematic) coloration** can itself provide protection from predators, which often learn from experience not to eat prey that are brightly colored.

Other prey species use **mimicry** as a defense: by resembling less palatable organisms or physical features of their environment, they cause potential predators to mistake them for something less desirable to eat. There are many forms of

mimicry. Some species have a shape or coloration that provides camouflage, allowing them to avoid detection by predators; this form of mimicry is called **crypsis** (from *cryptic*, or hidden). Examples include a gecko that blends in with leaves on the forest floor (**Figure 12.8C**), a flounder that is virtually indistinguishable from the ocean bottom on which it rests, and an octopus that changes its coloration rapidly to match its background. Other prey species use mimicry as a form of "false advertising": their shape and coloration mimic that of a species that is fierce or that contains a potent toxin. Examples include harmless flies that resemble yellow jacket wasps (**Figure 12.8D**) and viceroy butterflies that are good to eat but look like toxic monarch butterflies.

Finally, some species change their behavior to avoid predation. When predators are abundant, snowshoe hares forage less in open areas (where they are most vulnerable to attack). Meerkats also rely on behavioral defenses, with colony members keeping a close lookout for predators (**Figure 12.8E**). In some cases, there is a trade-off between behavioral and physical defenses, as the following example shows.

A CLOSER LOOK: SHELL SHAPE AND PREDATOR AVOIDANCE BEHAVIOR IN MARINE SNAILS   Crabs have extremely powerful claws that are capable of crushing the shells of many marine organisms. Over their long evolutionary history with crabs, snails have evolved defenses against them, including thicker shells and

a reduced shell aspect ratio (the ratio of shell height to shell width: **Figure 12.9A**) (Vermeij 1993). In addition, some species of snails can detect crabs and will retreat to safe havens when crab odors are present.

Peter Cotton and colleagues studied defensive adaptations in four species of marine snails eaten by the shore crab, *Carcinus maenas* (Cotton et al. 2004). There were no differences in the shell strengths of the four snail species. However, it took crabs only 47 seconds, on average, to crush *Littorina littorea* shells, compared with 562 seconds to crush *Gibbula umbilicalis* shells. This difference in handling time resulted from the fact that *L. littorea* shells had a higher aspect ratio (meaning that the shell was relatively tall and narrow), which made them easier for crabs to grip and manipulate. But *L. littorea* showed the largest behavioral response to predators: when crabs were present, they sought refuge more rapidly than the other three snail species did. In terms of their relative importance, these two defensive traits (shell aspect ratio and predator avoidance behavior) were exactly negatively correlated: the species whose shells required the least amount of handling time had the largest predator avoidance response, and vice-versa (**Figure 12.9B,C**). Thus, it appears that there may have been a trade-off between these two defenses.

Reducing herbivory: Avoidance, tolerance, and defenses

Some plants avoid herbivory by producing great numbers of seeds in some years and few or no seeds in other years, a phenomenon known as *masting*. Up to a hundred years may pass between bouts of seed production, as in the mass flowering of bamboos in China. Masting allows plants to hide (in time) from seed-eating herbivores, then overwhelm them by sheer numbers. Plants can also avoid herbivores in other ways, such as by producing leaves at times of the year when herbivores are scarce.

Other plants have adaptive growth responses that allow them to compensate for, and hence tolerate, the effects of herbivory—at least up to a point. **Compensation** occurs when removal of plant tissues stimulates the plant to produce new tissues, allowing for relatively rapid replacement of the material eaten by herbivores; when *exact compensation* occurs, herbivory causes no net loss of plant tissue. Compensation may occur, for example, when removal of leaf tissue decreases self-shading, resulting in increased plant growth, or when removal of apical buds (those at the end of

(A)

(B)

(C)

FIGURE 12.9 A Trade-off in Snail Defenses against Crab Predation
Marine snails have evolved several defenses against predation by crabs, including a reduced shell aspect ratio and predator avoidance responses. (A) The shell aspect ratio is the ratio of shell height to shell width. The higher the shell aspect ratio, the more easily crabs can manipulate the shell into a position from which a crushing force can be applied. (B) Time taken by crabs to manipulate and crush the shells of each of four snail species. (C) Index of the strength of predator avoidance response of each of four snail species; larger values indicate a more rapid response. (After Frid and Fordham 1994; Cotton et al. 2004.)

a branch or shoot) allows lower buds to open and grow. Beech trees respond to simulated herbivory (clipping) by increasing both leaf production and photosynthetic rate. Similarly, in field gentians, moderate to high levels of herbivory may benefit the plant under some circumstances (**Figure 12.10**). In this case, the timing of herbivory is critical: early in the growing season (up to July 20), compensation occurs, but later in the season (July 28), it does not. However, if the amount of material removed from any plant is large enough, or if insufficient resources are available for growth, the plant cannot compensate for the damage.

Finally, plants use an enormous array of structural and chemical defenses to ward off herbivores (Pellmyr et al. 2002; Agrawal and Fishbein 2006). A stroll through many plant communities makes this readily apparent: the leaves of many plants are tough, and many plant bodies are covered with spines, thorns, sawlike edges, or pernicious (nearly invisible)

hairs that can pierce the skin like miniature porcupine quills. The range of chemical defenses is just as impressive. Plants produce a wide variety of chemicals, called **secondary compounds**, that function to reduce herbivory. Some secondary compounds are toxic, protecting the plant from all but the relatively small number of herbivore species that can tolerate them. Others serve as chemical cues that attract predators or parasitoids to the plant, where they attack the herbivores (Schnee et al. 2006). Some plants produce secondary compounds constantly, regardless of whether herbivores have attacked the plant. In other cases, secondary compound production is an **induced defense** (stimulated by herbivore attack). Structural defensive mechanisms can also be induced; in some cactus species, for example, individual cacti increase their production of spines only after they have been grazed (Myers and Bazely 1991).

A CLOSER LOOK: INDUCED DEFENSES IN TOBACCO   The native tobacco *Nicotiana attenuata* is an annual plant species found in the Great Basin of the western United States. Populations of *N. attenuata* emerge after fires from seeds stored in the soil; these populations typically persist for 3 years or less. As a result, *N. attenuata* populations appear and disappear across the landscape, and they experience unpredictable levels of herbivory. At any given time, some populations may be hard hit, while others may experience little or no herbivory.

When it is attacked by herbivores, *N. attenuata* produces two induced defenses: toxic secondary compounds that deter herbivores directly, and compounds that deter herbivores indirectly by attracting predators and parasitoids. Acting together, these defenses are very effective in reducing losses to herbivores. In one experiment, the application of volatile compounds that are normally induced by herbivory to the stems of *N. attenuata* caused the numbers of a leaf-feeding herbivore on the plants to drop by more than 90% (Kessler and Baldwin 2001).

To evaluate the genetic basis for this phenomenon, André Kessler and colleagues (2004) used a "gene silencing" approach, in which particular genes are disabled. They developed three *N. attenuata* varieties in which one of three genes (*LOX3*, *HPL*, or *AOS*) was turned off. These genes are all part of a chemical pathway thought to control the induction of both direct (toxins) and indirect (predator and parasitoid attractants) defenses against herbivore attack. When grown in the field, the *not-LOX3* variety (in which the *LOX3* gene was silenced) suffered much more damage from herbivores than either control plants (in which no genes were silenced)

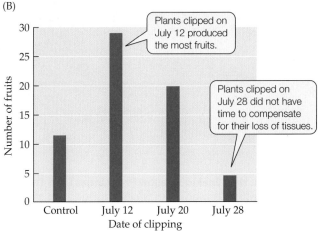

FIGURE 12.10 Compensating for Herbivory   Field gentians (*Gentianella campestris*) were clipped at different times during the growing season to simulate herbivory. (A) The shape and production of flowers in clipped and unclipped (control) plants. (B) Numbers of fruits produced by plants clipped on different dates. (After Lennartsson et al. 1998.)

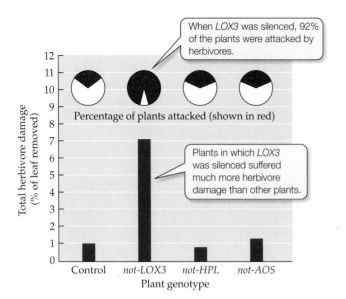

When *LOX3* was silenced, 92% of the plants were attacked by herbivores.

Percentage of plants attacked (shown in red)

Plants in which *LOX3* was silenced suffered much more herbivore damage than other plants.

FIGURE 12.11 Herbivores Damage Plants Lacking an Induced-Defense Gene Herbivore damage was measured in wild tobacco (*Nicotiana attenuata*) plants (controls) and in three experimental varieties of *N. attenuata*. In each experimental variety, one of three genes suspected to be involved in the plant's induced defenses (*LOX3*, *HPL*, or *AOS*) was silenced; these genotypes are referred to as *not-LOX3*, *not-HPL*, and *not-AOS*. The bar graphs show total herbivore damage (average percentage of leaf area removed). (After Kessler et al. 2004.)

or the other two experimental varieties (**Figure 12.11**). Kessler et al. also found that a greater number of herbivore species could feed on the *not-LOX3* plants than could feed on any of the other three plant types. These results showed that changes in a single gene can alter both the level of herbivory and the community of herbivores that attack a plant. More generally, this innovative study revealed the power of combining molecular genetic techniques with ecological field experiments: by doing so, the investigators were able to examine the effects of particular genes in a natural setting.

Predators and herbivores have adaptations to overcome the defenses of their food organisms

As we have just seen, predators and herbivores exert strong selection pressure on prey and food plants, leading to improvements in the defenses of those organisms. Those improvements, in turn, impose strong selection on predators and herbivores—organisms must eat, and a well-defended prey or plant species could provide an abundant food source for any species able to overcome its defenses. Evidence of increases in the ability of predators and herbivores to counter the defenses of their food organisms can be found throughout the broad sweep of the history of life on Earth; here we consider just a few specific examples.

Prey-capture adaptations: Physical features, toxins, mimicry, and detoxification For any given defensive mechanism used by prey, there is usually a predator that has a countervailing offensive mechanism. Cryptic prey, for example, may be detected by a predator that uses its sense of smell or touch.

Predators often have unusual physical features that help them capture prey. The body form of the cheetah, for example, enables great bursts of speed that allow it to catch rapidly fleeing gazelles. In another example, most snakes have the remarkable ability to swallow prey that are considerably larger than their heads (**Figure 12.12**). Unlike those of most terrestrial vertebrates, the bones of a snake's skull are not rigidly attached to one another. This unique feature allows the snake to open its jaws to a seemingly impossible extent; curved teeth mounted on bones that can move inward then help it to pull prey items down the throat. A person with the equivalent adaptations would be able to swallow a watermelon whole.

The bones shown in red can move, allowing the snake's mouth to open wide enough to eat large prey.

FIGURE 12.12 How Snakes Swallow Prey Larger Than Their Heads Snakes have movable skull bones that allow them to swallow prey larger than their heads, as shown in this photo of an egg-eater snake (*Dasypeltis scabra*) swallowing an egg.

While some predators depend primarily on their physical structure, others subdue prey with poison (e.g., venomous spiders). Still others use mimicry: ambush bugs, scorpionfishes, and many other predators blend into their environment so well that prey may be unaware of their presence until it is too late. Some predators have inducible traits that improve their ability to feed on specific prey species. The predatory ciliate *Lembadion bullinum* has such an inducible offense: individuals gradually adjust their size to match the size of the available prey (i.e., if they are small but the available prey are large, they increase in size—and vice versa). Finally, some predators detoxify or tolerate prey chemical defenses, as the following example shows.

A CLOSER LOOK: HOW A NONVENOMOUS SNAKE CAN EAT ITS DEADLY PREY
The garter snake, *Thamnophis sirtalis*, is the only predator known to eat the toxic rough-skinned newt, *Taricha granulosa* (**Figure 12.13**). In some of its populations, the skin of this newt contains unusually large amounts of tetrodotoxin (TTX), an extremely potent neurotoxin. TTX binds to sodium channels in nerve and muscle tissue, thus preventing nerve signal transmission and causing paralysis and death. A single newt can contain enough TTX to kill 25,000 mice—far more than enough to kill a person, as was tragically evident in 1979 when a 29-year-old man died after eating a rough-skinned newt on a dare.

Unlike many other snakes, garter snakes produce no poisons themselves, but some populations are resistant to the poisons of their prey. Garter snakes that can tolerate TTX do so by means of TTX-resistant sodium channels (Geffeney et al. 2005). Although resistant garter snakes are protected from the lethal effects of TTX, there are costs associated with the ability to eat toxic newts. Those garter snake individuals that tolerate the highest concentrations of TTX move more slowly than less resistant individuals—a trade-off between resistance to the poison and speed of locomotion. In addition, once they swallow a poisonous newt, the snakes are immobilized for up to 7 hours. During that time, the snakes are vulnerable to predation themselves and may also suffer from heat stress. It is possible (but not yet demonstrated) that the snake and the newt are locked in an evolutionary arms race: in populations where the newt has evolved to produce more TTX, the snake has evolved to tolerate the resulting higher concentrations of that poison (E. D. Brodie III, personal communication).

Overcoming plant defenses: Structural, chemical, and behavioral adaptations   The defenses used by plants are effective at preventing most herbivores from eating most plants. But for any given plant species, there are some herbivores that can cope with its defensive mechanisms. A plant covered with spines may be attacked by an herbivore that can avoid or tolerate those spines. Many herbivores have evolved digestive enzymes that enable them to disarm or tolerate plant chemical defenses. Such herbivores may gain a considerable advantage: they can eat plants that other herbivores cannot, and thus have access to an abundant food resource. Finally, some herbivores use behavioral responses to circumvent an otherwise effective defense. Let's consider how a group of tropical beetles copes with the defenses of its food plants.

A CLOSER LOOK: BEHAVIORAL AND CHEMICAL ADAPTATIONS IN TROPICAL BEETLES   Some tropical plants in the genus *Bursera* combine the production of toxic secondary compounds with a high-pressure delivery system: they store a toxic, sticky resin in a

FIGURE 12.13 A Nonvenomous Snake and Its Lethal Prey   The garter snake (*Thamnophis sirtalis*) is the only predator known that can eat the highly toxic, rough-skinned newt (*Taricha granulosa*).

FIGURE 12.14 Plant Defense and Herbivore Counterdefense Some plants in the genus *Bursera* store toxic resin under high pressure in leaf canals. (A) When herbivores eat the leaves, they chew through these canals, causing the resin to be squirted up to 2 m from the leaf. (B) Some beetles in the genus *Blepharida* can disable this defense by chewing slowly through the canals, releasing the pressure in a gradual and harmless way.

network of canals that runs through their leaves and stems (**Figure 12.14**). If an insect herbivore chews through one of these canals, the resin squirts from the plant under high pressure and may repel or even kill the insect (the resin hardens after it is exposed to air, so if an insect is drenched in resin, it can be entombed). Yet some tropical beetles in the genus *Blepharida* have evolved an effective defense (Becerra 2003). They chew slowly through the leaf veins where the resin canals are located, releasing the pressure so gradually that the resin does not squirt from the plant. It often takes a beetle more than an hour to "disarm" a leaf in this manner; once that job is done, the beetle eats the leaf in 10–20 minutes.

Plants in the genus *Bursera* typically make 1–5 defensive secondary compounds, most of which are of similar chemical composition. However, some *Bursera* species are protected from herbivores by a complex set of 7–12 toxins, some of which differ considerably in chemical composition. This set of noxious chemicals serves the plants well: only a small subgroup of *Blepharida* beetles can detoxify all of these compounds and hence eat the plants. These beetles diversified during the last 5–19 million years, roughly in synchrony with the plants they feed on.

Effects on Communities

CONCEPT 12.3
Predation and herbivory affect ecological communities greatly, in some cases causing a shift from one community type to another.

A general theme that runs through this book is that ecological interactions such as competition and predation affect the distributions and abundances of the interacting species. Such effects on populations in turn affect ecological communities. Although this is true by definition (because a community consists of interacting populations of different species that live in the same area), the community-level consequences of predation and herbivory can be profound—in some cases causing major shifts in the types of organisms found at a given location.

Exploitation influences the distributions and abundances of organisms

All exploitative interactions (e.g., predation, herbivory, parasitism) have the potential to reduce the growth, survival, or reproduction of the organisms that are eaten. These effects can be dramatic, as demonstrated in the case of a leaf-feeding beetle (*Chrysolina quadrigemina*) that rapidly reduced the density of Klamath weed, an introduced plant that is poisonous to livestock and that once infested about 4 million acres of rangeland in the western United States (**Figure 12.15**). Predators and parasitoids can also have dramatic effects. In six cases of successful biological pest control, introductions of wasps that preyed on crop-eating insects decreased their densities by 97.5% to 99.7%, thus reducing the economic damage caused by the pests.

The distributions and abundances of organisms can also be altered by indirect effects of predation or herbivory. As we've seen, predators and herbivores can change the outcome of competition (see Figure 11.14), thereby affecting the distributions or abundances of competitor species. In particular, inferior competitors may increase in abundance when they are in the presence of a predator or herbivore that decreases the performance of a dominant competitor. Paine (1974) found such a result: he showed that the removal of a sea star predator (*Pisaster*) led to the local extinction of all large invertebrates but one, a mussel. The mussel was a dominant competitor that, in the absence of the sea star, drove all the other large invertebrates to extinction.

(A)

FIGURE 12.15 A Beetle Controls a Noxious Rangeland Weed
Klamath weed (*Hypericum perforatum*), which poisons cattle, once covered about 4 million acres of rangeland in the western United States. (A) This photograph, taken in 1949, shows a field completely covered with Klamath weed at the flowering stage. (B) Abundances of the leaf-feeding beetle *Chrysolina quadrigemina* and Klamath weed in ¼ m² plots after the beetle's introduction in 1951. (After Hoffaker and Kennett 1957.)

We turn now to two other examples of how exploitation can affect the distributions and abundances of food organisms.

Predators can decrease the distributions and abundances of their prey   *Anolis* lizards are predators that eat a broad range of prey species, including spiders. Thomas Schoener and David Spiller studied the effects of lizard predators on their spider prey in the Bahamas. They selected twelve small islands, placing them in four groups of three islands that were similar in size and vegetation. Initially, each group of three islands contained one island with lizards and two without. One of the latter two islands was then chosen at random to have two male and three female adult *Anolis sagrei* lizards introduced to it; the other island was left as a control in which lizards were absent naturally.

The introduced lizards greatly reduced the distributions and abundances of their spider prey (Schoener and Spiller 1996). Before the experiment began, the number of spider species and the overall density of spiders were similar among the islands that lacked lizards. By the end of the experiment, however, the introduction of lizards to half of these islands had reduced the number and density of spider species to the levels found on the islands where lizards were present naturally. The proportion of spider species that went extinct was nearly 13 times higher on islands where lizards were introduced than on islands without lizards (**Figure 12.16**). Similarly, the density of spiders was about 6 times higher on islands without lizards

(B)

> When the beetle was introduced, it rapidly reduced the abundance of Klamath weed.

> Once Klamath weed densities had been reduced, beetle density dropped.

> The proportion of spider species that went extinct was nearly 13 times higher on islands where lizards were introduced than on islands without lizards.

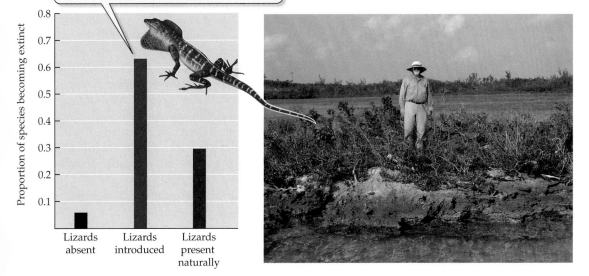

FIGURE 12.16 Lizard Predators Can Drive Their Spider Prey to Extinction   The experimental introduction of lizards to small islands in the Bahamas greatly increased the rate at which their spider prey became extinct. The photograph shows Thomas Schoener on one of the study islands. (After Schoener and Spiller 1996.)

than on islands that had lizards (either naturally or experimentally). The introduction of lizards reduced the density of both common and rare spider species, and most of the rare species went extinct. Similar results have been obtained for beetles eaten by rodents and grasshoppers eaten by birds.

Herbivores can decimate their food plants   Lesser snow geese (*Chen caerulescens*) migrate from overwintering grounds in the United States to breed in salt marshes that border Canada's Hudson Bay. During the summer, the geese graze on marsh grasses and sedges. Although the geese remove considerable plant matter, their presence can benefit the marshes. Nitrogen, a limiting resource for plant growth, is scarce in salt marshes. As they eat, the geese defecate once every 4 or 5 minutes, thereby adding nitrogen to the soil (nitrogen moves into the soil from goose feces more rapidly than it does from the decomposing leaves of marsh plants). The plants absorb the added nitrogen, which allows them to grow rapidly after being grazed. Overall, low to intermediate levels of grazing by geese lead to increased plant growth (Jefferies et al. 2003). For example, net primary productivity (NPP, measured as the annual amount of new aboveground plant growth) was higher in lightly grazed plots than in ungrazed plots (**Figure 12.17A**).

About 40 years ago, however, the situation described in the previous paragraph started to change. Beginning around 1970, lesser snow goose densities increased exponentially. This increase probably occurred because increased crop production near their overwintering sites provided the geese with a superabundant supply of food. The ensuing high densities of geese no longer benefited marsh plants. The geese completely removed the vegetation, drastically changing the distributions and abundances of marsh plant species (**Figure 12.17B**). Of an original 54,800 ha (135,400 acres) of intertidal marsh in the Hudson Bay region, geese are estimated to have destroyed 35% (19,200 hectares or 47,400 acres). An additional 30% (16,400 ha or 40,500 acres) of the original marsh has been badly damaged by the geese. Controlled hunts (from 1999 on) have slowed the population growth of the geese; this strategy may eventually lead to marsh recovery.

Exploitation can alter ecological communities

Schoener and Spiller's work on the effects of lizard predators on spiders shows that the direct effects of a predator can greatly reduce the diversity (i.e., the number) of prey species in a community. In other cases, a predator that suppresses a dominant competitor can (indirectly) cause the diversity of species in a community to increase (as in the sea star and mussel example). Indirect effects of predators can also alter ecological communities by affecting the transfer of nutrients from one ecosystem to another, as the following study on arctic foxes illustrates.

In the late nineteenth and early twentieth centuries, humans introduced arctic foxes (*Alopex lagopus*) to some of the

(A)

(B)

FIGURE 12.17 Snow Geese Can Benefit or Decimate Marshes (A) When lightly grazed (for a single 15–90-minute episode) by snow goose goslings, salt marsh plants increased their cumulative production of new biomass over a 60-day period. (B) Adult geese, however, can remove all plant matter from a square meter in an hour. Heavy grazing by high densities of snow geese can convert marshlands to mudflats, as seen by comparing the small remnant of marshland (protected from geese) with the surrounding mudflats (a former marsh that was grazed heavily by geese). (After Hik and Jeffries 1990.)

Aleutian Islands off the coast of Alaska. Other islands remained fox-free, either because foxes were never introduced there or because the introductions failed. Taking advantage of this inadvertent large-scale experiment, Croll et al. (2005) determined that, on average, the introduction of foxes to an island reduced the density of breeding seabird populations by nearly a hundredfold. The decrease in seabird numbers, in turn, reduced the input of guano (bird feces) to an island from roughly 362 grams per square meter to 6 grams per square meter. Seabird guano, which is rich in phosphorus and nitrogen, transfers nutrients from the ocean (where seabirds feed) to the land. By reducing the amount of guano that fertilized the (nutrient-limited) plant communities on the islands, the introduction of foxes caused dwarf shrubs and herbaceous plants other than grasses to increase in abundance at the expense of grasses. As a result, the introduction of foxes had the unexpected effect of transforming the community from grassland to tundra.

Herbivores can have equally large effects. Writing in *The Origin of Species*, Darwin (1859) noted the speed with which Scotch fir trees replaced heath after regions of heathland were enclosed to prevent grazing by cattle. When he observed heathlands grazed by cattle, "on looking closely between the stems of the heath, I found a multitude of seedlings and little trees, which had been perpetually browsed down by the cattle. In one square yard … I counted thirty-two little trees; and one of them, judging from the rings of growth, had during 26 years tried to raise its head above the stems of the heath, and had failed." Darwin concluded that seeds dispersed from trees located at the edge of the heath would germinate and overgrow the heath if not for grazing by cattle. Thus, the very existence of the heath community in that area depended on herbivory. Herbivores can also have pronounced effects in aquatic environments, as the following example shows.

Herbivores can alter aquatic communities The golden apple snail was introduced from South America to Taiwan in 1980 for aquaculture for local consumption and export. The snail escaped from cultivation and spread rapidly through Southeast Asia (**Figure 12.18**). Its spread caught the attention of researchers and government officials because it proved to be a serious pest of rice. The snail has also been found in Hawaii, the southern United States, and Australia, and is expected to reach Bangladesh and India in the near future (Carlsson et al. 2004).

Most freshwater snails eat algae, but the golden apple snail prefers to eat aquatic plants, including those that float on the water surface and those that attach themselves to the bottom. However, as mentioned on p. 264, golden apple snails are generalists, and if plants are not available, they can survive on phytoplankton (small, photosynthetic aquatic organisms, including algae and some bacteria), and detritus. As a result, these snails are resilient and hard to get rid of.

As a first step toward assessing how the snail had affected natural communities, Nils Carlsson and colleagues surveyed 14 wetlands in Thailand with varying densities of snails. They found that wetland communities with high densities of snails were characterized by few plants, high nutrient concentrations in the water, and high densities of algae and other phytoplankton (**Figure 12.19**).

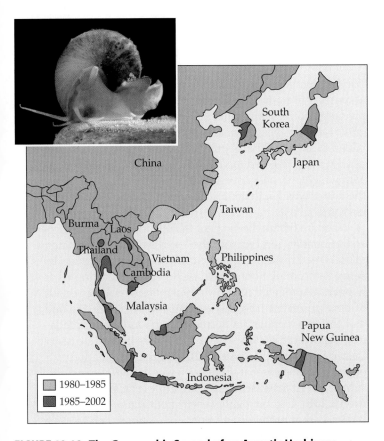

FIGURE 12.18 The Geographic Spread of an Aquatic Herbivore Since its introduction to Taiwan in 1980, the golden apple snail (*Pomacea canaliculata*) has spread rapidly across parts of Southeast Asia, threatening rice crops and native plant species. The map shows the regions the snail had occupied by 1985 and by 2002.

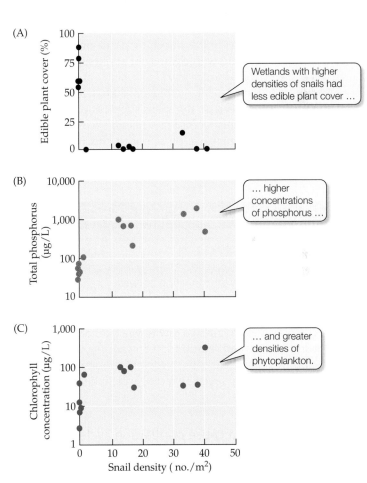

FIGURE 12.19 A Snail Herbivore Alters Aquatic Communities Nils Carlsson and colleagues measured several characteristics of 14 natural wetlands in Thailand that differed in their densities of golden apple snails (*Pomacea canaliculata*). (A) Percentage of the wetlands covered by edible plant species. (B) Concentrations of phosphorus in the water. (C) Chlorophyll concentrations (an indicator of phytoplankton densities). Note the log scale for parts B and C. Experiments conducted separately indicated that all of the trends shown here could have been caused by the snail. (After Carlsson et al. 2004.)

To test whether the trends observed in their survey could have been caused by the snail, Carlsson et al. (2004) constructed 24 1 × 1 × 1 m enclosures and placed them in a wetland in which snail densities were low. To each enclosure, they added about 420 grams of water hyacinth (*Eichhornia crassipes*), one of the most abundant plant species in many Southeast Asian wetlands. Next, they added 0, 2, 4, or 6 snails to the enclosures; there were six replicates of each of the four snail density treatments. Carlsson and colleagues then measured the effects of the snails on plant biomass and phytoplankton density. Water hyacinth biomass increased in the enclosures where no snails were present, but decreased in all the other enclosures. Phytoplankton densities increased, leading to increases in net primary productivity, at 6 snails per enclosure—far lower than the snail densities found in some invaded natural wetlands (see Figure 12.19).

The results of the survey and the experiment concur in suggesting that the golden apple snail can have an enormous effect on wetland communities, causing a complete shift from a wetland with clear water and many plants to a wetland with turbid water, few plants, high nutrient concentrations, and high densities of phytoplankton. It is likely that this shift occurs because the snails suppress plants directly (by eating them) and because they release the nutrients they obtain from the plants (e.g., phosphorus) into the water, thus providing improved growth conditions for algae and other phytoplankton.

Population Cycles

> **CONCEPT 12.4**
> Population cycles can be caused by feeding relations, such as a three-way interaction between predators, herbivores, and plants.

In the previous section, we saw how exploitation can alter the distributions and abundances of organisms, in some cases resulting in major changes to ecological communities. We turn now to a more specific effect of exploitation: population cycles.

Exploitation can cause population cycles

We introduced population cycles in Chapter 10 (e.g., Figure 10.8), and in this chapter's Case Study we described the most famous one of all, the hare–lynx cycle. Cyclic fluctuations in abundance are one of the most intriguing patterns in nature. After all, what could cause populations to change so considerably in size over time, yet in such a regular manner? We will return to the mechanisms that underlie the hare–lynx cycle on p. 277, but first we'll describe some insights into the causes of population cycles that have come from models, experiments, and field observations of predator–prey interactions.

Cycles in predator–prey models One way to evaluate possible causes of population cycles is to investigate the issue mathematically. In the 1920s, Alfred Lotka and Vito Volterra independently represented the dynamics of predator–prey interactions with what is now called the Lotka–Volterra predator–prey model:

$$\frac{dN}{dt} = rN - aNP$$

$$\frac{dP}{dt} = faNP - dP$$

(12.1)

In these equations, N represents the number of prey individuals and P represents the number of predator individuals. The equation for change in the prey population over time (dN/dt) assumes that when predators are absent ($P = 0$), the prey population grows exponentially (i.e., $dN/dt = rN$, where r is the population growth rate). When predators are present ($P \neq 0$), the rate at which they kill prey depends in part on how frequently predators and prey encounter one another. This frequency is expected to increase with the number of prey (N) and with the number of predators (P), so a multiplicative term (NP) is used in Equation 12.1. The rate at which predators kill prey also depends on the efficiency with which predators capture prey; this capture efficiency is represented by the constant a, so that the overall rate at which predators remove individuals from the prey population is aNP.

Predators starve when there are no prey. Thus, the equation for change in the predator population over time (dP/dt) assumes that in the absence of prey ($N = 0$), the number of predators decreases exponentially with a death rate of d (i.e., $dP/dt = -dP$). When prey are present ($N \neq 0$), individuals are added to the predator population according to the number of prey that are killed (aNP) and the feeding efficiency (f) with which prey are converted into predator offspring. Thus, the rate at which individuals are added to the predator population is $faNP$.

What are the implications of the relationships represented by Equation 12.1? As with the Lotka–Volterra competition model (see Equation 11.1), we can answer this question by using zero population growth isoclines to determine what happens to predator and prey populations over long periods of time (to review how this is done, see Box 11.2). We find that the prey population decreases if the number of predators is greater than r/a, whereas it increases if there are fewer than r/a predators (**Figure 12.20A**). Similarly, the predator population decreases if there are fewer than d/fa prey individuals, and it increases if there are more than d/fa prey (**Figure 12.20B**). Combining the results from Figures 12.20A and B reveals that predator and prey populations tend to cycle (**Figure 12.20C,D**).

The Lotka–Volterra predator–prey model thus yields an important result: it suggests that predator and prey populations have an inherent tendency to cycle. But the model also has a curious and unrealistic property: the amplitude of the cycle

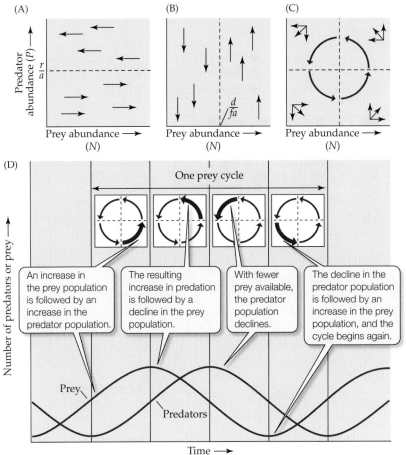

FIGURE 12.20 Predator–Prey Models Produce Population Cycles
(A) Considering the prey population first, the abundance of prey does not change when $dN/dt = 0$, which occurs when $P = r/a$ (see Equation 12.1). (B) Similarly, considering the predator population, the abundance of predators does not change when $dP/dt = 0$, which occurs when $N = d/fa$. Combining the results in parts A and B suggests that predator and prey populations have an inherent tendency to cycle. These cycles are shown here in two ways: (C) by plotting the abundance of predators versus the abundance of prey, and (D) by plotting the abundance of both predators and prey versus time; the four inset diagrams in (D) plot the abundance of predators vs. the abundance of prey.

released into the tray. After the predators were introduced, both prey and predator populations increased for a time, then declined to extinction (**Figure 12.21B**).

Huffaker also observed that the prey persisted longer if the oranges were widely spaced—presumably because it took the predators more time to find their prey. He test-

FIGURE 12.21 In a Simple Environment, Predators Drive Prey to Extinction (A) C. B. Huffaker constructed a simple laboratory environment to test for conditions under which predators and prey would coexist and produce population cycles. He placed oranges in a few of the positions of an experimental tray to provide food for the herbivorous six-spotted mite (*Eotetranychus sexmaculatus*); the remainder of the positions contained inedible rubber balls. (B) When a predatory mite (*Typhlodromus occidentalis*) was introduced to these simple environments, it drove the prey to extinction, causing its own population to go extinct as well. (After Huffaker 1958.)

(the magnitude by which predator and prey numbers rise and fall) depends on the initial numbers of predators and prey. If the initial numbers shift even slightly, the amplitude of the cycle will change (see **Web Extension 12.1** to find out what features of the model cause this change). Slightly more complex models (e.g., Rosenzweig and MacArthur 1963; Turchin 2003) still produce cycles, but do not show this unrealistic dependence on initial population sizes. Thus, a general conclusion emerges from predator–prey models: predator–prey interactions have the potential to cause population cycles.

Predator–prey cycles in the laboratory Turning from models to data, do real predator and prey populations tend to cycle? Laboratory experiments show that such cycles can be difficult to achieve. When prey are easy for predators to find, predators typically drive prey to extinction, then go extinct themselves. Such was the case in C. B. Huffaker's experiments with a predatory mite (*Typhlodromus occidentalis*) that eats the herbivorous six-spotted mite (*Eotetranychus sexmaculatus*) (Huffaker 1958). In an initial set of experiments, 20 six-spotted mites were released on a tray with 40 positions, a few of which contained oranges that these herbivorous mites could eat (**Figure 12.21A**). At first, the six-spotted mite population increased, in some cases reaching densities of 500 mites per orange. Eleven days after the start of the experiment, two predatory mites were

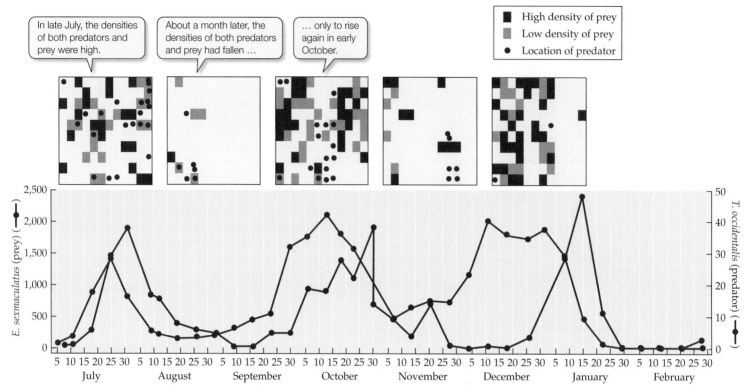

FIGURE 12.22 Predator–Prey Cycles in a Complex Habitat
Huffaker modified the simple experimental arena shown in Figure 12.21 to create a more complex habitat that aided the dispersal of the prey species, but hindered the dispersal of the predator. Under these conditions, predator and prey populations coexisted, and their abundances cycled over time. The top panels show the locations within the arena of prey (shaded regions) and predators (circles) at five different points in time. (After Huffaker 1958.)

ed this idea in a follow-up experiment in which he increased the complexity of the habitat in the following way. He added strips of Vaseline that partially blocked the predatory mites as they crawled from one orange to another, and he placed small wooden posts in an upright position on some of the oranges; these posts allowed the herbivorous mites to take advantage of their ability to spin a silken thread and float on air currents over the Vaseline barriers. Thus, he altered the experimental arena to favor dispersal of the six-spotted mite and impede dispersal of the predatory mite. Under these conditions, the prey and the predators both persisted, illustrating a form of "hide-and-seek" dynamics that produced population cycles (**Figure 12.22**). The six-spotted mites dispersed to unoccupied oranges, where their numbers increased. Once the predators found an orange with six-spotted mites, they ate them all, causing both prey and predator numbers on that orange to plummet. In the meantime, however, some six-spotted mites dispersed to other portions of the experimental habitat, where they increased in number until they too were discovered by the predators.

Population cycles and three-way feeding relationships The results of mathematical models, experiments, and observational studies indicate that predators influence population cycles in southern pine beetles, voles, collared lemmings, snowshoe hares, and moose (Turchin 2003; Gilg et al. 2003). But predation is not the only factor that causes population cycles in these species. The food supply of the herbivorous

prey can also play an important role, and in some cases, social interactions are important as well. Thus, reality is not as simple as implied by the results of predator–prey models (in which cycles are maintained purely by predator–prey interactions). Population cycles often seem to be caused by three-way feeding relationships—by the effects of predators and prey on each other, coupled with the effects of prey and their food sources (e.g., plants) on each other.

Coexistence and cycles of predators and prey in the field Natural populations of predators and prey can show dynamics similar to those of Huffaker's mites. Clumps of mussels off the coast of California, for example, can be driven to extinction by sea star predators. However, mussel larvae float in ocean currents and hence disperse more rapidly than the sea stars. As a result, the mussels continually establish new clumps that flourish until they are discovered by sea stars. Like the six-spotted mites, the mussels persist (and thus have the potential for population cycles) because portions of their population escape detection by predators for a period of time.

A variety of factors can prevent predators from driving prey to extinction. Such factors include habitat complexity and limited predator dispersal (as in Huffaker's mites), switching behavior in predators (see Figure 12.5), spatial refuges (i.e., areas in which predators cannot hunt effectively), and, as we will see next, evolutionary changes in the prey population.

Evolution can influence predator–prey cycles

In laboratory studies of an algal prey species and its rotifer predator, Nelson Hairston Jr. and colleagues obtained a puzzling result: they observed predator–prey cycles, but the predator populations peaked when the prey populations reached their lowest levels, and vice versa (**Figure 12.23**). The researchers knew that predator–prey interactions alone could not produce such a pattern, and they suggested four mechanisms that might: (1) rotifer egg viability might increase with prey density; (2) algal nutritional quality might increases with nitrogen concentrations; (3) accumulation of toxins alters algal physiology; and (4) the algae might evolve in response to predation.

Yoshida et al. (2003) tested these four hypotheses in two ways. First, they compared their data with results from four mathematical models (one for each mechanism). Only the model that included evolution in the prey population provided a good match to their data. Second, they performed an experimental test in which they manipulated the ability of the prey population to evolve; the idea was to see whether the puzzling results of Figure 12.23 would be duplicated when the prey could evolve, but not when the prey could not evolve. That is exactly what happened. In treatments in which the prey could not evolve (because only a single algal genotype was used), they observed typical predator–prey cycles; that is, predator abundance peaked shortly after prey abundance peaked (as in Figure 12.20D). In contrast, when the prey could evolve (because multiple genotypes were used), they observed cycles similar to those in Figure 12.23: predators were most abundant when prey were scarce.

Yoshida et al. (2003) found that the algal genotypes that were most resistant to predators were poor competitors. The puzzling results shown in Figure 12.23 appear to be due to the following scenario: When predator density is high, resistant algal genotypes have an advantage, and they increase in abundance. Eventually, the prey population consists mostly of resistant genotypes, and predator numbers drop and remain low, even though algae are abundant. Because there are now few predators, the nonresistant but competitively superior algal genotypes outcompete the resistant genotypes and increase in abundance. This increase in edible prey allows the predator population to increase, thus initiating another cycle. Hairston et al.'s results suggest an important lesson: ongoing evolutionary changes can have a powerful effect on predator–prey interactions.

CASE STUDY REVISITED
Snowshoe Hare Cycles

As we saw on pp. 260–261, neither the food supply hypothesis nor the predation hypothesis alone can explain hare population cycles. However, much of the variation in hare densities can be explained when we combine these two hypotheses—and add even more realism with a few new twists.

Charles Krebs and colleagues performed an experiment designed to determine whether food, predation, or their interaction caused population cycles in hares. The sheer scope of the experiment was impressive: the experimental treatments were performed in seven 1 × 1 km blocks of forest located in an isolated region of Canadian wilderness. Three blocks were not manipulated and were used as controls. Food was added to two blocks (the "+Food" treatment). In 1987, an electric fence (4 km in length) was constructed to

FIGURE 12.23 Evolution Causes Unusual Population Cycles In experiments performed in Hairston's laboratory, the abundances of a rotifer predator (*Brachionus calyciflorus*) and its algal prey (*Chlorella vulgaris*) fluctuated over time, but their population cycles differed from the typical predator–prey cycle (see Figures 12.2 and 12.20D). The researchers showed that these unusual population cycles were driven by evolutionary changes in the prey population. (After Yoshida et al. 2003.)

(A)

This conclusion was supported by results from a mathematical model that examined feeding relationships across three levels: vegetation (the hares' food), hares, and predators (King and Schaffer 2001). Field data were used to estimate the model parameters, and the model's predictions were compared with the actual results for Krebs et al.'s four treatments. Although the match was not exact, there was reasonably good agreement between the model and the results, again suggesting that both food and predators influence hare cycles (**Figure 12.25**).

While much progress has been made, some questions remain. We do not yet have a complete understanding of the factors that cause hare populations across broad re-

(B)

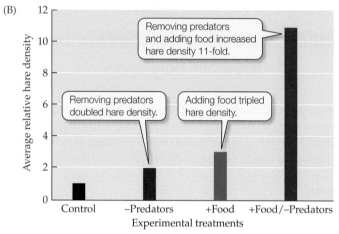

FIGURE 12.24 Both Predators and Food Influence Hare Density (A) This aerial photograph shows one of the 1 km² snowshoe hare study sites described in the text. (B) The number of hares in three experimental treatments increased relative to their densities in control blocks of forest. (B after Krebs et al. 1995.)

(A)

(B)

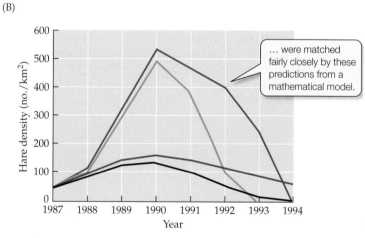

FIGURE 12.25 A Vegetation–Hare–Predator Model Predicts Hare Densities Accurately The predictions of a mathematical model were compared with the experimental results of Krebs et al. (1995). The model assumed that hare population densities were influenced by feeding relationships across three levels: vegetation, hares, and predators. Parameters for the model were estimated from field data. The investigators found a reasonably good match between (A) the experimental results and (B) the model's predictions. (After King and Schaffer 2001.)

exclude predators from one block of forest (the "–Predators" treatment). In the following year, a second 4 km fence was built; in the block of forest enclosed by this fence, food was added and predators were excluded (the "+Food/–Predators" treatment). The two fences (with a total length of 8 km) had to be monitored daily during the winter, when temperatures could plummet to –45°C (–49°F); this monitoring required so much time that it was impossible to replicate either fenced treatment. The survival rates and densities of hares in each blocks forest were monitored for an 8-year period.

Compared with the control blocks, hare densities were considerably higher in the +Food, –Predators, and +Food/–Predators blocks (**Figure 12.24**). The most pronounced effects were seen in the +Food/–Predators block, where, on average, hare densities were 11 times those in the control blocks. The strong effect of jointly adding food and removing predators suggests that hare population cycles are influenced by both food and predation.

gions of Canada to cycle in synchrony. Lynx can move from 500 to 1,100 km. If lynx move from areas with few prey to areas with abundant prey on a scale of hundreds of kilometers, their movements might be enough to cause geographic synchrony in hare cycles. In addition, large geographic regions in Canada experience a similar climate; within these regions, lynx cycles (which rise and fall with hare cycles) are similar to one another. The reason for this synchrony also remains to be determined.

Finally, the Krebs et al. experiment provided a test of whether the addition of food or the removal of predators (or both) could stop the hare population cycle. Although hare densities declined less in the +Food/–Predators block than in the control blocks, they did decline at the usual point in the hare cycle. Why did the +Food/–Predators treatment fail to stop the cycle? One possible reason is that the fences excluded lynx and coyotes, but did not exclude owls, goshawks, and other birds of prey. Collectively, these bird predators accounted for about 40% of snowshoe hare deaths, and thus could have contributed to the onset of the decline phase of the hare cycle in the +Food/–Predators block. Next, we'll explore another possible explanation: stress caused by the fear of predator attack.

CONNECTIONS IN NATURE

From Fear to Hormones to Demography

Predators not only affect their prey directly (by killing them), but also influence them indirectly—by altering prey behavior, for example (see p. 266). Boonstra et al. (1998) tested snowshoe hares for another possible indirect effect of predators: fear. Their results hint at a fascinating way that predation might influence the decline phase of the hare cycle.

When people are in a dangerous situation, we have a set of "fight-or-flight" responses that can produce rapid and sometimes astonishing results (such as the ability to move unusually heavy weights). Hares have a similar stress response. The stress response works by mobilizing energy and directing it to the muscles, and by suppressing body functions that are not essential for immediate survival, including growth, reproduction, and the immune system (**Figure 12.26**).

The stress response works well for immediate or acute forms of stress, such as an attack by a predator. Energy is provided to the muscles rapidly to help the animal deal with the threatening situation. Shortly thereafter, the response is shut down by a negative feedback process. The stress response works less well for long-term or chronic stress, however. In such cases, the negative feedback signals are weak, and the stress response is maintained for a long time. A failure to "turn off" the stress response can have harmful effects, including decreased growth and reproduction and increased susceptibility to disease. Collectively, such effects can reduce survival and reproduction rates.

When they are abundant, predators can cause up to 95% of hare deaths. At such times, hares are at increased risk of encountering predators; hares would also be likely to see or hear predators killing other hares, and to find the remains of hares that had been killed by predators. Reasoning that the fear provoked by such events could trigger chronic stress, Boonstra and colleagues measured the hormone and immune responses of hares exposed in the field to high versus low numbers of predators. In the decline phase of the hare cycle (when hares are exposed to many predators), cortisol levels increased, blood

FIGURE 12.26 The Stress Response When an animal is stressed, the hypothalamus releases a hormone called CRF, which stimulates a chain of reactions that affect a number of body processes. (After Boonstra et al. 1998.)

glucose levels increased, reproductive hormone levels decreased, and overall body condition worsened—as expected for hares experiencing chronic stress (see Figure 12.26).

Chronic stress can cause birth rates to drop, and this effect may be long-lasting: laboratory studies suggest that the conditions experienced by hares as they mature can influence their reproductive success for years to come. Overall, chronic stress induced by predation may help to explain some of the puzzling observations mentioned on p. 261, including why birth rates drop during the decline phase of the hare cycle and why hare numbers sometimes rebound slowly after predator numbers plummet. If future studies confirm the results of Boonstra et al. (1998), their work will provide a clear example of how predation risk can alter prey physiology, thereby changing prey demography and influencing predator–prey cycles.

SUMMARY

CONCEPT 12.1 Most predators have broad diets, whereas a majority of herbivores have relatively narrow diets.

- Over half of the organisms on Earth are predators, herbivores, or parasites that sustain themselves by feeding on other organisms.

- Many predators move about their habitat in search of prey; others, known as sit-and-wait predators, remain in one place and attack or trap prey that move within striking distance.

- Most predators do not specialize on particular prey species; instead, they typically eat prey in relation to their availability across a broad range of prey species.

- Many herbivores specialize on particular plant parts, such as leaves (the most common food source), roots, stems, seeds, or internal fluids.

- Herbivorous insects, which constitute a majority of herbivore species, tend to have relatively narrow diets, feeding on only one or a few plant species.

CONCEPT 12.2 Organisms have evolved a wide range of adaptations that help them capture food and avoid being eaten.

- In response to strong selection pressures exerted by predators and herbivores, prey species and plants have evolved a rich variety of defensive mechanisms.

- Prey may rely on physical defenses, toxins, mimicry, or behavioral responses to escape predation.

- Plants cope with herbivory via masting and other forms of avoidance, compensation (a form of tolerance), and secondary chemicals that deter herbivores.

- Predators have a wide range of adaptations for overcoming the defenses of their prey, including physical features, toxins, mimicry, and tolerance or detoxification of prey toxins.

- Although a plant's defensive mechanisms prevent most herbivores from eating it, typically some herbivores can overcome the plant's defenses via structural, chemical, or behavioral means.

CONCEPT 12.3 Predation and herbivory affect ecological communities greatly, in some cases causing a shift from one community type to another.

- Predators and herbivores may have direct effects on the distributions and abundances of their food organisms as well as indirect effects on other species in their communities.

- Predators can cause dramatic declines in the distributions and abundances of their prey; similarly, herbivores can decimate their food plants.

- Exploitation can also alter the composition of ecological communities, in some cases changing one community type to another.

CONCEPT 12.4 Population cycles can be caused by feeding relations, such as a three-way interaction between predators, herbivores, and plants.

- Results from mathematical models, laboratory experiments, and field observations suggest that population cycles can be caused by exploitation.

- Coexistence of predators and prey can be influenced by several factors, including dispersal and habitat complexity.

- Evolutionary change can affect predator–prey population dynamics, in some cases producing unusual population cycles.

REVIEW QUESTIONS

1. Compare and contrast the diet breadth of predators and herbivores.

2. Summarize the evolutionary effects that predators and herbivores have on their food organisms, as well as the effects that food organisms have on the predators and herbivores that eat them. Explain why these effects are pervasive and pronounced.

3. In this chapter, we claim that exploitative interactions such as predation and herbivory can have strong effects on ecological communities.

   a. Provide a logical argument to support this claim.

   b. Does the scientific evidence support or contradict this claim? Explain.

4. Based on the relationships shown in Equation 12.1, use the zero population growth isocline approach described in Box 11.2 to show that the prey population is expected to *decrease* if there are more than *r/a* predators.

HANDS-ON PROBLEM SOLVING

Predation Alters Behavior | This Web exercise (www.sinauer.com/ecology) explores the effect of predation on lizard behavior. You will interpret data from a recent paper that demonstrated changes in the behaviors of *Anolis* lizards after the introduction of a predatory lizard. This behavioral shift was manifest as a change in habitat use.

SUGGESTED READINGS

Clinchy, M., L. Zanette, R. Boonstra, J. C. Wingfield and J. N. M. Smith. 2004. Balancing food and predator pressure induces chronic stress in songbirds. *Proceedings of the Royal Society of London B* 271: 2473–2479. *In the wild, animals experience a constant tension between gathering food and avoiding predators; this tension can result in a physiological stress response that affects prey demography. This paper reports novel experiments conducted with birds and provides a general framework for understanding this critical effect of predators on their prey.*

Crawley, M. J. 1983. *Herbivory: The Dynamics of Animal–Plant Interactions.* Blackwell Scientific Publications. Oxford. *This mid-level but accessible text provides an excellent in-depth view of the effects that herbivores and plants have on each other.*

Forsyth, A. and K. Miyata. 1984. *Tropical Nature: Life and Death in the Rain Forests of Central and South America.* Scribner, New York. *A gripping description of life in the rain forest, including compelling examples of the predators and herbivores that live there.*

Lawton, J. H. and S. McNeill. 1979. Between the devil and the deep blue sea: On the problem of being an herbivore. *Symposium of the British Ecological Society* 20: 223–244. *This classic paper describes the tough life of an herbivore: it must evade predators (the "devil") while seeking nourishment from plants (the "deep blue sea"), a food source that is abundant but often well-defended and of poor nutritional quality.*

Myers, R. A., J. K. Baum, T. D. Shepherd, S. P. Powers and C. H. Peterson. 2007. Cascading effects of the loss of apex predatory sharks from a coastal ocean. *Science* 315: 1846–1850. *This paper described how decreases (due to overfishing) in the populations of great shark species have led not only to substantial increases in the populations of prey eaten by sharks, but also to the termination of a scallop fishery that had thrived for 100 years.*

Peterson, K. J., M. A. McPeek and D. A. D. Evans. 2005. Tempo and mode of early animal evolution: Inferences from rocks, Hox, and molecular clocks. *Paleobiology* 31 (2, Supplement): 36–55. *This thought-provoking paper illustrates how the strong selection imposed by exploitation on food organisms leaves a trace in the fossil record that can be used to shed light on key questions about the history of early animal life.*

Vermeij, G. J. 1987. *Evolution and Escalation: An Ecological History of Life.* Princeton University Press, Princeton, NJ. *This book provides a masterful description of how over long periods of time, predators such as crabs and shell-drilling snails evolved increasingly sophisticated ways to pry open, crush, or drill through their prey—and how prey responded by evolving increasingly effective means of defense.*

CHAPTER 13 Parasitism

CASE STUDY

Enslaver Parasites

In science fiction books and movies, villains sometimes use mind control or physical devices to break the will and control the actions of their victims. In these stories, a person may be forced to perform strange or grotesque actions, or to harm themselves or others—all against their will.

Real life can be just as strange. Consider the hapless cricket shown in the videoclip found in **Web Extension 13.1**. This cricket does something that a cricket ordinarily would never do: it walks to the edge of a body of water, jumps in, and drowns. Shortly afterward, a hairworm parasite begins to emerge from the body of the cricket (**Figure 13.1**). For the worm, this is the final step in a journey that begins when a terrestrial arthropod—such as a cricket—drinks water in which a hairworm larva swims. The larva enters the cricket's body and feeds on its tissues, growing from microscopic size into an adult that fills all of the cricket's body cavity except its head and legs. When fully grown, hairworms must return to the water, where adult males and females cluster in tight masses to mate. After the adult hairworms mate, the next generation of larvae are released to the water, where they will die unless they are ingested by a terrestrial arthropod host.

Has the hairworm "enslaved" its cricket host, forcing it to jump into the water—an act that kills the cricket, but is essential for the hairworm to complete its life cycle? The answer appears to be yes. Observations have shown that when crickets infected with hairworms are near water, they are much more likely to enter the water than are uninfected crickets (Thomas et al. 2002). Furthermore, in ten out of ten trials, when infected crickets were rescued from the water, they immediately jumped back in. Uninfected crickets do not do this.

Hairworms are not the only parasites that enslave their hosts. Maitland (1994) coined the term "enslaver parasites" for several fungal species that alter the perching behavior of their fly hosts in such a way that their spores can be dispersed more easily (**Figure 13.2**). Even vertebrates can be enslaved by parasites. Rats typically engage in predator avoidance behaviors in areas that show signs of cats. However, rats infected with the protozoan parasite *Toxoplasma gondii* behave abnormally toward cats: they do not avoid cats, and in some cases they are actually attracted to cats. While such a behavioral change can be a fatal attraction for the rat, it benefits the parasite because it increases the chance that the parasite will be transmitted to the next host in its complex life cycle—a cat.

FIGURE 13.1 Driven to Suicide The behavior of this wood cricket (*Nemobius sylvestris*) was manipulated by the hairworm (*Paragordius tricuspidatus*) exiting its body. By causing the cricket to jump into the water (where it drowns), the hairworm is able to continue its life cycle.

Hairworm

FIGURE 13.2 Enslaved by a Fungus
Shortly before they die from the infection, yellow dungflies infected by the fungus *Entomophthora muscae* move to the downwind side of a relatively tall plant and perch on the underside of one of its leaves. This position increases the chance that fungal spores released by *Entomophthora* will land on healthy yellow dungflies. (After Maitland 1994.)

Flies infected by the fungus perch in an atypical position, from which fungal spores can easily spread to healthy flies.

Wind

Spores

Healthy flies typically perch on the upper surfaces of low-lying vegetation—where they are exposed to fungal spores.

Finally, in what may be the most remarkable example of all, the tropical parasitoid wasp *Hymenoepimecis argyraphaga* manipulates its host, the orb-weaving spider *Plesiometa argyra*, so precisely that shortly before killing the spider, the wasp larva induces the spider to make a special "cocoon web." The cocoon web looks dramatically different from the spider's typical web, and its purpose has nothing to do with catching prey. As soon as the spider finishes making the cocoon web, the larva kills and eats it. The larva then spins a cocoon, in which it will complete its development, and hangs the cocoon from the web it induced the spider to make. The cocoon web serves as a strong support that protects the larva (in its cocoon) from being swept away by torrential rains.

How do some parasites enslave their hosts? Can the hosts fight back? More generally, what can these remarkable interactions tell us about host–parasite relationships?

More than half of the millions of species that live on Earth are **symbionts**, meaning that they live in or on other organisms. To begin to understand how many symbionts there are, we need look no further than our own bodies (**Figure 13.3**). Our faces are home to mites that feed on exudates from the pores of our skin and on secretions at the base of our eyelashes. There are bacteria and fungi that grow on our skin and under our toenails, and arthropods such as lice that live on our heads, pubic regions, and other parts of our bodies. Moving inward, our tissues, organs, and body cavities can be infested with a rich variety of organisms, from bacteria to worms to fungi to protozoans.

Although some symbionts are mutualists (as we will see in Chapter 14), the majority of them are parasites. A **parasite** consumes the tissues or body fluids of the organism on which it lives, its **host**; some parasites, called **pathogens**, cause diseases. As a group, parasites typically harm, but do not immediately kill, the organisms they eat (unlike predators). The negative effects of parasites on their hosts vary widely, however, from mild to lethal. We see this variation in our own species, for which some parasites, such as the fungus that causes athlete's foot, are little more than a nuisance. Others, such as the protozoan *Leishmania tropica*, can cause dis-

figurement, and still others, such as *Yersinia pestis*, the bacterium that causes the plague, can kill. There is similar variation in the degree of harm caused by parasites that infect other species. Parasites also vary in many other ways, as we'll see next as we examine their basic biology.

Parasite Natural History

CONCEPT 13.1
Parasites, which constitute roughly 50% of the species on Earth, typically feed on only one or a few host species.

Parasites vary in size from relatively large species (**macroparasites**), such as arthropods and worms, to species too small to be seen with the naked eye (**microparasites**), such as bacteria, protozoans, and unicellular fungi. But whether they are large or small, over the course of their lives, parasites typically feed on only one or a few host individuals. Thus, defined broadly, parasites include herbivores such as aphids or nematodes, that feed on one or a few host plants, as well as parasitoids (see p. 262), whose larvae feed on a single host, almost always killing it.

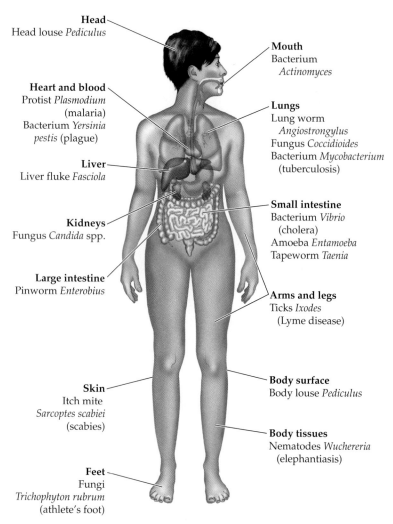

Head
Head louse *Pediculus*

Mouth
Bacterium
Actinomyces

Heart and blood
Protist *Plasmodium*
(malaria)
Bacterium *Yersinia
pestis* (plague)

Lungs
Lung worm
Angiostrongylus
Fungus *Coccidioides*
Bacterium *Mycobacterium*
(tuberculosis)

Liver
Liver fluke *Fasciola*

Small intestine
Bacterium *Vibrio*
(cholera)
Amoeba *Entamoeba*
Tapeworm *Taenia*

Kidneys
Fungus *Candida* spp.

Large intestine
Pinworm *Enterobius*

Arms and legs
Ticks *Ixodes*
(Lyme disease)

Skin
Itch mite
Sarcoptes scabiei
(scabies)

Body surface
Body louse *Pediculus*

Body tissues
Nematodes *Wuchereria*
(elephantiasis)

Feet
Fungi
Trichophyton rubrum
(athlete's foot)

FIGURE 13.3 The Human Body as Habitat Different parts of our bodies provide suitable habitat for a wide range of parasites.

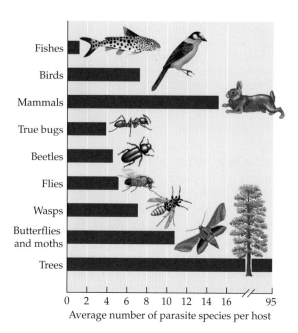

FIGURE 13.4 Many Species Are Host to More Than One Parasite Species In a study conducted in Britain, most host species were found to harbor more than one parasite species. The number of parasite species shown here for fishes, birds, and mammals includes only helminth worm parasites and hence is likely to underestimate the actual number of parasite species found in these vertebrates.

Most species are attacked by more than one parasite (**Figure 13.4**), and even parasites have parasites. Because parasites spend their lives feeding on one or a few host individuals, they tend to have a close relationship to the organisms they eat. For example, many parasites are closely adapted to particular host species, and many attack only one or a few host species. This specialization at the species level helps to explain why there are so many species of parasites—many host species have at least one parasite that eats *only* them. Overall, although the total number of parasite species is not known, a rough estimate is that 50% of the species on Earth are parasites (Windsor 1998).

Parasites are also specialized for living on or eating certain parts of the host's body. We'll focus next on this aspect of parasite specialization by describing both ectoparasites and endoparasites.

Ectoparasites live on the surface of their host

An **ectoparasite** lives on the outer body surface of its host (**Figure 13.5**). Plants that are ectoparasites include vines such as dodder (*Cuscuta* spp.) that grow on a host plant and obtain water and food by using modified roots called *haustoria* to penetrate the tissues of their host. Dodder cannot photosynthesize and hence depends on its host for both mineral nutrients and carbohydrates. Other plants, such as mistletoes (parasite species in the order Santalales that grow on the branches of trees or shrubs), are *hemiparasitic*: they extract water and mineral nutrients from their hosts, but since they have green leaves and can photosynthesize, they do not rely exclusively on their host for carbohydrates.

There are also many fungal and animal parasites that live on the surfaces of plants and animals, feeding on their hosts' tissues or body fluids. More than 5,000 species of fungi attack important crop and horticultural plants, causing billions of dollars of damage each year. Some fungi that attack plants, including mildews, rusts, and smuts, grow on the surface of the host plant and extend their hyphae (fungal filaments) within the plant to extract nutrients from its tissues. Plants are also attacked by numerous animal ectoparasites, including aphids, whiteflies, and scale insects, which are found on stems and leaves, and nematode worms, beetles, and (juve-

(A)

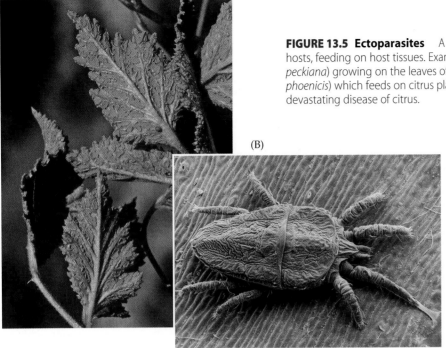

FIGURE 13.5 Ectoparasites A wide range of parasites live on the outer surfaces of their hosts, feeding on host tissues. Examples include (A) the orange rust fungus (*Gymnoconia peckiana*) growing on the leaves of its blackberry host, and (B) the flat mite (*Brevipalpus phoenicis*) which feeds on citrus plants and transmits the virus that causes citrus leprosis, a devastating disease of citrus.

(B)

Endoparasites live inside their host

If we ignore the details of their form, we can think of people and most other animals as doughnut-shaped: they have a ring of tissues (analogous to the part of the doughnut that you eat) surrounding a tube called the *alimentary canal* (analogous to the hole of the doughnut). The alimentary canal runs through the body from the mouth to the anus. Parasites that live inside their hosts, called **endoparasites**, include species that inhabit the alimentary canal as well as species that live within host cells or tissues (**Figure 13.6**).

nile) cicadas, which are found on roots. Animals like these that eat plants and live on their outer surfaces can be thought of both as herbivores (because they eat plant tissues) and as parasites (especially if they remain on a single host plant for much of their lives).

A similar array of fungal and animal ectoparasites can be found on the surfaces of animals. Familiar examples include *Trichophyton rubrum*, the fungus that causes athlete's foot, and fleas, mites, lice, and ticks, which feed on the tissues or blood of their hosts. Some of these parasites also transmit diseases to their hosts, including fleas that spread the plague and ticks that spread Lyme disease.

The alimentary canal provides excellent habitat for parasites. The host brings in food at one end (the mouth) and excretes what it cannot digest at the other (the anus). Parasites

FIGURE 13.6 Endoparasites Many parasites live within the body of their host, feeding on host tissues. (A) The tapeworm *Taenia taeniformis* uses the suckers and hooks shown here to attach to the intestinal wall of its mammalian host, often a rodent, rabbit, or cat. Once attached, an adult can grow to be over 5 meters in length. (B) The bacterium *Mycobacterium tuberculosis* causes the lung disease tuberculosis which kills two to three million people each year. (C) This section of a cactus shows the destruction wrought by *Erwinia corotovora*, a bacterium that causes soft rot. Affected areas become soft with decay and develop a distinctive foul odor.

(A)

(B)

(C)

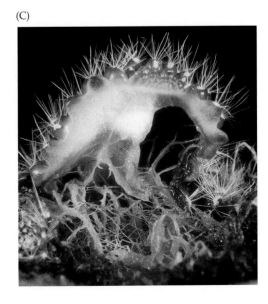

that live within the alimentary canal often do not eat host tissues at all—instead they rob the host of nutrients. Tapeworms, for example, have a scolex, a structure with suckers (and sometimes hooks) that they use to attach themselves to the inside of the host's intestine. Once it is attached, the tapeworm simply absorbs food that the host has already digested. Tapeworms that infect humans can grow up to 10–20 m (33–66 feet) long. Large tapeworms can block the intestines and cause nutritional deficiencies.

Many other endoparasites live within the cells or tissues of animal hosts, causing a wide range of symptoms as they reproduce within cells or consume host tissues. Examples in people include *Yersinia pestis*, the bacterium that causes the plague, and *Mycobacterium tuberculosis*, the bacterium that causes tuberculosis (TB). TB is a potentially fatal lung disease, aptly referred to as the "Captain of Death"; with the possible exception of malaria, it has killed more people than any other disease in human history. TB continues to kill 2 to 3 million adults each year (a number comparable to the 3 million that currently die each year from AIDS).

Plants too are attacked by a wide variety of endoparasites, including bacterial pathogens that cause soft rot in various plant parts, such as fruits (tomatoes or eggplants), fleshy storage tissues (potatoes), or succulent leaves (cabbage). Other plant pathogens include fungi that rot various plant parts from the inside out, and bacteria that invade plant vascular tissues, where they disrupt the flow of water and nutrients, causing wilting and often death.

Endoparasitism and ectoparasitism have advantages and disadvantages

There are advantages and disadvantages to living in or on a host (**Table 13.1**). Because ectoparasites live on the surface of their host, it is relatively easy for them or their offspring to disperse from one host individual to another. It is much more difficult for endoparasites to disperse to new hosts. Endoparasites solve this problem in a variety of ways. Some, like the

enslaver parasites discussed on p. 282, alter the physiology or behavior of the host in ways that facilitate parasite dispersal. Other examples include the bacterium (*Vibrio cholerae*) that causes cholera and the amoeba (*Entamoeba histolytica*) that causes amoebic dysentery. People with cholera and dysentery have diarrhea, a symptom that increases the chance that the parasite will contaminate drinking water and thereby spread to new hosts. Other endoparasites have complex life cycles that include stages that are specialized for dispersing from one host species to another (see Figure 13.9).

Although dispersal is relatively easy for ectoparasites, there are costs to life on the surface of a host. Compared with endoparasites, ectoparasites are more exposed to natural enemies such as predators, parasitoids, and parasites. Aphids, for example, are attacked by ladybugs, birds, and many other predators, along with parasites such as mites that suck fluids from their bodies and lethal parasitoids (see Figure 12.4). Endoparasites, in contrast, are safe from all but the most specialized predators and parasites. Endoparasites are also relatively well protected from the external environment, and they have relatively easy access to food—unlike an ectoparasite, an endoparasite does not have to pierce the host's protective outer surfaces to feed. But living within the host does expose endoparasites to a different sort of danger: attack by the host's immune system. Some parasites have evolved ways to tolerate or overcome immune-system defenses, as we shall see in the following section.

Defense and Counterdefense

CONCEPT 13.2
Hosts have adaptations for defending themselves against parasites, and parasites have adaptations for overcoming host defenses.

As we saw in Chapter 12, predators and herbivores exert strong selection on their food organisms, and vice versa. The prey species and plants eaten by predators and herbivores have adaptations that help them avoid being eaten; similarly, predators and herbivores have adaptations that help them to overcome the defenses of their prey or food plants. The same is true of parasites and their hosts: hosts have evolved ways to protect themselves against parasites, and parasites have evolved countermeasures to circumvent host defenses.

Host immune systems and biochemical defenses can protect against parasites

Host organisms have a wide range of defensive mechanisms that can prevent or limit the severity of parasite attacks. For example, a host may have a protective outer covering, such as the skin of a mammal or the hard exoskeleton of an insect, that can keep ectoparasites from piercing its body or

TABLE 13.1
Advantages and Disadvantages of Living In or On the Host

	Ectoparasitism	Endoparasitism
Advantages	Ease of dispersal	Ease of feeding
	Safe from host's immune system	Protected from external environment
		Safer from enemies
Disadvantages	Vulnerability to natural enemies	Vulnerability to host's immune system
	Exposure to external environment	
	Feeding more difficult	Dispersal difficult

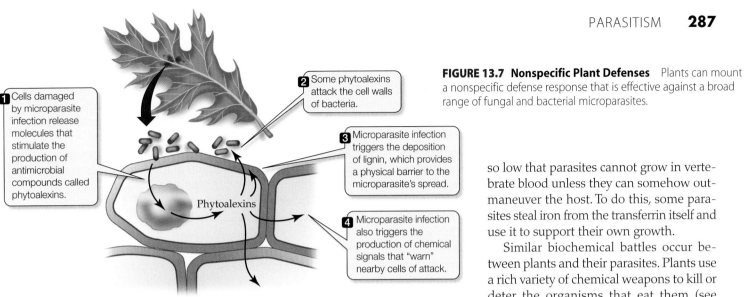

1 Cells damaged by microparasite infection release molecules that stimulate the production of antimicrobial compounds called phytoalexins.

2 Some phytoalexins attack the cell walls of bacteria.

3 Microparasite infection triggers the deposition of lignin, which provides a physical barrier to the microparasite's spread.

Phytoalexins

4 Microparasite infection also triggers the production of chemical signals that "warn" nearby cells of attack.

FIGURE 13.7 Nonspecific Plant Defenses Plants can mount a nonspecific defense response that is effective against a broad range of fungal and bacterial microparasites.

make it difficult for endoparasites to enter. Endoparasites that do manage to enter the host's body are often killed or rendered less effective by the host's immune system or biochemical defenses.

Immune systems The vertebrate immune system includes specialized cells that allow the host to recognize microparasites to which it has been previously exposed; in many instances, the "memory cells" of the immune system are so effective that the host has lifelong immunity against future attack by the same microparasite. Other immune system cells engulf and destroy parasites or mark them with chemicals that target them for later destruction.

Insects and plants can also mount highly effective responses to invasion by parasites. Some plants have resistance genes, the different alleles of which provide protection against microparasites that have a particular genotype; we will describe this defense system in more detail on pp. 290–291. Plants are not helpless, however, even when they lack alleles that provide resistance to a specific attacker. In such a case, the plant relies on a nonspecific immune system that produces antimicrobial compounds, including some that attack the cell walls of bacteria and others that are toxic to fungal parasites (**Figure 13.7**). The plant may also produce chemical signals that "warn" nearby cells of imminent attack, and still other chemicals that stimulate the deposition of lignin, a hard substance that provides a barricade against the invader's spread.

Biochemical defenses In addition to the defenses provided by their immune systems, hosts also have ways of regulating their biochemistry to limit parasite growth. Bacterial and fungal endoparasites, for example, require iron to grow. Vertebrate hosts—including mammals, birds, amphibians, and fishes—have a protein called transferrin that removes iron from their blood serum (where parasites could use it) and stores it in intracellular compartments (where parasites cannot get to it). Transferrins are so efficient that the concentration of free iron in mammalian blood serum is only 10^{-26} M—

so low that parasites cannot grow in vertebrate blood unless they can somehow outmaneuver the host. To do this, some parasites steal iron from the transferrin itself and use it to support their own growth.

Similar biochemical battles occur between plants and their parasites. Plants use a rich variety of chemical weapons to kill or deter the organisms that eat them (see Chapter 12). These defensive secondary compounds are so effective that some animals take advantage of them: they eat specific plants in order to treat or prevent parasite infections. For example, when parasitic flies lay eggs on their bodies, woolly bear caterpillars switch from their usual food plant (lupines) to a diet of poisonous hemlock (Karban and English-Loeb 1997). The new diet does not kill the parasites, but it does increase the chance that the caterpillar will survive the attack and metamorphose into an adult tiger moth (*Platyprepia virginalis*). Chimpanzees infected with nematodes (*Oesphogostomum stephanostomum*) specifically seek out and eat a bitter plant that scientists have learned contains compounds that kill or paralyze the nematodes and can also deter many other parasites (Huffman 1997) (**Figure 13.8**). Humans do essentially the same thing: we spend billions of dollars each year on pharmaceuticals that are based on compounds originally obtained from plants.

FIGURE 13.8 Using Plants to Fight Parasites A chimpanzee in Mahale National Park, Tanzania, chews on the bitter pith of *Vernonia amygdalina*, which produces a compound that is toxic to nematode parasites.

Effects of defenses on mate choice   Defense against parasites is so important that in some bird, mammal, and fish species, females select mates based on traits that indicate that a male has effective defenses. For example, a group of proteins known as the major histocompatibility complex (MHC) is a key component of the vertebrate immune system. Typically, the more different MHC proteins an individual has, the better protected it is against a wider range of parasites. Female stickleback fish prefer to mate with males that have many MHC proteins (Reusch et al. 2001). The females use their sense of smell to detect which males have many MHC proteins—and are thus likely to have few parasites.

Other species may use aspects of male performance to assess parasite loads. In *Copadichromis eucinostomus*, a cichlid fish from Lake Malawi, Africa, males court females by building a sand structure called a bower. Females prefer to mate with males that make large, smooth bowers. Researchers have found that such males have fewer tapeworm parasites than do males that make smaller bowers, perhaps because males with many tapeworms must spend more of their time eating to compensate for the nutrients they lose to the parasites, and hence cannot build large bowers. Similar results have been found in birds, in which traits such as beak color reflect parasite loads and are used by females as key criteria for mate choice.

Parasites have adaptations that circumvent host defenses

To survive and reproduce, a parasite must be able to tolerate or evade its host's defensive mechanisms. Aphids and other ectoparasites, for example, must be able to pierce the protective outer covering of the host, and they must be able to tolerate whatever chemical compounds are present in the host tissues or body fluids that they eat. Viewed broadly, the challenges faced by ectoparasites are similar to those faced by herbivores and predators as they attempt to cope with the toxins and physical structures that their food organisms use to defend themselves. We discussed such challenges in Chapter 12, so here we focus on how endoparasites cope with defenses found inside the host.

Counterdefenses against encapsulation   Endoparasites often face a formidable challenge from the host immune system and related aspects of host biochemistry. Host species typically have a number of ways to destroy parasite invaders. In addition to the strategies we have already described, some hosts can cover parasites or parasite eggs with capsules that kill them or render them harmless, a process called *encapsulation*.

Encapsulation is used by some insects to defend against macroparasites. Insect blood cells can engulf small invaders, such as bacteria, but they cannot engulf large objects, such as nematodes or parasitoid eggs. However, some insects have lamellocytes, which are blood cells that can form multicel-

lular sheaths (capsules) around large objects. When an insect mounts a vigorous encapsulation defense, most or all of the attacking parasites may be destroyed. As a result, the parasites are under strong selection to develop a counterdefense.

Drosophila fruit flies can mount an effective defense against wasp parasitoids by encapsulating (and hence killing) their eggs. Parasitoid wasps that attack fruit flies avoid encapsulation in several different ways. When wasps in the genus *Leptopilina* lay their eggs inside a fruit fly host, they also inject viruslike particles into the host. These particles infect the host's lamellocytes and cause them to self-destruct, thus weakening the host's resistance and increasing the percentage of wasp eggs that survive (Rizki and Rizki 1990). Other parasitoid wasps, such as *Asobara tabida*, lay eggs covered with filaments. These filaments cause the eggs to stick to and become embedded in fat cells and other host cells, where they are not detected by circulating lamellocytes.

Counterdefenses involving hundreds of genes   Some endoparasites have a complex set of adaptations that allows them to thrive inside their host. One such endoparasite is *Plasmodium*, the protozoan that causes malaria, a disease that kills 1 to 2 million people each year. *Plasmodium* has a complex life cycle with specialized stages that allow it to alternate between a mosquito and a human host (**Figure 13.9**). Infected mosquitoes contain one specialized *Plasmodium* stage, called a *sporozoite*, in their saliva. When an infected mosquito bites a human, sporozoites enter the victim's bloodstream and travel to the liver, where they divide to form another stage, called a *merozoite*. The merozoites penetrate red blood cells, where they multiply rapidly. After 48–72 hours, large numbers of merozoites break out of the red blood cells, causing the periodic chills and fever that are associated with malaria. Some of the offspring merozoites attack more red blood cells, while others transform into gamete-producing cells. If another mosquito bites the victim, it picks up some of the gamete-producing cells, which enter its digestive tract and form gametes. After fertilization occurs, the resulting zygotes produce thousands of sporozoites, which then migrate to the mosquito's salivary glands, where they await their transfer to another human host.

Plasmodium parasites face two potentially lethal challenges from their human host. First, red blood cells do not divide or grow, and hence they lack the cellular machinery needed to import nutrients necessary for growth. A *Plasmodium* merozoite inside a red blood cell would starve if it did not have a way to obtain essential nutrients. Second, after 24–48 hours, infection by *Plasmodium* causes red blood cells to have an abnormal shape. The human spleen recognizes and destroys such deformed cells, along with the parasites inside.

Plasmodium parasites address these challenges by having hundreds of genes whose function is to modify the host red blood cell in ways that allow the parasite to obtain food and escape destruction by the spleen (Marti et al. 2004; Hiller et

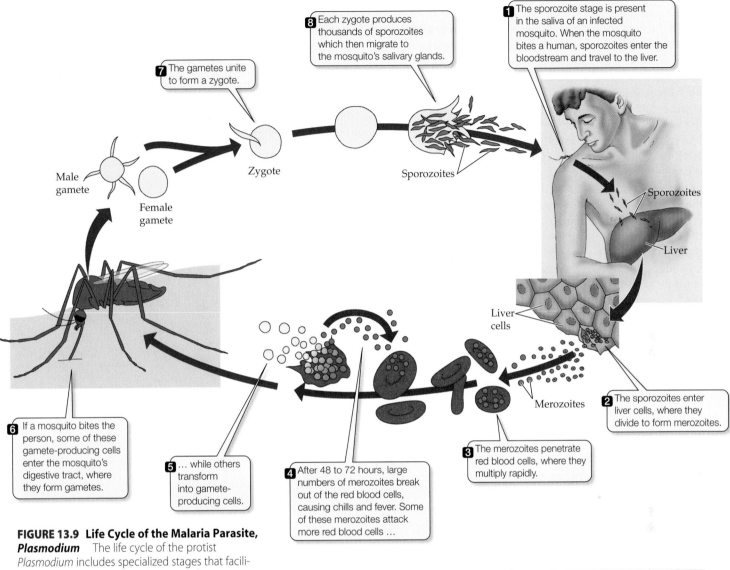

8 Each zygote produces thousands of sporozoites which then migrate to the mosquito's salivary glands.

7 The gametes unite to form a zygote.

1 The sporozoite stage is present in the saliva of an infected mosquito. When the mosquito bites a human, sporozoites enter the bloodstream and travel to the liver.

Male gamete

Female gamete

Zygote

Sporozoites

Sporozoites

Liver

Liver cells

2 The sporozoites enter liver cells, where they divide to form merozoites.

Merozoites

3 The merozoites penetrate red blood cells, where they multiply rapidly.

6 If a mosquito bites the person, some of these gamete-producing cells enter the mosquito's digestive tract, where they form gametes.

5 ... while others transform into gamete-producing cells.

4 After 48 to 72 hours, large numbers of merozoites break out of the red blood cells, causing chills and fever. Some of these merozoites attack more red blood cells ...

FIGURE 13.9 Life Cycle of the Malaria Parasite, *Plasmodium*   The life cycle of the protist *Plasmodium* includes specialized stages that facilitate the dispersal of this endoparasite from one host to another. The sporozoite stage, for example, enables the parasite to disperse from an infected mosquito to a human host, while the gamete-producing stage enables the parasite to disperse from the human host to a mosquito.

al. 2004). Some of these genes cause transport proteins to be placed on the surface of the red blood cell, thereby enabling the parasite to import essential nutrients into the host cell. Other genes guide the production of unique knobs that are added to the surface of the red blood cell. These knobs cause the infected red blood cell to stick to other human cells, thereby preventing it from traveling in the bloodstream to the spleen, where it would be destroyed. The proteins on these knobs vary greatly from one parasite individual to another, making it difficult for the human immune system to recognize and destroy the infected cells.

Coevolution

CONCEPT 13.3
Host and parasite populations can evolve together, each in response to selection imposed by the other.

As we have just seen, *Plasmodium* has specific adaptations that enable it to live inside a red blood cell. When a parasite and its host each possess such specific adaptations, that observation suggests that the strong selection pressure that hosts and parasites impose on each other has caused both of their populations to evolve. Such changes have been directly observed in Australia, where the *Myxoma* virus was introduced to control populations of the European rabbit (*Oryctolagus cuniculus*).

European rabbits were first introduced to Australia in 1859, when 24 wild rabbits were released at a ranch in Victoria. Within a decade, rabbit populations had grown so large, and were consuming so much plant material, that they posed a threat to cattle and sheep pasturelands and wool production. A series of control measures were enacted, including introductions of predators, shooting and poisoning of rabbits, and the building of fences to limit the spread of rabbits from one region to another (Fenner and Ratcliffe 1965). None of these methods worked. By the 1900s, hundreds of millions of rabbits had spread throughout much of the continent.

After years of investigation, Australian government officials settled on a new control measure: introduction of the *Myxoma* virus. A rabbit infected with this virus may suffer from skin lesions and severe swellings, which can lead to blindness, difficulty with feeding and drinking, and death (usually within 2 weeks of infection). The virus is transmitted from rabbit to rabbit by mosquitoes. In 1950, when *Myxoma* was first used to control rabbit populations, 99.8% of infected rabbits died. In the ensuing decades, millions of rabbits were killed by *Myxoma*, and the sizes of rabbit populations dropped dramatically throughout the Australian continent. Over time, however, the rabbit evolved resistance to the virus, and the virus evolved to become less lethal (**Figure 13.10**). *Myxoma* is still used to control rabbit populations, but doing so requires a constant search for new, lethal strains of the virus to which the rabbit has not evolved resistance.

The increased resistance of the rabbit and the reduced lethality of the virus illustrate **coevolution**, which occurs when populations of two interacting species evolve together, each in response to selection imposed by the other. The outcome of coevolution can vary greatly, depending on the biology of the interacting species. In the European rabbit, selection favored the evolution of increased resistance to viral attack, as you might expect. In addition, viral strains of intermediate lethality predominated, perhaps because such strains allowed rabbits to live long enough for one or more mosquitoes to bite them and transmit the virus to another host (mosquitoes do not bite dead rabbits). In other cases of host–parasite coevolution, the parasite evolves counterdefenses to overcome host resistance mechanisms, as the following examples illustrate.

Selection can favor a diversity of host and parasite genotypes

As mentioned earlier, plant defense systems include a specific response that makes particular plant genotypes resistant to particular parasite genotypes. Such **gene-for-gene interactions** are well documented in a number of plant species, including wheat, flax, and *Arabidopsis thaliana,* a plant in the mustard family that is often used in laboratory experiments and genetic studies. Wheat has dozens of different genes for

A rabbit infected with the *Myxoma* virus.

During the 1950s, the lethality of virus samples collected in the wild decreased.

By the early 1970s, viral lethality had stabilized at an intermediate level.

FIGURE 13.10 Coevolution of the European Rabbit and the *Myxoma* Virus (A) Periodically after the introduction of the *Myxoma* virus, rabbits were collected from wild populations and exposed to a standard strain of the virus; the standard strain killed 90% of naïve (unselected) laboratory rabbits. Over time, mortality declined as rabbits from wild populations evolved resistance to the virus. (B) The lethality of virus samples collected in the wild also declined when tested against a standard (unselected) line of rabbits. (A after Fenner and Ratcliffe 1965; B after May and Anderson 1983.)

resistance to fungi such as wheat rusts (*Puccinia*). Different wheat rust genotypes can overcome different wheat resistance genes, however, and periodically, mutations occur in

FIGURE 13.11 Adaptation by Parasites to Local Host Populations The graph shows the frequencies with which *Microphallus* trematodes from three lakes in New Zealand (Lake Mapourika [shown], Lake Wahapo, and Lake Paringa) were able to infect snails from the same three lakes. (After Lively 1989.)

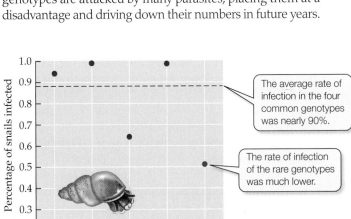

wheat rusts that produce new genotypes to which wheat is not resistant. Studies have shown that the frequencies of wheat rust genotypes vary considerably over time as farmers use different resistant varieties of wheat. For example, a rust variety may be abundant in one year because it can overcome the resistance genes of wheat varieties planted that year, yet less abundant the following year because it cannot overcome the resistance genes of wheat varieties planted that year.

Changes in the frequencies of host and parasite genotypes also occur in natural systems. In the lakes of New Zealand, a trematode worm, *Microphallus*, parasitizes the snail *Potamopyrgus antipodarum*. *Microphallus* has a serious negative effect on its snail hosts: it castrates the males and sterilizes the females. The parasite has a much shorter generation time than its host, and hence we might expect that it would rapidly evolve the ability to cope with snail defensive mechanisms. Lively (1989) tested this idea in an experiment that pitted parasites from each of three lakes against snails from the same three lakes. He found that parasites infected snails from their home lake more effectively than they infected snails from the other two lakes (**Figure 13.11**). This observation suggests that the parasite genotypes in each lake had evolved rapidly enough to overcome the defenses of the snail genotypes found in that lake.

The snails also evolved in response to the parasites, albeit more slowly. Dybdahl and Lively (1998) documented the abundances of different snail genotypes over a 5-year period in a New Zealand lake. The snail genotype that was most

abundant changed from one year to the next. Moreover, roughly a year after a genotype was the most abundant one in the population, snails of that genotype had a higher than typical number of parasites. Together with Lively's earlier study (1989), these results suggest that parasite populations evolve to exploit the snail genotypes found in their local environment. Refining this idea further, Dybdahl and Lively hypothesized that it should be easier for parasites to infect snails with a common genotype than it is for them to infect snails with a rare genotype. That is exactly what they found in a laboratory experiment (**Figure 13.12**). Hence, snail genotype frequencies may change from year to year because common genotypes are attacked by many parasites, placing them at a disadvantage and driving down their numbers in future years.

FIGURE 13.12 Parasites Infect Common Host Genotypes More Easily Than Rare Genotypes In a laboratory experiment, Dybdahl and Lively compared rates of *Microphallus* infection in four common snail genotypes (A–D) and a group of 40 rare snail genotypes (E). The parasites and snails in this experiment were taken from the same lake. (After Dybdahl and Lively 1998.)

There are costs to both host defenses and parasite counterdefenses

Parasites and hosts have such a powerful effect on each other that we might expect an ever-escalating "arms race" in which host resistance and parasite counterdefenses both get stronger and stronger over time. But such an outcome rarely occurs. In some cases—as in Dybdahl and Lively's snails and trematodes—host genotypes that are common decrease in frequency because they are attacked by many parasites, leading to an increase in the frequency of a previously rare genotype, so that the arms race continually begins anew. In other cases, the arms race eventually stops because a trait that improves either host defenses or parasite counterdefenses reduces some other aspect of the organism's growth, survival, or reproduction.

Such trade-offs have been documented in a number of host–parasite systems, including *Drosophila* fruit flies and the parasitoid wasps that attack them (described on p. 288). Alex Kraaijeveld and colleagues have shown that selection can increase both the frequency with which fruit fly hosts encapsulate wasp eggs (from 5% to 60% in 5 generations) and the ability of wasp eggs to avoid encapsulation (from 8% to 37% in 10 generations). But they have also shown that there are costs to these defenses and counterdefenses. For example, fruit flies from lineages that can mount an encapsulation defense have lower larval survival rates when they compete for food with flies of the same species from lineages that do not encapsulate wasp eggs. Similarly, wasp eggs that avoid encapsulation by becoming embedded in host tissues take longer to hatch than do other eggs (Kraaijeveld et al. 2001).

Similar results were obtained by Peter Thrall and Jeremy Burdon, who studied populations of wild flax, *Linum marginale*, and its rust pathogen, *Melampsora lini*. Some rust genotypes are more virulent than others, meaning that they can overcome more plant resistance genes. Thrall and Burdon (2003) showed that although the rust could disperse long distances, virulent rust genotypes were common only in host populations dominated by plants with many resistance genes. Here too, a trade-off appears to be at work. Virulent rust genotypes produce fewer spores than do rust genotypes that can overcome only a few resistance genes (**Figure 13.13**). Spores are the means by which rusts disperse and reproduce. In flax populations dominated by plants with few resistance genes, the ability of virulent genotypes to overcome many resistance genes provides a disadvantage (reduced spore production), but no advantage (since there are few or no resistance genes to overcome). Thus, virulent rust genotypes are not common in nonresistant plant populations.

The evolutionary changes in host and parasite populations that we've discussed in this section reflect the profound effects these organisms have on each other. Next, we'll focus on some of the ecological consequences of host–parasite interactions.

FIGURE 13.13 Virulent Rust Pathogens Reproduce Poorly Genotypes of a rust pathogen that attack flax show a trade-off between virulence (the ability to overcome many flax resistance genes) and spore production. (After Thrall and Burdon 2003.)

Ecological Effects of Parasites

CONCEPT 13.4
Parasites can reduce the sizes of host populations and alter the outcomes of species interactions, thereby causing communities to change.

As we've seen, parasites can reduce the survival or reproduction of their hosts—an observation that is illustrated clearly by the large drop in reproductive success that a sexually transmitted mite inflicts on its beetle host (**Figure 13.14**). At the population level, the harm that parasites cause host individuals translates into a reduction of the host population growth rate, λ (see Chapter 9). As we discuss in this section, the reduction in λ can be drastic: parasites may drive local host populations extinct or even reduce the geographic range of the host species. In other, less extreme cases, parasites may reduce host abundances or otherwise alter host population dynamics, but do not cause the extinction of host populations.

Parasites can drive host populations to extinction

The amphipod *Corophium volutator* lives in North Atlantic tidal mudflats. *Corophium* is small (about 1 cm long) and often very abundant, reaching densities of up to 100,000 individuals per square meter. *Corophium* builds tubular burrows in the mud, from which it feeds on plankton suspended in the water and on microorganisms found in sediments near the

(A) Females from control populations

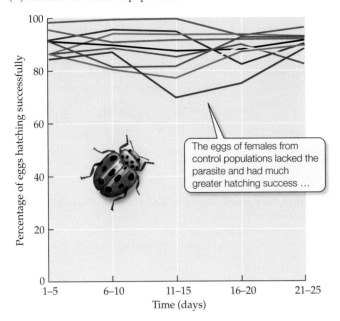

(B) Females from infected populations

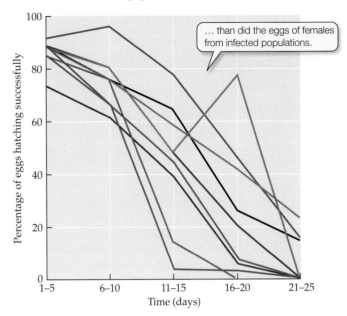

... than did the eggs of females from infected populations.

The eggs of females from control populations lacked the parasite and had much greater hatching success ...

FIGURE 13.14 Parasites Can Reduce Host Reproduction
Researchers infected experimental populations of the beetle *Adalia decempunctata* with a sexually transmitted mite parasite (*Coccipolipus hippodamiae*), then monitored the proportions of eggs laid by female beetles that hatched over the next 25 days. Each curve represents the eggs laid by a single female. (After Webberley et al. 2004.)

burrow opening. *Corophium* is eaten by a wide range of organisms, including migratory birds and trematode parasites. The parasites can reduce the size of *Corophium* populations greatly, even to the point of local extinction. For example, in a 4-month period, attack by trematodes caused the extinction of a *Corophium* population that initially had 18,000 individuals per square meter (Mouritsen et al. 1998). We'll return to this example on pp. 295–296.

Parasites can also drive host populations to extinction over a large geographic region. The American chestnut (*Castanea dentata*) once was a dominant member of deciduous forest communities in eastern North America (**Figure 13.15**), but the parasitic fungus *Cryphonectria parasitica* changed that

(A)

(B)

Some chestnut trees had a diameter up to twice the size of the one shown here.

FIGURE 13.15 Parasites Can Reduce Their Host's Geographic Range (A) The original distribution of the American chestnut (*Castanea dentata*) is shown in red. Although a few chestnut trees remain standing, this once-dominant species is now virtually extinct throughout its former range. (B) Chestnuts were once important timber trees (note the two loggers shown in the photograph).

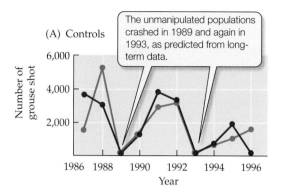

(A) Controls

The unmanipulated populations crashed in 1989 and again in 1993, as predicted from long-term data.

(B) Treated in 1989

(C) Treated in 1989 and 1993

Parasite removal did not stop the population cycles, but did reduce the fluctuation in grouse numbers.

FIGURE 13.16 Parasite Removal Reduces Host Population Fluctuations Hudson et al. studied the effects of parasite removal on the cycling of red grouse populations subjected to three treatments: (A) unmanipulated controls, (B) populations treated for nematode parasites in 1989, and (C) populations treated for parasites in 1989 and 1993. (After Turchin 2003.)

completely. This fungal pathogen causes chestnut blight, a disease that kills chestnut trees. The fungus was introduced to New York City from Asia in 1904 (Keever 1953). By mid-century, the fungus had wiped out most chestnut populations, greatly reducing the geographic range of this once-dominant species.

Isolated chestnut trees still can be found in North American forests, and some of these trees show signs of resistance to the fungus. But it is likely that many of the standing trees simply have not yet been found by the fungus. Once the fungus reaches a tree, it enters the tree through a hole or wound in the bark, killing the aboveground portion of the tree in 2–10 years. Before they die, infected trees may produce seeds, which may germinate and give rise to offspring that live for 10–15 years before they are killed by the fungus in turn. Some infected trees also produce sprouts from their roots, but these are usually killed a few years after they appear aboveground. Efforts are under way to breed resistant chestnut varieties, but at present it is not known whether chestnut populations will ever recover from the onslaught of the chestnut blight fungus.

Parasites can influence host population cycles

Ecologists have long sought to determine the causes of population cycles. As we saw in Chapter 12, such cycles may be caused by three-way feeding relationships—by the effects that predators and prey have on each other, coupled with the effects that prey and their food plants have on each other.

Population cycles can also be influenced by parasites. Consider the work of Peter Hudson and colleagues, who manipulated numbers of parasites in red grouse (*Lagopus lagopus*) populations on moors in northern England. Red grouse populations tend to crash every 4–8 years. Previous studies had shown that a parasitic nematode, *Trichostrongylus tenuis*, decreased the survival and reproductive success of individual red grouse. Hudson et al. (1998) investigated whether these parasites might also cause grouse populations to cycle.

To do this, the researchers studied changes in red grouse numbers in six replicate populations over the course of two population cycles. Long-term data indicated that red grouse

populations were likely to crash in 1989 and again in 1993. In an attempt to stop the crash from occurring in two of the six study populations, the researchers treated as many grouse as they could catch in 1989 and 1993 with a drug that killed the parasitic nematodes. In two of the other study populations, grouse were caught and treated for parasites in 1989 only; the remaining two populations served as unmanipulated controls. Because each replicate population covered a very large area (17–20 km^2), it was not possible to count red grouse numbers directly. Instead, Hudson et al. used the number of red grouse shot by hunters as an index of the actual population size.

In the control populations, red grouse numbers crashed as predicted in 1989 and 1993 (**Figure 13.16**). Although parasite removal did not completely stop the red grouse population cycle, it did reduce the fluctuation in grouse numbers considerably; this was particularly true for the populations that were treated for parasites in both 1989 and 1993. Thus, the experiment provided strong evidence that parasites influence—and may be the primary cause of—red grouse population cycles.

Parasites can alter ecological communities

The effects that parasites have on their hosts can have ripple effects: by reducing the performance of host individuals and the growth rates of host populations, parasites can change the outcome of species interactions, the composition of ecological communities, and even the physical environment in which a community is found.

Parasites can alter the outcomes of species interactions When two organisms interact with each other, the outcome of that interaction depends on many features of their biology. A predator that is young and healthy may be able to catch its prey—even though the prey organism literally "runs for its life"—whereas a predator that is old or sick may go hungry. Similarly, an individual that is in good condition may be able to compete effectively with others for resources, while an individual in poor condition may not.

Because they can affect host performance, parasites can also affect the outcome of interactions between their hosts and other species. Thomas Park conducted a series of experiments on factors that influenced the outcome of competition between flour beetle species. In one of those experiments, Park (1948) examined how the protozoan parasite *Adelina tribolii* affected the outcome of competition between two species of flour beetles, *Tribolium castaneum* and *T. confusum*. In the absence of the parasite, *T. castaneum* usually outcompeted *T. confusum*, driving it to extinction in 12 of 18 cases (**Figure 13.17**). The reverse was true when the parasite was present: *T. confusum* outcompeted *T. castaneum* in 11 of 15 cases. The outcome of competition was reversed because the parasite had a large negative effect on *T. castaneum* individuals, but virtually no effect on *T. confusum*. Parasites can also affect the outcome of competition in the field, as when the malaria parasite *Plasmodium azurophilum* reduced the competitive superiority of the lizard *Anolis gingivinus* over its smaller counterpart, *A. wattsi* (Schall 1992). Finally, parasites can alter the outcome of predator–prey interactions: by decreasing the physical condition of infected individuals, parasites may make predators less able to catch their prey, or prey less able to escape predation.

In the examples described in the previous paragraph, parasites affected the outcome of species interactions by altering the physical condition of their host. Parasites can also alter the outcome of species interactions by changing host behav-

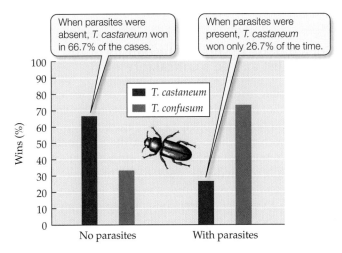

FIGURE 13.17 Parasites Can Alter the Outcome of Competition Thomas Park performed competition experiments between populations of the flour beetles *Tribolium castaneum* and *T. confusum* that were or were not infected with a protozoan parasite. (After Park 1948.)

ior. For example, when infected by a parasite, the host may behave in an unusual manner that makes it more vulnerable to predation. There are numerous examples of this phenomenon, including the protozoan parasites that make rats less wary of cats (see p. 282). Some worm parasites cause amphipods to move from sheltered areas to areas of relatively bright light, where the amphipods are more likely to be seen and eaten by fish or bird predators. In both of these cases, the parasite induces a change in host behavior that makes the host more likely to be eaten by a species (a cat in the first instance, a fish or bird in the second) that the parasite requires to complete its life cycle.

Parasites can change communities An ecological community can change when the abundance of one or more of the species in the community changes, when species interactions change, or when physical features of the environment change. Parasites can cause each of these types of community change.

We have seen several cases in which a parasites reduced the abundance of its host (*Myxoma* and rabbits; trematodes and amphipods; chestnut blight fungus and chestnuts), and hence altered the community in which it lived. We've also seen that parasites can change the outcome of species interactions. Such changes can have profound effects on communities. For example, a parasite that attacks a dominant competitor can suppress that species, causing the abundances of other species to increase (just as a predator or herbivore can do, as we saw in Chapter 12). Such a result was found in six stream communities studied by Kohler and Wiley (1997). Prior to recurrent outbreaks of a fungal pathogen, the caddisfly *Glossosoma nigrior* was the dominant herbivore in each of the six communities. The fungus devastated *Glossosoma* populations, reducing their densities nearly 25-fold, from an average of 4,600 individuals per square meter to an average of 190 individuals per square meter. This drastic reduction in the density of a dominant competitor allowed an increase in the abundances of dozens of other species, including algae, grazing insects that ate algae, and filter feeders such as blackfly larvae. In addition, several species that previously were extremely rare or absent were able to establish thriving populations, thus increasing the diversity of the community.

Parasites can also cause changes in the physical environment. This can happen when a parasite attacks a species that is an *ecosystem engineer*, a species whose actions change the physical character of its environment, as when a beaver builds a dam (see Chapter 15). The amphipod *Corophium* (see pp. 292–293) sometimes functions as an ecosystem engineer in its tidal mudflat environment: in some circumstances, the burrows it builds hold the mud together, preventing silt erosion and causing the formation of "mud islands" that rise above the surface of low-tide waters. As described earlier, trematode parasites can drive *Corophium* populations to ex-

(A)

(C)

Corophium burrows hold the mud together, causing the formation of "mud islands."

(B)

(D)

In the absence of *Corophium*, no mud islands are observed.

FIGURE 13.18 Parasites Can Alter the Physical Environment Infection of the amphipod *Corophium volutator* by a trematode parasite affects not only the host, but its entire tidal mudflat community. (A) The trematode can drive amphipod populations to local extinction. (B) In the absence of *Corophium*, the silt content of the mudflats decreases, and the erosion rate increases. (C, D) The overall physical structure of the mudflats also changes (compare C with D). (After Mouritsen and Poulin 2002.)

tinction (**Figure 13.18A**). When this happens, erosion rates increase, the silt content of the mudflats decreases, and the overall physical structure of the community changes noticeably (**Figure 13.18B–D**). Along with these physical changes, the abundances of ten large species in the mudflat community change considerably, including one species (a ribbon worm) that was driven to local extinction (K. N. Mouritsen, personal communication).

Dynamics and Spread of Diseases

CONCEPT 13.5
Simple models of host–pathogen dynamics suggest ways to control the establishment and spread of diseases.

As we've seen, parasites that cause diseases (pathogens) can greatly affect the population dynamics of both wild and domesticated plant and animal species. Pathogens also have a major effect on human populations—so much so that they are thought to have played a major role in the rise and fall of civilizations throughout the course of human history (Mc-Neill 1976; Diamond 1997). Examples include the European conquest of North America, where up to 95% (19 million of the original 20 million) of the native population was killed by new diseases brought to the continent by European trappers,

missionaries, settlers, and soldiers. Even with such massive mortality, the conquest took roughly 400 years; without it, the conquest would certainly have taken longer, and might have failed. Pathogens continue to be a major source of mortality today. Despite medical advances, millions of people die each year from diseases such as flu, AIDS, tuberculosis, and malaria.

For a disease to spread, the density of susceptible hosts must exceed a critical threshold

Considerable effort has been devoted to the development of mathematical models of host–pathogen population dynamics. These models often differ in three ways from those we have seen in earlier chapters. First, the host population is subdivided into different categories, such as susceptible individuals (S), infected individuals (I), and recovered and immune individuals (R). Second, it is often necessary to keep track of both host and pathogen genotypes because, as we have seen, host genotypes may differ greatly in their resistance to the pathogen, and pathogen genotypes may differ greatly in their ability to cause disease. Third, depending on the disease, it may be necessary to account for other factors that influence its spread, such as (1) differences in the likelihood that hosts of different ages will become infected, (2) a **latent period** in which an individual is infected but cannot spread the disease, and (3) **vertical transmission**, the spread of the disease from mother to newborn, as can occur in AIDS.

Models that include all of these factors can be very complicated. Here we'll consider a simple model that ignores most of these complicating factors, yet still yields an important insight: a disease will spread only if the density of susceptible hosts exceeds a critical, threshold density.

Modeling host–pathogen population dynamics   To develop a model that can be used to estimate the threshold density, we must determine how to represent the transmission of the disease from one host individual to the next. We'll denote the density of susceptible individuals by S, and the density of infected individuals by I. For a disease to spread, infected individuals must encounter susceptible individuals. Such encounters are assumed to occur at a rate that is proportional to the densities of susceptible and infected individuals; here, we'll assume that this rate is proportional to the product of their densities, SI. Diseases do not spread with every such encounter, however, so we multiply the encounter rate (SI) by a transmission coefficient (β) that indicates how effectively the disease spreads from infected to susceptible individuals. Thus, the key term in the model—disease transmission—is represented by the term βSI.

The density of infected individuals increases when the disease is transmitted successfully (at the rate βSI), and decreases when infected individuals die or recover from the disease. If we set the combined death and recovery rate equal to d, these assumptions yield the equation

$$\frac{dI}{dt} = \beta SI - dI \qquad\qquad \textbf{13.1}$$

where dI/dt represents the change in the density of infected individuals at each instant in time.

A disease will become established and spread when the density of infected individuals in a population increases over time. As explained in more detail in **Web Extension 13.2**, this occurs when dI/dt is greater than zero, which according to Equation 13.1, occurs when $\beta SI - dI > 0$ or $S > d/\beta$. Thus, a disease will become established and spread when the number of susceptible individuals exceeds the **threshold density**, S_T $= d/\beta$. For many diseases that affect people or animals, the transmission rate β and the death and recovery rate d are known, permitting estimation of the threshold density, S_T.

Strategies for controlling the spread of diseases   As Equation 13.1 suggests, to prevent the spread of a disease, the density of susceptible individuals must be less than the threshold density, S_T. There are several ways of achieving this goal. People sometimes slaughter large numbers of susceptible domesticated animals to reduce their density below S_T and hence prevent disease spread. This is typically done when the disease in question can spread to humans, as in highly virulent forms of bird flu. In human populations, if an effective and safe vaccine is available, the density of susceptible individuals can be reduced below S_T by a mass vaccination program. Such programs work, as illustrated by the dramatic results of a measles vaccination program in Romania (**Figure 13.19**).

In addition, public health measures can be taken to raise the threshold density, S_T, thereby making it more difficult for the disease to become established and spread. For example, the threshold density can be raised by taking actions that increase the rate at which infected individuals recover and become immune; one way to do this would be to improve the early detection and clinical treatment of the disease. The

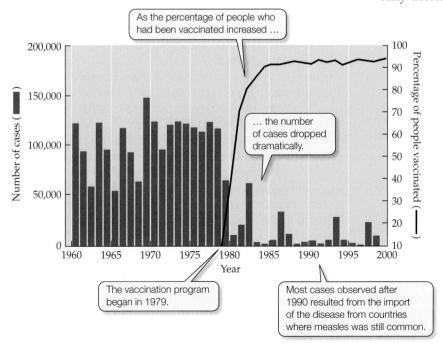

FIGURE 13.19 Vaccination Reduces the Incidence of Disease   The results of a measles vaccination program in Romania show how lowering the density of susceptible individuals can help efforts to control the spread of a disease. Measles often kills (especially in populations that are poorly nourished or that lack a history of exposure to the disease) and can cause severe complications in survivors, including blindness and pneumonia. (After Strebel and Cochi 2001.)

FIGURE 13.20 Determining Threshold Population Densities The percentage of bison that showed evidence of previous exposure to brucellosis was monitored in six national parks in the United States and Canada. Researchers monitored herds of different sizes in an effort to determine the threshold population density for establishment of the disease. (After Dobson and Meagher 1996.)

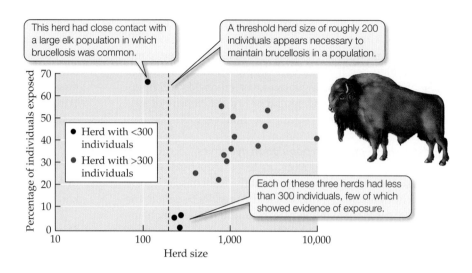

This herd had close contact with a large elk population in which brucellosis was common.

A threshold herd size of roughly 200 individuals appears necessary to maintain brucellosis in a population.

- Herd with <300 individuals
- Herd with >300 individuals

Each of these three herds had less than 300 individuals, few of which showed evidence of exposure.

threshold density can also be raised if β, the disease transmission rate, is decreased by quarantining infected individuals or by convincing people to engage in behaviors that make it more difficult for the disease to be transmitted from one person to the next.

Threshold densities are also important in wild populations. Dobson and Meagher (1996) studied bison populations to determine how best to prevent the spread of the bacterial disease brucellosis. Using data on previous exposure to the disease among bison herds in six national parks in Canada and the United States, they found that the threshold density (S_T) for disease establishment appeared to be a herd size of 200–300 bison (**Figure 13.20**). This field-based estimate of S_T was very similar to the estimated threshold density of 240 individuals calculated from a model similar to Equation 13.1. Many of the herds in the six national parks had 1,000–3,000 individuals, so reducing herd sizes below a threshold value of roughly 200 individuals would require implementing a vaccination program or killing large numbers of bison. An effective vaccine was not available, and killing many bison was not acceptable, either politically or ecologically (since herds as small as 200 individuals would face an increased risk of extinction). Thus, Dobson and Meagher concluded that it would be difficult to prevent the establishment of brucellosis in wild bison populations.

Enslaver Parasites

What are the biochemical and physiological mechanisms by which enslaver parasites manipulate the behavior of their hosts? Consider the parasitoid wasp *Hymenoepimecis* and its spider host (see p. 283). The larval stage of this wasp attaches to the exterior of the spider's abdomen and sucks the spider's body fluids. When the larva is fully grown, it induces the spider to build a cocoon web (**Figure 13.21**). Once the spider has built the cocoon web, the wasp larva kills and eats it. The larva then spins its cocoon and attaches it to the cocoon web.

A parasitized spider builds normal webs right up to the night when the wasp induces it to make a cocoon web. This sudden change in the spider's web-building behavior suggests that the wasp may inject the spider with a chemical that alters its behavior. To test this idea, William Eberhard (2001) removed wasp larvae from their hosts several hours before the time when a cocoon web usually would be made. Wasp removal sometimes resulted in the construction of a web that was very similar to a cocoon web, but more often resulted in the construction of a web that differed substantially from both normal and cocoon webs. In the days that followed the removal of the parasite, some spiders partially recovered the ability to make normal webs. These results are consistent with the idea that the parasite induces construction of a cocoon web by injecting a fast-acting chemical into the spider. The chemical appears to act in a dose-dependent manner; otherwise, we would expect spiders exposed to the chemical to build only cocoon webs, not webs that are intermediate in form. Spiders build cocoon webs by repeating the early steps of their normal web-building sequence a large number of times; thus, the chemical appears to act by interrupting the spiders' usual sequence of web-building behaviors.

Other enslaver parasites also appear to manipulate host body chemistry. On p. 282, we described hairworm parasites that cause crickets to commit suicide by jumping into water. Thomas et al. (2003) have shown that the hairworm causes biochemical and structural changes in the brain of its cricket host. The concentrations of three amino acids (taurine, valine, and tyrosine) in the brains of crickets that were parasitized differed from those in crickets that were not parasitized. Taurine, in particular, is an important neurotransmitter in insects, and it also regulates the brain's

FIGURE 13.21 Parasites Can Alter Host Behavior The parasitic wasp *Hymenoepimecis argyraphaga* dramatically alters the web-building behavior of the orb-weaving spider *Plesiometa argyra*. (A) The web of an uninfected spider. (B) The "cocoon" web of a parasitized spider.

ability to sense a lack of water. Hence, it is possible that the parasite induces its host to commit suicide by causing biochemical changes in its brain that alter the host's perception of thirst.

The papers by Eberhard (2001) and Thomas et al. (2003) suggest that some parasites enslave their hosts by manipulating them chemically. In other cases, however, little is known about how enslaver parasites alter the behavior of their hosts. For example, Laura Hechtel and colleagues

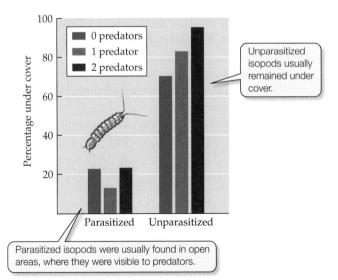

FIGURE 13.22 Making the Host Vulnerable to Predation Isopods that are not infected by worm parasites tend to remain under cover, where food is available and where they are not exposed to creek chubs, a predatory fish. Hechtel et al. exposed parasitized and unparasitized isopods to 0, 1, or 2 predator individuals and observed the influence of parasites on their behavior. (Data from Hechtel et al. 1993.)

studied isopods (*Caecidotea intermedius*) parasitized by the worm *Acanthocephalus dirus*. They observed that, compared with unparasitized isopods, those that were parasitized spent little time under cover, where food was available and where they could hide from a fish predator, the creek chub—the next host in the parasite's complex life cycle (**Figure 13.22**). Hechtel and colleagues also performed a choice experiment in which isopods could move either to one side of an aquarium, where a chub was present (but separated from the isopod by a net), or to the other side, where no chubs were present. They found that unparasitized isopods avoided the chubs, but parasitized isopods were actually drawn toward the chubs—a behavior that in the wild would benefit the parasite but be disastrous for the isopod.

The mechanism by which the parasite induces changes in isopod behavior is not known. Even in Eberhard's work, which suggests that the wasp injects a chemical into its spider host that induces it to make a cocoon web, the chemical in question has not been found. If this chemical were known, it could be injected into unparasitized spiders; if those spiders constructed cocoon webs, we would have a clear understanding of how the parasite manipulates the spider. Such a definitive experiment has yet to be performed for any parasite that enslaves its host.

CONNECTIONS IN NATURE

From Chemicals to Evolution and Ecosystems

Enslaver parasites that manipulate and harm their hosts exert strong selection pressure on host populations. As such, hosts might be expected to evolve resistance to the manipulations of enslaver parasites—for example, a host might evolve the ability to recognize and destroy chemicals that the parasite uses to alter its behavior. Likewise, enslaver parasites might be expected to evolve the ability to overcome host resistance. To date, we know of no such evidence of ongoing host–enslaver parasite coevolution.

However, interactions between enslaver parasites and their hosts provide evidence of previous evolutionary change. An enslaver parasite has many adaptations that allow it to cope with host defenses (otherwise it would not survive). A parasite that uses a chemical to manipulate a specific host behavior, for example, is beautifully adapted to take advantage of the body chemistry of its host. Such evolutionary links between enslaver parasites and their hosts illustrate a feature common to all host–parasite inter-

actions: it can be difficult to separate host–parasite ecological interactions from host–parasite evolutionary interactions. As we've seen in this chapter, the outcome of such ecological and evolutionary interactions can have profound effects on individuals, populations, communities, and ecosystems. As evolutionary change tips the balance back and forth, first in favor of the host, then in favor of the parasite, we can expect concomitant changes in the dynamics of other species, such as those that compete with or eat the host or the parasite. Viewed in this way, communities and ecosystems are highly dynamic, always shifting in response to the ongoing ecological and evolutionary changes that occur within them.

SUMMARY

CONCEPT 13.1 Parasites, which constitute roughly 50% of the species on Earth, typically feed on only one or a few host species.

- Parasites, including both macroparasites and microparasites, typically feed on only one or a few host individuals during the course of their lives; in addition, many parasites are closely adapted to particular host species.

- Some parasites are ectoparasites that live on the surface of their host; others are endoparasites the live within the body of their host.

- Endoparasitism and ectoparasitism each have advantages and disadvantages. It is easier for ectoparasites or their offspring to disperse from one host individual to another; however, ectoparasites are at greater risk from natural enemies than are endoparasites.

CONCEPT 13.2 Hosts have adaptations for defending themselves against parasites, and parasites have adaptations for overcoming host defenses.

- Many host organisms have immune systems that allow them to recognize and defend against endoparasites. Biochemical conditions inside the host's body can also provide protection against parasites.

- In some birds, mammals, and fishes, females select their mates based on traits that indicate whether a male is well defended from parasites.

- Parasites have a broad suite of adaptations that allow them to circumvent host defenses, from relatively simple counterdefenses against encapsulation to more complex counterdefenses that involve hundreds of parasite genes.

CONCEPT 13.3 Host and parasite populations can evolve together, each in response to selection imposed by the other.

- Host–parasite interactions can result in coevolution, in which populations of the host and parasite evolve

together, each in response to selection imposed by the other.

- Selection can favor a diversity of host and parasite genotypes. A rare host genotype may increase in frequency because few parasites can overcome its defenses; as a result, parasite genotypes that can cope with the now-common host genotype's defenses may also increase in frequency.

- Host–parasite interactions can exhibit trade-offs in which a trait that improves host defenses or parasite counterdefenses reduces other aspects of the organism's growth, survival, or reproduction.

CONCEPT 13.4 Parasites can reduce the sizes of host populations and alter the outcomes of species interactions, thereby causing communities to change.

- Parasites can reduce the abundances of host populations, in some cases driving local host populations to extinction or even changing the geographic distributions of host species.

- Evidence suggests that infection by parasites can cause host populations to cycle in abundance.

- Parasites can affect the outcomes of interactions between their hosts and other species; for example, a species that is a dominant competitor may become an inferior competitor when infected by a parasite.

- The effects of parasites can alter the composition of ecological communities and change features of the physical environment.

CONCEPT 13.5 Simple models of host–pathogen dynamics suggest ways to control the establishment and spread of diseases.

- Some models of host–pathogen population dynamics subdivide the host population into susceptible individuals, infected individuals, and recovered and immune individuals; track different host and pathogen genotypes; and take into account factors such as host age, latent periods, and vertical transmission.

- A simple mathematical model of host–pathogen dynamics (see Equation 13.1) yields an important insight: for a disease to become established and spread, the density of susceptible hosts must exceed a critical, threshold density.

- To control the spread of a disease, efforts may be made to lower the density of susceptible hosts (by slaughtering domesticated animals or undertaking vaccination programs) or to raise the threshold density (by increasing the recovery rate or decreasing the transmission rate).

REVIEW QUESTIONS

1. Define endoparasites and ectoparasites, giving an example of each. Describe some advantages and disadvantages associated with each of these two types of parasitism.

2. Given the effects that parasites can have on host individuals and host populations, would you expect that parasites could also alter the outcomes of species interactions and the composition of ecological communities? Explain.

3. a. Summarize the mechanisms that host organisms use to kill parasites or reduce the severity of their attack.

   b. With your answer to part (a) as background material, do you think the following statement from a news report could be true?

The parasite has a mild effect on a plant species in Australia, but after it was introduced for the first time to Europe, it had devastating effects on European populations of the same plant species.

Explain your reasoning, and illustrate your argument with an example of how a plant defensive mechanism might work—or fail to work—in a situation such as this.

4. a. What is meant by the concept of a threshold density for the establishment and spread of a disease? Why is this concept important?

   b. Explain the logic and show the algebraic steps by which Equation 13.1 can be used to calculate that the threshold density (S_T) has the value $S_T = d/\beta$.

HANDS-ON PROBLEM SOLVING

Evolutionary Interactions of a Fly and its Parasitoid | This Web exercise (www.sinauer.com/ecology) explores the evolutionary interactions between the fly *Drosophila melanogaster* and its wasp parasitoid. You will interpret data from a recent paper that examines the evolution of resistance of these flies to the parasitoids. Does resistance to parasitoids impose a cost?

SUGGESTED READINGS

Forsyth, A. and K. Miyata. 1984. *Tropical Nature: Life and Death in the Rain Forests of Central and South America*. Scribner, New York. *A gripping description of life in the rain forest, including compelling examples of the parasites that live there.*

Hoberg, E. P., N. L. Alkire, A. de Querioz and A. Jones. 2001. Out of Africa: Origins of the *Taenia* tapeworms in humans. *Proceedings of the Royal Society of London B* 268: 781–787. *The conventional wisdom has been that humans were first exposed to Taenia tapeworms (parasites that live in our intestines and that can reach up to 20 m (66 feet) in length) from domesticated animals such as dogs, pigs, and cattle. This paper turns that idea on its head, arguing instead that our domesticated animals "caught" their tapeworms from us. Taking advantage of the close relationship between parasites and their hosts, the paper also uses the evolutionary history of the tapeworm to shed light on open questions about human evolution. [See also Shipman (2002, A worm's view of human evolution. American Scientist 90: 508–510) for an informative popular account of the Hoberg et al. paper.]*

Jones, J. G. D. and J. L. Dangl. 2006. The plant immune system. *Nature* 444: 323–329. *An up-to-date review of how plants defend themselves against a broad suite of parasites, including an overview of how an understanding of plant immune systems is important for agriculture and pest control.*

May, R. M. 1983. Parasite infections as regulators of animal populations. *American Scientist* 71: 36–45. *An overview of the key roles parasites play in controlling the dynamics of animal populations, written by a leading researcher in the field.*

Pimentel, D., L. Lach, R. Zuniga and D. Morrison. 2000. Environmental and economic costs associated with nonindigenous species in the United States. *BioScience* 50: 53–65. *A highly cited review article that includes startling estimates of the economic impact of parasites (and other introduced species). The authors conclude, for example, that plant pathogens alone cause up to 30 billion dollars a year in damage to crop plants.*

Zimmer, C. 2000. *Parasite Rex: Inside the Bizarre World of Nature's Most Dangerous Creatures*. Free Press, New York. *A fascinating read, from the gruesome to the beautiful, about those most important and diverse of organisms—the parasites.*

14 Mutualism and Commensalism

CASE STUDY

The First Farmers

Humans first began to farm about 10,000 years ago. Agriculture was a revolutionary development that led to great increases in the size of our populations, as well as to innovations in government, science, the arts, and many other aspects of human societies. But people were far from the first species to farm. That distinction goes to ants in the tribe Attini, a group of 210 species, most of which live in tropical forests of South America. These ants, known informally as the attines or fungus-growing ants, started cultivating fungi for food at least 50 million years before the first human farmers (**Figure 14.1**).

Like human farmers, the ant farmers nourish, protect, and eat the species they grow, forming a relationship that benefits both the farmer and the crop. The attines cannot survive without the fungi they cultivate; many of the fungi depend absolutely on the ants as well. When a virgin queen ant leaves her mother's nest to mate and begin a new colony, she carries in her mouth some of the fungi from her birth colony. The fungi are cultivated by the colony in subterranean gardens (**Figure 14.2**). An ant colony may contain hundreds of gardens, each roughly the size of a football; these hundreds of gardens can provide enough food to support 2–8 million ants.

Some fungus-growing ants occasionally replace the fungi in their gardens with new, free-living fungi that they gather from surrounding soils. Other species, such as leaf-cutter ants in the genera *Atta* and *Acromyrmex*, never cultivate fungi found free-living in the environment. Instead, the fungi in their gardens always come from propagules passed from a parent ant colony to each of its descendant colonies.

As their name suggests, leaf-cutter ants cut portions of leaves from plants and feed them to the fungi in their gardens. Leaf-cutter ants are the largest of the attines. Soldier ants in the genus *Atta*, for example, may reach up to 14 mm in length (over 1/2 inch). They protect the colony with powerful mandibles that can cut half-moon sections out of leather shoes. *Atta* worker ants are much smaller than the soldiers, but they still have major effects on their surroundings: they maintain trails through the vegetation, and the workers of a single colony can harvest as much plant matter each day as it would take to feed a cow.

Back at the nest, the ants chew the leaves to a pulp, fertilize them with their own droppings, and "weed" the fungal gardens to help control bacterial and fungal invaders. In return,

FIGURE 14.1 Collecting Food for Their Fungi Fungus-growing ants (*Atta* sp.) in Costa Rica carrying leaf segments to their colony, where the leaves will be fed to the fungi the ants cultivate for food. The ant riding on a leaf is of the same species and is called a *minim* (in reference to its small size). Minims guard the ants carrying leaves from parasitic flies.

(A)

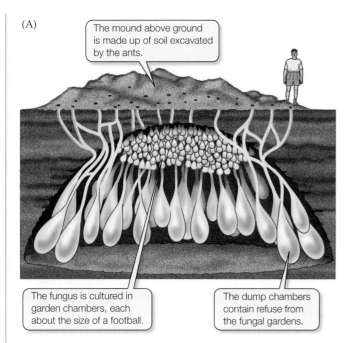

The mound above ground is made up of soil excavated by the ants.

The fungus is cultured in garden chambers, each about the size of a football.

The dump chambers contain refuse from the fungal gardens.

(B)

FIGURE 14.2 The Fungal Garden of a Leaf-cutter Ant
(A) A diagramatic representation of a large *Atta* leaf-cutter ant colony. (B) This photo shows a cut-away view of a garden chamber in a central Paraguay colony of the leaf-cutter ant, *Atta laegicata*. Several winged ants can be seen hiding from the disturbance created by excavating the garden: They have placed their heads into crevices of the garden, and they would be relatively motionless.

the fungi cultivated by the ants produce specialized structures, called *gongylidia*, on which the ants feed. The partnership between leaf-cutter ants and fungi has been called an "unholy alliance" because each partner helps the other to overcome the formidable defenses that protect plants from being eaten. The ant, for example, scrapes a waxy covering from the leaves that the fungi have difficulty penetrating, while the fungus digests and renders harmless the chemicals that plants use to kill or deter insect herbivores.

But all is not perfect in the gardens. Nonresident fungi, which themselves would benefit from ant cultivation, periodically invade leaf-cutter ant colonies. Furthermore, pathogens and parasites that attack the cultivated fungi occasionally outstrip the ants' ability to weed them out. What prevents such unwanted guests from destroying the garden?

The last three chapters emphasized interactions between species in which at least one member is harmed (competition, predation, herbivory, and parasitism). But life on Earth is also shaped by **positive interactions**, those in which one or both species benefit and neither is harmed.

Consider the vascular plants, a group that includes more than 90% of all extant plant species. These plants are named for their vascular tissues (xylem and phloem), which transport materials throughout the plant body. Most vascular plants form beneficial associations with fungi in which the fungus provides the plant with nutrients it absorbs from the soil, improving the plant's growth and survival. In fact, fossil evidence indicates that the earliest vascular plants formed similar associations with fungi more than 400 million years ago (Selosse and Le Tacon 1998). These early vascular plants lacked true roots, so their interactions with fungi may have increased their access to soil resources and aided their colonization of land.

As the vascular plant example suggests, positive interactions have influenced key events in the history of life—in this case, the colonization of land—as well as the growth and sur-

vival of organisms today. As we'll see, positive interactions can also influence the outcome of other interactions among organisms, thus shaping communities and influencing ecosystems. We will begin our study of positive interactions by defining some key terms and examining the scope of these interactions in ecological communities.

Positive Interactions

CONCEPT 14.1
Positive interactions occur when neither species is harmed and the benefits of the interaction are greater than the costs for at least one species.

There are two fundamental types of positive interactions: mutualism and commensalism. **Mutualism** is a mutually beneficial interaction between individuals of two species (a +/+ relationship). **Commensalism** is an interaction between two species in which individuals of one species benefit, while individuals of the other species do not benefit and are not harmed (a +/0 relationship). Many ecologists also refer to mutualism and commensalism collectively as *facilitation*, a synonym for positive interactions.

In some cases, the species involved in a positive interaction form a **symbiosis**, a relationship in which the two species live in close physiological contact with each other. Examples include the relationships between corals and algae (described at the end of this page) and between humans and bacteria (we have a diverse set of bacteria living in our guts, many of which are beneficial). However, parasites also form symbiotic associations with their hosts (see Figure 13.3). Thus, symbiotic relationships can range from parasitism (+/–) to commensalism (+/0) to mutualism (+/+).

In mutualism and commensalism, individuals benefit when their growth, survival, or reproduction is increased by their interaction with the other species. Such benefits can take a variety of forms. A species may provide its partner with food, shelter, or a substratum to grow on; it may transport its partner's pollen or seeds; it may reduce heat or water stress; or it may decrease the negative effects of competitors, herbivores, predators, or parasites. In mutualism, the organism that provides a benefit to its partner may do so at a cost to itself, as when supplying food to its partner reduces its own opportunity for growth. Nevertheless, the net effect of the interaction is positive because for each species, the benefits are greater than the costs.

In the remainder of this section, we will discuss some general observations that apply to both mutualism and commensalism. With this material under our belts, we will examine some characteristics that are unique to mutualism in Section 14.2, placing special emphasis on what can be learned from studies that document the costs and benefits of mutualistic interactions.

Mutualism and commensalism are ubiquitous

Mutualistic associations literally cover the land surface of Earth (**Figure 14.3**). Most plant species—including those that dominate terrestrial ecosystems—form **mycorrhizae**, symbiotic associations between plant roots and various types of fungi that are usually mutualistic. The fungi benefit the plants by increasing the surface area over which the plants can extract soil nutrients (over 3 m of fungal hyphae may extend from 1 cm of plant root); the fungi may also protect the plants from pathogens and help them take up water from the soil. The plants typically benefit the fungi by supplying them with carbohydrates.

There are two major types of mycorrhizae (**Figure 14.4**). **Ectomycorrhizae** are those in which the fungus typically grows between root cells and forms a mantle around the exterior of the root. **Arbuscular mycorrhizae**, in contrast, are those in which the fungus grows into the soil, extending some distance away from the root, and also grows between some root cells while penetrating others.* About 80% of angiosperms (flowering plants) and all gymnosperms (e.g., conifers, cycads, and the ginkgo) form mycorrhizal associations.

Mutualistic associations can also be found in many other organisms and habitats. In the seas, corals form a mutualism with symbiotic algae. The coral provides the alga with a home, nutrients (nitrogen and phosphorus), and access to

---

*Arbuscular mycorrhizae were previously known as "endomycorrhizae" (from the Greek *entos*, in), a term that referred to the fact that their hyphae can penetrate root cells. The term "endomycorrhizae" has been abandoned by many researchers, however, because the hyphae of some ectomycorrhizae can also penetrate plant root cells.

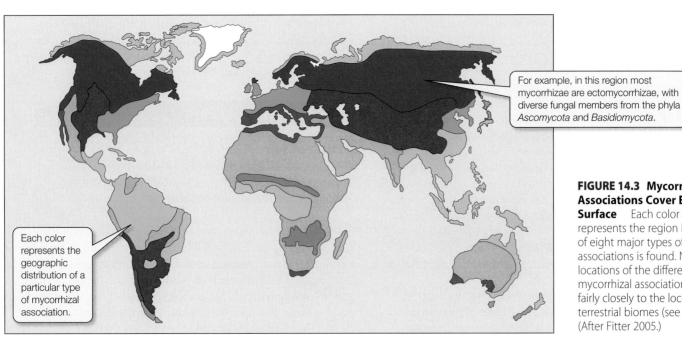

For example, in this region most mycorrhizae are ectomycorrhizae, with diverse fungal members from the phyla *Ascomycota* and *Basidiomycota*.

Each color represents the geographic distribution of a particular type of mycorrhizal association.

FIGURE 14.3 Mycorrhizal Associations Cover Earth's Land Surface Each color on the map represents the region in which one of eight major types of mycorrhizal associations is found. Notice that the locations of the different types of mycorrhizal associations correspond fairly closely to the locations of major terrestrial biomes (see Figure 3.5). (After Fitter 2005.)

(A) Ectomycorrhizae

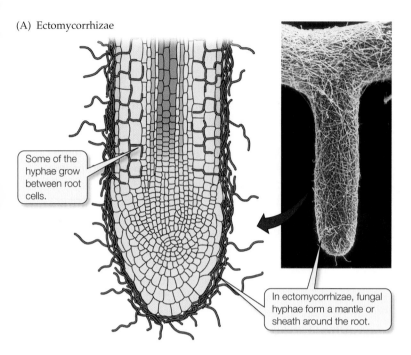

Some of the hyphae grow between root cells.

In ectomycorrhizae, fungal hyphae form a mantle or sheath around the root.

(B) Arbuscular mycorrhizae

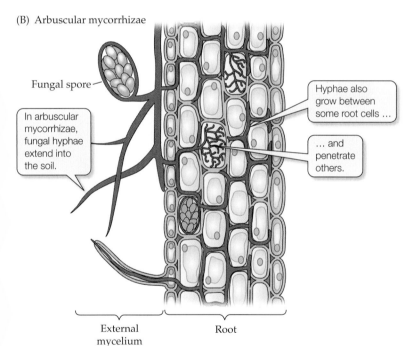

Fungal spore

In arbuscular mycorrhizae, fungal hyphae extend into the soil.

Hyphae also grow between some root cells …

… and penetrate others.

External mycelium

Root

FIGURE 14.4 Two Major Types of Mycorrhizae Mycorrhizae can be classified as (A) ectomycorrhizae or (B) arbuscular mycorrhizae. (A after Rovira et al. 1983; B after Mauseth 1988.)

gestible plant material, such as cellulose. Similarly, insects rely on mutualisms with a number of other species, including plants (e.g., pollination mutualisms, see p. 306), fungi (see p. 302), and protists and bacteria (**Figure 14.5**).

Commensalism, like mutualism, is everywhere—the ecological world is built on it. As we'll see in Chapter 15, millions of species form +/0 relationships with organisms that provide the habitat in which they live. In these relationships, a species that depends on the habitat provided by another species often has little or no effect on the species that provides the habitat. Examples include small species that grow on large species, such as lichens found on the bark of a tree or harmless bacteria that grow on the surface of your skin. Many algae, invertebrates, and fishes found in marine "kelp forests" go locally extinct if the kelp are removed (see p. 174); such species depend on the kelp for a home, but most of them do not harm or benefit the kelp. Likewise, although the numbers are quite uncertain, there may be more than a million insect species and thousands of understory plant species that

FIGURE 14.5 A Protist Gut Mutualist This wood-eating cockroach would starve if gut mutualists such as the protist shown here (a hypermastigote) did not help it to digest wood. The hypermastigote can break down cellulose, a major structural component of wood that the cockroach (and other wood-eating insects, such as termites) cannot digest on its own.

Hypermastigote (*Barbulanympha* sp.)

Wood-eating cockroach (*Cryptocercus punctulatus*)

sunlight; the alga provides the coral with carbohydrates produced by photosynthesis. Coral reefs are home to many different invertebrate and vertebrate species, all of which depend directly or indirectly on the coral–alga mutualism. On land, mammalian herbivores such as cattle and sheep, depend on bacteria and protists that live in their guts and help them metabolize otherwise indilive in tropical forests and nowhere else. These insects and small plants depend on the forest for their habitat, yet many have little or no effect on the trees that tower above them.

Mutualism and commensalism can evolve in many ways

Many different types of ecological interactions can evolve into commensalism or mutualism. For example, a lichen that grows on the surface of a tree's leaves may initially harm the tree by reducing its access to light. Over time, however, this relationship may become a commensalism if the tree species evolves the ability to tolerate the lichen's presence. Some trees have done so: The Australian palm *Calamus australis*, for example, increases the concentration of chlorophyll in portions of its leaves that are covered with lichens (Anthony et al. 2002). This response improves the efficiency with which the palm absorbs the low light levels that occur beneath the lichen.

It is also possible for a mutualism to emerge from what begins as a host–parasite interaction. In 1966, Kwang Jeon observed the spontaneous infection of a strain of *Amoeba proteus* by a rod-shaped bacterium. Large numbers of bacteria (ca. 40,000) infected each individual amoeba. Initially, these bacteria had a negative effect on their hosts: they often killed the host, and they caused infected hosts to be smaller, grow more slowly, and starve more easily than uninfected amoebas. As discussed in Section 13.3, however, parasites and hosts often coevolve, each in response to selection pressure imposed by the other. Five years later, the bacterium had evolved to become harmless to the amoeba, and the amoeba had evolved such that its nucleus depended on the bacterium for normal metabolic function. As Jeon (1972) showed in a series of experiments, neither species could survive without the other. Thus, what began as a parasitic relationship evolved into a mutualism in which each species provided the other with a clear benefit: the ability to survive.

Positive interactions can be obligate and coevolved or facultative and loosely structured

Mutualism and commensalism include a broad set of interactions, ranging from those that are species-specific, obligate (that is, not optional for either species), and coevolved to those that show none of these three characteristics. The leaf-cutter ant–fungus mutualism discussed in the Case Study illustrates one end of this spectrum: the ants and the fungi they cultivate have a highly specific, obligate relationship in which neither partner can survive without the other, and their interaction has led each partner to evolve unique features that benefit the other species.

Similarly, many tropical fig trees are pollinated by one or a few species of fig wasps. These relationships are mutually beneficial and obligate for both species in that neither species can reproduce without the other. Fig–fig wasp interactions also show clear signs of coevolution (Bronstein 1992). Fig flowers are contained within structures of fleshy stem tissue known as receptacles (**Figure 14.6**). In monoecious figs (those in which each tree has separate male and female flowers), the male and female flowers are located in different parts of the receptacle, and the male flowers mature after the female flowers. Female flowers range from those with short styles to those with long styles. A female fig wasp enters the receptacle, where she inserts her ovipositor through the styles to lay eggs in the ovaries (see Figure 14.6). The female then deposits pollen on the stigmas of both long- and short-styled flowers, and hence both flower types develop seeds. Perhaps because wasp ovipositors are not long enough to reach the ovaries of long-styled flowers, wasp larvae typically develop within short-styled flowers and feed on some of their seeds.

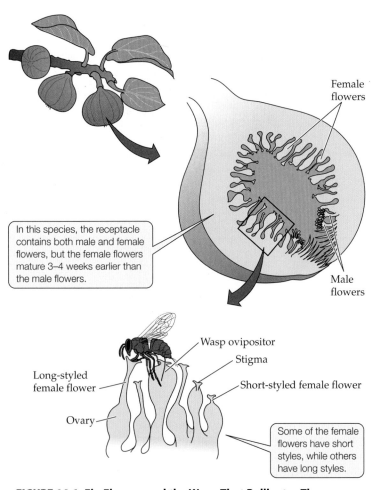

Female flowers

In this species, the receptacle contains both male and female flowers, but the female flowers mature 3–4 weeks earlier than the male flowers.

Male flowers

Wasp ovipositor

Stigma

Long-styled female flower

Short-styled female flower

Ovary

Some of the female flowers have short styles, while others have long styles.

FIGURE 14.6 Fig Flowers and the Wasp That Pollinates Them
The receptacle and flowers of a typical monoecious fig tree, *Ficus sycomorus*. (After Bronstein 1992.)

When the young wasps complete their development, they mate, the males burrow through the receptacle, and the females exit through this passageway. Before the females leave the receptacle, however, they visit male flowers (which are now mature), collect pollen from them, and store it in a specialized sac for use when they lay their eggs in another receptacle. The wasp's reproductive behavior is a remarkable example of a specialized behavior that provides a benefit to another species.

Unlike the ant–fungus and fig–fig wasp mutualisms, many mutualisms and commensalisms are facultative (not obligatory) and show few signs of coevolution. In desert environments, the soil beneath an adult plant is often cooler and moister than the soil of an adjacent open area. These differences in soil conditions may be so pronounced that the seeds of many plant species can germinate and survive only in the shade provided by an adult plant; such adults are called **nurse plants** because they "nurse" or protect the seedlings. A single species of nurse plant may protect the seedlings of many different species. Desert ironwood (*Olneya tesota*), for example, serves as a nurse plant for 165 different species, most of which can also germinate and grow under other plant species. This situation is typical of facultative interactions: a species that requires "nursing" may be found under a variety of nurse plant species (and hence has a facultative relationship with each of them), and the nurse plant and the beneficiary species may evolve little in response to one another. Once the beneficiary species reaches full size, it may in fact compete with the nurse plant for resources.

Facultative positive interactions that show little coevolution also occur in forest communities. For example, large herbivores such as deer or moose may inadvertently eat the seeds of small herbaceous plants whose leaves they feed on. The seeds may pass unharmed through the herbivore's di-

gestive tract and be deposited with its feces far from the parent plant (**Figure 14.7**). As we saw in Chapter 7, dispersal of offspring away from parents may be advantageous, so benefits may accrue to both the plant (whose seeds are dispersed) and the herbivore (which eats part of the plant). But such interactions are sporadic and facultative, and there is little evidence that the interacting species have evolved in response to each other.

Overall, mutualisms have the potential to be obligate for both species and highly coevolved, but not all mutualisms show these characteristics. In a commensalism, the relationship is always facultative for the species that does not benefit; in addition, coevolution does not occur because natural selection has little or no effect on the species that does not benefit from the relationship.

Positive interactions can cease to be beneficial under some circumstances

Interactions between two species can be categorized by determining for each species whether the outcome of the interaction is positive (benefits > costs), negative (costs > benefits), or neutral (benefits = costs). However, the costs and benefits experienced by interacting species can vary from one place and time to another (Bronstein 1994). Thus, depending on the circumstances, an interaction between two species may have either positive or negative outcomes.

Soil temperature, for example, influences whether a pair of wetland plants interact as commensals or competitors (Callaway and King 1996). Some wetland plants aerate anoxic soils by passively transporting oxygen through continuous air spaces in their leaves, stems, and roots. Oxygen leaked into the soil from the roots of such plants can become available to other plant species, thereby reducing the negative effects of the anoxic soil conditions. In a greenhouse experiment, Ragan Callaway and Leah King grew *Typha latifolia* (cattail), a species that has extensive air spaces, together with *Myosotis laxa* (small-flowered forget-me-not), a species that lacks continuous air spaces. They grew these plants under two different temperature regimes (11°C–12°C and 18°C–20°C) in pots filled with a mix of natural pond soil and peat, with the soil in the pots submerged under 1–2 cm of water to make it anoxic.

At the low soil temperatures, the dissolved oxygen content of the soil increased when *Typha* was present (**Figure 14.8A**);

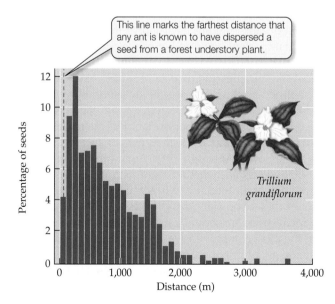

This line marks the farthest distance that any ant is known to have dispersed a seed from a forest understory plant.

Trillium grandiflorum

FIGURE 14.7 Deer Can Move Plant Seeds Long Distances These estimates of the distances that white-tailed deer disperse the seeds of the forest understory plant *Trillium grandiflorum* are based on observations of deer movements and of the length of time that deer retain plant seeds in their digestive tracts (from the time they eat the seeds until they defecate them). Although *T. grandiflorum* seeds are also dispersed by ants, deer move the seeds much farther. (After Vellend et al. 2003.)

but that did not happen at the high soil temperatures (**Figure 14.8B**). How did these different oxygen levels affect the outcome of the *Myosotis–Typha* interaction? At the low soil temperatures, the growth of *Myosotis* roots and shoots increased when *Typha* was present (**Figure 14.9A**). At the high soil temperatures, however, *Myosotis* root mass decreased when *Typha* was present (**Figure 14.9B**). Overall, these results suggest that at the low soil temperatures, *Typha* provided benefits to *Myosotis* (perhaps by aerating the soil), while at the high temperatures, *Typha* had a negative effect on *Myosotis*—just one example of how a change in environmental conditions can alter the outcome of an ecological interaction.

FIGURE 14.9 From Benefactor to Competitor The growth of *Myosotis laxa* under two temperature regimes in the presence and absence of *Typha latifolia* was measured by changes in three parameters: root length, root mass, and shoot mass. (After Callaway and King 1996.)

FIGURE 14.8 A Wetland Plant Aerates the Soil under Some Conditions Callaway and King measured the dissolved oxygen content of submerged soils in the presence and absence of *Typha latifolia* (cattail) under two soil temperature regimes. (After Callaway and King 1996.)

Positive interactions may be more common in stressful environments

In recent decades, studies have shown that positive interactions are important in a number of ecological communities, such as oak woodlands, coastal salt marshes, and marine intertidal communities. Many of these studies have focused on how individuals of a target species are affected by nearby individuals of one or more other species. These effects can be assessed by comparing the performance of the target species when neighbors are present with its performance when neighbors are removed. Although results from such studies cannot be used to determine whether mutualism, commensalism, or competition occur (because two-way interactions are not examined), they do provide a rough assessment of whether positive interactions are common in ecological communities.

In the most comprehensive study of this type that we are aware of, an international group of ecologists tested the effects that neighboring plants had on a total of 115 target plant species in 11 regions worldwide (**Figure 14.10**). In 8–12 repli-

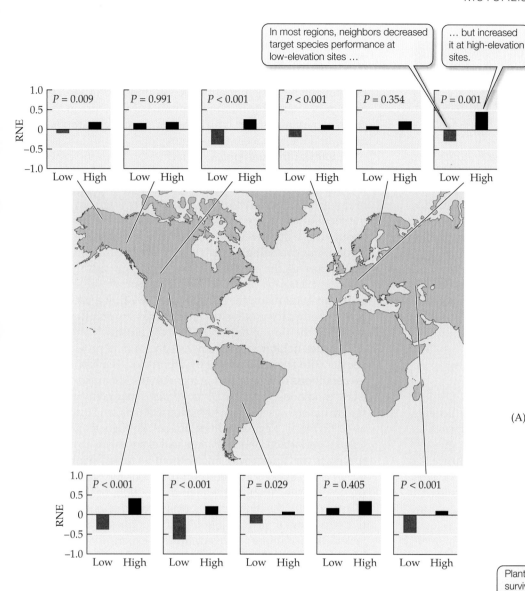

FIGURE 14.10 Neighbors Increase Plant Performance at High-Elevation Sites The relative neighbor effect (RNE, defined as the performance of the target plant species when neighbors are present minus its performance when neighbors are removed) was measured in experimental plots at high and low elevations in 11 regions. Performance was measured as change in biomass (most sites) or in leaf number. RNE values greater than zero (in black) indicate that neighboring plants increased the performance of target plant species; RNE values less than zero (in red) indicate that neighbors decreased the performance of target plant species. (After Callaway et al. 2002.)

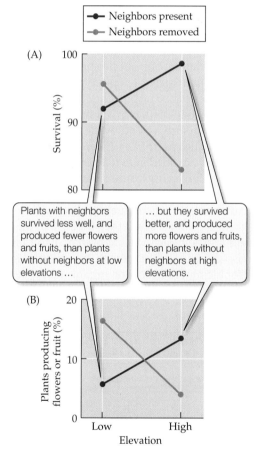

FIGURE 14.11 Negative Effects at Low Elevations, Benefits at High Elevations The proportions of target-species individuals with and without neighbors that (A) survived and (B) produced flowers or fruits differed between low and high elevations. (After Callaway et al. 2002.)

cate plots of each treatment for each target species, neighbors were either left in place or removed from the vicinity of the target species, and the "relative neighbor effect" (RNE, defined as the target species' performance with neighbors present minus its performance when neighbors were removed) was measured. The researchers found that RNE was generally positive at high-elevation sites—indicating that neighbors had a positive effect on the target species—but negative at low-elevation sites. In addition, neighbors tended to reduce the survival and reproduction of target-species individuals at low-elevation sites, but to increase their survival and reproduction at high-elevation sites (**Figure 14.11**). Since environmental conditions were generally more extreme at high-elevation sites (e.g., winter temperatures were colder), these results suggest that positive interactions may be more common in stressful environments. Similar results have been found in intertidal communities (Bertness 1989; Bertness and Leonard 1997).

Characteristics of Mutualism

CONCEPT 14.2
Each partner in a mutualism acts to serve its own ecological and evolutionary interests.

In the previous section we discussed some features that apply to both mutualism and commensalism: these two types of positive interactions are ubiquitous, they can evolve in many ways, and they can cease to be beneficial under some conditions. However, because mutualism is a reciprocal relationship in which both parties benefit, some of its characteristics differ from those of commensalism. A mutualism has costs as well as benefits, and if the costs exceed the benefits for one or both partners, their interaction will change. Before we describe the special characteristics of mutualism, however, we'll begin with a discussion of how mutualisms are classified.

Mutualisms can be categorized according to their effects

Mutualisms are often characterized by the types of benefits that the interacting species provide to each other, such as food, a place to live, or an ecological service. Although such categories can be helpful, they are not mutually exclusive. As we'll see, one partner in a mutualism may receive one type of benefit (such as food) while the other receives a different benefit (such as a place to live). In such cases, the mutualism could be classified in two different ways.

There are many **trophic mutualisms**, in which a mutualist receives energy or nutrients from its partner. In the leaf-cutter ant–fungus mutualism described in the Case Study, each partner feeds the other. (Recall that the ant and the fungus also help each other to overcome plant defenses, so each also provides the other with an ecological service.) In other trophic mutualisms, one organism may receive an energy source while the other receives growth-limiting nutrients. This is the case in mycorrhizae, in which the fungus receives energy in the form of carbohydrates and the plant gets help in taking up limiting nutrients, such as phosphorus. An exchange of energy for limiting nutrients also occurs in the coral–alga symbiosis, in which the coral receives carbohydrates and the alga receives nitrogen.

In **habitat mutualisms**, one partner provides the other with shelter, a place to live, or favorable habitat. Alpheid (pistol) shrimps form a habitat mutualism with some gobies (fish of the genera *Cryptocentrus* and *Vanderhorstia*) in environments with abundant food but little protective cover. The shrimp digs a burrow in the sediments, which it shares with a goby, thus providing the fish with a safe haven from danger. For its part, the goby serves as a "seeing-eye fish" for the shrimp, which is nearly blind. Outside the burrow, the shrimp keeps an antenna on the fish (**Figure 14.12**); if a predator or some other form of disturbance causes the fish to move suddenly, the shrimp darts back into the burrow.

In other habitat mutualisms, a species may provide its partner with favorable habitat by altering local environmental conditions or by improving its partner's tolerance of existing conditions. The grass *Dichanthelium lanuginosum* grows next to hot springs in soils whose temperatures can be as high as 60°C (140°F). Regina Redman, Russell Rodriguez, and colleagues performed laboratory and field experiments in which this grass was grown with and without its fungal symbiont, a *Curvularia* species that grows throughout the plant body (Redman et al. 2002). In the laboratory, 100% of the grass plants that had the fungal symbiont survived intermittent soil temperatures of 60°C, while 0% of the grass plants that lacked the symbiont survived. In field experiments, grass plants with symbionts had greater root and leaf mass than grass plants without symbionts in soils whose temperatures reached 40°C (104°F). In soils above 40°C, the grass plants with symbionts continued to grow well, but all of the grass plants without symbionts died. In a separate study, the transfer of the same fungal symbiont (*Curvularia*) to watermelons,

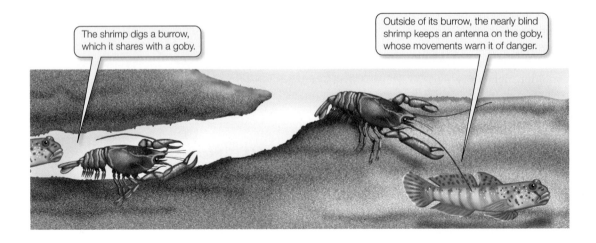

The shrimp digs a burrow, which it shares with a goby.

Outside of its burrow, the nearly blind shrimp keeps an antenna on the goby, whose movements warn it of danger.

FIGURE 14.12 A Seeing-Eye Fish In environments with little protective cover, a habitat mutualism between an alpheid (pistol) shrimp and goby fish benefits both partners.

tomatoes, and wheat improved the growth and survival of these crop species in high-temperature soils. Thus, *Curvularia* increased the ability of a wide range of plant species to tolerate high soil temperatures.

Our final category, **service mutualisms**, includes interactions in which one partner performs an ecological service for the other. Mutualists perform many such services, including pollination, dispersal, and defense against herbivores, predators, or parasites. We have already discussed several examples of service mutualisms (e.g., the fig–fig wasp pollination mutualism), and will consider others later in the chapter (e.g., the cleaner–client mutualism among coral reef fish described on p. 315). Many service mutualisms can also be viewed as trophic mutualisms; for example, a species may perform a service such as pollination in exchange for food.

Mutualists are in it for themselves

Although both partners in a mutualism benefit, that does not mean that a mutualism has no costs for the partners. In the coral–alga mutualism, for example, the coral receives benefits (an energy source), but incurs the costs of supplying the alga with nutrients and space; likewise, the alga gains limiting nutrients, but provides the coral with energy that it could have used to support its own growth and metabolism. The costs of mutualism may be especially clear when one species provides the other with a "reward" such as food for a service such as pollination. For example, during flowering, milkweeds use up to 37% of the energy that they gain from photosynthesis to produce the nectar that attracts insect pollinators such as honeybees.

Neither partner in a mutualism is in it for altruistic reasons. For an ecological interaction to be a mutualism, the net benefits must exceed the net costs for both partners. Should environmental conditions change so as to reduce the benefits or increase the costs for one of the partners, the outcome of the interaction may change. This is especially true if the interaction is not obligate. Ants, for example, often form facultative relationships in which they protect other insects from competitors, predators, and parasites. In one such case, ants protect treehoppers from predators, and the treehoppers secrete "honeydew" (a solution high in sugars), which the ants feed on (**Figure 14.13**). Treehoppers always secrete honeydew, so the ants always have access to this food source. However, in years when predator abundances are low, the treehoppers may receive no benefit from the ants. In such years, the outcome of the interaction may shift from +/+ (a mutualism) to +/0 (a commensalism).

Finally, under certain environmental conditions, a mutualist may withdraw the reward that it usually provides to its partner. In high-nutrient environments, some plants reduce the carbohydrate rewards that they usually provide to mycorrhizal fungi. In such environments, the plant can obtain ample nutrients on its own, and hence the fungus is of little

FIGURE 14.13 A Green Weaver Ant Guards Its Treehopper Mutualist To defend the treehopper larva that it is standing above, this Australian green weaver ant (*Oecophylla smaragdina*) has raised its abdomen, from which it can spray a toxic compound. Treehoppers secrete "honeydew," a liquid rich in carbohydrates that is used by the ants as a source of food.

benefit. Thus, when nutrients are plentiful, the plant may cease to reward the fungus because the costs of supporting fungal hyphae are greater than the benefits the fungus can provide.

Some mutualists have mechanisms to prevent overexploitation

As we've seen, there is an inherent conflict of interest between the partners in a mutualism: the benefit to one species comes at a cost to the other. In such a situation, natural selection may favor **cheaters**, individuals that increase their production of offspring by overexploiting their mutualistic partner. When one of the partners in a mutualism overexploits the other, it becomes less likely that the mutualism will persist. But mutualisms do persist, as the 50-million-year association between fungus-growing ants and the fungi they cultivate readily attests. What factors allow a mutualism to persist in spite of the conflict of interest between the partners?

One answer is provided by "penalties" imposed on individuals that overexploit a partner. If those penalties are high enough, they can reduce or remove any advantage gained by cheating. Olle Pellmyr and Chad Huth documented such a situation in an obligate, coevolved mutualism between the yucca *Yucca filamentosa* and its exclusive pollinator, the yucca moth *Tegeticula yuccasella* (Pellmyr and Huth 1994). Female yucca moths collect pollen from yucca plants with their unique mouthparts (**Figure 14.14A**). After collecting pollen, a female moth typically moves to another plant, lays eggs in

FIGURE 14.14 Yuccas and Yucca Moths *Yucca filamentosa* has an obligate relationship with its exclusive pollinator, the yucca moth *Tegeticula yuccasella*. (A) The female moth collects pollen from a yucca flower in specialized mouthparts. She may carry a load of up to 10,000 pollen grains, nearly 10% of her own weight. (B) The moth at the lower right of this photo is laying eggs in the ovary of a yucca flower; the moth at the top is placing pollen on the stigma.

(A)

(B)

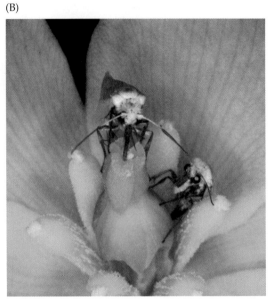

the ovary of a flower, and then walks up to the top of the style. There, the moth deliberately places some of the pollen she carries on the stigma, thus pollinating the plant (**Figure 14.14B**). The larvae that hatch from the moth's eggs complete their development by eating the seeds that develop in the ovary of the flower.

The moth and the plant depend absolutely on each other for reproduction. However, the mutualism is vulnerable to overexploitation by moths that lay too many eggs and hence consume too many seeds. Yuccas have a mechanism to prevent such overexploitation: they selectively abort flowers in which female moths have laid too many eggs (**Figure 14.15**). On average, yuccas retain 62% of the flowers that contain 0–6 moth eggs, but only 11% of the flowers that contain 9 or more eggs. When a flower is not retained, the yucca aborts it before the moth larvae hatch from their eggs. Although the cue that determines flower abortion is not known, it is clear that it is a powerful mechanism for reducing overexploitation: all the moth larvae in an aborted flower die.

Few other clear cases of a penalty for cheating have been documented, so we do not yet know whether such penalties are common in nature. Be that as it may, the yucca–yucca moth interaction illustrates the theme that runs throughout this section: the partners in a mutualism are not altruistic. Instead, the yucca takes actions that promote its own interests, and the yucca moth does the same. In general, a mutualism evolves and is maintained because its net effect is advantageous to both parties. If the net effect of a mutualism were to impair the growth, survival, or reproduction of one of the interacting species, the ecological interests of that species would not be served, and the mutualism might break down, at least temporarily. Should such a situation continue, the longer-term or evolutionary interests of that species might also fail to be served, and the mutualism might break down on a more permanent basis.

Although it is possible for a mutualism to break down, as we've seen, mutualism and commensalism are very common, and some of these interactions have been maintained for millions of years. We turn now to the ecological effects of these pervasive interactions.

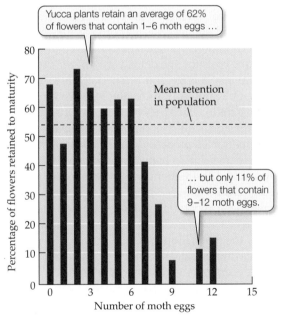

Yucca plants retain an average of 62% of flowers that contain 1–6 moth eggs …

Mean retention in population

… but only 11% of flowers that contain 9–12 moth eggs.

FIGURE 14.15 A Penalty for Cheating Yucca plants selectively abort flowers in which yucca moths have laid too many eggs. (After Pellmyr and Huth 1994.)

Ecological Consequences

> **CONCEPT 14.3**
> Positive interactions affect the distributions and abundances of organisms as well as the composition of ecological communities.

So far in this chapter, we've discussed features that are common to commensalism and mutualism as well as characteristics that are unique to mutualism. As various points in these discussions, we've mentioned ecological consequences of positive interactions, such as increased survival rates and the provision of habitats. In this section, we'll take a closer look at how positive interactions affect populations of organisms and the communities in which they are found.

Positive interactions influence the sizes and distributions of populations

Mutualism and commensalism can provide benefits that increase the growth, survival, or reproduction of individuals in one or both of the interacting species. As a result, these inter-actions have the potential to affect the abundances and distributions of the interacting species. To explore these issues further, we will first examine how an ant–plant mutualism affects demographic factors that influence the abundance of one of its members. We will then examine how mutualism and commensalism influence the distributions of organisms.

Mutualism can influence demographic factors that determine the abundances of organisms The effects of mutualism on demography can be seen in the obligate relationship that some ants in the genus *Pseudomyrmex* have with swollen-thorn acacias. These plants grow to be shrubs or small trees, and they have unusually large thorns, which house colonies of ants (**Figure 14.16A**). The thorns have a tough, woody covering but a soft, pithy interior that is easy for the ants to excavate. A queen ant of the species *Pseudomyrmex ferruginea* establishes a new colony on a bullhorn acacia (*Acacia cornigera*) by burrowing into a green thorn, removing some of its pithy interior, and laying eggs inside the thorn.

The queen feeds her larvae with nectar that the plant secretes from specialized nectaries and with modified leaflet tips called Beltian bodies, which are high in protein and fat (**Figure 14.16B**). As the colony grows, it eventually occupies

(A)

(B)

(C)

Nectary

Beltian body

FIGURE 14.16 An Ant–Plant Mutualism
(A) An ant colony within the thorn of a bullhorn acacia. (B) A nectary at the base of a leaf and Beltian bodies at the leaflet tips. (C) Ants have removed the plants that grew near the acacia in which they live, creating a competitor-free zone for their acacia.

all of the acacia's thorns, and its workers patrol the outside of the plant 24 hours a day. The ants aggressively attack insect and even mammalian herbivores (such as deer) that would eat the plant. The ants also use their mandibles to maul other plants that venture within 10–150 cm of their home acacia, thus providing the acacia with a competitor-free zone in which to grow (**Figure 14.16C**).

Do the services provided by the ants translate into benefits for the acacias? To find out, Dan Janzen removed ants from some acacia plants and compared the growth and survival of those plants with the growth and survival of plants that had ant colonies; he also compared the frequencies with which herbivorous insects were found on both groups of plants. The results were striking. On average, swollen-thorn acacias with ant colonies weighed over 14 times as much as did plants that lacked colonies; acacias with ants also survived better and were attacked by insect herbivores less frequently (Janzen 1966) (**Figure 14.17**).

If an acacia lacks an ant colony, the repeated loss of its leaves and growing tips to herbivores often kills the plant in 6 to 12 months. The ants, in turn, depend on the acacias for food and a home, and they cannot survive without these plants. Thus, the ant–acacia mutualism appears to be obligate for both partners. Furthermore, the ant and the plant have each evolved unusual characteristics that benefit their partners. For example, *Pseudomyrmex* ants that depend on acacias are highly aggressive, remain active for 24 hours a day (patrolling the plant surface), and attack vegetation that grows near their home plant; *Pseudomyrmex* species that do not form mutualisms with acacias show none of these traits. Similarly,

swollen-thorn acacias have enlarged thorns, specialized nectaries, and Beltian bodies on their leaves, and they produce leaves nearly year-round (thus providing reliable sources of food for the ants); most nonmutualistic acacia species do not show these traits. The striking coevolutionary changes that have occurred in the ants and acacias appear to have been selected for by the strong positive effects that each species has on the growth and survival of its partner.

Mutualism and commensalism can determine the distributions of organisms   There are literally millions of interactions in which one species provides another with favorable habitat and thus influences its distribution. Specific examples include corals that provide algal symbionts with a home and fungal mutualists that enable plants to live in environments they otherwise could not tolerate (such as the *Curvularia* fungi that enabled the grass *Dichanthelium* to live in high-temperature soils). Of course, obligate mutualisms, such as the fig–fig wasp mutualism discussed earlier, have a profound influence on the geographic distribution of the interacting species because neither can live where its partner is absent.

It is very common for a group of dominant species (such as the trees in a forest) to determine the distributions of other species by physically providing the habitat on which they depend. For example, many plant and animal species are found only in forests. Such "forest specialists" either cannot tolerate the physical conditions of more open areas (such as a nearby meadow), or they could live in those open areas but are prevented from doing so by competition with other species. Similarly, at low tide in marine intertidal communities, many species (e.g., crabs, snails, sea stars, sea urchins, barnacles) can be found under the strands of seaweeds that are attached to the rocks. The seaweeds provide a moist and relatively cool environment that enables some species to live at higher regions of the intertidal zone than they otherwise could. Finally, many sandy and cobblestone beaches are stabilized by grasses such as *Ammophila breviligulata* and *Spartina alterniflora*. By holding the substratum together, these species enable the formation of entire communities of plants and animals.

Many forest specialists have little direct effect on the trees under which they live; hence, they have a commensalism with the trees of the forest. The same is true of many marine species that seek shelter under seaweeds, and of many of the plants and animals that depend on substratum stabilization by grasses such as *Ammophila* or *Spartina*. In each of these cases, a positive interaction (often, a commensalism) allows one species to occupy a greater area than it otherwise could.

Positive interactions can alter communities and ecosystems

The effects that commensalism and mutualism have on the abundances and distributions of species can affect interactions among species, and that, in turn, can have a large effect

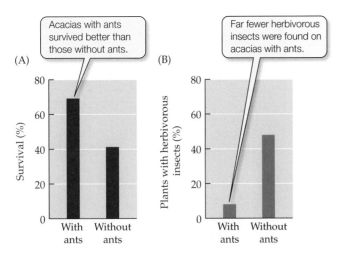

FIGURE 14.17 Effects of a Mutualism with Ants on Swollen-thorn Acacias   Removal experiments in which the performance of plants with ant colonies was compared to the performance of plants whose colonies were removed show that ants in the genus *Pseudomyrmex* provide clear benefits to swollen-thorn acacias. (A) The impact of ants on plant survival. (B) The impact of ants on the percentage of plants with insect herbivores. (After Janzen 1966.)

(A)

Cleaner fish

(B)

(C)

FIGURE 14.18 Ecological effects of the cleaner fish, *Labroides dimidiatus*   (A) As it looks for parasites, a *Labroides dimidiatus* cleaner fish places its head within the mouth of a much larger client fish. The experimental removal of *L. dimidiatus* from small reefs within the Great Barrier Reef of Australia led to (B) a drop in the number of fish species found on the reefs, and (C) a decrease in the total abundance of fishes on the reefs. (B,C after Grutter et al. 2003.)

on a community. For example, if a competitive dominant depends on a mutualist, loss of the mutualist may reduce the performance of the dominant species and increase the performance of other species—thus changing the mix of species in the community or their relative abundances. As we'll see, when the composition of a community changes, properties of the ecosystem may also change.

Removal of cleaner fish reduces the number of other fish species in a coral reef

Coral reefs are known for their astonishing beauty, and they are exceptional ecologically in that their fish communities are the most diverse vertebrate communities in the world. One of the most common interactions among these diverse coral reef fish is a service mutualism in which a small species (the cleaner) removes parasites from a larger fish (the client). The cleaner often ventures into the mouth of the client (**Figure 14.18A**). What prevents the client from simply eating the cleaner?

The answer appears to be that the benefit a client receives from cleaning (parasite removal) is greater than the energy benefit it could gain by eating the cleaner. In the Great Barrier Reef of Australia, individuals of the cleaner fish species *Labroides dimidiatus* were visited by an average of 2,297 clients each day, from which the cleaner fish removed (and ate) an average of 1,218 parasites per day (0.53 parasites per client). To determine whether the activities of cleaner fish translat-

ed into a reduction in the number of parasites found on client fish, Alexandra Grutter experimentally removed *L. dimidiatus* from three of five small reefs. On the reefs from which the cleaner fish were removed, after 12 days, there were 3.8 times more parasites on *Hemigymnus melapterus* fish than on the control reefs. In follow-up studies, Grutter and colleagues (2003) examined the effect of *L. dimidiatus* on the number of species and the total abundance of fish found on coral reefs. The results were dramatic: removal and exclusion of the cleaner fish for a period of 18 months caused large drops in both the number of species and the total abundance of fish found on the reefs (**Figure 14.18B**).

Grutter's work shows that a service mutualism can have a major effect on the diversity of species found in a community. Most of the loss of diversity came from species that typically move among reefs, including some large predators. Since large predators can themselves affect the diversity and abundance of species, the long-term effects of removing cleaner fish could be even greater than those shown in Figure 14.18.

Mycorrhizal fungi can alter species interactions and ecosystem properties

Barbara Hetrick and colleagues (1989) performed greenhouse experiments in which the presence of mycorrhizal fungi altered the outcome of competition between two prairie grasses, big bluestem (*Andropogon gerardii*) and junegrass (*Koeleria pyranidata*): big bluestem dominated when mycorrhizal fungi were present, and junegrass dominated when they were not. In a natural prairie community of which big bluestem was a dominant member, when David Hartnett and Gail Wilson (1999) suppressed mycorrhizal fungi with a fungicide, the performance of big bluestem decreased. At the same time, the performance of a variety of subordinate species, including the grasses *Poa pratensis* and *Dichanthelium oligosanthes* and the wildflowers *Aster ericoides*, *Salvia azurea*, and *Solidago missouriensis*, increased. Hartnett and Wilson suggested that big bluestem's dominance may have come from a competitive advantage caused by its association with mycorrhizal fungi, and that removal of these fungi removed that advantage and released the subordinate species from the negative effects of competition.

Mycorrhizal associations can also affect certain features of ecosystems, as shown in a 1998 study by Marcel van der Heijden, John Klironomos, and colleagues. In a large-scale field experiment, these scientists manipulated the number of species of mycorrhizal fungi found in soils in which identical

(A)

FIGURE 14.19 Mycorrhizal Fungi Affect Ecosystem Properties Researchers measured the effects of the number of mycorrhizal fungal species in the soil on (A) average shoot biomass, (B) average root biomass, and (C) phosphorus content of individuals of 15 species of plants grown from seed in a field experiment. (After van der Heijden et al. 1998.)

> Shoot and root biomass increased initially and then leveled off as the number of mycorrhizal fungal species increased.

(B)

(C)

> Plant phosphorus content rose steadily as the number of mycorrhizal fungal species increased.

evading host defenses. Are there any parasites that specialize in attacking fungal gardens?

Although you might expect that the answer would be yes, for more than a hundred years after the fungus-growing role of leaf-cutter ants was discovered (Belt 1874), no such parasites were known. That changed in 1999, when Cameron Currie and colleagues found that leaf-cutter ant gardens were plagued by a virulent, parasitic fungus of the genus *Escovopsis*. The parasite can be transmitted from one garden to the next, where it can rapidly destroy the gardens it invades, leading to the death of the ant colony.

Ants respond to *Escovopsis* by dramatically increasing the rate at which they weed their gardens (**Figure 14.20**). The ants also appear to enlist the help of other species. Portions of the underside of the ant's body are covered with a bacterium that produces chemicals that inhibit *Escovopsis*. The queen carries the bacteria on her body when she begins a new colony. These bacteria are good for the gardens in two ways (Currie et al. 1999). First, they secrete compounds that stimulate the growth of the cultivated fungi. Second, they produce fungicides that appear to be targeted specifically against *Escovopsis*—these fungicides are more effective against *Escovopsis* than against other fungi. While the ant clearly benefits from the use of these

mixtures of the seeds of 15 plant species had been sown. After one growing season, plant dry weights and phosphorus content were measured. Plant root and shoot biomass increased as the number of species of fungi increased (**Figure 14.19A,B**), as did the efficiency of phosphorus uptake by plants (**Figure 14.19C**). These results show that mycorrhizal fungi can influence key features of ecosystems such as net primary productivity (measured as the amount of new plant growth over one growing season) and the supply and cycling of nutrients such as phosphorus.

CASE STUDY REVISITED

The First Farmers

The fungal gardens described in the opening pages of this chapter represent an enormous food resource for any species able to overcome the ants' defenses. As we saw in Chapter 13, roughly half of the world's species are parasites, and many of them have remarkable adaptations for

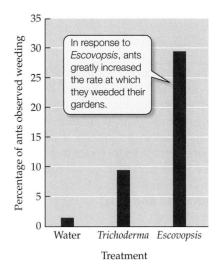

> In response to *Escovopsis*, ants greatly increased the rate at which they weeded their gardens.

FIGURE 14.20 A Specialized Parasite Stimulates Weeding by Ants Currie and Stuart measured the frequency with which the leaf-cutter ant *Atta colombica* weeded its fungal gardens after colonies were exposed to water, *Trichoderma viride* (a generalist fungal pathogen), and the specialized fungal pathogen *Escovopsis*. (After Currie and Stuart 2001.)

(A)

No incompatibility (score = 0): the two clones can grow into one another (no demarcation zone develops).

Mild incompatibility (score = 1): a slight but visible demarcation zone develops.

Moderate incompatibility (score = 2): a distinct demarcation zone develops.

Strong incompatibility (score = 3): a broad demarcation zone with brown coloration develops.

(B)

FIGURE 14.21 Resident Fungi Inhibit Foreign Fungi When grown in the laboratory, resident fungal clones cultivated by the leaf-cutter ant, *Acromyrmex octospinosus*, show varying degrees of incompatibility with foreign fungi introduced from nearby ant colonies. (A) Fungal incompatibility was measured on the four-point scale shown here. (B) Fungal incompatibility was strongly influenced by the genetic difference between resident and foreign fungi; genetic difference was observed to range from a low of 0.0 (no difference, a result found in all within-colony comparisons of resident fungi) to a high of approximately 0.4 (considerable difference). The size of the circles indicates the number of times a given result was obtained. (After Poulsen and Boomsma 2005.)

gardens to the ant and the bacterium. But Poulsen and Boomsma (2005) found otherwise: the cultivated fungus actively rejects fungi introduced from outside the ant colony. What's more, the strength of this reaction depends on the degree of genetic difference between the cultivated (resident) fungus and the introduced (foreign) fungi (**Figure 14.21**)—the more different the foreign fungi are from the resident fungus, the stronger the rejection. Poulsen and Boomsma conclude that, far from being a passive player, the fungus actively excludes nonresident clones and thereby imposes single-crop farming on the leaf-cutter ants!

CONNECTIONS IN NATURE

From Mandibles to Nutrient Cycling

While you have been reading this chapter, billions of pairs of leaf-cutter ant mandibles have been removing leaves from the forests of the Americas. People have long known that leaf-cutter ants are potent herbivores. Weber (1966) describes reports—the earliest from 1559—of how leaf-cutter ants destroyed the crops of Spanish colonists, and they still plague farmers today. In tropical regions, these ants tend to increase in abundance after a forest is cut down. Anecdotal evidence suggests that the thriving ant colonies found in deforested areas are one of the reasons that farms in some tropical regions are often abandoned just a few years after trees have been removed to make room for farmland (other reasons relate to a point described in Chapter 21—some tropical soils are nutrient poor).

In addition to their effects on human farmers, leaf-cutter ants introduce large amounts of organic matter into tropical forest soils. As a consequence, they affect the supply and cycling of nutrients in the forest. For example, nutrients in the leaf litter that falls to the forest floor enter the soil when the leaves decompose. Bruce Haines (1978) compared the amounts (grams/m^2) of 13 mineral nutrients contained in leaf litter with the amounts of the same nutrients found in areas where colonies of the leaf-cutter ant

fungicides, what of the bacterium? Recent work (Currie et al. 2006) indicates that the bacterium also benefits: the ant provides it with both a place to live (it is housed in specialized structures called *crypts* that are located on the ant's exoskeleton) and a source of food (glandular secretions). Thus, the bacterium appears to be a third mutualist that benefits from, and contributes to, these unique fungal gardens.

Let's step back from the ants and bacteria and focus instead on the fungi. As we saw in the Case Study, leaf-cutter ants are "single-crop farmers" in the sense that a colony (and its descendant colonies) rears a single fungal clone. From what we've described so far, it may seem likely that the cultivated fungal clone would leave the defense of the

Atta colombica deposit refuse. Averaged across the 13 nutrients, the ants' refuse areas contained about 48 times the nutrients found in the leaf litter. Plants respond to this concentration of nutrients by increasing their production of fine roots in the *Atta* refuse areas. Furthermore, the activities of leaf-cutter ants have the effect of tilling the soil near their nests, making it easier for plant roots to penetrate the soil (Moutinho et al. 2003). Moutinho and colleagues also found that the leaf material ants bring into colonies fertilizes the soil, causing soils beneath ant colonies to be 3 to 4 times richer in calcium and 7 to 14 times richer in potassium than are soils that are 15 m away from the nest.

While the ants clearly affect the ecosystems in which they are found, their overall effects are complex. In forest ecosystems, net primary productivity (NPP) is usually measured as new aboveground plant growth (see Chapter 19); root growth is often ignored, since it is difficult to measure in trees. Although leaf-cutter ants reduce NPP by harvesting leaves, some of the ants' other activities (e.g., tillage, fertilization) may increase NPP. As a result, the net effect of the ants on NPP is difficult to estimate. While it may prove possible to disentangle such effects in future studies, other puzzles remain. For example, ecologists have been known to fall through the soil in tropical forests, landing in what appear to be empty ant chambers several meters below the soil surface. Why are these chambers empty? If the chambers once contained resource-rich fungal gardens, why do plant roots fail to proliferate in their vicinity? If the chambers never housed fungal gardens, where do they come from? Here, as is often the case in science, as we learn more about these remarkable fungal gardens, fascinating new questions are raised.

SUMMARY

CONCEPT 14.1 Positive interactions occur when neither species is harmed and the benefits of the interaction are greater than the costs for at least one species.

- Mutualism and commensalism are ubiquitous interactions that are important in both terrestrial and aquatic communities.
- Mutualism and commensalism can evolve from other kinds of interactions in many ways.
- Mutualism may or may not be species-specific, obligate, and coevolved. Commensalism is always facultative for the species that does not benefit and is not coevolved.
- Positive interactions cease to be beneficial under some circumstances because the costs and benefits of an interaction may vary from one place and time to another.
- Positive interactions may be more common in stressful environments.

CONCEPT 14.2 Each partner in a mutualism acts to serve its own ecological and evolutionary interests.

- Mutualisms can be categorized by whether one partner provides the other with food (a trophic mutualism), a place to live (a habitat mutualism), or an ecological service (a service mutualism).
- The partners in a mutualism are in it for themselves—it is not an altruistic interaction.
- Some mutualists have mechanisms to prevent overexploitation by cheaters.

CONCEPT 14.3 Positive interactions affect the distributions and abundances of organisms as well as the composition of ecological communities.

- Positive interactions can provide benefits that increase the growth, survival, or reproduction of one or both of the interacting species.
- As a result of such demographic effects, positive interactions can determine the distributions and abundances of populations of the interacting species.
- Positive interactions can also affect interactions among organisms and hence the composition of ecological communities and the properties of the ecosystems of which those communities are a part.

REVIEW QUESTIONS

1. Summarize the key features of positive interactions described in Section 14.1.
2. Most researchers who study mutualism do not think of it as an altruistic interaction. Explain why.
3. High water temperature is one of several stressors that can cause *coral bleaching*, a process in which a coral expels its algal mutualists and hence loses its color. If bleaching occurs repeatedly, coral death may result. Some corals are more sensitive than others to high water temperatures. What do you think would happen if a reef containing a mixture of coral species was exposed to increasingly high water temperatures over a series of years?

HANDS-ON PROBLEM SOLVING

| **From a Parasite to a Mutualist** | This Web exercise (www.sinauer.com/ecology) explores the evolution of mutualism between bacteria that live inside *Drosophila* flies and their fly host. You will interpret data from a recent paper that demonstrates that, in the course of evolving with flies, the bacteria evolved from parasites into mutualists. |

SUGGESTED READINGS

Bruno, J. F., J. J. Stachowicz and M. D. Bertness. 2003. Inclusion of facilitation into ecological theory. *Trends in Ecology and Evolution* 18: 119–125. *This stimulating review article documents the importance of positive interactions and discusses how incorporating these interactions into ecological models may enrich our understanding of natural communities and yield unanticipated predictions.*

Clay, K., J. Holah and J. A. Rudgers. 2005. Herbivores cause a rapid increase in hereditary symbiosis and alter plant community compositions. *Proceedings of the National Academy of Sciences USA* 102: 12465–12470. *An experimental study demonstrating that in the presence of herbivores, the frequency with which plants harbor beneficial symbionts increases, leading to changes in the composition of the plant community.*

Hacker, S. D. and S. D. Gaines. 1997. Some implications of direct positive interactions for community species diversity. *Ecology* 78: 1990–2003. *An influential paper that emphasizes how positive interactions affect the composition of ecological communities and are especially common in environments with relatively high levels of physical disturbance, stress, or predation.*

Holland, J. N., J. H. Ness, A. Boyle and J. L. Bronstein. 2005. Mutualisms as consumer–resource interactions. In *Ecology of Predator–Prey Interactions*, P. Barbosa and I. Castellanos (eds.), 17–33. Oxford University Press, Oxford. *This chapter presents a new way of thinking about mutualism: as a consumer–resource interaction, a type of interaction more commonly associated with predator–prey, host–parasite, or plant–herbivore interactions. The authors first provide a well-reasoned argument for this view, then explore its implications for understanding populations.*

Pellmyr, O. 2003. Yuccas, yucca moths, and coevolution: A review. *Annals of the Missouri Botanical Garden* 90: 35–55. *Written by a leading researcher, this authoritative and interesting review describes one of the best-documented mutualisms in nature, the yucca–yucca moth interaction.*

Stadler, B. and T. Dixon. 2008. *Mutualism: Ants and Their Insect Partners*. Cambridge University Press, Cambridge. *This book looks at many different facets of mutualisms from a resource-based, cost–benefit perspective by focusing on ants as examples. Ants affect many ecological interactions and are themselves affected by a broad range of mutualisms. The book examines how these interactions affect not only the ants and their partners, but also the communities and ecosystems of which they are a part.*

UNIT 4 Communities

15 The Nature of Communities

KEY CONCEPTS

CONCEPT 15.1 Communities are groups of interacting species that occur together at the same place and time.

CONCEPT 15.2 Species diversity and species composition are important descriptors of community structure.

CONCEPT 15.3 Communities can be characterized by complex networks of direct and indirect interactions that vary in strength and direction.

CASE STUDY

"Killer Algae!"

I n 1988, a French marine biology student dove into the crystal clear water of the Mediterranean Sea and made an unusual discovery. On the seafloor, just below the cliffs of the palatial Oceanographic Museum of Monaco, grew an unusual alga, *Caulerpa taxifolia* (**Figure 15.1**), a native of the warm tropical waters of the Caribbean. The student told Alexandre Meinesz, a leading expert on tropical algae and a professor at the University of Nice, about the unusual species. Over the following year, Meinesz confirmed its presence and determined that its feathery fluorescent green fronds, interconnected by creeping "roots," literally carpeted an underwater area in front of the museum.

Meinesz was astonished by his observations because this species had never been seen in such cold waters, and had certainly never reached the high densities he recorded. As it later turned out, earlier sightings from 1984 allowed Meinesz to calculate a spread of more than 1 hectare in 5 years! Over the next few months, he asked himself and his colleagues some important questions. First, how did the alga get to the Mediterranean in the first place, and how could it survive in temperatures as cold as 12°C–13°C (given that its normal temperature range is 18°C–20°C)? Second, did this species occur anywhere else in the Mediterranean, and was it spreading beyond the soft-sediment habitats located in front of the museum? Most importantly, at such high densities, how was it interacting with native algae and seagrasses, both of which serve as critical habitat and food for fish and invertebrate species?

A definitive answer to the second question came in July 1990, when the alga was found 5 km east of the museum, at a popular fishing location. Evidently, fragments had been caught on the gear and anchors of fishing vessels and transported to new sites of colonization. The find generated media coverage that included information on the toxicity of the seaweed, which produces a peppery secondary compound to deter the fish and invertebrate herbivores that abound in the tropics. The press sensationalized *Caulerpa's* natural toxicity with headlines such as "Killer algae!" a misleading title that suggested the seaweed was toxic to humans (it is not). As the news spread, so did the sightings of the fluorescent green alga. By 1991, 50 sightings had been reported in France alone. The alga had indiscriminately colonized muddy, sandy, and rocky bottoms at 3 to 30 meters water depth. By 2000, the alga had moved from France to Italy, then to Croatia to the east and Spain to the west, eventually

FIGURE 15.1 Invading Algae *Caulerpa taxifolia* rapidly invaded and dominated native marine benthic communities in the Mediterranean Sea.

Caulerpa was unintentionally released by the Oceanographic Museum of Monaco in 1984.

- 1 to 1,000 m^2
- 1,000 to 20,000 m^2
- >3,000 ha

1 French Coast (1990)
2 Mallorca, Spain (1992)
3 Elba, Italy (1993)
4 Sicily, Italy (1993)
5 Island of Hvar, Croatia (1995)
6 Upper Adriatic Sea, Croatia (1996)
7 Sousse, Tunisia (2000)

By 2000, Caulerpa had spread as far as Tunisia.

FIGURE 15.2 Spread of *Caulerpa* in the Mediterranean Sea Densities of *Caulerpa* are highest in Monaco and France and lowest in the most recently invaded sites in Croatia and Tunisia.

spreading as far as Tunisia (**Figure 15.2**). It had invaded roughly 3,000 hectares, despite frantic but futile efforts to remove it using a variety of methods.

From the very beginning, Meinesz suspected that the answer to his first question lay with the museum. In 1980, a cold-resistant strain of *Caulerpa taxifolia* had been discovered and propagated in the tropical aquariums at the Wilhelmina Zoo in Stuttgart, Germany. Cuttings were sent to other aquariums, including the one in Monaco, to be grown as aesthetically pleasing backdrops to tropical fish displays. The museum admit-

ted to unintentionally releasing *Caulerpa* in the process of cleaning tanks, but believed the alga would die in the cold waters of the Mediterranean.

Given that *Caulerpa* did not die, but instead quickly invaded and overtook shallow areas of the Mediterranean, scientists and fisherman alike wanted to understand how this abundant and fast-spreading seaweed would affect marine habitats and the fisheries dependent on them. How do interactions with one very abundant species influence the hundreds of other species with which it shares a community?

We have emphasized throughout the book that species are connected with one another and their environment. Ecology is, at its very essence, a study of these interconnections. In the last unit, we formally considered species interactions as two-way relationships with one species either eating, competing, or facilitating another species. For ease of mathematical modeling, we quantitatively considered these pairwise interactions in isolation, even though we have emphasized that, in reality, species experience multiple interactions that largely shape the communities in which they live. In this chapter, we will explore multiple species interactions and how they shape the nature of communities. We will consider the various ways ecologists have defined communities, the metrics used to measure community structure, and the types of species interactions that characterize communities.

What Are Communities?

CONCEPT 15.1
Communities are groups of interacting species that occur together at the same place and time.

Ecologists define **communities** as groups of interacting species that occur together at the same place and time. Interactions that occur among multiple species give communities their character and function. They make communities into something more than the sum of their parts. For example, we all know that the human body, made up of various limbs and organs, assumes true structure and function only when all the parts interact. These interactions can be negative, positive, direct, or indirect, but they all work in a synergistic fashion to produce a functioning human individual. This synergistic and interactive view is highly applicable to communities as well. Whether we are dealing with a desert, a kelp forest, or the gut of an ungulate, the existence of the community is dependent on the individual species that are present and how they interact with one another and their physical surroundings.

The technical definition of a community given above is more theoretical than operational. In practical terms, delineating a community usually requires the use of physical or biological characteristics as a guide. A community may be defined by the physical characteristics of its environment; for example, a physically defined community might encompass all the species in a sand dune, a mountain stream, or a desert (**Figure 15.3A**). The biomes described in Chapter 3 are based largely on physical characteristics thought to be important in defining communities. Similarly, a biologically defined community might include all the species associated with a kelp forest, a freshwater bog, or a coral reef (**Figure 15.3B**). This method uses the presence and implied importance of com-

mon species, such as kelps, wetland plants, or corals, as the basis for community delineation.

Most ecologists use some combination of physical and biological attributes to delineate a community, but they still must contend with the difficult issue of counting all the species present. Merely creating a species list for a community is a huge undertaking, and one that is essentially impossible to complete, especially if small or relatively unknown species are considered. As we will see in Chapter 22, taxonomists have officially described about 1.8 million species, but we know from sampling studies of tropical insects and microorganisms that this value greatly underestimates the actual number of species on Earth, which could be up to 15 million species or even higher. For this reason, and because of the difficulty of studying many species at one time, ecologists usually consider a subset of species when they define and study communities.

FIGURE 15.3 Defining Communities Ecologists often delineate communities based on (A) physical attributes such as in sand dunes and mountain streams, or (B) biological attributes such as in kelp forests and freshwater bogs.

(A) Physically defined communities

(B) Biologically defined communities

FIGURE 15.4 Subsets of Species in Communities Ecologists may study subsets of communities based on (A) taxonomic affinity, (B) guilds, or species that use the same resources, and (C) functional groups, or species that function in similar ways, but use different resources.

Ecologists use subsets of species to define communities

One common way of subdividing a community is based on taxonomic affinity (**Figure 15.4A**). For example, a study of a community might be limited to all the bird, insect, or plant species within that community. Another useful subset of a community is a **guild**, a group of species that use the same resources even though they might be taxonomically distant (**Figure 15.4B**). For example, some birds, bees, and bats can feed on flower pollen, thus forming a guild of pollen-eating animals. Finally, a **functional group** is a subset of a community that includes species that function in similar ways, but do not necessarily use the same resources (**Figure 15.4C**). For example, mosquitoes and aphids both have stylet mouthparts, but one feeds on mammal blood and the other feeds on plant phloem.

There are other subsets of communities, such as food webs (**Figure 15.5A**), that allow ecologists to organize species based on their trophic or energetic interactions. Food webs can be further organized into **trophic levels**, or groups of species that have similar ways of obtaining energy. The lowest trophic level contains primary producers, which can be photosynthetic organisms such as plants or other autotrophic organisms. The primary producers are fed on by organisms at the second level, the primary consumers or herbivores. The third level contains the secondary consumers or carnivores, which are fed on in turn by tertiary consumers, also carnivores. Almost all food webs that have been studied in any detail appear to have two to five trophic levels, but some may have many as ten levels.

Traditionally, food webs have been used as a descriptive or idealized method of understanding the feeding relationships among the species in a community. Food webs tell us little, however, about the strength of interactions or their importance in the community. In addition, the use of trophic levels can create confusion because some species span two trophic levels (e.g., corals, because they have symbiotic algae, can be classified as herbivores and carnivores), some species change their feeding status as they mature (e.g.,

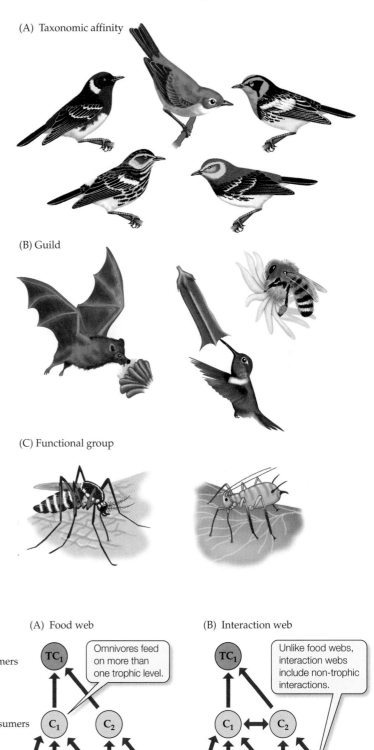

(A) Taxonomic affinity

(B) Guild

(C) Functional group

FIGURE 15.5 Four-Level Food and Interaction Webs (A) Food webs describe the trophic interactions among species. (B) Interaction webs include both trophic interactions (vertical arrows) and non-trophic (horizontal) competitive and facultative interactions.

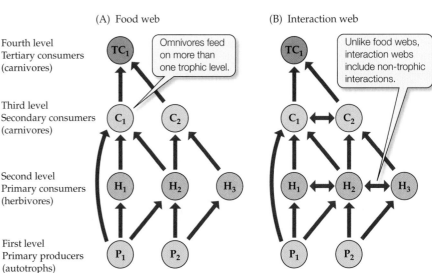

(A) Food web

Fourth level
Tertiary consumers
(carnivores)

Omnivores feed on more than one trophic level.

Third level
Secondary consumers
(carnivores)

Second level
Primary consumers
(herbivores)

First level
Primary producers
(autotrophs)

(B) Interaction web

Unlike food webs, interaction webs include non-trophic interactions.

amphibians can be herbivores as tadpoles, and carnivores as adults), and some species are **omnivores**, feeding on more than one trophic level (e.g., some fish that feed on both algae and invertebrates). Finally, food webs do not include non-trophic interactions (so-called **horizontal interactions**, such as competition and some positive interactions), which we know can influence community character. The concept of an **interaction web** has been introduced to more accurately describe both the trophic (vertical) and non-trophic (horizontal) interactions in a traditional food web (**Figure 15.5B**). Despite these drawbacks, the food web concept remains a strong one, if only for its visual representation of important consumer interactions within a community.

We will learn much more about food webs in Chapter 20. Let's move from defining what a community is to considering the important properties of communities that allow us to characterize and distinguish one from another.

Community Structure

CONCEPT 15.2
Species diversity and species composition are important descriptors of community structure.

We all know that communities vary significantly in the number of species they contain. A tropical rainforest has many more tree species than a temperate rainforest. A midwestern prairie has many more insect species than a New England salt marsh. Ecologists have devoted substantial effort to measuring this variation in different communities and at different spatial scales. Species diversity and species composition are important descriptors of **community structure**, or the set of characteristics that shape communities. Community structure is descriptive in nature, but provides the necessary scientific basis for generating hypotheses and experiments directed at understanding how communities work.

Species diversity is an important measure of community structure

Species diversity is the most commonly used measure of community structure. Even though the term is often used generally to describe the number of species within a community, it has a more precise definition. **Species diversity** is a measure that combines both the number of species (species richness) and their relative abundances compared with one another (species evenness). **Species richness** is the easiest metric to determine (one simply counts all the species of interest in the community one has delineated). **Species evenness**, which tells us about the commonness or rarity of species, requires knowing the abundance (either number, biomass, or percentage of cover) of each species relative to the other species within the community, a harder value to obtain.

The contributions of species richness and species evenness to species diversity can be illustrated using a hypothetical example (**Figure 15.6**). Let's imagine two mushroom communities, each containing four species. Both communities have the same species richness, so the community with the greatest diversity would be the one with the greatest evenness value. In community A, one mushroom species

Community A

In community A, the abundance of one species (the yellow mushrooms) is high relative to the other species, so this community has low species evenness.

Community B

In community B, each species has the same abundance, so this community has high species evenness.

FIGURE 15.6 Species Richness and Species Evenness   The two hypothetical mushroom communities shown here each have the same number of species (species richness), but different relative abundances (species evenness). Community A has lower evenness than community B; thus, species diversity is lower in community A (see Table 15.1).

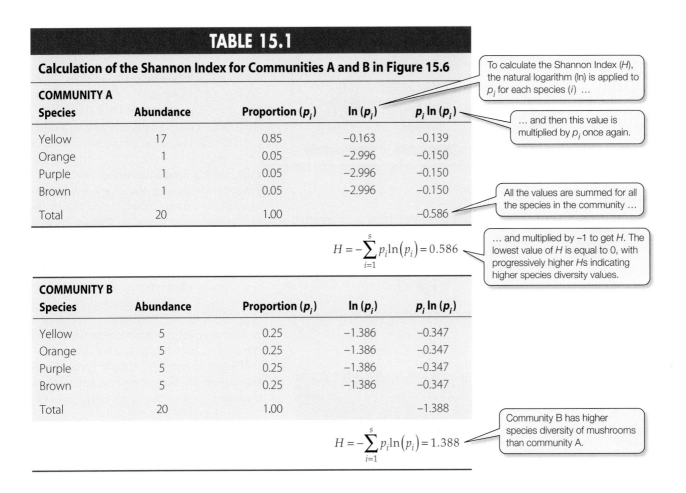

TABLE 15.1

Calculation of the Shannon Index for Communities A and B in Figure 15.6

To calculate the Shannon Index (H), the natural logarithm (ln) is applied to p_i for each species (i) ...

COMMUNITY A

Species	Abundance	Proportion (p_i)	ln (p_i)	p_i ln (p_i)
Yellow	17	0.85	−0.163	−0.139
Orange	1	0.05	−2.996	−0.150
Purple	1	0.05	−2.996	−0.150
Brown	1	0.05	−2.996	−0.150
Total	20	1.00		−0.586

... and then this value is multiplied by p_i once again.

All the values are summed for all the species in the community ...

$$H = -\sum_{i=1}^{s} p_i \ln(p_i) = 0.586$$

... and multiplied by −1 to get H. The lowest value of H is equal to 0, with progressively higher Hs indicating higher species diversity values.

COMMUNITY B

Species	Abundance	Proportion (p_i)	ln (p_i)	p_i ln (p_i)
Yellow	5	0.25	−1.386	−0.347
Orange	5	0.25	−1.386	−0.347
Purple	5	0.25	−1.386	−0.347
Brown	5	0.25	−1.386	−0.347
Total	20	1.00		−1.388

$$H = -\sum_{i=1}^{s} p_i \ln(p_i) = 1.388$$

Community B has higher species diversity of mushrooms than community A.

constitutes 85% of mushroom abundance, while the others constitute only 5% each; thus, species evenness is low. In community B, abundances are evenly divided among the four species (25% each), so species evenness is high. In this case, even though each community has the same species richness (four species), community B has the highest species diversity.

There are a number of species diversity indices that can be used to describe species diversity quantitatively. Anne Magurran (2004) outlines 13 indices that emphasize different aspects of species diversity. By far the most commonly used is the **Shannon index**,

$$H = -\sum_{i=1}^{s} p_i \ln(p_i)$$

where

H = the Shannon index value

p_i = the proportion of individuals found in the ith species

ln = the natural logarithm

s = the number of species in the community

We can calculate the Shannon index for our two mushroom communities in Figure 15.6 (**Table 15.1**). These calculations show that community A has the lower Shannon index value (H) of the two communities confirming mathematically that this community has lower species diversity than community B. The greater the H value, the greater the species diversity in a community.

As we mentioned earlier, the term "species diversity" is often used to describe the number of species in a community without regard to the relative abundances of species or even species diversity indices. For example, one commonly hears the assertion that "species diversity" is higher in tropical communities than in temperate communities without any accompanying information about the actual relative abundances of species in the two communities. Another term that is often used interchangeably with "species diversity" is "biodiversity." Technically, **biodiversity** is a term used to describe the diversity of important ecological entities that span multiple spatial scales, from genes to species to communities (**Figure 15.7**). Implicit in the term is the interconnectedness of individuals, populations, species, and even community-level components of diversity. That interconnectedness works something like this. As we saw in Chapter 10,

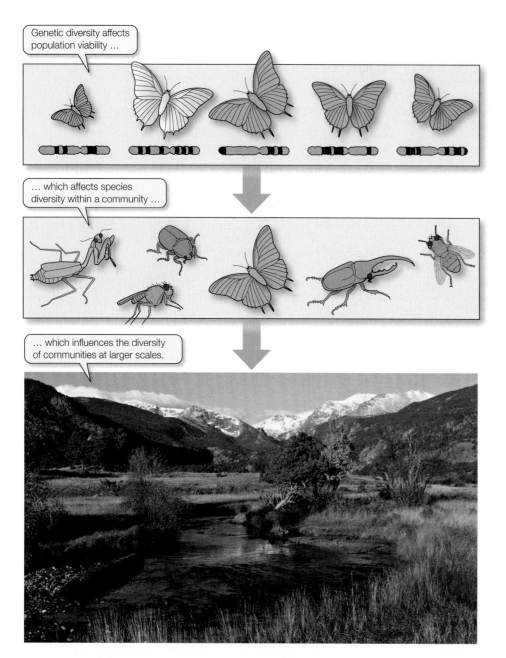

Genetic diversity affects population viability ...

... which affects species diversity within a community ...

... which influences the diversity of communities at larger scales.

FIGURE 15.7 Biodiversity Considers Multiple Spatial Scales Diversity can be measured at different spatial scales that range from genes to species to communities.

Species within communities differ in their commonness or rarity

Although species diversity indices allow ecologists to make comparisons among different communities, graphical representations of species diversity can give us a more explicit view of the commonness or rarity of the species in communities. Such graphs, called **rank abundance curves**, plot the proportional abundance of each species (p_i) relative to the others in rank order, from most abundant to least abundant (**Figure 15.8**). If we use rank abundance curves to compare our two mushroom communities from Figure 15.6 and Table 15.1, we can see that community A has one abundant mushroom species (i.e., the yellow species) and many rare species (i.e., the orange, purple, and brown species), while in community B all the species have the same abundance.

These two patterns could indicate something about the types of species interactions that occur in these two communities. For example, the dominance of the yellow mushroom species in community A might indicate that it has a strong negative effect on the three rare species in the community. In community B, where all the species have the same abundance, their interactions might be fairly equivalent, with no one species dominating the others. We can design manipulative experiments to explore the relationship between species abundances and the types of interactions that occur among the species in a community. As we will see in the next section, experiments of this kind typically involve adding or removing a species and measuring the responses of other species in the community to the manipulation.

For simplicity, we have considered a hypothetical example of species diversity patterns in mushroom communities. What do real communities reveal in this regard? Let's consider a case study in which species diversity and rank abundance curves were determined for two soil bacteria communities in pastures in Scotland. Allison McCaig and colleagues (1999) sampled bacteria from the soils of an undisturbed pasture and a pasture that had been fertilized regularly. How does one go about identifying different species of bacteria or determining their abundance? It is impractical to do this visually, but we can use molecular techniques that allow DNA sequences to be determined (in this study, a technique known as 16S ribosomal DNA sequence analysis was

the genetic variation among individuals within populations influences population viability. This population viability, in turn, has important consequences for species viability, and ultimately species diversity within communities. Moreover, the number of different kinds of communities in an area is critical to diversity at larger regional and latitudinal scales. We will discuss the importance of spatial scale and biodiversity in chapters to come, but it is worth defining here some of the ways in which the term "diversity" is used as a starting point for those later discussions.

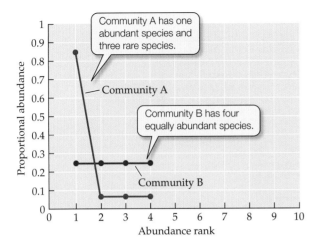

FIGURE 15.8 Are Species Common or Rare? Using rank abundance curves, we can see that the two hypothetical mushroom communities in Figure 15.6 differ in the commonness and rarity of the same four mushroom species.

used). Unique DNA sequences can then be grouped into groups of bacteria using phylogenetic analysis. Phylogenetic analysis involves methods to understand the evolutionary relationships and common ancestry of organisms. As is generally true for microbial communities, it is often difficult to identify bacteria at the species level, so microbiologists usually group them into higher taxonomic levels using phylogenetic analysis.

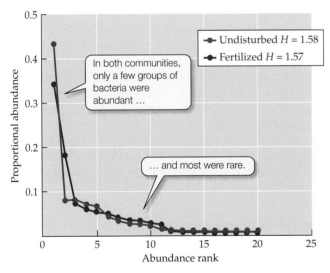

FIGURE 15.9 Bacterial Diversity in Pastures in Scotland
When Allison McCaig and colleagues calculated rank abundance curves and the Shannon index (*H*) for communities of bacteria in undisturbed and fertilized soils, they found that the two communities did not differ significantly by either measure. The composition of bacteria, however, did differ between the two communities. (After McCaig et al. 1999.)

McCaig and colleagues' molecular work identified 275 unique DNA sequences, which they grouped, using statistical techniques, into 22 phylogenetic groups of bacteria. Each sequence can be thought of as an "individual" within a group thus allowing abundance for each group to be determined. When they calculated rank abundance curves and diversity indices, they saw that there was little difference in the structure of the two bacterial assemblages (**Figure 15.9**). In addition, they found a pattern that is consistent with most communities in which larger organisms are the focus: a few species were abundant, and the majority of the rest were rare. One group of bacteria, the α-proteobacteria, was abundant (constituting 43% and 34% of the community in the undisturbed and fertilized pastures, respectively), and the other groups were rare. As mentioned earlier, whether this pattern of a few abundant species and many rare species tells us something about the importance of the species and their interactions is largely unknown, especially for microbial communities.

Species diversity estimates vary with sampling effort and scale

Let's imagine that you are sampling your backyard for insect species. It makes sense that the more areas you sample, the more species you are likely to find, until eventually every new area yields the same insect species over and over again. Essentially, you reach a point in your sampling effort at which any additional sampling will reveal so few new species that you could stop sampling and still have a good notion of the species richness of your backyard. That point of "no significant return" for your effort can be determined using **species accumulation curves** (**Figure 15.10**). These curves are calculated by plotting species richness as a function of the total number of individuals that have accumulated with each ad-

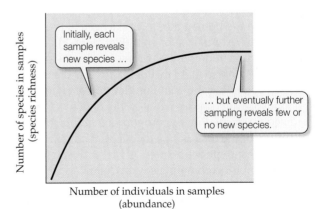

FIGURE 15.10 When Are All the Species Sampled? Species accumulation curves can help us determine when most or all of the species in a community have been observed. In this hypothetical example, the number of new species observed in each sample decreases after about half the individuals in the samples have accumulated.

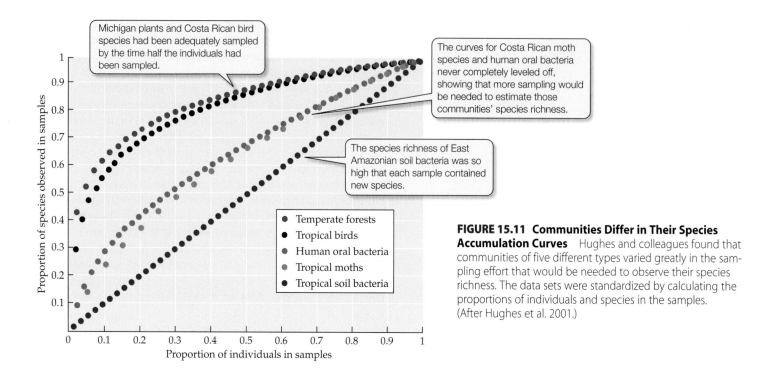

Michigan plants and Costa Rican bird species had been adequately sampled by the time half the individuals had been sampled.

The curves for Costa Rican moth species and human oral bacteria never completely leveled off, showing that more sampling would be needed to estimate those communities' species richness.

The species richness of East Amazonian soil bacteria was so high that each sample contained new species.

- Temperate forests
- Tropical birds
- Human oral bacteria
- Tropical moths
- Tropical soil bacteria

FIGURE 15.11 Communities Differ in Their Species Accumulation Curves Hughes and colleagues found that communities of five different types varied greatly in the sampling effort that would be needed to observe their species richness. The data sets were standardized by calculating the proportions of individuals and species in the samples. (After Hughes et al. 2001.)

ditional sample. The more samples taken, the more individuals will be added, and the more species will be found. At some point, however, the curve will reach a threshold at which no new species are added despite additional sampling.

Jennifer Hughes and colleagues (2001) used species accumulation curves to ask how communities differ in the relationship between species richness and sampling effort. Are there some very diverse communities in which we are unable to estimate species richness accurately despite intensive sampling? Hughes et al. calculated species accumulation curves for five different communities: a temperate forest community in Michigan, a tropical bird community in Costa Rica, a tropical moth community in Costa Rica, a bacterial community from the human mouth, and a bacterial community from tropical East Amazonian soils (**Figure 15.11**). Each data point on a curve represents the total number of individuals and species sampled up to that point. To compare the curves properly, given that the communities differed substantially in abundance and species richness, the data sets were standardized by calculating the proportions of individuals and species out of the total sampled up to that point. The results showed that the species richness of the Michigan forest and Costa Rican bird communities was adequately represented well before half the individuals were sampled. Human oral bacteria and Costa Rican moth communities had species accumulation curves that never completely leveled off, suggesting that their species richness was high and that additional sampling would be required to achieve an approximation of this richness. Finally, Amazon soil bacteria had a linear species accumulation curve, demonstrating that each new

sample resulted in the observation of many new bacterial species. Based on this analysis, it is clear that the sampling effort for tropical bacteria was well below that needed to adequately estimate species diversity in these hyperdiverse communities.

A comparison of species accumulation curves not only provides valuable insight into the differences in species richness among communities, but also demonstrates the influence of the spatial scale at which sampling is carried out. For example, if we were to sample the richness of bacteria in tropical soils at the same scale as Costa Rican moths, the bacterial diversity would be immense in comparison. But these comparisons do suggest that our ability to sample all the bacteria in the human mouth is roughly equivalent to our ability to sample all the moth species in a few hundred square kilometers of tropical forest. The work of Hughes et al. (2001) also reminds us how little we know about the community structure characteristics of rarely sampled assemblages, such as microbial communities.

Species composition tells us who is in the community

A final element of a community's structure is its **species composition**, or the identity of species present in the community. Species composition is an obvious but important characteristic that is not revealed in species diversity indices. For example, two communities might have the same species diversity values, but have completely different members. In the case of the bacterial communities in pastures that we considered above (McCaig et al. 1999), although diversity indices were nearly identical for the two communities (see Figure 15.9), their composition differed greatly. Five taxonomic groups out of the 22 the researchers found were present in one or the other pasture, but not in both.

As we will see in the next section and the next few chapters to come, knowing the identity of the species in a community makes all the difference in our understanding of the types of interactions they are having with one another and the significance of those interactions for the community as a whole. In many ways, community structure is the starting point for more interesting questions: How do species in the community interact with one another? Do some species play greater roles in the community than others? How is species diversity maintained? How does this information shape our view of communities in terms of conservation and the services they provide to humans? Let's move from the rather static view of communities as groups of species occurring together at the same place and time to a more active view of them as complex networks of species with connections or interactions that vary in strength and direction.

Interactions of Multiple Species

CONCEPT 15.3
Communities can be characterized by complex networks of direct and indirect interactions that vary in strength and direction.

The way we think about the interrelationships of species changes dramatically when we consider that they are embedded in a community of multiple interactors. Instead of a particular species experiencing a single, direct interaction with another species, we are now dealing with multiple species interactions that generate a multitude of connections, some direct, but many indirect (**Figure 15.12**). **Direct interactions** occur between two species and include traditional relationships such as competition, predation, and facilitation. **Indirect interactions** occur when the relationship between two species is mediated by a third (or more) species. The simple addition of a third species to a two-species interaction creates many more effects, both direct and indirect, that have the potential to change the outcome of the original interaction dramatically.

A social interaction analogy fits well here. Let's say you have a good friend whom you interact with well. Now, suppose this friend meets another person who dominates your friend's time. They go to dinner, the movies, and bowling—all things you and your friend enjoyed together—without you. At some point, this new friend might begin to significantly interfere with your friendship, possibly compromising it to the point where it becomes antagonistic. Sadly, the indirect effect of the third person changes your friendship irreparably. You might say that "the friend of my friend is my enemy." Likewise, one could imagine gaining a friend indirectly, if your foe had a foe (in this case, "the enemy of my enemy is my friend"). The point is that simply adding another person to the social circle can change the outcome of your relationship completely. The same is true of species interactions when we view them in the community context, rather than as isolated entities.

Indirect species interactions can have large effects

Charles Darwin was one of the first to convey the importance of indirect effects in *The Origin of Species*, published in 1859. Darwin set the scene by describing the role of bees in flower pollination, and hence seed production, among native plants in the region of England where he lived. In the book, he established the hypothesis that the number of bees is dependent on the number of field mice, which prey on the combs and nests of bees. Mice, in turn, are eaten by cats, leading Darwin to muse, "Hence it is quite credible that the presence of a feline animal in large numbers in a district might determine, through the intervention first of mice and then of bees, the frequency of certain flowers in that district!" (Darwin 1859, p. 59).

It is only recently that the sheer number and variation of indirect effects have been documented (Menge 1995). In many cases, indirect effects are discovered almost by accident when species are experimentally removed to study the strength of a direct negative interaction such as predation or competition. A good example of this type of indirect effect comes in the form of an interaction web called a **trophic cascade (Figure 15.13A)**. A trophic cascade is generated when

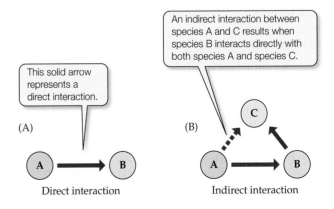

(A) This solid arrow represents a direct interaction.

(B) An indirect interaction between species A and C results when species B interacts directly with both species A and species C.

Direct interaction

Indirect interaction

FIGURE 15.12 Direct and Indirect Species Interactions
(A) A direct interaction occurs between two species. (B) An indirect interaction (dashed arrow) occurs when the direct interaction between two species is mediated by a third species.

(A) Trophic cascade

(B) Trophic facilitation

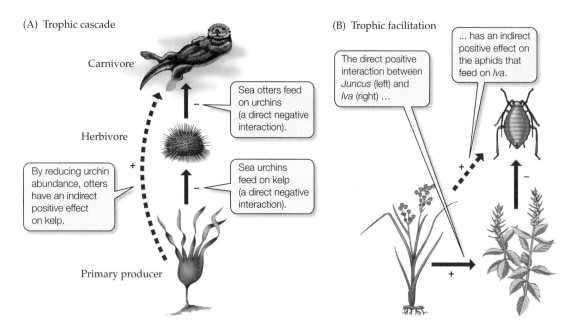

FIGURE 15.13 Indirect Effects in Interaction Webs   (A) A trophic cascade occurs when a carnivore feeds on an herbivore and thus has an indirect positive effect on a primary producer that is eaten by that herbivore. (B) A trophic facilitation occurs when a consumer is indirectly facilitated by a positive interaction between its prey and another species.

a carnivore eats an herbivore (having a direct negative effect on the herbivore) and decreases its abundance such that there is an indirect positive effect on a primary producer. One of the best-known examples is the indirect regulation of kelp forests by the sea otter (*Enhydra lutris*) through its direct interaction with sea urchins (*Strongylocentrotus* sp.) along the west coast of North America (see the Case Study in Chapter 8; Simenstad et al. 1978). Two direct negative interactions, those of sea otters feeding on sea urchins and sea urchins feeding on kelp, generate indirect positive interactions, including the effect of sea otters on kelp (via their reduction of urchin abundance) and kelp on sea otters (via the food provided for the urchins). Kelp, in turn, can positively affect the abundances of other seaweeds, which serve as habitat and food for marine invertebrates and fishes. The indirect effects generated in this simple food web are just as important as the direct effects in determining whether this system will be a kelp forest or a barren rock surface (see Figure 8.21). We will explore indirect effects in the context of species diversity (see Chapter 18) and food webs (see Chapter 20) in more detail later in the book.

Indirect effects can also emerge from direct positive interactions called trophic facilitations (**Figure 15.13B**). A **trophic facilitation** occurs when a consumer is indirectly facilitated by a positive interaction between its prey and another species. An example of this type of indirect effect was demonstrated by Sally Hacker (Oregon State University) and Mark Bertness (Brown University), who studied salt marsh plant and insect interaction webs in New England. Their research showed that a commensalistic interaction between two salt marsh plants—a rush, *Juncus gerardii*, and a shrub, *Iva frutescens*—has important indirect effects on aphids feeding on *Iva* (Hacker and Bertness 1996).

To explore these findings in greater detail, let's first consider the commensalistic interaction between the two plant species. When *Juncus* was experimentally removed, the growth rate of *Iva* decreased (**Figure 15.14A**), and it showed severe signs of stress. In contrast, removing *Iva* had no effect on *Juncus*. In the absence of *Juncus*, soil salinity increased and oxygen content decreased considerably around *Iva*, suggesting that the presence of *Juncus* ameliorated harsh physical conditions for *Iva*. *Juncus*, by shading the soil surface and thus decreasing water evaporation from the surface of the marsh, decreases salt buildup. It also has specialized tissue called aerenchyma, which allows it to deal with flood conditions. Oxygen can move down the aerenchyma tissue into the belowground parts of the plant, thus keeping it from suffocating during daily high tides. Some of the oxygen "leaks" out of the plant and can be used by other neighboring plants, such as *Iva*.

To understand the importance of this direct positive interaction, Hacker and Bertness (1996) measured the population growth rate of aphids on *Iva* with and without *Juncus* (**Figure 15.14B**). They found that aphids had a much harder time finding shrubs in the presence of the rush, but once they did, their population growth rates were significantly higher. Using the exponential growth equation, they predicted that aphids would become locally extinct in the salt marsh without the indirect positive effects of *Juncus* (**Figure 15.14C**). It is clear from this example that interactions in trophic facilitation webs can have both positive (as when *Juncus* improves soil conditions for *Iva*) and negative effects (as when *Juncus* facilitates aphids that feed on *Iva*), but it is the sum total of these effects that determine whether the interaction is beneficial or not. Given that the ultimate fate of *Iva* without *Juncus* is death, the positive effects greatly outweigh the negative ones.

(A) Removal of *Juncus* resulted in decreased growth (measured as photosynthetic rate) of *Iva* throughout the summer.

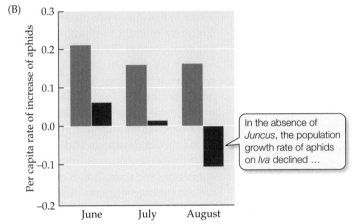

(B) In the absence of *Juncus*, the population growth rate of aphids on *Iva* declined …

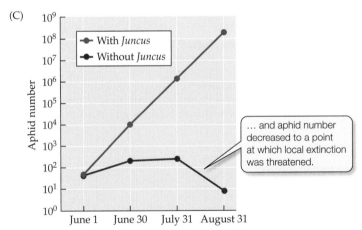

(C) … and aphid number decreased to a point at which local extinction was threatened.

FIGURE 15.14 Results of Trophic Facilitation in a New England Salt Marsh Removal experiments demonstrated that aphids are indirectly facilitated by the rush *Juncus gerardii*, which has a direct positive effect on the shrub *Iva frutescens*, on which the aphids feed. (After Hacker and Bertness 1996.)

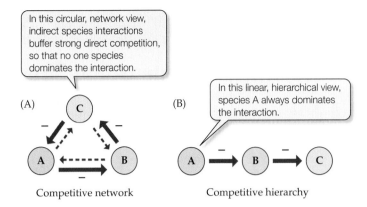

In this circular, network view, indirect species interactions buffer strong direct competition, so that no one species dominates the interaction.

In this linear, hierarchical view, species A always dominates the interaction.

(A) Competitive network

(B) Competitive hierarchy

FIGURE 15.15 Competitive Networks versus Competitive Hierarchies

Finally, important indirect effects can arise from multiple species interactions at one trophic level (i.e., the horizontal interactions in Figure 15.5B). Buss and Jackson (1979), looking for an explanation for the coexistence of competitors, hypothesized that competitive interactions that occur in a network fashion (i.e., every species negatively interacts with every other species) might be important in maintaining species richness. A *network*, as opposed to a hierarchy, is an interaction web that is circular rather than linear (**Figure 15.15A**). The idea is that networks of interacting species indirectly buffer strong direct competition, thus making competitive interactions weaker and more diffuse. So, for example, species A may have the potential to outcompete species B, and species B may have the potential to outcompete species C, but because species C also has the potential to outcompete species A, no one species dominates the interaction. This is clearly an example of the "enemy of my enemy is my friend" effect described earlier. All else being equal, a hierarchical view of competition, with species A outcompeting B and B outcompeting C (**Figure 15.15B**), always results in species A dominating the interaction.

Buss and Jackson (1979) tested this hypothesis using encrusting invertebrates and algae that live on the undersides of coral reefs in Jamaica (**Figure 15.16**). These species compete for space by overgrowing one another. The researchers collect-

FIGURE 15.16 Competitive Networks in Reef Organisms Encrusting invertebrates and algae compete for space on coral reefs.

BOX 15.1

Measurements of Interaction Strength

We can measure interaction strength by experimentally manipulating species interactions. The usual procedure involves either the removal or the addition of one of the species involved in the interaction (the interactor species) and measurement of the response of the other species (the target species). There are several different ways to measure interaction strength, but the most common is to calculate the per capita interaction strength using the following equation:

$$\text{Per capita interaction strength} = \ln\frac{\left(\dfrac{C}{E}\right)}{I}$$

where

C = the number of target individuals in the presence of the interactor

E = the number of target individuals in the absence of the interactor

I = the number of interactor individuals

ln = natural logarithm

Interaction strength can vary depending on the environmental context in which the interaction is measured. For example, Menge et al. (1996) measured the interaction strength of sea star (*Pisaster ochraceus*) predation on mussels (*Mytilus trossulus*) in wave-exposed versus wave-protected areas of the shoreline at Strawberry Hill on the coast of Oregon (see figure). Sea stars (I) were excluded from mussel beds using cages in both exposed and protected areas, and the number of mussels in the cages (E) was compared to the number of mussels that disappeared in the control plots (C) exposed to predation. (I) was determined by counting all the seastars at the site. An equation similar to that above was used to calculate the per capita interaction strength of sea star predation on mussels. The results showed that interaction strength was greater in wave-protected than in wave-exposed areas. Sea stars probably cannot feed as efficiently when subjected to the crashing waves characteristic of wave-exposed areas. This study points to the importance of environmental context (in this case, wave exposure) to the strength of species interactions and shows how those interactions can change over relatively small spatial scales (e.g., within different wave-exposed and wave-protected areas of the Strawberry Hill shoreline).

(i)

How Much Does Predation by Sea Stars Matter? It Depends (i) The per capita interaction strength of predation by sea stars (*Pisaster ochraceus*) on mussels (*Mytilus trossulus*) was measured at wave-exposed and wave-protected areas along the shoreline at Strawberry Hill, Oregon. Mussel populations in plots (ii) with and (iii) without cages that excluded sea stars were counted and the interaction strength was calculated. (iv) The results showed that interaction strength was greater in protected than in exposed areas. (After Menge et al. 1996.)

(ii)

(iii)

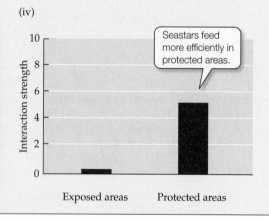

(iv)

Seastars feed more efficiently in protected areas.

ed samples at the margins between species, where one species overgrows another, for as many pairs of species as possible to determine the proportion of wins (species on the top) to losses (species on the bottom) for each species. Their results showed that every species either overgrew or was overgrown by another species, and that no one species consistently won the competition. The species interacted in a circular network rather than a linear hierarchy. Their observations demonstrate how competitive networks, by fostering diffuse and indirect interactions, can promote diversity in communities.

Species interactions vary greatly in strength and direction

You should realize by now that species interactions in a community vary greatly in strength and direction. Some species have a strong negative or positive effect on the community, while others probably have little or no effect. **Interaction strength**, the effect of one species on the number of individuals of another species, can be measured experimentally by removing one species (the *interactor species*) from the community and looking at the effect on the other species (the *target species*) (**Box 15.1**). If the removal of the interactor species results in a large decrease in the target species, we know that the interaction is strong and positive. However, if the abundance of the target species increases significantly, we know that the interactor species has a strong negative effect. The interaction strength "dynamic" (i.e., the relative proportion of strong to weak interactions or positive to negative interactions) is not well understood for any community because of the numbers of species involved and the many indirect interactions that emerge. As you will see in the following chapters, however, we can get an idea of what species are "in charge" of communities through both observations and experiments.

There are some large or abundant species, such as trees, that we know intuitively have large community-wide effects by virtue of providing habitat or food for other species. They might also be good competitors for space, nutrients, or light. These so-called **dominant species**, also known as **founda-**

FIGURE 15.18 Trees Are Dominant Species and Ecosystem Engineers Trees not only provide food for and compete with other species, but also act as ecosystem engineers by creating, modifying, or maintaining physical habitat for themselves and other species. (After Jones et al. 1997.)

tion species (Dayton 1971), have a large effect on other species, and thus species diversity, by virtue of their considerable abundance and biomass (**Figure 15.17**).

Some dominant species act by "bioengineering" their environment. These species, known as **ecosystem engineers** (Jones et al. 1994), are able to create, modify, or maintain physical habitat for themselves and other species. Consider the simple example of the tree mentioned above. Trees provide food for other organisms and compete for resources just like any other species. However, trees also engineer their environment in subtle but important ways (**Figure 15.18**; Jones et al. 1997). The trunk, branches, and leaves provide habitat for a multitude of species, from birds to insects to lichens. The physical structure of the tree reduces sunlight, wind, and rainfall, influencing temperature and moisture levels in the forest. The roots of the tree can increase weathering and aeration of the soil, and they can anchor and bind rocks that stabilize surrounding substrata. The tree's leaves fall to the forest floor, where they add moisture and nutrients to the soil and provide habitat for soil-dwelling invertebrates, seeds, and microorganisms. If the tree falls, it can become a "nurse log," providing space, nutrients, and moisture for tree seedlings. Thus, trees can have a large physical influence on the struc-

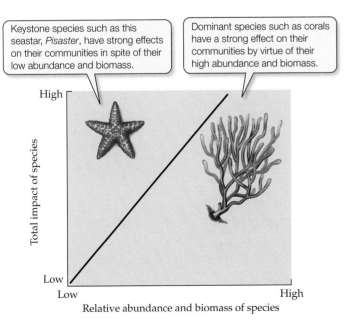

FIGURE 15.17 Dominant versus Keystone Species Species that have large effects on their communities may or may not do so by virtue of their high abundance or biomass. (After Power et al. 1996b.)

ture of a forest community, which obviously changes through time as trees grow, mature, and die.

Other strongly interacting species, so-called **keystone species**, have a strong effect not because of their abundance, but because of the roles they play in their communities. They differ from dominant species in that their effect is large in proportion to their abundance and biomass (see Figure 15.17). Keystone species usually influence community structure indirectly, via trophic means, as we saw in the case of sea otters (see Figure 15.13A and Chapter 8). Sea otters are considered keystone species because, by preying on sea urchins, they indirectly control the presence of kelp, which provides important habitat for many other species. We will consider the role of keystone species in more detail in Chapter 20.

There are also keystone species that act as ecosystem engineers. A great example is the beaver, a species in which just a few individuals can have dramatic effects on the landscape. Beavers dam streams with cut trees and woody debris. Very quickly, flooding ensues and sediment accumulates as the increasing number of woody obstacles slows the water flow. Eventually, the once swiftly flowing stream is replaced by a marsh, containing wetland plants that can deal with flood conditions. At the landscape level, by creating a mosaic of wetlands within a larger forested community, beavers can increase regional biodiversity significantly (**Figure 15.19**). Naiman et al. (1988) showed that there was a 13-fold increase in wetland area in one region of Minnesota (roughly 200 ha to 2,600 ha) when beavers were allowed to recolonize areas where they had been nearly hunted to extinction some 60 years previously.

Environmental context can change the outcome of species interactions

As we have seen, species interactions can vary in strength and direction, and their outcome is highly dependent on the influence of other species in the community. As we saw in Chapter 14 and Box 15.1, another important factor in the outcome of species interactions is the environmental context in which they occur. For example, under benign environmental conditions that are favorable for population growth, it makes sense that species will thrive and limit resources, thus engaging in more antagonistic interactions such as competition or predation. Under harsh environmental conditions, species will naturally be more limited by physical factors, thus interacting either weakly or cooperatively with other species.

This view of species interactions as **context-dependent**, or changeable under different environmental conditions, is relatively new to ecology, but a number of important examples exist. These examples usually involve keystone or dom-

FIGURE 15.19 Beavers Are Keystone Species and Ecosystem Engineers By damming streams, beavers created networks of different types of wetlands (in red) in a 45 km^2 watershed on the Kabetogama Peninsula in Minnesota, thus increasing biodiversity within the region. (After Naiman et al. 1988.)

In 1940, beavers were nearly extinct in this region, and there were few wetlands (in red).

By 1986, beavers had recolonized the region, and wetland area (in red) had increased 13-fold.

1940

1961

1986

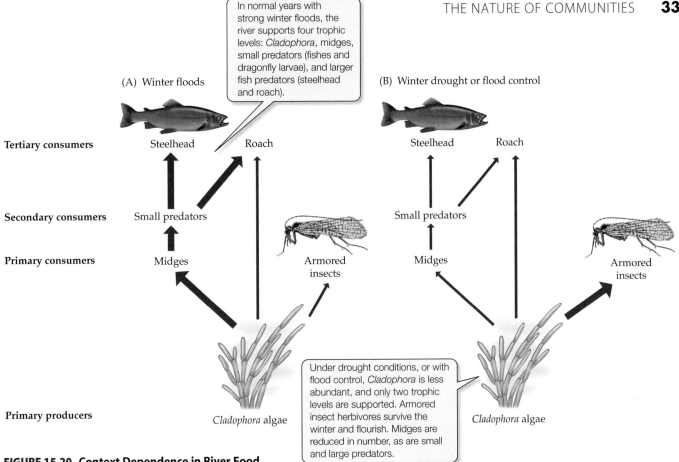

FIGURE 15.20 Context Dependence in River Food Webs Drought and flood control changes the relative importance of different trophic levels in the Eel River of northern California. Wider arrows represent stronger interactions than narrower arrows. (After Power et al. 1996a.)

inant species that play important roles in their communities in one context, but not in another. Mary Power, a professor at the University of California, Berkeley, who works on stream communities in northern California, has shown that the role of fish predators (roach, *Hesperoleucas symmetricus*, and steelhead, *Oncorhynchus mykiss*) changes from year to year. The role of these predators shifts from that of keystone species following winters of scouring floods to one of weak interactors during years with winter droughts or where flood control is operating (**Figure 15.20**; Power et al. 1996).

In the northern California rivers where Power works, there is a natural winter flood regime that produces dramatic population cycles of a green filamentous alga (*Cladophora glomerata*). In most years, scouring winter floods remove most of the inhabitants—particularly armored herbivorous insects—from the river bottom. In the following spring, there are large algal blooms. Increased light and nutrients and the lack of invertebrate herbivores all allow *Cladophora* mats to grow profusely over the rocks, producing filaments up to 8 meters long. By midsummer, these mats detach from the rocks and cover large portions of the river, at which time midge larvae, which feed on the floating alga and use it to weave small

homes, increase in number. The midges are fed on by small fishes and damselfly larvae, which in turn are eaten by steelhead and roach (a trophic cascade with four trophic levels; **Figure 15.20A**). The steelhead and roach are able to decrease the size of the algal mats through the indirect positive effects of eating small fishes and damselfly larvae, which feed on midge larvae.

During drought years, however, and in rivers where flood control is operating, scouring of the river bottom does not occur. In these years, *Cladophora* persists, but does not form large, lush mats. Power and colleagues showed that this change was due to the presence of more armored herbivorous insects that were not removed by the floods, and which ate the *Cladophora* while it was still attached to the rocks. This led to chronic declines in *Cladophora* and the loss of the detachment phase of the alga. The armored insects are much less susceptible to predation than the midges and thus are not controlled by higher trophic levels. In essence, the typical food web with four trophic levels is converted to a two-level food web during drought years, and the steelhead and roach that once were keystone predators are now minor players in the food web (**Figure 15.20B**).

In upcoming chapters, we will consider in much more detail the role of disturbance, stress, and predation, as well as physical factors, on the outcome of species interactions and ultimately on species diversity.

"Killer Algae!"

The introduction of *Caulerpa taxifolia* to the Mediterranean Sea in the early 1980s set in motion a series of unfortunate events that resulted in large carpets of fluorescent green algae dominating formerly species-rich marine benthic communities. *Caulerpa* thrived because humans facilitated its dispersal and its physiological tolerance as a consequence of the aquarium trade. Even in the early stages of the invasion, Meinesz documented the seaweed in at least three types of communities, with different species compositions, on rocky, sandy, and muddy substrata. Together, these communities are home to several hundred species of algae and three marine flowering plants, as well as a number of animal species. Once *Caulerpa* arrived, native competitive or predator species were unable to deter the alga from spreading.

The invasion of *Caulerpa* has dramatically changed the ways in which the native species interact with one another, and thus the structure and function of the native communities. One obvious consequence of the presence of *Caulerpa* is the overgrowth and decline of seagrass meadows dominated by *Posidonia oceanica* (**Figure 15.21**). This seagrass has been likened to an "underwater tree" because of its long life and slow growth (patches grow to 3 meters in diameter in 100 years). Just like forests, seagrass meadows support a multitude of species that use the vegetation as habitat. Research showed that *Posidonia* and *Caulerpa* have different growth cycles: *Posidonia* loses blades in the summer, when *Caulerpa* is most productive. Over time, this asynchronous growth pattern results in *Caulerpa* overtopping the existing seagrasses and establishing itself as the dominant species. Additional research has shown that *Caulerpa* acts as an ecosystem engineer, accumulating sediments around its roots more readily than *Posidonia*, which dramatically changes the species composition of the small invertebrates that live on the seafloor. Surveys have revealed a significant drop in the numbers and sizes of fish that utilize the *Caulerpa*-invaded communities, suggesting that these habitats are no longer suitable for commercially important species.

Many of the changes to Mediterranean seagrass meadows, and species dependent on them, are difficult to predict, given the sheer number of species that are potentially affected by *Caulerpa*, the indirect effects that will be generated by changing interactions, and the relatively short time that has elapsed since the invasion began. A scientific approach, guided by a combination of theory and empirical data, will be necessary if future predictions are to be made about the ultimate effect of *Caulerpa* on this vanishing underwater community.

FIGURE 15.21 A Mediterranean Seagrass Meadow Native communities like this one, dominated by the seagrass *Posidonia oceanica*, are quickly overcome by invasive *Caulerpa*.

Stopping Invasions Requires Commitment

Even though it may be too late to stop the invasion of *Caulerpa* in the Mediterranean, the lessons learned there have been important in other regions of the world. In 2000, just as Meinesz was making progress in banning international trade of the alga, he received an e-mail from an environmental consultant in San Diego, California. While surveying eelgrass in a lagoon, she had noticed a large patch of what was later identified as *Caulerpa taxifolia*. Acting on Meinesz's recommendation, a team of scientists and managers from county, state, and federal agencies immediately assembled to design an eradication plan. This plan involved treating the alga with chlorine gas injected under tarps placed on top of algal patches. More than $1 million was initially budgeted for the project in 2000, but it eventually took 6 years and $7 million to eradicate the alga. The invasion was widely publicized, resulting in the discovery of another patch of *Caulerpa* in another lagoon near Los Angeles, which was also eradicated. The U.S. experience is a rare success story only because immediate action was taken by scientists, managers, and politicians to deal with the invasion before eradication became an ecological and fiscal impossibility. It is estimated that control of invasive species in the United States costs $22 billion per year (Pimentel et al. 2000).

To determine the origin of the *Caulerpa* that invaded California, molecular evidence was needed. Specimens of the alga were sent to geneticists at two universities, who quickly determined that the sequences of its ITS rDNA were identical to those from the Mediterranean, the Wilhelmina Zoo in Germany (where the strain was first cultivated), and many other public aquariums around the world (Jousson et al. 2000). Unfortunately, it is still unknown how the species was introduced into the two California lagoons, but hypotheses range from amateur aquarists cleaning

their tanks in the lagoons to an accidental release from aquariums onboard the Prince of Saudi Arabia's yacht, which was being repainted in San Diego at about the time the alga probably arrived. Through the use of DNA analysis, it has been determined that *Caulerpa* involved in subsequent invasions in Australia and Japan are genetically identical to the original German strain. The molecular evidence makes it clear that the trade of this alga in aquarium circles poses a global threat to nearshore temperate marine environments. Legislation is now in place to ban the "killer alga" from a number of other countries where it has a good chance of invading successfully.

SUMMARY

CONCEPT 15.1 Communities are groups of interacting species that occur together at the same place and time.

- Communities can be delineated by the characteristics of their physical environment or by biological characteristics such as the presence of common species.

- Ecologists use subsets of species to define and study communities because it is impractical to count all the species within a community, especially if they are small or undescribed species.

- Subsets of species used to study communities include taxonomic groups, guilds, functional groups, and food and interaction webs.

CONCEPT 15.2 Species diversity and species composition are important descriptors of community structure.

- Species diversity, the most commonly used measure of community structure, is a combination of the number of species (species richness) and the abundances of those species relative to one another (species evenness).

- Communities differ in the commonness or rarity of their species, such that even in communities with the same numbers of species, one community might have a few very abundant species and many rare species, while another might have species of equal abundance.

- Species richness estimates for communities improve

with increased sampling effort up to a certain point, at which additional samples reveal few or no new species.

- Species composition—the identity of the species present in a community—is an obvious but important characteristic of community structure that is not revealed in measures of species diversity.

CONCEPT 15.3 Communities can be characterized by complex networks of direct and indirect interactions that vary in strength and direction.

- Indirect species interactions, in which the relationship between two species is mediated by a third (or more) species, can have large effects on the outcomes of species interactions.

- Some species have a strong negative or positive effect on their communities, but others probably have little or no effect.

- Species that have strong effects by virtue of their large abundance and biomass are known as dominant species while those that have strong effects due to their interactions, rather than size are known as keystone species.

- Ecosystem engineers such as trees create, modify, or maintain physical habitat for themselves and other species.

- The environmental context of species interactions can modify them enough to change their strength or direction.

REVIEW QUESTIONS

1. What is the formal definition of a community? Why is incorporating species interactions important to the definition? What kinds of characteristics of communities can be used as a guide to define a community in practical terms?

2. Species diversity measurements take into account both species richness and species evenness. Why would

these measurements be preferred to species richness alone? What do rank abundance curves add to one's knowledge about community structure?

3. Species vary in their strength and function within communities. Species that interact strongly with other species include dominant species, keystone species, and ecosystem engineer species. Describe the differ-

ences between these three types of species. Can dominant and keystone species also be ecosystem engineers?

4. The following interaction web is common in the rocky intertidal zone of Washington and Oregon. The arrows and their signs (+ or –) represent interactions that occur between species in the rocky intertidal zone community.

a. Suppose seagulls were removed from this community. Determine the two direct effects this manipulation would have on limpets and gooseneck barnacles. Would it increase, decrease, or not change the abundances of those species, and through what type of interaction (i.e., by predation, competition, or facilitation)?

b. Determine the four indirect effects of removing seagulls from the community. How would each species respond (i.e., would it increase, decrease, or not change in abundance), and through what type of interaction (i.e., predation, competition, or facilitation)?

Seagulls

Limpet (*Lotia strigatella*) Limpet (*Lotia digitalis*) Gooseneck barnacles (*Pollicipes polymerus*) Mussels (*Mytilus edulis*)

Large seaweed Phytoplankton

HANDS-ON PROBLEM SOLVING

Measuring Marine Species Diversity | This Web exercise (www.sinauer.com/ecology) explores various methods of measuring species diversity. You will interpret data from a recent paper that measured marine species diversity, with emphasis on the benthic fauna of the continental shelf of Norway, using the Shannon index as well as other indices of diversity.

SUGGESTED READINGS

Jones, C. G., J. H. Lawton and M. Shachak. 1994. Organisms as ecosystem engineers. *Oikos* 69: 373–386. *The first paper, with many good examples, that describes the importance of organisms that create, modify, or maintain physical habitat for themselves and other species.*

Magurran, A. E. 2001. *Measuring Biological Diversity.* Blackwell Science Publishing. *This general and accessible book provides a comprehensive overview of how to measure biodiversity.*

Meinesz, A. 2001. *Killer Algae.* University of Chicago Press, Chicago. *A book that describes the ecological and political ramifications of a serious marine invasion.*

Pimentel, D., L. Lach, R. Zuniga and D. Morrison. 2000. Environmental and economic costs of nonindigenous species in the United States. *BioScience* 50: 53–65. *A paper that describes the surprising and hidden costs of non-native species introductions.*

Power, M. E. and 9 others. 1996. Challenges in the quest for keystones. *BioScience* 46: 609–620. *A thoughtful analysis of the importance of keystone species especially to conservation biology.*

Strauss, S. Y. 1991. Indirect effects in community ecology: Their definition, study and importance. *Trends in Ecology and Evolution.* 6: 206–210. *A paper that describes the diversity of indirect effects observed in nature.*

CASE STUDY

A Natural Experiment of Mountainous Proportions

The eruption of Mount St. Helens was a defining moment for ecologists interested in catastrophic change. The once frosty-topped mountain in the geologically active Cascade Range, located in the Pacific Northwest region of North America (**Figure 16.1**), once had a rich diversity of ecological communities. If you visited Mount St. Helens in the summer, you could have seen alpine meadows filled with colorful wildflowers and grazing elk. At lower elevations, you could have hiked across the cool fern- and moss-covered forest floor under massive old-growth trees. You could have swum in the blue, clear water of Spirit Lake, or fished along its shores. But a few minutes after 8:30 A.M. on May 18, 1980, all that was living on Mount St. Helens would be gone. On the north side of the mountain, a huge magma-filled bulge had been forming for months, and it gave way that morning in an explosive eruption and the largest avalanche in recorded history.

Photos of the eruption show that mud and rock flowed down the face of Mount St. Helens and was deposited tens of meters deep in some areas (**Figure 16.2**). The wave of debris that passed over Spirit Lake was 260 meters deep and increased the lake's elevation by 60 meters. The bulk of the avalanche traveled 23 km in about 10 minutes to the North Fork Toutle River, where it scoured the entire valley, from floor to rim, with material from the volcano and left a truly massive pile of tangled vegetation at its tail end. In addition to the avalanche, the blast produced a cloud of hot air that burned forests to ash near the mountain, blew down trees over a large area, and left dead but standing trees stretching for miles away from the mountain. Ash from the explosion blanketed forests, grasslands, and deserts located hundreds of kilometers away.

The destruction that ensued on that day created whole new habitats on Mount St. Helens, some of which were completely devoid of any living organisms. At one extreme, there was the Pumice Plain, a large, gently sloping moonscape of a place below the volcano that was pelted with hot, sterilizing pumice (see Figure 16.2). This harsh and geologically monotonic environment lacked life, and even organic matter, of any form. All life in Spirit Lake was extinguished, and huge amounts of woody debris were deposited there, some of which still floats on top of the lake today. But not surprisingly, given the large forests that had surrounded the mountain, the majority of the landscape now consisted of downed or denuded trees, covered with rock, gravel, and

FIGURE 16.1 Once a Peaceful Mountain Before the eruption shown here on May 18, 1980, Mount St. Helens, in southwestern Washington State, had a diversity of communities, including alpine meadows, old-growth forests, and lakes and streams.

FIGURE 16.2 A Transformed Mount St. Helens Organisms on Mount St. Helens were scorched, pounded by pumice, covered in mud, and blown down by the eruption. The eruption had different effects on the geology of the mountain at different locations, creating many new habitats. (After Dale et al. 2005.)

Blowdown zone

Scorch zone

5km

Debris avalanche

- ■ Lava dome
- ■ Crater walls
- ■ Mudflow zone
- ■ Pumice Plain
- ■ Debris avalanche
- ■ Blowdown zone
- ■ Scorch zone
- □ Lakes
- ⌒ Streams

Toutle River

Spirit Lake

Mount St. Helens

Pumice Plain

mud tens of meters deep in some places (see Figure 16.2). Compared with the Pumice Plain, these blowdown areas had some hope of biological legacy buried under the piles of trees and ash.

Shortly after the eruption, helicopters delivered the first scientists to the mountain to begin studying what was essentially a natural experiment of epic proportions.

A few lucky ecologists recorded the first observations of the sequence of biological changes that began soon after the eruption. Field excursions in the summers of 1980 and 1981 were organized, and valuable baseline data were collected. Now, nearly 30 years later, hundreds of ecologists have studied the reemergence of life on Mount St. Helens. For many, the experience has been life-changing, and their careers have been consumed by research on this fascinating study system. Much of what has been learned has been unexpected and has changed the way we view the recovery of communities and the persistence of life on Earth.

One constant that all ecologists can agree on is that communities are always changing. Some communities show more dynamism than others. For example, it is hard to imagine how deserts, with their large, stoic cacti, have changed much over time. This is espe-

cially true if you compare deserts with, for example, high mountain streams or rocky intertidal zones, where species are coming and going on a regular basis. But community change is relative, and there is no question that even deserts change, though at a much slower pace than we might realize on the basis of one visit, or even one ecological study. Unfortunately, we humans cannot deny that our actions are becoming one of the strongest forces of change in communities, and that we are taking those actions with an imperfect understanding of their consequences. In this chapter, we will consider the agents of change in communities, from subtle to catastrophic, and their effects on community structure over time.

Agents of Change

> **CONCEPT 16.1**
> Agents of change act on communities across multiple temporal and spatial scales.

Let's imagine for a moment that you have the ability to look back in time and follow the change in a typical coral reef community in the Indian Ocean (**Figure 16.3**). Over the last few decades, you might have seen considerable change, both subtle and catastrophic. Subtle changes might include the slow rise to dominance of certain coral species, and the slow decline of others, due to the effects of competition, predation, and disease. More catastrophic changes might include massive deaths of corals in the last decade due to coral bleaching events (i.e., loss of symbiotic algae) and the great tsunami of 2004, resulting in the replacement of some coral species with other species, or no replacement at all. Taken together, these changes make the community what it is today: a reef that has many fewer coral species than it did a few decades ago due to a combination of natural and anthropogenic (human-caused) agents of change.

Succession is the change in species composition in communities over time. Succession is the result of a variety of abiotic (physical and chemical) and biotic agents of change. In the following sections, we will consider the theory behind succession and examples that illustrate how it works in a variety of systems. But first, we need to identify and define the important agents of change that are responsible for driving succession.

Agents of change can be abiotic or biotic

Communities, and the species contained within them, change in response to a number of abiotic and biotic factors (**Table 16.1**). We have considered many of these factors in previous chapters. In Unit 1, we learned that abiotic factors, in the form of climate, soils, nutrients, and water, vary over daily, seasonal, decadal, and even 100,000-year time scales. This variation has important implications for community change. For example, in our Indian Ocean coral reef community (see Figure 16.3), unusually high water temperatures driven by large-scale climate change have been implicated in recent losses of symbiotic algae in corals, resulting in coral bleaching. If the symbiotic algae do not return, the corals will eventually die, thus creating the conditions for species replacement. Likewise, increases in sea level can decrease the light that reaches the corals. If light decreases beyond the physiological limits of some coral species, they could slowly be replaced by more tolerant species, or even by macroalgae. Because abiotic conditions are constantly changing, communities are doing the same, at a pace consistent with their environment.

FIGURE 16.3 Change Happens Coral reef communities in the Indian Ocean have experienced significant changes over the last few decades. The agents of change have been both subtle and catastrophic, natural and human-caused.

> Species interactions, such as competition, predation, and disease, can cause the gradual replacement of species over time.

> Changes in abiotic conditions, such as sea level and water temperature, can cause physiological stress, coral bleaching, and eventually mortality.

> Catastrophic disturbances, such as tsunamis, can cause massive injury and death in coral reefs.

TABLE 16.1

Examples of Abiotic and Biotic Agents of Change and Their Effects on Organisms

Agent of change	Effects	Habitats	Examples
ABIOTIC FACTORS			
Waves, currents	Organisms are detached, injured, or killed	Terrestrial, marine, freshwater	Storms, hurricanes, floods, tsunamis, ocean upwelling
Wind	Organisms are detached, injured, or killed	Terrestrial, marine, freshwater	Storms, hurricanes, wind-driven sediment scouring
Water supply	Organisms grow slowly, are injured, or are killed	Terrestrial, marine, freshwater	Droughts, floods, mudslides
Chemical composition	Organisms grow slowly, are injured, or killed	Terrestrial, marine, freshwater	Pollution, acid rain, high or low salinity, high or low nutrients
Temperature	Organisms grow slowly, are injured, or are killed	Terrestrial, marine, freshwater	Freezing, snow and ice, avalanches, excessive heat, fire, sea level rise or fall
Volcanic activity	Organisms are injured or killed	Terrestrial, marine	Lava, hot gas, mudslides, flying rocks and debris, floods
BIOTIC FACTORS			
Negative interactions	Organisms grow slowly, are injured, or are killed	Terrestrial, marine, freshwater	Competition, predation, herbivory, disease, parasitism
Positive interactions	Organisms grow rapidly, less injury and death	Terrestrial, marine, freshwater	Mutualisms, commensalisms

Source: Adapted, with additions, from Sousa 2001.

Abiotic agents of change can be placed into two categories, disturbance and stress, which differ in the effects they have on species. A **disturbance** is an abiotic event that injures or kills some individuals and creates opportunities for other individuals to grow or reproduce. In our coral reef example, the 2004 tsunami can be viewed as a disturbance because the force of water passing over the reef injured and killed many coral individuals. Examples of other disturbances are given in Table 16.1. **Stress**, on the other hand, occurs when some abiotic factor reduces the growth or reproduction of individuals and creates opportunities for other individuals. A stress in our coral reef might be the effect of water temperature or sea level rise on the growth and reproduction of corals. Examples of other stresses are included in Table 16.1. Both disturbance and stress are believed to play critical roles in driving succession.

How do biotic factors influence community change? In Unit 3, we saw that species interactions, both negative and positive, can result in the replacement of one species with another (see Table 16.1). In our coral reef (see Figure 16.3), change might be driven by competition between, for example, platelike corals and branched forms, with the platelike forms eventually dominating over time. Coral diseases are another example of species interactions that can initiate change in communities by causing particular species to grow

more slowly or eventually die. Equally common agents of change are the actions of ecosystem engineers or keystone species (see Figures 15.17 and 15.18). Both types of species have strong effects on other species that result in community change. Beavers, for example, act both as ecosystem engineers and as keystone species when they create dams in a forested riparian community (see Figure 15.19). Their initial effect is tree removal, which leads to the damming of the stream, flooding of the surrounding area, and the death of additional streamside vegetation. The resulting community is a wetland, created through the actions of the beavers and the physical modifications of the environment that stem from their actions.

Finally, it is important to realize that abiotic and biotic factors often interact to produce change in communities. We see this in the case of ecosystem engineers, where a single species causes changes in abiotic conditions that can cause species replacement. Beavers creating a wetland (see p. 336) is a perfect example. Similarly, abiotic factors such as wind, waves, or temperature can act by modifying species interactions, either positively or negatively, thus creating opportunities for other species. We saw examples of this kind of effect for sea palms in the rocky intertidal zone (see p. 253), plants in alpine regions (see Figure 14.10), and stream insects in northern California (see Figure 15.20).

Agents of change vary in their intensity, frequency, and extent

As you might guess, how often, and at what magnitude and areal extent, agents of change act largely determines the tempo of succession. For example, when the avalanche produced by Mount St. Helens ripped through the alpine community back in 1980, it produced a disturbance that was larger and more severe than any others that had occurred that year, that decade, or that century. The intensity of that disturbance—the amount of damage and death it caused—was huge, both because of the massive physical force involved and because of the area covered. In contrast, the frequency of the disturbance was low because these eruptive episodes are so rare (occurring once every few centuries). Extremely intense and infrequent events, such as the eruption of Mount St. Helens, are at the far end of the spectrum of disturbances species experience in communities (**Figure 16.4**). In this case, the entire community is affected, and recovery involves the reassembly of the community over time. We will consider the mechanisms of that reassembly in the following sections.

At the other end of the spectrum are weak and frequent disturbances that may have more subtle effects or affect a smaller area (see Figure 16.4). Prior to the eruption of Mount St. Helens, such disturbances might have included wind blowing down old trees living in the Douglas fir forests surrounding the mountain. These more frequent disturbances form patches of available resources that can be used by individuals of the same or different species. A mosaic of disturbed patches can promote species diversity in communities over time, but might not lead to much successional change. We will learn more about smaller disturbances and their effects on species diversity in Chapter 18.

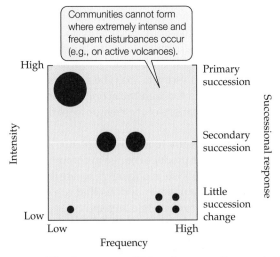

FIGURE 16.4 The Spectrum of Disturbance How much biomass is removed (its intensity) and how often it is removed (its frequency) can influence the amount of disturbance (represented by the red circles) that occurs and the type of succession that is possible afterward.

Let's turn our attention now from the agents of change to their consequences for community succession.

Basics of Succession

CONCEPT 16.2
Succession is the change in species composition over time as a result of abiotic and biotic agents of change.

At the most basic level, succession describes a process in which the species composition in a community changes over time. Mechanistically, it involves the colonization and extinction of species in a community due to abiotic and biotic agents of change. Even though studies of succession often focus on vegetative change, the roles of animals, fungi, bacteria, and other microbes are equally important.

Theoretically, succession progresses through various stages that include a **climax stage** (**Figure 16.5**). The climax is thought to be a stable end point that experiences little change until a particularly intense disturbance sends the community back to an earlier stage. As we will see in the following sections, there is some argument about whether succession can ever lead to a stable end point.

Ecologists recognize two types of succession that differ in their initial stage. The first type, **primary succession**, involves the colonization of habitats devoid of life (see Figure 16.5), either as a result of catastrophic disturbance (see Figure 16.4), as we see on the Pumice Plain at Mount St. Helens, or because they are newly created habitats, such as volcanic rock. As you can imagine, primary succession can be very slow because the first arrivals typically face extremely inhospitable conditions. Even the most basic resources needed to fuel life, such as soil, nutrients, and water, may be lacking. The first colonizers, then, tend to be species that are capable of withstanding great physiological stress and transforming the habitat in ways that benefit their further growth and expansion (and that of others, as we will see).

The other type of succession, known as **secondary succession**, involves the reestablishment of a community in which most, but not all, of the organisms have been destroyed (see Figure 16.5). Agents of change that can create such conditions include fire, hurricanes, logging, and herbivory. Despite the catastrophic effect of the eruption on Mount St. Helens, there were many areas where some organisms survived and constitute examples where secondary succession took place. As you might expect, the legacy of the preexisting species and their interactions with colonizing species play larger roles in the trajectory of succession than in primary succession.

The early history of ecology is a study of succession

The modern study of ecology had its beginnings at the turn of the twentieth century. At that time, it was dominated by

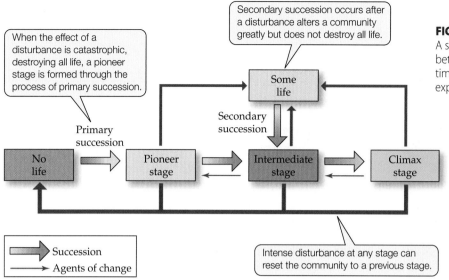

When the effect of a disturbance is catastrophic, destroying all life, a pioneer stage is formed through the process of primary succession.

Secondary succession occurs after a disturbance alters a community greatly but does not destroy all life.

FIGURE 16.5 A Theoretical Model of Succession
A simple model of succession involves transitions between stages driven by species replacements over time, which ultimately result in a climax stage that experiences little change.

Intense disturbance at any stage can reset the community to a previous stage.

where new sand was being deposited, were the youngest. As you walked from the lake to the back of the dune, he believed, you were traveling forward in time and able to imagine what the community would look like in centuries to come. The first stages were dominated by a hardy ecosystem engineer, American beach grass (*Ammophila breviligulata*). *Ammophila* (whose name literally means "sand lover") is excellent at trapping sand and creating hills that provide refuge on their leeward side for plants less tolerant of the constant burial and scouring experienced on the beachfront.

scientists fascinated with plant communities and the changes they undergo over time. One pioneer was Henry Chandler Cowles (1899), who studied the successional sequence of vegetation in sand dunes on the shore of Lake Michigan (**Figure 16.6**). In this system, the dune is growing with new sand deposition occurring at the shoreline. Cowles was able to infer the successional pattern of plant communities along a dune by assuming that the plant assemblages farthest from the lake's edge were the oldest and the ones nearest the lake,

Cowles made the assumption that the different plant assemblages—or "societies," as he called them—that he saw in different positions on a dune represented different successional stages. That assumption allowed him to predict how a community would change over time without actually waiting for the pattern to unfold, which would have taken decades to centuries. This assumption, known as the "space for time substitution" (Pickett 1989), is used frequently as a practical way to study communities over time scales that exceed the life span of an ecologist. It is assumed that time is the main factor causing communities to change and that unique conditions in particular locations are inconsequential. These are big assumptions, and they have fueled a current debate about the predictability of community dynamics over time. We will discuss this debate in more detail later in the chapter when we deal with alternative stable state theory.

Henry Cowles was not alone in his interest in plant succession. His peers included Frederick Clements and Henry Gleason, two men who had completely different and contentious views on the mechanisms driving succession (Kingsland 1991). Clements (1916), the first to write a formal book in 1907 on the new science of ecology, believed that plant communities were like "superorganisms," groups of species that worked together in a mutual effort toward some deterministic end. Succession was similar to the development of an organism, complete with a beginning (em-

Cowles assumed that the communities closest to the lake (dominated by *Ammophila breviligulata*) were early successional stages...

and that those farther back were later successional stages.

Front of dune (windward)

Back of dune (leeward)

Sand deposition

FIGURE 16.6 Space for Time Substitution   (A) Dune nearest the shoreline on Lake Michigan is covered in *Ammophila*. (B) When Henry Chandler Cowles studied succession on these dunes he assumed that the earliest successional stages occurred on the newly deposited sand at the front of the dune, and the later successional stages occurred at the back of the dune.

(A)

(B)

FIGURE 16.7 Elton's Trajectory of Pine Forest Succession
(A) Charles Elton at the age of 25, a year before the publication of his first book, *Animal Ecology* (1927). (B) Elton's book contained this diagram of succession in pine forests after logging. The successional trajectory differs depending on the moisture conditions in a particular area and can result in new pine woods or mixed woods.

bryonic stage), middle (adult stage), and end (death). Clements thought that each community had its own predictable life history and, if left undisturbed, ultimately reached a stable end point. This "climax community" was composed of species that dominated and persisted over many years and provided the type of stability that could potentially be maintained indefinitely.

Henry Gleason (1917) thought that viewing a community as an organism, with various interacting parts, ignored the responses of individual species to prevailing conditions. Communities were not the predictable and repeatable result of coordinated interactions among species, but rather the random product of fluctuating environmental conditions acting on individual species. Each community was the product of a particular place and time and thus unique in its own right.

Looking back, it is clear that Gleason and Clements had extreme views of succession. As we will see in Section 16.3, we can find elements of both theories in the results of studies that have accumulated over the last century. First, though, it is important to mention one last ecologist, Charles Elton (**Figure 16.7A**), whose perspective on succession was not only shaped by those of the botanists who came before him, but by his interest in animals. He wrote his first book, *Animal Ecology* (1927), in 3 months' time at the age of 26. The book addresses many important ideas in ecology, including succession. Elton believed that both organisms and the environment interact to shape the direction succession will take. He gave an example from pine forests in England that were being subjected to deforestation. After felling of the pines, the trajectory of succession varied depending on the mois-

ture conditions of the environment (**Figure 16.7B**). Wetter areas developed into sphagnum bogs, while drier areas developed into grass and rush marshes. Eventually, these communities converged into birch scrub, but then ultimately diverged into two types of forest. Through these observations, Elton emphasized that the only way to predict the trajectory of succession was to understand the biological and environmental context in which it occurred.

Elton's greatest contribution to the understanding of succession was his acknowledgment of the role of animals. Up to that point, most ecologists believed that plants drove succession, while animals were passive followers. Elton provided many examples showing how animals could create successional patterns by eating, dispersing, trampling, and destroying vegetation in ways that greatly affected the sequence and timing of succession. We will review some examples of animal-driven succession in Section 16.3, but it is clear that the observations and conclusions Elton made 80 years ago still hold today.

Multiple models of succession were stimulated by lack of scientific consensus

Fascination with the mechanisms responsible for succession, and attempts to integrate the controversial theories of Clements, Cowles, and Elton, led ecologists to use more scientifically rigorous methods to explore succession, including comprehensive reviews of the literature and manipulative experiments. Joseph Connell, along with his collaborator Ralph Slatyer (1977) surveyed the literature and proposed

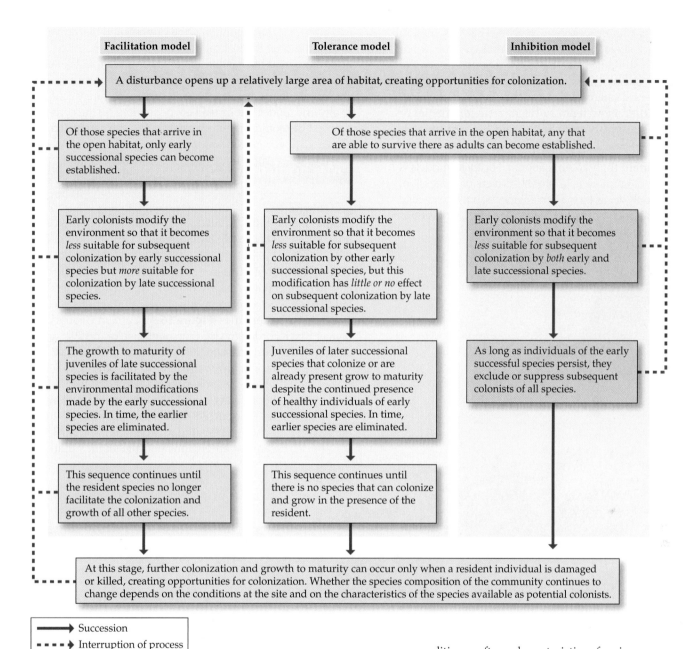

| Facilitation model | Tolerance model | Inhibition model |

A disturbance opens up a relatively large area of habitat, creating opportunities for colonization.

Facilitation model: Of those species that arrive in the open habitat, only early successional species can become established.

Tolerance & Inhibition model: Of those species that arrive in the open habitat, any that are able to survive there as adults can become established.

Facilitation: Early colonists modify the environment so that it becomes *less* suitable for subsequent colonization by early successional species but *more* suitable for colonization by late successional species.

Tolerance: Early colonists modify the environment so that it becomes *less* suitable for subsequent colonization by other early successional species, but this modification has *little or no* effect on subsequent colonization by late successional species.

Inhibition: Early colonists modify the environment so that it becomes *less* suitable for subsequent colonization by *both* early and late successional species.

Facilitation: The growth to maturity of juveniles of late successional species is facilitated by the environmental modifications made by the early successional species. In time, the earlier species are eliminated.

Tolerance: Juveniles of later successional species that colonize or are already present grow to maturity despite the continued presence of healthy individuals of early successional species. In time, earlier species are eliminated.

Inhibition: As long as individuals of the early successful species persist, they exclude or suppress subsequent colonists of all species.

Facilitation: This sequence continues until the resident species no longer facilitate the colonization and growth of all other species.

Tolerance: This sequence continues until there is no species that can colonize and grow in the presence of the resident.

At this stage, further colonization and growth to maturity can occur only when a resident individual is damaged or killed, creating opportunities for colonization. Whether the species composition of the community continues to change depends on the conditions at the site and on the characteristics of the species available as potential colonists.

⟶ Succession
┈┈▶ Interruption of process

FIGURE 16.8 Three Models of Succession Connell and Slatyer (1977) proposed three conceptual models—the facilitation, tolerance, and inhibition model—to describe succession. (After Connell and Slatyer 1977.)

three models of succession that they believed to be important (**Figure 16.8**):

■ The *facilitation model,* inspired by Clements, describes situations in which the earliest species modify the environment in positive ways that ultimately benefit later species, but hinder their own continued dominance. These early species are usually stress tolerant and are good at engineering the habitat and ameliorating the harsh physical conditions often characteristic of primary succession. Eventually, though, the sequence of species facilitations leads to a climax community composed of species that no longer facilitate other species and are displaced only by disturbances.

■ The *tolerance model* also assumes that the earliest species modify the environment, but in neutral ways that neither benefit nor inhibit later species. These early successional species have life history strategies that allow them to grow and reproduce quickly. Later species persist merely because they have life history strategies that allow them to tolerate environmental or biological stresses that would normally hinder early successional species.

■ The *inhibition model* assumes that early successional species modify conditions in negative or inhibitory ways that hinder later successional species. This suppression of the next stage of succession is broken only when stress or disturbance decreases the abundance of the inhibitory species. As in the tolerance model, later species persist merely because they have life history strategies that allow them to tolerate environmental or biological stresses that would normally hinder early successional species.

One important thing to note about all of these models is the inclusion of animals as important drivers of succession. Animals not only act as ecosystem engineers, but also change plants in ways that we discussed earlier, leading to transitions from one successional stage to another.

Thirty years have gone by since Connell and Slatyer wrote their influential theoretical paper on succession. Since that time, there have been a number of experimental tests of their three models. As we will see in the next section, these studies show that the mechanisms driving succession rarely conform to any one model, but instead are dependent on the community and the context in which experiments are conducted.

Mechanisms of Succession

CONCEPT 16.3
Experimental work on succession shows its mechanisms to be diverse and context-dependent.

One of the best-studied examples of primary succession occurs in Glacier Bay, Alaska, where the melting of glaciers has led to a sequence of community change that reflects succession over many centuries (**Figure 16.9**). Captain George Vancouver first recorded the location of glacial ice in 1794, while exploring the west coast of North America. Over the last 200 years, those glaciers have retreated up the bay, leaving behind bare, broken rock. John Muir, (1915) in his book "Travels in Alaska," first noted how much the glaciers had melted since Vancouver's time. When he visited Glacier Bay in 1879, he camped among ancient tree stumps that had

(A)

(B)

FIGURE 16.9 Glacial Retreat in Glacier Bay, Alaska Over more than 200 years, the melting of glaciers has exposed rock for community formation and succession. The oldest communities are at the mouth of the bay (A), and the youngest, pioneer communities are closest to existing glaciers (B). (A after Chapin et al. 1994.)

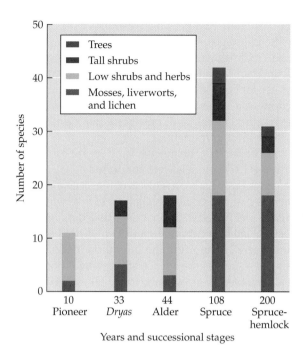

FIGURE 16.10 Successional Communities at Glacier Bay, Alaska Plant species richness has generally increased over the 200-year period of glacial retreat. It took a century for trees to colonize these communities. (Data from Reiners et al. 1971.)

est (the spruce stage) was in place, which fostered a diverse array of lichens, low shrubs, and herbs. Two hundred years after exposure, Reiners et al. (1971) documented that species richness decreased somewhat as Sitka spruce forests were transformed into forests of longer-living Western hemlocks.

The mechanisms underlying succession in this system have been studied extensively by F. Stuart Chapin and colleagues. Chapin et al. (1994) wondered, given the harsh physical conditions experienced by most species in the pioneer stage, whether the facilitation model could explain the pattern of succession observed by Cooper (1923b) and Reiners et al. (1971). First, they analyzed the soils of the different successional stages. They found significant changes in soil properties that were coincident with the increases in plant species richness (**Figure 16.11**). Not only were there increases in soil organic matter and soil moisture in later stages of succession, but nitrogen increased more than fivefold from the alder stage to the spruce stage. Chapin hypothesized that the assemblage

once been covered in ice and saw forests that had grown up in previously glaciated areas. He was impressed with the dynamic nature of the landscape and how the plant community responded to the changes.

Muir's book spiked the interest of William S. Cooper (1923a), who began his studies of Glacier Bay in 1915. A former student of Henry Chandler Cowles, Cooper saw Glacier Bay as an example of the "space for time substitution" so well documented by his advisor in the Michigan dunes. He established permanent plots (Cooper 1923b) that have allowed researchers to characterize the pattern of community change along the bay from Vancouver's time to today. This pattern is characterized by an increase in plant species richness and a change in plant species composition with time and distance from the melting ice front (**Figure 16.10**). In the first years after new habitat was exposed, a primary or **pioneer stage** developed, dominated by a few species that included lichens, mosses, horsetails, willows, and cottonwoods. Roughly 30 years after exposure, a second community developed, named the *Dryas* stage after the small shrub (*Dryas drummondii*) that dominates this community. In this stage, species richness increased, with willows, cottonwoods, alders (*Alnus sinuata*), and Sitka spruce (*Picea sitchensis*) sparsely distributed among the carpets of *Dryas*. After about 50 years (or some 20 km from the ice front), alders dominated, forming the third community, referred to as the alder stage. Finally, a century after glacial retreat, a mature Sitka spruce for-

FIGURE 16.11 Soil Properties Change with Succession Chapin and colleagues studied the properties of the soils in each of the four successional stages at Glacier Bay. (After Chapin et al. 1994.)

of species at each stage of succession was having effects on the physical environment that largely shaped the pattern of community formation. The question remained, however whether those effects were facilitative or inhibitory, and how they varied across the different successional stages.

To test their facilitation hypothesis, Chapin et al. (1994) conducted manipulative experiments. They added spruce seeds to each of the successional stages and observed their germination, growth, and survival over time. These experiments, along with observations of unmanipulated plots, showed that neighboring plants had both facilitative and inhibitory effects on the spruce seedlings, but that the directions and strengths of those effects varied with the stages of succession (**Figure 16.12**). For example, in the pioneer stage, spruce seedlings had a low germination rate, but a higher survival rate than in later successional stages. In the *Dryas* stage, spruce seedlings had weak germination and survival rates due to increases in seed predators, but those individuals that did survive grew better due to nitrogen fixing by bacteria associated with *Dryas*. In the alder stage, a further increase in nitrogen (alders also have nitrogen-fixing bacteria) and an increase in soil organic matter produced positive effects for spruce seedlings, but shading and seed predators led to overall poor germination and survival rates. In this stage, spruce seedlings that germinated before alders were able to

dominate experienced a net positive effect of the alders. Finally, in the last stage of succession, the effects of large spruce on spruce seedlings were mostly negative and long-lasting. Growth and survival rates were low due to competition with adult spruce for light, space, and nitrogen. Interestingly, seed production was enhanced, which led to relatively high seedling numbers merely as a consequence of the many more seeds available for germination.

Thus, in Glacier Bay, the mechanisms outlined in Connell and Slatyer's (1977) models were supported in at least some stages of succession. Early on, aspects of the facilitation model were seen as plants modified the habitat in positive ways for other plants and animals. Species such as alders had negative effects on later successional species unless they were able to colonize early, supporting the inhibition model. Finally, some stages—such as the spruce stage, where dominance was an artifact of slow growth and long life—were driven by life history characteristics, a signature of the tolerance model.

FIGURE 16.12 Both Positive and Negative Effects Influence Succession   The relative contributions of positive and negative effects on spruce seedling establishment changed across successional stages in Glacier Bay, Alaska. Positive effects outweighed negative effects in early and middle successional stages, but the opposite was seen in the latest stage. (After Chapin et al. 1994.)

In the pioneer stage, spruce seedlings had a low germination rate, but a higher survival rate than in later successional stages.

In the *Dryas* stage, seed predators kept germination and survival rates low, but a small increase in nitrogen had a positive effect on the survivors.

In the alder stage, shading and seed predators had negative effects, but increases in nitrogen and soil organic matter had positive effects on those seedlings that survived.

In the spruce stage, adult spruce had negative effects on spruce seedlings.

Successional stages	Pioneer	*Dryas*	Alder	Spruce
Positive effects	Higher survival rate	Higher nitrogen level Higher growth rate	Higher soil organic matter Higher nitrogen level More mycorrhizae Higher growth rate	Higher germination rate
Negative effects	Lower germination rate	Lower germination rate Lower survival rate Higher seed predation and mortality rate	Lower germination rate Lower survival rate Higher seed predation and mortality rate Root competition Light competition	Lower survival rate Higher seed predation and mortality rate Root competition Light competition Lower nitrogen level

Studies of succession show that no one model fits any one community

What do other studies show with regard to the three models of Connell and Slatyer? Mark Bertness and Scott Shumway studied the relative importance of facilitative versus inhibitory interactions in controlling secondary succession in a New England salt marsh. Salt marshes are characterized by different species compositions and physical conditions at different tidal elevations. The shoreline border of the marsh is dominated by the cordgrass *Spartina patens*, whereas dense stands of the black rush *Juncus gerardii* are found at the terrestrial border. A common natural disturbance in salt marsh habitats is the deposition of tidally transported dead plant material known as wrack (**Figure 16.13**). The wrack smothers and kills plants, creating bare patches where secondary succession takes place. These bare patches are places of high soil salinity because, without shading by plants, water evaporation increases, leaving behind salt deposits. Patches are initially colonized by the spike grass *Distichlis spicata*, an early successional species that is eventually outcompeted by both *Spartina* and *Juncus* in their respective zones.

Bertness and Shumway (1993) hypothesized that *Distichlis* could either facilitate or inhibit the later colonization of *Spartina* or *Juncus* depending on the salt stress experienced by the plants. To test this idea, they created bare patches in two zones of the marsh and manipulated plant interactions

FIGURE 16.13 Wrack Creates Bare Patches in Salt Marshes A tidal deposit of wrack at Rumstick Cove, Rhode Island, where Shumway and Bertness conducted their research on secondary succession. This dead plant material smothers living plants, creating bare patches with high soil salinity.

shortly after the patches had been colonized (**Figure 16.14**). In the low intertidal zone (the *Spartina* zone, close to the shoreline), they removed *Distichlis* from half the newly colonized patches, leaving *Spartina*, and removed *Spartina* from

(A) Low intertidal (*Spartina patens*) zone

In all the treatments in which *Spartina* was present, it inhibited colonization by *Distichlis*.

(B) Middle intertidal (*Juncus gerardii*) zone

Under high salt stress conditions, *Distichlis* facilitated colonization by *Juncus*.

Under low salt stress conditions, *Juncus* outcompeted *Distichlis*.

■ Neighbors removed
■ Neighbors present

FIGURE 16.14 New England Salt Marsh Succession Is Context-Dependent The trajectory of succession in salt marshes depends on soil salinity and the physiological tolerances of plant species. The kinds of interactions observed differed between the low intertidal zone (A) and the middle intertidal zone (B). (Data from Bertness and Shumway 1993.)

the other half, leaving *Distichlis*. In the middle intertidal zone (the *Juncus* zone, close to the terrestrial border of the marsh), they performed similar manipulations, but *Juncus* and *Distichlis* were the target species. Control patches, in which the colonization process was not manipulated, were maintained in both zones. Then, to manipulate salt stress, they watered half the patches in each treatment group with fresh water to alleviate salt stress, and left half as controls.

After observing the patches for 2 years, Bertness and Shumway (1993) found that the mechanisms of succession differed depending on the level of salt stress experienced by the plants and the species interactions involved. In the low intertidal zone, *Spartina* always colonized and dominated the plots, whether or not *Distichlis* was present or watering occurred (Figure 16.14A). *Distichlis* was able to dominate only if *Spartina* was removed from the plots, so it was clearly inhibited by *Spartina*, the competitive dominant. In the middle intertidal zone, *Juncus* was able to colonize only if *Distichlis* was present or watering occurred (Figure 16.14B). Measurements of soil salinity confirmed that the presence of *Distichlis* helped to shade the soil surface, thus decreasing salt accumulation and reducing stress for *Juncus*. *Distichlis*, however, was able to colonize plots with *Juncus* only when salt stress was high; that is, under the control conditions. If plots were watered, *Distichlis* was easily outcompeted by *Juncus*.

The experimental manipulations of Bertness and Shumway (1993) confirmed that the mechanisms important to succession are context-dependent. No single model is sufficient to explain the underlying causes of succession. In the middle intertidal zone, facilitation by *Distichlis* was a key interaction for the colonization of *Juncus*. Once this facilitation occurred, the interaction was tipped in favor of *Juncus*, which outcompeted *Distichlis* (see Figure 16.14B). In the low intertidal zone, *Distichlis* and *Spartina* were equally able to colonize and grow in salty patches. If *Spartina* arrived first, one would expect it to inhibit *Distichlis* colonization. If *Distichlis* arrived first, one would expect it to persist only if *Spartina* did not arrive and displace it (see Figure 16.14A).

Our final examples come from a community where succession has been studied extensively: that of the rocky intertidal zone. Here, disturbances are created mainly by waves, which can rip out organisms during large storm events or propel objects such as logs or boulders into them. In addition, stresses caused by low tides that expose organisms to high or low air temperatures can easily kill organisms or cause them to lose their attachment to the rocks. The resulting bare rock patches then become active areas of colonization and succession. Some of the first experimental work on succession in the rocky intertidal zone was done on boulder fields in southern California by Wayne Sousa, a graduate student at the time. Sousa (1979) noticed that the algae-dominated communities on these boulders experienced disturbance every time the boulders were overturned by waves. When he cleared some patches on the boulders and followed community formation in those patches over time, he found that the first species to colonize and dominate a patch was always the bright green seaweed, *Ulva lactuca* (**Figure 16.15A**). It was followed by the perennial red seaweed species, *Gigartina canaliculata*. To understand the mechanisms controlling this successional sequence, Sousa

FIGURE 16.15 Algal Succession on Southern California Boulders Is Driven by Inhibition   (A) The successional sequence of algae in bare patches on boulder fields in the intertidal zone of southern California. (B) Sousa performed removal experiments on concrete blocks to understand the mechanisms of succession in this system. (Data from Sousa 1979b.)

performed removal experiments on concrete blocks that he had allowed *Ulva* to colonize. He found that colonization by *Gigartina* could be accelerated if *Ulva* was removed (**Figure 16.15B**). This result favored inhibition as the main mechanism controlling succession, but a question remained: if *Ulva* is able to inhibit other seaweed species, why doesn't it always dominate? Through a series of further experiments, Sousa (1979) found that grazing crabs preferentially fed on *Ulva*, thus initiating a transition from the early *Ulva* stage to the mid-successional red algal species. In turn, the mid-successional species were more susceptible to stress and epiphytes (parasitic algae) than the late successional *Gigartina*. *Gigartina* dominated because it was the least susceptible to stress or consumer pressures.

This view of rocky intertidal succession as driven by inhibition was the accepted paradigm for many years. Facilitation and tolerance were thought to be much less important in a system where competition for space was the main driving factor. More recent work by Terence Farrell (1991) and others (e.g., Berlow 1997) has demonstrated that the relative importance of inhibition is probably much more context-dependent than previously thought. In the more productive rocky intertidal zone of the Oregon coast, the communities include many more sessile invertebrates, such as barnacles and mussels, than Sousa's communities of the southern California coast, where macroalgae dominate. In the high intertidal zone of Oregon, Farrell (1991) found that the first colonizer of bare patches was a barnacle, *Chthamalus dalli*, which was replaced by another, larger barnacle species, *Balanus glandula*, which was then replaced by three species of macroalgae, *Pelvetiopsis limitata*, *Fucus gardneri*, and *Endocladia muricata*. A series of removal experiments showed that *Chthamalus* did not inhibit colonization by *Balanus*, but that *Balanus* was able to outcompete *Chthamalus* over time, thus supporting the tolerance model. Likewise, *Balanus* did not hinder macroalgal colonization, but in fact facilitated it, lending credibility to the facilitation model.

But why and how would *Balanus* facilitate macroalgal colonization? Farrell suspected that *Balanus* provided protection for the algae in some way, possibly from desiccation stress or grazing by limpets (a common herbivorous gastropod). To test this idea, Farrell created experimental plots from which *Balanus*, limpets, or both were removed, then observed macroalgal colonization in those plots. He found that macroalgae colonized all of the plots without limpets, but had a much higher density in the plots with barnacles than in those without barnacles (**Figure 16.16A**). These results suggested that *Balanus* did indeed act to impede limpets from grazing on newly settled macroalgal sporelings.

You might be asking yourself, why doesn't *Chthamalus* have the same facilitative effect on macroalgae that *Balanus* does? Ferrell suspected that the reason was *Balanus*'s larger size (it is nearly three times wider than *Chthamalus*). By using plaster casts to mimic barnacles that were slightly larger than

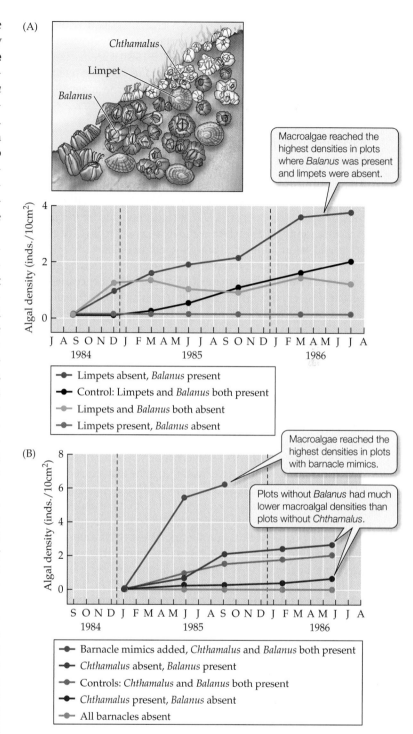

FIGURE 16.16 Algal Succession on the Oregon Coast Is Driven by Facilitation (A) Changes in macroalgal densities were measured in plots from which *Balanus* barnacles, limpets, or both had been removed. The results suggested that *Balanus* facilitates macroalgae by regulating limpet grazing. (B) To understand the mechanisms of facilitation, large barnacle mimics were added to some plots, which were compared with plots from which the real barnacle species—*Balanus*, *Chthamalus*, or both—had been removed. The results suggested that the larger the barnacle species, the better it protects macroalgae against limpet grazing and desiccation. (Data from Farrell 1991.)

Balanus, Ferrell found that these barnacle mimics had a more positive effect on macroalgal colonization than smaller-sized live barnacles of either species (**Figure 16.16B**). It seems likely that the smaller and smoother *Chthamalus* does not retain as much moisture, or block as many limpets, as the larger and more sculpted *Balanus*—or the mimics, for that matter.

Experiments show facilitation to be important in early stages

A number of experimental studies like the ones we have just described, initially stimulated by Connell and Slatyer's thought-provoking paper (1977), suggest that succession in any community is driven by a complex array of positive, negative, and neutral mechanisms. No one model fits any one community; instead, each community is characterized by elements of all the models. In most successional sequences, especially those that have a physically stressful pioneer stage, facilitative interactions are important drivers of early succession. Organisms that can tolerate and modify these physically challenging environments will thrive and facilitate other organisms that lack those capabilities. As succession progresses, slow-growing and long-lived species begin to dominate. Those species tend to be larger and more competitively dominant than early successional species. For this reason, one might expect competition to play a more dominant role than facilitation later in succession.

As succession proceeds, species richness typically increases (see Figure 16.10); thus, we must recognize that a vast array of both positive and negative interactions are operating in mid- to late successional stages. We will learn more about the mechanisms responsible for controlling species diversity in Chapter 18, but let's turn our attention next to the question of whether succession always takes one predictable path, as Clements believed, or whether other paths are possible.

Alternative Stable States

CONCEPT 16.4
Communities can follow different successional paths and display alternative states.

Up to this point, we have assumed that the trajectory of succession will be repeatable and predictable. But what if, for example, a boulder in the rocky intertidal zone of southern California turns over and, instead of a seaweed community forming, as Sousa (1979) observed, a sessile invertebrate community forms instead? Or what if *Dryas* never colonizes the till left behind by a glacier at Glacier Bay, but is replaced by a grass that competes with, rather than facilitates, later successional species such as Sitka spruce? Might spruce forests never develop? Possibly. There are cases in which different communities develop in the same area under similar envi-

ronmental conditions. Ecologists refer to such alternative scenarios as **alternative stable states**. Richard Lewontin (1969) was one of the first to formally define and model alternative stable states in natural communities. A community is thought to be **stable** when it returns to its original state after some perturbation.

How stable are communities? This is a question that has perplexed ecologists for some time, partially because the notion of stability depends on variability in space and time. At a small spatial scale, such as a 1 m^2 plot in a midwestern prairie, there might be considerable change or instability over time. If the plants were removed from the plot, it is unlikely that all the same species would recolonize that particular plot, and certainly not in the exact same locations. Similarly, if one followed the plot for a short time, the chance that its species composition would change would be low. But the more time that elapsed, the more likely it would be that the community would change and thus appear unstable. With these caveats in mind, let's take a closer look at examples of communities that, once disturbed, do not revert to previous states, but instead show alternative stable states.

Alternative states are controlled by strongly interacting species

John Sutherland (1974) studied the marine fouling communities (sponges, hydroids, tunicates, and other invertebrates) that encrust ships, docks, and other hard surfaces in bays and estuaries. He suspended ceramic tiles in the early spring from the dock at Duke Marine Laboratory in Beaufort, North Carolina, and allowed them to be colonized by invertebrate larvae that passively floated by (**Figure 16.17A**). At the end of 2 years, even though a handful of species had colonized, most of the tiles were dominated by *Styela*, a solitary tunicate species. The dominance of *Styela* was not universal, however. It actually declined on Sutherland's tiles during the first winter and was replaced by the hydroid species *Tubularia*. This effect was due to the annual nature of *Styela*, which dies off in winter, and it quickly regained dominance in the next spring, when larvae started to settle.

By placing new tiles out periodically, Sutherland also concluded that *Styela* was able to persist despite the existence of other potential colonizers. These colonizers fouled the new tiles, but were unable to colonize those dominated by *Styela*. As such, Sutherland viewed this fouling community as stable. Within a few months, Sutherland also identified what he believed to be another stable fouling community, this one dominated by *Schizoporella*, an encrusting bryozoan (see Figure 16.17A). This community developed on new tiles suspended from the dock in late summer and was also impervious to colonization by other species, including *Styela*.

To understand what might be controlling these two alternative outcomes of succession, Sutherland (1974) submerged new tiles at the same spot on the dock, but excluded fish

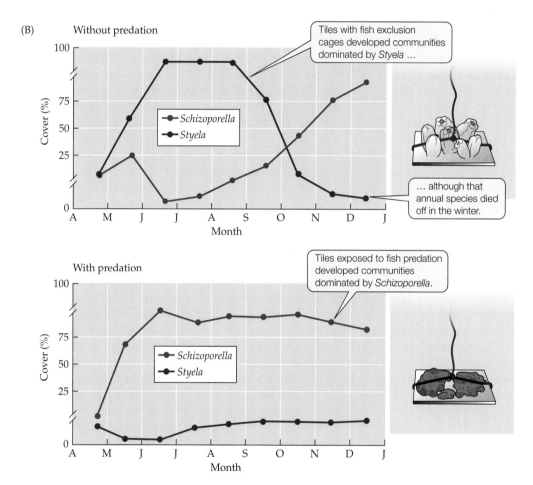

FIGURE 16.17 Fouling Communities Show Alternative States John Sutherland studied fouling communities by suspending ceramic tiles from a dock in North Carolina and allowing invertebrates to colonize them. (A) Two types of communities developed on the tiles over time, one dominated by *Styela* and another by *Schizoporella*. (B) Different communities developed on the tiles depending on whether they were protected from fish predation. (After Sutherland 1974.)

predators from half of the tiles by surrounding them with cages (**Figure 16.17B**). After a year, Sutherland found that the tiles protected from fish predation had formed communities dominated by *Styela*, while those exposed to fish predation had formed communities dominated by *Schizoporella*. He also noticed that the abundances of both species in the predator exclusion cages were reversed when *Styela* began to die off in the winter. These results suggested that *Styela* is competitively dominant if left undisturbed, but is outcompeted by *Schizoporella* when disturbed. Sutherland explained his original observations of *Styela* dominance by suggesting

that fish predation was spotty and that the tunicates themselves, once they reached a certain large size, might have acted as a natural "cage" or predator exclusion mechanism.

Lewontin (1969) and Sutherland (1974) believed that multiple stable states existed in communities and could be driven by the addition or exclusion of particularly strongly interacting species. If those species were missing or ineffective, communities could follow alternative successional trajectories that might never lead back to the original community type, but might instead form a new community type. We can visualize the theory behind alternative stable states by imag-

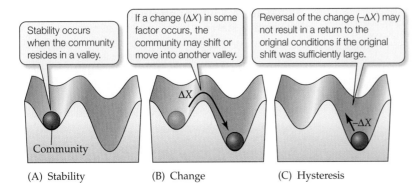

Stability occurs when the community resides in a valley.

If a change (ΔX) in some factor occurs, the community may shift or move into another valley.

Reversal of the change (−ΔX) may not result in a return to the original conditions if the original shift was sufficiently large.

Community

(A) Stability

(B) Change

(C) Hysteresis

FIGURE 16.18 A Model of Alternative Stable States   (A) A community is represented by a ball that moves along a landscape of community types (valleys). (B) Notice that the valleys can be shallower or deeper, suggesting the magnitude of change needed to shift the community from one state to another. (C) Reversal of the change may not be possible if the original change is too large. (After Beisner et al. 2003.)

ining a topographic surface where the valleys represent different community types and a ball represents a community (**Figure 16.18A**). The ball can move from one valley to another, depending on the presence or absence of strongly interacting species and how they shape the community (**Figure 16.18B**). For example, it might take only a slight change in one or more dominant species to force the ball into a new valley, or it might require complete removal of a species to cause this change. If we use Sutherland's (1974) work as an example, we can think of the *Styela* and *Schizoporella* communities as two different valleys. Whether the ball resides in the *Schizoporella* or *Styela* valley depends on the presence of fish predators. Interestingly, in this system, the ball might not simply move back to the *Schizoporella* valley if access by fish predators is restored (**Figure 16.18C**). As Sutherland noted, *Styela* was able to escape predation once it reached a certain large size. Thus, this system might show **hysteresis**, or an inability to shift back to the original community type, even when the original conditions are restored.

Connell and Sousa (1983) were skeptical that Sutherland (1974) had demonstrated the existence of alternative stable states. First, they thought the tile communities did not persist long enough, or have a spatial scale large enough, to be considered stable. If the tiles could be followed over multiple years, they asked, would they not all end up being dominated by one or the other species? In addition, they wondered whether the fouling communities could have been sustained outside of an experimental setting in which predators were removed. Finally, although it was not a criticism of Sutherland's (1974) study per se, they believed that alternative stable states could be driven only by species interactions and not by physical changes in the community. They argued that many of the examples Sutherland (1974) had used to bolster the importance of alternative stable states fell into the latter category. Their requirement that the physical environment not change is problematic because it excludes as drivers of succession all species that interact with other species by modifying their physical environment—that is, all ecosystem engineers. We know that ecosystem engineers can have strong effects on communities, so excluding them is unrealistic to most ecologists.

Human actions have caused communities to shift to alternative states

The stringent requirements suggested by Connell and Sousa (1983) had the effect of delaying alternative stable state research for two decades. Recently, however, there has been renewed interest in alternative stable states, spurred by the increasing evidence that human activities, such as habitat destruction, non-native species introductions, and overharvesting of wild species, are shifting communities to alternative states. Some examples we have already seen include the change from kelp forests to urchin barrens due to the decline of sea otters (see Chapter 8), the crash of the anchovy fishery in the Black Sea due to the introduction of *Mnemiopsis* (see Chapter 10), and the invasion of the aquarium strain of *Caulerpa taxifolia* in the Mediterranean, Australia, Japan, and North America (see Chapter 15). These so-called regime shifts are caused by the removal or addition of key species that maintain one community type over others. Once communities have been "manipulated" by human activities and a new regime is in place, ecologists are uncertain whether the results can be reversed. Will the reintroduction of sea otters rejuvenate kelp forests? Will the cessation of nutrient enrichment in the Black Sea revitalize the anchovy fishery? And will the removal of *Caulerpa* restore seagrass meadows? These are all questions whose answers may be found in a better understanding of the factors that drive alternative stable states and the role restoration of the original conditions can play in reversing their effects.

CASE STUDY REVISITED

A Natural Experiment of Mountainous Proportions

On the twentieth anniversary of the eruption of Mount St. Helens, in 2000, a group of ecologists gathered on the once smoking and ash-covered volcano to participate in a week-long field camp. They gathered their gear together, including tape measures, quadrat frames, and maps, and visited the same sites they had explored two decades earlier. This

TABLE 16.2

Surviving Organisms Found on Mount St. Helens within a Few Years after the Eruption

Disturbance zone	Mean vegetation cover (%)	Average number of plant species/m²	Animal species					
			Small mammals	Large mammals	Birds	Lake fish	Amphibians	Reptiles
Pumice Plain	0.0	0.0	0	0	0	0	0	0
Mudflow zone (central flow path)	0.0	0.0	0	0	0	—	0	0
Blowdown zone			8	0	0	4	11	1
Pre-eruption clear-cut	3.8	0.0050						
Forest without snow	0.06	0.0021						
Forest with snow	3.3	0.0064						
Scorch zone	0.4	0.0039		0	0	2	12	1

Source: Adapted from Crisafulli et al. 2005.

visit, termed a "pulse," was an opportunity to establish a 20-year benchmark of data comparable to those first collected in 1980 and 1981. Many of the participants had spent the past 20 years —for some their entire careers—studying recolonization and succession patterns in those once-devastated landscapes. When they departed, they agreed to write a book, the chapters of which would contain all that was known about the extraordinary ecology of this system, with the hope that young ecologists would be motivated to continue the research and carry on their legacy. The book, *Ecological Responses to the 1980 Eruption of Mount St. Helens* (Dale et al. 2005) was published 5 years later.

What does the book tell us about succession on Mount St. Helens? First, the eruption created disturbances that varied in their effects depending on the distance from the volcano and the habitat type (e.g., aquatic versus terrestrial) (**Table 16.2**). Although areas close to the summit, such as the Pumice Plain, were literally sterilized by the heat of the eruption, ecologists were surprised to discover how many species actually survived on the mountain. Because the eruption occurred in the spring, many species were still dormant under the winter snows. Survivors included plants with underground buds or rhizomes, animals such as rodents and insects with burrows, and fish and other aquatic species in ice-covered lakes. In the blowdown zone, large trees and animals perished while smaller organisms survived in the protection of their larger neighbors. The opposite was true in areas outside the blowdown zone, where falling rocks and ash smothered smaller plants and animals, but not larger organisms.

A second important research discovery from Mount St. Helens is the role survivors have played in controlling the pace and pattern of succession. In many cases, these species were thrust into novel physical habitats and species assemblages without time to adapt over evolutionary time scales. Some species thrived, while others fared poorly, but their adaptability and unpredictability were surprising. Unlikely alliances were formed that hastened succession in particular habitats. For example, newly formed and isolated ponds and lakes were colonized by amphibians much faster than was thought possible (**Figure 16.19**). Scientists discovered that frogs and salamanders were using tunnels created by northern pocket gophers (*Thomomys talpoides*) to make their way from one pond to another

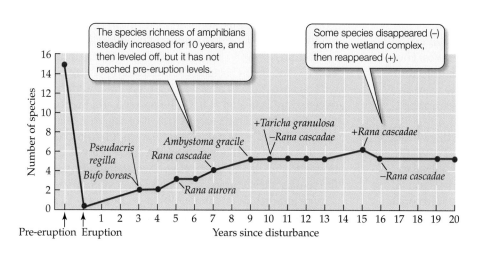

FIGURE 16.19 Rapid Amphibian Colonization   Frog and salamander species rapidly colonized a wetland complex in the Pumice Plain on Mount St. Helens. (Data from Crisafulli et al. 2005.)

FIGURE 16.20 Pocket Gophers to the Rescue The burrowing activity of northern pocket gophers, some of which survived the eruption underground, brought organic soil, seeds, and fungal spores to the surface, creating microhabitats like this one in the Pumice Plain, where plants could grow.

across the arid landscape (Crisafulli et al. 2005). Gophers were particularly successful on Mount St. Helens, both because they survived the eruption in their tunnels and because grassy meadows, their preferred habitat, greatly expanded after the eruption. Interestingly, gophers were also responsible for facilitating plant succession; their burrowing activity brought to the surface organic soil, seeds, and fungal spores buried deep under the volcanic rock and ash (Crisafulli et al. 2005; **Figure 16.20**).

A third important discovery was the realization that multiple mechanisms were responsible for primary succession on Mount St. Helens. Facilitation on the Pumice Plain was exemplified by dwarf lupines (*Lupinus lepidus*), which trapped seeds and detritus and increased soil nitrogen content though their association with nitrogen-fixing bacteria (del Moral et al. 2005). The lupines, in turn, were inhibited by multiple insect herbivores, which essentially controlled the pace of primary succession (Bishop et al. 2005). Tolerance was evident in some primary successional habitats, where Douglas fir lived in concert with annual herbs. The diversity of strategies species used, and the resulting communities, never ceased to amaze ecologists, who were guided, up to that point, mostly by the models of Connell and Slatyer (1977).

Despite 30 years of data and a trove of novel discoveries, research on Mount St. Helens has only just begun. Will the communities there follow paths of succession that lead to predictable and repeatable outcomes? Or will they form alternative states that are highly dependent on their historical legacies? Geological work suggests that Mount St. Helens erupts roughly every 300 years. The life span of community succession thus greatly exceeds our own life span by hundreds of years. As such, we must be content with the limited knowledge we have gained from studying what is arguably the most interesting phase of succession on Mount St. Helens, and with the hope that ecologists will continue the strong research tradition for years to come.

CONNECTIONS IN NATURE

Primary Succession and Nitrogen-Fixing Bacteria

All the examples of primary succession in terrestrial environments presented in this chapter involve plants with symbiotic nitrogen-fixing bacteria. These bacteria form nodules in the roots of their plant hosts, where they convert nitrogen gas from the atmosphere (N_2) into a form that is usable by plants (ammonia; NH_4^+). The plants provide the bacteria with sugars produced by photosynthesis. This interaction appears to be extremely important for plants and animals colonizing completely sterile environments. We saw that *Dryas* and alders, both species that form tight mutualisms with bacteria, were some of the first species to colonize the till left behind by glaciers at Glacier Bay, Alaska. Similarly, the legume *Lupinus lepidus* was able to use the nitrogen produced by its bacterial symbionts to colonize the sterile Pumice Plain of Mount St. Helens after the eruption. Lupines were the major source of nitrogen for subsequent plants and herbivorous insects for many years. Thus, lupines and their symbiotic bacteria play a large role in controlling the rate of primary succession on Mount St. Helens.

The nitrogen-fixing bacteria involved in symbioses are extremely diverse. Only a couple of groups of bacteria live in root nodules formed by the plant; all the rest are associated either with the surfaces of roots or the guts of ruminants. The nodule-forming bacteria include the Rhizobia, a taxonomic group associated with legumes, and the Actinomycetes of the genus *Frankia*, which are associated with woody plants such as alders and gale (*Myrica*). Nodule formation involves a complex series of chemical and cellular interactions between the root and the bacteria (**Figure 16.21**). Free-living bacteria are attracted to root exudates, which cause them to attach to the roots and multiply. Sets of genes are activated in both the bacteria and the root cells that allow the bacteria to enter the root, the root cells to divide, and the nodule to be formed. A vascular system develops that supplies sugars to the bacteria and carries fixed nitrogen to the plant. The enzymes involved in nitrogen fixing (known as nitrogenases) are highly sensitive to oxygen and require anaerobic conditions. Wherever nitrogen-fixing symbioses occur, there is some structural component to the interaction, such as membranes within the root nodule, that

FIGURE 16.21 Dwarf Lupines and Nitrogen-fixing Bacteria (A) Dwarf lupines, *Lupinus lepidus*, are an example of a legume with nitrogen-fixing bacteria, and were the first colonizers on Mount St. Helens. (B) Root nodule development is controlled by a strong interaction between the plant and bacteria.

(A)

(B)

Cortical cells   Root hair

Rhizobia

Free living bacteria are attracted to and infect the root.

Developing bacteriod region

A special vascular system develops in the root, supplying food for the bacteria in the nodules and carrying fixed nitrogen products to the plant.

Infection thread

Vascular system   Root hairs

Cells of the nodule primordium now infected and differentiating.

Root tip

Nodule

The bacteria form a nodule by actively dividing within the confines of the root membranes.

Bacteroids

produces anaerobic conditions. The bacteria, however, need oxygen to metabolize, and so a hemoglobin protein known as leghemoglobin, that has a high oxygen affinity, is produced in the nodules to deliver oxygen to the bacteria in an essentially anaerobic environment. Nodules often have an eerie pink color that is associated with the leghemoglobin.

Maintaining a symbiosis with nitrogen-fixing bacteria is costly to plants. Estimates suggest that creating and maintaining the nodules alone costs a plant 12–25% of its total photosynthetic output. Plants may be able to shoulder this cost, especially if it allows them to live in environments free of competitors and herbivores. But as they increase the nitrogen in the soils of their isolated environments, they also make conditions better for other species that are likely to be competitors. Thus they face a trade-off between improving the environment for themselves versus other species, which makes their role in early successional environments important, but also somewhat ironic.

SUMMARY

CONCEPT 16.1 Agents of change act on communities across multiple temporal and spatial scales.

■ Agents of change include both abiotic (physical and chemical) and biotic factors.

■ Abiotic agents of change can act as disturbances (injuring or killing organisms) or stresses (reducing the growth or reproduction of organisms).

■ Biotic agents of change include both positive and negative species interactions. Ecosystem engineers and keystone species are common agents of change.

■ Agents of change vary in their intensity, frequency, and areal extent.

CONCEPT 16.2 Succession is the change in species composition over time as a result of abiotic and biotic agents of change.

■ Theoretically, succession involves a series of stages that include a stable end point, or climax stage.

■ Primary succession involves the colonization of habitats that are devoid of life.

■ Secondary succession involves the reestablishment of a community in which most, but not all, of the organisms have been destroyed.

- Early ecologists were fascinated with succession, but disagreed about whether it proceeded in deterministic or indeterministic ways.

- Connell and Slatyer proposed three models of succession in 1977, known as the facilitation model, tolerance model, and inhibition model.

CONCEPT 16.3 Experimental work on succession shows its mechanisms to be diverse and context-dependent.

- Multiple studies of succession have shown that no one model fits any one community. Aspects of the facilitation, tolerance, and inhibition models can be seen in almost all systems studied.

- Generally, experiments show that facilitation tends to be important in early stages of succession and competition in later stages of succession.

CONCEPT 16.4 Communities can follow different successional paths and display alternative states.

- Alternative stable states occur when different communities develop in the same area under similar environmental conditions.

- In communities that experience alternative states, succession is typically controlled by strongly interacting species.

- Human activities have caused "regime shifts" in communities that may or may not be reversible.

REVIEW QUESTIONS

1. List some abiotic and biotic agents of change in communities. Describe the intensities and frequencies at which they are likely to act. Describe some ways in which abiotic and biotic agents of change interact with one another.

2. Describe the differences between primary and secondary succession and what they mean for colonizing species.

3. Connell and Slatyer (1977) proposed three separate models of succession: the facilitation model, the tolerance model, and the inhibition model. Choose a hypothetical community and describe the different circumstances that would be required to support each one of the models.

4. Why is it hard to determine whether a community is stable? Do you think Sutherland (1974) was able to demonstrate alternative stable states on his tiles? Why or why not?

HANDS-ON PROBLEM SOLVING

Soil Invertebrates and Succession | This Web exercise (www.sinauer.com/ecology) explores how invertebrates that live in the soil could affect patterns of succession. You will interpret data from a recent paper that shows the effects of soil invertebrates on the growth of early- and mid-succession plants.

SUGGESTED READINGS

Beisner, B. E., D. T. Haydon and K. Cuddington. 2003. Alternative stable states in ecology. *Frontiers in Ecology and the Environment* 1: 376–382. *A good introductory paper for general audiences on the basics behind alternative stable state theory.*

Dale, V. H., F. J. Swanson and C. M. Crisafulli, editors. 2005. *Ecological Responses to the 1980 Eruption of Mount St. Helens.* Springer, New York. *A broad overview of all that has been learned about the recovery of life on Mount St. Helens since the eruption 20 years earlier.*

Elton, C. 1927. *Animal Ecology.* Sidgwick and Jackson, Ltd. London. *A surprisingly insightful book on the then new science of ecology. Many of the questions Elton addresses are still relevant today.*

Hughes, T. P., D. R. Bellwood, C. Folke, R. S. Steneck and J. Wilson. 2005. New paradigms for supporting the resilience of marine ecosystems. *Trends in Ecology and Evolution* 20: 380–386. *A short overview of the regime-shifting effects of humans on marine ecosystems.*

Muir, J. 1915. *Travels in Alaska.* Houghton Mifflin, Boston. *An early account of one man's travels to the largely unexplored Alaskan wilderness. The chapter "Stickeen" is a dog story not to be missed.*

Sousa, W. P. 2001. Natural disturbance and the dynamics of marine benthic communities. In M. D. Bertness, S. D. Gaines, and M. E. Hay (eds.), *Marine Community Ecology,* (pp. 85-130). Sinauer Associates, Sunderland, MA. *An extensive review of disturbances—both what they are—and how they affect marine communities, in particular.*

CHAPTER 17 Biogeography

FIGURE 17.1 Diversity Abounds in the Amazon Freshwater fish caught in the Amazon River on display in a market in Manaus, Brazil.

CASE STUDY

The Largest Ecological Experiment on Earth

There is probably only one place on Earth where a person could hear the songs of a hundred species of birds or smell the fragrances of a thousand species of flowers or see the leaf patterns of 750 species of trees all in 1 hectare (2.5 acres) of land. That place is the Amazon, where half the world's remaining tropical rainforests and species reside. One hectare of rainforest in the Amazon contains more plant species than all of Europe! Of course, not all of those species are confined to the rainforest itself. The Amazon Basin contains the largest watershed in the world; one-fifth of all the fresh water on Earth rages down the slopes of the Andes through over a thousand forested tributaries and eventually flows into the Amazon River and out to sea. A trip to a fish market in Manaus, Brazil, would reveal the amazing aquatic diversity in these rivers (**Figure 17.1**). The number of fish species in the Amazon River exceeds the total number found in the entire Atlantic Ocean.

Ironically, with this incredible species diversity can come devastating species losses when these ecosystems are disturbed. The main destructive force has been deforestation, which began in earnest with the building of roads in the 1960s (Bierregaard et al. 2001). Before then, most of the Amazon had no roads and was completely isolated from modern society. In 50 years' time, 15% of the rainforest has been converted to pastureland, towns, roads, and mines. Although this percentage might seem modest, it is deceivingly so, both because of the sheer number of species involved and because of the pattern of deforestation. Logging practices have caused extreme habitat fragmentation, sometimes resulting in a "fishbone" pattern in which thin linear fragments of rainforest are surrounded by strips of nonforested land (see Figure 3.6). As we will see, habitat fragmentation can have serious consequences for the maintenance of species diversity.

It was this fragmentation of the Amazon rainforest that motivated Thomas Lovejoy and his colleagues to initiate the largest and longest-running ecological experiment ever conducted. When the Biological Dynamics of Forest Fragments Project (BDFFP) began in 1979, Lovejoy seized a unique opportunity to find out what was happening to the amazing species diversity of the Amazon as more and more forests were being logged. He was guided by an elegant model in Robert MacArthur and Edward O. Wilson's 1967 book, *The Theory of Island Biogeography*, which presents an explanation for the observation that more species are found on large islands than on small islands. By taking advantage of a Brazilian law requiring landowners to leave half of their land as forest, Lovejoy was able to arrange for the designation of different-sized forest plots that would be surrounded by either forested (controls) or deforested land (fragments) (**Figure 17.2**). The control plots and the fragments

(A)

Plots of four sizes—1, 10, 100, and 1,000 hectares—were designated before logging took place.

Control plots remained surrounded by forested land.

Experimental fragments were surrounded by deforested land.

(B)

FIGURE 17.2 Studying Habitat Fragmentation in Tropical Rainforests The Biological Dynamics of Forest Fragments Project (BDFFP) near Manaus, Brazil, was designed to study the effects of habitat fragment size on species diversity. (A) Plots of different sizes were isolated by logging or left as controls. (B) Aerial photo of a 10 ha and 1 ha fragment isolated in 1983. (A after Bierregaard et al. 2001.)

were of four sizes: 1, 10, 100, or 1,000 hectares. Baseline data collected at the start of the experiment showed little difference between control plots and fragments. By the mid-1980s, the ecologists had a fully replicated experiment at a scale unimaginable in the past.

Over the last 25 years, the BDFFP has evolved from a study that asks the simple question, "What is the minimum area of rainforest needed to maintain species diversity?" to one that asks, "What role does the isolation, shape, and configuration of forest fragments play in maintaining species diversity? How does the surround-

ing matrix habitat influence that diversity? And what is the future prognosis for the Amazon rainforest, one of the most deforested but species-rich terrestrial biomes on Earth?"

ooking out over a community such as a rocky intertidal zone on the northern California coast, it is obvious that the locations of species on the shoreline are controlled by important physical factors, such as tide height and wave action, but also by a variety of biological interactions. Sea stars eat sessile mussels in the low intertidal zone, thus limiting them to the higher intertidal zones. In those zones, the crevices between mussels provide habitat for many species that otherwise would be absent. Local conditions such as these are important regulators of species distributions. However, as important as these conditions appear to us, we must always be cognizant of the influence of processes operating at larger geographic scales. Oceanographic processes, such as currents or upwelling, regulate the delivery of species to rocky shorelines. At a global scale, oceanic circulation patterns control current direction and can isolate species over ecological and evolutionary time. As a result, a distinct biota comprises the pool of possible recruits to the

northern California coast, and the assemblage of species there is ultimately based on a foundation of global and regional processes. In this chapter, we will consider the effects of these large-scale geographic processes on one of the most recognizable ecological patterns known, that of the distribution and diversity of species on Earth.

Biogeography and Spatial Scale

CONCEPT 17.1
Patterns of species diversity and distribution vary at global, regional, and local spatial scales.

One of the most obvious ecological patterns on Earth is the variation we see in species composition and diversity among geographic locations. The study of this variation is known as **biogeography**. Pretend for a moment that you have a lifelong desire to see all the forest biomes on Earth. In this imaginary scenario, you have the ability to move from one geographic region on Earth to another. Think Google Earth,* but with the ability to fly down into a community and see species up close. You start in the tropics at 4° S latitude and 60° W longitude and fly into the Amazon rainforest, the most species-rich forest on Earth (**Table 17.1**). At 20 meters altitude, you fly through the middle of the humid forest, and every hectare brings new tree species (**Figure 17.3A**). You may have encountered half of them in the previous hectare, but at least half are completely new. The more area you cover, the more tree species you see. The richness is almost mind-numbing, and the heat and humidity are stifling, so you decide to head north to drier climes.

You arrive at 35° N, 125° W. This is the southern coast of California, where the forests are oak woodland—a dry biome, as we learned in Chapter 3, with mostly nonconiferous evergreen trees and shrubs interspersed with grassland (**Figure 17.3B**). Flying down through the vegetation, you notice the many kinds of trees and shrubs, all with small leaves and thick bark. The

*See http://earth.google.com

woodland is aromatic because of the volatile oils contained in the local shrubs and herbaceous plants. Plant species richness is high, but just a fraction of that in the Amazon (see Table 17.1).

It's still warm, so you decide to head north to 45° N, 123° W, where the forest is cool and very wet. This is the Pacific Northwest region of North America, where the forests are dominated by large conifers. As you fly through, you notice the incredible lushness of the forest, with its lichen-filled canopy and fern-covered forest floor (**Figure 17.3C**). Tree species richness in these lowland temperate forests is a fraction of that in the two previous forests you've visited (see Table 17.1). There are only a handful of tree species: Douglas fir, western hemlock, western red cedar, red alder, and big leaf maple. What these forests lack in species richness, however, is made up in the huge biomass produced in these communities.

FIGURE 17.3 Forests around the World Different forest biomes vary greatly in their species composition and species richness. (A) A tropical rainforest in Brazil. (B) Oak woodland in southern California. (C) Lowland temperate forest in the Pacific Northwest. (D) Boreal spruce forest in northern Canada.

(A)

(B)

(C)

(D)

TABLE 17.1

Tree Species Richness in Different Forests around the World

Forest location/ forest type	Latitude, longitude	Approximate tree species richness	Source
Amazon, Brazil	4°S, 60°W	1,300	Laurance 2001
Southern California, USA	35°N, 125°W	57	Allen et al. 2007
Pacific Northwest, USA			Franklin and Dyrness 1988
Douglas fir forest	45°N, 123°W	7	
Garry oak forest	45°N, 123°W	4	
Boreal forest, Canada	64°N, 125°W	2	Kricher 1998
New Zealand			Dawson and Lucas 2000
Beech forest	45°S, 170°E	20	
Flowering tree forest	35°S, 170°E	100	

You want to see the extremes in species richness, so your next stop is 60° N, 125° W, in the boreal forests of Canada. Flying over the cold landscape, you notice rows and rows of identical spruce trees, broken once in a while by large wetlands (**Figure 17.3D**). Dipping down into the canopy, you are struck by the dense and monotonic nature of the forest. It's dark down under those fir boughs, but low-lying berry bushes are a reminder that light does penetrate this northern forest, es-pecially in the summer months. You continue to fly north, and the forests thin until the landscape is one long expanse of treeless tundra.

Your trip could end here, but because you have always wanted to visit New Zealand, you take the time to fly back to the Southern Hemisphere. New Zealand was separated from the ancient continent of Gondwana roughly 80 million years ago, and since that time evolution has produced forests that are completely unique (**Figure 17.4**). Roughly 80% of the species in New Zealand are **endemic**, meaning that they occur nowhere else on Earth. Dialed into 45° S, 170° E, you fly through the Southern Hemisphere equivalent of the Pacific Northwest. Instead of conifers,

FIGURE 17.4 Forests of North and South Island, New Zealand
The two islands of New Zealand span a large latitudinal gradient (35°– 47° S) and have different forest types. (A) The forests of South Island are dominated by beech tree species. (B) The forests of the warmer North Island have greater tree species diversity and different species composition than those on South Island (see Table 17.1).

(A)

The forests of South Island are dominated by southern beech trees bearing small leaves.

Tasman Sea

Auckland

North Island

Wellington

Christchurch

South Island

Pacific Ocean

The branches of divaricating understory shrubs have a zigzag appearance.

(B)

North Island has kauris, one of the largest tree species on Earth.

The ancient tree ferns of North Island have "fiddleheads" similar to those of more modern ferns.

the forests are dominated by four species of southern beech trees with billowy layers of twisted branches (see Figure 17.4A). Below the canopy are incredible "divaricating" shrubs, found in highest abundance in New Zealand, with multiple-angled branches that give them a zigzag appearance. Although temperate forests in the Northern and Southern hemispheres each have fairly low tree species richness, they are made up of completely different species assemblages with very different evolutionary histories.

Even within New Zealand, over a distance that extends from 35° to 47° S (a latitudinal distance identical to that from southern California to British Columbia in the Northern Hemisphere), there are big differences in tree species richness and composition. North Island is warmer (closer to the equator) than South Island and has more diverse forests, consisting of many flowering tree species with a few tall emergent conifers (**Figure 17.4B**). These forests have a tropical feel to them because of all the flowering trees and the multitude of vines and epiphytes (a plant or lichen that lives on a larger plant). Tree ferns growing here are similar to those that were dominant 100 million years ago, during the age of the dinosaurs. One of the most extraordinary trees is the kauri (*Agathis australis*), among the largest tree species on Earth (interestingly, the largest are the California redwoods, which occur at roughly the same latitude in the Northern Hemisphere). Some kauri trees are 60 m (200 ft) high and 7 m in diameter (23 ft). Unfortunately, like redwoods, kauris have been extensively logged, and they exist in a forest community in only two small reserves, 100 km^2 in total size. Given that old-growth stands of kauris take 1,000–2,000 years to generate, these forests are irreplaceable to modern society. If we contrast the tree species richness of the forests characteristic of North Island with those on South Island, we find that there are more than 100 tree species in the northern forests, compared with the 10–20 species in the less diverse beech forests characteristic of the temperate south (see Table 17.1).

With our world forest tour at its end, what can we conclude about biogeographic patterns on Earth, assuming that forest communities are good global representatives? First, species richness and composition vary with latitude: the lower tropical latitudes have many more, and different, species than the higher temperate and polar latitudes. Second, species richness and composition vary from continent to continent, even where longitude or latitude is roughly similar. Third, the same community type or biome can vary in species richness and composition depending on its location on Earth. As we will see in the rest of this chapter, these are reliable patterns that have been demonstrated over and over again for many regions of the world and many community types. What has puzzled ecologists for centuries is just what processes control these biogeographic patterns. Why are more species found in some areas than in other areas? Why do some regions harbor species assemblages that are not found anywhere else on Earth?

A number of hypotheses have been posed to explain biogeographic variation in species composition and diversity. These hypotheses are highly dependent on the spatial scale at which they are applied. Let's deal with the issue of spatial scale before we go any further.

Patterns of species diversity at different spatial scales are interconnected

On our world forest tour, we saw that patterns of species diversity and composition varied at global, regional, and local spatial scales. We can think of these spatial scales as interconnected in a hierarchical way, with the patterns of species diversity and composition at one spatial scale setting the conditions for patterns at smaller spatial scales (Whittaker et al. 2001). Let's start with the largest spatial scale and work down.

As the term suggests, the global scale includes the entire world, a huge geographic area over which there are major variations with latitude and longitude (**Figure 17.5**). Species have been isolated from one another, possibly on different continents or in different oceans, by long distances and over long periods. As such, differences in the rates of three process-

Global patterns of species diversity and composition are driven by variation in speciation, extinction, and migration rates across latitudes and longitudes.

Within regions, patterns of species diversity and composition are driven by migration and extinction rates across the landscape.

The local and regional scales are connected by turnover, the difference in species number and composition as one moves across the landscape from one community type to another.

Local patterns of species diversity and composition are driven by physical conditions and species interactions.

FIGURE 17.5 Interconnected Spatial Scales of Species Diversity The arrows represent the relationships between, and processes important to, species diversity and composition at global, regional, landscape, and local scales.

es—speciation, extinction, and migration—help determine differences in species diversity and composition. We will consider these processes in more detail in Section 17.2.

The **regional scale** encompasses a smaller geographic area where the climate is roughly uniform and the species contained therein are bound by dispersal to that region (see Figure 17.5). All the species contained within a region are known as the **regional species pool** (sometimes called the **gamma diversity** of the region). Earth's regions differ in species diversity and composition due to the global-scale processes mentioned above. Obviously, a region in the Amazon will have many more species, and thus a larger species pool, than a region in Antarctica.

The topographic and environmental features of a region, such as the number, area and distance from one another of mountains, valleys, deserts, islands, and lakes—referred to collectively as the *landscape*—are all critical to within-region biogeography (see Figure 17.5). Species composition and diversity vary within a region depending on how the landscape shapes the rate of migration and extinction to and from critical local habitats. The **local scale**, which is essentially equivalent to a community, reflects the suitability of particular abiotic and biotic characteristics of habitats for species once they reach those habitats through dispersal (see Figure 17.5). Species physiology and interactions with other species weigh heavily in the resulting species diversity (sometimes called **alpha diversity**). The important connection between local and regional scales of species diversity comes in the form of a measurement known as **beta diversity**. Beta diversity tells us the change in species number and composition, or *turnover* of species, as one moves from one community type to another.

Knowing how spatial scales are related to one another in a hierarchical way is important, but are there actual area values one could apply to local and regional spatial scales? For example, how much area does a region or locality encompass? The answer is highly dependent on the species and communities of interest. For example, Shmida and Wilson (1985) suggest that terrestrial plants might have a local scale of 10^2–10^4 m^2 and a regional scale of 10^6–10^8 m^2. But for phytoplankton, the local scale might be something more like 10^2 cm^2. As we will see, the actual area we use to define species diversity measurements can be critical to our interpretation of the processes controlling biogeographic patterns.

Local and regional processes interact to determine local species diversity

It is obvious from Figure 17.5 that patterns of species diversity, and the processes that control them, are interconnected across spatial scales. Given this interconnection, ecologists are interested in knowing just how much variation in species diversity at the local scale is dependent on larger spatial scales. We know that the regional species pool provides the raw material for local assemblages and sets the theoret-

FIGURE 17.6 What Determines Local Species Diversity? The relative influences of local and regional processes in a community can be determined by plotting local species diversity against regional species diversity.

ical upper limit on species diversity for communities. But is that limit also set by local conditions, including species interactions and the physical environment, as suggested in Figure 17.5? One way we can consider this question quantitatively is by plotting local species diversity against the corresponding regional species diversity (**Figure 17.6**). Three basic types of relationships can be seen in such plots. First, if local species diversity and regional species diversity are equal (slope = 1), then all the species within a region will be found in all the communities of that region. This pattern is a theoretical possibility, but an empirical impossibility, for the simple reason that regions will always have landscape and habitat features that exclude some species from some communities (e.g., lowland tree or fish species will not be found in alpine forests or lakes). Second, if local species richness is simply proportional to regional species richness (i.e., local species richness increases with increasing regional species richness, but the relationship is not 1:1), then we can assume that community species richness is largely determined by the regional species pool, with local processes playing some role. Finally, if local species richness levels off despite a large regional species pool, then local processes can be assumed to limit local species diversity. The degree to which local richness levels off can tell us something about how important species interactions and physical conditions are in setting a *saturation point*—a limit on diversity—for communities.

Let's move away from these theoretical constructs and look at what real data show us about the relationship between local and regional species richness. Witman et al. (2004) considered this relationship for marine invertebrate communities living on subtidal rock walls at a variety of locations throughout the

(A)

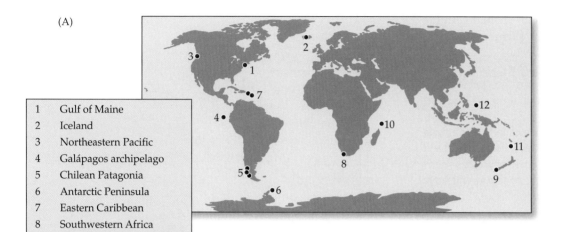

1	Gulf of Maine
2	Iceland
3	Northeastern Pacific
4	Galápagos archipelago
5	Chilean Patagonia
6	Antarctic Peninsula
7	Eastern Caribbean
8	Southwestern Africa
9	Southwestern New Zealand
10	Seychelles
11	Norfolk Island
12	Palau

FIGURE 17.7 Marine Invertebrate Communities May Be Limited by Regional Processes Among shallow subtidal marine invertebrate communities, regional species richness explains approximately 75% of the local species richness. (A) The 12 regions of the world where the 49 sampling sites were located. (B) A plot of local species diversity against regional species diversity. Each dot represents one of the 49 sampling sites. (After Witman et al. 2004.)

(B)

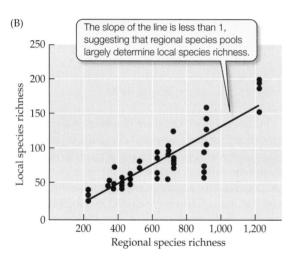

The slope of the line is less than 1, suggesting that regional species pools largely determine local species richness.

world (**Figure 17.7**). At 49 local sites in 12 regions of the world, they surveyed species richness in 0.25 m² plots on rock walls at a 10–15 m water depth. Local species richness values were then compared with regional species richness values from published lists of invertebrate species capable of living on hard substrata at shallow depths (< 20 m). A plot of all the local sites showed that local species richness was always proportionally lower than regional species richness and that it never leveled off, or became saturated, at high regional richness values. Instead, regional species richness explained approximately 75% of the variation in local species richness. As such, this study suggests that regional species pools largely determine the number of species present in the local marine invertebrate communities.

Does the lack of saturation detected in this study and others indicate that local processes are unimportant in determining local community species diversity? The answer is no,

for at least two reasons. First, there is still considerable unexplained variation among local communities within regions that could be attributable to the effects of local processes. Moreover, the effects of species interactions, in particular, are likely to be highly sensitive to the local spatial scale chosen. Although the small spatial scale of the Witman et al. (2004) study is probably appropriate for species interacting on marine rock walls, some studies have used inappropriate (usually too large) spatial scales that are unlikely to detect local effects. Still, the strong influence of regional-scale processes on local species diversity suggests that both marine and terrestrial communities are likely to be much more susceptible to, for example, species invasions from outside regions than previously thought.

In the remainder of this chapter, we will explore the factors controlling variation in species diversity at global and regional biogeographic scales. We will wait until the next chapter to delve in more detail into the causes and consequences of species diversity differences at the local scale.

Global Biogeography

CONCEPT 17.2
Global patterns of species diversity and composition are controlled by geographic area and isolation, evolutionary history, and global climate.

It must have been incredible to be a scientific explorer 200 years ago. You would have left the safety of your home in Europe to travel by ship to a destination largely unknown. You would have had to endure seasickness, disease, accidents of all kinds, and years away from your family, friends, and colleagues. You might have had many years of financial debt to repay unless you were independently wealthy or could sell your collections. But you would have been the first scientist to document and collect animal and plant species of incredible beauty, oddity, and rarity. It was under these cir-

(A)

(B)

FIGURE 17.8 Alfred Russel Wallace and His Collections (A) A photograph of Wallace taken in Singapore in 1862, during his expedition to the Malay Archipelago. (B) Some of Wallace's rare beetle collections from the Malay Archipelago, found in an attic by his grandson in 2005.

cumstances that the science of biogeography was born and many important discoveries were made. Up to that point, European scientists had very little information about the natural history and ecology of other parts of the world; most was second-hand or anecdotal. What these early naturalists were able to bring back were specimens, journals, and most of all, theories.

Though not the first in his field, Alfred Russel Wallace (1823–1913) rightly earned his place as the father of biogeography (**Figure 17.8A**). Inspired by naturalists such as Alexander von Humboldt, Charles Darwin, and Joseph Hooker, Wallace came on the scene with considerably less wealth or education, but his intellectual motivation more than made up for these insufficiencies. Wallace is best known, along with Charles Darwin, as the codiscoverer of the principles of natural selection, although he has always stood in the shadow of Darwin in that regard. But his main contribution was the study of species distributions across large spatial scales.

Wallace left England for Brazil in 1848 and explored the Amazon rainforest for 4 years. On his way back, the ship he was traveling on burned in the middle of the Sargasso Sea, destroying all his samples and most of his notes. After 10 days in a lifeboat, he was rescued and made his way back to England, where he published an impressive six papers on his observations.

Even though he had vowed never to travel again, Wallace left, in 1852, for the Malay Archipelago (present-day Indonesia) (**Figure 17.8B**). It was here that he made the puzzling observation described in Chapter 8 and in his book *The Malay Archipelago* (Wallace 1869a): that the mammals of the Philippines were more similar to those in Africa (5,500 km away) than they were to those in New Guinea (750 km away) (see Figure 8.10). Wallace was the first to notice the clear demarcation between these two faunas, which came to be known as Wallace's line. It turns out, as we'll see shortly, that these separate groups of animals evolved on two different continents that have come into close proximity only within the last 15 million years. Wallace's biogeographic research culminated in the publication of a two-volume work called *The Geographical Distribution of Animals*, published in 1876. In this book, Wallace overlays species distributions on top of geographic regions and reveals two important global patterns:

- Earth's land mass can be divided into six recognizable biogeographic regions that differ markedly in species composition and diversity.

- There is a gradient of species diversity with latitude; species diversity is greatest in the tropics and decreases toward the poles.

These two patterns are necessarily interrelated; the latitudinal gradient is superimposed over the biogeographic regions. For ease of explanation, let's first spend some time learning

FIGURE 17.9 Six Biogeographic Regions
Wallace identified six biogeographic regions using the distributions of terrestrial animals. These six regions roughly correspond to the major tectonic plates.

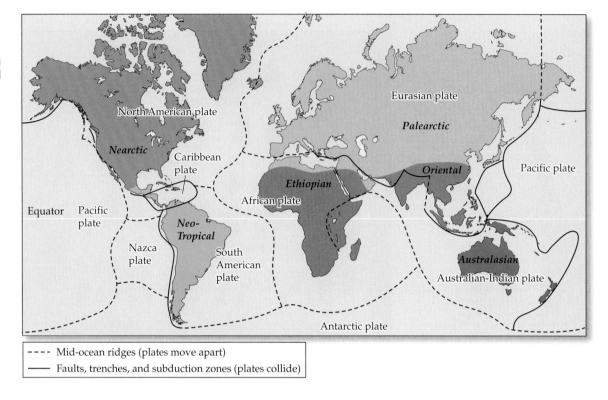

--- Mid-ocean ridges (plates move apart)
— Faults, trenches, and subduction zones (plates collide)

about the biogeographic regions described by Wallace and the underlying forces that created them. We will then consider some of the processes likely to be responsible for the latitudinal gradient in species diversity.

The biotas of biogeographic regions reflect evolutionary isolation

Earth's six biogeographic regions are the Nearctic (North America), Neotropical (Central and South America), Palearctic (Europe and parts of Asia and Africa), Ethiopian (most of Africa), Oriental (India, China, and Southeast Asia), and Australasian (Australia, the Indo-Pacific, and New Zealand) (**Figure 17.9**). It is no coincidence that these regions correspond roughly to Earth's six major tectonic plates. These plates are sections of Earth's crust that move or drift (thus the name **continental drift** given to this process) across Earth's surface through the action of currents generated deep within its molten rock mantle (**Figure 17.10**). In areas known as *mid-ocean ridges*, this molten rock flows out of the seams between plates and cools, creating new crust and forcing the plates to

spread. In some areas where two plates meet, known as *subduction zones*, one plate is forced downward under the second plate. These areas are associated with strong earthquakes, volcanic activity, and mountain range formation. In other areas where two plates meet, the plates slide sideways past each other, forming a *fault*.

As a result of these processes, the positions of the plates, and of the continents that sit on them, have changed dramatically over geologic time. For our purposes here, let's consider

FIGURE 17.10 Mechanisms of Continental Drift
Over geologic time, currents generated deep within Earth's molten rock mantle (magma) move sections of Earth's crust across its surface.

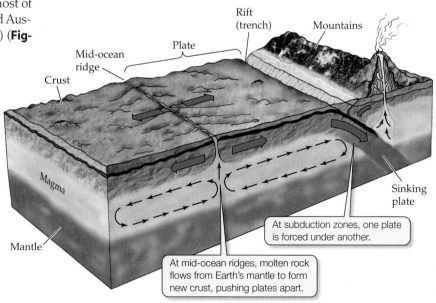

At subduction zones, one plate is forced under another.

At mid-ocean ridges, molten rock flows from Earth's mantle to form new crust, pushing plates apart.

(A)

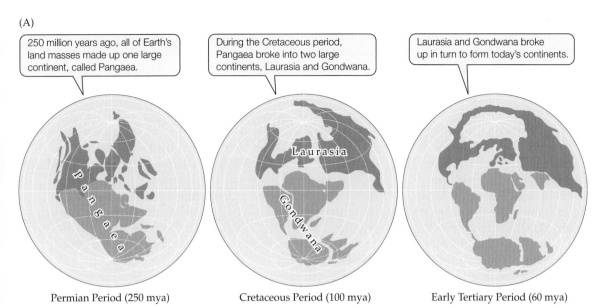

250 million years ago, all of Earth's land masses made up one large continent, called Pangaea.

During the Cretaceous period, Pangaea broke into two large continents, Laurasia and Gondwana.

Laurasia and Gondwana broke up in turn to form today's continents.

Permian Period (250 mya) Cretaceous Period (100 mya) Early Tertiary Period (60 mya)

FIGURE 17.11 The Positions of Continents and Oceans Have Changed over Geologic Time The locations of continents and oceans have changed dramatically over the last 250 million years due to continental drift. (A) The breakup of Pangaea. (B) A summary of the movements that led to the configuration of the continents we know today. Red arrows show the time (in millions of years) since land masses joined; black arrows show the time since land masses separated.

(B)

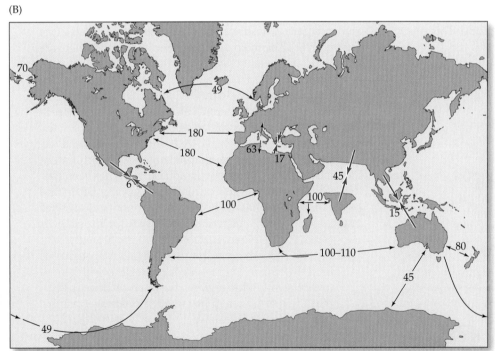

continental drift since the end of the Permian period (250 million years ago), when all of Earth's land masses made up one large continent, called Pangaea (**Figure 17.11A**). About 144 million years ago, during the Cretaceous period, Pangaea began to split into two land masses, Laurasia to the north and Gondwana to the south. By the early Tertiary period (60 million years ago), Gondwana had separated into the present-day continents of South America, Africa, India, Antarctica, and Australia. Laurasia eventually split up into the North American, European, and Asian continents. Most of these movements resulted in the separation of continents from one another, but

some continents were brought together (**Figure 17.11B**). For example, North and South America joined at the Isthmus of Panama, India collided with Asia to create the Himalayas, Africa and Europe united at the Mediterranean Sea, and a land bridge was formed between North America and Asia at the Bering Strait.

Continental drift thus separated one large group of terrestrial species on Pangaea, united by geography and phylogeny, into six biogeographically distinct groups of species by isolating them on different continents. The sequence and tempo of the continental movements has resulted in some biogeographic regions having very different flora and fauna than others. For example, the Neotropical, Ethiopian, and Australian regions, all once part of Gondwana, have been isolated for quite some time and have very distinctive forms of life. In other cases, however, distinct groups of species have been united. For example, the biota of the Nearctic region differs substantially from that of the Neotropical region despite their modern-day proximity. Because North America was part of Laurasia and South America was part of Gondwana, they had no contact until about 6 million years ago. Within that time, however, many species have moved from one continent to the other (for example, mountain lions, wolves, and the precursors of llamas moved to South America, while armadillos and opossums moved to North America), somewhat homogenizing the biota of the two regions. Interestingly, there is also evidence that some terrestrial mammal families (24%) went extinct once the two continents merged, suggesting that ecological coexistence was not possible for some species (Flessa 1975). Finally, the Nearctic and Palearctic, both part of ancient Laurasia, have similarities in biota across what is now Greenland as well as across the Bering Strait, where a land bridge has allowed exchanges of species over the last 100 million years.

The legacy of continental drift can be found in a number of existing taxonomic groups as well as in the fossil record. The evolutionary separation of species due to a barrier such as continental drift is known as **vicariance**. A good example of the effects of vicariance can be seen among the large flightless birds known as ratites, all of which had a common ancestor from Gondwana (**Figure 17.12**). Over the millennia, as Gondwana broke up, the rheas of South America, the ostriches of Africa, the cassowaries and emus of Australia, and the moas of New Zealand became isolated from one another. They all evolved unique characteristics in isolation, but still retained their large size and inability to fly. Interestingly, the kiwis of New Zealand are more closely related to ostriches, cassowaries, and emus than they are to moas, despite their co-occurrence with moas on New Zealand. This ob-servation suggests that the kiwis evolved elsewhere and im-migrated to New Zealand sometime after the breakup of Gondwana.

Tracing the threads of vicariance over large geographic areas and long time periods provided important evidence for early theories of evolution. For example, as Wallace began to amass more species and make geographic connections between them, his ideas about the origin of species started to solidify. In an 1855 paper titled "On the law which has regulated the introduction of new species," he wrote, "Every species has come into existence coincident both in space and time with a pre-existing closely allied species." Despite the biogeographic evidence of evolutionary connections among species, it took a few more years for the mechanism of evolution (i.e., natural selection) to be formally proposed by both Wallace (1858) and Darwin (1859).

Before we move on, it is important to consider whether the biogeographic regionalism first identified by Wallace is present in the oceans as well as on land. After all, the oceans make up 75% of Earth's surface area and, just as we have seen for continents, oceans "drift," in the sense that they are created, merged, or destroyed by the movements of plates (see Figure 17.11). The main question, then, is whether there are barriers to dispersal between oceans as there are between continents. Despite what you might think, oceans do have significant impediments to the exchange of biota, in the form of continents, currents, thermal gradients, and differences in water depth. Oceanographic discontinuities have isolated species from one another, allowed for evolutionary change, and created unique biogeographic regions. Unfortunately, identification of marine biogeographic regions has been hindered by the extra complicating factor of water depth and by the basic lack of natural history and taxonomic knowledge of the deep oceans. One recent model by Adey and Steneck (2001) shows 24 recognizable biogeographic regions for intertidal benthic marine macroalgae. Although it is hard to compare these macroalgal regions with terrestrial biogeographic regions, the analysis does suggest that the marine realm has much more biogeographic variation than we realized.

FIGURE 17.12 Vicariance among the Ratites The pattern of evolutionary relationships among the ratites shown here corresponds to the pattern of continental drift as Gondwana broke up. These large flightless birds share a common ancestor that once lived on Gondwana, but they evolved differently after their populations were isolated by continental drift. (After Van Tyne and Berger 1959; Haddrath and Baker 2001.)

South America | Africa | Australia and New Guinea | Australia | New Zealand

Rhea | Ostrich | Cassowary | Emu | Kiwi | Moa (extinct)

Common ancestor

The rhea became isolated from the other ratites when South America separated from Africa.

The kiwi appears to have evolved in Australia and migrated to New Zealand later.

Latitudinal gradients have multiple, interrelated causes

If you recall our Google Earth-style tour of the globe in Section 17.1, it was clear that tree species diversity and composition changed dramatically with latitude: species diversity was highest at tropical latitudes and decreased toward the poles. Wallace and the other nineteenth-century European scientific explorers became keenly aware of this pattern as they collected thousands of species in the tropics and compared them with their more meager European collections. As more data have accumulated over the last 200 years, the latitudinal gradient in species diversity has been more firmly established (**Figure 17.13**). Willig et al. (2003) tallied the results of 162 studies on a variety of taxa that considered whether diversity and latitude showed a negative relationship (with diversity decreasing toward the poles), a positive relationship (increasing toward the poles), a unimodal relationship (increasing toward mid-latitudes and then declining toward the poles), or no relationship. At broad spatial scales (> 20° latitude), negative relationships were by far the most often seen, but unimodal relationships were also evident.

In addition to this undeniably strong latitudinal gradient, biogeographers have observed an important pattern of longitudinal variation. Gaston et al. (1995) measured the numbers of families along multiple transects running north to south and

(A) Global seabird diversity

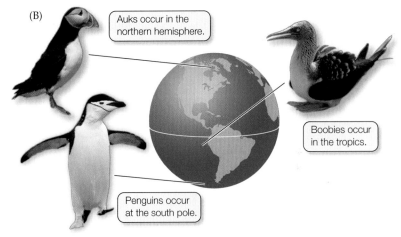

(B)

Auks occur in the northern hemisphere.

Boobies occur in the tropics.

Penguins occur at the south pole.

FIGURE 17.14 Seabirds Go against Conventional Wisdom Global seabird species richness shows a latitudinal pattern opposite to that of most fauna. (A) Species richness among seabirds is high in temperate and polar regions and much lower in the tropics. (B) The composition of major seabird groups shows strong latitudinal differences. (Data from Harrison 1987.)

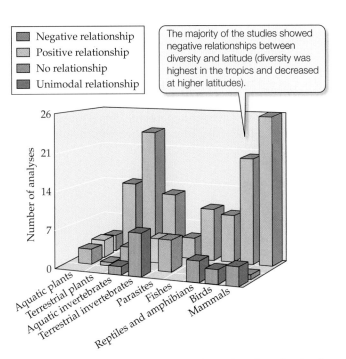

Negative relationship
Positive relationship
No relationship
Unimodal relationship

The majority of the studies showed negative relationships between diversity and latitude (diversity was highest in the tropics and decreased at higher latitudes).

FIGURE 17.13 Studies of Latitude and Species Diversity Confirm Conventional Wisdom Willig and colleagues tallied the results of a number of studies of relationships between species diversity and latitude. The studies included a variety of taxa, but all were done at broad latitudinal scales (> 20°). (After Willig et al. 2003.)

separated by 10° longitude. Families of seed plants, amphibians, reptiles, and mammals all increased in number toward the equator and declined at higher latitudes in both the Northern and Southern hemispheres. These researchers determined, however, that the number of families also depended on the longitude chosen. Their observations point to the importance of so-called *hot spots* or areas of high species richness that occur at particular longitudes, sometimes secondary to latitude.

Of course, as Figure 17.13 indicates, not all groups of organisms show decreases in species richness at higher latitudes. Some species display the opposite pattern. Seabirds, for example, have their highest diversity at temperate and polar latitudes (Harrison 1987) (**Figure 17.14A**). Seabirds of the Antarctic and subantarctic include penguins, albatrosses, petrels, and skuas (**Figure 17.14B**). In the Arctic and subarctic, auks replace penguins as ecological equivalents, and gulls, terns, and grebes are common. In the tropics and subtropics, seabird diversity declines; the seabird community here is composed mostly of pelicans, boobies, cormorants, and frigatebirds. The pattern of seabird diversity correlates well with marine productivity, which is substantially higher in temperate and polar oceans (see Figure 19.10). The same pattern has been observed in marine benthic communities, which experience much higher productivity at higher latitudes. As we will see, productivity differences are one possible explanation for latitudinal gradients in species diversity. Let's turn now to some of those other possible explanations.

Large-scale, global patterns of species richness should ultimately be controlled by the rates of three processes: speciation, extinction, and migration. Let's assume here for simplicity's sake that the migration rate is roughly the same worldwide. We can then predict that the number of species at any particular location on the globe will reflect a balance, or equilibrium, between the rates of two fundamental processes: speciation and extinction (**Figure 17.15A**). According to this model, we would expect both speciation and extinction rates to increase with species richness. So, as speciation increases, the number of species increases, which means that we would expect more species to evolve from them (this is a classic positive feedback loop for species diversification). Additionally, as the number of species increases, we would expect the probability of extinction to increase by a simple law of numbers (the more species, the more extinctions), but also because more species would presumably cause more resource depletion and thus extinctions. If we plot the rates of these two processes at any location against numbers of species, the two curves should intersect at an equilibrium point, representing the number of species (S) present at that location. We can then use this model to make predictions about the variation in species richness at different latitudes (**Figure 17.15B**). For example, we can assume, based on this model, that the highest speciation and extinction rates should occur in the trop-

ics and the lowest rates in the polar regions. Temperate-zone speciation and extinction rates should be somewhere in between the tropical and polar extremes.

Before we leave the model, it is interesting to speculate whether there is some upper limit to speciation because of the sheer number of species already in existence. In other words, is there an upper limit to the number of species that can coexist on Earth? Some ecologists (e.g., Rohde 2005) have suggested that the number of ecological niches, or roles, that species might assume are endless, and that in the absence of major global disturbances (e.g., climate change, deforestation, or catastrophic events such as meteorite impacts or volcanic explosions), there is no reason why global species diversity could not continue to increase indefinitely.

Nonetheless, if speciation and extinction rates contribute to large-scale patterns of species diversity, we then have to ask, what ultimately controls those rates? How do speciation and extinction change with geographic location? As mentioned earlier, there are dozens of hypotheses, but very little agreement among ecologists. Part of the reason lies in the fact that there are multiple and confounding gradients in geographic area, evolutionary age, and climate that are correlated with species diversity gradients. In addition, the global scale makes it impossible to conduct a manipulative experiment to isolate various factors and separate correlation from causation. Finally, although it is likely that some factors are more important than others, it is clear that they are not mutually exclusive, but instead interact to produce the gradients we see. But let's take a look at some of the more convincing hypotheses.

One important hypothesis relates latitudinal gradients in species diversity to geographic area and temperature. John Terborgh (1973) and Michael Rosenzweig (1992) proposed that terrestrial species diversity is highest in the tropics because the tropics have the largest land area (**Figure 17.16A**). Rosenzweig (1992) calculated that the region between 26° N

(A)

The point where the speciation and extinction curves intersect is the equilibrium point, representing the number of species (S) present.

(B)

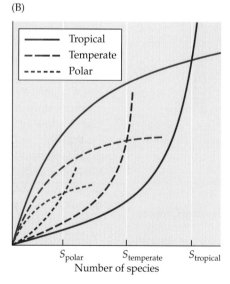

FIGURE 17.15 How Many Species?
(A) An equilibrium between the speciation rate and extinction rate controls the number of species (S) at any particular location.
(B) Tropical, temperate, and polar regions differ in their speciation and extinction rates at equilibrium thus explaining why tropical regions have the most species compared to temperate and polar regions. (A after MacArthur 1969; B after Rosenzweig 1992.)

(A)

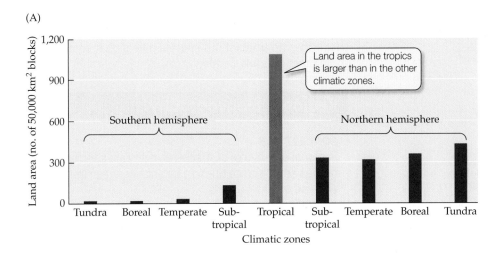

Land area in the tropics is larger than in the other climatic zones.

FIGURE 17.16 Do Land Area and Temperature Influence Species Diversity? Michael Rosenzweig hypothesized that two characteristics of the tropics lead to high speciation rates and low extinction rates: (A) their land area and (B) their stable temperatures. (After Rosenzweig 1992.)

(B)

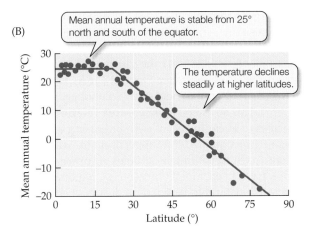

Mean annual temperature is stable from 25° north and south of the equator.

The temperature declines steadily at higher latitudes.

and S has 2.5 times more land area than any other latitude range on Earth. This makes intuitive sense, given that this latitude range is at the middle, and thus at the widest part, of the planet. Equally interesting are data showing that this very large area is also the most thermally stable location on Earth (**Figure 17.16B**). A plot of mean annual temperature against latitude by Terborgh (1973) showed that temperatures are remarkably uniform over a wide area between 25° N and S, but then drop off rapidly at higher latitudes.

Why would a larger land area and more stable temperatures foster greater species diversity? Rosenzweig (1992) suggested that area and temperature combine to decrease extinction rates and increase speciation rates in tropical regions. He argued that a larger and more thermally stable area should decrease extinction rates in two ways: first, by increasing the population sizes of species (assuming that their densities are the same worldwide), and thus decreasing their chance of extinction due to chance events, and second, by increasing the geographic ranges of species, and thus decreasing their chances of extinction by spreading the risk over a larger geographic area (see Chapter 10). He further suggested that speciation should increase in larger areas because

species should have larger geographic ranges, and thus should have a greater chance of genetic isolation and speciation (see Chapter 6).

A second important hypothesis, first suggested by Wallace (1878), supports the notion that tropical regions have longer evolutionary histories than temperate or polar regions. Because the tropics are perceived as more climatically stable (see Figure 17.16B), they should accumulate more species over time. In contrast, we would expect higher latitudes to have shorter evolutionary histories because severe climatic conditions, such as ice ages, would increase extinction rates and hinder speciation. In other words, severe climatic conditions would decrease the evolutionary time that species had to diversify. With these ideas in mind, we can expand our theory a bit and consider the possibility that most species actually originate in the tropics and then move to temperate regions during times of climatic stability.

The idea that the tropics serve as a "cradle" for diversity was originally proposed by Stebbins (1974). Jablonski et al. (2006) recently examined this hypothesis by comparing modern-day marine bivalve (e.g., clams, mussels, and oysters) fauna with marine bivalve fossils from as far back as 11 million years. They found that the majority of extant marine bivalve taxa originated in the tropics (**Figure 17.17A**) and spread toward the poles (**Figure 17.17B**), but without losing their tropical presence. Thus, in this particular case, we can think of the tropics as a cradle of species diversity because the majority of extant taxa originated there. But, as Jablonski et al. (2006) also pointed out, the tropics can serve as a "museum" as well as a "cradle." If extinction rates in the tropics are low, then species that diversify there will tend to stay there "on display," if you will. Jablonski and colleagues suggested that the current loss of biodiversity in the tropics is likely to have profound effects because it not only compromises species richness today, but could also conceivably cut off the supply of new species to higher latitudes in the future.

The final hypothesis for the latitudinal gradient in species diversity that we will consider is based on productivity. The

(A)

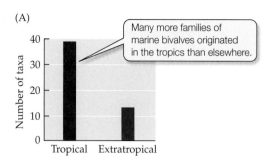

Many more families of marine bivalves originated in the tropics than elsewhere.

(B)

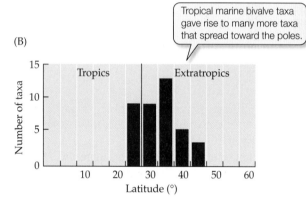

Tropical marine bivalve taxa gave rise to many more taxa that spread toward the poles.

FIGURE 17.17 The Tropics Are a Cradle and Museum for Speciation Jablonski et al. examined extant and fossil marine bivalve taxa to evaluate the hypothesis that longer evolutionary histories in the tropics contribute to the latitudinal gradient in species diversity. (A) Climatic zones of first occurrence of marine bivalve taxa (based on families of fossils). (B) Range limits of modern-day marine bivalve taxa with tropical origins. (After Jablonski et al. 2006.)

productivity hypothesis, proposed as far back as Hutchinson (1959), posits that species diversity is higher in the tropics because that is where productivity is highest, at least for terrestrial systems (see Figure 19.8). The thought is that higher productivity will promote larger population sizes, which will lead to lower extinction rates and overall higher species richness as a result. The productivity hypothesis might also explain why we see a reversal in the latitudinal gradient for some marine organisms, such as seabirds (see Figure 17.14). Productivity is generally higher in temperate coastal marine habitats than in tropical regions (see Figure 19.10). But we also know that some of the most productive habitats on Earth, such as estuaries, typically have very low species diversity. Suffice it to say, the productivity hypothesis is complex and unsatisfactory in many cases. In Chapter 18, we will consider how productivity influences diversity at local scales, where manipulative experiments can give us more insight into its effects.

As we have seen, biogeographic patterns have motivated and inspired some of the best and brightest scientists of modern times. Their fascination with the differences in the numbers and kinds of species at large geographic scales and their overwhelming drive to understand why these differences exist have resulted in some of the most influential scientific

theories of all time, including that of the origin of species. In the next section, we will consider another important theory that strives to understand species diversity at smaller and more practical spatial scales.

Regional Biogeography

CONCEPT 17.3
Regional differences of species diversity are controlled by area and distance due to a balance between immigration and extinction rates.

An important thread that runs through this chapter, and through biogeography generally, is the relationship between species number and geographic area. We saw in the Case Study that larger fragments of Amazon rainforest had greater species richness than smaller fragments. In our global tour of the world's forests, we saw that species diversity was greatest in the tropics (see Table 17.1), the climatic zone whose geographic area is also largest (see Figure 17.16A). We further learned that this pattern has held over evolutionary time (see Figure 17.17). This so-called **species–area relationship**, in which species richness increases with increasing area sampled, has been documented at a variety of spatial scales, from small ponds to whole continents. Studies of species–area relationships have been mostly targeted at regional spatial scales, where they tend to be good predictors of species diversity differences.

Species richness increases with area and decreases with distance

In 1859, H. C. Watson plotted the first curve showing a quantitative species–area relationship—in this case, for plants within Great Britain (**Figure 17.18**; Williams 1943). The curve

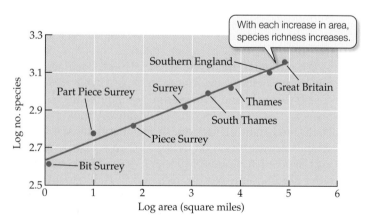

FIGURE 17.18 The Species–Area Relationship The first species–area curve, for British plants, was constructed by H. C. Watson in 1859. (After Williams 1943.)

BOX 17.1

Species–Area Curves

Species–area curves are the result of plotting the species richness (S) of a particular sample against the area (A) of that sample. A linear regression equation estimates the relationship between S and A in the following manner:

$$S = zA + c$$

where z is the slope of the line and c is the y-intercept of the line.

Because species–area data are typically nonlinear, ecologists transform S and A into logarithmic values so that the data fall along a straight line and conform to a linear regression model. Nonlinear regression models, such as polynomial or quadratic equations can be hard to visualize and understand.

The graph below shows species–area curves for plants on the Channel Islands (off the coast of France) and on the French mainland (Williams 1964). Log transformations were con-

ducted on both the island and mainland data, they were plotted separately, and a linear model was used to estimate the best fit to the data.

An important characteristic of species–area curves for islands versus mainlands is evident in this figure. The steeper the slope of the line (that is, the greater the z value), the greater the difference in species richness among the sampling areas. The Channel Islands have a much steeper slope than the French mainland areas, for the reasons outlined on p. 382.

Species–Area Relationships of Island versus Mainland Areas   Species–area curves plotted for plant species on the Channel Islands and in mainland France show that the slope of a linear regression equation (z) is greater for the islands than for the mainland areas. (After Williams 1964).

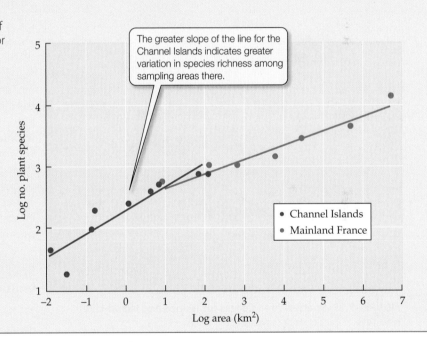

The greater slope of the line for the Channel Islands indicates greater variation in species richness among sampling areas there.

starts with a small "bit" of Surrey County and expands to ever-increasing areas that eventually encompass all of Surrey County, southern England, and finally Great Britain. With each increase in area, species richness increases until it reaches a maximum number bounded by the largest area considered. (**Box 17.1** describes how species–area curves are plotted and interpreted.)

Most species–area relationships have been documented for islands (**Figure 17.19**). Islands, in this case, include all kinds of isolated areas surrounded by dissimilar habitat (referred to as *matrix* habitat). So "islands" could include real islands surrounded by ocean, lake islands surrounded

by land, or mountain islands surrounded by valleys. They could also include habitat fragments, like those produced by the deforestation of the Amazon (see Figure 17.2). Nonetheless, these islands and island-like habitats all display the same basic pattern: large islands have more species than small islands.

In addition, because of the isolated nature of islands, species diversity on islands shows a strong negative relationship to distance from the main source of species. For example, Lomolino et al. (1989) found that mammal species richness on mountaintops in the American Southwest decreases as a function of the distance from the main species sources—

(A) Reptiles

(B) Mammals

(C) Fishes

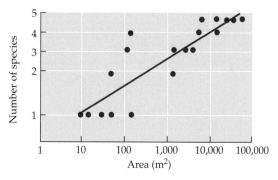

FIGURE 17.19 Species–Area Curves for Islands and Island-Like Habitats Species–area curves plotted for (A) reptiles on Caribbean islands, (B) mammals on mountaintops in the American Southwest, and (C) fish living in desert springs in Australia all show a positive relationship between area and species richness. (A after Wright 1981; B after Lomolino et al. 1989; C after Kodric-Brown and Brown 1993.)

in this case, two large mountain ranges in the American Southwest. This and other examples generally show that islands more distant from source populations, such as those in mainland areas or unfragmented habitats, have fewer species than islands closer to them.

Almost always, though, island isolation and size are confounded. Robert MacArthur and Edward O. Wilson (1963) illustrated this problem by plotting the relationship between

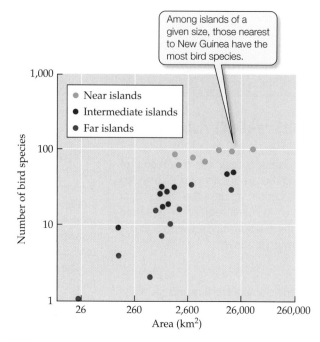

FIGURE 17.20 Area and Isolation Influence Species Richness on Islands MacArthur and Wilson plotted species–area relationships for birds on islands of different sizes and at different distances from New Guinea. (Data from MacArthur and Wilson 1963.)

bird species richness and island area for a group of islands in the Pacific Ocean off New Guinea (**Figure 17.20**). Here, the islands varied both in size and in degree of isolation from the mainland, but some patterns were evident. For example, if we compare islands of equivalent size, the island farthest from New Guinea has fewer bird species than the island nearest the source.

Let's turn now to the question of how island area and isolation from species sources could together act to produce these commonly observed species diversity patterns.

Species richness is a balance between immigration and extinction

The Theory of Island Biogeography (1967) was the first breakthrough in the science of biogeography since Wallace's time. The book was born out of the common interests of two scientists: a mathematical ecologist, Robert MacArthur, and a taxonomist and biogeographer, Edward O. Wilson. Wilson, who had studied the biogeography of ants for his Ph.D. thesis work, had made a few key observations about islands in the South Pacific, which he found himself discussing with MacArthur when they met at a scientific meeting (Wilson 1994). The first observation was that for every tenfold increase in island area, there was a rough doubling of ant species number. The second was that as ant species spread from mainland areas to islands, the new species tended to replace the existing species, but there was no net gain in species richness. There appeared to be an equilibrium number of species on the islands, which was dependent on their size and distance from the mainland, but species composition on the islands could, and did, change over time.

MacArthur, a gifted mathematical ecologist, was just 31 years old when he and Wilson developed these observations into the beginnings of a simple but elegant theoretical regional biogeographic model. The model, published in their book 5 years later, became more commonly known as the **equilibrium theory of island biogeography**. The theory is based on the idea that the number of species on an island or in an island-like habitat is dependent on a balance between immigration rates and extinction rates. The theory works something like this: Imagine an empty island open for colonization by species from mainland or source populations. As new species arrive on the island, by whatever means necessary, the island starts to fill up. The rate of immigration (the number of new species arriving) will decrease over time as more and more species are added, eventually reaching zero when the entire pool of new species that could reach the island is exhausted. But as the number of species on the island increases, there should also be an increase in the rate of extinction. This assumption makes sense according to the simple law of numbers mentioned above: as the number of species increases, there should also be an increase in the number of extinctions. Additionally, as the number of species increases, the population size of each species should get smaller. Conceivably, this could occur for two reasons. First, competition may increase, thus decreasing the population sizes of species as they vie for the same space and resources. Second, predation may increase as more consumer species are added to the island. The result of either species interaction is smaller population sizes and thus a greater risk of species extinction. If we plot the immigration rate against the extinction rate, the actual number of species on the island should fall where the two curves intersect, or where species immigration and extinction are in balance (**Figure 17.21**). This equilibrium number is the number of species that should theoretically "fit" on the island, irrespective of the **turnover,** or replacement of one species with another, that occurs on the island over time.

To understand the influence of island size and isolation on island species richness, MacArthur and Wilson (1967) simply adjusted their curves up or down to reflect their effects (see Figure 17.21). They assumed that island size mainly controls the extinction rate. They reasoned that small islands should have higher extinction rates than large islands, for the same two reasons described above, resulting in an extinction curve for small islands that is higher than that for large islands. Likewise, they reasoned that the distance of an island from the mainland mainly controls the immigration rate. Distant islands should have lower rates of immigration than islands near the mainland, resulting in an immigration curve for distant islands that is lower than that for islands near the mainland.

To test their theory, MacArthur and Wilson (1967) applied it to data from the small volcanic island of Krakatau, near

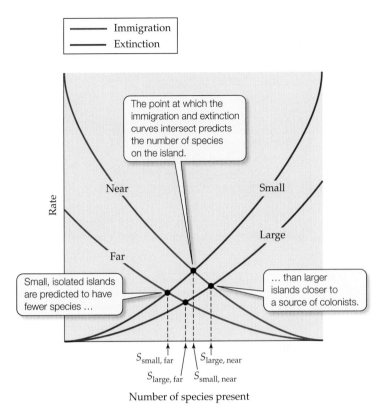

FIGURE 17.21 The Equilibrium Theory of Island Biogeography
MacArthur and Wilson's theory emphasized the balance between species immigration and species extinction rates for islands of different sizes and at different distances from a source of colonizing species. (After MacArthur and Wilson 1967.)

Sumatra and Java, which erupted violently in 1883, wiping out all life on the island (**Figure 17.22**). Surprisingly, scientists began observing the return of animal and plant species within a year of the explosion. MacArthur and Wilson calculated the immigration and extinction rates of birds on the island from data from three surveys at various times since the eruption. Based on these rates, they predicted that the island should sustain roughly 30 species at equilibrium, with a turnover of 1 species. The data showed that bird species richness on the island had indeed reached 30 species within 40 years after the eruption and had remained close to that number thereafter. Unfortunately, they also found that turnover was much higher, at 5 species. Whether this difference was due to a sampling error or a problem with the model is unknown, but this example motivated Wilson and others (for example, the researchers whose work is described in the Case Study) to start testing the model using manipulative experiments.

One of the best-known experiments to test the equilibrium theory of island biogeography was conducted by Daniel Simberloff and his advisor, E. O. Wilson, on mangrove islands and their insect inhabitants in the Florida Keys (Wil-

(A)

(B)

(C)

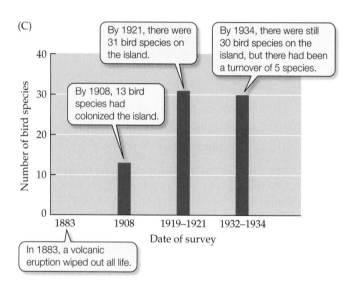

By 1921, there were 31 bird species on the island.

By 1934, there were still 30 bird species on the island, but there had been a turnover of 5 species.

By 1908, 13 bird species had colonized the island.

In 1883, a volcanic eruption wiped out all life.

FIGURE 17.22 The Krakatau Test (A) The eruption of the small volcanic island of Krakatau near Sumatra and Java in 1883, provided a natural test of the equilibrium theory of island biogeography. (B) A drawing based on a photo taken from a ship a few hours before the major eruption. (C) By 1921, the number of bird species reached 31, and in 1934, it was at 30—the equilibrium number predicted by MacArthur and Wilson's theory. Turnover, however, was five times higher than the theory had predicted. (Data from MacArthur and Wilson 1967.)

the least (**Figure 17.23B**). Interestingly, the farthest island had not quite regained its original species richness even after 2 years, but all the islands showed considerable turnover of species, as might be expected for small islands (Simberloff and Wilson 1969).

The equilibrium theory of island biogeography holds true for mainland areas

Do the effects of area and isolation influence differences in species richness in mainland areas as well as on islands? As we saw in Watson's graph of plant species richness in Great Britain, the species–area relationships observed on islands also hold for mainland areas (see Figure 17.18). How, then, does the biogeography of mainland areas differ from that of islands?

Let's consider a plot of plant species richness on the Channel Islands versus that in mainland areas of France (see Box 17.1). Williams (1964) showed that plant species richness increases with area in both locations, but the slope of the line representing the increase was steeper for the Channel Islands than for the French mainland (i.e., the z value was greater on islands; see Box 17.1). How can we interpret this difference? In mainland areas, just as on islands, the number of species is controlled by rates of immigration and extinction. In mainland areas, however, these rates are very different from those on islands. Immigration rates are greater in mainland areas because the barriers to dispersal are lower. Species can move from one area to the next, presumably through continuous,

son and Simberloff 1969; Simberloff and Wilson 1969). These small mangrove islands were scattered at various distances from large "mainland" mangrove stands (**Figure 17.23A**). Simberloff and Wilson manipulated a handful of these islands by spraying them with an insecticide to remove all of their insects and spiders. They then surveyed the defaunated islands over a year-long period. By the end of the year, species numbers on the islands were similar to those before the defaunation; furthermore, the island closest to a source of colonists had the most species, and the farthest island had

(A)

(B)

The island closest to the source of colonists recovered its predefaunation species richness in 140 days.

The island farthest from the source colonists still had not recovered its predefaunation species richness after a year.

Number of species

Days since defaunation

- - - - Predefaunation species number on near island
- - - - Predefaunation species number on far island
──●── Near island
──●── Far island

FIGURE 17.23 The Mangrove Experiment
To test the equilibrium theory of island biogeography, Simberloff and Wilson defaunated mangrove islands (A) located at different distances from larger mangrove stands and recorded the number of insect species that recolonized the islands. (B) Results for two islands, one near and one far from a source of colonists, are shown here. (After Simberloff and Wilson 1969.)

non-island habitat. In addition, extinction rates are much lower in mainland areas because of the continual immigration of new individuals from the larger mainland population. Species will always have a good chance of being "rescued" from local extinction by other population members. The end result of these higher immigration and lower extinction rates in mainland areas is a lower rate of increase in species richness with increasing area, and thus a gentler slope, than in island areas.

We have seen over and over again in this chapter that geographic area has a large influence over species diversity at global and regional spatial scales. This effect takes on heightened significance as more habitats become "island-like" due to human influences. As we will see in the Case Study Revisited, the theory and practice of island biogeography is timely and relevant to the issues of conservation that we deal with today.

The Largest Ecological Experiment on Earth

One goal of ecologists is to understand the science behind the conservation of species threatened by habitat destruction and fragmentation. As we set aside more and more reserves to protect species diversity, the areas around those reserves continue to be changed by human activities, leaving many of these islands in a sea of degraded habitat that is unsuitable for the species they contain. Thus, it is critical that we understand reserve design if we are to maximize our conservation goals. When Lovejoy and colleagues embarked, nearly 30 years ago, on the Biological Dynamics of Forest Fragments Project (BDFFP) in the Amazon, one of their goals was to study the effects of reserve design on the maintenance of species diversity

(Bierregaard et al. 2001). What they found surprised them. Unfortunately, as it turned out, they learned that habitat fragmentation had more negative and complicated effects than they had originally anticipated.

One of the first things they learned was that forest fragments needed to be quite large and close together to effectively maintain their original species diversity. For example, in a study of forest understory birds, Ferraz et al. (2003) found that even the largest fragments they surveyed (100 hectares) lost 50% of their species within 15 years. Given the regeneration time for these tropical rainforests, which ranges from several decades to a century, they projected that even the largest fragments in the experiment would be ineffective at maintaining bird species richness until forest regeneration could "rescue" species within the fragments. The ecologists calculated that over 1,000 hectares would be needed to maintain bird species richness until the forests could be regenerated, an area far greater than the average Amazon rainforest fragment in existence today (Gascon et al. 2000). If forest regeneration did not occur, as is likely when the land around a forest fragment is developed or used for agriculture, the fragment would have to contain 10,000 hectares or more to maintain most of its bird species over more than 100 years of isolation (although even a fragment of that size could not sustain them all).

The researchers of the BDFFP were also quite surprised at the short distances between fragments that resulted in almost complete isolation of species. Clearings even 80 meters wide hindered the recolonization of fragments by birds, insects, and arboreal mammals (Laurance et al. 2002). It seemed that animals avoid entering clearings for a number of interrelated reasons, the most obvious of which is that they have no innate reason to do so. They have evolved within large, continuous, and climatically stable habitats that lack the fragmentation imposed on them by deforestation. This may be one reason why almost none of the endemic species in the Amazon are migratory. Moreover, even if some animals were inclined to venture into the clearings, specific requirements for their movement, such as trees for arboreal mammals, would not be present to facilitate their travel to another forest patches.

A second major finding of the BDFFP was that habitat fragmentation exposes the species within a fragment to a wide variety of potential hazards, including harsh environmental conditions, fires, hunting, predators, diseases, and invasive species. These **edge effects**, which occur at the transition between forest and nonforested habitat, can act synergistically to increase local species extinctions. Trees, for example, can be killed or damaged by their sudden exposure to brighter light, higher temperatures, wind, fire, and diseases (**Figure 17.24**). Over time, depending on the surrounding matrix habitat, the ultimate influences of edge effects are revealed. If the surrounding habitat is allowed to regenerate, secondary succession occurs in the edge habi-

tats, mediating edge effects. However, if the surrounding habitat continues to be degraded, then the area subjected to edge effects may increase in size. For example, Gascon et al. (2000) describes forest fragments in the southern Amazon that are embedded in huge non-native sugarcane

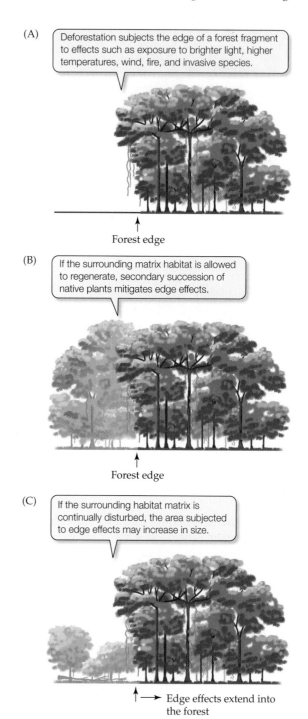

(A) Deforestation subjects the edge of a forest fragment to effects such as exposure to brighter light, higher temperatures, wind, fire, and invasive species.

Forest edge

(B) If the surrounding matrix habitat is allowed to regenerate, secondary succession of native plants mitigates edge effects.

Forest edge

(C) If the surrounding habitat matrix is continually disturbed, the area subjected to edge effects may increase in size.

Edge effects extend into the forest

FIGURE 17.24 Tropical Rainforests on the Edge Research on the BDFFP forest fragments showed that deforestation subjects the remaining forest fragments to negative edge effects. (After Gascon et al. 2000.)

and *Eucalyptus* plantations where burning is used regularly for crop rotation. The burning maintains the forest edges in a constant state of disturbance. Fire-tolerant plant species, many of them non-native, become more common at the edges and act as conduits for more fires. This positive feedback loop ends up decreasing the effective size of the forest fragment and continually increasing the area subjected to edge effects. Some edge habitats can extend a kilometer or more into a fragment, essentially influencing the entire area of a 1,000-hectare fragment.

The results of the BDFFP have contributed immensely and soberly to our understanding of forest fragmentation. As Laurance et al. (2002) point out, the BDFFP is a controlled experiment that probably provides a conservative estimate of species losses. The reasons have to do with fragment shape, location, and protection from human disturbance. For example, the fragments in the BDFFP are primarily square, which makes for a smaller amount of edge area compared with, say, a rectangular fragment. In addition, the fragments in the BDFFP are close to continuous tracts of pristine forest, thus increasing the chance of rescue effects. Finally, the matrix habitat in the BDFFP is being managed for restoration, thus decreasing the chances of edge effects such as fire, invasive species, or hunting. Although it is unlikely that a single "minimum critical size" for forest fragments will ever be identified given the complexities presented here, the BDFFP has shown us that most forest fragments are too small to maintain all their original species and that conservation will be most effective if we err on the side of larger, closer, and more numerous fragments.

CONNECTIONS IN NATURE
Human Benefits of Tropical Rainforest Diversity

Why do we care when species go extinct in a rainforest far away? Such extinctions raise ethical and aesthetic concerns, similar to those when great works of art or antiquities are lost to society. There are economic concerns about the loss of important products such as timber. But there are also concerns about losses of other products and services that are important to human health and well-being. Many of our foods and medicines, for example, have their origins in rainforests. At least 80% of the developed world's diet originated in tropical rainforests, including corn, rice, potatoes, squash, yams, oranges, coconuts, lemons, tomatoes, and nuts and spices of all kinds. Twenty-five percent of all commercial pharmaceuticals are derived from tropical rainforest plants, but less than 1% of tropical rainforest plants have been tested for their potential uses.

These statistics beg the question, "What is the relative economic value of tropical rainforest plants used for non-timber purposes?" It turns out that there have been very few economic analyses of this type. A few studies come from the Millennium Ecosystem Assessment (2005), a synthesis of studies on the use of the environment and its relationship to human needs created by leading scientists from around the world. An example comes from Cambodia, where the total economic value of traditional forest uses (e.g., fuelwood, rattan and bamboo, malva nuts, and medicines) was compared with that of unsustainable forest harvesting. The value of traditional forest uses is four to five times greater ($700–$3,900 per hectare) than that of unsustainable forest harvesting ($150–$1,100 per hectare).

Recognition of the economic benefits of changing our resource management practices has only just begun. Why is this? Part of the answer lies in our not formally recognizing the economic value of the services provided by species or whole communities. Tropical rainforests provide food, medicine, fuel, and a destination for tourists, all of which can be obtained without complete deforestation. Rainforests also regulate water flow, climate, and atmospheric CO_2 concentrations. Assigning a value to any of these important services is difficult compared with setting the market price of timber or agricultural products. For that reason, it is easier to justify the use of rainforest timber and land (or even some of the sustainable forest products) for private profit than the conservation of rainforests for the ecological services that benefit society in general. If private landowners are not given incentives to value the larger social benefits of ecological services, then maximizing personal gains will often drive their decisions. Given the importance of ecological services to our planet, we cannot afford to ignore these economic trade-offs any longer.

SUMMARY

CONCEPT 17.1 Patterns of species diversity and distribution vary at global, regional, and local spatial scales.

■ Biogeography is the study of species distribution and diversity at different geographic locations and spatial scales (global, regional, and local).

■ The global spatial scale includes the entire world, a huge geographic area over which there are major differences in climate and species distributions.

■ The regional spatial scale encompasses a smaller geographic area in which the climate is roughly uniform and the species contained therein are bound by dispersal to that region.

- The local spatial scale encompasses the smallest geographic area and is essentially equivalent to a community.

- Different spatial scales of species diversity are connected to one another in a hierarchical way, with one dependent on another.

- Studies show that the regional species pools can largely determine the number of species present in communities but local conditions are also important and cannot be discounted.

CONCEPT 17.2 Global patterns of species diversity and composition are controlled by geographic area and isolation, evolutionary history, and global climate.

- Earth's land mass can be divided into six biogeographic regions that vary markedly in species diversity and composition.

- The biota of the six biogeographic regions reflect a history of isolation due to continental drift.

- Tracing the threads of vicariance over large geographic areas and long time periods provided important evidence for early theories of evolution.

- Species diversity is greatest in the lower-latitude tropical zones and declines at higher latitudes.

- A number of hypotheses, involving geographic area and climatic stability, evolutionary history, and productivity, have been proposed to explain the latitudinal gradient in species diversity.

CONCEPT 17.3 Regional differences of species diversity are controlled by area and distance due to a balance between immigration and extinction rates.

- Species richness tends to increase with the area sampled and tends to decrease with distance from a source of species.

- Most species–area relationships have been documented for islands, which include all kinds of isolated areas surrounded by dissimilar habitat.

- The equilibrium theory of island biogeography predicts that a balance between immigration and extinction rates controls species diversity on islands or island-like areas.

- Larger islands closer to a source of species have more species than smaller islands that are more distant from a source of species, possibly due to higher immigration rates and lower extinction rates.

- The same species–area relationship observed on islands also holds for mainland areas, but the rate of increase in species richness with increasing area is lower than on islands.

REVIEW QUESTIONS

1. Spatial scale is important to the biogeographic patterns of species diversity and composition we see on Earth. Define the various spatial scales that are important to biogeography and describe how they are related to or interconnected with one another.

2. Describe the factors believed to have created biogeographic regions on land and in the oceans.

3. Latitudinal gradients in species diversity and composition are strong global features of biogeography. Describe three hypotheses proposed to explain why species diversity is higher in the tropics and decreases toward the poles for the majority of taxa on Earth.

4. Imagine that you work for your government's park service. You are presented with three proposed designs for a new park that is intended to preserve an important type of rare habitat. All of the designs consist of

fragments of rare habitat surrounded by a matrix of less desirable habitat. You must choose the park design that will best preserve the species richness of the rare habitat. Using MacArthur and Wilson's (1967) equilibrium theory of island biogeography, justify which of the three proposed park designs you would most recommend to the government.

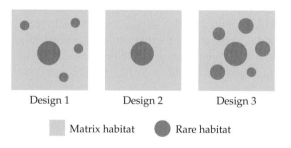

Design 1   Design 2   Design 3

Matrix habitat ● Rare habitat

HANDS-ON PROBLEM SOLVING

Island Biogeography   This Web exercise (www.sinauer.com/ecology) explores the factors that determine the number of species that can live on different islands according to the theory of island biogeography. In a series of simulations, you will manipulate the size of an island and the distance from the island to the mainland to demonstrate how these factors affect the equilibrium number of species on that island.

SUGGESTED READINGS

Bierregaard, R. O. Jr., C. Gascon, T. E. Lovejoy and R. C. G. Mesquita. 2001. *Lessons from Amazonia: The Ecology and Conservation of a Fragmented Forest.* Yale University Press, New Haven, CT. *An overview of all that has been learned from the BDFFP in the 25 years since its inception.*

Lomolino, M. V., B. R. Riddle and J. H. Brown. 2006. *Biogeography.* 3rd ed. Sinauer Associates, Sunderland, MA. *A comprehensive introductory text on biogeography.*

Quammen, D. 1996. *The Song of the Dodo: Island Biogeography in an Age of Extinctions.* Scribner, New York. *An engaging account of Quammen's travels with modern biogeographers seeking to discover the causes, and document modern-day losses, of species diversity around the world.*

Rosen, J. 2007. Missing link. *New Yorker.* February 12. *A critical account of Alfred Russel Wallace's work and the reasons why he never received the recognition that Charles Darwin did for his theory of speciation.*

Rosenzweig, M. L. 1995. *Species Diversity in Space and Time.* Cambridge University Press, Cambridge. *A comprehensive review of the patterns and processes influencing species diversity at regional and global spatial scales.*

Wilson, E. O. 1994. *Naturalist.* Island Press, Washington, DC. *Wilson's entertaining and well-written autobiography describes the origin and meaning of the equilibrium theory of island biogeography.*

Wilson, E. O. 1999. *The Diversity of Life.* 2nd ed. W. W. Norton, New York. *An entertaining narrative account of global species diversity past, present, and future.*

CHAPTER **18** Species Diversity in Communities

CASE STUDY

Powered by Prairies? Biodiversity and Biofuels

A hundred and twenty years ago, as the last covered wagons crossed the vast and beautiful prairies of the North American Great Plains on their way out west (**Figure 18.1**), two German inventors toiled in a laboratory to develop the first gasoline-powered car. In 1889, Gottlieb Daimler, along with his design partner, Wilhelm Maybach, built the first automobile from the ground up; it had a four-speed transmission and could reach speeds of 10 mph. By the turn of the twentieth century, automobiles were being produced by the thousands, and a century later, by the millions. Although automobiles dominate our lives today, in large part because of their convenience and affordability, their negative environmental impacts cannot be denied. Internal combustion engines create nearly 80% of the anthropogenic atmospheric CO_2 emissions that are contributing to global warming (IPCC 2007). Furthermore, the world's supplies of fossil fuels are not endless, and their costs will no doubt continue to rise. These circumstances have led to the development of alternative sources of energy, one of which is biofuels.

Biofuels are liquid or gas fuels made from plant material (biomass). Two basic kinds of biofuels are being made in the United States today: ethanol and biodiesel, produced mostly from corn and soybeans, respectively. Biofuels have two potential advantages over fossil fuels. First, they aim to be **carbon neutral**, meaning that the amount of CO_2 produced by burning them is matched by the amount taken up by the plants from which they are made. Second, their supply is potentially limitless as long as crops can be grown to fulfill production needs. Unfortunately, their downsides are also numerous. The growing of corn and soybeans for biofuels competes for land and water that could be used for food production. Furthermore, fossil fuels, in the form of fertilizers, are required to grow these

FIGURE 18.1 Powered by Prairies? A native prairie on the American Great Plains, USA. Could they be used to produce biomass more efficiently for biofuels?

crops. Large amounts of fossil fuels are also required to produce and transport biofuels. For these reasons, biofuels, as they are currently produced, have their own negative environmental impacts and only modestly reduce fossil fuel use.

One promising advance in biofuel technology involves the use of more sustainable alternatives to food crops. These include the by-products of food or lumber production, such as straw, corn stalks, or waste wood. This is also where those vast prairies, once an inspiration to the settlers of western North America, come into play. Most of the land that was once prairie has been converted to agriculture. Years of cultivation have degraded some of this land, making it unsuitable for high-yield food crops. This situation inspired David Tilman, a community ecologist at the University of Minnesota, to consider how prairie communities, once symbols of the limitless potential of the land and its resources, might be used to make biofuels in a more sustainable manner.

Tilman had been studying prairie communities for some time and was interested in how plant species diversity influenced the productivity of these communities. Through a set of experiments on abandoned agricultural land at Cedar Creek, Minnesota (**Figure 18.2**), he and his colleagues showed that plots with more plant species (up to 16 species in some plots) produced greater biomass for a given amount of water or nutrients than plots with fewer species (Tilman and Downing 1994; Tilman et al. 1996). Tilman reasoned that if diverse assemblages of prairie plants could be grown on degraded land, they might be a great source of biomass for biofuel production, given the low fossil fuel input required by the prairie system. Can prairie plants be used in biofuel production, and are they a viable substitute for traditional crops such as corn and soybeans? As we will see, prairie plant species diversity makes all the difference in the answers to these questions.

FIGURE 18.2 Plant Diversity Matters Experimental plots used to investigate the relationship between plant species richness and productivity at Cedar Creek, Minnesota.

ommunities vary tremendously in the numbers and kinds of species they contain. In the last chapter, our worldwide tour of forest communities demonstrated the wide variation in species diversity that occurs both globally and regionally. We saw that communities in the tropics (such as the Amazon rainforest) had many more tree species than those at higher latitudes (such as the forests of the Pacific Northwest or New Zealand). Moreover, we found that regional species pools had an important, but not an exclusive, influence over the number of species within a community. In New Zealand, for example, regional tree diversity is generally high, but there are many more species in the northern flowering tree communities than in the southern beech communities (see Table 17.1). This difference exists because local abiotic and biotic conditions have an important role in mediating species diversity. In this chapter, we will focus on species diversity at the local scale. We will ask two important questions: First, what are the factors that control species diversity within communities? Second, what is the function of this species diversity within communities?

Community Membership

CONCEPT 18.1
Species richness differs among communities due to variation in regional species pools, abiotic conditions, and species interactions.

If you looked across a landscape from the top of a mountain, you would see a patchwork of different communities that might consist of, say, forests, meadows, lakes, streams, and marshes (**Figure 18.3**). You can bet that each of those communities would have a different species composition and a different species richness. The meadow would probably be dominated by a variety of grasses, herbs, and terrestrial insects. The lake might be filled with different species of fish, plankton, and aquatic insects, possibly harboring as many species as the meadow. Even though some species might be able to move from one community to another (such as amphibians), the two communities would still be highly distinct from each other.

FIGURE 18.3 A View from Above   Looking at the Grand Tetons, it is easy to see that the landscape is made up of a patchwork of communities of different types.

How do collections of species end up coming together and forming communities with different species compositions and richnesses? More simply put, what controls species diversity in communities? One way to answer this question is to consider the factors that control species membership in communities. If you think about the sheer number of species that coexist within one community, it is clear that no one process is responsible for all the species we find there. As we saw in Chapters 8 and 17, the distributions and abundances of organisms within communities are dependent on three factors: (1) regional species pools and dispersal ability (species supply), (2) abiotic conditions, and (3) species interactions. We can think of these three factors as "filters" that act to exclude species from (or include them in) particular communities (**Figure 18.4**). Let's briefly consider each of them in more detail.

Species supply is the "first cut" to community membership

In Chapter 17, we saw that the regional pool of species provides an absolute upper limit on the numbers and types of species that can be present within communities (see Figure 17.6). Not surprisingly, we saw that regions of high species richness tend to have communities of high species richness

FIGURE 18.4 Community Membership: A Series of Filters
Species end up in a local community by passing through a series of "filters" that help determine community membership. Species are lost at each filter, so that local communities contain a fraction of the species in the regional pool. In practice, the filters work at the same time, rather than in series as the figure suggests. (After Lawton 2000.)

(see Figure 17.7). This relationship is due to the role of the regional species pool and, more specifically, dispersal in "supplying" species to communities (see Figure 18.4A). Nowhere is the controlling effect of dispersal on community membership more evident than in the invasion of communities by non-native species.

As ecologists are beginning to learn, humans have greatly expanded the regional species pools of communities by serving as vectors of dispersal. For example, we know that many aquatic species travel to distant, and naturally unreachable, parts of the world in the ballast water carried by ships (**Figure 18.5A**). Seawater is pumped into and out of ballast tanks, which serve to balance and stabilize cargo-carrying ships, all over the world. Most of the time, the water—along with the pelagic organisms it contains (from bacteria to planktonic larvae to fish)—is taken up and released close to ports, where some of the organisms have the opportunity to colo-

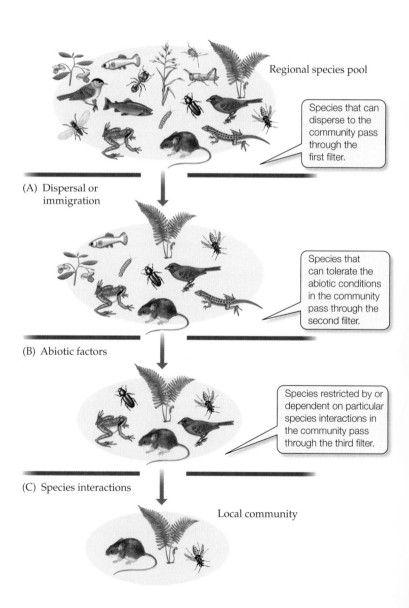

Regional species pool

Species that can disperse to the community pass through the first filter.

(A) Dispersal or immigration

Species that can tolerate the abiotic conditions in the community pass through the second filter.

(B) Abiotic factors

Species restricted by or dependent on particular species interactions in the community pass through the third filter.

(C) Species interactions

Local community

(A)

(B)

FIGURE 18.5 Humans Are Vectors for Invasive Species (A) Large and fast oceangoing ships are carrying marine species to all parts of the world in their ballast water. (B) The zebra mussel (*Dreissena polymorpha*), a destructive invader of the inland waterways of the United States, was carried there from Europe in ballast water.

nize nearshore communities. As many as 5,000 freshwater and marine species are transported in the ballast water of oceangoing vessels each day. Ballast water introductions have increased substantially over the past few decades because ships are larger and faster, so that more species can be taken along and survive the trip. In 1993, Carlton and Geller listed 46 known examples of ballast-mediated invasions in the previous 20 years alone. One species, the zebra mussel (*Dreissena polymorpha*), arrived in North America in the late 1980s in ballast water discharged into the Great Lakes (**Figure 18.5B**). It spread rapidly throughout the St. Lawrence River, into Lake Champlain, and into the Mississippi River drainage. It has had community-changing effects on inland waterways by fouling infrastructure and dramatically changing water properties. Zebra mussels have been recorded at densities as high as 700,000 per square meter, and their voracious filter feeding has been known to decrease phytoplankton populations by

80%–90% compared with preinvasion values (Benson and Raikow 2007). Another example of a ballast water introduction, which we learned about in Chapter 10, is the release of the comb jelly *Mnemiopsis leidyi* into the Black Sea and the negative ecological consequences of that introduction.

The mounting evidence from studies of damaging species invasions has made it clear that ecologists cannot ignore the role of large-scale processes of dispersal in determining species richness at the local scale. But let's turn our attention now to the role of local conditions, particularly the abiotic and biotic characteristics of communities that help determine their structure.

Abiotic conditions play a strong role in limiting community membership

A species may be able to get to a community but be physiologically unable to tolerate the abiotic conditions that are imposed by its environment (see Figure 18.4B). These physiological constraints can be quite obvious. For example, if we return to our thought experiment of viewing the landscape from the top of a mountain, it is reasonable to assume that the abiotic attributes of lakes make them good places for fishes, plankton, and aquatic insects, but not for terrestrial plants. Similarly, lakes might be good habitat for certain species of fish, plankton, and aquatic insects, but not for all of them. Some of these species depend on fast-flowing water and are thus restricted to streams rather than lakes. These differences among abiotic environments are obvious constraints (or requirements, depending on how you look at it) that largely determine where particular species can and cannot occur within a region.

There are many examples throughout this book that demonstrate how physiological constraints can control the distributions and abundances of species [see, for example, the aspen example (p. 84 and Figure 4.4), the creosote bush and saguaro cactus example (p. 179 and Figure 8.7), and the barnacle example (p. 180 and Figure 8.9)]. In our earlier discussion of species introductions by ballast water, it was clear that humans could transport many more species than could actually survive in the new locations to which they were carried. For example, the majority of organisms released with ballast water find themselves in coastal waters that may not have the temperature, salinity, or light regimes they need to survive or grow. Luckily, many of these individuals die before they can become a threat to the native community. But ecologists know, based on examples such as the *Caulerpa* invasion in the Mediterranean (see the Case Study in Chapter 15), that it is not wise to rely on physiological constraints as a mechanism to exclude potential invaders from a community. It may be that, with multiple introductions, particular individuals with slightly different physiological capabilities could survive and reproduce in an environment once thought uninhabitable by their species. Or it is possible that species require particular interactions with other species that help ameliorate the unfavorable conditions. We turn next to species interactions, the final factor determining community membership.

Who you interact with makes all the difference in community membership

Even if species are able to get to a community and cope with its potentially restrictive abiotic conditions, the final cut requires coexistence with other species (see Figure 18.4C). Clearly, for species that depend on other species for their growth, reproduction, and survival, those other species must be present if they are to gain membership in a community. Equally important, species may be excluded from a community by competition, predation, parasitism, or disease. For example, returning to our thought experiment, we might assume that lakes are great habitats for many fish species, but how could those species all live together in one lake, given that resources are limiting? A simple view suggests that the best competitors or predators should dominate in the lake, thus excluding weaker species and resulting in a low-diversity ecosystem. But we know that most communities are full of species that are actively interacting and coexisting. So what allows this coexistence? There are a number of important mechanisms that allow species to coexist, and we will spend the next two sections considering them. But first let's ask how species might be excluded from communities by biological interactions, a question that is a bit different, but equally relevant.

The best examples of exclusion of species from communities by other species again come from the invasive species literature. The failure of some non-native species to be incorporated into communities has been attributed to interactions with native species that exclude or slow the growth of the non-native species—a phenomenon that ecologists call **biotic resistance**. Multiple studies in a variety of communities have shown that native herbivores have the ability to reduce the spread of non-native plants in substantial ways. Maron and Vila (2001) found that mortality of non-native plants due to native herbivores can be quite high (about 60%), especially at the seedling stage (up to 90% in some studies). But while native herbivores can kill individual non-native plants, it is still unknown how important native species are in completely excluding non-native species from a community. For example, Faithfull (1997) found that in Australia, adults and larvae of the native Lucerne seed web moth (*Etiella behrii*) breed and feed on the seed pods of the invasive gorse shrub (*Ulex europaeus*), but the plant still continues to spread (**Figure 18.6**). This lack of knowledge about biotic resistance may be an artifact of ecologists more often studying why a particular non-native species does or does not spread once it becomes a provisional member of the community than studying all the cases in which it is unable to gain a foothold because of interactions with native species. It may also be true that most failed introductions of non-native species go completely undetected.

Studying invasions gives us valuable insights into the question of how species are excluded by other species from communities. But the flip side of that question, which asks why species are included in communities, or better yet, how they coexist, can be more complicated. In the next two sections, we will consider theories that seek to answer these questions. There are two schools of thought, most easily classified as **equilibrium theory** and **nonequilibrium theory**. Equilibrium theory relies on ecological and evolutionary

Etiella behrii

FIGURE 18.6 Stopping Gorse Invasion?   Herbivory by adults and larvae of the native Lucerne seed web moth (*Etiella behrii*) has slowed, but has not stopped, an invasion of the non-native gorse shrub (*Ulex europaeus*) in Australia.

"compromises" that result in partitioning of resource use as the mechanism of relatively stable coexistence. Nonequilibrium theory relies on fluctuating conditions that maintain coexistence by keeping dominant species from monopolizing resources. Let's consider the main equilibrium theory, that of resource partitioning, next.

Resource Partitioning

> **CONCEPT 18.2**
> Resource partitioning among the species in a community reduces competition and increases species richness.

In our first discussion of species interactions in Chapter 11, we used the Lotka–Volterra competition model to explore the outcome of competition between two species. A key assumption of the model was that as two species interact, they reach an equilibrium state defined by their individual carrying capacities and competition coefficients. The outcome of the interaction (coexistence or competitive exclusion) is determined when an equilibrium state is reached (see Figure 11.13). We used the model to show that competing species are more likely to coexist when they use resources in different ways (see Figure 11.13D). Let's consider how this concept, known as **resource partitioning**, has been used both in theory and in practice to explain species diversity within communities.

Resource partitioning allows more species to coexist along a resource spectrum

A simple model of resource partitioning envisions each type of resource available in a community along a resource spectrum (**Figure 18.7**). The resource spectrum could represent, for example, different nutrients, prey sizes, or habitat types (note that the spectrum represents the variability of an available resource, but not its amount). We can assume that each species' resource use falls somewhere along this spectrum and overlaps with the uses of other species to varying degrees (Figure 18.7A). The more overlap, the more competition between species, with the extreme being complete overlap and competitive exclusion. The less overlap, the more specialized species have become on that particular part of the spectrum, and the less strongly they will compete with one another.

Using this guiding principle, we can consider some of the possible ways that resource partitioning might result in higher species richness in some communities than in others. First, species richness could be high in some communities because species show a high degree of specialization along the resource spectrum (Figure 18.7B). More species can be "packed" into a community if the overlap in resource use among the species is low, leading to less competition and ultimately higher species richness. Second, species richness could be high in some communities because the resource spectrum is broad (Figure 18.7C). A broader resource spectrum would make more kinds of resources available to be used by a diversity of species, resulting in higher species richness. Finally, species richness could be high in some communities if species were generalists with highly overlapping patterns of resource use (Figure 18.7D). In this scenario, species would be subject to more competition for the same resources,

(A)

Resource use

Different species

Each curve represents the resource use of a different species in the community.

(B) High specialization, little overlap

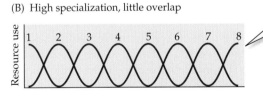

Species in this community show a high degree of specialization (little overlap) in their resource use.

(C) Broad resource spectrum, little overlap

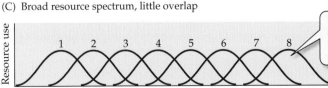

In this community, the resource spectrum is broad, making more kinds of resources available to support more species.

(D) High generalization, high overlap

Resource spectrum

Species in this community are generalists (their resource use overlaps) and compete with one another, but that competition keeps their population sizes small, allowing more species in the community.

FIGURE 18.7 Resource Partitioning
Species coexistence within communities may depend on how the species divide resources. (A) The principle of resource partitioning along a resource spectrum. (B–D) Three characteristics of communities that can result in higher species richness.

and thus would maintain smaller population sizes, but those smaller populations would allow more species to be packed into the community.

At this point, let's turn our attention away from models and take a look at some real communities to see how resource partitioning might work in practice.

Studies of resource partitioning show that specialization is the main mechanism of coexistence

As we learned earlier from the two-species studies of Gause (1934a) on *Paramecium* (p. 248) and Connell (1961b) on barnacles (p. 246), species that compete with each other can coexist by using slightly different resources. Robert MacArthur, whose work on the equilibrium theory of island biogeography we described in Chapter 17, played a pioneering role in understanding how this mechanism might be applied to whole communities, where multiple species interactions are occurring all at once.

MacArthur studied warblers, small and brightly colored birds that co-occur in the forests of northern North America. The idyllic New England forests that MacArthur studied are home to an array of warbler species (*Dendroica* spp.) that migrate from the tropics each spring to breed and feed on insects in the north. Through a series of detailed natural history observations in the summers of 1956 and 1957 in Maine and Vermont, MacArthur (1958) recorded the feeding habits, nesting locations, and breeding territories of five species of warblers to find out how they might coexist in the face of very similar resource needs.

MacArthur began mapping the locations of warbler activity in tree canopies and found that the warblers were using different parts of the habitat in different ways (**Figure 18.8**). For example, yellow-rumped (*D. coronata*) warblers fed from the tops of trees to the forest floor, while bay-breasted (*D. castanea*) and black-throated green (*D. virens*) warblers fed more in the middle of a tree, both inside and toward the outside of the tree canopy. Blackburnian (*D. fusca*) and Cape May (*D. tigrina*) warblers both fed on the outside tops of trees, more commonly catching their prey

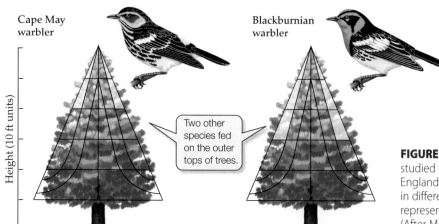

FIGURE 18.8 Resource Partitioning by Warblers Robert MacArthur studied the habitat and food choices of five species of warblers in New England forests. He found that the warblers partition resources by feeding in different parts of the same trees. The shaded areas in each tree diagram represent the parts of trees where each warbler species fed most often. (After MacArthur 1958.)

in midflight. MacArthur found that the nesting heights of the five warbler species also varied, as did their use of breeding territories. Taken together, these observations supported his hypothesis that the warblers, although using the same habitat and food resources, were able to coexist by partitioning those resources in slightly different ways. His work, which was part of his Ph.D. thesis, earned him the Mercer Award, a prestigious annual award for the best paper in ecology.

MacArthur, along with John MacArthur (MacArthur and MacArthur 1961), extended these ideas about resource partitioning in a study of the relationship between bird species diversity (calculated using the Shannon index; see p. 327) and foliage height diversity (a measure of the number of vegetation layers in a community that serves as an indication of habitat complexity, also calculated using the Shannon index). They found a positive relationship between the two in 13 tropical and temperate bird habitats from Panama to Maine (**Figure 18.9**). Interestingly, bird species diversity was not related to plant diversity per se, beyond the effects of foliage height diversity, suggesting that tree species identity was less important than the structural complexity of the habitat.

Another important resource partitioning study comes from phytoplankton communities. In Chapter 11, we learned about David Tilman et al.'s (1981) study of two species of diatoms that competed for silica, which diatoms use to build their cell walls. When the two species were grown together in a laboratory environment with limited supplies of silica, one outcompeted and excluded the other (see Figure 11.4). How, then, do diatom species coexist in nature? Tilman (1977) proposed what has become known as the **resource ratio hy-**

FIGURE 18.10 Resource Ratio Hypothesis Two competing diatoms (*Asterionella* and *Cyclotella*) can coexist by using slightly different proportions of phosphorus and silica. (Data from Tilman 1977.)

pothesis, which posits that species coexist by using resources in different proportions. He predicted that diatoms would be able to coexist, despite using the same set of limiting nutrients, by acquiring these nutrients in different proportions (**Figure 18.10**). By growing two diatom species, *Cyclotella* and *Asterionella*, in laboratory environments that differed in their ratios of silica (SiO_2) to phosphorus (PO_4), Tilman found that *Cyclotella* was able to dominate only when the ratio of silica to phosphorus was low (approximately 1:1). When the ratio of silica to phosphorus was high (more like 1,000:1), *Asterionella* outcompeted *Cyclotella*. Only when the proportion of silica and phosphorus were limiting to both species (in the range of 100:1 to 10:1) could they coexist with one another. Even though both species needed the same set of nutrients, it was the way in which they partitioned these resources that allowed them to coexist. Outside of a laboratory setting, this partitioning would work best if nutrients naturally varied within the environment. What is the support for this possibility in the field?

In a detailed survey, Robertson et al. (1988) mapped resource distribution in an abandoned agriculture field in Michigan that had been colonized by grassland plants. They found considerable variation in soil moisture and nitrogen concentrations at spatial scales of a meter or less (**Figure 18.11**).

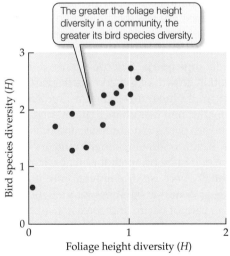

FIGURE 18.9 Bird Species Diversity Is Higher in More Complex Habitats MacArthur and MacArthur plotted bird species diversity against foliage height diversity (a measure of habitat complexity) for 13 different communities. Both kinds of diversity were calculated for each community using the Shannon index (*H*). (Data from MacArthur and MacArthur 1961.)

(A) Soil nitrogen levels

Nitrate (NO_3^-) concentration ($\mu g/cm^2$)

■ ≥ 8.9   ■ 6.4–8.8   ▨ 3.9–6.3   ☐ 0.5–3.8

Both nitrogen concentrations and soil moisture showed great variation over short distances.

(B) Soil moisture levels

Moisture ($\mu g/cm^2$)

■ ≥ 1.06   ■ 0.76–1.05   ▨ 0.46–0.75   ☐ 0.10–0.45

FIGURE 18.11 Resource Distribution Maps   Mapping of (A) nitrogen (nitrate, NO_3^-) concentrations and (B) soil moisture levels in an abandoned agricultural field revealed considerabe small-scale variation. (From Robertson et al. 1988.)

investigated, but it does suggest that resource partitioning could occur in plants (see Figure 21.9). Some of the best evidence of resource partitioning in plants comes from experiments that manipulate species richness and measure productivity, as in the Case Study on prairie diversity. We will explore this evidence in more detail in Section 18.4.

The theory of resource partitioning relies on the assumption that species have reached a stable population size (the carrying capacity) and that resources are limiting. Some ecologists have argued that this assumption is unrealistic because species' populations fluctuate considerably in space and time. For example, we saw in Chapter 16 that processes such as disturbance, stress, or predation can act to lower the abundances of some species, thus freeing up resources for other species and allowing coexistence. In the next section, we will consider how nonequilibrium conditions might control species diversity at the local scale.

Nonequilibrium Theories

CONCEPT 18.3
Nonequilbrium processes such as disturbance, stress, and predation can mediate resource availability, thus affecting species interactions and coexistence.

We have seen in previous chapters that disturbance, stress, and predation can modify species interactions and allow for species coexistence. We saw that when two species are competing with each other for the same resource, as in the case of the sea palms and mussels competing for space in the rocky intertidal zone (see p. 258), coexistence can be achieved if the population growth of the dominant species is disrupted. In that example, the mussels are the dominant competitors, and only through disturbance of the mussels via wave action can the sea palms acquire space and thus coexist with the mussels. In this and many other examples in this book, as long as the dominant competitor is unable to reach its own carrying capacity because of a reduction in its population numbers due to disturbance, stress, or predation, competitive exclusion cannot occur, and coexistence will be maintained (**Figure 18.12**).

On the same theme of coexistence and variability, we have explored the effect of nonequilibrium conditions on the formation of positive interactions. For example, we saw in the cases of salt marsh plants (Figure 16.14) and plants at high elevations (Figure 14.11) that stress can shift what might be a negative interaction under more benign conditions to a positive one. These studies show that species that might nor-

These patches of water and nitrogen resources did not necessarily correspond to topographic differences, and they were not correlated with each other. If we were to overlay the water map on the nitrogen map, we would find even smaller patches corresponding to different proportions of these two resources. Whether this spatial distribution of resources affected or reflected the structure of the plant community was not

(A)

(B)

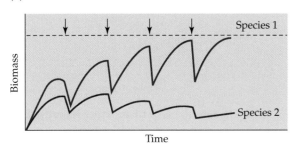

FIGURE 18.12 The Outcome of Competition under Equilibrium versus Nonequilibrium Conditions (A) Under equilibrium conditions, species 1 (the dominant competitor) outcompetes species 2 when it reaches its own carrying capacity (K) (B) If nonequilibrium processes such as disturbance, stress, or predation (represented by the arrows) reduce the population of species 1, it will never reach its carrying capacity and will not outcompete species 2. (After Huston 1979.)

mally be unable to tolerate extreme variability can maintain viable populations under these conditions because of the facilitative effects of other species.

Let's expand these ideas about modification of species interactions to whole communities and ask how nonequilibrium conditions, and the population fluctuations they create, might be responsible for the maintenance of species diversity.

Nonequilibrium theories are alternatives to resource partitioning

There is an old adage among ecologists that goes something like this: "If you think it's a new idea, check out Darwin. He probably discovered it first." In fact, when it comes to nonequilibrium theories, Darwin was the first to formally recognize disturbance as a mechanism for the maintenance of species diversity. In *The Origin of Species* (1859, p. 55), he noted the following results after an impromptu experiment in which he left a meadow on his property undisturbed by mowing: "Out of twenty species growing on a little plot of mown turf (three feet by four) nine species perished, from the other species being allowed to grow freely" with mowing, the dominant species in this community were competitively excluding weedy plants and cutting species richness nearly in half. Darwin used this

example, along with a multitude of others, to support the argument that nature applies limits to the tendency of species to increase in abundance and outcompete other species. His hypothesis was that species struggle for existence, a necessary first piece to his theory of natural selection.

In 1961, G. E. Hutchinson revived the idea of nonequilibrium processes in a paper titled "The Paradox of the Plankton." Hutchinson, an influential community ecologist from Yale University (and major professor to Robert MacArthur), provided one of the first mechanistic descriptions of nonequilibrium conditions by focusing on phytoplankton communities in temperate freshwater lakes (**Figure 18.13**). The simple idea behind Hutchinson's model was the seeming paradox between the presence of 30–40 species of phytoplankton and the relatively limited number of resources at their disposal. He reasoned that all of the phytoplankton compete for the same array of resources, including CO_2, nitrogen, phosphorus, sulfur, trace elements, and vitamins, that are likely to be evenly distributed in lakes. How could so many species manage to coexist with so few resources and in such a structurally simple environment as a lake? Hutchinson hypothesized that the conditions in the lake changed seasonally and over longer periods, which kept any one species from outcompeting the others. As long as conditions in the lake changed before the competitively superior species reached its carrying capacity, coexistence would be possible.

Hutchinson's model has two components that interact to control coexistence among species. One is the time required for one species to competitively exclude another species (t_c), which depends on the population growth rates of the two competing species. The second is the time it takes for envi-

FIGURE 18.13 Paradox of the Plankton How could so many freshwater phytoplankton species coexist in a lake using the same set of basic resources? G. E. Hutchinson suggested the influence of environmental variation over time.

ronmental variation to act on the population growth of the two competing species (t_e). Hutchinson predicted that when competitive exclusion occurs more rapidly than environmental conditions can change ($t_c \ll t_e$), coexistence cannot be achieved. One could imagine this occurring in communities where there is little environmental change or where the dominant competitor has very rapid growth rates. Conversely, in a fluctuating environment to which the competitors are adapted (where $t_c \gg t_e$), environmental variation does not affect the competitive interactions, and competitive exclusion occurs. One could imagine this pattern in environments with frequent, low-intensity environmental fluctuations and long-lived species. Hutchinson argued that it is only when the time it takes for competitive exclusion to occur is roughly equal to the time it takes for environmental variation to interrupt the competitive interaction (when $t_c = t_e$) that competitive exclusion is thwarted and coexistence occurs. Hutchinson argued that this condition is likely to be met quite often in phytoplankton communities, otherwise very few species, rather than tens of species, would coexist.

The idea that competitive exclusion is rare in nature was proposed, but not tested, in Hutchinson's "Paradox of the Plankton" paper. It was Robert Paine's work in the rocky intertidal zone of the west coast of North America in the late 1960s that provided some of the most rigorous and convincing evidence that nonequilibrium forces were actually at work in nature. Paine (1966) manipulated population densities of *Pisaster*, a predatory sea star that feeds preferentially on the mussel *Mytilus californianus*. In plots from which *Pisaster* was removed, species richness decreased as mussels outcompeted barnacles and other competitively inferior species. In plots where *Pisaster* was present, species diversity was enhanced. There are a number of important aspects to Paine's work, including the keystone species concept and the effects of indirect interactions, but we will consider these aspects in more detail in Chapter 20 when we discuss food webs. For now, let's concentrate on an idea that sprung from the work of Darwin, Hutchinson, and Paine: the intermediate disturbance hypothesis.

The intermediate disturbance hypothesis explains diversity under variable conditions

Robert Paine's work (1966, 1969) had a transformative effect on community ecology and stimulated one of the best-known and best-documented nonequilibrium theories. That theory, known as the **intermediate disturbance hypothesis**, considers how gradients in disturbance (although we can easily include stress and predation in this model) affect community species diversity (**Figure 18.14**). The theory was first formally proposed by Joseph Connell, a contemporary of Paine's and author of the classic work on barnacle competition (see Chapter 11). Connell (1978) recognized that the level of disturbance a particular community experiences could

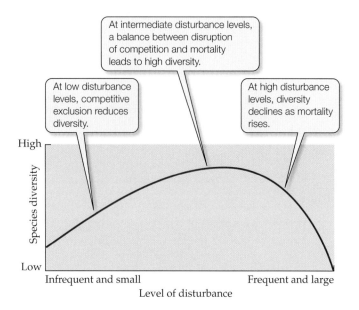

FIGURE 18.14 The Intermediate Disturbance Hypothesis
Species diversity is expected to be greatest at intermediate levels of disturbance. (After Connell 1978.)

have dramatic effects on species diversity. He hypothesized that species diversity would be greatest at intermediate levels of disturbance and lowest at high and low levels of disturbance. Why would this be the case? At low levels (low frequencies or low intensities) of disturbance, competition would regulate species diversity because dominant species would be free to exclude competitively inferior species. At high levels (high frequencies or high intensities) of disturbance, on the other hand, species diversity would decline because many individuals would die and some species would become locally extinct as a result. At intermediate levels of disturbance, species diversity would be maximized simply due to the balance between disruption of competition and mortality due to disturbance.

The intermediate disturbance hypothesis is highly amenable to empirical testing. Researchers have successfully applied both observational and experimental data to the theory. One interesting example comes from Wayne Sousa (1979a), who studied succession in intertidal boulder fields in southern California (Sousa 1979b) (see Figure 16.15). In a different but related study, Sousa measured the rate of disturbance of communities growing on the boulders and documented the resulting species richness (**Figure 18.15**). Small boulders were rolled over frequently by waves, and thus constituted a highly disturbed environment for the marine algae and invertebrate species that lived on them. The opposite was true for large boulders, which rarely experienced wave forces large enough to dislodge them. Intermediate-sized boulders, of course, were rolled over at intermediate frequencies. After 2 years, Sousa found that the highest percentage of the small boulders had only one species (early successional species: the macroalga *Ulva* or the barnacle *Chthamalus*), while the highest percentage of the large boulders had two species (late suc-

(A) Large boulders/low level of disturbance

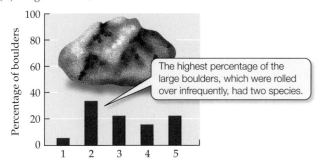

The highest percentage of the large boulders, which were rolled over infrequently, had two species.

(B) Intermediate boulders/intermediate level of disturbance

The intermediate-sized boulders, which were rolled over at intermediate frequencies, had the highest species richness.

(C) Small boulders/high level of disturbance

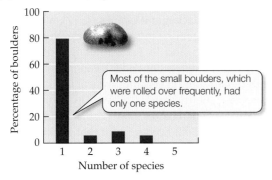

Most of the small boulders, which were rolled over frequently, had only one species.

FIGURE 18.15 A Test of the Intermediate Disturbance Hypothesis
Wayne Sousa conducted community surveys on marine intertidal boulders that differed in the level of disturbance they experienced from being rolled over by wave action. (Data from Sousa 1979a.)

turbance on competition, but reasoned that a second process, competitive displacement, could be an important mediating factor. Competitive displacement can be thought of as the growth rate of the strongest competitors in a community and is dependent on the productivity of the community. Huston's (1979) **dynamic equilibrium model** considers how the frequency or intensity of disturbance and the rate of competitive displacement combine to determine species diversity (**Figure 18.16**). Like Hutchinson's ideas, Huston's model predicts maximum species diversity when the level of disturbance and the rate of competitive displacement are roughly equivalent (thus the term "equilibrium" in the model name). Species diversity will be highest when the frequency or intensity of disturbance and the rate of competitive displacement are both at low to intermediate levels (Figure 18.16, point AC). Moreover, the lowest species diversity will occur either when the frequency or intensity of disturbance is high and competitive displacement is low (point AD), or compet-

cessional species: the macroalga *Gigartina canaliculata* and others). The highest percentage of the intermediate-sized boulders had four species, but some had up to seven species (a mixture of early-, mid-, and late-successional species). Sousa's work is just one of many studies that have demonstrated the highest diversity at intermediate disturbance levels. Others include studies of the effects of hurricanes on coral reefs (Connell 1978), of snail herbivores on tide pool algae (Lubchenco 1978), of logging in tropical rainforests (Molino and Sabatier 2001), and of flooding in streams (Townsend et al. 1997).

There have been several elaborations on the intermediate disturbance hypothesis

The intermediate disturbance hypothesis is a simple model that relies on variation in disturbance levels to explain species diversity in communities. A handful of ecologists have used it as a foundation for adding more complexity and realism to their theories. One of the first to elaborate on the theory was Michael Huston (1979), who acknowledged the effect of dis-

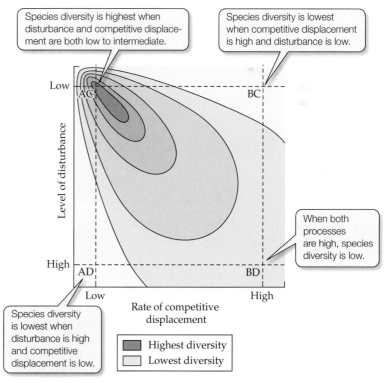

Species diversity is highest when disturbance and competitive displacement are both low to intermediate.

Species diversity is lowest when competitive displacement is high and disturbance is low.

When both processes are high, species diversity is low.

Species diversity is lowest when disturbance is high and competitive displacement is low.

Highest diversity
Lowest diversity

FIGURE 18.16 The Dynamic Equilibrium Model   This model predicts that species diversity will be highest when the level of disturbance and the rate of competitive displacement are low to intermediate. (After Huston 1979.)

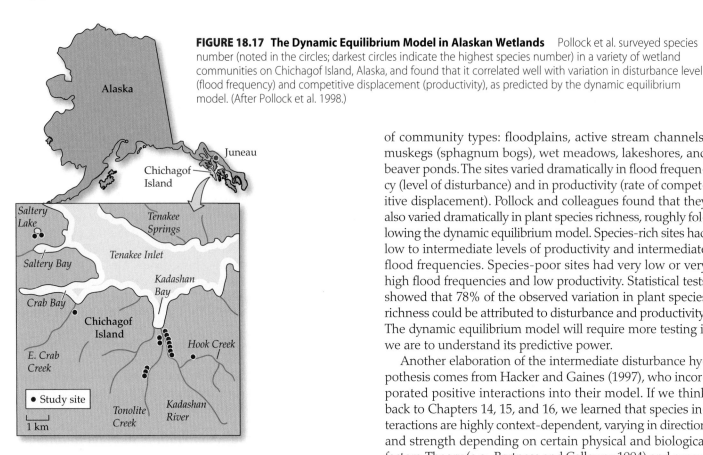

FIGURE 18.17 The Dynamic Equilibrium Model in Alaskan Wetlands Pollock et al. surveyed species number (noted in the circles; darkest circles indicate the highest species number) in a variety of wetland communities on Chichagof Island, Alaska, and found that it correlated well with variation in disturbance level (flood frequency) and competitive displacement (productivity), as predicted by the dynamic equilibrium model. (After Pollock et al. 1998.)

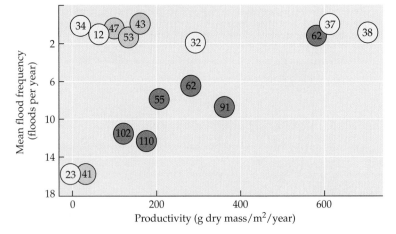

of community types: floodplains, active stream channels, muskegs (sphagnum bogs), wet meadows, lakeshores, and beaver ponds. The sites varied dramatically in flood frequency (level of disturbance) and in productivity (rate of competitive displacement). Pollock and colleagues found that they also varied dramatically in plant species richness, roughly following the dynamic equilibrium model. Species-rich sites had low to intermediate levels of productivity and intermediate flood frequencies. Species-poor sites had very low or very high flood frequencies and low productivity. Statistical tests showed that 78% of the observed variation in plant species richness could be attributed to disturbance and productivity. The dynamic equilibrium model will require more testing if we are to understand its predictive power.

Another elaboration of the intermediate disturbance hypothesis comes from Hacker and Gaines (1997), who incorporated positive interactions into their model. If we think back to Chapters 14, 15, and 16, we learned that species interactions are highly context-dependent, varying in direction and strength depending on certain physical and biological factors. Theory (e.g., Bertness and Callaway 1994) and experiments (e.g., Bertness and Hacker 1994; Bertness and Leonard 1997; Callaway et al. 2002) both suggest that positive interactions are more common under relatively high levels of disturbance, stress, or predation, all circumstances in which associations among species could increase their growth and survival. Hacker and Gaines (1997) reasoned that positive interactions might be particularly important in promoting

itive displacement is high and the frequency and intensity of disturbance is low (point BC). When both processes are high and roughly similar (point BD), we expect species diversity to be relatively low because both high mortality and competitive displacement will be acting to reduce species diversity.

Perhaps because of the added complexity, there have been few empirical studies of the dynamic equilibrium model. One example comes from an observational study of riparian wetlands in Alaska. Pollock et al. (1998) surveyed 16 sites in and near the Kadashan River basin on a coastal island in Southeast Alaska (**Figure 18.17**). The sites represented a number

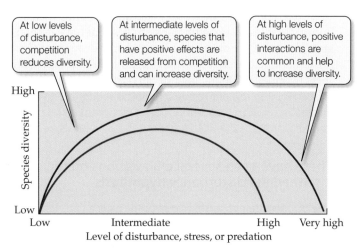

FIGURE 18.18 Positive Interactions and Species Diversity The intermediate disturbance hypothesis (blue curve) has been elaborated to include positive interactions (red curve). (After Hacker and Gaines 1997.)

species diversity at intermediate to high levels of disturbance, stress, or predation for two reasons (**Figure 18.18**). First, under high levels of disturbance, stress, or predation, positive interactions should increase the survival of individuals of the interacting species, through both amelioration of harsh habitat conditions and associational defenses. Second, even under intermediate levels of disturbance, stress, or predation, species involved in positive interactions should be released from competition, an effect that should further increase species diversity.

Hacker and Gaines (1997) used a New England salt marsh case study to support their theory. In this community, there is a strong gradient in physical stress from saltwater inundation. The highest stress occurs closest to the shoreline, where the tides inundate the plants most frequently. A survey of plants, insects, and spiders across the marsh revealed three distinct zones and showed that the middle intertidal zone had a higher species richness than the high or low intertidal zones (**Figure 18.19A**). The researchers then conducted transplant experiments in which all the plant species were moved to all three zones with or without the most abundant plant of their own zone: the tall shrub *Iva frutescens* in the high intertidal zone and the rush *Juncus gerardii* in the middle and low intertidal zones (Bertness and Hacker 1994; Hacker and Bertness 1999). The results revealed that competition with *Iva* in the high intertidal zone led to the competitive exclusion of most plant species transplanted there. In the low intertidal zone, physiological stress was the main factor in controlling population numbers, as many individuals died whether *Juncus* was present or absent. However, in the middle intertidal zone, *Juncus* facilitated other plant species. Without *Juncus*, mortality was 100% for most species by the end of the summer. The mechanism of facilitation, described in Chapter 15, was amelioration of both salt stress and hypoxia by *Juncus* (see p. 332). Additionally, we learned in that example that *Juncus* indirectly facilitates an aphid herbivore that depends on *Iva* for survival. It turns out that these indirect interactions hold for a number of insect herbivores that feed on a variety of other plants facilitated by *Juncus* in the marsh. Hacker and Gaines (1997) predicted, based on these studies, that positive interactions were critically important in maintaining species diversity, especially at the intermediate levels of physical stress that characterized the middle intertidal zone (**Figure 18.19B**). They recognized that physical stress in the middle intertidal zone both decreases the competitive effect of *Iva* and increases the facilitative effect of *Juncus* (and its indirect effects on insects), thus providing a "perfect storm" for enhanced species coexistence and diversity.

The intermediate disturbance hypothesis assumes that there is an underlying competitive hierarchy among species—that is, that some species are much stronger competitors than others and thus dominate if not kept in check by nonequilibrium processes. What happens if we assume that there is

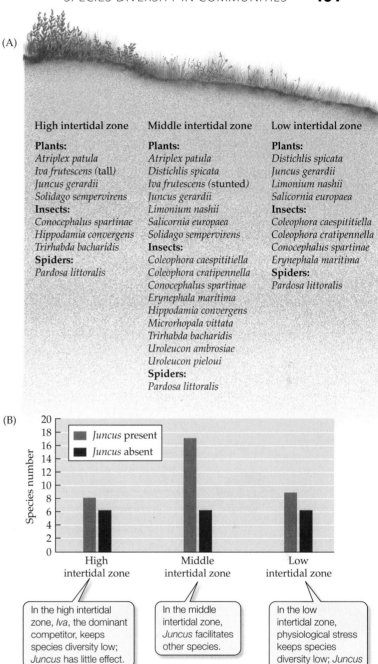

FIGURE 18.19 Positive Interactions: Key to Local Diversity in Salt Marshes? (A) Surveys of plant and insect species diversity in a New England salt marsh show diversity to be greatest in the middle intertidal zone. (B) Experiments suggest that this high diversity is controlled by the direct and indirect effects of the facilitating rush species *Juncus gerardii* and by the absence of the dominant competitor, *Iva frutescens*, due to physical stress. (After Hacker and Gaines 1997.)

no competitive hierarchy among species? If species have equivalent interaction strengths, then the ability of any one species to live in a community will depend more on chance than on "conflict resolution." Let's spend a moment discussing this unconventional theory of species diversity.

The lottery model relies on equality and chance

A final type of nonequilibrium model is the so-called **lottery model**. As the name suggests, this kind of model emphasizes the role of chance in the maintenance of species diversity. Lottery models assume that resources in a community, made available by the effects of disturbance, stress, or predation, are captured at random by recruits from a larger pool of potential colonists. For this mechanism to work, species must have fairly similar interaction strengths and population growth rates, and they must have the ability to respond quickly, by dispersing, to disturbances that free up resources. If there is a large disparity in competitive abilities among species, the dominant competitor will have a greater chance of obtaining resources and eventually forming a monopoly. In the lottery model, it is the equal chance of all species to obtain resources that allows species coexistence and results in a multiple-species system.

Lottery models have most often been applied to species-rich communities. Peter Sale (1977, 1979) conducted one of the earliest and best-known tests of the lottery model using fishes on the Great Barrier Reef of Australia. Fish diversity on this reef ranges from 1,500 species in the north to 900 species in the south. On any one small patch reef (3 meters in diam-

eter), up to 75 species might be recorded. In the reef ecosystem, there is strong habitat fidelity and severe space limitation, such that many individual fish spend their entire lives in roughly the same spot on the reef. Given these conditions, Sale asked the obvious question: how could so many species coexist in such a small space for so long?

Sale (1977) reasoned that only a portion of the coexistence among these fishes could be explained by resource partitioning because the species tended to have very similar diets. Instead, vacant sites or territories were highly desirable and made available rather unpredictably by the deaths of individual occupants (due, for example, to predation, disturbance, starvation, or disease). To look at this system in more detail, Sale (1979) followed losses and recruitment to newly vacated sites among three species of territorial pomacentrid fishes (*Eupomacentrus apicalis*, *Plectroglyphidodon lacrymatus*, and *Pomacentrus wardi*). He found the pattern of occupation to be random (**Figure 18.20**): the species that had previously occupied and lost a site had no bearing on which species was recruited to that site. One species, *P. wardi*, both lost and occupied sites at a greater rate than the other species, but this had no effect on its overall ability to coexist with the other two species. Sale noted that one important component of this lottery system is that fishes produce many, highly mobile juveniles that can saturate a reef and take advantage of open space made available. As Sale put it, "the species of a guild are competing in a lottery for living space in which larvae are tickets and the first arrival at a vacant site wins that site" (Sale 1977, p. 351).

The role of chance in maintaining species diversity, especially in unpredictable environments, has intuitive appeal. As long as species win the lottery every once in a while, they will continue to reproduce (i.e., buy more tickets) and be able to enter the lottery once again. It is easy to see how this mechanism might be particularly relevant in highly diverse communities such as tropical rainforests and grasslands, where so many species overlap in their resource requirements. Its relevance decreases, however, in communities where species have large disparities in interaction strength (i.e, in which one species has a much stronger effect than another). In those communities, it appears that the "great equalizers" might be processes that decrease competitive exclusion, such as disturbance, stress, or predation. Ecologists are a long way from agreeing on any one theory to explain why certain species end up coexisting in space and time. Instead, they continue

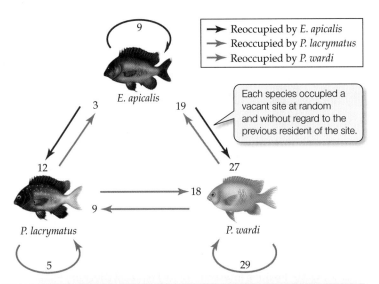

FIGURE 18.20 The Lottery Model In an ecological lottery, when resources become available, all species in the community have an equal chance to "win" them, but they must be in the right place at the right time. Sale tested this model using coral reef fishes living on the Great Barrier Reef of Australia. By counting the individuals of three fish species (*Eupomacentrus apicalis*, *Plectroglyphidodon lacrymatus*, and *Pomacentrus wardi*) that occupied vacated sites, he found that occupation was random and unrelated to the species that had previously occupied the site. (Data from Sale 1979.)

to strive for generalities while recognizing that the relative importance of different mechanisms of species diversity may depend on the characteristics of the community in question.

Up to this point in the chapter, we have focused on the causes of species diversity at the community level. We have asked, "Why and how does species diversity differ among communities?" In the next section, we will shift gears and instead ask what might be considered the flip side of that question. We want to know, given the variation in species diversity among communities (and the current losses of species diversity due to human activities), whether diversity matters. In other words, what do species do in communities? Does species diversity have functional significance?

The Consequences of Diversity

CONCEPT 18.4
Experiments show that species diversity is positively related to community function.

In the Case Study at the opening of this chapter, we considered a scenario in which communities (in this case, patches of prairie) with higher species richness produced greater biomass than those with lower species richness. These results support the notion that species diversity can control certain functions of the community. These functions are numerous, but include plant productivity, soil fertility, water quality and availability, atmospheric chemistry, and even resistance to disturbance (and the speed of recovery afterward, also known as **resilience**). Many of these functions provide valuable services to humans; for example, food and fuel production, water purification, O_2 and CO_2 exchange, and protection from catastrophic events such as floods or tsunamis. The Millennium Ecosystem Assessment (2005), a publication produced under the auspices of the United Nations, details the importance of these ecosystem services to humans. The assessment predicts that if the current losses of species diversity continue, the world's human populations will be severely affected by the loss of the services species provide. These are fairly dire predictions, prompting the question, "What evidence underlies these findings?" Recent research has attempted to look at the connections between species diversity and community function, both because of these concerns and to seek basic insights into community ecology.

Relationships between species diversity and community function are positive

David Tilman and colleagues, who study prairie communities in Minnesota, performed the first experimental test of a long-standing theory, proposed by both Robert MacArthur (1955) and Charles Elton (1958), that species richness is positively related to community stability. **Stability**, in this case, is defined as the tendency of the community to remain the same in struc-

ture and function. The diversity–stability theory remained "conventional wisdom" until the mid-1970s, when it was tested mathematically using food web models that varied in species richness and complexity. We will consider these models in more detail in Chapter 20. But when Tilman and Downing (1994) noticed that some of their experimental plots at Cedar Creek (see Figure 18.2) seemed to be responding to drought differently, they were reminded of these models and of MacArthur and Elton's conventional wisdom. A survey of their experiment showed that plots with higher species richness (but the same density of plants) were better able to withstand drought than plots with lower species richness (**Figure 18.21A**). Total plant biomass decreased less due to the drought

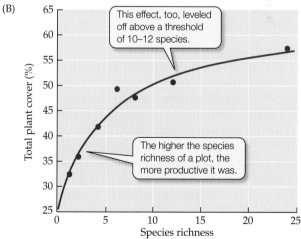

FIGURE 18.21 Species Diversity and Community Function Tilman and colleagues performed two separate experiments using prairie plants at the Cedar Creek site in Minnesota (see Figure 18.2) to test the effects of species richness on community function. (A) They measured the effects of drought on plant biomass in plots that varied in species richness. (B) They then created plots that varied in species richness, but all had the same density of individual plants, and measured biomass in those plots after 2 years of growth. (A after Tilman and Downing 1994; B after Tilman et al. 1996.)

in species-rich plots than in species-poor ones, resulting in a positive, curvilinear relationship between species richness and drought resistance (measured as the difference between biomass before and after the drought). Tilman and Downing reasoned that a curvilinear relationship would be expected if additional species beyond some threshold (the point at which the curve levels off; roughly 10–12 species in this study) had little additional effect on drought resistance. These species could be considered redundant in the sense that they have essentially the same effects on drought resistance as other species. Tilman and Downing suggested, however, that once the number of species in a plot declined below that threshold, each additional species lost from the plot meant a progressively greater negative effect of drought on the community.

To test this idea more rigorously, Tilman et al. (1996) conducted a well-replicated experiment in which species diversity was directly manipulated in the same prairie ecosystem. A series of plots that differed in species richness, but not in the number of individuals, was created by randomly selecting sets of species from a pool of 24 species. Each plot was provided with the same amount of water and nutrients, and biomass in the plots was measured after 2 years of growth. The results confirmed the curvilinear effect of species richness on productivity and additionally showed that nitrogen was more efficiently used as species richness increased (**Figure 18.21B**). Similarly designed studies have shown species richness to have positive effects on productivity (e.g., Hector et al. 1999), community respiration (e.g., Naeem et al. 1994), biotic resistance (e.g., Tilman 1997; Levine 2000), resilience (e.g., Allison 2004), and trophic interactions (e.g., Duffy et al. 2007).

Hypotheses explaining diversity–function relationships are numerous

Although experiments documenting the relationship between species diversity and community function continue to increase in their sophistication, the mechanisms underlying these relationships have been difficult to resolve. Despite the intense debate among ecologists over these mechanisms (e.g., Grime 1997; Loreau et al. 2001), there are at least four hypotheses, outlined by Peterson et al. (1998), that might explain the positive relationships between species diversity and community function. These hypotheses differ in their assumptions concerning two variables: the degree of overlap in the ecological functions of species, and variation in the strengths of the ecological functions of species (**Figure 18.22A**). These different assumptions result in diversity–function curves with differing shapes.

The first hypothesis, known as the **complementarity hypothesis**, proposes that as species richness increases, there will be a linear increase in community function (**Figure 18.22B**). In this case, each species added to the community will have a unique and equally incremental effect on community function. We might expect this type of pattern if we assume that species are equally partitioning their functions within a community. For example, as more and more species are added to the community, each of their individual unique functions will contribute to an ever-increasing community function value.

The second model, known as the **redundancy hypothesis**, assumes something similar to the complementarity hypothesis, but provides an upper limit on the effect of species richness on community function (**Figure 18.22C**). This model best fits the work of Tilman and colleagues cited above (see Figure 18.21), in which the functional contribution of additional species reaches a threshold. This threshold is reached because as more species are added to the community, there is overlap in their function—essentially, there is redundancy among species. In this model, species can be thought of as belonging to certain functional groups (see Figure 15.4). As long as all the important functional groups are represented, the actual species composition of the community is of little importance to its overall function.

The third model, known as the **driver and passenger hypothesis**, proposes that the strengths of the effects of species' ecological functions vary dramatically among species (**Figure 18.22D**). "Driver" species have a large effect on community function, while "passenger" species have a minimal effect. The addition of driver and passenger species to a community will therefore have unequal effects on community function, producing a "stair step" curve as shown in Figure 18.22D. The shape of the curve—the rise and run of the stair steps—will depend on the numbers of driver and passenger species added and on how much their effects differ. If communities are assembled in such a way that there are only a few driver species (e.g., keystone or dominant species; see Figure 15.17) but many passenger species, then one would expect community function values to vary dramatically with species richness; that is, the stair steps will vary both in rise and run, depending on whether driver species are present or not. However, as species richness increases, the chance that the driver species will be present becomes very high, so that the variation in community function values, or the stair steps, should decrease.

The fourth model is a variation on the driver and passenger hypothesis. It assumes that there could be overlap in the function of drivers and passengers, and thus predicts a curvilinear relationship, with a threshold at high species richness (**Figure 18.22E**).

Although these models differ in the functional roles they assign to species, the underlying ecological interactions among species are not well understood. For example, is competition driving species to partition resources and act in unique and complementary ways? Or are positive interactions influencing community function? Equally important, what kinds of trophic interactions are occurring among species, and how do they affect community function? Conceivably, these questions can be answered through experi-

(A)

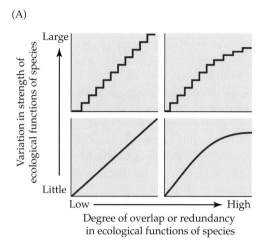

FIGURE 18.22 Hypotheses on Species Richness and Community Function A number of hypotheses have been proposed to explain the positive relationship between species richness and community function. Two variables distinguish these hypotheses: the degree of overlap in the ecological functions of species and variation in the strength of the ecological functions of species. (After Peterson et al. 1998.)

(B) Complementarity hypothesis

(C) Redundancy hypothesis

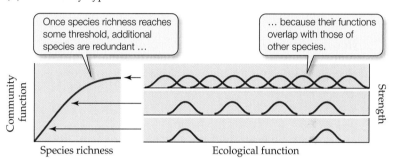

ments that manipulate species and measure interaction strengths (e.g., see Box 15.1), but such experiments are logistically challenging. In many ways, these experiments and models are at the heart of modern ecology, not only because they tell us something about how communities work, but also because they may be able to tell us what the future holds for communities that are both losing (by extinction) and gaining (by invasions) species through human influence.

(D) Driver and passenger hypothesis

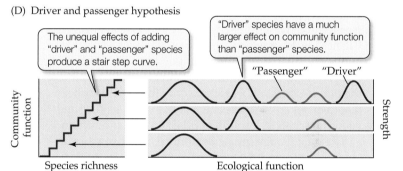

CASE STUDY REVISITED

Powered by Prairies? Biodiversity and Biofuels

The potential value of understanding how species diversity controls community function is limitless when we consider the services natural communities provide to humans. Although the functions of communities are as basic as biomass production, atmospheric gas exchange, water regulation, and resistance to disturbance, their services to humans can be numerous and diverse. One potential service that has been overlooked until recently is biofuel production. Tilman's prairie studies had already shown that biomass production is greater in high-diversity plots than in single-species plots (*monocultures*). Tilman and colleagues (2006) established that their high-diversity plots produced nearly 238% more biomass per input of energy than the single-species

(E) Driver and passenger hypothesis with overlap

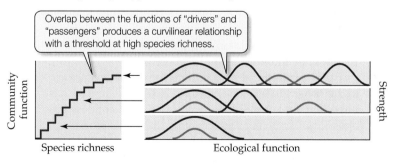

plots. They used these experimental results to compare the biofuel that could be produced from traditional crops and that of high-diversity prairie plants. In particular, they compared three biomass types—soybeans, corn, and low-input, high-diversity (LIHD) biomass from their prairie plots—and three types of biofuel products—biodiesel, ethanol,

FIGURE 18.23 Biofuel Comparisons Net energy balance (NEB), a measure of the biofuel produced minus the fossil fuels needed to produce them, for corn ethanol, soybean biodiesel, and two biofuel types made from LIHD prairie biomass. (After Tilman et al. 2006.)

and synfuel (a type of synthetic gasoline)—produced from them (**Figure 18.23**). When the products created from LIHD prairie biomass were compared to the traditional crops, they found that synfuel production from prairie biomass netted the most energy (measured as "net energy balance" or NEB, which is the amount of biofuel produced minus the amount of fossil fuels used to produce them) compared to the other crops. A comparison of NEB among the biomass types showed that corn ethanol and soybean biodiesel netted 51% less energy than LIHD synfuel, and equivalent or even lower than LIHD ethanol production.

These higher energy yields resulted from two main differences between LIHD biomass and the two traditional biofuel crops. First, energy inputs were lower for LIHD crops because they are perennial plants and require little water, fertilizer, or pesticides (**Figure 18.24A,B**). Second, LIHD crops had a very high yield of biomass due to diversity effects on biomass and the fact that all of the above-ground plant material (rather than just the seeds, as in the case of corn and soybeans) can be used.

Prairie plants are also superior to corn or soybean crops in their ability to take up and store CO_2. This difference has important implications for the reduction of greenhouse gas emissions. Tilman et al. (2006) found that LIHD plots sequestered 160% more CO_2 in plant roots and soil than single-species prairie plots. When they calculated the amounts of CO_2 and other greenhouse gases produced and consumed by the different biofuels, they found that the traditional biofuels are net carbon sources, while LIHD biofuels are carbon neutral. Greenhouse gas emission reductions relative to the burning of fossil fuels were 6 to 16 times greater for LIHD fuels than for corn ethanol or soybean biodiesel (**Figure 18.24C**). In addition, because LIHD prairie biomass

can be grown on degraded former agricultural land, it need not take up land that could be used for growing food.

The results of this study are promising, but can LIHD biomass provide enough fuel to make a difference? Tilman et al. (2006) estimate that about 5×10^8 hectares of abandoned agricultural land, supporting a biomass of 90 gigajoules (GJ) per hectare per year, could produce enough fuel to substitute for about 13% of global petroleum and 19% of global electricity consumption. This strategy could also eliminate 15% of global CO_2 emissions. Even if these are optimistic values, it is clear that this new twist on an emerging technology deserves more attention.

By applying basic principles of ecology to the biofuels debate, Tilman et al. (2006) show elegantly that we cannot

FIGURE 18.24 Environmental Effects of Biofuels Tilman and colleagues compared the environmental effects of biofuels made from corn, soybeans, and LIHD prairie biomass. (A) Fertilizer requirements. (B) Pesticide requirements. (C) Reduction of greenhouse gas emissions relative to the combustion of an equal amount of fossil fuel, taking into account the uptake and release of CO_2 and other greenhouse gases through the entire life cycle of the biofuel, from crop production to combustion. (After Tilman et al. 2006.)

merely compare the attributes of natural systems with those of agricultural systems. We must consider what might seem like inconsequential and esoteric details, such as the number of species that coexist within communities. In this case, species richness makes all the difference, not only in terms of the beauty and productivity that inspired North American settlers who crossed the prairies back in the late 1880s, but also in the promise of energy independence those prairies may provide for the future.

CONNECTIONS IN NATURE

Barriers to Biofuels: The Plant Cell Wall Conundrum

Biofuels vary considerably in the biomass needed to produce them and the energy required to refine them. Biodiesel, for example, is easily produced from oil seed crops such as soybeans (most common in the United States) and palms (most common in Malaysia and Indonesia). The main limitation in production occurs at the crop stage. Even though soybeans are legumes, and thus have symbiotic nitrogen-fixing bacteria in their roots (see p. 361), they can promote soil erosion and require lots of water for irrigation. Biodiesel production has the potential to compete with food production as well.

Ethanol is most commonly made from corn grains that are fermented and distilled. The energy costs associated with growing the grain and producing the ethanol are high, so there is only a slight energy gain in ethanol production (see Figure 18.23). Additionally, ethanol production from corn competes with the food needs of humans. For example, an acre of corn produces about 440 gallons of ethanol, which represents 4–5 months of driving for the average individual in the United States. The same amount of corn could feed one person for 20–27 years.

For these reasons, and the reasons given in the Case Study, scientists are turning to nonfood biomass, such as crop residues, logging wastes, and prairie plants, to produce what is called cellulosic ethanol. Unlike corn ethanol, which is made of sugars broken down from starch, cellulosic ethanol is made of sugars liberated from cellulose. Breaking down cellulose—the major component of plant cell walls—is extremely difficult and requires special enzymes that are still in the research and development stage. Plants have evolved many protective mechanisms to resist digestion by both bacteria and herbivorous animals, including epidermal waxes, thick and reinforced cell walls, enzyme-insoluble surfaces, and inhibitors of fermentation.

Himmel et al. (2007) have outlined some of the techniques that molecular biologists are developing to overcome this recalcitrance, most of which involve genetically engineered enzymes that work on the plant both externally and internally. The thought is that, one day, plant cells might even be genetically programmed to self-destruct before harvest. It is clear that for biofuel production to be a viable alternative to fossil fuels, ecologists and molecular biologists will have to work together to break down the barriers to biofuels that currently exist.

SUMMARY

CONCEPT 18.1 Species richness differs among communities due to variation in regional species pools, abiotic conditions, and species interactions.

- The regional species pool and the dispersal abilities of species play important roles in supplying species to communities.

- Humans have greatly expanded the regional species pools of communities by serving as vectors for the dispersal of non-native species.

- Physiological constraints imposed by abiotic conditions act as a strong "filter" for community membership.

- When species depend on other species for their growth, reproduction, and survival, those other species must be present if they are to gain membership in a community.

- Species may be excluded from communities by competition, predation, parasitism, or disease.

CONCEPT 18.2 Resource partitioning among the species in a community reduces competition and increases species richness.

- Resource partitioning is an equilibrium theory predicting that species must use slightly different resources along a resource spectrum if they are to avoid competitive exclusion.

- One model of resource partitioning states that the less overlap between the species in a community in their resource use, the more species can coexist in the community.

- The resource ratio hypothesis posits that species that use the same set of resources are able to partition them by using them in different proportions.

CONCEPT 18.3 Nonequilbrium processes such as disturbance, stress, and predation can mediate resource availability, thus affecting species interactions and coexistence.

- Nonequilibrium theories state that disturbance, stress, or predation can cause species' populations to fluctuate in space and time, making resources available and increasing species coexistence.

- The intermediate disturbance hypothesis states that intermediate levels of disturbance promote species diversity by reducing competitive exclusion. Species diversity is low at low levels of disturbance due to competitive exclusion and at high levels of disturbance due to high mortality.

- The dynamic equilibrium model predicts that species diversity will be highest when the level of disturbance and the rate of competitive displacement are roughly equivalent.

- Positive interactions can promote species diversity, particularly at intermediate to high levels of disturbance, stress, or predation.

- Lottery models assume that resources made available by disturbance, stress, or predation are captured at random by recruits from a larger pool of colonists, all of which have an equal chance to do so.

CONCEPT 18.4 Experiments show that species diversity is positively related to community function.

- Species can control numerous functions of communities, including productivity, soil fertility, water quality and availability, atmospheric chemistry, and responses to disturbance.

- Many manipulative experiments in different communities have shown that as species diversity increases, so does community function.

- Hypotheses proposed to explain the positive relationship between species diversity and community function fall into four general categories that include varying assumptions about the degree of overlap in the ecological functions of species and about variation in the strengths of the effects of the ecological functions of species.

REVIEW QUESTIONS

1. Suppose you are an ecologist studying prairie grassland communities in Minnesota. As you are doing your fieldwork, grass seeds with hooked spines attach to your shoes. You then travel to New Zealand to study the grasslands on South Island. When you enter the Customs area in the Auckland airport, the officers in charge ask if you have visited a natural area or farm recently. You say "yes," and they tell you to take off your shoes and wait while they disinfect them with bleach. Given what you know about the mechanisms important to community membership, is it worth the time and money required to clean all that footwear before allowing it into New Zealand?

2. We know that species diversity varies greatly among communities. Describe how equilibrium and nonequilibrium theories differ in their explanations of the mechanisms controlling species diversity within communities.

3. Suppose you are studying a tropical rainforest community in Panama. You obtain a 50-year data set for the forest that records both the mortality of adult trees and the emergence of new tree seedlings. As you analyze the data, you try to determine whether there is a pattern of species replacement, with individuals of one species generally occupying the same sites in preference to individuals of other species. After much work, you are convinced that no pattern of replacement exists in this forest—instead, sites are colonized in an entirely random fashion, with no one species having an advantage. What model of species diversity best describes your observations and why?

4. Recent experimental work in communities has shown positive relationships between species diversity and community function. We learned that there is considerable debate as to the mechanisms controlling these relationships and that at least four hypotheses have been developed to explain them. Below are four graphs (A, B, C, and D) of species diversity–community function relationships that vary in the shapes of their curves. Describe which hypothesis best fits each curve, and why.

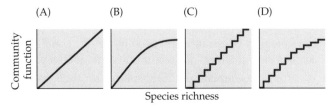

HANDS-ON PROBLEM SOLVING	
Periodic Disturbance and Its Effect on Species	This Web exercise (www.sinauer.com/ecology) demonstrates how periodic disturbance can maintain species in a community that otherwise could not coexist. You will manipulate the frequency and intensity of disturbances to investigate this effect.

SUGGESTED READINGS

DeLuca, T. H. 2007. Environmental benefits and consequences of biofuels development in the United States. The Wilderness Society, Washington, DC. *A comprehensive assessment of the costs and benefits of developing biofuels.*

Elton, C. S. 1958. *The Ecology of Invasions by Animals and Plants.* Chapman and Hall, London. *The classic book on the relationship between species diversity and species invasions.*

Huston, M. A. 1994. *Biological Diversity: The Coexistence of Species on Changing Landscapes.* Cambridge University Press, Cambridge. *The definitive book on nonequilibrium theories of species diversity.*

Millennium Ecosystem Assessment. 2005. *Ecosystems and Human Well-Being: Biodiversity Synthesis.* World Resources Institute, Washington, DC. *A complete review of the ecosystem services provided by species diversity.*

Tilman, D., D. Wedin and J. Knops. 1996. Productivity and sustainability influenced by biodiversity in grassland ecosystems. *Nature* 379: 718–720. *A report on one of the first manipulative experiments undertaken to understand the role of species diversity in community function.*

Tilman, D., J. Hill and C. Lehman. 2006. Carbon-negative biofuels from low-input high-diversity grassland biomass. *Science* 314: 1598. *The quantitative evidence behind the suggestion that prairies are good sources of biomass for biofuel production.*

UNIT 5 | Ecosystems

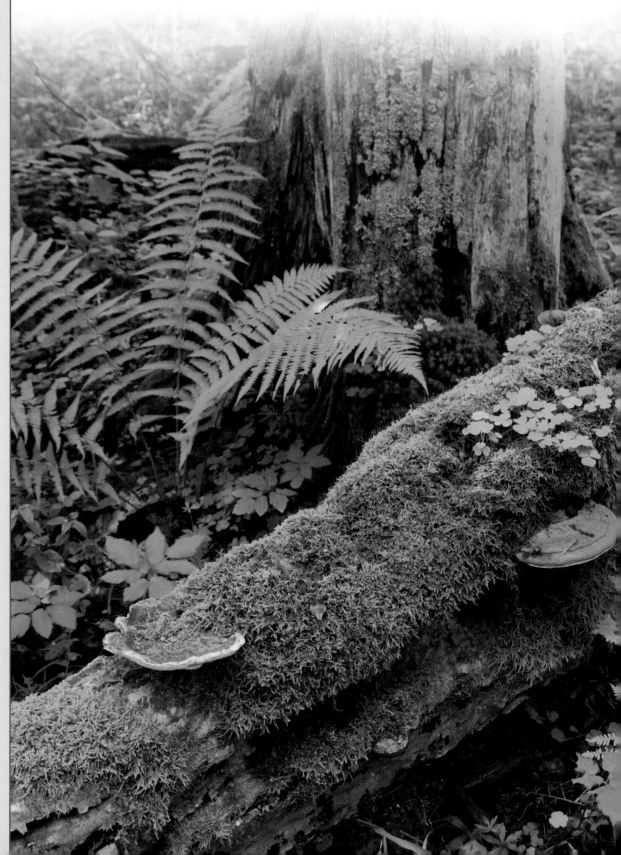

CHAPTER 19 Production

FIGURE 19.1 *Alvin* in Action The deep-sea submersible craft *Alvin* was instrumental in locating and exploring the first known hydrothermal vent site in 1977. The *Alvin* can carry two scientists, and is equipped with video cameras and robotic arms for collecting specimens from the seafloor.

CASE STUDY

Life in the Deep Blue Sea, How Can It Be?

Ecologists once considered the deep sea to be the marine equivalent of a desert. The physical environment between 1,500 and 4,000 m depth did not seem conducive to life as we knew it. It is completely dark, so photosynthesis is not possible. The water pressure reaches values 300 times greater than at the surface of the ocean, enough to crush terrestrial animals like roadkill. Organisms living on the bottom of the deep sea were thought to obtain energy exclusively from the sparse rain of the bodies of dead algae and zooplankton falling from the sunlit zone in the upper layers of the ocean, where phytoplankton carry out photosynthesis. The only known deep-sea organisms included detritus feeders such as echinoderms (e.g., sea stars), mollusks, crustaceans, and polychaete worms.

Our view of deep-sea life was changed dramatically in 1977 when an expedition using the submersible craft *Alvin*, led by Robert Ballard of the Woods Hole Oceanographic Institution, visited a mid-ocean ridge near the Galápagos archipelago (**Figure 19.1**). The *Alvin* was in search of deep-sea hot springs thought to occur along mid-ocean ridges. These ridges lie at the junctions of tectonic plates, where the seafloor spreads as the plates are pulled apart by tectonic forces. The ridges are volcanically active, and geologists and oceanographers had hypothesized that heat from Earth's crust would be released there by hot springs as well as by emerging lava. Hot springs were also considered potential sources of chemicals for the ocean system. Despite their hypothesized existence, no such hot springs had ever been located.

Ballard's group did indeed find hot springs, known as hydrothermal vents, but from an ecological perspective, this geochemical finding paled in comparison to their biological discovery: the areas around the hydrothermal vents were teeming with life. Dense assemblages of tube worms (e.g., genus *Riftia*), giant clams (e.g., genus *Calyptogena*), shrimps, crabs, and polychaete worms were found in the areas surrounding the vents (**Figure 19.2**). The density of organisms was unprecedented in the deep, dark ocean. The discovery of these hydrothermal vent communities posed an immediate question: how did the organisms

FIGURE 19.2 Life around a Hydrothermal Vent

Tubeworms over 2 m in length surround a "black smoker" hydrothermal vent. The vent is spewing superheated water as hot as 400°C, which contains high concentrations of dissolved metals and chemicals, particularly hydrogen sulfide.

obtain the energy needed to sustain themselves in such abundance? The rate at which dead algae and animals from the upper zones of the ocean accumulate on the bottom is very low (0.5 to 0.01 mm/year). The seafloors where the vents are located are only decades old, and thus the amount of food that would have accumulated would not be enough to sustain these high densities of hydrothermal vent organisms. Photosynthesis in the surface waters therefore did not appear to be the energy source supporting these hydrothermal vent communities.

In addition to the lack of an apparent energy source, the water being emitted from the hydrothermal vents posed a problem: its chemical composition would be toxic to most organisms. Hydrothermal vents result when seawater seeping into cracks in the ocean floor near the mid-ocean ridges is superheated by pockets of magma, chemically transformed, and then ejected. The water emitted by the vents is rich in sulfides as well as heavy metals such as lead, cobalt, zinc, copper, and silver, which inhibit metabolic activity in most organisms. Hydrothermal vent communities thus pose two mysteries: first, what is the source of energy that sustains these diverse and productive communities, and second, how do the organisms tolerate the high concentrations of potentially toxic sulfides in the water? As we shall see, these two problems are intimately related.

n 1942, a controversial paper authored by Ray Lindeman, describing energy flow in a Minnesota bog, was published in the journal *Ecology*. Lindeman had studied the energy relationships among the organisms and nonliving components of the bog system (Lindeman 1942). Rather than grouping its component plants, animals, and bacteria according to their taxonomic categories, Lindeman grouped them into functional categories, based primarily on how they obtained their energy (**Figure 19.3**). His views on the importance of the energy base of the system ("ooze"—particulate and dissolved organic matter) and on the efficiency of energy transfer among the biological components of the system were groundbreaking. Lindeman's treatment of energy flow in the bog was considered too theoretical at the time, and his paper was initially rejected, although the publishers later recanted after Lindeman's mentor, G. Evelyn Hutchinson, advocated its acceptance.

Lindeman's pioneering paper was among the first in the area of ecosystem science, and it is now considered a foundation paper in the discipline. The term **ecosystem** was first coined by A. G. Tansley, a plant ecologist, to refer to all of the components of an ecological system, biotic and abiotic, that influence the flow of energy and elements (Tansley 1935). The "elements" considered in ecosystem studies are primarily nutrients, but also include pollutants; the movements of those elements through ecosystems are the topic of Chapter 21. The

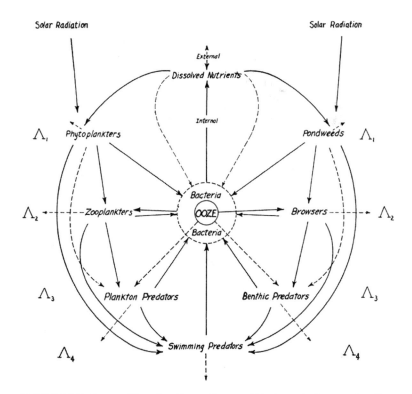

FIGURE 19.3 Energy Flow in a Bog Raymond Lindeman's diagram describes the movement of energy among groups of organisms at Cedar Bog, Minnesota. Note the general functional categories of organisms Lindeman used, as well as the central position of "ooze" (organic matter) in the diagram. The subscripts next to the upper case Greek lambdas represent trophic levels. (From Lindeman 1942.)

ecosystem concept is now well established and has become a powerful tool for integrating ecology with other disciplines such as geochemistry, hydrology, and atmospheric science.

In Chapter 5, we described the physiological basis for the capture of energy through photosynthesis and chemosynthesis by autotrophs, as well as how heterotrophs obtain that energy by consuming autotrophs. In this chapter, we return to the topic of energy as we review how energy enters ecosystems, how it is measured, and what controls rates of energy flow. In the next chapter, we will describe the movements of energy through the biotic components of ecosystems and the importance of those movements to the dynamics of communities and ecosystems.

Primary Production

CONCEPT 19.1
Energy in ecosystems originates with primary production by autotrophs.

The chemical energy generated by autotrophs, known as **primary production**, is derived from the fixation of carbon during photosynthesis and chemosynthesis. (Chemosynthesis accounts for a very small amount of this energy globally, although it can be important locally, as we will see.) Primary production represents an important energy transition: the conversion of light energy from the sun into chemical energy that can be used by autotrophs and consumed by heterotrophs. Primary production is the source of energy for all organisms, from bacteria to humans. The energy assimilated by autotrophs is stored as carbon compounds in plant tissues; therefore, carbon is the currency used for the measurement of primary production. The rate of primary production is sometimes referred to as *primary productivity*.

Gross primary production is total ecosystem photosynthesis

The total amount of carbon fixed by the autotrophs in an ecosystem is called **gross primary production** (**GPP**). The GPP in most terrestrial ecosystems is equivalent to the total of all plant photosynthesis.

The amount of carbon fixed by plants is controlled by the influence of climate on rates of photosynthesis (see Chapter 5) and by the leaf area of the plants per unit of ground area (the **leaf area index** [**LAI**], a metric that lacks units, since it is an area divided by an area). The leaf area index varies among biomes, from less than 0.1 in Arctic tundra (i.e., less than 10% of the ground surface has leaf cover) to 12 in boreal and tropical forests (i.e., on average, there are 12 layers of leaves between the canopy and the ground). Shading of the leaves below the uppermost layer increases with the addition of each new leaf layer, so the incremental gain in photosynthesis for each added leaf layer decreases (**Figure 19.4**). Eventually, the respiratory costs associated with adding leaf layers outweigh the photosynthetic benefits. Plants general-

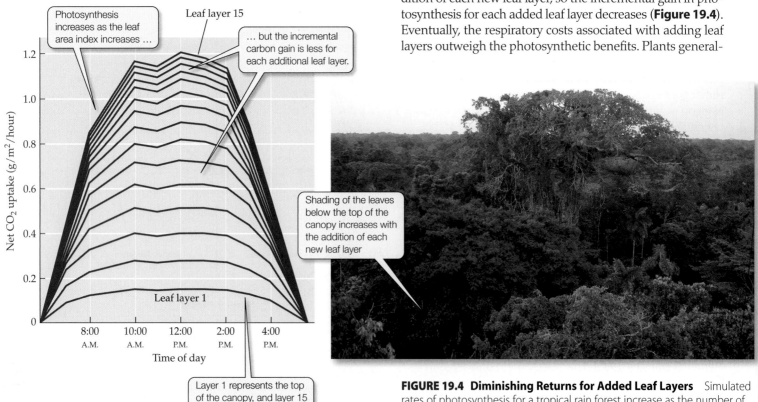

FIGURE 19.4 Diminishing Returns for Added Leaf Layers Simulated rates of photosynthesis for a tropical rain forest increase as the number of leaf layers, or leaf area index, increase. (After Larcher 1980.)

ly optimize their leaf area index in relation to the climatic conditions and the supply of resources, particularly water and nutrients, available to them.

A plant uses approximately half of the carbon it fixes by photosynthesis in cellular respiration to support biosynthesis and cellular maintenance. All living plant tissues lose carbon via respiration, but not all of them acquire carbon via photosynthesis. Thus, plants that have a large proportion of nonphotosynthetic stem tissue, such as trees and shrubs, tend to have higher overall respiratory carbon losses than herbaceous plants. Plant respiration rates increase with increasing temperatures, and as a result, respiratory carbon losses are higher in tropical forests than in temperate and boreal forests.

Net primary production is the energy remaining after respiratory losses

The balance between GPP and respiration is called **net primary production** (**NPP**):

$$NPP = GPP - respiration$$

NPP represents the biomass gained by the plant as a result of carbon fixation. NPP is the energy left over for plant growth and consumption by detritivores and herbivores. It also represents short-term (annual) to long-term storage of carbon in ecosystems.

Carbon not used in respiration can be allocated to growth and reproduction, storage, and defense against herbivory. Furthermore, plants can respond to environmental conditions by allocating carbon to the growth of different tissues. The allocation of carbon within a plant varies considerably according to the species, the availability of resources, and the climate. Allocation of carbon to photosynthetic tissues results in greater future NPP, but the demands of the plant for other resources, particularly water and nutrients, as well as biological interactions such as herbivory, influence whether that investment pays off.

The allocation of NPP to the growth of leaves, stems, and roots is generally balanced so that plants can maintain supplies of water, nutrients, and carbon to match their requirements. For example, plants growing in the soils of desert, grassland, and tundra ecosystems are regularly exposed to shortages of water and nutrients. Plants in these ecosystems may allocate a greater proportion of NPP to root growth, relative to the growth of shoots (leaves and stems), than plants growing in ecosystems with higher soil resource availability (**Figure 19.5**). This greater allocation to roots facilitates their acquisition of the resources that are in short supply. In contrast, plants growing in dense communities with neighbors that may shade them may allocate NPP preferentially to stems and leaves in order to capture more sunlight for photosynthesis. In other words, plants tend to allocate the most NPP to those tissues that acquire the resources that limit their growth.

Allocation of NPP to storage compounds such as starch and carbohydrates provides insurance against losses of tissues to herbivores, disturbances such as fire, and climatic events such as frost. These compounds are usually stored in stems of woody plants or in belowground stems and roots of herbaceous plants. Where levels of herbivory are high, plants may allocate a substantial amount of NPP (up to 20%) to defensive secondary compounds such as tannins or terpenes.

NPP changes during ecosystem development

As ecosystems develop during primary or secondary succession (see Chapter 16), NPP changes as the leaf area index, the ratio of photosynthetic to nonphotosynthetic tissue, and plant species composition all change. Most ecosystems have their highest NPP during intermediate stages of succession. Several factors contribute to this pattern, including the tendency for photosynthetic tissues, plant diversity, and nutrient supply to be highest at intermediate stages of succession. In forest ecosystems, the leaf area index and the photosynthetic rates of leaves decrease in old-growth stands, lowering GPP and thus NPP (**Figure 19.6**). In some grasslands, such as the tallgrass prairies of the central United States, the accumulation of dead leaves near the ground surface and the

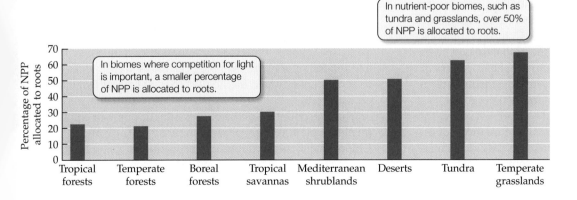

FIGURE 19.5 Allocation of NPP to Roots The proportion of their NPP that plants allocate to roots varies with the resources available to them. (After Chapin et al. 2002.)

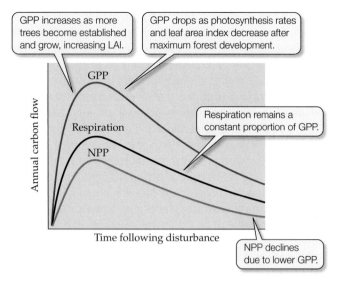

GPP increases as more trees become established and grow, increasing LAI.

GPP drops as photosynthesis rates and leaf area index decrease after maximum forest development.

Respiration remains a constant proportion of GPP.

NPP declines due to lower GPP.

FIGURE 19.6 NPP Changes during Forest Succession Carbon flux in forests ecosystems changes as succession proceeds after a disturbance. (After Ryan 1997.)

development of a closed upper canopy of leaves decrease light penetration for short plants, lowering the photosynthetic carbon gain of the ecosystem. However, the decrease in NPP over time is far less pronounced in grasslands than in forest ecosystems. Although NPP may decrease in later successional stages, these old-growth ecosystems have large pools of stored carbon and nutrients and provide habitat for late successional animal species.

NPP can be estimated by a number of methods

There are a number of reasons why it is important to be able to measure NPP in an ecosystem. As we have seen, NPP is the ultimate source of energy for all organisms in an ecosystem, and thus determines the amount of energy available to support that ecosystem. It varies tremendously over space and time, at a number of scales. Year-to-year variation in NPP provides a metric for examining ecosystem health, since changes in primary productivity can be symptomatic of stresses such as drought or acid rain. In addition, because the movement of carbon from the atmosphere into ecosystems is an important control over atmospheric CO_2 concentrations, NPP is intimately associated with the global carbon cycle, and thus with global climate change (see Chapter 24). For all these reasons, there have been many improvements in techniques for estimating NPP over the past two decades.

Terrestrial ecosystems There are multiple methods of estimating NPP in terrestrial ecosystems. Methods for estimating NPP in forest and grassland ecosystems are the best established due to their economic importance for wood and

forage production. Traditional techniques measure the increase in plant biomass during the growing season by harvesting plant tissues in experimental plots and scaling up to the ecosystem level. For example, in temperate grassland ecosystems, the aboveground biomass (living plant mass) can be harvested from plots at the start of the growing season and again when the amount of plant biomass reaches its maximum. The difference in plant biomass between the two harvests is used as an estimate of NPP. In forests, the radial growth of wood must be included in estimates of NPP. In the tropics, plants may continue to grow throughout the year, and tissues that die decompose rapidly, making the use of harvest techniques problematic. Despite these shortcomings, harvest techniques still provide reasonable estimates of aboveground NPP, particularly if corrections are made for tissue loss to herbivory and mortality.

Measuring belowground NPP is more difficult because root growth is more dynamic than the growth of leaves and stems. As noted earlier, the proportion of NPP in roots exceeds that in aboveground tissues in some ecosystems. In grassland ecosystems, root growth may be twice that of the aboveground leaves, stems, and flowers combined. Roots *turn over* more quickly than shoots; that is, more roots are "born" and die during the growing season than stems and leaves. In addition, roots may exude a significant amount of carbon into the soil, and they may transfer carbon to mycorrhizal or bacterial symbionts. Therefore, harvests for measuring root biomass must be more frequent, and additional correction factors must be used when estimating belowground NPP. Algorithms relating aboveground to belowground NPP have been developed for some forest and grassland ecosystems, so that measurements of aboveground NPP can be used to estimate whole-ecosystem NPP. The use of *minirhizotrons*, underground viewing tubes outfitted with video cameras, has led to significant advances in the understanding of belowground production processes (**Figure 19.7**).

The labor-intensive and destructive nature of harvest techniques makes them impractical for estimating NPP over large areas or in biologically diverse ecosystems. Several nondestructive techniques have been developed that allow more frequent estimation of NPP over much larger spatial scales, although with lower precision than harvest techniques. Some of these techniques, which include remote sensing and atmospheric CO_2 measurements, provide a quantitative index rather than an absolute measure of NPP. Some techniques require a combination of data collection and modeling of plant physiological and climatic processes to accurately estimate the actual fluxes of carbon associated with NPP.

The concentration of the photosynthetic pigment chlorophyll in a plant canopy provides a proxy for GPP and NPP. Chlorophyll concentrations can be estimated using remote sensing techniques that rely on the reflection of solar radiation. Since chlorophyll absorbs visible solar radiation in blue

FIGURE 19.7 A Tool for Viewing Belowground Dynamics (A) Minirhizotrons allow researchers to observe the dynamics of root growth and death belowground. (B) A view of roots through a minirhizotron.

and red wavelengths, it has a characteristic "spectral signature." In addition, plants have higher reflectance in the infrared wavelengths than do bare soil or water. Therefore, researchers can measure the reflection of several specific wavelengths from a land or water surface and estimate NPP using several indices that have been developed. One of the most common indices is the normalized difference vegetation index, or NDVI, which uses differences between visible-light and near-infrared reflectance to estimate the absorption of light by chlorophyll. That estimate is then coupled with estimates of the efficiency of light absorption to estimate CO_2 uptake. NDVI can be measured frequently, at spatial scales up to the entire globe, using satellite-based sensors (**Figure 19.8**). Indices such as NDVI that are based on chlorophyll concentrations can overestimate NPP if the vegetation is not physiologically active, as in boreal forests in winter.

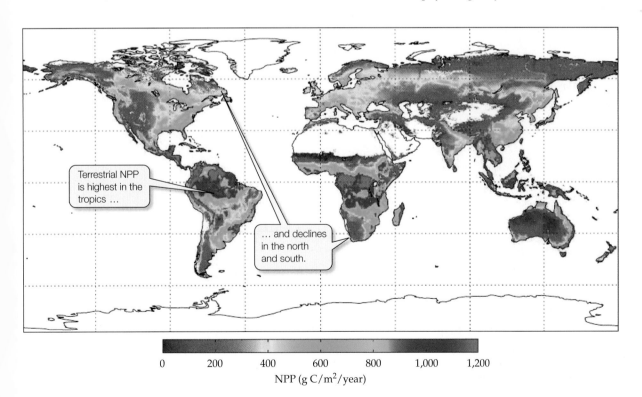

FIGURE 19.8 Remote Sensing of Terrestrial NPP Terrestrial NPP for the globe for the period 2000–2005, estimated using a satellite-based sensor [Moderate resolution Imaging Spectroradiometer (MODIS)]. Note the latitudinal patterns in NPP corresponding to climatic regimes. (From Zhao et al. 2006.)

NPP can also be estimated from direct measurements of its components: GPP and plant respiration. This approach typically involves measuring the change in CO_2 concentration in a closed system, which is created by placing a chamber around stems and leaves, whole plants, or whole stands of plants. For example, Howard Odum estimated NPP for a tropical forest in Puerto Rico by enclosing a stand of trees inside a 200 m^2 × 20 m tall clear plastic "tent" (Odum and Jordan 1970). The exchange of CO_2 with the atmosphere was measured. The sources of CO_2 added to the atmosphere are respiration by the plants and heterotrophs, including microorganisms in the soil and animals in the forest. Uptake of CO_2 from the atmosphere results from photosynthesis. The net change in CO_2 inside the tent results from the balance between GPP uptake and total respiratory release by the plants and the heterotrophs. This net exchange in CO_2 is called **net ecosystem production** or **net ecosystem exchange** (**NEE**). Heterotrophic respiration must be subtracted to obtain NPP. Over long periods, carbon lost through leaching in the soil and disturbances (e.g., fire) must also be incorporated into estimates of NPP.

Another noninvasive approach to estimating NEE uses frequent measurements of CO_2 and microclimate at various heights through a plant canopy and into the open air above the canopy. This technique, known as eddy covariance or eddy correlation, takes advantage of the gradient in CO_2 concentration between the plant canopy and the atmosphere that develops because of photosynthesis and respiration. During the day, when plants are photosynthetically active, the concentration of CO_2 is lower in the plant canopy than in the air above the plant canopy. At night, when photosynthesis shuts down but respiration continues, the CO_2 concentration in the canopy is higher than that in the atmosphere. Instrument-bearing towers established in forest, shrubland, and grassland canopies are used to measure the net ecosystem exchange of CO_2 over long periods (**Figure 19.9**). Depending on the tower height, eddy covariance can provide an integrated NEE for up to several square kilometers of the surrounding area. A network of eddy covariance sites in the Americas (Ameriflux: http://public.ornl.gov/ameriflux/) has been established to better understand the uptake and fate of carbon in terrestrial ecosystems and how carbon uptake is influenced by climate.

Aquatic ecosystems   The dominant photosynthesizers in both freshwater and marine ecosystems are phytoplankton. Phytoplankton turn over much more rapidly than terrestrial plants, so the biomass present at any given time is very low compared with NPP; therefore, harvest techniques are not used in aquatic ecosystems. Instead, photosynthesis and respiration rates are measured in water samples collected in bottles and incubated at the site with light (for photosynthesis)

(A)

(B)

CO$_2$ concentrations are highest at night without photosynthesis.

Photosynthesis during the day draws canopy CO$_2$ concentration down to the same level or lower than the atmosphere above it.

FIGURE 19.9 Eddy Covariance Estimates of NPP   (A) A tower projecting above a subalpine forest on Niwot Ridge, Colorado. Attached to the tower are instruments for measuring microclimate (temperature, wind speed, radiation) and atmospheric CO_2 concentrations at frequent intervals. These measurements are used for estimating net ecosystem exchange of CO_2. (B) Concentration of CO_2 (parts per million) from the ground surface to above the canopy in a boreal forest in Siberia, measured over the course of a 24-hour period in the summer. (B after Hollinger et al. 1998.)

NPP (g C/m²/year)

FIGURE 19.10 Remote Sensing of Marine NPP Primary production in the oceans, estimated using a satellite-based sensor [Sea-viewing Wide Field-of-view Sensor (SeaWiFS)].

and without light (for respiration). The difference in the rates is equal to NPP. Although there are errors associated with the artificial environment of the bottles, as well as the inclusion of heterotrophic respiration of bacteria and zooplankton in the bottles, this technique is used widely in freshwater and marine ecosystems.

Remote sensing of chlorophyll concentrations in the oceans using satellite-based sensors provides good estimates of marine NPP (**Figure 19.10**). Indices similar to NDVI are used to indicate how much light is being absorbed by chlorophyll, which is then related to NPP by using a light utilization coefficient to translate light absorption into CO_2 uptake.

Environmental Controls on NPP

> **CONCEPT 19.2**
> Net primary productivity is constrained by both physical and biotic environmental factors.

As we have seen, NPP varies substantially over space and time. For example, there is as much as a 50-fold difference in NPP between Arctic and tropical ecosystems. Much of this variation is associated with differences in climate, such as latitudinal gradients in temperature and precipitation. In this section, we explore the factors that constrain rates of NPP.

NPP in terrestrial ecosystems is controlled by climatic factors

Variation in terrestrial NPP at the continental to global scale correlates well with variation in temperature and precipitation. NPP increases as mean annual precipitation increases, up to a maximum of about 2,400 mm per year, after which it decreases (**Figure 19.11A**). At very high precipitation levels, NPP may decrease because of greater cloud cover and lower sunlight, leaching of nutrients from soils, and soil saturation, which results in anoxic conditions that cause stress for both plants and decomposers. NPP increases with annual average temperature (**Figure 19.11B**). These correlations suggest that NPP is directly linked to water availability and temperature. Such links makes sense when we consider the direct influence of water availability on photosynthesis via the opening and closing of stomates and the influence of temperature on the enzymes that facilitate photosynthesis (see Chapter 5). In deserts and in some grassland ecosystems, water availability has a clear, direct influence on NPP. In other ecosystems, the causal connection between precipitation and NPP is less clear.

The links between climate and NPP may also be indirect, mediated by factors such as nutrient availability or the particular plant species found within an ecosystem. How can we detect whether the influence of climate on NPP is direct or indirect? Several approaches, both observational and exper-

(A)

(B)

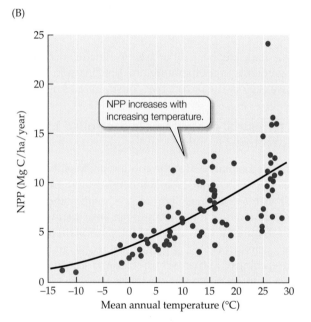

FIGURE 19.11 Global Patterns of Terrestrial NPP Are Correlated with Climate These graphs show the relationships between NPP and (A) precipitation and (B) temperature in terrestrial ecosystems worldwide (Mg = 10^6g). (After Schuur 2003.)

imental, have been used. William Lauenroth and Osvaldo Sala examined how NPP in a short-grass steppe ecosystem changed in response to year-to-year variation in precipitation (Lauenroth and Sala 1992). They also examined the relationship between NPP and precipitation across several grassland ecosystems of different types at different locations in the central United States. When they compared the correlations between NPP and precipitation in their two analyses,

they found a steeper relationship between increases in precipitation and increases in NPP for the site-to-site comparison among the different grassland types than for the comparison among years in the short-grass steppe (**Figure 19.12**). In other words, the rate at which NPP changed with variation in precipitation over space in the comparison of grassland ecosystems was greater than the rate at which it changed in the short-grass steppe over time. They attributed this difference in the responsiveness of NPP to precipitation to differences in the species that made up the different grasslands. Different grass species have different inherent abilities to increase their growth in response to changes in water availability, including differences in their ability to produce new shoots and in vegetative and seed reproduction. Lauenroth and Sala also suggested that there was a time lag in the response to increased precipitation in the short-grass steppe ecosystem; that is, the increase in NPP with an increase in precipitation did not occur in the same year, but was delayed one to several years. Thus, within the grassland biome, differences in species' characteristics can contribute to site-to-site variation in NPP, and the correlation between climate and NPP among sites is related in part to differences in species composition.

Experimental manipulations of water, nutrients, CO_2, and plant species composition have been used to examine the direct influence of those factors on NPP. The results of numerous experiments indicate that nutrients, particularly nitrogen, control NPP in terrestrial ecosystems. For example, William Bowman, Terry Theodose, and their colleagues used a fertilization experiment in alpine communities of the southern

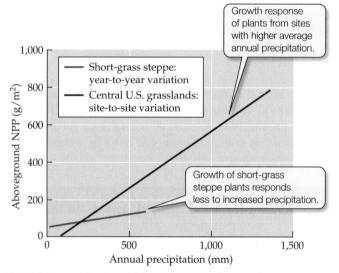

FIGURE 19.12 The Sensitivity of NPP to Changes in Precipitation Varies among Grassland Ecosystems The relationship between aboveground NPP and precipitation is shown for a short-grass steppe ecosystem and for several grassland ecosystems of different types at different sites. (After Lauenroth and Sala 1992.)

Rocky Mountains to determine whether the supply of nutrients limits NPP (Bowman et al. 1993). As in the grassland ecosystems described above, spatial differences in NPP among alpine communities are correlated with differences in soil water availability. Bowman et al.'s fertilization experiment was performed in two communities, dry and wet meadows, that differed in soil moisture and nutrient availability. They sought to determine whether these belowground resources and the species composition of the communities influenced their responses to added nutrients. They added nitrogen, phosphorus, and both nitrogen and phosphorus to plots in both communities, and maintained plots with no nutrient additions as controls. Their results indicated that the supply of nitrogen limited NPP in the dry meadow (which had lower soil moisture and nutrient availability), while nitrogen and phosphorus both limited NPP in the wet meadow (which had higher soil moisture and nutrient availability) (**Figure 19.13**). An additional experiment indicated that the addition of water to the dry meadow did not increase NPP, despite the positive relationship between NPP and soil moisture across the communities. These results suggest that the correlation between soil moisture and NPP in these alpine communities does not indicate a direct causal relationship, but rather is determined by the effect of soil moisture on nutrient supply through its effects on decomposition and movement of nutrients in the soil (described in Chapter 21).

Closer examination of the data in Figure 19.13 shows that the increase in NPP was not uniform across all plant species groups. The dominant plant of the alpine dry meadow (*Ko-*

bresia) did not increase its biomass as much as the less common sedge and grass species. The change in NPP in the dry meadow occurred largely as a result of a change in plant species composition within the experimental plots. This was not the case in the wet meadows, where the dominant sedges increased their growth more than the subdominant forb species. These results are consistent with the general trend of results from many fertilization experiments, which indicate that plant species from resource-poor communities have lower growth responses to fertilization than species from resource-rich communities. This apparent contradiction is the result of differences in the capacity of plant species to respond. Plants of resource-poor communities tend to have low intrinsic growth rates, a characteristic that lowers their resource requirements. Plants of resource-rich communities tend to have higher growth rates, which make them better able to compete for resources, particularly light. Although NPP increases in nutrient-poor communities when they are fertilized, the change in plant species composition that occurs in many such experiments indicates that plant species composition can determine the intrinsic capacity of an

FIGURE 19.13 Nutrient Availability Influences NPP in Alpine Communities (A) Fertilization plots in a dry meadow alpine community in the Colorado Rocky Mountains, dominated by sedges, forbs, and grasses (see Figure 3.11). (B) Fertilization of a resource-poor dry meadow and a resource-rich wet meadow with nitrogen (N), phosphorous (P), and both N and P in an alpine ecosystem indicated nutrient limitations of NPP. (B after Bowman et al. 1993.)

(A)

(B)
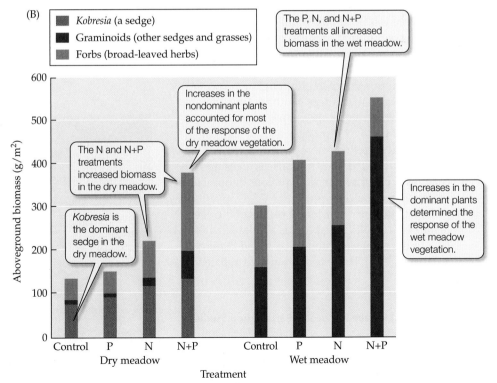

FIGURE 19.14 Growth Responses of Alpine Plants to Added Nitrogen Subjecting plants to low and high nitrogen, while keeping all other nutrients at optimal concentrations, indicates substantial variation exists in the ability of alpine species to respond to variation in nutrient availability. (After Bowman and Bilbrough 2001.)

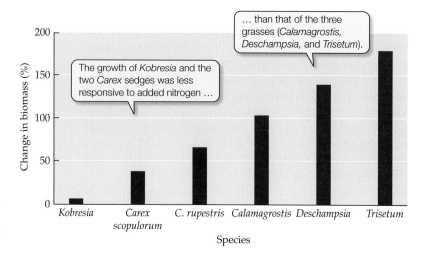

ecosystem to increase its NPP when resources are increased (**Figure 19.14**).

NPP is often limited by nutrients in nondesert terrestrial ecosystems. Some general differences among terrestrial ecosytems have emerged from resource manipulation experiments and measurements of plant and soil chemistry. NPP of lowland tropical rainforests is often limited by the supply of phosphorus, since the relatively old, leached tropical soils in which they grow are low in available phosphorus relative to other nutrients. Other nutrients, such as calcium and potassium, can also limit production in lowland tropical ecosystems. Montane tropical ecosystems, and most temperate and Arctic ecosystems, are limited by the supply of nitrogen, and occasionally by phosphorus. Even in some desert ecosystems, NPP is co-limited by water and nitrogen.

NPP in aquatic ecosystems is controlled by nutrient availability

The primary producers in lake ecosystems are phytoplankton and rooted vascular plants at the shallow margins of lakes, called *macrophytes* (*macro*, "large"; *phyte*, "plant"). NPP in lake ecosystems is often limited by the supply of phosphorus, as we know not only from the results of experimental manipulations, but also from unintentional "experiments" set in motion by wastewater discharges into lakes (see p. 470). A common approach to determining the responses of NPP in lakes to changes in nutrient supply is to incubate translucent or open-top containers, sometimes referred to as "limnocorrals," of lake water, amended with one or more nutrients, in the lake (**Figure 19.15**). The NPP response is measured by changes in chlorophyll concentrations or numbers of phytoplankton cells.

One of the most convincing studies of the effect of nutrients on NPP in lakes was a series of whole-lake fertilization experiments by David Schindler (Schindler 1974). The experiments were initiated in

1969 in the Experimental Lakes Area in Ontario, a series of 58 small lakes set aside for experimental manipulations. Concern over declining water quality in the lakes of North America and Europe motivated Schindler and his colleagues to establish several experiments to determine whether inputs of nutrients in wastewater were involved in the dramatic increases in the growth of phytoplankton that had been observed. They added nitrogen, carbon, and phosphorus to all or half of several individual lakes. The results of these experiments provided strong evidence for phosphorus limitation of NPP (**Figure 19.16**). Massive increases in the abundances of cyanobacteria were responsible for the increase in NPP with phosphorus addition. Evidence for N limitation of NPP in high elevation lakes also exists based on small-scale fertilization experiments and measurements of the ratio of nitrogen to phosphorus in the water.

NPP in streams and rivers is often low, and the majority of the energy in those ecosystems is derived from terrestrial

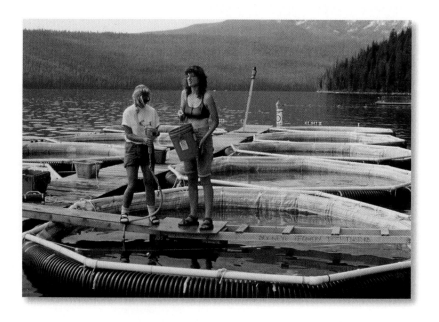

FIGURE 19.15 Lake Mesocosm Fertilization Studies Student assistants add nutrients to an experimental enclosure in Redfish Lake, Idaho. The experiment tests whether nutrients stimulate NPP in the lake to assist recovery of endangered Snake River sockeye salmon.

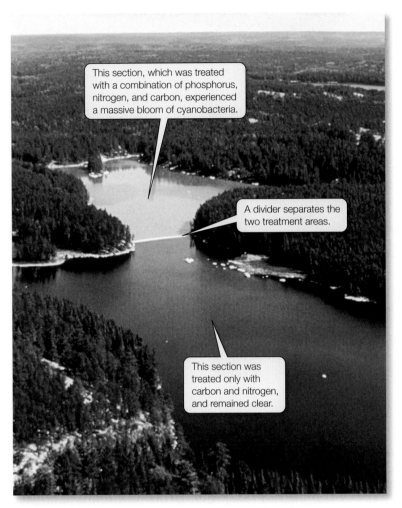

FIGURE 19.16 Response of a Lake to Phosphorus Fertilization
Experimental Lake 226 was divided into two sections as part of David Schindler's experiments on the effect of nutrient supply on NPP.

organic matter (see Chapter 20). Water movement limits the abundance of phytoplankton, except where the water velocity is relatively low. In Chapter 3, we introduced the *river continuum concept*, which describes the increasing importance of in-stream NPP as the river flows downstream. Most of the NPP in streams and rivers comes from photosynthesis by macrophytes and algae attached to the bottom in shallow areas where there is enough light for photosynthesis. The amount of suspended sediment in rivers can limit light penetration; thus, water clarity often controls NPP. Nutrients, particularly nitrogen and phosphorus, can also limit NPP in streams in rivers.

Marine NPP is usually limited by nutrient supply, but the specific limiting nutrients vary among marine ecosystems. Estuaries, the zones where rivers empty into the ocean (see Chapter 3), are rich in nutrients relative to other marine ecosystems. Variation in NPP among estuaries is correlated with variation in nitrogen inputs from rivers. Agricultural and industrial activities have increased riverine inputs of nitrogen into estuaries, which have caused periodic "blooms" of algae. These blooms have been implicated in the development of "dead zones"—areas of high fish and zooplankton mortality—in the Gulf of Mexico. When the algae die, they are consumed by growing bacterial populations, which deplete the oxygen supply in the water, killing the other organisms.

NPP in the open ocean is derived primarily from phytoplankton, including a group referred to as the *picoplankton*, with cells smaller than 1 μm. Picoplankton contribute as much as 50% of the total marine NPP. Smaller contributions come from floating mats of seaweeds such as *Sargassum*. Near the coast, kelp forests may have leaf area indices and rates of NPP as high as those of tropical forests. "Meadows" of seagrasses such as eelgrass (genus *Zostera*) are also important contributors to NPP in shallow near-shore zones.

In much of the open ocean, NPP is limited by nitrogen. In the equatorial Pacific Ocean, however, detectable concentrations of nitrogen can be found in the water even when peak NPP occurs, suggesting that some other factor limits NPP. John Martin and colleagues measured the concentrations of nutrients in the waters of the open ocean and performed bottle incubation experiments with iron additions. They found that adding iron to the bottles increased NPP (Martin et al. 1994). Based on this evidence that iron limits NPP in some ocean regions, Martin suggested that aeolian (windblown) dust from Asia, a source of iron for the open ocean, could play an important role in the global climate system through its influence on marine NPP, and thus on atmospheric CO_2 concentrations. During glacial periods, large areas of the continents could have contributed aeolian dust that would have fertilized the ocean. As NPP in marine ecosystems increased, they might have taken up more CO_2 from the atmosphere, reducing its atmospheric concentration and serving as a positive feedback to cool the climate further. Martin suggested that these findings might be applied to address global warming, saying at the time, "Give me half a tanker-load of iron, and I'll give you an Ice Age." He recommended the use of large-scale experiments to investigate the influence of iron on ocean NPP. Unfortunately, Martin died in 1993, before his ambitious experiment could be carried out.

Martin's colleagues subsequently performed the first of several experiments in 1993, adding iron sulfate to surface waters of the equatorial Pacific west of the Galápagos archipelago. This experiment was alternatively referred to as IronEx I or the "Geritol solution"* to global climate change. During

*Geritol was a supplement once widely believed to help cure "iron-poor, tired blood."

(A)

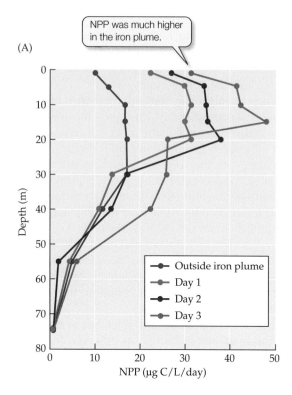

NPP was much higher in the iron plume.

- Outside iron plume
- Day 1
- Day 2
- Day 3

(B)

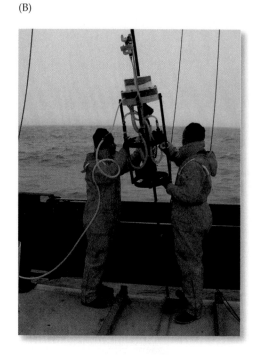

FIGURE 19.17 Effect of Iron Fertilization on Marine NPP IronEx I released a plume of iron into the equatorial Pacific Ocean to study the effects of iron fertilization on NPP. (A) This vertical profile shows primary production at various depths outside the iron plume and inside the plume on three specific days: 1, 2, and 3 days following the release of the iron. (B) Researchers deploy a pump to add iron to the ocean. (A after Martin et al. 1994.)

IronEx I, a 64 km^2 area was fertilized with 445 kg of iron, which resulted in a doubling of phytoplankton biomass and a fourfold increase in NPP (**Figure 19.17**). Three other iron fertilization experiments were subsequently performed, one in 1995 (IronEx II), which produced a tenfold increase in phytoplankton biomass, a second in 1999 in the Southern Ocean, and the last in 2002, also in the Southern Ocean. While the iron limitation hypothesis has been strongly supported by these experiments, fertilizing large areas of the ocean is unlikely to provide a solution to increasing atmospheric CO_2 concentrations and global climate change. Some of the CO_2 taken up by phytoplankton is eventually reemitted to the atmosphere via respiration by zooplankton and bacteria that consume the phytoplankton. In addition, the iron is lost relatively quickly from the surface photic zone, sinking to deeper layers where it is unavailable to support phytoplankton photosynthesis and growth.

Global Patterns of NPP

CONCEPT 19.3
Global patterns of net primary production reflect climatic controls and biome types.

The development of remote sensing and eddy covariance techniques has improved our ability to estimate global patterns of NPP. Initial estimates of global NPP were based on compilations of plot-level measurements from different bio-mes, scaled up using estimates of the spatial distributions of those biomes. These estimates were subject to error associated with the uncertainty of the actual area covered by each biome type, as well as with the potential for overestimating NPP if undisturbed, old-growth study plots were selected to represent a biome. Remote sensing data now provide direct measurements of NPP independent of estimates of biome type or landscape condition.

Total terrestrial and oceanic NPP are nearly equal

Chris Field and colleagues used remote sensing data collected over multiple years to estimate NPP for the entire planet at 105 petagrams (1 Pg = 10^{15} g) of carbon per year (Field et al. 1998). They determined that 54% of this carbon is taken up by terrestrial ecosystems, while the remaining 46% is taken up by primary producers in the oceans. Their estimate of oceanic NPP (which comes to 48 Pg C/year) is considerably higher than previous estimates. Despite their similar contributions to total global NPP, the average rate of NPP for the land surface (426 g C/m^2/year) is higher than that for the oceans (140 g C/m^2/year). The lower rate in the ocean is compensated for by the greater percentage (70%) of Earth's surface it covers. As we will see in Chapter 24, terrestrial NPP has been increasing since the 1980s in a manner consistent with the climate changes associated with greenhouse gas emissions.

Most of the surface of both oceans and land is dominated by areas with relatively low NPP (see Figures 19.8 and 19.10).

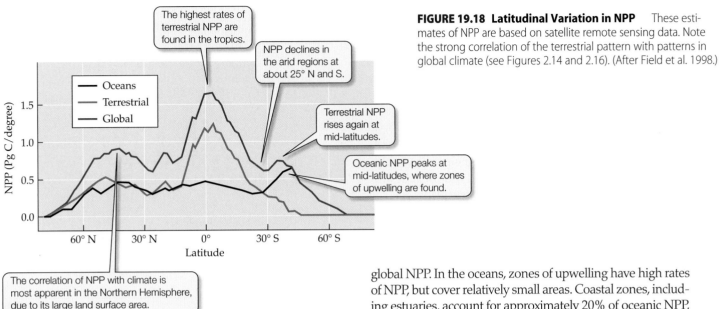

FIGURE 19.18 Latitudinal Variation in NPP These estimates of NPP are based on satellite remote sensing data. Note the strong correlation of the terrestrial pattern with patterns in global climate (see Figures 2.14 and 2.16). (After Field et al. 1998.)

The highest rates of NPP on land are found in the tropics (**Figure 19.18**). This pattern results from latitudinal variation in climate and in the length of the growing season. Higher latitudes have shorter growing seasons, and low temperatures constrain nutrient supply by lowering decomposition rates, which in turn limits NPP. Tropical zones have long growing seasons and high rates of precipitation, promoting high rates of NPP. NPP declines to the north and south of the tropics at about 25° N and S, reflecting the increasing aridity at those latitudes (see Chapter 2). Another peak in terrestrial NPP occurs at the northern mid-latitudes, where temperate forest biomes are found. NPP in the mid- to high latitudes shows strong seasonal trends, with peaks in summer and declines in winter. In contrast, seasonal trends in the tropics are often slight and are associated with dry–wet cycles.

Oceanic NPP peaks at the mid-latitudes between 40° and 60° N and S (see Figure 19.18). These peaks are associated with zones of upwelling, areas where ocean currents bring nutrient-rich deep water to the surface (see Chapter 3). High NPP is also associated with estuaries at these latitudes. Seasonal trends in NPP are not as apparent for the oceans as for the land surface.

Differences among biomes in NPP reflect climatic and biotic variation

It is not surprising that NPP varies among biomes, since biomes are associated with latitudinal climatic variation. For example, the high NPP in the tropics is associated with tropical ecosystems, including forests, grasslands, and savannas. The low NPP at high latitudes is associated with boreal forests and tundra. Tropical forests and savannas contribute approximately 60% of terrestrial NPP (**Table 19.1**), or about 30% of total

global NPP. In the oceans, zones of upwelling have high rates of NPP, but cover relatively small areas. Coastal zones, including estuaries, account for approximately 20% of oceanic NPP, or about 10% of total global NPP. Despite its low mean rate of NPP, the vast area of the open ocean accounts for the majority of oceanic NPP, and approximately 40% of total global NPP.

Much of the variation in NPP among terrestrial biomes is associated with differences in leaf area index (see Figure 19.4). In addition, the length of the growing season varies markedly among biomes, from year-round in some tropical ecosystems to 100 days or less in Arctic tundra. Variation associated with different plant growth forms (e.g., C_3 versus C_4 plants; grasses versus shrubs versus trees) is significant, but contributes less to variation among biomes than do growing season and leaf area index. Variation in NPP among aquatic ecosystems is primarily related to variation in inputs of nutrients (see Chapter 21).

TABLE 19.1

Variation in NPP among Terrestrial Biomes

Biome	NPP (Pg/year)	Terrestrial NPP (%)
Tropical forest	17.8	32
Broad-leaved deciduous forest	1.5	3
Broad-leaved and needle-leaved forest	3.1	5
Needle-leaved evergreen forest	3.1	5
Needle-leaved deciduous forest	1.8	3
Savanna	16.8	30
Grassland	2.4	4
Shrubland	1.0	2
Tundra	0.8	1
Desert	0.5	1
Crops	8.0	14

Source: Field et al. 1998.

Secondary Production

> **CONCEPT 19.4**
> Secondary production is generated through the consumption of organic matter by heterotrophs.

Energy that is derived from the consumption of organic compounds produced by other organisms is known as **secondary production**. Organisms that obtain their energy in this manner are known as heterotrophs, and they include animals, archaea, bacteria, fungi, and even a few plants (see the Case Study in Chapter 11). In this section, we will introduce some of the concepts associated with secondary production. We will cover energy flow among organisms and its consequences for population growth, community dynamics, and ecosystem function in the next chapter.

Heterotrophs are categorized by their diet

Heterotrophs are classified according to the type of food they consume. The most general categories, introduced in Chapter 5, are herbivores, which consume plants and algae; carnivores, which consume other live animals; and detritivores, which consume dead organic matter (detritus). Organisms that consume live organic matter from both plants and animals are called omnivores. Further refinement of feeding preferences is sometimes incorporated into the terminology used to describe autotrophs; insect eaters, for example, are referred to as *insectivores*.

Heterotroph diets can be determined from the isotopic composition of food sources

Determining what organisms eat may be as simple as watching them feed. Such observations, however, may be a time-consuming and imprecise exercise. An alternative method involves analyzing the isotopic composition of the organism and its potential food sources. The concentrations of naturally occurring rare stable (i.e., nonradioactive) isotopes of carbon (^{13}C), nitrogen (^{15}N), and sulfur (^{34}S) differ among potential food items. Measuring the isotopic composition of the feeding organism and its potential food sources provides an indication of the identity of the food sources making up its diet (Peterson and Fry 1987). Determining the principal food source of a large colony of ants, for example, is often difficult. Ants may consume a variety of food items, including plant tissues, phloem sap, fungi tended by aphids, as well as other animals. Comparing the isotopic composition of the ants' potential food sources with that of a sample of ants may indicate what they are feeding on.

Diane Davidson and colleagues used an isotopic composition approach to examine the diets of ants in tropical rain forest canopies. They addressed the question of why ants are

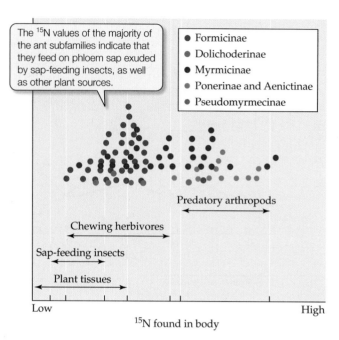

FIGURE 19.19 Nitrogen Isotopic Composition of Ants and Their Diets Each dot represents a different ant subfamily collected from tropical forest canopies in Brunei. Its position on the *x*-axis indicates the ^{15}N values found in its body. Arrows indicate the ^{15}N values associated with various potential ant foods. Similar data were obtained from ant communities in tropical forest canopies of Peru. (After Davidson et al. 2002.)

so abundant (representing 80%–90% of the arthropod biomass) (Davidson et al. 2002). The abundance of suitable prey is low compared to the ants' abundance. Davidson and colleagues therefore hypothesized that the ants must be obtaining most of their food directly or indirectly from plant sources, including plant tissues as well as phloem sap exuded by aphids and other sap-feeding insects. Davidson and colleagues examined the diets of ants in tropical rain forest canopies in two locations, Peru and Brunei. They evaluated the use of potential ant food sources using the nitrogen isotopic composition of plants, sap-feeding insects, herbivores, and predatory arthropods (**Figure 19.19**). The nitrogen isotopic composition of the ants indicated that the majority of their nitrogen, and thus their diet, came from sap exuded by sap-feeding insects. Many groups of ants tend these insects, protecting them from predators and parasites and feeding on the sap they produce. Some of the ant groups that tended sap-feeding insects also contained nitrogen from predation on insects. There was also some evidence that nitrogen-fixing bacterial symbionts in the guts of some ant groups provided them with nitrogen. However, Davidson et al.'s hypothesis was supported: most of the diet of the ants was derived from plant sources, helping to explain how ants could make up such a high proportion of the canopy arthropod community.

Net secondary production is equal to heterotroph growth

Not all of the organic matter consumed by heterotrophs is incorporated into biomass. Some is used in respiration, and some is egested (lost in urine and feces). **Net secondary production** is therefore the balance between ingestion, respiratory loss, and egestion:

$$\text{net secondary production} = \text{ingestion} - \text{respiration} - \text{egestion}$$

Net secondary production by a heterotroph depends on the "quality" of the heterotroph's food, in terms of its digestibility and nutrient content. In addition, the physiology of the heterotroph influences how effectively its food intake is channeled into growth. Animals with high respiration rates (e.g., endotherms) have less energy left over to allocate to growth.

Net secondary production in most ecosystems is a small fraction of NPP. Net secondary production in aquatic ecosystems represents a greater fraction of NPP than it does in terrestrial ecosystems. The majority of net secondary production in most ecosystems is associated with detritivores, primarily bacterial and fungi.

CASE STUDY REVISITED

Life in the Deep Blue Sea, How Can It Be?

In this chapter, we have emphasized the importance of photosynthetic autotrophs as the source of energy for ecosystems, since the vast majority of the energy that enters ecosystems is derived from visible solar radiation. Here and in Chapter 5, however, we have alluded to another source of energy for ecosystems: chemosynthesis. Some bacteria can use chemicals such as hydrogen sulfide (H_2S and related chemical forms, HS^- and S^{2-}) as electron donors to take up carbon dioxide and convert it to carbohydrates:

$$S^{2-} + CO_2 + O_2 + H_2O \rightarrow SO_4^{2-} + (CH_2O)_n$$

Bacteria that provide energy for ecosystems via chemosynthesis are known as chemoautotrophs. The existence of these bacteria was known for at least a century before the discovery of hydrothermal vents. However, their role in providing energy for the vent communities was uncertain.

Initially, hypotheses suggested that the high velocity of water flow around the hydrothermal vents helped focus organic matter derived from the photic zone toward the filter-feeding invertebrates. However, several lines of evidence suggested that chemoautotrophs were the major source of energy for these ecosystems. First, the ratios of carbon isotopes ($^{13}C/^{12}C$) in the bodies of the vent invertebrates were different from those of phytoplankton in the photic zone. Second, tube worms collected from the vents (genus *Riftia*) were found to lack mouths and digestive systems. These gutless tube worms also have structures called trophosomes, highly vascularized tissues with specialized cells containing large amounts of bacteria (**Figure 19.20**). Elemental sulfur was found in the trophosomes, suggesting that sulfides were being chemically transformed in the tube worm's body. Finally, enzymes associated with the Calvin cycle, the biochemical pathway used by autotrophs to synthesize carbohydrates (see Chapter 5), as well as enzymes involved in sulfur metabolism, were found in the trophosomes. Clams and other mollusks found in the vent communities were initially thought to be filter feeders on suspended organic matter. However, samples collected from the vents indicated that these bivalves lacked some of the critical tissues for filter feeding, and that they also had large amounts of bacteria in specialized tissues, as well as enzymes associated with the Calvin cycle.

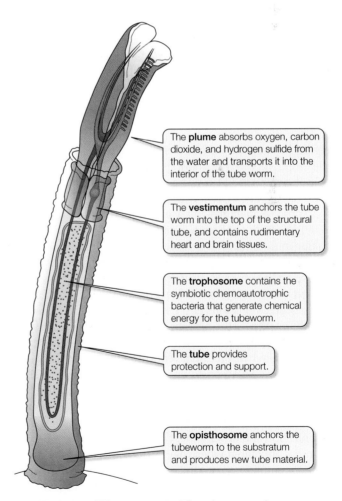

The **plume** absorbs oxygen, carbon dioxide, and hydrogen sulfide from the water and transports it into the interior of the tube worm.

The **vestimentum** anchors the tube worm into the top of the structural tube, and contains rudimentary heart and brain tissues.

The **trophosome** contains the symbiotic chemoautotrophic bacteria that generate chemical energy for the tubeworm.

The **tube** provides protection and support.

The **opisthosome** anchors the tubeworm to the substratum and produces new tube material.

FIGURE 19.20 *Riftia* Anatomy *Riftia* tube worms have a number of specialized structures that are well adapted to their hydrothermal vent environment.

All of this evidence pointed to the conclusion that deep-sea hydrothermal vent communities derive their energy from chemoautotrophic bacteria. These bacteria also aid in detoxifying the sulfides in the water, which would normally inhibit aerobic respiration. Furthermore, many of the abundant organisms have symbiotic relationships with the bacteria— that is, they house the autotrophs in their bodies, often in specialized structures. Is this a mutualistic symbiosis (see Chapter 14)? The tube worms and clams housing the bacteria benefit by obtaining carbohydrates to fuel their metabolic processes, growth, and reproduction, as well as from detoxification of the sulfides. Do the bacteria derive any benefit from the invertebrates? The answer is yes: the invertebrates provide them with a chemical environment unlike that found in the surrounding water, supplying them with carbon dioxide, oxygen, and sulfides at higher rates than what they could obtain for themselves if they were free-living in the water or the sediments surrounding the vent. The symbiosis between the bacteria and the invertebrates therefore results higher productivity than if the organisms lived separately.

Bacterially generated sediments cloud the water within weeks of the initial eruption of a vent (April 1991).

The site has been colonized by tube worms in the genus *Tevnia* (bottom right) (March 1992).

CONNECTIONS IN NATURE

Energy-Driven Succession and Evolution in Hydrothermal Vent Communities

Larger tubeworms in the genus *Riftia* dominate the site (December 1993).

Hydrothermal vent environments are dynamic, born with the eruption of new hot springs, which eventually cease to emit their sulfide-laden water as the subsurface water channels are altered and heat exchange with the geothermal source equilibrates the heat differential (Van Dover 2000). When the hot springs no longer emit water, and the sulfide in the water has been consumed, the communities surrounding the vents collapse as their energy source disappears and the physical substratum falls apart. The life span of vent communities varies from approximately 20 to 200 years. Studies of colonization and development in these communities over the past three decades have provided insights into succession in marine communities in general. Succession in hydrothermal vent communities is relatively rapid and can be observed by periodic revisitation of specific vents (**Figure 19.21**). The logistic difficulty and expense of such investigations has limited the number of observations, although some general trends have emerged. The rates of colonization and development of hydrothermal vent communities are higher when they are closer to other existing vent communities. Since their energy is derived from chemosynthesis, colonization begins with chemoautotrophic bacteria, sometimes in numbers large enough to cloud the water. Tube worms are often the first invertebrates to arrive. Clams and other mollusks are thought to be better

Continued dominance by *Riftia* (October 1994).

A decrease in the temperature of the vent water has increased the iron concentration in the water, resulting in iron oxide precipitation that has given the *Riftia* individuals a rusty appearance (November 1995).

FIGURE 19.21 Succession in Hydrothermal Vent Communities Composition and abundance of hydrothermal vent communities change through time following the emergence of a hot spring.

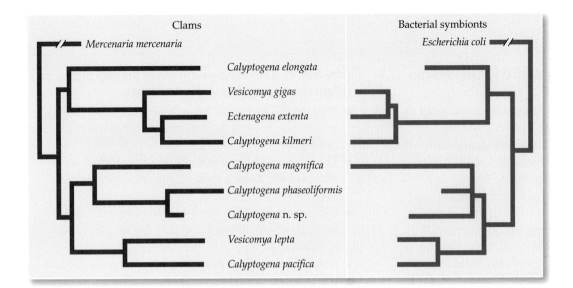

Clams

Mercenaria mercenaria

Calyptogena elongata

Vesicomya gigas

Ectenagena extenta

Calyptogena kilmeri

Calyptogena magnifica

Calyptogena phaseoliformis

Calyptogena n. sp.

Vesicomya lepta

Calyptogena pacifica

Bacterial symbionts

Escherichia coli

FIGURE 19.22 Coevolution of Vent Clams and Their Symbiotic Bacteria The phylogenetic trees of vesicomyid clams collected from hydrothermal vents and their accompanying chemoautotrophic bacterial symbionts show remarkable parallels, suggesting coevolutionary development of these species. (After Peek et al. 1998.)

competitors for sites with optimal temperatures and water chemistry, and over time they increase in abundance at the expense of the tube worms. A few scavengers and carnivores, such as crabs and lobsters, are found in the developing community, although at low abundances. As the tube worm and bivalve populations decline with the decline in sulfide inputs when water flow from the vent decreases, the abundance of scavenger organisms increases until the energy available in the form of detritus is gone.

The pattern of succession in hydrothermal vent communities is subject to the same stochastic or random factors found in terrestrial and aquatic habitats: the order of arrival of organisms at a site can influence the long-term dynamics of the community (Chapter 17). Neighboring vent communities found in the same area of a mid-ocean ridge may show different stages of succession, associated with the stages of hot spring development, as well as different "paths" of succession due to differences in the organisms present. Thus, collections of hydrothermal vents within the same general area are a mosaic of communities at different successional stages, similar to those in terrestrial forest patches, albeit separated by greater distances than the patches within a forest.

The unique nature of the energy supply in hydrothermal vent communities would suggest strong evolutionary divergence between the organisms that inhabit the vents and their nearest non-vent relatives. Where phylogenetic relationships between the vent organisms and their non-vent relatives have been worked out, the divergence is indeed deep, usually at the level of genus, family, or order. Since the discovery of hydrothermal vents, approximately 500 new vent species have been described; of these species, about 90% are endemic to hydrothermal vents. However, large areas of mid-ocean ridges potentially containing hydrothermal vents have yet to be explored.

The close association between the chemoautotrophic bacteria and their invertebrate hosts suggests the potential for a coevolutionary relationship. Have the invertebrates and the sulfide-oxidizing bacteria that they host evolved in concert? To address this question, Andrew Peek and colleagues compared the evolutionary relationships (phylogenetic trees) of vent-dwelling clams in the family Vesicomyidae and their symbiotic bacteria (Peek et al. 1998). Clams in this family transfer bacteria to their offspring in the cytoplasm of their eggs. Peek et al. collected eight species of clams in three genera from hydrothermal vent communities at latitudes ranging from 18° to 47° N and depths ranging from 500 to 6,370 m. Ribosomal DNA taken from the clams and the bacteria was used to construct the phylogenetic trees. The two trees showed remarkable congruence (**Figure 19.22**), providing strong evidence that speciation in the clams and in their bacterial symbionts has occurred synchronously. Other vent groups lack this apparent coevolutionary relationship, however. For example, three different species of tube worms found in different geographic locations have been found to contain the same species of sulfur-oxidizing bacteria.

Recently, it has been suggested that hydrothermal vents are a potential site for the origin of life on Earth. The reducing (i.e., electron donating) geochemical environment of hydrothermal vents is conducive to the abiotic synthesis of amino acids, required for the development of living systems. Although amino acids are not stable in ocean water under the high pressures or temperatures found in some deep hydrothermal vents, there are vents with lower temperatures at shallower depths where amino acid genesis could (and does) occur. As Cyndy Lee Van Dover (2000) so eloquently stated, "Vent water may be the ultimate soup in the sorcerer's kettle."

SUMMARY

CONCEPT 19.1 Energy in ecosystems originates with primary production by autotrophs.

- Gross primary production (GPP) is the total amount of carbon fixed by the autotrophs in an ecosystem.

- The amount of GPP in an ecosystem is determined by photosynthetic rates per unit leaf area and the leaf area index.

- Net primary production is equal to gross primary production minus respiration.

- Net primary productivity changes during succession due to changes in leaf area index and in the balance between photosynthetic and nonphotosynthetic tissues.

- Researchers have developed diverse approaches to measuring net primary production at different spatial and temporal scales.

CONCEPT 19.2 Net primary productivity is constrained by both physical and biotic environmental factors.

- Variation in terrestrial net primary productivity is associated with variation in temperature and precipitation, which affects resource availability and the types and abundances of plants.

- The intrinsic growth rate of plants influences temporal variation in net primary productivity and ecosystem response to climatic variation.

- Net primary productivity in aquatic ecosystems is controlled by the supply of nutrients, particularly phosphorus and nitrogen.

CONCEPT 19.3 Global patterns of net primary production reflect climatic controls and biome types.

- Terrestrial and oceanic net primary production contribute nearly equal proportions of global net primary production

- The majority of terrestrial net primary production occurs in the tropics, including tropical forests and savannas.

- Differences in NPP among terrestrial biomes reflect differences in leaf area index and in the length of the growing season.

- Although nearshore ecosystems are the most productive per unit of area, the majority of oceanic net primary production is derived from the open ocean due to its larger area.

CONCEPT 19.4 Secondary production is generated through the consumption of organic matter by heterotrophs.

- Heterotrophs derive energy from the consumption of live or dead organic matter.

- Net secondary production is the energy ingested by heterotrophs minus the energy used in respiration and egested in feces and urine.

- Heterotroph diet can be estimated by comparing the natural abundances of C, N, or S isotopes in potential food sources with isotopic composition of the tissues of the consumer.

REVIEW QUESTIONS

1. Why is it important to know how much primary production occurs in ecosystems?

2. Plants allocate the energy they acquire through photosynthesis to different functions, including growth and metabolism. The allocation to growth can go preferentially to different tissues, such as leaves, stems, roots, or flowers. How would you expect the allocation of growth among plant tissues to change as the amount of terrestrial ecosystem NPP increased? Explain why you would expect the allocation pattern you describe.

3. Ecologists interested in the underlying factors that control variation in NPP in Arctic tundra measured the growth of plants at several locations over multiple years. The researchers also measured air and soil temperatures, wind speed, solar radiation, and soil mois-

ture. When they analyzed all of their data, they found that the best correlation between NPP and the physical environmental factors they measured is with soil temperature. The researchers concluded that NPP in Arctic tundra is controlled by the effect of soil temperature on root growth. Is this conclusion correct?

4. What are some of the benefits and drawbacks associated with measuring NPP using (a) harvest techniques and (b) remote sensing?

5. Stable isotopes are often used as indicators of diet preference in animals. Would you expect the isotopic composition of ^{15}N in a population of an omnivore to vary more or less among individuals than that in a population of an herbivore?

HANDS-ON PROBLEM SOLVING

| **Drought Reduces Productivity Across Europe** | This Web exercise (www.sinauer.com/ecology) examines the effect of the 2003 European drought on primary productivity. You will interpret data from a recent paper that demonstrated that the drought did substantially reduce primary productivity at various sites in Europe. |

SUGGESTED READINGS

Chapin, F. S. III, P. A. Matson and H. A. Mooney. 2002. *Principles of Terrestrial Ecosystem Ecology*. Springer, New York. *An excellent text covering the basics of terrestrial ecosystem science.*

The following two syntheses cover estimates of global primary production and the underlying environmental controls on biome and latitudinal patterns.

Field, C. B., M. J. Behrenfeld, J. T. Randerson and P. Falkowski. 1998. Primary production of the biosphere: Integrating terrestrial and oceanic components. *Science* 281: 237–240.

Lieth, H. and R. H. Whittaker. 1975. *Primary Productivity of the Biosphere*. Springer-Verlag, New York.

Sala, O. E., R. B. Jackson, H. A. Mooney and R. W. Howarth. 2000. *Methods in Ecosystem Science*. Springer, New York. *This book provides detailed overviews of the theory and methods involved in measuring ecosystem processes, including four chapters dedicated to primary production.*

Van Dover, C. L. 2000. *The Ecology of Deep-Sea Hydrothermal Vents*. Princeton University Press, Princeton, NJ. *This book provides an entertaining account of all aspects of hydrothermal vent science, from the geology of mid-ocean ridges to the successional ecology of vent communities.*

CHAPTER **20** Energy Flow and Food Webs

KEY CONCEPTS

CONCEPT 20.1 Trophic levels describe the feeding positions of groups of organisms in ecosystems.

CONCEPT 20.2 The amount of energy transferred from one trophic level to the next depends on food quality and consumer abundance and physiology.

CONCEPT 20.3 Changes in the abundances of organisms at one trophic level can influence energy flow at multiple trophic levels.

CONCEPT 20.4 Food webs are conceptual models of the trophic interactions of organisms in an ecosystem.

FIGURE 20.1 Subsistence Hunting Inuit hunters peel layers of skin and fat off of a slaughtered seal in a remote, very sparsely populated Arctic region.

CASE STUDY

Toxins in Remote Places

The Arctic is considered one of most pristine and remote regions on Earth. Human effects on its environment are thought to be slight relative to the temperate and tropical zones where the vast majority of humans live. Thus, the Arctic is one of the last places one would expect to find high levels of pollutants in living organisms.

In the mid-1980s, Eric Dewailly was studying concentrations of polychlorinated biphenyls (PCBs) in the breast milk of mothers in southern Quebec. PCBs belong to a group of chemical compounds called persistent organic pollutants (POPs) because they remain in the environment for a long time. POPs originate from industrial and agricultural activities and from the burning of industrial, medical, or municipal wastes. Exposure to PCBs has been linked to increased incidence of cancer, impairment of the ability to fight infections, decreased learning ability in children, and lower birth weights in newborns.

Dewailly was seeking a human population from a pristine area that could be used as a control in his study. He enlisted the help of some Inuit mothers from northern Canada. The Inuit are primarily subsistence hunters of the Arctic, and they have no developed industry or agriculture that would expose them to POPs (**Figure 20.1**). Dewailly therefore assumed that they would have few or no PCBs in their breast milk, providing a benchmark against which to compare populations in more industrialized areas.

What Dewailly found was startling: the Inuit women had concentrations of PCBs in their breast milk that were seven times higher than those in women to the south (**Figure 20.2**) (Dewailly et al. 1993). These alarming findings were reinforced by the work of Harriet Kuhnlein, who at about the same time found that approximately two-thirds of the children from an Inuit community in northeastern Canada had PCB levels in their blood that exceeded Canadian health guidelines (Kuhnlein et al. 1995). More extensive surveys found that the presence of POPs was widespread in Inuit populations. As many as 95% of the people in Inuit communities of Greenland had blood levels of PCBs that exceeded health standards (Pearce 1997).

How were these toxins finding their way into the Arctic environments where the Inuit live? The POPs that were found in the tissues of Inuit populations occur in gaseous form at most environmental temperatures. Produced

FIGURE 20.2 Persistent Organic Pollutants in Canadian Women The breast milk of Inuit mothers from northern Canada was found to contain substantially higher concentrations of polychlorinated biphenyls (PCBs) and two other POPs—dichloro-diphenyldichloroethylene (DDE, a pesticide similar to DDT), and hexachlorobenzene (HCB, an agricultural fungicide)—than that of mothers from southern Quebec. (After Dewailly et al. 1993.)

in lower-latitude industrial areas, these compounds enter the atmosphere under warmer conditions, but when carried by atmospheric circulation patterns into the colder atmosphere of the Arctic, they condense into liquid forms and fall from the atmosphere, sometimes in snowflakes. The manufacture and use of most POPs has been banned in North America since the 1970s. Some developing countries continue to produce POPs, however, and they are important sources of the compounds found in Arctic regions. Although emissions of POPs have decreased, these compounds may remain in Arctic snow and ice for many decades, being released slowly during snowmelt every spring and summer.

Why are concentrations of these compounds so high in the Inuit? The concentrations of POPs in their drinking water are not high enough to explain this phenomenon. One hint comes from the correlation between the levels of the toxins in people and their preferred diets. Communities that have traditionally relied on marine mammals for their food tend to have the highest levels of POPs, while communities that consume herbivorous caribou tend to have lower levels. We will discover the ecological basis for this difference as we trace the flow of energy through ecosystems in this chapter.

To begin our discussion of energy flow in ecosystems, we move from the Arctic to a much warmer desert. Despite their aridity, deserts contain diverse assemblages of plants, animals, and microorganisms. This diversity is reflected in the variation in size, shape, and physiology of the animals making up desert fauna, from nematodes in the soil to grasshoppers in the plant canopy to hawks in the sky. What links these animals together in the context of ecological functioning isn't necessarily their physical appearances or their phylogenetic relationships. Rather, their ecological roles are determined by what they eat and by what eats them—that is, by their feeding, or *trophic*, interactions. In other words, the influence an organism has on the movement of energy and nutrients through an ecosystem is determined by the type of food it consumes, as well as by what consumes it. For example, grasshoppers and scorpions are both arthropods, with similar morphology and physiology, yet their ecological effects on energy flow through the desert ecosystem are quite different. In the context of energy flow, grasshoppers are more similar to mule deer than to scorpions. Grasshoppers and mule deer are both generalist herbivores that consume a variety of desert plant species. The scorpion, by contrast, is a carnivorous insect, feeding primarily on other insects, and has an ecological role more similar to that of a hawk than to that of a grasshopper.

In this chapter, we will continue our discussion of energy, describing its flow through ecosystems and the factors that control its movement through different trophic levels. We will also look at the feeding relationships in an ecosystem as an intricate web of interactions between species, a view that has important implications for energy flow, species interactions, and community and ecosystem dynamics.

Feeding Relationships

CONCEPT 20.1
Trophic levels describe the feeding positions of groups of organisms in ecosystems.

In Chapter 19, we introduced Ray Lindeman's simplified approach to categorizing groups of organisms in an ecosystem according to how they obtain energy (see Figure 19.2). Rather than grouping them by their taxonomic identity, he grouped them into functional categories based on their roles in moving energy through the ecosystem. In this section, we'll take a closer look at these functional categories.

Organisms can be grouped into trophic levels

Each feeding category, or **trophic level**, is based on the number of feeding steps by which it is separated from autotrophs (**Figure 20.3**). The first trophic level consists of the autotrophs, the primary producers that generate chemical energy from sunlight or inorganic chemical compounds. The first trophic level also generates most of the dead organic matter in an ecosystem. In our desert ecosystem, the first trophic level includes all of the plants, which we lump together to form a single component, regardless of their taxonomic identity. In Lindeman's bog ecosystem, the first trophic level was associated primarily with dead organic matter, which Lindeman poetically referred to as "ooze," as well as with autotrophs such as phytoplankton and pondweeds. The second trophic level is composed of the herbivores that consume the autotroph biomass—which in our desert ecosystem would include grasshoppers and mule deer— as well as the detritivores that consume the dead organic matter. The remaining trophic levels (third and up) contain the carnivores that consume the animals at the trophic level below them. The primary carnivores constituting the third trophic level in our desert ecosystem would include small birds and scorpions, while examples of the secondary carnivores making up the fourth trophic level would include foxes and birds of prey. Most ecosystems have four or fewer trophic levels.

Some organisms do not conveniently fit into the trophic levels we have defined here. Coyotes, for example, operate as opportunistic feeders, consuming vegetation, mice, other carnivores, and old leather boots. Omnivores* defy our attempt to group organisms into simple feeding categories. However, their diets can be partitioned to reflect how much energy they consume within a given trophic level (Pimm 2002).

All organisms are either consumed or end up as detritus

All organisms in an ecosystem are either consumed by other organisms at higher trophic levels or enter the pool of dead organic matter, or detritus (Lindeman's ooze) (see Figure 20.3). In most terrestrial ecosystems, only a relatively small

*In trophic studies, the term "omnivory" may be used to refer to feeding at multiple trophic levels, in contrast to our earlier definition of omnivores in Chapter 19 as heterotrophs consuming both plants and animals.

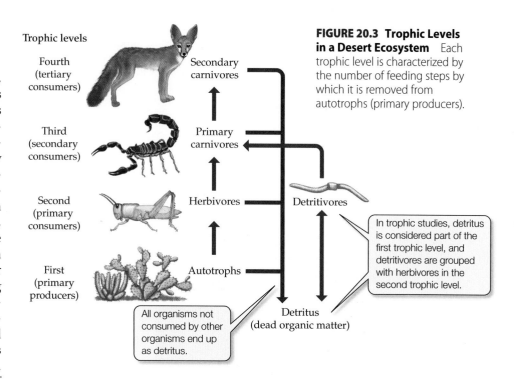

FIGURE 20.3 Trophic Levels in a Desert Ecosystem Each trophic level is characterized by the number of feeding steps by which it is removed from autotrophs (primary producers).

In trophic studies, detritus is considered part of the first trophic level, and detritivores are grouped with herbivores in the second trophic level.

All organisms not consumed by other organisms end up as detritus.

proportion of the biomass is consumed, and most of the energy flow passes through detritus (**Figure 20.4**). Because most of this energy flow occurs in the soil, we are not always aware of its magnitude and importance. Dead plant, microbial, and animal matter, as well as feces, are consumed by a multitude of organisms, known as detritivores (primarily bacteria and fungi), in a process known as *decomposition*. We will describe decomposition in more detail in Chapter 21 in the context of nutrient cycling. Detritus is part of the first trophic level, while detritivores join herbivores in the second trophic level. While live plant–based and detritus-based trophic levels are sometimes considered separately, they are tightly linked through primary production, nutrient cycling, and the many organisms that acquire energy from both plants and detritus.

Energy flow through detritus is important in both terrestrial and aquatic ecosystems. Detritus in terrestrial ecosystems comes almost exclusively from plants within the ecosystem. On the other hand, a large proportion of the input of detritus into stream, lake, and estuarine ecosystems is derived from terrestrial organic matter, considered external to the aquatic ecosystem. External energy inputs are referred to as **allochthonous** inputs, while energy produced by autotrophs within the system is known as **autochthonous** energy. Allochthonous inputs include plant leaves, stems, wood, and dissolved organic matter that fall in from adjacent terrestrial ecosystems or flow in via ground water. These inputs tend to be more important in stream and river ecosystems than in lake and marine ecosystems. For example, Bear Brook, a headwater stream in New Hampshire, receives 99.8% of its energy as allochthonous inputs; the rest is NPP derived from benthic algae and mosses in the stream

(A)

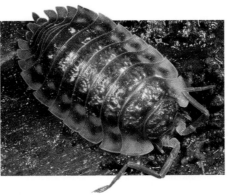

FIGURE 20.4 Ecosystem Energy Flow through Detritus (A) Detritus is consumed by a multitude of organisms, incuding fungi (*Marasmius* sp. in Khao Yai National Park, Thailand) and crustaceans such as the common wood louse, *Oniscus* sp., native to Europe, but introduced to North America. Most of the NPP in terrestrial and aquatic ecosystems ends up as detritus (B), as indicated by the percentage of detritus reported in studies. Similarly, relatively little NPP is consumed by herbivores (C). (After Cebrian and Lartigue 2004.)

(B)

(C)

(Fisher and Likens 1973). In contrast, autochthonous energy accounts for almost 80% of the energy budget in nearby Mirror Lake (Jordan and Likens 1975). As described in our discussion of the river continuum concept (see Chapter 3), the importance of autochthonous energy inputs usually increases from the headwaters toward the middle reaches of a river, in concert with decreases in water velocity and increases in nutrient concentrations.

As this aquatic example shows, grouping organisms into trophic levels makes it easier to trace the flow of energy through an ecosystem. That flow is the topic to which we'll turn next.

Energy Flow among Trophic Levels

CONCEPT 20.2
The amount of energy transferred from one trophic level to the next depends on food quality and consumer abundance and physiology.

The second law of thermodynamics states that during any transfer of energy, some energy is lost due to the tendency toward an increase in disorder (entropy). Thus, we can expect that energy will decrease with each trophic level as we move from the first trophic level upward. We know from our discussion of primary production in Chapters 5 and 19 that autotrophs lose chemical energy through cellular respiration,

lowering the amount of energy available to the second trophic level. In this section, we will examine more closely the factors influencing energy movement between trophic levels.

Energy flow between trophic levels can be depicted using energy or biomass pyramids

A common approach to conceptualizing trophic relationships in an ecosystem is to construct a stack of rectangles, each of which represents the amount of energy or biomass within one trophic level. When assembled from lower to higher trophic levels, these rectangles form a **trophic pyramid**. By portraying the relative amounts of energy or biomass at each trophic level, these pyramids show us how energy flows through the ecosystem.

As we have noted, a proportion of the biomass at each trophic level is not consumed, and a proportion of the energy at each trophic level is lost in the transfer to the next trophic level. Therefore, the sizes of the energy rectangles always decrease as we move from one trophic level to the one above it. In terrestrial ecosystems, energy and biomass pyramids are usually similar because biomass is typically closely associated with energy production (**Figure 20.5A**). In aquatic ecosystems, however, the high consumption rate and the relatively short life spans of the primary producers, mainly phytoplankton, result in an unusual situation in which the biomass pyramid is inverted relative to the energy pyramid (**Figure 20.5B**). In other words, the biomass of heterotrophs

(A) Terrestrial ecosystems

(B) Aquatic ecosystems

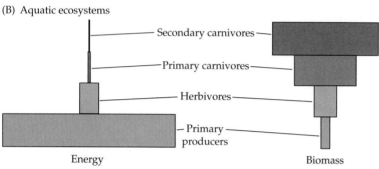

(C)

FIGURE 20.5 Trophic Pyramid Schemes Representative energy and biomass pyramids for (A) terrestrial ecosystems and (B) aquatic ecosystems. (C) Inverted pyramids in aquatic ecosystems are more common in nutrient-poor waters with low autotrophic biomass. (C after Gasol et al. 1997.)

Energy flow between trophic levels differs among ecosystem types

What factors determine the amount of energy flow from one trophic level to the next? In Chapter 19, we evaluated the factors that influence the amount of primary production in terrestrial and aquatic ecosystems, emphasizing abiotic factors such as climate and nutrient availability as well as differences in the inherent ability of autotroph species to produce biomass. It would be reasonable to assume that the flow of energy to higher trophic levels is associated with the amount of primary production at the base of the food web. However, we will see that the situation is not quite so simple. The proportion of each trophic level consumed by the one above it, the nutritional content of autotrophs, detritus, and prey, and the efficiency of energy transfers also play roles in determining the flow of energy between trophic levels.

A comparison of consumption of autotrophs between terrestrial and aquatic ecosystems provides some insight into the factors that influence energy flow between trophic levels. When viewed from outer space, some parts of Earth's terrestrial surface appear green, while the ocean appears blue. Why is the land surface green and the sky blue? In Chapter 19, we saw that very productive lakes (e.g., those that are experimentally fertilized; see Figure 19.15) can appear green. What these green areas have in common is productivity that far exceeds rates of herbivory. Herbivores on land consume a much lower proportion of autotroph biomass than herbivores in most aquatic ecosystems. On average, about 13% of terrestrial NPP is consumed (range 0.1%–75%), while in aquatic ecosystems, an average of 35% of NPP is consumed (range 0.3%–100%) (Cebrian and Lartigue 2004).

There is a positive relationship between net primary production and the amount of biomass consumed by herbivores (**Figure 20.6**). This relationship, which holds across most ecosystem types, would seem to suggest that herbivore production is limited by the amount of food available. Why, then, is the proportion of autotroph biomass consumed in terres-

may be greater at any given time than the biomass of autotrophs. However, the energy produced by the autotrophs is still greater than that produced by the heterotrophs. This tendency toward inverted biomass pyramids is greatest where productivity is lowest, such as in nutrient-poor regions of the open ocean (**Figure 20.5C**). The higher proportion of primary consumer biomass relative to producer biomass in these nutrient-poor regions results from a more rapid turnover of phytoplankton, associated with higher growth rate and shorter life-span compared with phytoplankton of more nutrient rich waters. Phytoplankton in nutrient-poor regions thus provide a greater energy supply per unit of time (Gasol et al. 1997). In addition, detritus makes a higher proportional contribution to energy flow in these nutrient-poor waters than in nutrient-rich waters.

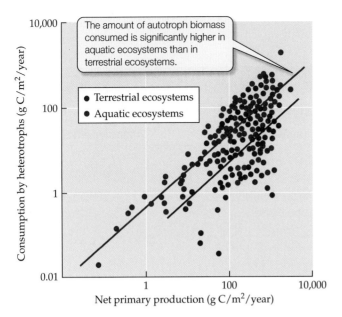

FIGURE 20.6 Consumption of Autotroph Biomass Is Correlated with NPP The amount of autotroph biomass consumed increases with increasing NPP in both terrestrial and aquatic ecosystems. (After Cebrian and Lartigue 2004.)

trial ecosystems relatively low? If herbivore production is limited by the supply of energy and nutrients from plants, why don't terrestrial herbivores consume a greater proportion of the biomass that is available?

Several hypotheses have been proposed to explain the lower proportion of autotroph biomass consumed in terrestrial ecosystems. First, Nelson Hairston and others (1960) have argued that the population growth of herbivores is constrained by predation, keeping population sizes below their potential carrying capacities. Predator removal experiments, described in Chapter 12 and later in this chapter, have provided support for this hypothesis in some ecosystems by demonstrating that significant increases in prey populations follow removal of their predators. Second, defenses against herbivory, including secondary compounds and structural defenses, lower the amount of consumption of some autotrophs. Plants of resource-poor environments, such as desert and tundra ecosystems, tend to be more strongly defended against herbivory than plants from resource-rich environments. This greater allocation to defense may explain why the proportion of plant biomass consumed is lower in resource-poor environments. Third, terrestrial plants contain nutrient-poor structural materials such as stems and wood, which are typically absent in aquatic autotrophs. The chemical composition of phytoplankton makes them more nutritious for herbivores than terrestrial plants. Herbivores typically have high requirements for nutrients such as nitrogen and phosphorus to meet their demands for structural growth, metabolism, and protein synthesis. The ratio of carbon to

nutrient in a food is thus an important measure of its quality, with carbon representing energy. Carbon:nutrient ratios differ markedly between autotrophs in terrestrial and in freshwater ecosystems: freshwater phytoplankton have carbon:nutrient ratios closer to those of herbivores than those of terrestrial plants (Elser et al. 2000), better meeting the nutritional demands of the herbivores. Each of these factors—predation, plant defenses, and food quality—contribute to differences in the proportion of NPP consumed among ecosystems, in particular the greater consumption of autotroph biomass in aquatic ecosystems.

The efficiency of energy transfer varies among consumers

Not all of the energy consumed by a heterotroph gets incorporated into heterotroph biomass. We can consider the transfer of energy between trophic levels using the concept of energy *efficiency*, defined as the output of energy per unit of energy input. In studies of energy transfer in trophic systems, the concept of **trophic efficiency** is used, defined as the amount of energy at one trophic level divided by the amount of energy at the trophic level immediately below it. Trophic efficiency incorporates the proportion of available energy that is consumed (consumption efficiency), the proportion of ingested food that is assimilated by the consumer (assimilation efficiency), and the proportion of assimilated food that goes into producing new consumer biomass (production efficiency) (**Figure 20.7**).

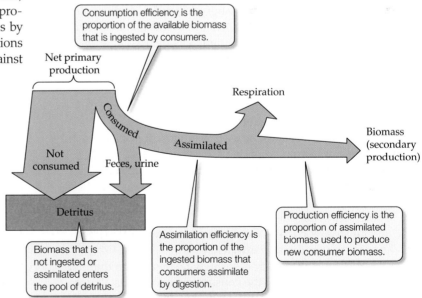

FIGURE 20.7 Energy Flow and Trophic Efficiency The proportion of energy transferred between trophic levels depends on efficiencies of consumption, assimilation, and production.

As we saw above, not all of the biomass available at one trophic level is consumed by the next trophic level. The proportion of the available biomass that is ingested is the **consumption efficiency**. We have seen that, in general, consumption efficiency was higher in aquatic ecosystems than in terrestrial ecosystems. Consumption efficiencies also tend to be higher for carnivores than for herbivores, although a systematic survey comparing the two groups has not been done.

Once biomass is ingested by the consumer, it must be assimilated by the digestive system before the energy it contains can be used to produce new biomass. The proportion of the ingested food that is assimilated is the **assimilation efficiency**. Food that is ingested but not assimilated is lost as feces to the environment, entering the pool of detritus. Assimilation efficiency is determined by the quality of the food (its chemical composition) and the physiology of the consumer.

The quality of the food available to herbivores and detritivores is generally lower than that of the food available to carnivores. Plants and detritus are composed of relatively complex carbon compounds, such as cellulose, lignins, and humic acids, that are not easily digested. In addition, plants and detritus have low concentrations of nutrients such as nitrogen and phosphorus. Animal bodies, on the other hand, have a carbon:nutrient ratio that is usually very similar to that of the animal consuming them, and so are assimilated more readily. Assimilation efficiencies of herbivores and detritivores vary between 20%–50%, while those of carnivores are about 80%.

The digestive capacity of consumers is associated with their thermal physiology. Endotherms (animals that generate heat internally; see Chapter 4) tend to digest food more completely than ectotherms (animals that rely on heat exchange with the environment for thermoregulation), and therefore have higher assimilation efficiencies. Additionally, some herbivores have mutualistic symbionts that help them digest cellulose. For example, ruminants such as cattle, deer, and camels have a modified foregut that contains bacteria and protists that increase the breakdown of cellulose-rich foods. This mutualistic symbiosis, coupled with a longer period of digestion, gives ruminants higher assimilation efficiencies than nonruminant herbivores.

Assimilated food can be used to produce new biomass in the form of consumer growth and production of new consumer individuals (reproduction). However, a portion of the assimilated food must be used for respiration associated with construction of new biomass and maintenance of existing molecules and tissues (see Chapter 5). The proportion of the assimilated food that is used to produce new consumer biomass is the **production efficiency**.

Production efficiency is strongly related to the thermal physiology and size of the consumer. Endotherms allocate much of their assimilated food to metabolic production of heat, and therefore have less energy left over to allocate to growth and reproduction than ectotherms do (**Table 20.1**). Thus, ectotherms have considerably higher production effi-

TABLE 20.1

Production Efficiencies of Consumers

Consumer group	Production efficiency (%)
Endotherms	
Birds	1.3
Small mammals	1.5
Large mammals	3.1
Ectotherms	
Fishes and social insects	9.8
Nonsocial insects	40.7
Herbivores	38.8
Detritivores	47.0
Carnivores	55.6
Non-insect invertebrates	25.0
Herbivores	20.9
Detritivores	36.2
Carnivores	27.6

Source: Chapin et al. 2002; data from Humphreys 1979.

ciencies than endotherms. Body size in endotherms is also an important determinant of production efficiency. If morphology (i.e., the relative sizes of trunk and appendages) and insulation (fat, feathers, and fur) are held constant, then as body size increases, the surface area-to-volume ratio decreases. Thus, a small endotherm, such as a shrew, will lose a greater proportion of its internally generated heat across its body surface than a large endotherm, such as a grizzly bear, and will tend to have a lower production efficiency.

Trophic efficiencies can influence population dynamics

Changes in food quantity and quality, and the resulting changes in trophic efficiency, can determine the consumer population sizes that can be sustained and the health of the individuals in consumer populations. Here we'll examine the influence of food quality on population changes of Steller sea lions (*Eumetopias jubatus*) in Alaska.

From the late 1970s into the 1990s, the total population of Steller sea lions in the Gulf of Alaska and the Aleutian Islands decreased by about 80%, from approximately 250,000 sea lions in 1975 to 50,000 in 2000 (**Figure 20.8**). Andrew Trites and C. P. Donnelly reviewed the available information to try to determine possible causes for this decline (Trites and Donnelly 2003). They found that individual sea lions within the same age classes collected during the period of decline were smaller than those collected before the start of the decline. There was also a reduction in the number of pups born

FIGURE 20.8 Steller Sea Lion Population Decline in Alaska
The population of sea lions in the Gulf of Alaska and the Aleutian Islands decreased by about 80% over 25 years. (After Trites and Donnelly 2003.)

during this period, which resulted in a shift in the age structure toward older individuals. No evidence was found for outbreaks of disease or parasites. The lower body sizes and declining birth rates suggested that there were fewer prey available or that the available prey were not providing sufficient nourishment to sustain the sea lions— in other words, that trophic efficiency had declined. Additional data indicated that the sea lions were obtaining prey, primarily fish, as regularly as before the decline. Nursing females in the declining population were actually spending less time hunting for the same amount of fish as nursing females in other populations that were not declining. Therefore, it did not appear that the availability of prey, or the sea lions' ability to capture it, was limiting growth and reproduction.

Trites and Donnelly considered the possibility that changes in the species of prey fish available had contributed to the decline of the Steller sea lions. They and others suggested that the decline might be related to declining prey quality, an idea they referred to as the "junk food hypothesis." Prior to the de-

cline, the diet of the sea lions was primarily herring, a fish that is relatively rich in fats, along with small amounts of pollock, cod, salmon, and squid. During the period of the population decline, the sea lion diet shifted away from herring toward a greater proportion of pollock and cod (**Table 20.2**). This change in diet reflected a shift toward cod dominance of the fish community from the 1970s through the 1990s. The causes of the change in fish community composition are uncertain, but may be associated with long-term climate change. The fat and energy contents per fish mass of pollock and cod are approximately half the proportions for herring. Captive Steller sea lions raised on a diet of herring, and then switched to a diet of pollock, lose significant amounts of body mass and fat, even with an unlimited supply of pollock.

Based on their review of the available information, Trites and Donnelly concluded that nutritional stress was the most likely cause of the decline in the Steller sea lion population. The amount of prey available to the sea lions did not appear to change, but changes in the quality of that prey, and associated changes in trophic efficiency, contributed to the decline in the population through their effects on individual growth rates and birth rates. The decline has continued into the twenty-first century, but at a slower pace than in the 1980s.

TABLE 20.2

Proportion of Steller Sea Lion Scats and Stomachs Containing Five Prey Categories

	Gadids (cod, pollock, hake)	Salmon	Small schooling fish (herring, capelin, eulachon, sand lance)	Cephalopods (squid)	Flatfish (flounder, sole)
1990–1993	85.2	18.5	18.5	11.1	13.0
1985–1986	60.0	20.0	20.0	20.0	5.0
1976–1978	32.1	17.9	60.7	0.0	0.0

Source: Trites and Donnelly 2003; data from Merrick et al. 1997.

Trophic Cascades

CONCEPT 20.3
Changes in the abundances of organisms at one trophic level can influence energy flow at multiple trophic levels.

There are two possible ways to look at the control of energy flow through ecosystems. First, the amount of energy that flows through trophic levels may be determined by how much energy enters an ecosystem via net primary production (NPP), which in turn is related to the supply of resources (see Chapter 19). The greater the NPP entering the ecosystem, the more energy can be passed on to higher trophic levels. This view, which is often referred to as "bottom-up" control of energy flow, holds that the resources that limit NPP determine energy flow through an ecosystem (**Figure 20.9A**). Alternatively, energy flow may be governed by rates of consumption by predators at the highest trophic levels, which influence abundances and species composition of multiple trophic levels below them. This view is often referred to as

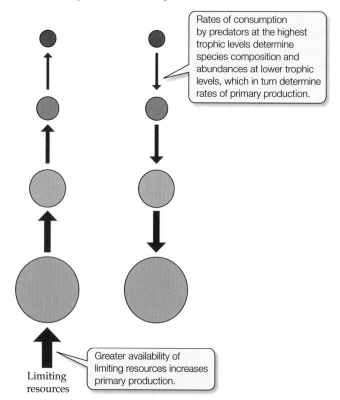

(A) Bottom-up control   (B) Top-down control

Rates of consumption by predators at the highest trophic levels determine species composition and abundances at lower trophic levels, which in turn determine rates of primary production.

Greater availability of limiting resources increases primary production.

Limiting resources

FIGURE 20.9 Bottom-up and Top-down Control of Productivity The amount of production in an ecosystem can be viewed as being controlled by (A) limiting resources or by (B) controls exerted on the composition and abundance of autotrophs due to consumption by higher trophic levels.

"top-down" control of energy flow (**Figure 20.9B**). In reality, both bottom-up and top-down controls are operating simultaneously in ecosystems, but top-down control has important implications for the effects of trophic interactions on energy flow in ecosystems.

Trophic interactions can trickle down through multiple trophic levels

Changes in the composition or abundances of organisms at one trophic level can lead to important, somewhat unpredictable changes in the amount of energy and species composition at other trophic levels. For example, increases in the rate of predation by a carnivore at the fourth trophic level on carnivores at the third trophic level would lead to a lower rate of consumption of herbivores at the second trophic level. Less herbivory would result in higher rates of NPP, and possibly changes in abundances and species composition at the first trophic level. Such a series of changes in energy and species composition is referred to as a **trophic cascade**.

Trophic cascades have been described primarily in aquatic ecosystems, although there are examples from terrestrial ecosystems as well. They are most often associated with changes in the abundance of top specialist predators. Omnivory in food webs may act to buffer the effects of trophic cascades. Finally, trophic cascades have been hypothesized to be most important in relatively simple, species-poor ecosystems. However, several recent experiments have demonstrated trophic cascades in ecosystems with relatively high species diversity.

An aquatic trophic cascade Many examples of trophic cascades come from unintended experiments associated with introductions of non-native species or near-extinctions of native species. A classic example of the latter involves the interaction of sea otters (*Enhydra lutris*), sea urchins, and orcas on the west coast of North America, which was discussed in the Case Study in Chapter 8. Unfortunately, there is no shortage of examples of trophic cascades associated with the intentional or unintentional release of non-native species. One such example resulted from the introduction of brown trout (*Salmo trutta*), a popular sport fish, into streams and lakes of New Zealand. The stocking of Kiwi waters by European settlers began in the 1860s, and by 1920 an estimated 60 million fish had been released throughout New Zealand. Native fish populations have declined as a result, and some species have disappeared from streams now dominated by trout.

Alexander Flecker and Colin Townsend (1994) investigated the influence of brown trout on the composition of prey species (primarily stream insects) and associated effects on primary production in the Shag River. Brown trout were originally released into the Shag River in 1869 by the "Otago Acclimatisation Society" to make settlers feel more at home. The Shag River is one of a small number of streams in New

Zealand that still holds both native fish and trout in the same sections. The native fish include the common river galaxias (*Galaxias vulgaris*). The morphology and feeding behavior of galaxias are similar to those of trout, as indicated by their common name, Maori trout.

Flecker and Townsend compared the effects of brown trout and galaxias on stream invertebrate composition and abundance as well as on primary production by algae. To manipulate fish presence and absence, they constructed artificial stream channels adjacent to the natural channel, made of 5 m lengths of half cylinders of PVC pipe. The PVC channels had mesh on the ends that kept the fish in or out, but allowed free movement of stream invertebrates and algae. The researchers placed clean gravel and stone cobbles in the bottoms of the channels to provide a substratum for the invertebrates and algae. The channels were allowed to accumulate algae and invertebrates for 10 days before the fish were added. Three treatments were initiated: channels with introduced trout, channels with galaxias, and channels with no fish (controls). Eight fish of similar size and mass were used for both fish species additions. The experiment was run for 10 days, after which samples were collected to determine invertebrate species composition and abundance and algal biomass.

Flecker and Townsend had expected brown trout to decrease invertebrate diversity more than the native galaxias, but the effect of fish on invertebrate diversity was relatively small and did not differ between the two fish species. The brown trout, however, reduced total invertebrate density by approximately 40% relative to the control channels, while galaxias resulted in a smaller reduction (**Figure 20.10A**). The abundance of algae increased with both fish, but the effect was greater in the channels with trout (**Figure 20.10B**). Flecker and Townsend suggested that the effect on algal biomass was the result of a trophic cascade in which fish predation not only reduced the density of stream invertebrates, but also caused them to spend more time in refugia on the stream bottom rather than feeding on algae. The trout had a much greater effect on invertebrate density, and thus on primary production, than the native galaxias. These results suggested that trophic cascades associated with stocking of non-native fish for sport may have consequences not just for native biodiversity, but for the functioning of stream ecosystems as well.

A terrestrial trophic cascade As mentioned earlier, trophic cascades have been most commonly observed in aquatic ecosystems, and their existence in terrestrial ecosystems is less certain. Terrestrial ecosystems were thought to be more complex than aquatic ecosystems. In addition, it was believed that a decrease in the abundance of one species in a terrestrial ecosystem was more likely to be compensated for by an increase in the abundance of similar species that were not being consumed as heavily. Thus, it was considered unlikely that trophic cascades would be apparent in diverse terrestrial ecosystems such as tropical forests.

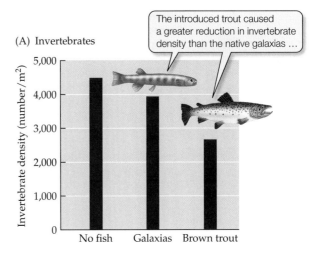

(A) Invertebrates

The introduced trout caused a greater reduction in invertebrate density than the native galaxias …

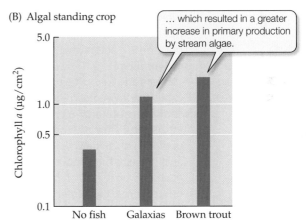

(B) Algal standing crop

… which resulted in a greater increase in primary production by stream algae.

FIGURE 20.10 An Aquatic Trophic Cascade Flecker and Townsend used artificial stream channels to study the effects of non-native brown trout and a native fish (galaxias) on stream invertebrates and algae in the Shag River, New Zealand. (A) Effects on invertebrate density. (B) Effects on algal biomass, as estimated using chlorophyll concentrations in stream water. (After Flecker and Townsend 1994.)

Lee Dyer and Deborah Letourneau (1999a) tested the effects of a potential trophic cascade on the production of *Piper cenocladum* trees in the understory of a lowland wet tropical forest in Costa Rica. *Piper cenocladum* is a relatively common component of the understory in these forests and is eaten by dozens of different herbivore species. Ants of the genus *Pheidole* live in chambers of the petioles of the *Piper* trees. The ants eat food bodies provided by the trees, but they also consume herbivores that attack the trees. These ants, in turn, are eaten by beetles of the genus *Tarsobaenus*. Thus, four distinct trophic levels exist in this system (**Figure 20.11**). Dyer and Letourneau had previously noted that production was lower, and rates of herbivory were higher, when densities of *Tarsobaenus* beetles were high. They performed experiments to test whether a top-down trophic cascade involving the beetles, ants, and herbivores influenced the production of the

Tarsobaenus beetles

Pheidole ants

Herbivores

Piper cenocladum

FIGURE 20.11 A Terrestrial Trophic Cascade Trophic interactions in the understory ecosystem of a lowland wet tropical rainforest in Costa Rica are indicated by the arrows. *Piper cenocladum* trees are consumed by herbivores, but provide shelter for *Pheidole* ants, which consume herbivores attacking the trees. *Pheidole* ants are consumed by *Tarsobaenus* beetles. Both ants and beetles also consume food bodies produced by the trees. (After Dyer and Letourneau 1999a.)

Piper trees, and how strong that influence was compared with that of bottom-up factors such as light and soil fertility.

Dyer and Letourneau established experimental plots in the understory containing uniform-sized cuttings of *Piper* trees. They treated two groups of plots with an insecticide to kill any ants present, then added *Tarsobaenus* beetle larvae to half of those plots. In other words, they established three groups of plots, one group of insecticide-treated plots with beetles and two control groups, one group of insecticide-treated plots without beetles, and one group of untreated plots. The insecticide treatment improved the establishment of the beetles by preventing ant attacks on the beetle larvae. Half of the plots were on a relatively fertile soil type, and the other half were on a relatively infertile soil type. Natural light levels in the plots varied such that half of the plots were assigned to a high-light treatment and half to a low-light treatment. Dyer and Letourneau maintained the treatments for 18 months and measured herbivory and leaf production within each of the plots.

If production of the *Piper* trees was limited primarily by resource supply, then the addition of the ant-predator *Tarsobaenus* beetles would be expected to have little effect on *Piper* leaf production. Soil fertility and light levels would be expected to have greater effects on leaf production if these bottom-up effects were more important than the influence of the trophic cascade associated with beetles, ants, and herbivores. Dyer and Letourneau found, however, that the trophic cascade was the only significant influence on leaf production. The addition of the predatory beetles decreased ant abundance fivefold, increased rates of herbivory threefold,

and decreased leaf area per tree to half that in the control plots (**Figure 20.12**). The results of this experiment provide convincing evidence of a top-down trophic cascade affecting the production of the *Piper* trees. However, it should be noted that the lack of an effect of soil fertility and light in the control treatments, which had low rates of herbivory, indicate that the resource(s) that actually limit production may not have been manipulated in this experiment. An additional ex-

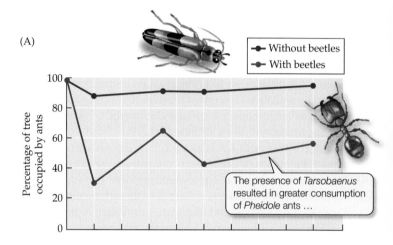

(A)

— Without beetles
— With beetles

The presence of *Tarsobaenus* resulted in greater consumption of *Pheidole* ants ...

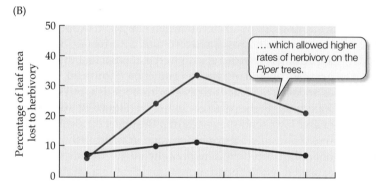

(B)

... which allowed higher rates of herbivory on the *Piper* trees.

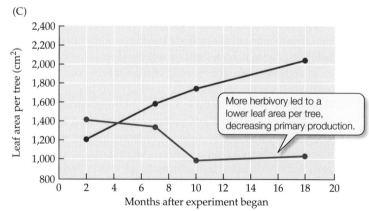

(C)

More herbivory led to a lower leaf area per tree, decreasing primary production.

FIGURE 20.12 Effects of a Trophic Cascade on Production A trophic cascade in a tropical rainforest understory ecosystem (see Figure 20.11) was shown to have important effects on (A) predation, (B) herbivory, and (C) production. (After Dyer and Letourneau 1999a.)

periment using more controlled manipulation of light levels and soil nutrients, rather than relying on variation in natural background levels, found significant effects of these resources on *Piper* production, but also found a continued strong effect of herbivory (Dyer and Letourneau 1999b). Thus, it is clear that top-down trophic cascades do occur in diverse terrestrial ecosystems, although they may require strong interactions between specialist predators and their prey.

What determines the number of trophic levels?

Ecologists have wondered what determines the variation among ecosystems in their numbers of trophic levels and why so few ecosystems have five or more trophic levels. This question is not simply an academic one. Through the effect of trophic cascades, the number of trophic levels in an ecosystem can influence movements of energy and nutrients, as well as the potential for toxins in the environment to become concentrated at higher trophic levels.

There are three basic, interacting controls on the number of trophic levels in an ecosystem (Post 2002). First, dispersal ability may constrain the ability of higher-level predators to enter an ecosystem. Such constraints would be most important in geographically isolated areas such as islands or mountains. Second, the amount of energy entering an ecosystem through primary production may determine how many trophic levels can be sustained. Because a relatively large amount of energy is lost in the transfer from one trophic level to the next, the more energy entering a system, the more is potentially available to support viable populations of higher-level predators. Finally, the frequency of disturbances or other agents of change, such as disease outbreaks, can determine whether populations of higher-level predators can be sustained. Following a disturbance, there is a time lag before the community returns to its original state. Because lower trophic levels are required to sustain higher trophic levels, there is a longer time lag for the reestablishment of the highest trophic levels. If disturbance is frequent, then higher trophic levels may never become established, no matter how much energy is entering the system (**Figure 20.13**). For example, despite having as much as an order of magnitude of variation in NPP (from 100 to 1,900 mg $C/m^2/day$), grassland ecosystems from the Arctic to the tropics have very similar food chain lengths, with three trophic levels and occasionally a fourth (Pimm 2002). This invariance in the number of trophic levels appears to be related primarily to disturbance frequencies, which vary little among grasslands. The constraints imposed on energy transfer to higher trophic levels by trophic efficiency and disturbance dynamics are manifested in a rarity of "big, fierce animals" (Colinvaux 1978). These constraints also explain why carnivores are the most common threatened and endangered mammals.

We turn our attention next to a more detailed investigation of trophic relationships in ecosystems, bridging the dis-

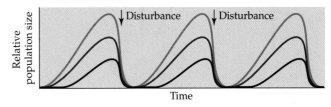

FIGURE 20.13 Disturbance Influences the Number of Trophic Levels in an Ecosystem (A) The recovery of higher trophic levels following a disturbance is delayed until lower trophic levels, containing the prey species needed to sustain the predators at those higher trophic levels, can become established. (B) If disturbances occur frequently, predators at higher trophic levels may never become established.

ciplinary boundaries of ecosystem and community ecology to examine how energy movement can influence the diversity and stability of communities and ecosystems.

Food Webs

CONCEPT 20.4
Food webs are conceptual models of the trophic interactions of organisms in an ecosystem.

Ever since Charles Darwin (1859) described "a tangled bank, clothed with many plants of many kinds, with birds singing on the bushes, with various insects flitting about, ... dependent upon each other in so complex a manner" in his *Origin of Species*, the interdependence of species has been a central concept in ecology. When these links among species focus on feeding relations, they can be described by a **food web**, a diagram showing the connections between organisms and the food they consume. Food webs offer an important tool for modeling the ecological interactions that define energy flows and predator–prey relations, and as a result, the dynamics of associated populations and communities. For the desert ecosystem we considered at the start of this chapter, we can

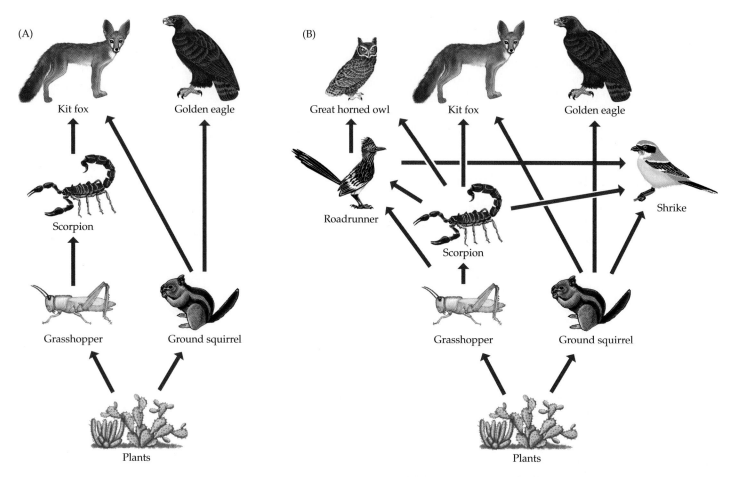

(A)

Kit fox   Golden eagle

Scorpion

Grasshopper   Ground squirrel

Plants

(B)

Great horned owl   Kit fox   Golden eagle

Roadrunner

Shrike

Scorpion

Grasshopper   Ground squirrel

Plants

FIGURE 20.14 Desert Food Webs Food webs may be simple or complex depending on their purpose. (A) A simple six-member food web for a representative desert grassland. (B) Addition of more participants to the food web adds realism, but the inclusion of additional species adds complexity.

construct a simplified food web showing that plants are consumed by insects and ground squirrels, and that these herbivores are food for scorpions, eagles, and foxes (**Figure 20.14A**). In this way, we can begin to understand qualitatively how energy flows from one component of this ecosystem to another, and how that energy flow may determine changes in population sizes and in the composition of communities.

Food webs are complex

The desert food web in Figure 20.14A is far from complete. We can add many other organisms and links to the food web, providing additional complexity. For example, the scorpion consumes insects such as the grasshopper, but like the grasshopper, it may be food for birds such as shrikes and owls (**Figure 20.14B**). As we continue to add more and more organisms to the food web, we add significant complexity, such that the food web may take on the appearance of a "spaghetti diagram" (**Figure 20.15**). In order to add greater realism, it is important to recognize that the feeding relationships of animals can span multiple trophic levels and may even include cannibalism (circular arrows in Figure 20.15) (Polis 1991).

Although food webs are useful conceptual tools, it is important to remember that even a simplified food web is only one static description of energy flow and trophic interactions

in an ecosystem. Actual trophic interactions can be dynamic over time (Wilbur 1997). Some organisms may change their feeding patterns as they age; maturing frogs, for example, make the transition from omnivorous aquatic tadpoles to carnivorous adults. Some animals, such as migratory birds, are relatively mobile and are thus components of multiple food webs. Most food webs fail to account for additional biological interactions among organisms that influence population and community dynamics, such as pollination mutualisms. (In community studies, this problem may be addressed by the use of interaction webs, as described in Chapter 15.) The critically important roles of microorganisms are often ignored, despite their processing of a substantial amount of the energy moving through an ecosystem. What are food webs good for, then? Despite these apparent shortcomings, food webs are important conceptual tools for understanding the dynamics of energy flow in ecosystems, and hence the community and population dynamics of their component organisms.

FIGURE 20.15 Complexity of Desert Food Webs In this desert food web, complexity overwhelms any interpretation of interactions among the members. Even this food web, however, lacks the majority of the trophic interactions in the ecosystem. (From Polis 1991.)

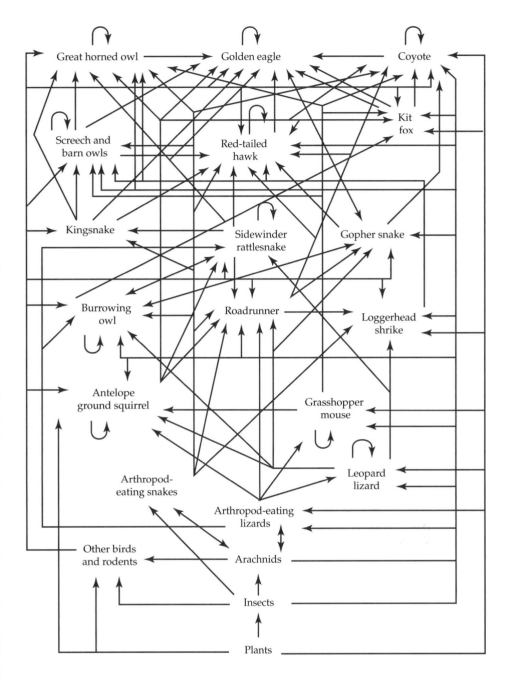

The strengths of trophic interactions are variable

As indicated in the quote from Darwin above and in earlier chapters, a core concept of ecological thought is that "everything is connected to everything else." However, the links among the species in an ecosystem vary in their importance to energy flow and species population dynamics; in other words, not all connections are equally important. Some trophic relationships may play a larger role than others in dictating how energy flows through the ecosystem. **Interaction strength** is a measure of the effect of one species' population on the size of another species' population. Determining interaction strengths is an important goal of ecologists because it helps us simplify the "spaghetti" in a complex food web by focusing attention those links that are most important for basic research and conservation.

How are interaction strengths determined? Several approaches have been used. Removal experiments, like those described in Chapter 15 to determine competition or facilitation, can be employed, but it is logistically overwhelming to consider performing such experimental removals to quantify every link in a web. Therefore, much current ecological research is devoted to discovering simpler, less direct measures that still give us a reliable estimate of the relative importance of different links. For example, simple food webs can be coupled with observations of the feeding preferences of predators and of changes in the population sizes of predators and prey over time to provide an estimate of which interactions are the strongest. Similarly, comparisons of two or more food webs when a predator or prey is present in one but absent in the other may provide evidence for the importance of links. Predator and prey body sizes have been used to predict the strengths of predator–prey interactions because feeding rate is known to be related to metabolic rate, which in turn is governed by body size. The best estimates of interaction strengths in food webs often come from a combination of these approaches.

A series of classic studies examining interaction strengths in food webs was performed in rocky intertidal zones of the Pacific Northwest by Robert Paine. Paine (1966) had observed that the diversity of organisms in rocky intertidal zones declined as the number of predators in those ecosystems decreased. He reasoned that some of these consumers might be playing a greater role than others in controlling the numbers of species coexisting in these communities. One of Paine's critical observations was that one mussel species (*Mytilus californianus*) had the ability to overgrow and smother many of the other sessile invertebrate species that compete with it for space. Paine hypothesized that sea stars might

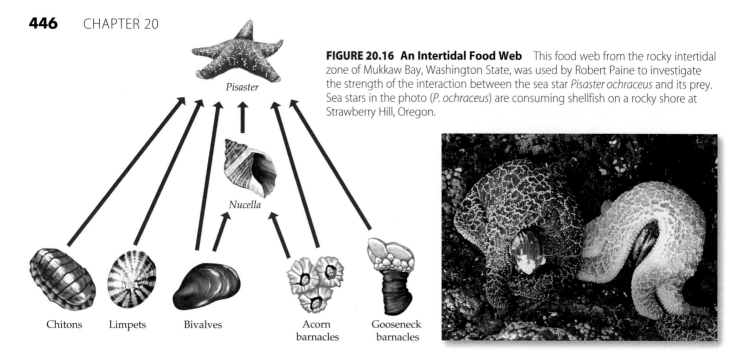

FIGURE 20.16 An Intertidal Food Web This food web from the rocky intertidal zone of Mukkaw Bay, Washington State, was used by Robert Paine to investigate the strength of the interaction between the sea star *Pisaster ochraceus* and its prey. Sea stars in the photo (*P. ochraceus*) are consuming shellfish on a rocky shore at Strawberry Hill, Oregon.

Pisaster

Nucella

Chitons   Limpets   Bivalves   Acorn barnacles   Gooseneck barnacles

play a key role in maintaining diversity in this community by acting as a strong consumer of these mussels and preventing them from competitively excluding other species.

To test these hypotheses, Paine conducted an experiment in Washington State, in which he removed the top predator in the system, the sea star *Pisaster ochraceus*. *Pisaster* feeds primarily on bivalves and barnacles, and to a lesser extent on other mollusks, including chitons, limpets, and a predatory gastropod (genus *Nucella*) (**Figure 20.16**) Following the continuous manual removal of *Pisaster* from 16 m² plots, acorn barnacles (*Balanus glandula*) became more abundant, but with time, they were crowded out by mussels (*Mytilus*) and gooseneck barnacles (genus *Pollicipes*). After 2½ years, the number of species in the community had decreased from 15 to 8. Even 5 years after the initiation of the experiment, when sea stars were no longer removed, dominance by the mussels continued, as individual mussels had grown to sizes that prevented predation by sea stars, and diversity remained lower in the experimental plots than in adjacent plots (Paine et al. 1985). Experimental removals of higher-level predators in other intertidal zones, including one in New Zealand, which shares no species with the intertidal zone of the Pacific Northwest, resulted in similar reductions in diversity. Predators in these intertidal ecosystems are thus key to maintaining species diversity by preventing one competitively dominant species from excluding the rest.

Some species are more important in food webs than their numbers would indicate

The experimental research of Paine and others was an encouraging advance in ecology because it demonstrated that, despite the potential complexity of trophic interactions among species, patterns of energy flow and community structure might be governed by a small subset of those species. Paine called animals such as *Pisaster* **keystone species**, defining them as species that have a greater influence on energy flow and community composition than their abundance or biomass would predict (see Figure 15.17). The keystone species concept has become an important focus in ecology and conservation because it implies that protecting a keystone species may be critical for protection the many other species that depend on it. Examples of keystone species include organisms from almost all ecosystem types. Keystone species tend to be predators at higher trophic levels, but they are not limited to this group. Top predators tend to have a large effect on prey species populations relative to their own abundance.

Some species behave as keystone species in only part of their geographic range, suggesting that interaction strengths are dependent on the environmental context of the habitat (Chapter 15). For example, Bruce Menge and his colleagues found that while *Pisaster ochraceus*, the sea star made famous by Paine's research, behaved as a keystone species in wave-exposed rocky intertidal zones, its effect on other species, particularly mussels, was much weaker in wave-sheltered sites (Menge et al. 1994). The population dynamics of mussels in the wave-sheltered sites were determined more by the sparse recruitment of young individuals than by sea star predation. Other studies have found similar context-dependent variation in the degree to which species behave as keystone species. Thus, while the keystone species concept is intuitively simple, predicting when and where a particular species will behave as a keystone species remains a challenge.

Direct and indirect effects determine net interaction strength

One reason it remains difficult to predict a strong keystone effect is that the ecological importance of a keystone species, such as *Pisaster*, manifests itself not only through one strong link, such as that between *Pisaster* and mussels, but also through strong indirect effects (see Figure 15.12), such as the effects *Pisaster* has on other species by reducing the abundance of mussels. If *Pisaster* instead consumed only the species that are inferior competitors for space (such as barnacles), it would not play a keystone role in the rocky intertidal community. Thus, predicting the effects of species loss on the remaining community requires an understanding not only of the strengths of individual links, but also the strengths of chains of indirect effects.

Removal experiments can provide estimates of the *net effect* of an interactor on a target species. This net effect includes the sum of the direct effect and all possible indirect effects mediated through the other species present in the community (**Figure 20.17**). The net effect of a predator on its prey, for example, includes not only the direct effect of prey consumption, but also the effects the predator has on other species that compete with, facilitate, or modify the environment of the target prey species. For example, *Pisaster* has a negative direct effect on barnacles by consuming them, but the positive indirect effect by consuming mussels (and thus freeing barnacles from competition for space) is much stronger, leading to a net positive effect of *Pisaster* on this inferior competitor. If the negative direct effect of a predator on the abundance of a target prey species is offset by its positive indirect effects on the target prey species, the net effect may appear weak. Alternatively, the net effect of a pred-

ator on a target prey species may appear very strong if both the direct and indirect effects are in the same direction.

The potential for indirect effects to offset or reinforce the direct effects of predators on prey was demonstrated experimentally by Eric Berlow. Berlow (1999) hypothesized that the potential for indirect effects to offset or reinforce the direct effect of a predator is greatest when the direct effect is weak. Berlow tested his hypothesis by manipulating the direct and indirect effects of predatory whelks (snails of the genus *Nucella*) on a target prey species, small mussels (*Mytilus trossulus*). The whelks also prey on acorn barnacles (*Balanus glandula*), which compete with the mussels for space. While whelks have a negative direct effect on mussels (by eating them), they can have either a positive or a negative indirect effect on mussels by eating barnacles. Barnacles generally facilitate mussels by providing safe "nooks and crannies" for mussel larvae to settle in. However, if barnacles settle on a rock in very dense clusters, they grow thin and are not well attached to the rock. When this happens, the barnacles, and any mussels that have settled on them, are more easily knocked off the rocks by waves. Therefore, when barnacles are at low densities, whelk predation on barnacles has a negative indirect effect on mussels by removing the mussels' preferred settlement substratum. However, when barnacles are at high densities, some thinning of the barnacles by whelks has a positive indirect effect on mussels by providing them with a more stable settlement substratum (**Figure 20.18A**).

Berlow controlled the strength of the direct effect of whelk predation by using high and low densities of whelks. He then measured the effect of whelks on mussel settlement rates in the presence (with indirect effects) and absence (no indirect effects) of barnacles. Whelk predation, without the indirect effects mediated by barnacles, had a consistent negative direct effect on the settlement rate of mussels, regardless of whelk density (**Figure 20.18B**). In the presence of barnacles, however, the effect of whelk density on mussel settlement rates changed. When whelks were at low densities (i.e., when their direct effect was "weak"), whether their net effect on mussel settlement was positive or negative depended on the density of the barnacles, as described above. At high whelk densities (i.e., when the direct effect of whelks was "strong"), the whelks had a consistently negative net effect on mussel settlement, regardless of the densities of barnacles.

The results of Berlow's experiment supported his hypothesis, showing that weak direct effects could show up as variable net effects under the influence of indirect effects. This variation reflects a combination of a significant direct effect of predation with indirect effects of predation, mediated by other prey species, that may offset, or in some cases accentuate, the direct effect. A mix of positive and negative indirect effects contributes to variation in the outcomes of species removal experiments when the direct effect is too weak to overcome the indirect effects. Why is this variation important? If a predator has varying (positive or negative) effects

FIGURE 20.17 Direct and Indirect Effects of Trophic Interactions The net effect of a consumer (C) on a target prey species (P₁) population includes direct consumption as well as all possible indirect effects of the consumer on other species (P₂) that interact with the prey species population in the community.

FIGURE 20.18 Strong and Weak Interactions Produce Variable Net Effects The indirect effects of a predator have the potential to offset or reinforce its direct effects. (A) Whether the influence of whelks (a predator) on the settlement rate of mussels (a prey species) is positive or negative depends on the density of barnacles (another prey species). (B) The indirect effect of whelks was apparent, however, only at low whelk densities (when the direct effect of whelks is "weak"). (After Berlow 1999.)

on a particular prey species depending on the presence or absence of another species, the potential for the predator to eliminate that prey species throughout its range is less. The variation associated with weak interactions may therefore promote coexistence of multiple prey species because different prey are facilitated in different places.

Does complexity enhance stability in food webs?

Ecologists have pondered whether more complex food webs—those with more species and more links among them—are more stable than simpler food webs with lower diversity and fewer links. Stability in this context is usually gauged by the magnitude of changes in the population sizes of the organisms in the food web over time. As we saw in Chapter 10, greater oscillations in population size increase the susceptibility of species to local extinction. Thus, a less stable food web means a greater potential for extinction for its component species. The question of stability has taken on ever greater importance with increasing rates of biodiversity loss and non-native species invasions worldwide. How an ecosystem responds to species loss or gain is strongly related to the stability of food webs.

Early proponents of the idea that food web complexity increased stability based their arguments on observations of real trophic interactions and on intuition. Ecologists such as

Charles Elton and Eugene Odum argued that simpler, less diverse food webs should be more easily perturbed and experience greater species losses as a result of larger changes in population densities. More rigorous mathematical analyses of food webs, however, provided a contrary view. Robert May (1973) used food webs made up of random assemblages of organisms to demonstrate that food webs with higher diversity are less stable than those with lower diversity. The instability in May's models resulted from accentuation of population fluctuations by strong trophic interactions: the more interacting species there were, the more likely that their population fluctuations would reinforce one another, leading to extinction of one or more of the species.

May's work overturned the notion that more complex systems are inherently more stable than simpler ones. Yet anyone visiting a tropical rainforest or a coral reef can attest to the fact that highly diverse and complex communities do persist in nature. Therefore, much ecological research has been devoted to discovering the factors that allow naturally complex food webs to be stable. More recent models, for example, have incorporated distributions of interaction strengths among organisms more closely resembling those observed in nature. In addition, there is a greater realization that weak interactions can stabilize trophic interactions, as demonstrated by the results of Eric Berlow's work described above. These

FIGURE 20.19 Diversity and Stability in a Food Web
Increasing the number of protozoan species in laboratory micro-cosms decreased the stability of food webs, as indicated by increases in the percentage of species going extinct. (After Lawler 1993.)

models and experiments suggest that, while more complex systems are not necessarily more stable, natural food webs may have a particular structure or organization that allows increased species diversity to have a stabilizing effect.

Sharon Lawler tested the relationship between species diversity and food web stability in an experiment using laboratory microcosms (small closed-system experimental containers) containing protozoan food webs of different diversities (Lawler 1993). She established microcosms in glass jars containing two, four, or eight protozoan species, all with equal numbers of prey and predator species. The prey species consumed bacteria, which in turn fed on a nutritive medium that was added to the microcosms. Lawler evaluated the stability of each of the component protozoan species by their persistence (whether their populations went extinct), the variation in their population sizes over time, and their mean population sizes.

Increasing the number of species in Lawler's microcosms resulted in more population extinctions (**Figure 20.19**), but no changes in variation in population sizes over time. The increase in extinctions corresponded with decreased popula-

tion sizes, although the total abundance of all individuals of all species remained relatively constant among the microcosms. Lawler suggested that the decrease in population sizes with increasing diversity may have increased the susceptibility of the populations to extinction.

Does Lawler's experiment indicate that diverse food webs are inherently more unstable? No, such a conclusion would be too simplistic. Lawler cautioned that the species composition of the food web was also an important influence in her experiment. In other words, the *identity* of the species that made up the food web was important in determining its stability. In Lawler's experiment, some species were more likely than others to become extinct, and some showed greater variation in their population sizes in the presence of other species than others; furthermore, some predators were more likely to cause the extinction of prey species than others. Thus, both species diversity and species composition appeared to be important in determining the stability of the protozoan food webs. Additional studies suggest that the buffering influence of weak interactions (Neutel et al. 2002) and adaptive changes in prey choice (Kondoh 2003) help to reduce the population fluctuations associated with complex food webs.

CASE STUDY REVISITED

Toxins in Remote Places

Knowledge of energy flow through the trophic levels of ecosystems is key to understanding the environmental effects of persistent organic pollutants. Some chemical compounds taken up by organisms, either directly from the environment or by consumption with their food, can become concentrated in their tissues. These compounds may not be metabolized or excreted for a variety of reasons, so they become progressively more concentrated in the body over the organism's lifetime, a process known as **bioaccumulation**. Bioaccumulation can lead to increasing tissue concentrations of these compounds in animals at higher trophic levels as animals at each trophic level consume prey with higher concentrations of the compounds. This process is known as **biomagnification (Figure 20.20)**. The POPs we

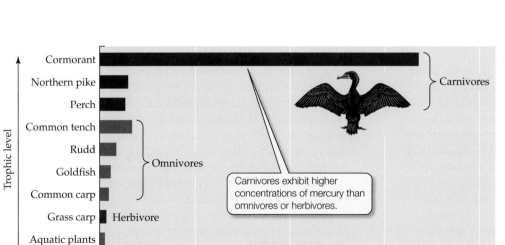

FIGURE 20.20 Bioaccumulation and Biomagnification Levels of mercury (a toxic heavy metal) show bioaccumulation and biomagnification in a Czech pond ecosystem. (From Houserová et al. 2007.)

discussed at the beginning of this chapter are particularly subject to these processes.

The potential dangers associated with bioaccumulation and biomagnification of POPs were well publicized by Rachel Carson's book *Silent Spring*, published in 1962, in which she described the devastating effects that pesticides, particularly DDT, were having on nontarget bird and mammal populations. DDT was thought of as a "miracle" insecticide during the 1940s and 1950s, when it was widely used to control a variety of crop and garden pests and disease vectors. However, DDT was also accumulating in higher-level predators as a result of biomagnification, and it contributed to the near-extinction of some birds of prey, including the peregrine falcon and the bald eagle. In *Silent Spring*, Carson described the persistence of DDT in the environment, its accumulation in the tissues of consumers, including humans, and its health hazards. Because of Carson's careful documentation and her ability to convey her message in a well-crafted manner that could be appreciated by the general public, *Silent Spring* led to increased scrutiny of the use of chemical pesticides, eventually resulting in a ban on the manufacture and use of DDT in the United States.

The concept of biomagnification led researchers to suspect that the high concentrations of POPs found in the Inuit resulted from their position at the highest trophic levels of the Arctic ecosystem. This suspicion was reinforced by comparisons of the concentrations of toxins among different Inuit communities. The highest concentrations of toxins were found in communities that consumed marine mammals such as whales, seals, and walruses—animals that occupy the third, fourth, or fifth trophic levels. On the other hand, inhabitants of communities where herbivorous caribou (at the second trophic level) were a more important part of the diet had lower concentrations of toxins. The Inuit preference for foods rich in fatty tissues, such as whale blubber (muktuk), poses a significant problem as well, because many POPs are preferentially stored in the fatty tissues of prey animals.

Although emissions of some POPs and other pollutants are declining globally as awareness of their effects increases and regulations are put in place, the potential for long-term storage of these compounds in the Arctic environment means that their effects are unlikely to disappear any time soon (Pearce 1997). Concentrations of PCBs and DDT in Arctic lake sediments have continued to increase over time, while concentrations of these POPs in lake sediments to the south have tended to decrease. The cold temperatures and relatively low light levels in the Arctic limit the chemical breakdown of POPs. Unfortunately, the long-term effects of these POPs on Inuit populations is uncertain. While switching to alternative food sources might seem to be a potential solution to the problem, the cultural identity of the Inuit is strongly associated with their hunt-ing traditions and their diet, and they would be unlikely to make such a switch easily.

Biological Transport of Pollutants

Anthropogenic pollutants have been reported in almost all environments on Earth—even Antarctic ice holds trace amounts of DDT and lead emitted from the burning of leaded gasoline. Animals in many remote areas have been reported to have elevated levels of industrial and agricultural toxins in their tissues. Fish in isolated alpine lakes of the Canadian Rockies, for example, show elevated concentrations of POPs, which have been associated with condensation of these compounds in snowfields and glaciers above the lakes (Blais et al. 1998). As suggested above, the concentrations of these pollutants are related to the trophic positions of the animals, with consumers at the highest trophic levels, such as polar bears, seals, and birds of prey, containing the highest amounts of pollutants.

The movement of POPs and other anthropogenic toxins is usually associated with atmospheric transport from low to high latitudes. However, the behaviors of animals can also influence the movement of POPs to remote places. Salmon, for example, have been shown to be important in the movement of nutrients from marine to freshwater and terrestrial ecosystems during their spawning runs. At reproductive maturity, salmon leave the ocean and move up rivers in large numbers. From the rivers, they move into freshwater lakes and streams, where they spawn and then die. The potential exists for salmon to move toxins, as well as nutrients, from the oceans to freshwater ecosystems via this spawning behavior.

E. M. Krümmel and colleagues studied the potential for spawning sockeye salmon (*Oncorhynchus nerka*) to act as a "fish pump" for pollutants, moving PCBs from the ocean to remote lakes in Alaska (Krümmel et al. 2003). Salmon occupy the fourth trophic level and thus, through bioaccumulation and biomagnification, accumulate PCBs in their body fat at concentrations more than 2,500 times higher than those found in seawater. Krümmel et al. sampled eight lakes in southwestern Alaska that had different densities of spawning salmon (ranging from 0 to 40,000 spawners/km²). The researchers also collected sediment cores from the lakes and measured PCBs in the sediments. They found that the concentrations of PCBs were strongly correlated with the density of spawners (**Figure 20.21**). Lakes that did not have visits from spawning fish had concentrations of PCBs similar to expectations based on atmospheric transport alone. The lake with the highest density of spawning fish (40,000 per km²) had PCB concentrations that were six times higher than the "back-

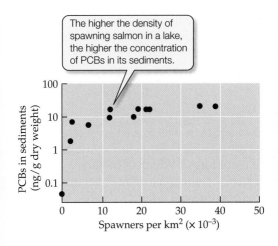

The higher the density of spawning salmon in a lake, the higher the concentration of PCBs in its sediments.

FIGURE 20.21 Biological Pumping of Pollutants
Spawning salmon act as biological pumps, concentrating toxins from the oceans in their bodies and transporting them en masse to freshwater ecosystems. (After Krümmel et al. 2003.)

ground" levels associated with atmospheric transport. A similar study found that DDT, other POPs, and mercury had been transported by northern fulmars (*Fulmarus glacialis*, pelagic fish-eating seabirds) from the ocean to small ponds near their nesting colonies (Blais et al. 2005).

These examples demonstrate how the behaviors of some species (spawning in fish, colonial nesting in birds) can exacerbate problems of pollution associated with biomagnification in ecosystems.

SUMMARY

CONCEPT 20.1 Trophic levels describe the feeding positions of groups of organisms in ecosystems.

- An organism's trophic level is determined by the number of feeding steps by which it is removed from the first trophic level, which contains autotrophs and detritus.

- Omnivores feed at multiple trophic levels, although their diets can be partitioned to reflect their consumption at each level.

- All organisms eventually end up as food for other organisms or as detritus.

CONCEPT 20.2 The amount of energy transferred from one trophic level to the next depends on food quality and consumer abundance and physiology.

- Energy and biomass pyramids portray the relative amounts of energy or biomass at different trophic levels.

- The high turnover of autotroph biomass in aquatic ecosystems can result in biomass pyramids that are inverted relative to energy pyramids.

- The proportion of autotroph biomass consumed in terrestrial ecosystems tends to be lower than that in aquatic ecosystems.

- The efficiency of energy transfer from one trophic level to the next is determined by food quality and the physiology of consumers.

CONCEPT 20.3 Changes in the abundances of organisms at one trophic level can influence energy flow at multiple trophic levels.

- Changes in the numbers and types of consumers at higher trophic levels can influence primary production through influences on the consumption of herbivores.

- Trophic cascades tend to be more apparent in aquatic ecosystems than in terrestrial ecosystems, but they have been demonstrated in complex terrestrial ecosystems as well.

- The number of trophic levels that can be sustained in an ecosystem is determined by the dispersal ability of organisms, the amount of energy entering the ecosystem through primary production, and the frequency of disturbance.

CONCEPT 20.4 Food webs are conceptual models of the trophic interactions of organisms in an ecosystem.

- Food webs are diagrams that portray the diverse trophic interactions among species in an ecosystem.

- Although trophic interactions are extremely complex, food webs can be simplified by focusing on the strongest interactions among the component organisms.

- Keystone species have greater effects on energy flow and species composition in their communities than their abundance or biomass would indicate.

■ Indirect effects of a consumer on a target prey species, including effects on other species that compete with, facilitate, or modify the environment of the target species, can negate or reinforce the direct effects of consumption. These indirect effects may have important stabilizing effects on inherently unstable food webs.

REVIEW QUESTIONS

1. Suppose one population of coyotes (population A) demonstrates a greater degree of omnivory than another population (population B). Population A relies on a diet that includes road-killed animal carcasses, plants, and rotten food from dumpsters, while population B has a steady diet of small rodents. Which population should have a higher assimilation efficiency, and why?

2. Mammals in temperate terrestrial and temperate marine ecosystems occupying similar trophic levels may have different production efficiencies. Assuming similar food quality, abundance, and capture rates, explain why the production efficiencies of these animals would differ between a marine ecosystem and a terrestrial ecosystem. (Hint: Consider how the animals maintain their body heat, and the temperature variation of their environments covered in Chapter 2.)

3. Which ecosystem would you expect to have a greater total amount of energy passing through its trophic levels: a lake or an adjacent forest? Which of these ecosystems would have a higher *proportion* of net primary production moving through different trophic levels, the forest or the lake?

4. Generalist herbivores consume a greater number of plant species than specialist herbivores. If a trophic cascade results in a reduction in the consumption of the predators of an herbivore, would there be a greater or a lesser effect on primary production if the herbivore were a specialist?

HANDS-ON PROBLEM SOLVING

Are Complex Food Webs More Stable? This Web exercise (www.sinauer.com/ecology) explores whether or not food webs that are more complex are also more stable. You will interpret data from a recent paper that experimentally addressed the relationship between species diversity in grassland food webs and the stability of the food web during the decade-long study.

SUGGESTED READINGS

Hairston, N. G. Sr., F. E. Smith and L. B. Slobodkin. 1960. Community structure, population control, and competition. *American Naturalist* 94: 421–425. *A classic article that is credited with first suggesting that predation kept the world green by limiting the abundance of herbivores.*

Pace, M. L., J. J. Cole, S. R. Carpenter and J. F. Kitchell. 1999. Trophic cascades revealed in diverse ecosystems. *Trends in Ecology and Evolution* 14: 483–488. *This paper documents the widespread nature of trophic cascades in marine, freshwater, and terrestrial ecosystems, touching on the effects of changes in trophic cascades due to introductions of organisms and the implications for conservation.*

Pimm, S. L. 2002. *Food Webs.* University of Chicago Press, Chicago. *This book provides a readable introduction to the concepts of food webs in ecosystems, including models, complexity, and stability—topics that continue to be controversial in the ecological literature.*

Polis, G. A. 1991. Complex trophic interactions in deserts: An empirical critique of food web theory. *American Naturalist* 138: 123–155. *A critique of the development of food web theory, employing consideration of real desert food webs to evaluate the predictions and assumptions of prior models using simplified food webs.*

Power, M. E. and 9 others. 1996b. Challenges in the quest for keystones. *BioScience* 46: 609–620. *This article reviews the keystone concept, clarifies some of the confusing issues and sloppy usage of the term, and provides examples of approaches and case studies identifying keystones in a variety of ecosystems.*

CHAPTER **21** Nutrient Supply and Cycling

KEY CONCEPTS

CONCEPT 21.1 Nutrients enter ecosystems through the chemical breakdown of minerals in rocks or through fixation of gases in the atmosphere.

CONCEPT 21.2 Chemical and biological transformations in ecosystems alter the chemical form and supply of nutrients.

CONCEPT 21.3 Nutrients cycle repeatedly through the components of ecosystems.

CONCEPT 21.4 Freshwater and marine ecosystems receive nutrient inputs from terrestrial ecosystems.

CASE STUDY

A Fragile Crust

The Colorado Plateau in western North America includes vast expanses of isolated mountains, intricately folded sandstone formations, and deeply cut, multicolored canyons. One of the most unusual features found in this rugged and beautiful region, however, occurs at a very small scale: its patchy cover of dark, bumpy soil (**Figure 21.1**). On closer examination, the soil looks like a miniature landscape of hills and valleys, covered with black, dark green, and white splotches resembling lichens. The comparison is apt, because this crust on the soil surface, known simply as a **biological crust** (or a *cryptobiotic* crust), is composed of a mix of hundreds of species of cyanobacteria, lichens, and mosses (Belnap 2003). Approximately 70% of the soils on the Colorado Plateau, covering parts of Utah, Arizona, Colorado, and New Mexico, have some biological crust development. Similar crusts, containing a surprisingly similar suite of species, are found in many other arid and semiarid regions throughout the world, . The crusty nature of the soil is largely the work of filamentous cyanobacteria, which create a sheath of mucilaginous material as they move through the soil after a rain. When the soil dries out, the cyanobacteria withdraw to deeper soil layers, leaving behind the sheathing material, which helps bind the coarse soil particles together (**Figure 21.2**).

The soils of the Colorado Plateau are exposed to tremendous climatic variation and strong erosive forces (Belnap 2003). Surface temperatures can range from –20°C (–4°F) in winter to 70°C (158°F) in summer. High evapotranspiration rates often dry out the soils, and the sparseness of the vegetation allows the strong surface winds to carry away fine soil particles. Precipitation in spring and summer often comes in the form of brief, intense thunderstorms. Biological crusts are critical for anchoring the soil in place in the face of high winds and torrential rains.

FIGURE 21.1 Biological Crust on the Colorado Plateau Biological crusts are a common feature in the deserts of the Colorado Plateau. The surface topography and coloration of the crust are clearly visible in this photo.

FIGURE 21.2 Cyanobacterial Sheaths Bind Soil into Crusts
(A) Cyanobacterial strands surround themselves with a sheath of mucilaginous material as they move through the soil. (B) The sheaths left behind by the cyanobacteria help to bind soil particles together and protect soils from erosional loss.

Although the Colorado Plateau is sparsely populated, humans have had a large and lasting effect on its landscape. Livestock grazing has been an important use of public lands in the region since cattle were introduced there in the 1880s. Most of the land has been affected to some degree by grazing, which has resulted in the trampling of biological crusts and overgrazing of vegetation. Until recently, grazing was the most important human-associated disturbance in the region. Recently, however, a proliferation of off-road vehicles has invaded the region. During the 2005 Moab Jeep Safari, for example, an estimated 30,000–40,000 participants descended on a town with a year-round population of 5,000. All-terrain vehicle use is also increasing dramatically, joining the motorcycle, mountain bike, and hiking traffic in the

wilds. The majority of these users of the desert backcountry obey federal and local laws, staying on designated trails and roads. However, a minority of users drive their vehicles off designated roads and across soils covered with biological crusts.

While the spatial extent of soil surface disturbance associated with off-road vehicle use and livestock grazing has not been well quantified, it is clear that a large part of the landscape has been disturbed to some degree during the past 150 years, and that the rate of disturbance is increasing. The recovery of biological crusts following disturbance is extremely slow in arid environments: decades are required for the reestablishment of the cyanobacteria and up to centuries for recolonization by lichens and mosses (Belnap and Eldridge 2001).

What are the implications of the loss of biological crusts for the functioning of desert ecosystems? How important are they to nutrient inputs and retention? Given the long-term nature of disturbances associated with livestock across the Colorado Plateau, can we still find areas that can serve as controls for studies of the disturbance that has already occurred?

In addition to energy, all organisms require specific chemical elements to meet their fundamental biochemical requirements for metabolism and growth. Organisms get these elements by absorbing them from the environment or by consuming other organisms, living or dead. Iron, for example, is needed for several important metabolic functions by all organisms, but how those organisms get their iron, and where it comes from, varies substantially. Phytoplankton in the Atlantic Ocean may take up iron that came from dust that blew in from the Sahara. Lions on an African savanna get their iron from the prey they kill and consume. Aphids get their iron in the sap they suck from a plant, while the plant obtains its iron when it takes up water containing dissolved iron from the soil. The ultimate source of all of this iron is

solid minerals in Earth's crust, which are subjected to chemical alteration as they move through the different physical and biological components of ecosystems.

The study of the physical, chemical, and biological factors that influence the movements and transformations of elements is known as **biogeochemistry**. Understanding the biogeochemistry of elements is important for determining their availability as **nutrients**, defined as the chemical elements an organism requires for its metabolism and growth. Nutrients must be present in certain chemical forms to be available for uptake by organisms. The rate at which physical and chemical transformations occur determines the supply of nutrients. Biogeochemistry also involves the study of non-nutrient elements that serve as tracers in ecosystems or pollutant compounds that cause environmental damage. Biogeochemistry is an integrative discipline involving contributions from soil science, hydrology, and atmospheric science as well as ecology.

In this chapter, we will consider the biological, chemical, and physical factors that control the supply and availability of nutrients in ecosystems. We will emphasize the roles of autotrophs (photosynthetic and chemosynthetic plants and bacteria) because they are the principal source of nutrients for heterotrophs. We will examine which nutrients are important for organismal functioning, describe the sources of those nutrients and how they enter ecosystems, and review some of the important chemical and biological transformations that constitute the cycling of nutrients in ecosystems. In Chapter 24, we will consider the global-scale cycling of some of these elements.

Nutrient Requirements and Sources

CONCEPT 21.1
Nutrients enter ecosystems through the chemical breakdown of minerals in rocks or through fixation of gases in the atmosphere.

All organisms, from bacteria to blue whales, share similarities in their nutrient requirements. How those nutrients are obtained, the chemical forms of those nutrients that are taken up, and the relative amounts of those nutrients that are required vary significantly among organisms. All of these nutrients, however, come from a common source: inorganic mineral forms that are present in Earth's crust or as gases in the atmosphere.

Organisms have specific nutrient requirements

An organism's nutrient requirements are related to its physiology. The amounts and specific nutrients needed therefore vary according to the organism's mode of energy acquisition (autotrophs versus heterotrophs), mobility, and thermal physiology (ectotherms versus endotherms). Mobile animals, for example, generally have higher rates of metabolic activity than plants and therefore have higher requirements for nutrients such as nitrogen (N) and phosphorus (P) to support the biochemical reactions associated with movement. This difference is reflected in the composition of the dry mass of animals and plants (**Table 21.1**). Carbon is often associated with structural compounds in plant cells and tissues, while nitrogen is largely tied up in enzymes. Accordingly, the ratios of carbon to nitrogen (C:N) in organisms can indicate the relative concentrations of biochemical machinery in cells. Animals typically have a lower C:N ratio than plants: for example, humans have a C:N ratio of 6.0, whereas that of plants ranges from 10 to 40. This difference is one reason

TABLE 21.1

Elemental Composition (as Percentage of Dry Mass) of a Plant and an Animal

Element (symbol)	Plant (corn, *Zea mays*)	Animal (human, *Homo sapiens*)
Oxygen (O)	44.43	14.62
Carbon (C)	43.57	55.99
Hydrogen (H)	6.24	7.46
Nitrogen (N)	1.46	9.33
Silicon (Si)	1.17	0.005
Potassium (K)	0.92	1.09
Calcium (Ca)	0.23	4.67
Phosphorus (P)	0.20	3.11
Magnesium (Mg)	0.18	0.16
Sulfur (S)	0.17	0.78
Chlorine (Cl)	0.14	0.47
Iron (Fe)	0.08	0.012
Manganese (Mn)	0.04	—
Sodium (Na)	—	0.47
Zinc (Zn)	—	0.01
Rubidium (Rb)	—	0.005

Source: Epstein and Bloom 2005.
Note: Dashes indicate a neglible amount of an element.

why herbivores must consume more food than carnivores to acquire enough nutrients to meet their nutritional demands.

The nutrients essential for all plants, and the functions associated with them, are presented in **Table 21.2**. Some plant species have specific requirements for other nutrients not found in Table 21.2. For example, many, but not all, C_4 and CAM plants (see Chapter 5 for discussion of these photosynthetic pathways) require sodium. (In contrast, sodium is an essential nutrient for all animals, critical for maintaining pH and osmotic balances.) Cobalt is required by some plants that host nitrogen-fixing symbionts (discussed later in this chapter). Selenium is toxic to most plants, but a small number of plants growing on soils rich in selenium may require it (in contrast, selenium is an essential nutrient for animals and bacteria).

Plants and microorganisms usually take up nutrients from their environment in relatively simple, soluble chemical forms, from which they synthesize the larger molecules needed for their metabolism and growth. Animals, on the other hand, typically take up their nutrients through the consumption of living organisms or detritus, obtaining their nutrients in larger, more complex chemical forms. Animals break down some of these compounds and resynthesize new molecules; others are absorbed intact and used directly in biosynthesis. For example, 9 of the 20 amino acids that are essen-

TABLE 21.2

Plant Nutrients and Their Principal Functions

Nutrients	Principal functions
Carbon, hydrogen, oxygen	Components of organic molecules
Nitrogen	Component of amino acids, proteins, chlorophyll, nucleic acids
Phosphorus	Component of ATP, NADP, nucleic acids, phospholipids
Potassium	Ionic/osmotic balance, pH regulation, regulation of guard cell turgor
Calcium	Cell wall strengthening and functioning, ionic balance, membrane permeability
Magnesium	Component of chlorophyll, enzyme activation
Sulfur	Component of amino acids, proteins
Iron	Component of proteins (e.g., heme groups), oxidation–reduction reactions
Copper	Component of enzymes
Manganese	Component of enzymes, activation of enzymes
Zinc	Component of enzymes, activation of enzymes, component of ribosomes, maintenance of membrane integrity
Nickel	Component of enzymes
Molybdenum	Component of enzymes
Boron	Cell wall synthesis, membrane function
Chlorine	Photosynthesis (water splitting), ionic and electrochemical balance

Sources: Salisbury and Ross 1992; Marschner 1995.

tial for metabolism in humans and other mammals must be absorbed intact, since we cannot synthesize them ourselves.

Minerals and atmospheric gases are the ultimate sources of nutrients

All nutrients are ultimately derived from abiotic sources: minerals in rocks and gases in the atmosphere. Over time, as they are taken up and incorporated by organisms, they accumulate in ecosystems in organic forms (i.e., in association with carbon and hydrogen molecules). Nutrients may be cycled within an ecosystem, repeatedly passing through organisms and the soil or water in which they live. They may even be cycled internally within an organism, stored or mobilized for use as its needs for specific nutrients change. Here we describe the inputs of nutrients into ecosystems from minerals and the atmosphere. In the following sections, we will complete the steps that constitute nutrient cycling within an ecosystem.

Mineral sources of nutrients The breakdown of minerals in rock supplies ecosystems with nutrients such as potassium, calcium, magnesium, and phosphorus. *Minerals* are solid substances with characteristic chemical properties, derived from a multitude of geological processes. *Rocks* are collections of different minerals. Nutrients and other elements are released from rock minerals in a two-step process known as **weathering**. The first step, **mechanical weathering**, is the physical breakdown of rocks. Expansion and contraction processes, such as freezing–thawing and drying–rewetting cycles, act to break rock into progressively smaller particles. Gravitational mechanisms (such as landslides) and the growth of plant roots also contribute to mechanical weathering. Mechanical weathering exposes the surfaces of mineral fragments to the second step of the process, called **chemical weathering**, in which the minerals are subjected to chemical reactions that release soluble forms of nutrients.

Weathering is one of the processes involved in soil development. **Soil** is formally defined as a mix of mineral particles, solid organic matter (detritus, primarily decomposing plant matter), water containing dissolved organic matter, minerals, and gases (the *soil solution*), and organisms. Soils have several important properties that influence the delivery of nutrients to plants and microorganisms. One such property is their texture, which is defined by the sizes of the particles that make up the soil. The coarsest soil particles (0.05–2 mm) are referred to as **sand**. Fine soil particles (< 2 μm), known as **clays**, have a semicrystalline structure and weak negative charges on their surfaces that can hold onto cations and exchange them with the soil solution. As a result, clay particles serve as a reservoir of nutrient cations such as Ca^{2+}, K^+, and Mg^{2+}. A soil's ability to hold these cations and exchange them with the soil solution, referred to as its **cation exchange capacity**, is determined by the amounts and types of clay the soil contains. Soil texture also influences the soil's water-holding capacity and thus the movement of nutrients in the soil solution. Soils with a high proportion of sand have a large volume of spaces between particles. These spaces (called *macropores*) allow water to drain through the soil and limit the amount of water it can hold.

The **parent material** of a soil is the rock or mineral material that was broken down by weathering to form that soil. It is usually the underlying bedrock, but may also include thick layers of sediment deposited by glaciers (known as **till**), wind (**loess**), or water. The chemistry and structure of the parent material is important for determining the rate of weathering and the amount and type of nutrients released, and thus influences the fertility of the soil. Limestone, for example, is high in the nutrient cations Ca^{2+}, K^+, and Mg^{2+}. Soils derived from more acidic parent material, such as granites, have lower concentrations of these elements. In addition, the higher acidity (lower pH) of soils derived from granites lowers the availability of nitrogen and phosphorus to plants.

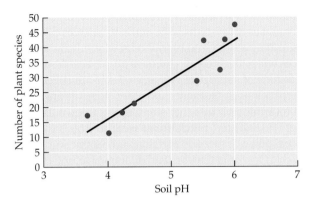

FIGURE 21.3 Plant Species Richness Decreases with Increasing Soil Acidity   Increases in soil acidity (decreases in pH) result in lower richness of vascular plant species in Alaskan Arctic tundra. The gradient in soil acidity is primarily due to differences in parent material: less acidic soils are associated with greater loess deposits. (After Gough et al. 2000.)

The chemistry and pH of the parent material exerts an important influence on the abundance, growth, and diversity of plants in ecosystems. For example, Laura Gough and colleagues (2000) demonstrated that variation in the acidity of the parent material is associated with differences in plant species richness among Arctic ecosystems in Alaska. They surveyed Arctic vegetation across natural gradients in soil acidity associated with the differential distribution of calcium-rich glacial loess (areas with more loess had lower acidity than areas with less loess.) They found that the diversity of plant species increased as acidity decreased (**Figure 21.3**). This variation in diversity was attributed to the negative ef-

fects of soil acidity on nutrient availability as well as the inhibitory effects of low soil pH on plant establishment.

Over time, soils undergo changes associated with weathering, accumulation of organic matter, and chemical alteration and movement (*leaching*) of dissolved organic matter and fine mineral particles from upper to lower layers. These processes form **horizons**, layers of soil distinguished by their color, texture, and permeability (**Figure 21.4**). Soil horizons are an important tool used by soil scientists to characterize different soil types.

Climate influences the rates of many of the processes associated with soil development, including weathering, biological activity (such as the input of organic matter and its processing in the soil), and leaching. In general, these processes occur most rapidly under warm, wet conditions. Thus, the soils of lowland tropical forest ecosystems, which have experienced high rates of weathering and leaching for a long time, are poor in mineral-derived nutrients. In contrast to terrestrial ecosystems at higher latitudes and elevations, a high proportion of the nutrients in lowland tropical forest ecosystems are found in the living biomass of trees. When the forests are cleared and burned to make way for pastures or cropland, most of the nutrients are lost in smoke and ash, as well as through soil erosion following the fires. As a result, these ecosystems may become severely nutrient impoverished, and it may take them centuries to return to their previous state. Soils in higher-latitude ecosystems have lower leaching rates and are richer in mineral-derived nutrients.

Organisms, primarily plants, bacteria, and fungi, influence soil development by contributing organic matter, which is an important reservoir of nutrients such as nitrogen and phosphorus. Organisms also increase rates of chemical weathering through the release of organic acids from plants and detritus and the release of CO_2 by plants and microorganisms. Thus, rates of biological activity have a strong influence on the development of soils.

Forest vegetation

Organic matter accumulates in the top soil horizon.

Mineral nutrients leached from the top soil horizon accumulate in the next soil horizon.

Clay accumulation.

Physical weathering breaks bedrock into ever smaller particles.

Relatively unweathered bedrock

Fresh unweathered bedrock

Well-developed soil

Time

FIGURE 21.4 Development of Soil Horizons   Soils develop over time as parent material is weathered and broken up into ever finer soil particles, increasing amounts of organic matter accumulate in the soil, and materials are leached and deposited in deeper soil layers. The rate of soil development is dependent on climate, the parent material, and the organisms associated with the soil.

Atmospheric sources of nutrients   The atmosphere is composed of 78% nitrogen (as dinitrogen gas, N_2), 21% oxygen, 0.9% argon, increasing amounts of carbon dioxide (0.038%, or 380 parts per million, in 2007), and other trace gases, some natural, others pollutants derived from human activities. The atmosphere is the ultimate source of carbon and nitrogen. These nutrients become biologically available when they are taken up from the atmosphere and transformed, or *fixed*, by organisms. They may then be transferred from organism to organism before returning to the atmosphere.

Carbon is taken up by autotrophs as CO_2 through photosynthesis. Photosynthesis was described in Chapter 5, and the global carbon cycle is discussed in Chapter 24. Carbon compounds store energy in chemical bonds, and they are important structural compounds (e.g. cellulose) as well.

Although the atmosphere is a huge reservoir of nitrogen, it is in a chemically inert form (N_2) that cannot be used by most organisms due to the high energy required to break the triple bond between the two molecules. The process of taking up N_2 and converting it into chemically available is known as **nitrogen fixation** (see Connections in Nature, Chapter 16). Biological nitrogen fixation is accomplished with the aid of the enzyme **nitrogenase**, which occurs only in certain bacteria. Some of these nitrogen-fixing bacteria are free-living; others are partners in symbiotic relationships. Nitrogen-fixing symbiotic relationships include associations between plant roots and soil bacteria, most notably between legumes and bacteria in the family Rhizobiaceae. Leguminous plants "host" nitrogen-fixing bacteria in special root structures called nodules and supply them with carbon compounds as an energy source (room and board!) to meet the high energy demands of nitrogen fixation (**Figure 21.5**). In return, the plant gets nitrogen fixed by the bacteria. Other examples of nitrogen-fixing symbioses include associations between woody plants such as alders and bacteria in the genus *Frankia* (called actinorhizal associations), associations

between the fern *Azolla* and cyanobacteria, lichens that include fungal and nitrogen-fixing cyanobacterial symbionts, and termites with nitrogen-fixing bacteria in their guts. Humans also fix atmospheric nitrogen when they manufacture synthetic fertilizers using the Haber–Bosch process, in which ammonia is produced from atmospheric nitrogen under high pressure using an iron catalyst.

Nitrogen fixation requires a large amount of energy, consuming as much as 25% of the photosynthetic energy fixed by plants with nitrogen-fixing symbiotic partners. Thus, nitrogen fixation provides symbiotic plants a source of nitrogen, but represents a trade-off with other energy demanding processes such as growth and reproduction. Allocation of energy to nitrogen fixation rather than to growth lowers the ability of nitrogen-fixing plants to compete for resources other than nitrogen. Nitrogen fixation is particularly important during the early stages of primary succession, as we saw in Chapter 16.

In addition to carbon and nitrogen, the atmosphere contains fine soil particles (dust) and a collection of suspended solid, liquid, and gaseous particles known as **aerosols**. Some of this particulate matter enters ecosystems when it falls from the atmosphere due to gravity or in precipitation, a process known as **atmospheric deposition**. Atmospheric deposition represents an important natural source of nutrients for some ecosystems. Aerosols containing cations derived from sea spray, for example, may be an important source of nutrients in coastal areas. Atmospheric deposition of dust originating in the Sahara Desert is an important input of iron into the Atlantic Ocean and of phosphorus into the Amazon Basin. On the other hand, some ecosystems have been negatively affected by atmospheric deposition associated with industrial and agricultural activities. Acid rain is an atmospheric dep-

FIGURE 21.5 Legumes Form Nitrogen-Fixing Nodules
(A) These swollen nodules on the roots of a soybean (*Glycine max*) plant contain nitrogen-fixing *Rhizobium* bacteria. (B) *Rhizobia* are visible in the root cells inside a nodule.

(A)

(B)

Rhizobia

osition process that has been associated with declines in forest ecosystems in the eastern United States and Europe (as we will see in Chapter 24).

Now that we've seen how nutrients enter ecosystems, let's follow their movements within ecosystems as they are taken up and transformed. The next two sections will focus on terrestrial ecosystems; we will take a closer look at nutrient cycling in aquatic ecosystems in the final section.

Nutrient Transformations

CONCEPT 21.2
Chemical and biological transformations in ecosystems alter the chemical form and supply of nutrients.

Following their input into ecosystems, nutrients are subjected to further modifications as a result of uptake by organisms and other chemical reactions that alter their form and influence their movement and retention within ecosystems. Foremost among these transformations is the decomposition of organic matter, which releases nutrients back into the ecosystem.

Decomposition is a key nutrient recycling process

As detritus (dead plants, animals, and microorganisms and egested waste products) builds up in an ecosystem, it becomes an increasingly important source of nutrients, particularly nitrogen and phosphorus. Those nutrients are made available by decomposition, the process by which detritivores break down detritus to obtain energy and nutrients (**Figure 21.6**). **Decomposition** releases nutrients as simple, soluble organic and inorganic compounds that can be taken up by other organisms.

Organic matter in soil is derived primarily from plant matter, which comes from above and below the soil surface. Fresh, undecomposed organic matter on the soil surface is known as **litter** and is typically the most abundant substrate for decomposition. The litter is used by animals, protists, bacteria, and fungi in the soil as a source of energy and nutrients. As animals such as earthworms, termites, and nematodes consume the litter, they break it up into progressively finer particles. This physical fragmentation enhances the chemical breakdown of the litter by increasing its surface area.

The chemical conversion of organic matter into inorganic nutrients (i.e., nutrients that are not associated with carbon molecules) is known as **mineralization**. It is the result of the breakdown of organic macromolecules in the soil by enzymes released by heterotrophic microorganisms. Because plants often rely on inorganic nutrients, ecologists use measurements of mineralization to estimate rates of nutrient supply. Understanding of the abiotic and biotic controls on decomposition and mineralization is key to understanding nutrient availability to autotrophs.

Rates of decomposition and mineralization are greatly influenced by climate. Decomposition, like other biologically mediated processes, proceeds more rapidly at warmer temperatures. Soil moisture also controls rates of decomposition

Litter input includes leaves, stems, roots, and dead animals.

The litter is broken up by small animals into progressively smaller fragments with greater surface area.

Small organic compounds and inorganic nutrients are released into the soil solution, from which they can be taken up by plants and microorganisms.

Soluble nutrients

Fragmentation

Mineralization

Bacteria and fungi release enzymes that act on the exposed surfaces of the fragments to convert organic macromolecules into inorganic nutrients.

FIGURE 21.6 Decomposition Decomposition of organic matter in the soil provides an important input of nutrients to terrestrial ecosystems. Similar steps occur in freshwater and marine ecosystems.

FIGURE 21.7 Climate Controls the Activity of Decomposers
Changes in soil microbial respiration, used as an estimate of decomposition, are plotted as a function of soil moisture at different temperatures.

by influencing the availability of water and oxygen to detritivores. Dry soils may not provide enough water for these organisms, and wet soils have low oxygen concentrations, which lower aerobic respiration and the rate of biological activity. Therefore, the activity of detritivores is highest at intermediate soil moistures and warm temperatures (**Figure 21.7**).

Since some nutrients are consumed by detritivores, the majority of which are microorganisms during the process of decomposition, not all of the nutrients released during mineralization become available for uptake by autotrophs. The amount of nutrients in organic matter that are released during its decomposition depends on the nutrient requirements of the organisms carrying out the decomposition and the amount of energy the organic matter contains. These factors can be approximated by the ratio of carbon (representing energy) to nitrogen (since nitrogen is the nutrient most often in short supply for detritivores) in the organic matter. Organic matter with a high C:N ratio will result in a low net release of nutrients, since heterotrophic microbial growth is more limited by N supply than by energy. For example, heterotrophic microorganisms require approximately 10 molecules of carbon for every molecule of nitrogen they take up. About 60% of the carbon they take up is lost through respiration. Therefore, the optimal C:N ratio of organic matter for microbial growth is about 25:1; after a 60% loss of carbon, such material would yield a C:N ratio of 10:1, identical to that of the microbes. Organic matter with a C:N ratio greater than 25:1 would result in all of the nitrogen being taken up by the mi-

crobes during decomposition, while organic matter with a C:N ratio less than 25:1 would result in some nitrogen being released into the soil.

The chemistry of the carbon in litter determines how rapidly the material can be decomposed. **Lignin**, a structural carbon compound that strengthens plant cell walls, is difficult for soil microorganisms to degrade and thus decomposes very slowly (**Figure 21.8**). The rate of release of nutrients from plant material containing high concentrations of lignin, such as oak or pine litter, is lower than that from material from species with low concentrations of lignin, such as maple and aspen litter. In addition, plant litter may contain secondary compounds (chemicals not associated with growth, including those produced as defenses against herbivores and parasites; see Chapter 12, as well as compounds that protect against excess light energy; see Chapter 5). High concentrations of secondary compounds in litter can lower nutrient release during decomposition, either by inhibiting the activity of heterotrophic microorganisms or, in some cases, by stimulating their growth, leading to greater microbial uptake of nutrients.

By varying the chemistry of their litter, as well as the amount of litter they produce, plants can influence decomposition rates in the soil. Lowering decomposition rates lowers the fertility of the soil. What is the consequence for a plant of decreasing its own nutrient supply? For plants that have inherently slow growth rates, lowering soil fertility may protect them from competitive exclusion by neighbors with higher growth and resource uptake rates. Low soil nutrient concentrations can therefore be perpetuated through plant

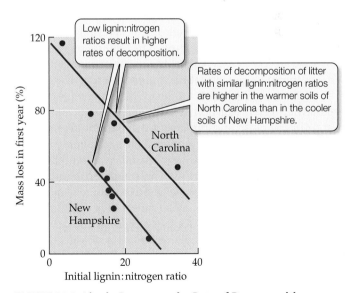

FIGURE 21.8 Lignin Decreases the Rate of Decomposition
The rate of decomposition of leaf litter, expressed as the percentage of mass lost during the first year, decreases as the ratio of lignin to nitrogen in the litter increases. This ratio varies among forest tree species. Note, however, that climate also has an important influence on decomposition rates. (After Melillo et al. 1982.)

chemistry in a way that benefits the plants themselves (Van Breemen and Finzi 1998).

Microorganisms modify the chemical form of nutrients

Microorganisms in soil (as well as in freshwater and marine ecosystems) transform some of the inorganic nutrients released during the process of mineralization. These transformations are particularly important in the case of nitrogen, since they can determine its availability to plants and the rate at which it is lost from the ecosystem (see Figure 21.10B). Certain chemoautotrophic bacteria, known as nitrifying bacteria, convert ammonia (NH_3) and ammonium (NH_4^+) to nitrate (NO_3^-), by a process called **nitrification**. Nitrification occurs under aerobic conditions, so it is limited primarily to terrestrial environments. Under anoxic conditions, some bacteria use nitrate as an electron acceptor, converting it into N_2 and nitrous oxide (N_2O) by a process known as **denitrification**. These gaseous forms of nitrogen cannot be easily used by organisms and are lost to the atmosphere.

If plants are to acquire nitrogen from organic matter, then the proteins contained in detritus must be converted by microbial processes during decomposition into smaller, soluble forms of nitrogen that can be taken up by plants. Plant ecologists and physiologists have long held that nitrogen availability to plants was dependent on the supply of inorganic nitrogen, principally nitrate (NO_3^-), and in some cases ammonium (NH_4^+), particularly in wet soils, because nitrification rates are low under anoxic conditions. Therefore, soil fertility has traditionally been estimated using measurements of the concentrations and supplies of these inorganic forms of nitrogen.

During the 1990s, much effort was invested toward understanding what controls nitrogen mineralization rates, particularly in ecosystems where fertilization experiments indicated that nitrogen availability limited primary production and influenced community diversity. Measurements of inorganic nitrogen production in forest and grassland soils generally came close to estimates of the amount taken up by plants. However, rates of inorganic nitrogen supply in Arctic and alpine ecosystems were substantially lower than what plants were actually taking up. These apparent shortfalls in nitrogen supply led to the realization that some plants were using organic forms of nitrogen to meet their nutritional requirements. Earlier work in marine ecosystems had shown that phytoplankton could take up amino acids directly from water, and mycorrhizae had been shown to take up organic nitrogen from the soil and supply it to plants. However, Terry Chapin and colleagues (1993) and Ted Raab and colleagues (1996) demonstrated that some species, primarily sedges, take up organic forms of nitrogen without mycorrhizae. Arctic sedges may take up as much as 60% of their nitrogen in organic form. Organic nitrogen uptake has been observed in plants in other ecosystems as well, including boreal forests,

salt marshes, savannas, and rainforests. Thus, the mineralization step in decomposition may not be as necessary for nitrogen supply in plants as has been commonly thought (Schimel and Bennett 2004).

The use of soluble organic nitrogen by plants has important implications for competition among plants and between plants and soil microorganisms. There is evidence to support the hypothesis that plants in some Arctic and alpine communities avoid competition through the preferential uptake of specific forms of nitrogen. Robert McKane and colleagues (2002) examined the forms of nitrogen taken up by several plant species growing together in the Arctic tundra of northern Alaska. They measured each species' uptake of inorganic and organic forms of nitrogen, as well as the depth in the soil at which nitrogen was taken up and the time of year when it was taken up. They found that all three factors (form of nitrogen, depth of uptake, and timing of uptake) differed among species. The use of different forms of N by these tundra plants is a rare example of resource partitioning in plants, as discussed in Chapter 18. Furthermore, they found that dominance in the community was related to uptake of the form of nitrogen that was most abundant in the soil (**Figure 21.9**). Thus, a species' ability to dominate a community where nitrogen limits growth may be determined in part by its ability to take up a specific form of nitrogen that is most abundant.

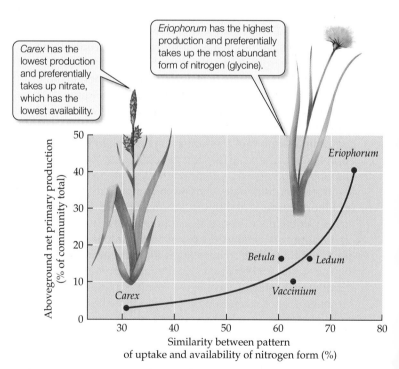

FIGURE 21.9 Community Dominance and Nitrogen Uptake Dominance (measured by proportional contribution to community production) of a plant community in the Alaskan Arctic tundra is related to the similarity between a species' preferred form of nitrogen (ammonium, nitrate, or glycine, a small amino acid) and the availability of that form in the soil. (After McKane et al. 2002.)

Plants can recycle nutrients internally

Leaves and flowers contain the highest nutrient concentrations of any plant organ. During seasonal leaf senescence, nutrients and nonstructural carbon compounds (such as starch and carbohydrates) in perennial plants are broken down into simpler, more soluble chemical forms and moved into stems and roots, where they are stored. This phenomenon is most obvious in mid- to high-latitude ecosystems, when chlorophyll molecules in the leaves of deciduous species are broken down to recover their nitrogen and other nutrients, while other pigments, such as carotenoids, xanthophylls, and anthocyanins, remain, providing the autumnal splendor we humans enjoy. When growth resumes in spring, the nutrients are transported to the growing meristems for use in biosynthesis. Plants may resorb as much as 60%–70% of the nitrogen and 40%–50% of the phosphorus in their leaves before they fall. This recycling effort reduces their need to take up "new" nutrients in the following growing season.

Nutrient Cycles and Losses

CONCEPT 21.3
Nutrients cycle repeatedly through the components of ecosystems.

In the previous section, we saw how nutrients undergo biological, chemical, and physical transformations as they are taken up by organisms and released through decomposition, ultimately returning to their original inorganic form, or a similar form, over time. This movement of nutrients within ecosystems is known as **nutrient cycling** (**Figure 21.10**). For example, we've traced the path of nitrogen into and through an ecosystem, starting with nitrogen-fixing microorganisms, as it is converted into chemical forms that can be used by plants. The plants incorporate the nitrogen into organic compounds (e.g., proteins and enzymes), which may end up being consumed by herbivores. Eventually plants, animals, and microorganisms all end up as detritus. Inorganic and organic nitrogen released from the detritus by decomposition is taken up again by plants, thereby completing the nitrogen cycle (**Figure 21.11**).

Nutrients cycle at different rates according to identity and ecosystem type

The time it takes a nutrient molecule to go from uptake to release to subsequent uptake can vary substantially, depending on the element in question and the ecosystem(s) where the cycle is occurring. In general, nutrients that limit primary

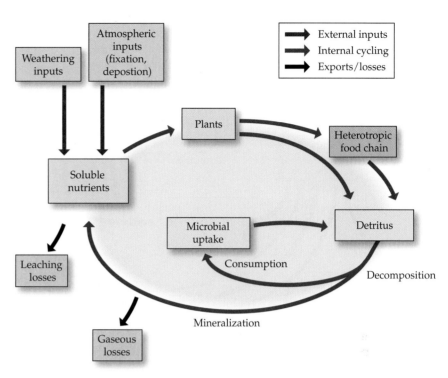

FIGURE 21.10 Nutrient Cycles   A generalized nutrient cycle, showing the movements among the ecosystem components, and the potential pathways for inputs and losses.

production are cycled more rapidly than nonlimiting nutrients. For example, nitrogen and phosphorus may cycle through the photic zone of the open ocean over a period of hours or days, while zinc may cycle over geological time scales associated with sedimentation, mountain building, and erosional processes. Nutrient cycling rates also vary with climate due to the effects of temperature and moisture on the metabolic rates of the organisms associated with production, decomposition, and chemical transformations of nutrients.

Biogeochemists quantify rates of nutrient cycling by estimating the **mean residence times** (also known as **turnover rates**) of elements:

mean residence time = total pool of element/rate of input

In other words, the mean residence time is the amount of time an average molecule of an element spends in a pool before leaving it. The **pool** of an element is the total amount found within a component of the ecosystem, such as soil or biomass. Inputs include all possible sources of the element for that ecosystem component. This approach to estimating mean residence time assumes that pools of nutrients do not change over time and that the mean residence time reflects the overall rate of nutrient cycling. Mean residence time is most commonly used for estimating rates of nutrient turnover

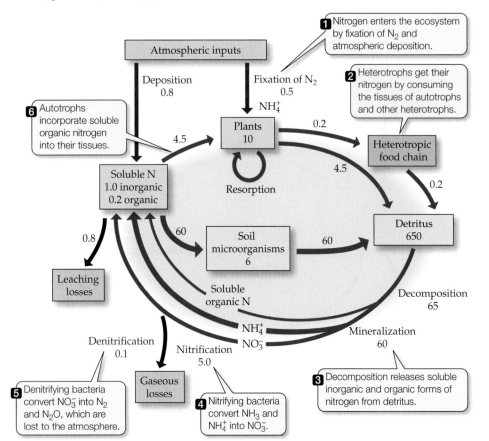

FIGURE 21.11 Nitrogen cycle for an alpine ecosystem, Niwot Ridge, Colorado Boxes represent pools of nitrogen, and arrows represent flows of nitrogen. Amounts of nitrogen are given in units of grams of N per square meter for the pools and grams of N per square meter per year for the flow. Note the large amount of nitrogen passing through soil microorganisms, indicating a high turnover rate for nitrogen in this relatively small pool. (Data from Bowman and Seastedt 2001.)

nutrients? A comparison of mean residence times for organic matter and several nutrients indicates that nutrient pools in the soils of tropical forests are much smaller than those in boreal forests (**Table 21.3**). The turnover rates of nitrogen and phosphorus are more than 100 times faster in tropical forest soils than in boreal forest soils. Temperate forests and chaparral have turnover rates that fall in between, but are closer to those in the tropics.

The main reason for this trend in residence times of nutrients and organic matter is that the influence of climate on rates of decomposition is greater than its influence on primary productivity. Boreal forest soils often have permafrost layers underneath, which cool the soils and lower rates of biological activity. The permafrost also blocks the percolation of water through the soil, creating wet, anoxic soil conditions. Furthermore, the litter produced by boreal forest trees is rich in secondary compounds that slow rates of decomposition in the soil. The variation in mean residence times among specific nutrients is related to their chemical properties (e.g., solubility). Some nutrients (e.g., K) occur in more soluble forms, and thus are lost from soil organic matter more quickly, than others that are found as insoluble organic compounds (e.g., N).

in soil organic matter, which reflect constraints associated with inputs of nutrients and subsequent decomposition. Decomposition rates, as we have seen, are related to climate and the chemistry of plant litter.

Given that both inputs of plant litter and decomposition rates control the mean residence times of nutrients in soil, and that both are subject to climatic control, what differences would we expect to see among ecosystems with similar plant growth forms (e.g., forests) in different climates? Tropical forests have higher net primary productivity, and therefore higher litter input rates, than boreal forests. Does this difference result in differences in the mean residence times of

TABLE 21.3

Turnover Rates of Soil Organic Matter and Nutrients in Forest and Shrubland Ecosystems

Ecosystem type	Soil organic matter	Mean residence time (years)				
		N	P	K	Ca	Mg
Boreal forest	353	230	324	94	149	455
Temperate coniferous forest	17	18	15	2	6	13
Temperate deciduous forest	4	5	6	1	3	3
Chaparral	4	4	4	1	5	3
Tropical rainforest	0.4	2	2	1	1.5	1

Source: Schlesinger 1997.

FIGURE 21.12 Catchments Are Common Units of Ecosystem Study A drainage basin (known as a catchment or watershed) associated with a single stream system (blue lines), with boundaries determined by topographic divides (outlined in white), is a unit commonly used in terrestrial ecosystem studies to measure inputs and outputs of nutrients. This catchment is the upper Hunters Creek basin, draining the south side of Longs Peak in Rocky Mountain National Park.

In Chapter 24, we will return to nutrient cycling at a much larger spatial scale as we consider global cycles of carbon, nitrogen, phosphorus, and sulfur in the context of human alterations of these cycles.

Catchment studies measure losses of nutrients from ecosystems

What determines how long nutrients remain in an ecosystem? The retention of nutrients within an ecosystem is related to their uptake into its biological and physical pools (e.g., organic matter and soil) and to the stability of their forms. Nitrogen, for example, is more stable as part of an insoluble organic molecule, such as a protein, than as nitrate (NO_3^-), which is more easily leached from the soil. Nutrients are lost from an ecosystem when they move below the rooting zone by leaching, and from there into groundwater and streams, as well as by gaseous losses and conversion into chemical forms that cannot be used by organisms.

In our discussion of nutrient inputs into and losses from ecosystems, we have been referring to ecosystems as if they were definitive spatial units, but what defines the boundaries of an ecosystem? Ecologists studying terrestrial ecosystems commonly focus on a single drainage basin. This unit of study, which is called a **catchment** or **watershed**, includes the terrestrial area that is drained by a single stream (**Figure 21.12**). By measuring the inputs and outputs of elements in a catch-

ment and calculating the balance between them, ecologists can make inferences about the use of nutrients in the ecosystem and their importance to ecosystem processes such as primary production.

Figure 21.13 presents a conceptual model of a catchment. Nutrient inputs into the catchment include atmospheric deposition and nitrogen fixation. Nutrients that enter the catchment may be stored in the soil (on cation exchange sites or in the soil solution) or taken up by organisms. They are transferred within and between these ecosystem components by herbivory and predation, decomposition, and weathering processes. Nutrients are assumed to be lost from the catchment primarily in stream water, so measurements of dissolved and particulate matter in streams draining the catchment are often used to quantify these losses. In reality, the situation is often more complicated, as nutrients are also lost to the atmosphere in gaseous forms (e.g., N_2 and N_2O from denitrification). However, measurement of the input–output balance of different nutrients is instructive for determining their biological importance and subsequent retention.

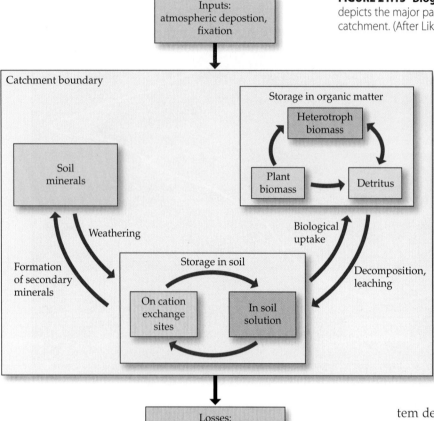

FIGURE 21.13 Biogeochemistry of a Catchment This conceptual model depicts the major pathways of nutrient movement into, through, and out of a catchment. (After Likens and Bormann 1995.)

His hypothesis was based on observations of multiple watersheds at different stages of succession. Vitousek's results showed that losses of nitrogen (a limiting nutrient) as nitrate in stream water from forests at intermediate successional stages were much less than those from old-growth forests, while losses of nonlimiting nutrients, such as potassium, magnesium, and calcium, showed less sensitivity to forest successional stage (**Figure 21.14**).

Long-term ecosystem development affects nutrient cycling and limitations on primary production

As terrestrial ecosystems develop on new substrates (e.g., in primary succession following glacial retreat or on new volcanic flows), soil weathering, nitrogen fixation, and the buildup of organic matter in the soil determine the supply of nutrients available to plants. Early in ecosystem development, there is little organic matter in the soil, so nitrogen supplies derived from decomposition are low. Supplies of mineral-derived nutrients are also low, but higher than the supply of nitrogen. Accordingly, nitrogen availability should be an important limit on primary production and plant community composition early in primary succession (see Chapter 16). As the pool of nitrogen in soil organic matter increases, its limitation of primary production should decrease.

Phosphorus originates from the weathering of a single rock mineral (apatite), and its supply is high relative to that of nitrogen early in succession. As the supply of phosphorus from weathering is exhausted over time, however, decomposition becomes increasingly important in supplying phosphorus to plants. In addition, soluble phosphorus may combine with iron, calcium, or aluminum to form insoluble compounds that are unavailable as nutrients, a process known as **occlusion**. The amount of phosphorus in occluded forms increases over time, further reducing its availability. As a result, phosphorus should become more limiting to primary production during later stages of succession (Walker and Syers 1976).

These observations provide a hypothetical framework for considering how changes in nutrient cycling during ecosystem development should influence the specific nutrients limiting primary production. Nitrogen should be most important in limiting primary production early in succession, nitrogen and phosphorus should both be important at intermediate stages of succession, and phosphorus should be

The best-known catchment studies have been performed at the Hubbard Brook Experimental Forest in New Hampshire (Likens and Bormann 1995), which is considered to be representative of the northern deciduous forests of the United States. Continuous monitoring of the Hubbard Brook catchment began in 1963 under the direction of Herb Bormann and Gene Likens, whose studies there have served as models for a number of other catchment-level studies. These studies are providing information about the roles of organisms and soils in nutrient retention, how ecosystems respond to disturbances such as logging and fire, and long-term trends in nutrient flows associated with acid rain and climate change.

An excellent example of the utility of catchment studies comes from an examination of the effect of disturbance on nutrient retention in ecosystems. Peter Vitousek (1977) studied nutrient retention by spruce–fir forests at different stages of secondary succession using a catchment approach. He proposed that retention of nutrients would be related to the rate of forest growth, predicting that the high rates of primary production usually observed during the intermediate stages of succession would result in the highest retention of nutrients, and that those nutrients most limiting to primary production would be retained more tightly than nonlimiting nutrients.

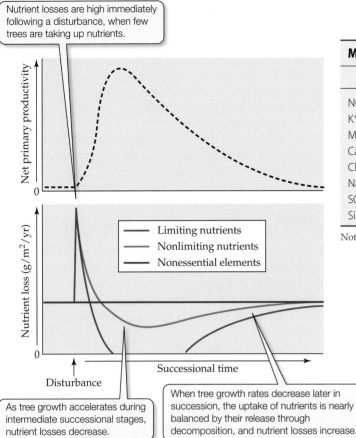

Nutrient losses are high immediately following a disturbance, when few trees are taking up nutrients.

As tree growth accelerates during intermediate successional stages, nutrient losses decrease.

When tree growth rates decrease later in succession, the uptake of nutrients is nearly balanced by their release through decomposition, and nutrient losses increase.

- Limiting nutrients
- Nonlimiting nutrients
- Nonessential elements

Mean Growing Season Streamwater Concentrations (µeq/liter)

	Old-growth	Successional	Ratio of concentrations
NO_3^-	53 (5)	8 (1.3)	6.63
K^+	13 (1)	7 (0.5)	1.81
Mg^{++}	40 (4.9)	24 (1.6)	1.67
Ca^{++}	56 (4.5)	36 (2.5)	1.56
Cl^-	15 (0.3)	13 (0.3)	1.15
Na^+	29 (2.6)	28 (0.9)	1.04
SO_4^-	119 (4.6)	123 (6.5)	0.97
Si	75 (7)	86 (5)	0.87

Note: Standard error in parentheses.

FIGURE 21.14 Retention of Nutrients Is Highest at Intermediate Stages of Forest Succession Loss of nutrients from forest catchments is high following disturbances, due to lack of biological uptake. As forest growth accelerates, the loss of nutrients decreases as tree uptake increases. Limiting nutrients have a greater degree of retention than nonlimiting nutrients. When tree growth rates decrease later in succession, uptake of nutrients is nearly balanced by release from decomposition, and nutrient loss increases in proportion to external inputs. The table shows changes in streamwater outputs from old growth and younger successional stands of spruce–fir forests that support this hypothesis, with larger changes (higher ratios) in output of limiting nutrients (N, K) relative to nonlimiting nutrients (Ca, Cl, S). (Data from Vitousek 1977.)

FIGURE 21.15 Nutrient Limitation of Primary Production Changes with Ecosystem Development (A) Fertilization experiments were conducted in three ecosystems with soils of different ages in the Hawaiian Islands: Thurston (300 years old), Laupahoehoe (20,000 years old), and Kokee (4.1 million years old). Vegetation at all sites is dominated by a single species of tree, Ohi'a (*Metrosideros polymorpha*). (B) Ohi'a growth rates in response to fertilization treatments with nitrogen (N), phosphorus (P), and both N + P in the three ecosystems. The more an added nutrient increased tree growth, the more limiting it was assumed to be. Note the differences in the ranges of the y-axes. (A after Crews et al. 1995; B after Vitousek and Farrington 1997.)

most important late in succession. This hypothesis was tested in the Hawaiian Islands by Peter Vitousek and colleagues. The movement of the Pacific tectonic plate over millions of years has given rise to the chain of volcanoes that formed these Islands. The oldest islands are in the northwestern part of the chain, the youngest in the southwest (**Figure 21.15A**).

Vitousek's group studied Hawaiian ecosystems with ages ranging from 300 years to over 4 million years to determine what nutrients were most limiting to primary production. The study was aided by the similarity of the vegetation and climate in each of the study sites. Vitousek and colleagues added nitrogen, phosphorus, and both nitrogen and phosphorus to plots in three ecosystems of different ages and measured the effects of these treatments on the growth of the dominant tree, Ohi'a (*Metrosideros polymorpha*). Consistent with their hypothesis, nitrogen was most limiting to tree growth in the youngest ecosystem, while phosphorus was most important in the oldest ecosystem (Vitousek and Farrington 1997) (**Figure 21.15B**). Nitrogen and phosphorus in combination increased tree growth in the intermediate-aged ecosystem. In contrast to these tropical soils, the soils of ecosystems in temperate, high-latitude, and high-elevation zones are often subjected to major disturbances (e.g., large-scale glaciation, landslides) and are less likely to reach ages at which phosphorus becomes limiting.

Nutrients lost from terrestrial ecosystems often end up in streams, lakes, and oceans. They are a critical source of nutrients for those aquatic ecosystems, but they can have negative effects as well, as we'll see in the next section.

Nutrients in Aquatic Ecosystems

> **CONCEPT 21.4**
> Freshwater and marine ecosystems receive nutrient inputs from terrestrial ecosystems.

In freshwater and marine ecosystems, nutrient transfers have the added complexity of occurring in a moving aqueous medium. Inputs of nutrients from outside the ecosystem are much more important than in terrestrial ecosystems. In addition, oxygen concentrations are often lower than in terrestrial ecosystems, constraining aerobic respiration and the biogeochemical processes that influence the chemical forms of nutrients and their rates of supply.

Nutrients in streams and rivers cycle in spirals

Nutrient supplies in streams and rivers are very dependent on external inputs from terrestrial ecosystems. Terrestrial inputs of organic matter, dissolved nutrients derived from chemical weathering and decomposition in surrounding soils, and particulate minerals are the primary sources of nutrients for riverine organisms. Rivers and streams carry all of these materials to the ocean, but they are not just conduits for the movement of material between terrestrial and marine ecosystems. Biogeochemical processing in moving stream water can significantly affect the forms and concentrations of the elements it contains. For example, denitrification and biological uptake in streams and rivers may result in significant losses of nitrogen during transport in stream water. These processes may explain why rivers export less nitrate from regions receiving high amounts of anthropogenic nitrogen pollution than would be expected (**Figure 21.16**).

Nutrients in rivers and streams are cycled repeatedly as the water flows downstream. Nutrients are transferred between dissolved inorganic forms and organisms, primarily bacteria and phytoplankton, that take them up and incorporate them into organic molecules. The nutrients taken up by these organisms may be consumed by other organisms and pass through a food web, eventually entering the pool of stream detritus. Following decomposition, the mineralized nutrients are released back into the water in dissolved inorganic forms. This repeated uptake and release in association with the movement of water can be thought of as nutrient "spiraling" (Newbold et al. 1983) (**Figure 21.17**). The time it takes for a full nutrient spiral to occur (i.e., from uptake and incorporation into organic forms to release in inorganic forms) is related to the amount of biological activity in the stream, the water velocity, and the chemical form of the nutrient. Retention of nitrate and phosphate in rivers increases downstream due to increasing spiral lengths; thus, higher-order streams (see Chapter 3) are particularly important in buffering the effects of nutrient pollutants on estuarine and marine ecosystems (Ensign and Doyle 2006).

Nutrients cycle tightly in lake ecosystems

Lake ecosystems receive inputs of nutrients from streams by atmospheric deposition, and litter falling from adjacent terrestrial ecosystems. Biological demand for nutrients is highest in the photic zone (the zone with enough light for photosynthesis; see Chapter 3), where phytoplankton are suspended in the water column, and in the shallow zones at the margins of the lake, where rooted aquatic plants are found. Phosphorus commonly limits primary production in lakes, although nitrogen may also be limiting in some lakes. Nutrient transfers between trophic levels, like energy transfers (see Chapter 20), are very efficient in lakes. Some detritus is decomposed and mineralized in the water column and in sediments in the shallow zones, providing an internal input of nutrients. Nitrogen fixation by cyanobacteria occurs in the photic zone, usually when that process is favored by a low ratio of dissolved nitrogen to phosphorus (N:P < 10). Rates of nitrogen fixation in lake ecosystems are similar to those in terrestrial ecosystems.

Over time, nutrients are progressively lost from the photic zone of a lake. Dead organisms sink through the water column, passing through the nonmixing layers in a stratified lake (see Chapter 2), and are deposited in the sediments of the benthic zone (at the bottom of the lake). Lake sediments are characterized by anoxic conditions that limit biological activity, including decomposition, and a reducing chemical environment that may change the chemical form of some nutri-

(A)

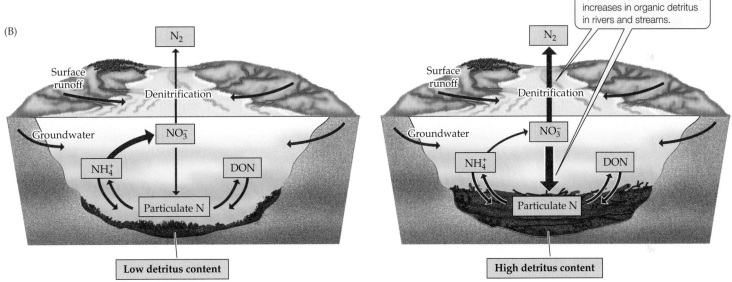

(B)

FIGURE 21.16 Rivers Are Important Modifiers of Nitrogen Exports   Nitrogen that enters rivers from terrestrial ecosystems is not simply carried to the ocean. (A) The rates of nitrogen exports to the North Atlantic Ocean from major drainage basins are correlated with rates of nitrogen inputs into rivers by human activities. The export rates, however, are substantially lower than the input rates due to biogeochemical processing of the nitrogen in the rivers. (B) Denitrification and biological uptake are two of the main processes that lower the export of nitrogen from drainage basins and are enhanced when benthic detritus is high. DON, dissolved organic nitrogen. (A after Howarth et al. 1996; B after Bernhardt et al. 2005.)

ents. Iron, for example, is often reduced from Fe^{3+} to Fe^{2+}, contributing to the dark color of lake sediments. Denitrification is also promoted by the low oxygen concentrations in the sediments, and bacteria may reduce sulfate (SO_4^{2-}) to hydrogen sulfide (H_2S).

Decomposition in the benthic sediments cannot provide nutrients to the photic zone unless there is mixing of the water column. As we saw in Chapter 2, this mixing occurs in temperate-zone lakes during fall and spring, when the water becomes isothermal and wind facilitates its turnover. This seasonal turnover brings dissolved nutrients from the bottom water to the surface layers, along with detritus that may be subsequently decomposed by bacteria. Mixing of the water layers is less common in tropical

FIGURE 21.17 Nutrient Spiraling in Stream and River Ecosystems   Cycling of nutrients as the water moves downstream results in repeated spirals of nutrient uptake and release.

lakes, so external inputs of nutrients may be more important for maintaining production in those lakes.

Lake ecosystems are often classified according to their nutrient status (that is, whether nutrient availability is high or low). Nutrient-poor waters with low primary productivity are referred to as **oligotrophic**, while nutrient-rich waters with high primary productivity are referred to as **eutrophic**. **Mesotrophic** waters are intermediate in nutrient status between oligotrophic and eutrophic waters. The nutrient status of a lake is the result of natural processes associated with climate and lake size and shape. For example, lakes in high mountain areas are typically oligotrophic due to their short growing season, low temperatures, and tendency to be deep with a low surface area-to-volume ratio, which limits atmospheric deposition of nutrients. In contrast, shallow lakes at lower elevations or in the tropics tend to be eutrophic, with warmer temperatures, higher primary productivity, and higher nutrient availability.

The nutrient status of a lake tends to shift naturally from oligotrophic to eutrophic over time. This process, known as **eutrophication**, occurs as sediments accumulate on the lake bottom. As the lake becomes shallower, its summer temperatures become warmer, more decomposition occurs, nutrient pools and their tendency to mix increase, and the lake becomes more productive (**Figure 21.18**). Human activities have accelerated the process of eutrophication through discharges of sewage, agricultural fertilizers, and industrial wastes containing high concentrations of nitrogen and phosphorus into rivers and lakes. The water of Lake Tahoe, on the border between Nevada and California, has lost much of its clarity due to increased inputs of phosphorus and nitrogen from streams, groundwater, and surface runoff from neighboring communities. Water clarity, which is used as an indicator of a lake's nutrient status, is primarily determined by the density of plankton in the water column. It can be measured using a Secchi disk, a black and white circular plate that is lowered gradually into the water; the maximum depth at which the disk can be seen is referred to as the *depth of clarity*. Over the past three decades, the average depth of clarity in Lake Tahoe has risen by 10 m (Murphy and Knopp 2000).

Anthropogenic eutrophication can be reversed if the discharge of wastes into surface waters is decreased. A classic example of such a reversal occurred in the 1960s and 1970s in Lake Washington, near Seattle. Treated sewage containing high phosphorus concentrations from Seattle was released into Lake Washington beginning in the late 1940s as neighborhoods and accompanying sewage treatment plants were built near the shore of the lake. Significant decreases in water clarity were noted during the 1950s, corresponding to increases in phytoplankton densities and blooms of cyanobacteria. Public concern grew, and local governments debated what action to take. Based on advice from a prominent local limnologist, W. T. Edmondson, who believed that the problem was associated with phosphorus inputs from treated sewage, including wastewater from washing machines containing phosphorus-laden detergents, Seattle quit dumping its sewage into Lake Washington. Between 1963 and 1968, the amount of sewage discharged into Lake Washington was progressively dropped to zero. Increases in water clarity were soon noted, and by 1975, it was considered recovered from eutrophication

FIGURE 21.18 Lake Sediments and Depth   Organic sediments accumulate at the bottom of a lake over time, making it progressively shallower and leading to eutrophication. Changes in the depth contours of Mirror Lake in New Hampshire show the accumulation of sediments over the past 14,000 years. (After Davis et al. 1985.)

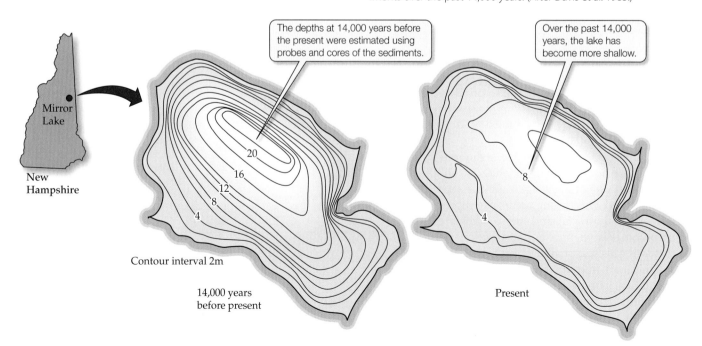

The depths at 14,000 years before the present were estimated using probes and cores of the sediments.

Over the past 14,000 years, the lake has become more shallow.

Mirror Lake

New Hampshire

Contour interval 2m

14,000 years before present

Present

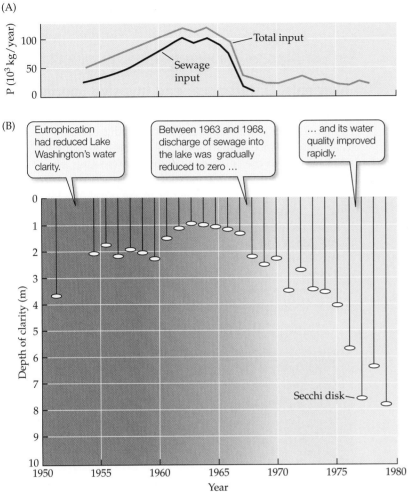

(A)

(B)

Eutrophication had reduced Lake Washington's water clarity.

Between 1963 and 1968, discharge of sewage into the lake was gradually reduced to zero …

… and its water quality improved rapidly.

Secchi disk

FIGURE 21.19 Lake Washington: Reversal of Fortune
Inputs of treated sewage between the 1940s and the 1960s caused eutrophication in Lake Washington; cessation of sewage inputs between 1963 and 1968 increased lake clarity. (A) Phosphorus inputs. (B) Measurements of water clarity made with a Secchi disk. (A after Edmondson and Litt 1982.)

marshes, which are rich in nutrients because they trap both riverine and ocean sediments. Like benthic sediments in lakes, estuarine and salt marsh sediments have low oxygen concentrations that limit decomposition.

As we saw in Chapter 19, primary production in the open ocean is limited by several nutrients, including nitrogen, phosphorus, and in some areas, iron and silica. Seawater has relatively high concentrations of magnesium, calcium, potassium, chloride, and sulfur. Sources of nitrogen in marine ecosystems include inputs from rivers and atmospheric deposition as well as internal cycling through decomposition. Rates of nitrogen fixation by cyanobacteria in the oceans are lower than those in freshwater lakes, possibly because they are limited by molybdenum, which is a component of the nitrogenase enzyme. Phosphorus, iron, and silica enter the marine system primarily in dissolved and particulate form in rivers, with a smaller but important component coming from atmospheric deposition of dust. Inputs from both of these terrestrial sources are increasing as a result of human activities, including large-scale desertification and deforestation.

Deep deposits of sediments (up to 10 km, or 6 miles!) have accumulated in the benthic zones of the open ocean. These deposits, which consist of a mix of ocean-derived detritus and terrestrial erosional sediments, are important potential sources of nutrients. Sulfate reduction and denitrification are important processes in these anoxic sediments, but some decomposition and mineralization of organic matter occurs. Bacteria have been found as deep as 500 m in these sediments. Mixing of deep, nutrient-rich waters with nutrient-poor surface waters occurs in zones of upwelling (see Chapter 2), where ocean currents bring deep waters to the surface. These zones of upwelling are highly productive (see Chapter 19) and thus are important areas for commercial fisheries.

(**Figure 21.19**). Edmondson's recommendation was crucial to this recovery, and the case contributed to the current restrictions on the use of phosphates in detergents.

Imports and upwelling are important to nutrient supply in marine ecosystems

Rivers join marine ecosystems in estuaries (see Chapter 3). In these zones where fresh water meets seawater, salinity, and thus water density, is variable. This variation influences the mixing of waters and the chemical forms of some nutrients. For example, the phosphorus bound to soil particles may be released in a form more available to phytoplankton as a result of changes in pH and water chemistry when river water mixes with seawater.

As the velocity of a river decreases, suspended sediments begin to settle out and are deposited at the mouth of the river. These sediments are resuspended by upwelling of saltier, denser seawater into the fresher, less dense river water, providing detritus for detrivores and nutrients for phytoplankton in the estuary. Estuaries are often associated with salt

CASE STUDY REVISITED

A Fragile Crust

We've seen that nutrient supplies for plants in terrestrial ecosystems are dependent on the weathering of rock minerals and the decomposition of detritus in the soil, as well as on the fixation of atmospheric nitrogen. How might the loss of biological crusts from desert soils influence these processes? As we have seen, the crusts prevent erosional losses of soil by helping to bind soil particles together. The activity of the organisms that make up the crust may also influence nutrient inputs, and in turn the productivity of the desert ecosystem, as well as its capacity to withstand the desert climate.

Jason Neff and colleagues conducted a study to evaluate the effects of cattle grazing on soil erosion and nutrient availability on the Colorado Plateau (Neff et al. 2005). They used a comparative approach, selecting three study sites in Canyonlands National Park: one that had never been grazed and two that had been grazed historically, but were closed to grazing after 1974. Cattle grazing was initiated in the 1880's, and most of the soil surface in the area has been impacted. The ungrazed site was surrounded by rock formations that prevented the movement of cattle into the area. The study sites all had the same parent material, similar plant communities, and were located within 10 km of one another. Biological crusts were present at all three sites, although those at the historically grazed sites had clearly been damaged, as they appeared less well developed than those at the site that had never been grazed.

Samples of soil and bedrock were collected from each of the sites, and the textures and nutrient contents of the soils were compared. In addition, the retention of fine dust from the atmosphere was estimated by measuring the magnetic properties of the soil. Dust blown in from distant areas contains higher amounts of iron oxides than the native soil, so the more dust present, the stronger the magnetic signal. Retention of this dust is important because it is a source of mineral nutrients; in addition, loss of this dust indicates the potential for erosional loss of the native soil as well.

Neff et al. found that the historically grazed soils had less fine textured soil and substantially less magnesium and phosphorus (**Figure 21.20**), than the ungrazed soils. They attributed these differences to greater retention of dust and lower rates of erosion in the soils with better-developed biological crusts. The crusts may also enhance rates of weathering by altering pH, increasing the rates of chemical reactions that release mineral nutrients, and increasing water retention in the soil. Soils in the historically grazed sites also had 60%–70% less carbon (from organic matter) and nitrogen than the ungrazed sites. These differences were also related to losses of biological crust. Although a crust had begun to recover at the historically grazed sites, the cumulative loss of carbon and nitrogen from the soils during the period of grazing was high. The cyanobacteria in biological crusts fix significant amounts of atmospheric N_2 (Belnap 2003), which represents an important input of a nutrient that may limit plant growth in the absence of water limitation during the spring growing season. In addition, the dark surface of the crust absorbs more solar radiation, and crust-covered soils retain more water,

than in soils without crusts, creating conditions more conducive to decomposition and mineralization.

Nutrients, Disturbance, and Invasive Species

By increasing nutrient supplies and stabilizing soils, biological crusts enhance primary production. Plants growing in association with the crusts have higher growth rates, and contain more nutrients, than plants growing in soils without crusts. Plant cover also increases in the presence of biological crusts. Finally, biological crusts have been shown to lower the germination and survival rates of invasive plants (Mack and Thompson 1982).

Are the negative effects of cattle grazing on soil stability and nutrient availability that Neff et al. observed in Canyonlands National Park common in other areas? The answer lies in part with the long-term history of grazing

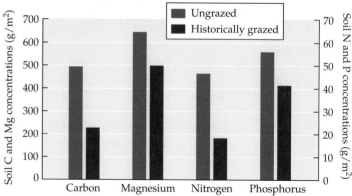

FIGURE 21.20 Loss of Biological Crusts Results in Smaller Nutrient Pools Historically grazed soils in Canyonlands National Park contained less carbon, magnesium, nitrogen, and phosphorus than soils that had never been grazed. (After Neff et al. 2005.)

FIGURE 21.21 Scourge of the Intermountain West Large areas of the intermountain West of North America are now dominated by cheatgrass (*Bromus tectorum*), an invasive species that increases fire frequencies, outcompetes native plants for resources, and spreads rapidly across the landscape.

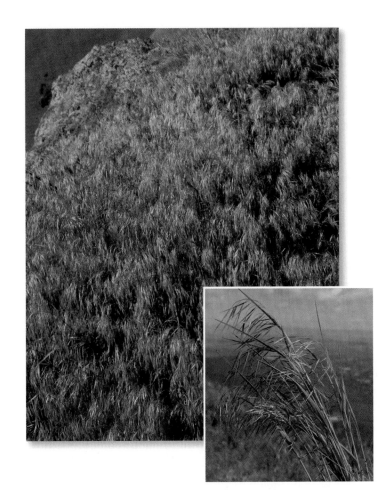

and climate in North America. Prior to Euro-American settlement, soils in much of the intermountain West did not experience the amount of grazing by native animals that occurred in other areas, such as the Great Plains, where large herds of bison roamed. A combination of aridity and long-term development of biological crusts may have given the soils of the Colorado Plateau an especially low tolerance for heavy grazing.

In the grasslands of the intermountain West, the combination of soil disturbance and loss of biological crust has created a situation conducive to the spread of non-native species, most notably cheatgrass (*Bromus tectorum*; **Figure 21.21**), a native of Eurasia. Cheatgrass has had profound effects on the ecology of much of western North America. Cheatgrass is a spring annual that sets seed, dies, and dries out by early summer. This life cycle increases the amount of dry, combustible vegetation that is present during the summer. As a result, cheatgrass has increased the frequency of fires, which now have return intervals of about 3–5 years, compared with more natural fire frequencies of 60–100 years. Since native grasses and shrubs cannot recover from such frequent fires, cheatgrass increases its dominance under these conditions. Cheatgrass is an effective competitor for soil resources, and it also lowers rates of nitrogen cycling by producing litter with a higher C:N ratio relative to native species (Evans et al. 2001). The combination of in-

creasing fire frequency, increasing competition, and changing nutrient cycling has led to decreases in native species richness in many parts of the intermountain grasslands.

SUMMARY

CONCEPT 21.1 Nutrients enter ecosystems through the chemical breakdown of minerals in rocks or through fixation of gases in the atmosphere.

- The nutrient requirements of organisms are specific to their physiological demands and thus differ between autotrophs and heterotrophs.

- Autotrophs absorb nutrients in soluble forms from their environment, while heterotrophs obtain them by consuming prey or detritus.

- The physical and chemical breakdown of rock minerals (weathering) releases soluble nutrients.

- Soils are made up of mineral particles, detritus, dissolved organic matter, water containing dissolved minerals and gases, and organisms.

- Carbon and nitrogen enter ecosystems through fixation of atmospheric gases by autotrophs and by prokaryotes, respectively.

CONCEPT 21.2 Chemical and biological transformations in ecosystems alter the chemical form and supply of nutrients.

- Decomposition of organic matter releases the nutrients it contains in soluble forms that can be reused by plants and microorganisms.

- Modification of the chemical forms of nutrients, particularly nitrogen, by microorganisms influences their availability as nutrients or loss from the ecosystem.

- Plants recycle nutrients by reabsorbing them from senescing tissues and remobilizing them when growth commences again.

CONCEPT 21.3 Nutrients cycle repeatedly through the components of ecosystems.

- Nutrient cycling rates are controlled primarily by the rate of decomposition, which in turn is controlled by climate and the chemistry of plant litter.

- Losses of nutrients from terrestrial ecosystems can be estimated by measuring nutrient outputs in stream water.

- Changes in the relative amounts of nutrients supplied by weathering and decomposition determine the specific nutrients that limit primary production at different stages of ecosystem development.

CONCEPT 21.4 Freshwater and marine ecosystems receive nutrient inputs from terrestrial ecosystems.

- Nutrient cycling in streams and rivers can be thought of as a spiral of repeated biological uptake and incorporation into organic forms followed by release in inorganic forms.

- In lakes, nutrients are cycled between the water column and the bottom sediments.

- Imports of nutrients from rivers and terrestrial ecosystems support production in marine ecosystems.

REVIEW QUESTIONS

1. Describe the processes involved in the transformation of solid minerals in rock into soluble nutrients in soil. What biological factors can influence the rate of this transformation?

2. Why is nitrogen often in short supply relative to other nutrients required by plants, despite being the most abundant element in the atmosphere? How does the source of nitrogen change during terrestrial ecosystem development?

3. Which factor is more important in controlling the mean residence times and pools of nutrients in soil organic matter in terrestrial ecosystems, the rate of input (i.e., net primary productivity) or the rate of decomposition? Would you expect to find greater nutrient pools in the soils of a tropical forest than in those of a boreal forest, given that primary productivity is higher in the tropics?

4. Why might you expect nutrient input from terrestrial and stream ecosystems to be more important in tropical lakes than in temperate-zone lakes?

HANDS-ON PROBLEM SOLVING

Dry Decomposition | This Web exercise (www.sinauer.com/ecology) explores how plant litter decomposes in a dry climate. You will interpret data from a recent paper demonstrating the factors responsible for litter decomposition in a semi-arid ecosystem in Patagonia.

SUGGESTED READINGS

The following two texts provide excellent introductions to nutrient cycling in terrestrial ecosystems, including environmental controls on nutrient supply to plants, feedbacks to other ecosystem functions such as production, and how human influences are affecting nutrient availability.

Aber, J. D. and J. M. Melillo. 2001. *Terrestrial Ecosystems.* Harcourt Academic Press, San Diego, CA.

Chapin, F. S. III, P. A. Matson and H. A. Mooney. 2002. *Principles of Terrestrial Ecosystem Ecology.* Springer, New York.

Terry Chapin's classic review paper, one of the most heavily cited in the field of ecosystem science and plant physiological ecology, and the follow-up paper coauthored with Rien Aerts, provide thorough coverage of the feedbacks between plants and soil biogeochemistry.

Chapin, F. S. III. 1980. The mineral nutrition of wild plants. *Annual Review of Ecology and Systematics* 11: 233–260.

Aerts, R. and F. S. Chapin III. 2000. The mineral nutrition of wild plants revisited: A re-evaluation of processes and patterns. *Advances in Ecological Research* 30: 1–67.

Newbold, J. D., J. W. Elwood, R. V. O'Neill and A. L. Sheldon. 1983. Phosphorus dynamics in a woodland stream ecosystem: A study of nutrient spiralling. *Ecology* 64: 1249–1265. *This paper presents one of the earliest empirical determinations of nutrient spiraling in a stream ecosystem, using radioactively labeled phosphorus to trace the movements of the element through stream water and into and out of stream biota and detritus.*

The following two works emphasize the importance of nutrient ratios to determining specific nutrient limitations on production. Redfield observed that organic material in deep ocean water had a constant ratio of C:N:P, and proposed that biological activity was key to maintaining this ratio (actually described in an earlier 1934 paper). Deviations from the ratio of N:P dictate the nutrients limiting pro-

NUTRIENT SUPPLY AND CYCLING

duction. *Sterner and Elser's book expands the theme of nutrient ratios and their relationships to energy, nutrient cycling, and other ecological functions.*

Redfield, A. C. 1958. The biological control of chemical factors in the environment. *American Scientist* 46: 205–221.

Sterner, R. W. and J. J. Elser. 2002. *Ecological Stoichiometry: The Biology of the Elements from Molecules to the Biosphere.* Princeton University Press, Princeton, NJ.

Vitousek, P. M. and W. A. Reiners. 1975. Ecosystem succession and nutrient retention: A hypothesis. *BioScience* 25: 376–381. *Vitousek and Reiners's paper questioned the belief that nutrient retention in forest ecosystems increased during ecosystem development, reaching a maximum during the later stages of succession. Although controversial when first published, their hypothesis of higher nutrient retention during intermediate stages of succession has been supported by the majority of studies since that time.*

Applied and Large-Scale Ecology

CHAPTER 22 Conservation Biology

CASE STUDY

Can Birds and Bombs Coexist?

Could it be that using land as a bombing range has been the secret to conservation success? Although it seems strange, decades of bombing on the Fort Bragg military base in North Carolina have inadvertently protected thousands of acres of longleaf pine savanna, aiding efforts to save the endangered red-cockaded woodpecker (**Figure 22.1**).

For 90 years, the forests of Fort Bragg in the North Carolina Sandhills have been used for military training exercises, degraded by off-road vehicles and earth-moving equipment, and set on fire by explosives. These destructive activities take place in the midst of a vibrant but now uncommon ecosystem, one that, ironically, survives in large part as a result of the military presence. How can this be? First, the longleaf pine savanna ecosystem depends on fire for its persistence. Second, the designation of large blocks of forest land for military use has kept them from being converted to farmland, forestry, and residential uses.

The longleaf pine savanna ecosystem has been reduced to only 3% of its original coverage of more than 30 million hectares (75 million acres), making it one of the most threatened ecosystems in the United States (**Figure 22.2**). Various factors contributed to this decline, including rapid growth of the human population, the clearing of land for large plantations where other tree species, such as loblolly pine, are grown, and fire suppression. With the decline of the longleaf pine ecosystem, several plant, insect, and vertebrate species that depended on it have also undergone substantial declines.

One of these species is the red-cockaded woodpecker (*Picoides borealis*), a small insectivorous bird specifically adapted to large tracts of open pine savanna. Once far more abundant, the species currently stands at about 6,100 breeding pairs and their associated helpers. While other woodpeckers nest in dead snags, red-cockaded woodpeckers require mature, living pines, especially the longleaf pine (*Pinus palustris*), for their nesting cavities. In addition, like few other woodpecker species, they are cooperative breeders. Hatchlings are raised by a breeding pair along with two to four nonbreeding helpers, generally males born to the pair in previous years.

FIGURE 22.1 The Red-Cockaded Woodpecker: An Endangered Species A male red-cockaded woodpecker (*Picoides borealis*) perched by its nest hole. This species was once abundant throughout the pine forests of the United States, but has been severely reduced in numbers due to the loss of its required habitat.

FIGURE 22.2 Decline of the Longleaf Pine Ecosystem (A) The estimated area covered by longleaf pine forests at different points in time. (B) As seen in this photograph from the southeastern USA, longleaf pine (*Pinus palustris*) savanna consists of open forest with a grass understory. (A after Van Lear et al. 2005.)

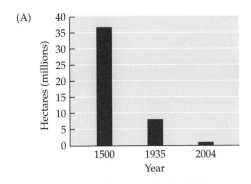

Periodic fires historically helped to maintain longleaf pine savanna. Without those fires, the longleaf pine community soon undergoes succession. As an understory of young oaks and other hardwoods grows up, red-cockaded woodpeckers abandon their nesting cavities, apparently due to a decrease in food resources. In the past, the birds would move to parts of the forest that had been more recently burned, but as the area of suitably mature longleaf pines declines, there are few places for the birds to go.

This loss of habitat has reduced the woodpeckers' populations, making them vulnerable to the problems of small, isolated populations that we discussed in Chapter 10. There is evidence of inbreeding among the birds, and in 1989, Hurricane Hugo killed 70% of the birds in one population. West Nile virus has not yet been detected in this species, but it is highly lethal to other bird species, so if it emerged, the results could be devastating.

The recent history of the red-cockaded woodpecker reflects that of thousands of other imperiled species around the world that have gone from being a vital component of an extensive ecosystem through a gradual decline with loss of habitat to critically low numbers. When organisms rely on a specific habitat that is lost to the effects of human activities, they are found in diminishing numbers until, in some cases, they vanish. But the history of the red-cockaded woodpecker continues. Legal protection and extraordinary human effort have resulted in a stabilization and slow recovery of numbers.

What can people do to rescue such species from extinction? Do we have a responsibility to protect existing biodiversity and to restore some of what has been lost? If so, how can we best allocate our limited resources to be most effective in our conservation efforts?

Over the last few centuries, as the human population has grown, and as humans have cut, plowed, drained, and dammed, the species that evolved in the ecosystems around us have lost their habitats through outright destruction or through changes in their physical properties. These changes have given rise to a biodiversity crisis. The most recent Red List of Threatened Species, compiled by the World Conservation Union, lists 16,118 species as threatened with extinction—about 1% of all species worldwide (**Table 22.1**). This number is certainly an underestimate, as only the best-studied taxonomic groups have been assessed for their conservation status in any comprehensive way.

Ecologists play an important role in observing, measuring, and communicating the changes in species abundances, distributions, and biological traits that have resulted from anthropogenic alterations of the planet's ecosystems. They are the eyes, the ears, and the calculators that record and report the dramatic declines in biological diversity that have come

about in the twentieth and twenty-first centuries. And ecologists and other biologists are also one component of a diverse team working to find ways of reversing these declines. This chapter and the next will show how ecologists fit into the complex picture of biodiversity conservation.

Conservation Biology

CONCEPT 22.1
Conservation biology is an integrative discipline that applies the principles of ecology to the conservation of biodiversity.

The stabilization of red-cockaded woodpecker populations required expertise from biological disciplines such as demography, genetics, and pathology, but has also included contributions from other arenas outside biology, including law, economics, political science, communications, and sociolo-

TABLE 22.1

Global Summary of the Number of Documented Imperiled Species

Group	Number of described species	Percentage of group evaluated in 2006	Number of species evaluated in 2006	Number of threatened species in 2006	Number threatened in 2006, as percentage of described species
VERTEBRATES					
Mammals	5,416	89.7	4,856	1,093	20
Birds	9,934	100.0	9,934	1,206	12
Reptiles	8,240	8.1	664	341	4
Amphibians	5,918	100.0	5,918	1,811	31
Fishes	29,300	9.9	2,914	1,173	4
Subtotal	*58,808*	*41.3*	*24,286*	*5,624*	*10*
INVERTEBRATES					
Insects	950,000	0.1	1,192	623	0.07
Mollusks	70,000	3.1	2,163	975	1.39
Crustaceans	40,000	1.3	537	459	1.15
Others	130,200	0.1	86	44	0.03
Subtotal	*1,190,200*	*0.3*	*3,978*	*2,101*	*0.18*
PLANTS					
Mosses	15,000	0.6	93	80	0.53
Ferns and allies	13,025	1.6	212	139	1
Gymnosperms	980	92.7	908	306	31
Dicotyledons	199,350	4.8	9,538	7,086	4
Monocotyledons	59,300	1.9	1,150	779	1
Subtotal	*287,655*	*4.1*	*11,901*	*8,390*	*3*
OTHERS					
Lichens	10,000	> 0.1	2	2	0.02
Mushrooms	16,000	> 0.1	1	1	0.01
Subtotal	*26,000*	*> 0.1*	*3*	*3*	*0.01*
TOTAL	**1,562,663**	**2.6**	**40,168**	**16,118**	**1**

Source: Data from World Conservation Union Red List Summary 2007.

Note: Some groups have been completely evaluated (amphibians, birds) for conservation status, but for many groups, only a small percentage of described species have been evaluated. For those groups that have been only partially evaluated, there may be a bias toward completing assessments of imperiled species, with assessments of more common species a lower priority. That only 1% of species are shown as imperiled is an artifact of incomplete evaluation, as the percentage is believed to be much higher.

gy. It has required working with farmers, landowners, the U.S. military, and the business community. Arriving at a successful management approach required not only data collection and analysis, but also creativity and the ability to work with a wide variety of stakeholders with varying interests in the ecosystem. Such an integrative approach is characteristic of conservation biology.

Conservation biology is the scientific study of phenomena that affect the maintenance, loss, and restoration of biodiversity. It applies many of the ecological principles and tools that you have studied in this book to the halting or reversal of biodiversity declines. Later in this chapter, we will look at the reasons why biodiversity is declining and at the tools conservation biologists use to address conservation problems. But first let's look at why preventing and reversing declines in biodiversity is an important task for the current generation of ecologists.

Protecting biodiversity is important for both practical and moral reasons

People rely on nature's diversity. In addition to the hundreds of domesticated species that sustain us, we make abundant use of wild species for food, fuel, and fiber. We harvest wild species for medicines, building materials, spices, and decorative items. Many people rely on these natural resources for their livelihoods. All of us are dependent on a wide range of **ecosystem services**, such as water purification, generation and maintenance of soils, pollination of crops, climate regulation, and flood control (Ehrlich and Wilson 1991). These natural processes that sustain human life depend on the functional integrity of natural communities and ecosystems. For our emotional health, most of us require time spent surrounded by nature's beauty and complexity. Spiritually, we go to natural ecosystems for solace, wonder, and insight.

But beyond our physical dependence on biodiversity, do we have some moral obligation to the other species that inhabit Earth? For many people, biodiversity has inherent value and warrants protection simply for that reason. For others, religious or spiritual beliefs lead to a sense of stewardship, or to the view that other species have a right to exist just as we do. Many others, however, see natural resources as simply there for the taking, as commodities awaiting human extraction.

The field of conservation biology arose in response to global biodiversity losses

Scientists have long been aware that human activity negatively affects the abundances and distributions of organisms. In the nineteenth century, Alfred Russel Wallace, the "father of biogeography" whose work we described in Chapter 17, foresaw current biodiversity crisis, warning in 1869 that humanity was at risk of obscuring the record of past evolution by bringing about extinctions. In the United States, there was a rising public outcry over the rapid decline of bison in the West, the stunning harvest to extinction of the passenger pigeon (**Figure 22.3**), the flagrant use of bird feathers in ladies' hats, and other assaults on animal populations.

Throughout the early history of ecology in the United States, ecologists were divided over how strongly they could advocate for the preservation of nature while still maintaining scientific objectivity (Kinchy 2006). Before 1945, the Ecological Society of America frequently lobbied Congress for the establishment of national parks or for better management of existing parks. In 1948, however, the Ecological Society of America decided to separate "pure" science from advocacy, and the Ecologists' Union branched off as an independent entity focused on the preservation of nature. In 1950, this offshoot organization changed its name to The Nature Conservancy, rising in prominence as a nonprofit organization that integrates science with advocacy and on-the-ground conservation work (Burgess 1977).

Conservation biology emerged as a scientific discipline distinct from ecology only in the early 1980s, as ecologists saw the need to apply their science to the preservation of species and ecosystems. The Society for Conservation Biology, founded in 1985, arose in response to the biodiversity crisis. The emergence of a growing number of professional journals dedicated to conservation biology during the 1980s and 1990s, and a growing number of academic programs for the training of graduate students and professionals, indicate the growing acceptance of and need for this specialized discipline.

FIGURE 22.3 The Passenger Pigeon: From Great Abundance to Extinction The passenger pigeon (*Ectopistes migratorius*; museum specimen shown here), once one of the most abundant birds in North America, was subject to massive hunts in the nineteenth century. The last passenger pigeon died in the Cincinnati Zoo in 1914. The ecological effects of its extinction on the eastern deciduous forest, coincident with the loss of the American chestnut (see Chapter 13), are difficult to estimate, but are presumed to be considerable.

Conservation biology is a value-based discipline

The methods of science call for objectivity—an assurance that the collection and interpretation of data are unbiased by preconceived ideas. Yet science is not free of human values, and it inevitably takes place within a larger social context. Conservation biologists have had to come to terms with the implicit and explicit values that are part of their work. From the founding of the Society for Conservation Biology, the designation of the discipline as "mission-driven" (Soulé and Wilcox 1980; Meine et al. 2006) and "crisis-oriented" (Soulé 1985) explicitly revealed the values behind the science.

Many ecologists have chosen to speak up, and even to refocus their research programs, as they have come to understand the irreversible biological consequences of the changes taking place on the planet. In 1986, Dan Janzen, a tropical biologist who had largely committed himself until then to studying tropical plant–insect interactions, wrote that "if biologists want a tropics in which to biologize, they are going to have to buy it with care, energy, effort, strategy, tactics, time, and cash." Edward O. Wilson, who had primarily distinguished himself in ant biology, island biogeography, and sociobiology, began to write in the mid-1980s about biodiversity loss and has rallied others to the conservation cause ever since. Such motivation does not necessarily detract from the solidity or objectivity of the scientific studies done by conservation biologists, as they understand that addressing the problems posed by global change will require decisions based on sound and credible analysis.

In the next section, we'll meet one ecologist who put these values into practice. Then we'll look at what ecologists know about the extent and causes of the current declines in biodiversity.

Declining Biodiversity

CONCEPT 22.2
Biodiversity is declining globally, and Earth's biota is becoming increasingly homogenized.

One of the greatest Neotropical botanists of our era, Alwyn Gentry devoted his life to identifying, classifying, and mapping the immense diversity of plants found in Central and South America. Gentry played a vital role in cataloging the diversity and distribution of Neotropical trees. He was a tireless collector who had a host of tricks for identifying mammoth tropical forest trees from the ground. He also became an eyewitness to plant species extinctions as the region underwent rapid deforestation. It was not uncommon for him to identify a new endemic plant species during an expedition into Ecuador or Peru, only to return to the same spot a few years later to find the forest cleared (Dodson and Gentry

1991) (**Figure 22.4**). Gentry worked with a growing sense of urgency to identify rare species in order to protect them from this fate. His untimely death in a plane crash in the Ecuadorian forest in 1993, while doing an aerial survey of land proposed for conservation, cut this work short, and was an enormous loss to tropical botany.

Gentry was just one of many ecologists who have been finding and describing species on the one hand and watching their destruction on the other. Extinctions of barely known tropical plant species (and most likely of other species that we have yet to discover) continue throughout the tropics at a staggering pace despite our decades-long recognition of the problem. Through greater efforts to explore Earth's ecosystems and a deeper understanding of systematics, ecologists are gaining knowledge of the world's biota and tabulating new species at an accelerating rate, but threats to those species are keeping pace with such gains in our knowledge about them.

The rate at which Earth is losing species is accelerating

How do ecologists determine how rapidly species are being lost? Because of the uncertainly over how many species exist, and because we know the biology of only a fraction of them, we cannot know precisely the number of extinctions taking place. Yet biologists have found ways to estimate rates of extinction using indirect measures (May et al. 1995).

For example, estimates of extinction rates from the fossil record can be used to establish a "background" extinction rate with which current rates can be compared. For the best-known taxonomic groups, the mammals and birds, paleontologists have estimated that the background extinction rate is on the order of one extinction every 200 years, which is equivalent to an average species life span of 1 million to 10

(A) Before European colonization

(B) 1958

(C) 1988

FIGURE 22.4 Loss of Forest Cover in Western Ecuador Between 1958 and 1988, a growing human population and policies that served to stimulate rapid economic development led to rapid deforestation in western Ecuador. Green indicates forest cover. The extensive loss of forest habitat in this region is estimated to have resulted in the loss of more than 1,000 endemic species (Brooks et al. 2002).

million years. By contrast, we have seen about one extinction per year among the mammals and birds over the past century, or an average species life span of only 10,000 years. Thus, overall, the rate of extinction in the twentieth century was 100 to 1,000 times higher than the background rate estimated from the fossil record (May 2006).

One method for estimating current extinction rates uses the species–area relationship (see Chapter 17)—or, more specifically, the relationship between the number of endemic species and area—to estimate the number of species lost based on the extent of habitat loss (Kinzig and Harte 2000). Biologists have also used changes over time in the threat statuses of species (for instance, a shift from endangered to critically endangered) to forecast rates of extinction (Smith et al. 1993). A third approach is based on the rates of population decline or range contraction of common species (Balmford et al. 2003).

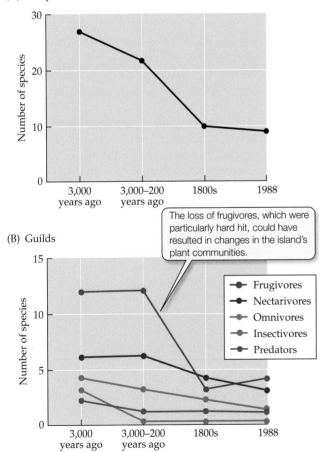

(A) All species

(B) Guilds

The loss of frugivores, which were particularly hard hit, could have resulted in changes in the island's plant communities.

• Frugivores
• Nectarivores
• Omnivores
• Insectivores
• Predators

FIGURE 22.5 Anthropogenic Extinctions Have Been Occurring for Millennia Trends over time in (A) total number of bird species and (B) the number of species by feeding guild found on the Pacific island nation of Eua (Tonga). Prehistoric extinctions (3000–200 years ago) occurred on many Pacific islands as a result of hunting and the introduction of rats, dogs, and pigs. (After Steadman 1995.)

All of these methods are fraught with uncertainty, yet they are the best ways we have devised to document losses of biodiversity—an inherently challenging task. It can be difficult to ascertain when a species is definitely, irrevocably extinct. Many species are known from a single specimen or location, and the logistics of relocating them may be insurmountable. Even an exhaustive hunt for a very rare species can fail to detect some remnant populations. Declaring a species extinct, however, has been known to stimulate biologists' search efforts. Since the publication of a flora of Hawaiian plants in 1990, for example, 35 species listed there as extinct have been relocated, though only a few individuals were found. The joy of their rediscovery is compromised by the realization that these extremely small populations cannot serve the same ecological functions as more substantial populations, and that 8% of Hawaii's native flora of 1,342 species is now considered extinct (Wagner et al. 1999).

While humanity's growing ecological footprint has accelerated the rates of biodiversity loss over the last century, humans have had substantial effects on Earth's biota for millennia. Steadman (1995) describes how bones found on Pacific islands reveal the prehistoric extinction of up to 8,000 species of birds (of which perhaps 2,000 species were endemic flightless rails) following colonization by Polynesians. Most of these species were island endemics, and the extinctions encompassed entire ecological guilds, which are now absent from these islands (**Figure 22.5**). Ecologists can only speculate about the roles the lost frugivores and nectarivores played in maintaining endemic tree populations. Steadman's grim chronicle reminds us that extinctions do not just eliminate individual species, but can also bring about cascading shifts in ecological communities.

Extinction is the end point of incremental biological decline

An ecological observation by Andrewartha and Birch (1954) states that "there is no fundamental distinction to be made between the extinction of a local population and the extinction of a species other than this: that the species becomes extinct with the extinction of the last local population." Sometimes species gradually erode away, population by population, and sometimes they vanish in a spectacular collapse, as in the case of the passenger pigeon.

Conservation biologists have viewed the process of biological decline and extinction in numerous ways. Initial research focused on the problems of small populations. As we saw in Chapter 10, small populations are particularly vulnerable to genetic, demographic, and environmental events that reduce their chances of persistence. As a result, if a population drops below a certain size, it may become vulnerable to processes that act to reduce it even further (a pattern referred to as an **extinction vortex**). Caughley (1994) contrasted this focus on the process of incremental decline with what he believed was a more pragmatic approach: determining the causes of population declines in particular species, with the aim of identifying actions that would counteract these declines before the extinction vortex was invoked.

Ecologists can also take a spatial approach by tracking changes in species' ranges. Ceballos and Ehrlich (2002) examined patterns of range contraction in 173 declining mammal species worldwide. They found that, collectively, these species had lost half of their range area. In Africa, for example, the cheetah occupies only 56% of the land it once did. Channell and Lomolino (2000) examined the patterns of range contraction in 309 declining species. They found that a decline tends to move through the species' historic range like a wave, from one end to the other. This pattern contrasts with a retreat from all edges of the range into its center, which would probably occur if small population effects were prevailing. The ranges of most of the species they examined were shrinking as a result of a particular disturbance, and these disturbances—whether due to invasive species or anthropogenic land use changes—tended to enter the species' range at one edge and move through, eliminating it population by population.

The decline of a species does not take place in a vacuum. When populations are lost from an ecological community, there are consequences not only for the declining species, but also for that species' predators, prey, or mutualistic partners. The resulting changes at the community level may bring about secondary extinctions and ultimately affect ecosystem function (Christianou and Ebenman 2005). Using models, ecologists have found that while food webs are generally quite resilient to species removal, the deletion of certain species can trigger a cascade of secondary extinctions. Generally, the stronger a species' interactions in the food web, the greater the effect of its removal (Solé and Montoya 2001). Models of plant–pollinator interaction webs in which pollinators were selectively removed found that the effect on plant species depended on whether the removed pollinators were specialists or generalists (Memmott et al. 2004) (**Figure 22.6**). Both the food web and plant–pollinator models demonstrate the importance of considering the broader ecological consequences of incremental species loss.

Earth's biota is becoming increasingly homogenized

Organisms move about. They always have, and they always will. Over the last century, however, people have moved over Earth's surface at an unprecedented rate, carrying organisms with them and greatly enhancing rates of introductions of new species to all parts of the globe (**Figure 22.7**). These range expansions coincide with the range contractions of many native species as they are pushed back by habitat loss and declining numbers. Typically, the greatest "losers" among the native species tend to be specialists—those with morphological, physiological, or behavioral adaptations that result from evolution in a particular place—while the "winners" tend to be generalists with less stringent habitat requirements. The spread of introduced species and native generalists, coupled with declining abundances and distributions of native specialists, are part of a growing **taxonomic homogenization** of Earth's biota (Olden et al. 2004).

Island biotas are particularly vulnerable to both extinction and invasion, and often the extinction of island endemics is coupled with, if not driven by, the introduction of more cosmopolitan species. In a recent survey of American Samoa, Robert Cowie (2001) found just 19 of the 42 species of land snails historically known from the island group, plus 5 species not previously found there, but which he presumed were native. He also found that there were 12 non-native species present on the islands. These non-natives occurred in high abundances, representing about 40% of the individuals collected (there was also one abundant native species). Cowie concluded that most native species were declining in abundance, while many non-natives were increasing. Furthermore, the predators contributing to the declines of native land snail species were also non-natives, such as the predatory snail *Euglandina rosea* and the house mouse (*Mus musculus*). Cowie has found this trend toward homogenization of land snail faunas to be widespread among Pacific islands.

Legend:
— Generalist pollinators removed first
— Pollinators removed at random
— Specialist pollinators removed first

1 If generalist pollinators (those that visited the most plant species) were removed first, there was a more rapid reduction of pollinator visits to plants …

2 … than if the removal started with the most specialized pollinators.

FIGURE 22.6 Effects of Pollinator Losses on Plant Species Depend on Pollinator Specialization In a model of a plant–pollinator interaction web from the Rocky Mountains (Pike's Peak, Colorado, U.S.), pollinator species were removed one by one and the effects of their losses on plant visitation rates were recorded. (After Memmott et al. 2004.)

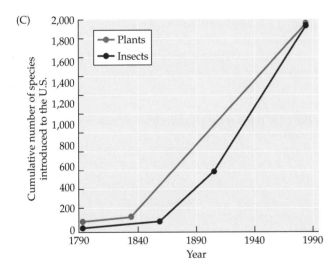

FIGURE 22.7 Species Introductions Have Become a Growing Problem   The number of non-native species present in the United States has increased about fivefold over the past century for (A) plant pathogens and terrestrial vertebrates, (B) mollusks and fishes, and (C) plants and insects. A similar pattern is seen in many other countries. (After U.S. Congress, OTA 1993.)

west as fish bait about 50 years ago. Riley and colleagues found that all the ponds in their study area contained hybrid animals, and that the rare *A. californiense* was under threat of extinction through genetic swamping by *A. tigrinum* (Riley et al. 2003).

A third level of homogenization is suspected to be occurring at the level of ecosystem function, given that with lower biodiversity and a greater preponderance of generalists will

Homogenization has also been observed among the freshwater fishes of the United States, largely as a result of widespread introductions of game fishes. Rahel (2000) quantified the homogenization of U.S. fish faunas by examining the change in the number of species shared between all possible pairs of the lower 48 states. He found that, on average, pairs of states shared 15 more species than they did at the time of European colonization (Rahel 2000) (**Figure 22.8**).

In addition to taxonomic homogenization at the species level, we are seeing homogenization within taxa at the genetic level and within ecosystems at the level of functional groups (Olden et al. 2004). Interspecific genetic homogenization is occurring through hybridization between native and nonnative species. An example is the California tiger salamander (*Ambystoma californiense*), a threatened California endemic that has hybridized with another species of tiger salamander (*Ambystoma tigrinum*) introduced from the Mid-

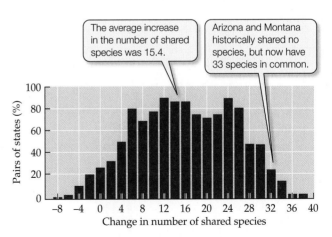

FIGURE 22.8 U.S. Fish Faunas Are Undergoing Taxonomic Homogenization   The numbers of fish species shared by pairs of the lower 48 U.S. states have increased by an average of 15 since European settlement. (After Rahel 2000.)

come a lower number of functional groups in natural communities. While this possibility has been little explored empirically, theoretical considerations lead to serious questions about how it might affect the ecosystem services on which human society depends (Olden et al. 2004).

We have seen that biodiversity is being lost at the levels of species diversity, genetic diversity, and ecosystem function as a result of humanity's increasing ecological footprint on the planet. Let's look in more detail at the reasons for these losses, and then consider what role ecologists can play in counteracting them.

Threats to Biodiversity

CONCEPT 22.3
The primary threats to biodiversity are habitat loss and degradation, invasive species, and overexploitation.

Understanding the threats that are bringing about biodiversity losses is the first step toward reversing them. In any particular case, multiple factors are likely to contribute to a species' decline and eventual extinction. For example, the last Pyrenean ibex (*Capra pyrenaica pyrenaica*) was killed in 2000 by a falling tree. This mountain goat, which was endemic to the Pyrenees in Spain and France, was abundant in the fourteenth century, but its numbers declined gradually due to hunting, climate change, disease, and competition with domesticated livestock as well as with non-native ungulate species such as chamois (Perez et al. 2002). For the past century, no more than 40 animals had been counted, and the

problems of small populations seem to have contributed to the ibex's ultimate extinction. This scenario, in which multiple types of threats contribute to the decline and extinction of a taxon, is common.

Habitat loss and degradation are the most important threats to biodiversity

As you fly in an airplane over Earth's surface, look down and ask yourself, "What species lived here before these farms, woodlots, and cities were here?" "Where do the species native to this place live, and how do they or their propagules move about?" From 30,000 feet above the landscape, you will find yourself face to face with the source of the biodiversity crisis: the ecological footprint of humanity on Earth is large and rapidly increasing. Earth has been modified to some extent over 83% of its land surface (Sanderson et al. 2002). One species, *Homo sapiens*, is now appropriating between 10% and 55% of Earth's primary production (Rojstaczer et al. 2001), and has appropriated an astonishing 98% of the area where wheat, corn, or rice can be grown.

This human footprint is by far the most important factor contributing to global declines in biodiversity (Sax and Gaines

FIGURE 22.9 Habitat Loss Results from a Growing Human Footprint A growing percentage of Earth's land surface is being substantially altered by human activities, resulting in losses of habitat for many species. The human footprint shown here is a quantitative measure of the overall human impact on the environment based on geographic data describing human population size, density, land development, human infrastructure, and resource use. (After Sanderson et al. 2002.)

| 0–1 |
| 1–10 |
| 10–20 |
| 20–30 |
| 30–40 |
| 40–60 |
| 60–80 |
| 80–100 |
| No data |

2003). There are areas of extreme human influence, such as urban areas and agricultural fields, and areas of little influence, such as deserts and boreal forests, but overall, 60% of the terrestrial surface lies in areas of moderate human influence (**Figure 22.9**) (Sanderson et al. 2002). It is little wonder, then, that addressing the degradation, fragmentation, and loss of habitat is central to conservation work. **Habitat degradation** refers to changes that reduce the quality of the habitat for many, but not all, species. **Habitat fragmentation** refers to the breaking up of once continuous habitat into a complex matrix of habitat patches amid a human-dominated landscape. Chapter 23 will address habitat fragmentation and its effects in detail. **Habitat loss** refers to the outright conversion of an ecosystem to another use.

The Atlantic Forest of Brazil is an example of a habitat that has suffered large losses. This moist tropical forest has many endemic species because it has been isolated from the Amazon rainforest for millions of years. Of South America's 904 mammal species, 73 are endemic to this forest, and 25 of these are threatened with extinction. Its location also coincides with 70% of Brazil's human population. As a result, more than 92% of this habitat has been cleared to make room for agriculture and urban development (**Figure 22.10**), and what remains has been highly fragmented, pushing many species to endangerment. There is currently a push for conservation investment in this region to preserve what remains of the Atlantic Forest habitat and the species it contains (Fonseca 1985; Ranta et al. 1998).

How serious has this loss of habitat been to the loss of biodiversity? Brooks et al. (1999) asked why there have been no reports of extinctions among birds of this region. They offered three possible suggestions, which might apply to patterns of biological decline in other ecosystems as well. First, the birds might be adapting to inhabiting forest fragments. Second, the vulnerable species might have gone extinct before they were known to biologists. Their third suggestion, which they see as the most plausible, is that the time lag between deforestation and extinction of species has not yet played out. While there may have been no extinctions yet, populations have been reduced to such an extent that the birds are no longer capable of maintaining their populations. Unless drastic measures are taken, such species are doomed for extinction.

Habitat degradation is even more widespread than habitat loss, and its effects have only begun to be clarified by ecological studies. It has diverse causes and takes many forms. For example, in the Sinai Peninsula of Egypt, Omar Attum and his colleagues compared sand dune habitat that had been degraded through grazing, vegetation harvesting, and small-scale agriculture with undisturbed habitat. The degradation had lowered the percentage of plant cover and the height of the vegetation. Lizard communities in these habitats had declined as a result of the degradation: there were fewer individual lizards as well as a lower diversity of lizard species in the degraded habitats (Attum et al. 2006).

FIGURE 22.10 The Atlantic Forest of Brazil Has Been Significantly Reduced in Area   The Atlantic Forest has been reduced to less than 8% of its original extent since European colonization. The smallest forest fragments are not visible on this map. (After Morellato and Haddad 2000.)

Let's look at one particular way in which habitat is degraded: through the presence of invasive species.

Invasive species can displace native species and alter ecosystem properties

The second most important factor leading to declines in biodiversity is the change in biotic environments that results from **invasive species**, which are non-native, introduced species that sustain growing populations and have large effects on communities. Worldwide, 20% of endangered vertebrates, particularly those on islands, are imperiled as a result of invasive species (MacDonald et al. 1989). Coastal marine and freshwater environments have been particularly affected.

Invasive species are of particular concern where they compete with, prey on, or change the physical environment of endangered native species. The effect of the Eurasian zebra mussel (*Dreissena polymorpha*) on the freshwater mussel species of North America is a prime example. North America is the center of diversity for freshwater mussels (bivalves of the order Unionoida), with 297 species, a third of those in the world. Prior to the invasion of the zebra mussel in the late 1980s (see Chapter 18), North American freshwater mussels were already of conservation concern. Most of these species are globally imperiled, many are endemic and thus naturally rare, and all are threatened by compromised water quality and river channelization (Ricciardi et al. 1998). Competition with zebra mussels has brought about steep declines in populations of native freshwater mussels (60%–90%), including some regional extinctions (Strayer and Malcom 2007). Whether the invasion will bring some species to complete extinction remains to be seen.

In many ecosystems, habitat degradation following fragmentation has increased vulnerability to invasion by non-native species, which in turn may lead to ecosystem consequences that further degrade the system. In Hawaii, the area of tropical dry forest has been reduced by 90% since human settlement, and this ecosystem harbors more than 25% of Hawaii's threatened plant species. So the addition of an invasive species of fountain grass (*Pennisetum setaceum*) has made a bad situation worse. In addition to outcompeting and displacing local plants, it has increased fire frequency in this ecosystem. The presence of grazing animals, both feral and domesticated, facilitates invasion by fountain grass. Removal of the grass and exclusion of grazers has permitted the recovery of a number of rare plants (Cabin et al. 2000), but such measures are not feasible on a landscape-wide scale.

Invasive species are of particular concern where they compete with, prey on, or change the physical environment of endangered native species or change their interactions with other species. We have seen a number of examples of such introductions in previous chapters of this book. Another is the introduction of the Nile perch (*Lates niloticus*) into Lake Victoria in Africa. This predator has had a dramatic negative effect on the diversity and abundance of the native cichlid fishes, a family that had been a textbook case of adaptive radiation (see Chapter 6). Historically, more than 500 species of cichlids had been recorded, of which many were endemic to the lake. The Nile perch, a preferred game fish (**Figure 22.11**), was introduced into the lake in the early 1960s, and was present at first at relatively low abundances. After about 15 years, however, its populations began to increase substantially, coincident with a decrease in the abundances and diversity of cichlids. It is uncertain how many cichlid species have gone extinct, but the number is thought to be as high as 200. Before the introduction, the cichlids made up 80% of the biomass in the lake; the Nile perch now accounts for 80% of the lake's biomass. As in most cases, more than one factor is driving the cichlids' decline; pollution and overfishing augment the negative effect of predation by the Nile perch (Ogutu-Ohwayo 1993; Seehausen et al. 1997).

Control or eradication of invasive species is difficult, labor-intensive, and expensive, but at times may be warranted in the interest of protecting economically or culturally valuable native species or natural resources. The best strategy for combating this threat is to prevent it altogether through careful

FIGURE 22.11 Invasive Species Can Reduce Native Fish Populations The Nile perch was introduced into Lake Victoria in the early 1950s. The presence of this large predator has resulted in dramatic declines of a remarkably diverse family of native cichlid fishes.

screening of biological materials at international borders. But once potentially invasive species are present, control measures are best implemented immediately; constant vigilance and quick action are key to minimizing their effects (Simberloff 2003).

Overexploitation of species has large effects on ecological communities

While many North Americans get their food from a supermarket, for many of the world's people, food comes, at least in part, directly from a natural ecosystem. The problem is that as the human population increases and natural habitat shrinks, the harvesting of many species from the wild has become unsustainable. Globally, overexploitation is contributing to the imperilment of many species, including many mammals, birds, and fishes, but also reptiles, and many plants. Overexploitation has been the cause of the probable extinction of at least one primate subspecies, Miss Waldron's Red Colobus monkey (*Procolobus badius waldroni*), endemic to Ghana and Côte d'Ivoire that was last seen in the 1970s (Oates et al. 2000).

The effects of overhunting on forests, particularly tropical forests, have been substantial, resulting in what Kent Redford (1992) has called an "empty forest." He points to forests that look healthy in satellite images, but which are in fact devoid of their large vertebrate faunas. The ecological consequences of the loss of frugivores and top predators from these biological communities have been examined for only a few

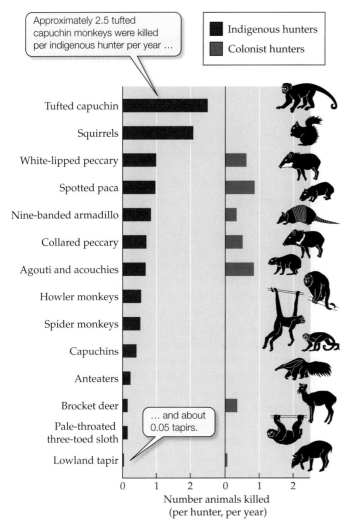

FIGURE 22.12 Unsustainable Hunting Is Defaunating Tropical Forests Wildlife hunting is an important source of protein for people in many parts of the world. The numbers of different mammal species killed per hunter by indigenous people and colonists in the Amazon rainforest are shown here. (After Redford 1992.)

and be released back to the sea; the rest comprises bycatch. The bycatch of prominent species of conservation concern, such as marine mammals, seabirds, and marine turtles, has received attention from fisheries managers, and in some cases, losses have been reduced through changes in gear design (see Chapter 9). But concern has been raised about the ecological effects of this unnecessary mortality on marine food webs (Lewison et al. 2004). In addition, repeated trawling on the coastal sea bottom has significantly affected benthic species such as corals and sponges and has thereby degraded habitat for many other species. Although few studies of recolonization in these benthic habitats have been conducted, early results indicate that habitat recovery following trawling is very slow (National Research Council 2002).

Selective overharvesting of other resources from natural ecosystems, such as wood, fibers, medicines, and oils, has brought many species to a threatened status. All three species of mahogany (*Swietenia* spp.) have become imperiled from overharvesting as well as from habitat loss. Many of the wild palm species used for making rattan furniture are of conservation concern. The global growth in interest in herbal medicine has threatened wild populations of many medicinal plants. The premier wood for making violin bows comes from a tree (*Caesalpinia echinata*) that is globally endangered.

Whenever a species has market value, it is likely to be overharvested. And, in an unfortunate confluence between human behavior (i.e., greed) and declining animal and plant populations, we see that for some desirable species, an "anthropogenic Allee effect" is invoked, as the increased economic value of a rare specimen drives more aggressive search and collection missions (Courchamp et al. 2006). The best approach to protecting overexploited species is to determine the levels of harvest that will be biologically sustainable and to establish regulatory mechanisms to permit only those levels to be taken. Some species may be maintained by developing systems to propagate them in captivity, though "cheaters" who harvest from the wild will always be rewarded.

Pollution, climate change, and disease erode the viability of populations

More insidious anthropogenic factors, such as air and water pollution, climate change, and diseases, are also chipping away at populations of many species. We are also seeing the emergence of new diseases and the crossing over of diseases from domesticated animals into wildlife. The effects of all these factors exacerbate declines in species already reduced in numbers by habitat loss or overexploitation.

Pollutants are omnipresent in air and water. Pollutants become a contributor to habitat degradation and biodiversity loss where they are present at levels that cause physiological stress. We will see in Chapter 24 how some of these pollutants degrade habitats, reduce populations, and threaten the persistence of species.

species and in a few places, but as we have seen, they are likely to be important. Increased accessibility of forests as roads are built through them facilitates this overharvesting of wildlife, as does the widespread availability of guns. The enormous quantity of "bushmeat" being taken from tropical forests is sobering. Redford has calculated that 13 million mammals are killed each year in the Amazon rainforests of Brazil by rural hunters (**Figure 22.12**), and it is estimated that in western and central Africa, 1 million tons of forest animals are taken annually for food (Wilkie and Carpenter 1999).

In the oceans, what characterizes overfishing is widely debated, but rapid and steep declines, particularly in the abundances of top-level marine predators, are clearly taking place (Myers and Worm 2003). For every ton of fish caught by commercial trawlers, 1 to 4 tons of other marine life may be brought aboard. Some organisms may survive the experience

(A)

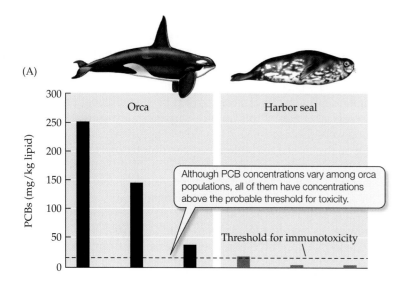

Although PCB concentrations vary among orca populations, all of them have concentrations above the probable threshold for toxicity.

Threshold for immunotoxicity

(B)

Concentrations of flame-retardant PBDEs in the orcas were so high that they were referred to as "fireproof killer whales."

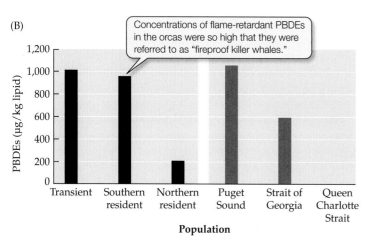

Population

FIGURE 22.13 Persistent Synthetic Chemicals that Disrupt the Endocrine System Are a Growing Threat to Marine Mammals In British Columbia, the concentrations of PCBs (A) and PBDEs (B) found in orcas (*Orcinus orca*) and harbor seals (*Phoca velutina*) are very high. (After Ross 2006.)

Such problems for species already at low numbers do not improve the outlook for their future.

Biologists have only recently detected the "fingerprints" of global warming in the shifting of species' distributions to higher latitudes or elevations. Specific cases of changed conservation status due to climate change, however, have been few in number (Root et al. 2003). The extinction of the golden toad (*Bufo periglenes*), a former inhabitant of the cloud forest of Costa Rica, has been attributed in part to shifts in fog regimes due to climate change (Pounds et al. 1999). It is anticipated that continued warming will result in changed ecological configurations and additional physiological stresses, which in turn will lead to local and global extinctions. Of utmost concern to conservation biologists is the possibility that the pace of warming will exceed the capacity of species to migrate to new ranges or adapt to changing conditions, and that the protected areas we are establishing today will prove less effective over time as their physical environments become less suitable for the species living there. We will explore climate change in greater depth in the next two chapters.

Disease has also contributed to the decline of many endangered species. The final decline to extinction of the thylacine, or Tasmanian wolf (*Thylacinus cynocephalus*), in the 1930s was hastened by an undetermined disease (McCallum and Dobson 1995), and now the Tasmanian devil (*Sarcophilus harrisii*) appears to be similarly threatened due to the spread of a facial tumor disease (Hawkins et al. 2006). In the North American prairie, the threatened status of the black-footed ferret (*Mustela nigripes*) has been exacerbated by canine distemper (Woodroffe 1999). Zoologists are also seeing newly emerging diseases infecting wild species; canine parvovirus, for example, emerged as a mutation of feline panleukopenia virus and subsequently spread to canid species worldwide (Holmes 1996). Increased contact between domesticated and wild animals has contributed to the crossing over of diseases such as rabies and rinderpest into wild populations (Daszak et al. 2000). Increased mortality from disease can negatively affect both wild and captive populations of threatened organisms.

The world's biomes are all affected to some extent by the threats we have just described as the human population climbs toward the 7 billion mark, but the importance of these threats varies among biomes (**Figure 22.14**). Habitat loss is greater in the tropics than in the polar zones, for example, but climate change is having more of an effect in the polar zones than in the tropics. What can conservation biologists offer as solutions to these threats from so many fronts?

An emerging pollution threat is the growing concentration of persistent endocrine-disrupting contaminants (EDCs), particularly in the marine environment. As we saw in Chapter 20, persistent organic pollutants such as DDT, PCBs, flame retardants, and organophosphates from agricultural pesticides end up in marine food webs, where they are bioaccumulated and biomagnified, particularly in top predators. The number of chemicals found, the number of animals affected, and the concentrations found in marine mammals have risen markedly in the last 40 years (Tanabe 2002). Peter Ross refers to the orcas of British Columbia as "fireproof killer whales" due to the extremely high levels of flame-retardant chemicals (polybrominated diphenyl ethers, or PBDEs) found in their bodies (**Figure 22.13**). These EDCs have been observed to interfere with reproduction, neurological development and immune function in mammals (Ross 2006). EDCs have also interfered with reproduction—basically by turning males into females—in the endangered pallid sturgeon (*Scaphirhynchus albus*) in the Mississippi River downriver from Saint Louis (Harshbarger et al. 2000).

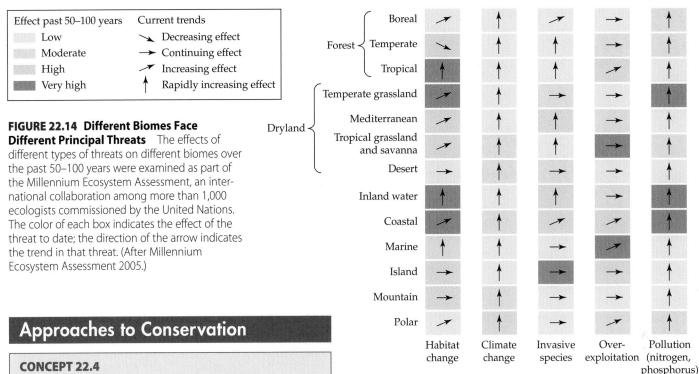

| | Habitat change | Climate change | Invasive species | Over-exploitation | Pollution (nitrogen, phosphorus) |

FIGURE 22.14 Different Biomes Face Different Principal Threats The effects of different types of threats on different biomes over the past 50–100 years were examined as part of the Millennium Ecosystem Assessment, an international collaboration among more than 1,000 ecologists commissioned by the United Nations. The color of each box indicates the effect of the threat to date; the direction of the arrow indicates the trend in that threat. (After Millennium Ecosystem Assessment 2005.)

Approaches to Conservation

CONCEPT 22.4
Conservation biologists use many tools and work at multiple scales to manage declining populations.

Species or habitat? Where should we put our focus? Conservation biologists have debated this question back and forth, and have generally concluded that protecting habitat is of primary importance, but that understanding species is also important. There is no real dichotomy here, as we must understand a threatened species' biology to understand its decline and recovery, but we must also identify and preserve its habitat if we are to conserve it. The U.S. Endangered Species Act functions through the listing of particular species, but for each endangered species, it mandates the identification and protection of critical habitat. Worldwide, many other laws protecting biodiversity take a similar approach.

These contrasting emphases have been labeled **fine-filter** (genes/populations/species) and **coarse-filter** (landscape/ecosystem/habitat) approaches to conservation planning. While we now recognize that the coarse-filter emphasis on maintaining ecosystem processes may be more efficient in the long run because it protects many species at once, the choice of where on the landscape to focus our efforts must be dictated by a certain amount of fine-filter information (Schwartz 1999).

Chapter 23 will describe how the principles of ecology are applied to protecting habitat and how conservation biologists work to manage ecosystems and landscapes. In this section, we will look at the variety of ways in which conservation biologists work to understand biodiversity at the level of genes, populations, and species and how conservation solutions are generated using a fine-filter approach.

Conserving current genetic diversity enables future evolution

Conservation biologists are concerned that as populations become smaller, their future evolutionary potential will be diminished. As we saw in Chapters 6 and 10, small populations are particularly vulnerable to the effects of genetic drift and inbreeding, which can result in the loss of genetic variability and the fixation of deleterious alleles. Insight into the importance of genetics in conservation first emerged when a molecular genetic analysis of the cheetah (*Acinonyx jubatus*) revealed extremely low genetic variability, low sperm counts, and poor reproductive success, both in the wild and in captivity. Unrelated animals were so genetically similar that reciprocal skin grafts between them were not rejected. Biologists attributed this phenomenon to a genetic bottleneck in the Pleistocene (an event that reduced the population to only a few individuals), and the cheetah stood as the poster child of conservation genetics for some years (O'Brien and Johnson 2005).

While the genetic study of threatened species remains an important component of our multipronged conservation efforts, just how central a role genetics plays in their downfall remains undetermined. Derek Spielman and his colleagues analyzed a number of studies to show that threatened species are generally characterized by lower heterozygosity than their more common relatives (Spielman et al. 2004). Whether this reduced genetic variability has contributed to the conservation status of the threatened species remains unclear, though it is generally believed that such a reduction could lead to a greater chance of extinction in the future. A primary genetic concern with small populations is that inbreeding or a loss of genetic variability due to genetic bottlenecks will result in a loss of the species' ability to evolve in response to environmental change. Examination of many species has shown that the loss of genetic variability is less of a problem than inbreeding (Amos and Balmford 2001).

Genetic information and modeling can inform conservation decisions by revealing the genetic diversity present in a species as well as gene flow between populations. It can also improve our taxonomic understanding. Genetic analyses are also being implemented to determine the appropriate targets of management within species (e.g., subspecies, populations) through the identification of **evolutionarily significant units** or *management units* (Waples 1998). In the Pacific Northwest, for example, genetic studies of salmonid fishes (salmon and their relatives; *Oncorhynchus* spp.) have had both pragmatic aims—to distinguish populations warranting protection under the Endangered Species Act—and theoretical aims—to help biologists understand the evolutionary history of the Pacific salmonids.

Molecular genetic techniques can also be used in forensic applications, such as the identification of the source of wildlife products confiscated from the illegal trade (**Figure 22.15**). For example, the use of molecular markers has permitted the identification of illegally harvested whale species in Japanese meat that was labeled as either dolphin or (southern hemisphere) minke whale, which are legal to hunt (Baker et al. 2002). Cycads, too, have been genetically "fingerprinted," allowing tracking of these highly valuable and frequently poached plants (Little and Stevenson 2007).

The availability of molecular genetic tools has enhanced our ability to understand some of the genetic problems faced by small populations, and has also given us tools to help address some of those problems. Let's turn next to the some of the ways we approach conservation at the population level.

Demographic models can guide management decisions

Is the growth rate of the Yellowstone grizzly bear population high enough to allow it to persist? At what life stages are loggerhead sea turtles most vulnerable to predation, and what management decisions would be most expedient to ensure their continued viability? How much old-growth forest habitat must be preserved to ensure the persistence of the northern spotted owl? Such questions arise with nearly any species of conservation concern, and demographic models offer approaches to answering them.

There are hundreds of quantitative demographic models in use, tailored to the specific biological traits of particular species. The quantitative approach most widely used for projecting the potential future status of populations in a species of interest is referred to as **population viability analysis** (**PVA**). This approach allows ecologists to assess extinction risks and evaluate management options for populations of rare or threatened species (Morris and Doak 2002). PVA is a process by which biologists can calculate the likelihood that a population will persist for a certain amount of time under various scenarios. It encompasses a whole suite of models, ranging from relatively simple stage- or age-based demographic models like those described in Chapter 9 to more complex, spatially explicit models that can take actual landscape features and dispersal of individuals from multiple populations into account (Norris 2004). These models serve as an assessment tool, providing conservation biologists with the probabilities that certain outcomes may occur, given certain assumptions about future conditions (e.g., changes in threats or in management efforts). Thus, PVA is a tool with which ecologists can synthesize data collected in the field, assess the risk of extinction of a population or populations, identify particularly vulnerable age or stage classes, determine how many animals to release or how many plants to propagate to ensure the establishment of a new population, or determine what might be a safe number of animals to harvest (Beissinger and Westphal 1998).

You read in Chapter 9 about the use of a demographic model for loggerhead sea turtles, from which ecologists con-

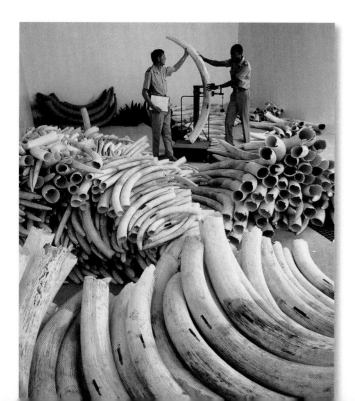

FIGURE 22.15 Molecular Genetic Techniques Can Identify the Origin of Organisms Intercepted in Illegal Trade   As in human forensic cases, DNA evidence can be used in the prosecution of criminals involved in the trade of protected organisms. In one case, DNA fingerprinting was used to show that a large illegal shipment of ivory (similar to that shown here) originated in Zambia.

cluded that protection of mature individuals would be key to slowing declines in populations (see Box 9.1). PVA has been used to make a wide variety of decisions about how best to manage rare species. In Florida, the fire regime that would best serve population growth in a rare plant (*Chamaecrista keyensis*) was determined through PVA simulations of burns at different times of year and at different intervals (Liu et al. 2005). In Australia, the forest cutting practices that would best serve the perseverance of two endangered arboreal marsupial species, the greater glider and Leadbeater's possum, were determined through extensive PVA modeling coupled with long-term monitoring data to verify the accuracy of the information going into the model (Lindenmayer and McCarthy 2006).

While such analyses have played a critical role in management decisions for a number of species, Some conservation biologists, however, caution against excessive reliance on conclusions based solely on the results of PVA, particularly when they regard time to extinction or probability of extinction. They point to the high level of uncertainty in the dynamics of small populations, the paucity of demographic and envi-

ronmental data for many endangered species, and the high probability that a model will leave some critical factor unaccounted for. To be used effectively, such models need to be constantly refined and revisited by different researchers to check their validity against field observations, just as management strategies must be checked and adjusted for effectiveness (Beissinger and Westphal 1998).

Ex situ conservation is a last-resort measure to rescue species on the brink of extinction

When remaining populations of a species fall below a certain number, demographic models become inappropriate, and direct, hands-on action is called for. The only hope for the species' survival at that point may be to take the remaining individuals out of their habitat—ex situ—and allow them to multiply in sheltered conditions under human care with the hope of later returning some individuals to the wild. The rescue of the California condor (*Gymnogyps californianus*) is a leading example of this strategy (**Figure 22.16**). In the late

FIGURE 22.16 Ex Situ Conservation Efforts Can Rescue Species from the Brink of Extinction
Ex situ efforts to save the California Condor (*Gymnogyps californianus*) involve multiple steps, including those shown here. (A) To reduce inbreeding and increase the number of eggs that hatch successfully, a USFWS biologist removes eggs from the wild (to be taken to a zoo) and replaces them with one egg from the San Diego Zoo. (B) In the San Diego Zoo, condor chick "Hoy" is being fed by a condor-feeding puppet. (C) Two condors at the time of their release (spring 2000). The instrument in the right foreground is a weighing scale from which condor mass can be read by telescope when a bird perches on it. (D) This adult, with a wingspan of 9 feet, was bred in captivity and later released.

Pleistocene, this great bird ranged throughout much of North America, and by the nineteenth century, it was still distributed from British Columbia to Baja California. The condor population declined steeply between the 1960s and 1980s, however, due to multiple causes, reaching a low of 22 birds by 1982. The species became extinct in the wild in 1987 when the last birds were captured and brought to an ex situ facility in California for breeding (Ralls and Ballou 2004). The species now numbers more than 200 birds, some released to the wild and some remaining in captivity. Increasing the population to this point has required careful genetic analysis, hand rearing of some chicks, and wide cooperation among zoos, managers of natural areas, hunters, and ranchers. Principal barriers to the condor's recovery include lead poisoning from ammunition found in the carrion condors feed on, negative health effects of ingesting plastic and other trash, West Nile virus, and genetic drift. Given all of these risks and costs, is the recovery of the California condor worth all the effort that has gone into making it happen? Without that effort, the species would now be extinct.

Ex situ conservation programs are taking place in zoos, special breeding facilities, botanical gardens, and aquariums all over the world. Such programs have allowed many species at risk of extinction to build their numbers sufficiently to permit reintroduction into the wild. While ex situ programs play important roles in keeping our most threatened species from extinction, as well as in publicizing the plight of these species, they are expensive, and they can introduce a host of problems, such as disease exposure, genetic adaptation to captivity, and behavioral changes (Snyder et al. 1996). Furthermore, they do not have a very high rate of success in terms of restoration of self-sustaining populations in the wild. Could the funds dedicated to ex situ efforts be better spent on managing species in the wild or on securing land for the establishment of new protected areas—that is, for in situ conservation? Sometimes the answer is no, usually when populations have been reduced to critical levels or when not enough suitable habitat is available. But the question must always be asked.

Legal and policy measures support biological methods of protecting species and habitat

Conservation biologists seek to gain the best scientific information possible to inform decisions that will benefit the welfare of species or ecosystems. The process of making many of these decisions, however, is more of a societal than a scientific one and falls into the realm of public policy and communications. Policy that reflects society's collective decisions takes the form of national and state laws, policies set by natural resource agencies, and the work and policies of nongovernmental organizations. At times, these decisions are guided by global processes in the form of international treaties, agreements, and conventions. At other times, they

are driven by grassroots efforts. The interplay between science and this composite of human decision-making entities is complex, but it is an integral part of any successful conservation effort.

In the United States, the most prominent legislation protecting species, the Endangered Species Act (ESA), has played a vital role in protecting many of the country's most threatened species. It was passed by Congress in 1973 to "provide a means whereby the ecosystems on which endangered and threatened species depend may be conserved, and to provide a program for the conservation of these species." The U.S. Fish and Wildlife Service and the National Marine Fisheries Service are charged with listing federally threatened and endangered species, identifying critical habitat for each species, drafting recovery plans, and carrying out actions necessary to increase abundances to target numbers.

The ESA currently protects more than 1,300 species native to the United States and another 570 from other countries. The ESA extends its influence beyond U.S. borders by regulating trade in endangered species as a result of an international treaty called the Convention on International Trade in Endangered Species of Wild Fauna and Flora (CITES). This treaty, which has been in place for more than 35 years, regulates international trade in listed organisms and their parts. It mandates a virtual prohibition on trade in some species recognized to be endangered, while others, deemed to be less threatened, must be monitored in their home countries for indications that trade should be restricted. Currently, 167 countries have agreed to adhere to CITES regulations, and about 33,000 species receive some protection. But this agreement goes only so far. International trade in wild species is worth many billions of dollars, and much of this trade is illegal. While enforcement of CITES regulation remains a difficult task for governments, the treaty has been a key instrument in protecting species worldwide.

The Convention on Biological Diversity, which most nations have signed (the most notable exception being the United States), is the outcome of the Earth Summit held in Rio de Janeiro in 1992. This agreement acknowledges declining biodiversity as a problem shared by all the world's people and establishes goals for actions to counteract it. For example, the agreement urges nations to document the biodiversity contained within their borders, identifies the rights of nations to benefit from patents derived from their biodiversity, and calls for mechanisms to protect biodiversity from genetically modified organisms. It has provided a fruitful structure for nations developing their own conservation plans and has served as a framework for bringing nations together to address global biodiversity losses.

The protection of biodiversity is also dependent on a wide array of national, state, and local regulations and policies that set restrictions on land development, call for environmental review prior to land disturbances, and manage harvest levels. There is a limit to the percentage of the landscape we will be able to place under some protected status, so much of the world's biodiversity will continue to reside on private lands and working landscapes. Therefore, it is vital that we maintain a legal framework that will serve to protect the most critical components of biodiversity in these places, and that these laws and policies be crafted with the best available science to support them.

Ranking Species for Protection

CONCEPT 22.5
Prioritizing species helps maximize the biodiversity that can be protected with limited resources.

How do we allocate the limited resources that are available for species conservation? Do we protect those species that are most threatened, or do we focus on those that play a substantial ecological role? How can public policymakers and conservation organizations know which land areas are the most critical to protect? If a species is rare, is it necessarily at risk of extinction? Do we determine how threatened an insect or a lichen is in the same way that we do for a bird, a mammal, or a conifer?

The rarest and the most rapidly declining species are priorities for protection

Many species have become rare only recently as a result of the threats we outlined earlier in this chapter. Other species may have always been rare. In either case, having a measure of how threatened a species is permits us to focus our efforts on those species that are most threatened: the rarest and the most rapidly declining. We may be able to postpone attending to species that are naturally low in abundance but not particularly threatened.

What do we mean by rarity, and how do we determine just how rare something is? To clarify the different concepts of rarity, we can use a matrix that sorts out whether a species has a wide or a narrow geographic range, whether it is broad or restricted in its habitat specificity, and whether local populations tend to be small or large (**Table 22.2**). There are some rare species, for example, that exist over a wide geographic area, are relatively broad in their habitat requirements, yet tend to occur in very small populations; other rare species inhabit specific habitats in a narrow geographic range, but may have large populations in those specific locations (Rabinowitz et al. 1986). Conservation of these different types of rare species requires different approaches, at times justifying small reserves to protect well-established populations, at other times calling for management practices that create habitat conditions suitable for a rare but geographically widespread species.

Assessment of the conservation status of species in an objective, scientific way began in 1963 with the red-listing process of the World Conservation Union. A parallel effort was developed in the United States by The Nature Conservancy, which established the Natural Heritage Program (now NatureServe) in the early 1970s in order to assess the conservation status of American species. Both organizations have developed a ranking structure that indicates how threatened a species is and an assessment protocol to determine a species' rank. The assessment takes into account not only numbers of populations or individuals, but also the total geographic area that the species occupies, the rate of its decline, and its risk of extinction. Because of the challenge of creating a system that can be applied equally well to a skipper butterfly, a cycad, or a cheetah, and because the information available on rare species is often incomplete, assessors for both systems can choose among different sets of criteria to decide whether a species is vulnerable, endangered, or critically imperiled, or under some lesser level of

TABLE 22.2

Seven Forms of Rarity

Local population size	Geographic distribution: WIDE		Geographic distribution: NARROW	
	Habitat specificity BROAD	Habitat specificity RESTRICTED	Habitat specificity BROAD	Habitat specificity RESTRICTED
SOMEWHERE LARGE	Common	RARE: Widely distributed in an uncommon specific habitat	RARE: Narrow endemic, but with a broad ecological tolerance and locally abundant	RARE: Endemic, with narrow ecological requirements, but locally abundant
EVERYWHERE SMALL	RARE: Never abundant, but distributed over a wide geographic and habitat range	RARE: Small populations in a specific habitat, but over a wide geographic area	RARE: Endemic, broad ecological tolerance, small populations	RARE: Endemic, narrow ecological tolerance, small populations

Source: Rabinowitz, et al. 1986.

threat. The criteria must be flexible while still permitting consistent conclusions by different evaluators. Both systems come up with a global ranking, but are also applicable on a regional level. The effort to evaluate species is a large and ongoing process, and most taxonomic groups have had only a small proportion of their members evaluated (de Grammont and Cuarón 2006).

These assessments of conservation status play a vital role in conservation planning. They can be are used to locate clusters of threatened species and thus identify areas that are critical to protect (**Figure 22.17**). They are frequently consulted when development projects are planned, and they are also important for keeping the public aware of the degree of threat faced by Earth's biota. They are dynamic databases, changing as scientific information is updated, allowing for both downgrading of conservation status when a species has been found to be more common than previously thought and upgrading as species decline in number.

Protection of surrogate species can provide protection for other species with similar habitat requirements

If we protect the habitat necessary for the red-cockaded woodpecker, will we simultaneously provide protection for the gopher tortoise, Bachman's sparrow, Michaux's sumac, and other rare species that are dependent on the longleaf pine savanna ecosystem? Species may become conservation priorities not only because of their own conservation status, but also because of their capacity to serve as **surro-**

gate species whose conservation will serve to protect many other species with overlapping habitat requirements. We may choose surrogate species as a shortcut when we lack sufficient information on the distributions and abundances of many species of conservation concern in a region. We may be interested in a species that will help us garner public support for a conservation project; examples of such **flagship species** include charismatic animals such as the giant panda (**Figure 22.18**). An **umbrella species**, by contrast, is one that we select with the assumption that protection of its habitat will serve as an "umbrella" to protect many other species with similar habitat requirements. Umbrella species are typically species with large area requirements, such as grizzly bears, or habitat specialists, such as the red-cockaded woodpecker. But they may also include animals that are relatively easy to count, such as butterflies (Fleishman et al. 2000). Some researchers think it better to choose not just one species, but several **focal species**, selected for their different ecological requirements or susceptibility to different threats, with the realization that by thus casting a broader net, we improve our chances of better covering regional biodiversity with protection (Lambeck 1997).

Methods have been devised and criteria established to allow for strategic selection of the one or several surrogate species that will best serve conservation aims (Andelman and Fagan 2000; Favreau et al. 2006). Conservation biologists recognize, however, that surrogate species approaches are not without problems, and that the distribution or habitat requirements of any one species cannot capture all the conservation targets we may have.

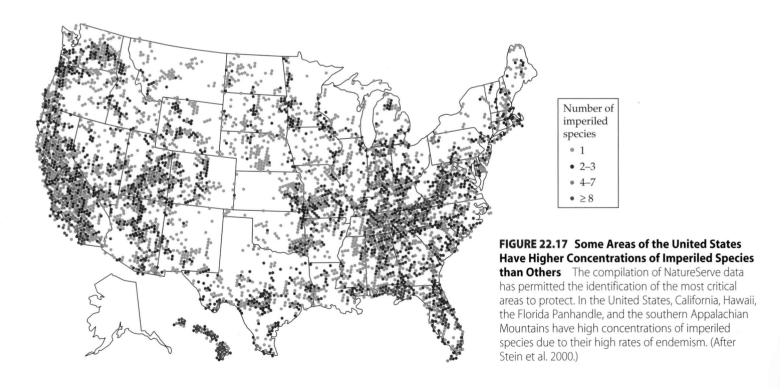

FIGURE 22.17 Some Areas of the United States Have Higher Concentrations of Imperiled Species than Others   The compilation of NatureServe data has permitted the identification of the most critical areas to protect. In the United States, California, Hawaii, the Florida Panhandle, and the southern Appalachian Mountains have high concentrations of imperiled species due to their high rates of endemism. (After Stein et al. 2000.)

Number of imperiled species
- 1
- 2–3
- 4–7
- ≥ 8

FIGURE 22.18 A Flagship Species, the Giant Panda   This giant panda (*Ailuropoda melanoleuca*) is eating bamboo, which constitutes 99% of its diet. Pandas are endangered, primarily due to habitat loss. Native to China, about 2,000 pandas remain in the wild; another 300 pandas live in zoos or breeding centers, mostly in China.

CASE STUDY REVISITED

Can Birds and Bombs Coexist?

As the longleaf pine ecosystem lost its functional integrity and 97% of its area over 200 years, the biological traits of the red-cockaded woodpecker that had worked well in the extensive pine forests of the past turned out to be detrimental in its changing environment. Prime habitat became fragmented, consisting of islands of usable habitat in an unsuitable landscape. The cooperative breeding system of red-cockaded woodpeckers evolved in response to a shortage of suitable breeding and nesting sites (Walters 1991). This problem was a result of the woodpeckers' peculiar habit of excavating nests in living trees—a process that usually takes a year or more to complete, and which made cavity availability a limiting factor for woodpecker populations. The chance of a young dispersing bird finding a vacant cavity is very small. Under these circumstances, it is advantageous for young birds to remain with their parents and help them rear their younger siblings until one of them inherits a breeding site in their natal territory, rather than dispersing into territory that is already occupied to capacity (Koenig et al. 1992).

Jeff Walters and his colleagues tested the hypothesis that a lack of high-quality habitat was limiting population growth in the red-cockaded woodpecker by constructing artificial nest cavities, putting them in clusters, and observing woodpecker behavior. They tried this for a few reasons.

First, each bird in a cooperative breeding group must have its own cavity for roosting; this explains why cavities are found in clusters. Second, cavity clusters are typically abandoned after several years' use, primarily due to cavity entrance enlargement by other species or mortality of cavity trees, so there is a continual demand for cavity clusters (Harding and Walters 2002). Third, trees suitable for cavity excavation must be mature enough (at least 90 years old) to be sufficiently large and to have developed some heart rot fungus, which facilitates excavation, and such trees are in particularly short supply. The cavity clusters constructed by the researchers were rapidly colonized, mostly by helper birds from the vicinity and young dispersing birds (Walters et al. 1992; Copeyon et al. 1991).

From these results came the realization that people could help the red-cockaded woodpecker increase its numbers by going out with drill, wood, wire, and glue and building clusters of nesting cavities within living longleaf pines. This realization has proved a boon to woodpecker recovery. When Hurricane Hugo hit the South Carolina coast in 1989, the population of red-cockaded woodpeckers in Francis Marion National Forest, just north of Charleston, was severely reduced. It had been the second largest extant population of red-cockaded woodpeckers, with 344 breeding groups. When the hurricane came, 87% of their cavity trees and 60% of their forage trees were toppled by wind overnight. Some 63% of the birds did not survive the storm, and another 18% died the following winter (Hooper et al. 2004). Within 2 years of the storm, however, National Forest workers had installed 443 artificial cavities. This strategy averted what seemed likely to be a severe population decline; by 1992 the population had recovered to 332 breeding groups (**Figure 22.19**).

But now that managers have identified cavity construction and maintenance as a critical factor for the species' recovery, they are obliged to continue doing it. This strategy is labor-intensive and expensive, but it is necessary for the red-cockaded woodpecker's continued existence. How long can we sustain this effort? Will we reach a point at which there is enough longleaf pine forest that the woodpeckers will be able to maintain their own numbers without human assistance? What effect will artificial nest cavity construction have on the bird's behavior and evolution? We just don't know the answers to these questions.

In the decades that Walters and others have been researching the red-cockaded woodpecker, they have used almost all of the kinds of tools described in this chapter. Models of population dynamics have facilitated the identification of vulnerable stages in the woodpecker's life cycle. Genetic studies and modeling have focused attention on the threat of inbreeding. Field studies have demonstrated the need for prescribed burning and for protection of mature longleaf pine individuals to maintain the community structure favored by the woodpeckers. Economic and soci-

(A)

(B)

(C)

FIGURE 22.19 Construction and Installation of Artificial Nest Cavities Has Allowed Populations of Red-Cockaded Woodpeckers to Increase (A) An artificial nest cavity built for a red-cockaded woodpecker. (B) Cutting a hole for the artificial nest. (C) Installing the artificial nest.

ological analyses have led to the development of a "safe-harbor" program that makes endangered species management more palatable to private landowners. And managers are reaching into an actual toolbox to build nesting cavities. Much of this work has been dictated by the creation more than 30 years ago of the U.S. Endangered Species Act.

CONNECTIONS IN NATURE

Some Burning Questions

The key to managing the longleaf pine system is fire—either naturally ignited or intentionally set under controlled conditions. Fire affects ecosystems at multiple scales, from the cellular and biochemical to the atmospheric. Variations in fire regimes regulate species abundances, alter the cycling of nutrients and water, and emit carbon dioxide to the atmosphere. Prescribed burning is used as a management tool for conserving species in numerous ecosystems where fire has been a regular disturbance (**Figure 22.20**).

But there can be unintended and undesirable ecological outcomes when fire is used as a management tool where non-native invasive species are present. In some longleaf pine forests in Florida, openings resulting from burning have provided favorable habitat for the establishment of cogongrass (*Imperata cylindrica*), an invasive plant from Asia. The presence of this grass, in turn, causes fires to burn hotter, higher, and more evenly on a horizontal plane. The consequences of these hotter fires are increased mortality of longleaf pine seedlings and native wiregrass, favorable conditions for further infiltration of cogongrass, and a resulting threat to the astounding native plant diversity in the understory of the longleaf pine savanna (Lippincott 2000). Land managers in this situation are faced with a dilemma: to burn or not to burn? The right question is more likely to be when to burn, and how often.

In New Mexico, a burning question arose in the Animas Mountains, where fire maintains the native grassland that serves as habitat for a number of rare and endangered species. However, fire poses a threat to another threatened species that lives there, the New Mexico ridge-nosed rattlesnake (*Crotalus willardi obscurus*). If fires burn too hot, as they can when fuel loads have built up after decades of fire suppression, they can kill this rare snake (Smith et al. 2001). Again, the question seems to be what cycle and intensity of burning will minimize rattlesnake mortality while still maintaining the grassland.

Adding people to the burning landscape further complicates matters. Throughout the southeastern United States, prescribed burns are taking place in a complex landscape where patches of forest are adjacent to peoples' homes and businesses. Convincing the public that these fires are necessary has required considerable outreach and public education. In the North Carolina Sandhills, the days for prescribed burns are chosen not only for safe conditions, but

FIGURE 22.20 Prescribed Burning Is a Vital Management Tool in Some Ecosystems   In the southeastern United States, regular burning is used to maintain the high plant biodiversity characteristic of the understory in the pine savanna ecosystems. Many threatened species, including the red-cockaded woodpecker, rely on regular burning for their persistence. Here, USFWS firefighters monitor a prescribed burn intended to preserve habitat for the endangered Florida panther.

also with regard to wind direction so as to minimize the amount of smoke in population centers.

Here, as elsewhere, recognition of people as an integral component of the landscapes that must harbor all of nature's diversity has been a vital piece of the conservation picture. Establishing protected natural areas as sanctuaries for wildlife is an important part of the solution, but we must also do what we can to ensure that the vast majority of Earth's surface outside of protected areas is able to sustain both people's livelihoods and habitat for all the other species. This is a very difficult challenge that will involve constant education, negotiation, legislation, and many creative approaches.

SUMMARY

CONCEPT 22.1 Conservation biology is an integrative discipline that applies the principles of ecology to the conservation of biodiversity.

- Conservation biology is the scientific study of phenomena that affect the maintenance, loss, and restoration of biodiversity.

- Biodiversity is important to human society because of our reliance on resources and ecosystem services that derive from intact natural communities and ecosystems.

- With the growing awareness of accelerating losses of global biodiversity, ecologists saw a need for a separate discipline that would apply the principles of ecology to the preservation of species and ecosystems.

- Conservation biology is a scientific discipline instilled with the value of biodiversity.

CONCEPT 22.2 Biodiversity is declining globally, and Earth's biota is becoming increasingly homogenized.

- Earth is losing species at an accelerating rate, largely due to humanity's growing footprint on the planet.

- Extinction is the end point of incremental biological decline as species lose individuals and populations and become increasingly vulnerable to the problems of small populations.

- Earth's biota is becoming increasingly homogenized due to a rise in generalist species and a decline in specialist species, losses of genetic diversity within some taxa, and probable losses of ecosystem function.

CONCEPT 22.3 The primary threats to biodiversity are habitat loss and degradation, invasive species, and overexploitation.

- Habitat degradation, fragmentation, and loss are the most important threats to biodiversity.

- Invasive species degrade local habitats by preying on or competing with native species and altering ecosystem properties.

- Overexploitation of selected species has large effects on communities and ecosystems.

- Other factors that erode the viability of populations and contribute to losses of biodiversity include air and water pollution, global climate change, and diseases.

CONCEPT 22.4 Conservation biologists use many tools and work at multiple scales to manage declining populations.

- Genetics has been used as a tool to understand and manage genetic diversity within rare species, to identify appropriate management units, and in forensic analyses of illegally harvested organisms.

- Population viability analysis (PVA) is an approach that uses demographic models to assess extinction risks and evaluate proposed management actions.

- Ex situ conservation, which involves taking organisms from the wild into human care, is a last-resort measure to rescue species on the brink of extinction.

- Laws, policies, and international treaties are vital supplements to biological methods of protecting species and habitat.

CONCEPT 22.5 Prioritizing species helps maximize the biodiversity that can be protected with limited resources.

- Conservation biologists identify those species of the highest priority for protection—the rarest and the most rapidly declining species—by assessing numbers of individuals and populations, rates of decline, and the degree of threat faced.

- Identification of surrogate species can make conservation planning more efficient by providing protection for other species with similar habitat requirements.

REVIEW QUESTIONS

1. Think about what motivates your own interest in ecology, and contemplate your own feelings as you witness destruction of natural places. Is your interest in part motivated by a desire to do something about it? How would you separate out your values as you consider conducting research in conservation? Are your research questions "value-laden"?

2. Describe the three components of biodiversity that are being homogenized due to human actions.

3. What are the principal threats to biodiversity? Describe some examples in which multiple threats have contributed to a species' decline.

4. What is the difference between a species determined to be endangered by the Natural Heritage/NatureServe program and one that is listed as endangered under the Endangered Species Act? What are the consequences for management of each?

5. Identify five imperiled species that live in your region, including a plant, a mammal, a bird, a fish, and an invertebrate. Are any of these species endemic to your region? For each species you have identified, try to find out what whether it was rare prior to human settlement of the region. What threats does this species face today? What is being done to protect this species? Based on the ecological knowledge you have gained, what questions do you think should be researched to aid in the species' recovery? (Much of this information is available at http://www.natureserve.org/.)

HANDS-ON PROBLEM SOLVING

Evolution of an Invasive Species | This Web exercise (www.sinauer.com/ecology) explores hypotheses for why invasive plant species are often better competitors in their introduced ranges than in their native ranges. You will interpret data from a recent paper that was designed to test whether the success of an invasive tree species in its introduced range has resulted from evolutionary change.

SUGGESTED READINGS

Brooks, T. M. and 9 others. 2002. Habitat loss and extinction in the hotspots of biodiversity. *Conservation Biology* 16(4): 909–923.

Caughley, G. 1994. Directions in conservation biology. *Journal of Animal Ecology* 63 (2): 215–244. *In this lively critique of a relatively young discipline, Caughley contrasts the small population approach and the declining population approach to conservation biology and discusses the need for integration of theoretical and applied approaches to conservation.*

Gomez-Pompa, A. and A. Kaus. 1992. Taming the wilderness myth. *BioScience* 42(4): 271. *This paper vividly describes the biases that many bring to considerations of wilderness and argues for making conservation decisions considering people that rely on the land as a vital part of the system.*

Groom, M. J., G. K. Meffe and C. R. Carroll. 2006. *Principles of Conservation Biology.* 3rd ed. Sinauer Associates, Sunderland, MA. *A comprehensive and engaging text covering all aspects of conservation biology.*

Redford, K. H., 1992. The empty forest. *BioScience* 42 (6): 412–422. *Redford describes the necessity not only of saving rainforests, but of ensuring that hunting is regulated within these forests. He details the problem of the bushmeat trade and subsistence hunting.*

Simberloff, D. 1998. Flagships, umbrellas, and keystones: Is single-species management passé in the landscape era? *Biological Conservation* 83 (3): 247–257. *Simberloff examines the use of surrogate species and contrasts the utility of focusing on species with that of doing landscape-level planning for conservation.*

Stein, B. A., L. S. Kutner and J. S. Adams, eds. 2000. *Precious Heritage: The Status of Biodiversity in the United States*. Oxford University Press, Oxford, New York. *A compilation of NatureServe data, with many graphs and maps showing the state of nature in the United States.*

Wilson, E. O. 2003. *The Future of Life*. Alfred A Knopf, New York. *An accessible and interesting discussion of biodiversity and the threat that it currently faces by one of conservation's great voices.*

CONCEPT 23.1 Landscape ecology examines spatial patterns and their relationship to ecological processes and changes.

CONCEPT 23.2 Habitat fragmentation decreases habitat area, isolates populations, and alters conditions at habitat edges.

CONCEPT 23.3 Biodiversity can best be sustained by large reserves connected across the landscape and buffered from areas of intense human use.

CONCEPT 23.4 Ecosystem management is a collaborative process with the maintenance of long-term ecological integrity as its core value.

FIGURE 23.1 A Top Predator Returns A gray wolf (*Canis lupus*) in Yellowstone National Park, with an elk (*Cervus canadensis*) kill. Wolves were reintroduced to Yellowstone in 1995 after nearly 70 years of absence, where they are now the main predators of elk.

CASE STUDY

Wolves in the Yellowstone Landscape

I magine that you have trekked by snowshoe to a wildlife blind in the North Range of Yellowstone National Park. You have your spotting scope trained on a herd of elk as they move in a loose cluster across an open icy meadow, scraping away the snow with their hooves to reach clumps of bunchgrass and forbs. As you watch, you notice that some animals are lifting their heads and casting their attention to the east. You look up and see an approaching pack of wolves. The herd draws together, and the elk begin to move, fleeing down the meadow. A young male in poor health drops behind, is separated from the herd, and is surrounded with wolves leaping at his hindquarters and neck. He falls, and is quickly killed.

In Yellowstone, such predation is an ancient scene, and it is now a regular occurrence as wolves, reintroduced to the northern Rockies in 1995 and 1996 after 70 years of absence, hunt among a smorgasbord of ungulates and other prey (**Figure 23.1**). The reintroduction of wolves was the culmination of years of research effort and hotly contested policy discussion, with vociferous objection from some residents of the region. Within a decade, its ecological consequences have proved to be multifaceted and profound, and public opinion has become generally more favorable.

The Greater Yellowstone Ecosystem (GYE) is an area that both symbolizes the soul of the American wilderness and encapsulates the challenges of managing public lands. But how "wild" is it? Larger in area than the state of West Virginia, the GYE includes two national parks and seven national forests as well as other public and private lands (**Figure 23.2**). The region is actively managed by more than 25 different state and federal agencies as well as private corporations, nongovernmental organizations, and private landowners. Decisions about the use of its land and natural resources are complex and often uncoordinated, yet when considered together, these decisions determine which species will or will not be sustained by the ecosystem (Parmenter et al. 2003).

Nevertheless, the GYE is often perceived as one of the most biologically intact ecosystems in North America, with seven species of native ungulates and five large carnivore species. Understanding how these predator and prey populations interact, and what effects their relative numbers have on the

FIGURE 23.2 The Greater Yellowstone Ecosystem The Greater Yellowstone Ecosystem falls under a diverse array of management agencies. It contains Yellowstone and Grand Teton national parks, seven different national forests, and land managed by the Bureau of Land Management, as well as private lands. (After Parmenter et al. 2003.)

Map labels: Bozeman · Livingston · Montana · Wyoming · Yellowstone National Park · Cody · Grand Teton National Park · Rexburg · Jackson · Wind River Indian Reservation · Lander · Idaho | Wyoming

Legend:
— Ecological boundary
National forest boundary
National Park boundary

50 km

ecosystem as a whole, has been a persistent challenge to ecologists who study the GYE, particularly in light of a century of management of wildlife populations. After wolves were eradicated in the mid 1920s, there were concerns that elk were overgrazing meadows in the northern part of the park. The elk population was regulated from the 1920s to the late 1960s by exporting animals to elk farms and by culling. In 1968, a new policy of "natural regulation" was implemented, and the elk population nearly quadrupled over a 30-year period, with subsequent suppression of the plants they feed on. The reintroduction of wolves has not only reduced the elk population, but has also affected the populations of many other species. How?

To start to answer that question, let's go back to the 1950s, when ecologists noticed that beavers had become scarce in Yellowstone National Park. Gradually, it became clear that the cause was elevated elk herbivory on the beavers' preferred food plants, willow and aspen. But a whole suite of other species depend on beaver ponds for their own persistence, and their abundances had declined along with the beavers'. The decision to eradicate wolves did not anticipate these ecological changes to the Yellowstone ecosystem. How can ecologists of today help managers of nature reserves make decisions that will foresee future consequences?

n this chapter, we will take a big step back, expand the scope of our view, and look at ecology from a landscape perspective. The emergence of this perspective has been accompanied by, or perhaps driven by, a powerful assemblage of tools that permits us to view and to sense the environment in all dimensions and at many scales. For example, the emergence of aerial photography gave ecologists a ready means of looking at "the big picture." More recently, our access to space has vastly expanded our ability to acquire images of Earth through remote sensing satellites and has permitted the interpretation of many large-scale ecological patterns, such as global patterns of net primary production. The use of geographic information systems (GIS) has become standard for landscape planning efforts, whether for conservation or for urban development (**Figure 23.3**). In the field, handheld global positioning systems (GPS) have permitted ecologists to document precise locations and integrate them with other variables through GIS. Radiotelemetry has greatly enhanced our ability to follow animal movements and migration patterns, again with the help of GIS. And our ability to analyze all this information is constantly growing thanks to better computers and new statistical methods of spatial analysis.

We saw in Chapter 22 that habitat degradation, fragmentation, and loss are the primary causes of the current declines in biodiversity. In this chapter, we will see how the tools and methods of landscape ecology are used to address this problem at the landscape and ecosystem scales. Because protected areas are at the heart of conservation strategies, we will see how conservation biologists identify and design them to maximize their conservation effectiveness. Finally, we will see how ecosystem management integrates ecological principles with social and economic information to arrive at decisions about land use.

Landscape Ecology

CONCEPT 23.1
Landscape ecology examines spatial patterns and their relationship to ecological processes and changes.

Landscape ecology is a sub-discipline of ecology that emphasizes the causes and consequences of spatial variation across a range of scales. As such, landscape ecologists docu-

Use a computer to produce data layers:

(A)

Satellite for gathering imaging data

Data gathered on the ground

Aerial photography

Satellite for GPS location (data gathered through telemetry)

Population density

Road density

Protected land

Final image:

(B)

Each colored region on the map is a watershed (drainage area) for which different GIS layers were combined to estimate the overall Landscape Condition.

Population density

Road density

Protected land

This watershed has a Landscape Condition score of 10, which was calculated by combining the scores for six GIS layers. Three of those layers are shown above.

Better   Worse
1 2 3 4 5 6 7 8 9 10 11 12 13
Landscape condition

0   50   100
Miles

FIGURE 23.3 Geographic Information Systems Integrate Spatial Data from Multiple Sources   Geographic Information Systems (GIS) are computer-based systems that allow the collection, storage, analysis, and display of geographically explicit data. (A) Data are derived from many sources, including aerial photographs, satellite imagery, and ground-based field studies. Layers of mapped data can be put together in ways that help to address particular questions. (B) In New York State, information on human land use patterns (human population density, road density, protected lands, dams) was combined with natural variables (natural land cover and interior forest areas) to estimate landscape condition and identify areas with the greatest conservation opportunities.

ment observed spatial patterns—including those that occur across broad geographic regions—and they study how those patterns affect and are affected by ecological processes. The patterns of interest concern the spatial arrangement of different *landscape elements* across Earth's surface. Examples of landscape elements include patches of forest surrounded by pasture, or lakes scattered across a large region of northern forest. At a smaller spatial scale, the elements of a landscape could

consist of individual creosote bushes in a desert. But whatever these elements are, they are arranged a certain way in space. As we will see, the spatial pattern of landscape elements can influence what species live in an area, as well as the dynamics of ecological processes such as disturbance and dispersal.

A landscape is a heterogeneous area composed of a dynamic mosaic of interacting ecosystems

A **landscape** is an area in which at least one element is spatially heterogeneous (varies from one place to another) (**Figure 23.4**). Landscapes can be heterogeneous either in what they are composed of—is the landscape composed of twelve

(A)   (B)   (C)

FIGURE 23.4 Landscape Heterogeneity Landscapes can be heterogeneous in many different kinds of elements, which may be arranged in ways independent of one another. (A) An aerial photograph of Michigan's Upper Peninsula. (B) A map of six different soil types in the same area. (C) A map of seven different landscape elements in the same area. (After Delcourt 2002.)

Lake
Dry acid sand without hardpan
Dry acid sand with hardpan
Moist calcareous loam
Moist acid loam
Wet calcareous loam
Wet acid sand and peat

Lake
Urban or quarry
Upland deciduous forest
Upland evergreen forest
Mixed deciduous/evergreen forest
Forested wetland
Nonforested wetland
Cropland or pasture

different vegetative cover types or only three?—or in the way their elements are arranged—are there many small patches arranged regularly over the landscape, or are there a few large patches? Ecologists often refer to this composite of heterogeneous elements as a **mosaic**.

The different ecosystems that make up a landscape are dynamic and continually interacting with one another. These interactions may occur through the flow of water, energy, nutrients, or pollutants between ecosystems, as from a forest ecosystem into an adjacent lake. There is also biotic flow between adjacent patches in the mosaic as animals, seeds, pollen, and other biological emissaries move between them (Forman 1995). For such movement to occur, the patches must either be directly connected to one another or the surrounding habitat (the *matrix*) must be of a type through which dispersal is possible (**Figure 23.5**). In Australia, for example, rats regularly leave forest habitat to forage in adjacent macadamia nut plantations, resulting in greater nut predation along edges that abut forest habitats than along edges with grasslands or agricultural fields (White et al. 1997). We will examine these interpatch dynamics in greater detail below.

Landscape heterogeneity is studied at a broad range of spatial scales, though most landscape studies are at the scale of hectares to square kilometers. The question of scale is central to landscape ecology studies, as the scale selected influences the conclusions reached (**Box 23.1**).

(A)   (B)

Patch
Matrix 1
Matrix 2
Corridor

FIGURE 23.5 Interpatch Dynamics Interactions between adjacent landscape elements may occur frequently (thicker arrows) or rarely (thinner arrows). (A) Exchange between a patch and the surrounding matrix types occurs frequently. (B) Exchange between patches of the same type occurs frequently, but exchange with the matrix occurs only rarely. (After Hersperger 2006.)

BOX 23.1

Thinking about Scale

Considerations of scale cannot be ignored in landscape ecology. A landscape may be heterogeneous at a scale important to a tiger beetle, but homogeneous to a warbler or a moose. The scale at which we choose to study a landscape determines the results we will obtain. Part of landscape ecology, therefore, is dedicated to understanding the implications of scale.

Scale, the spatial or temporal dimension of an object or process, is characterized by both grain and extent. **Grain** is the size of the smallest homogeneous unit of study (such as a pixel in a digital image) and determines the resolution at which we view the landscape. The selection of grain will affect the quantity of data that must be manipulated in analysis: using a large-grained approach may be appropriate when one is looking at patterns at a regional to continental scale (**Figures A–C**). **Extent** refers to the area or time period encompassed by our study. Consider how differently we might describe the composition of a landscape depending on how we define the spatial extent (**Figures D–F**). Figure D, for example, shows little late-succession whitebark pine, while Figure F contains a considerable area of it (Turner et al. 2001). There may be natural or anthropogenic boundaries that determine the extent of a study, or they may be defined by the researcher.

In examining questions of scale, landscape and ecosystem studies must also determine how processes scale up or down. For example, someone studying carbon exchange at the landscape level needs to know how leaf-based measurements of CO_2 exchange scale up to the whole plant, the ecosystem, and ultimately to the mosaic of ecosystems that make up the landscape. Ecologists have sought to elucidate general principles that can be applied to problems of scaling, the consequences of aggregating data or simplifying across levels of scale, and how patterns and phenomena at one scale affect those occurring either higher or lower in the scaling hierarchy (Levin 1992).

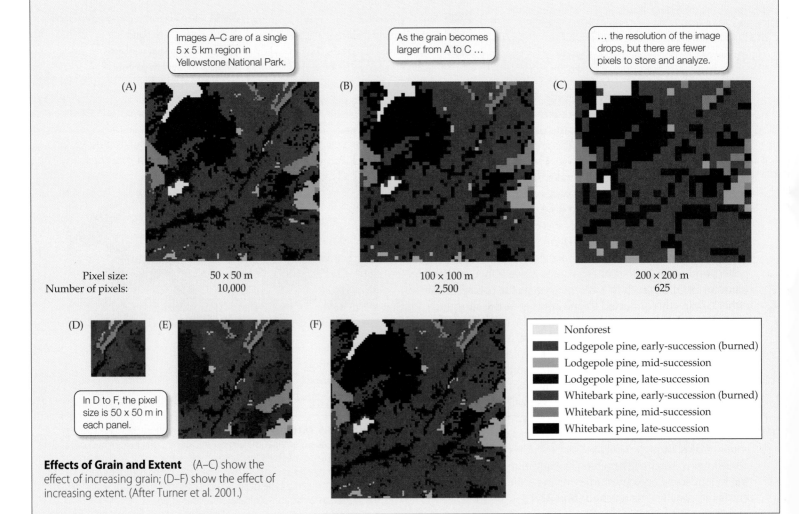

Images A–C are of a single 5 × 5 km region in Yellowstone National Park.

As the grain becomes larger from A to C …

… the resolution of the image drops, but there are fewer pixels to store and analyze.

(A) (B) (C)

Pixel size:	50 × 50 m	100 × 100 m	200 × 200 m
Number of pixels:	10,000	2,500	625

(D) (E) (F)

In D to F, the pixel size is 50 × 50 m in each panel.

Nonforest
Lodgepole pine, early-succession (burned)
Lodgepole pine, mid-succession
Lodgepole pine, late-succession
Whitebark pine, early-succession (burned)
Whitebark pine, mid-succession
Whitebark pine, late-succession

Effects of Grain and Extent (A–C) show the effect of increasing grain; (D–F) show the effect of increasing extent. (After Turner et al. 2001.)

Only remnants of older forest remain following recent fires.

Large stands of older trees are found in this unburned region.

Yellowstone Lake

Stand age classes
- 1–25
- 26–158
- 159–247
- 248–323
- 324–560

0   16
km

FIGURE 23.6 Landscape Composition and Structure In this 1985 map of lodgepole pine (*Pinus contorta* var. *latifolia*) forest in Yellowstone National Park, we can see that the landscape is composed of five different age classes of forest. Structural complexity varies across the landscape, as seen in the varying degree of natural fragmentation. (After Tinker et al. 2003.)

Landscape ecologists describe the heterogeneity that we see in landscapes in terms of composition and structure. **Landscape composition** refers to the kinds of elements or patches in a landscape, as well as to how much of each kind is present. These elements are defined by the investigator and influenced by the source of the data used (**Figure 23.6**). In an example from Yellowstone National Park, researchers designated five different age classes of lodgepole pine forest using ground-based fieldwork, aerial photographs, and GIS (Tinker et al. 2003). The composition of this landscape can be quantified by counting the kinds of elements in the mapped area (five in this case), by calculating the proportion of the mapped area covered by each element, or by measuring the diversity and dominance of the different landscape elements much as one does for species, using a measure such as the Shannon index (see Chapter 15).

In Figure 23.6, we can also see variation in the way that the different forest age classes are arranged across the landscape. When we note that one area is more fragmented than another, we are comparing **landscape structure**, or the physical configuration of the different compositional elements. We can see that in some parts of the landscape, there are large contiguous blocks of older forest (area B), while other parts of the landscape are more fragmented, with a mosaic of dif-

ferent forest age classes made up of smaller patches (area A), largely as a result of past fire history. How do landscape ecologists quantify these differences? Hundreds of different quantitative measures have been developed for measuring, analyzing, and interpreting landscape patterns, many involving complex analysis. Overall, they address whether the landscape is characterized by large or small patches, how aggregated or dispersed the patches are, whether the patches are simple or complicated in their shape, and how fragmented the landscape is (Turner et al. 2001). Tinker and colleagues were able to use the landscape metrics that they derived for Yellowstone to compare the natural, fire-caused fragmentation within the park with fragmentation caused by clear-cutting in adjacent national forests. There is no single measure that captures all the aspects of landscape structure that might apply in different situations. Whatever quantitative measures are used, this approach allows us to compare one landscape with another and to relate landscape patterns to ecological processes and to the dynamics of landscape change.

Landscape patterns affect ecological processes by regulating movement of elements and organisms

The spatial configuration of landscape elements plays important roles in ecological dynamics. For example, it can affect whether and how animals move, and can thus indirectly influence rates of pollination, dispersal, or predation. In French Guiana, Mickael Henry and his associates studied the movements of a frugivorous bat (*Rhinophylla pumilio*) in a mature tropical forest following fragmentation that resulted from the construction of a reservoir. By using landscape metrics that quantified the degree of landscape connectivity at their sampling points, they found that connectivity was an important determinant of bat density. More isolated habitat fragments were less likely to be visited by bats, even if they contained abundant food resources (Henry et al. 2007). Thus, the landscape structure affected bat foraging behavior, and thereby the dispersal patterns of the plants the bats fed on.

Landscape patterns also modulate biogeochemical cycling. Ecosystem ecologists have pointed to biogeochemical "hot spots" where chemical reaction rates are higher than in the surrounding landscape. Water plays a key role in this variation, as the interfaces between terrestrial and aquatic ecosystems are typically places of high rates of biogeochemical turnover, but other factors may also play a part (McClain et al. 2003). For example, Kathleen Weathers and her colleagues found that inputs of sulfur, calcium, and nitrogen from atmospheric deposition were higher at forest edges than in forest interiors, primarily as a result of greater interception of airborne particles by the denser canopies and greater physical complexity typically found at a forest edge. The fragmented forests typically surrounding urban areas may therefore be substantially influenced by atmospheric inputs of pollutants and nutrients—a finding that has implications for soil microbial dynamics,

FIGURE 23.7 Movement Patterns of the Bog Fritillary Butterfly The travel patterns of this butterfly (*Proclossiana eunomia*) are influenced by features of the surrounding landscape. Butterflies will hesitate to leave the patches they inhabit if there is not another suitable habitat patch nearby, but will traverse a surrounding matrix of unsuitable habitat when the next patch is close.

herbaceous vegetation growth, and animal communities in the edges of these fragments (Weathers et al. 2001).

In general, landscape patches vary in both their quality as habitat and their resource availability. This variation can affect the population densities of species inhabiting each patch, the time animals spend foraging in a patch, and the movement of organisms between patches. Patch boundaries, connections between patches, and the matrix between patches also can affect population dynamics, both within and among patches. For example, Schtickzelle and Baguette studied the movement patterns of the bog fritillary butterfly (*Proclossiana eunomia*) across fragmented landscapes in Belgium (**Figure 23.7**). Where patches of suitable habitat were aggregated, female butterflies crossed readily from patch to patch. However, where the habitat was more fragmented and there was a wider distance of matrix to cross, the butterflies were more hesitant to leave a patch (Schtickzelle and Baguette 2003).

The shape and orientation of landscape elements can also be important in physically intercepting organisms and may thus be instrumental in determining species composition. Gutzwiller and Anderson found that northward-migrating cavity-nesting birds were more likely to nest in forest patches in the Wyoming grasslands that were oriented along an east–west axis, such that the habitat patch effectively serves as a net, intercepting birds as they migrate north. They found no such association for resident bird species. Here landscape structure in part determines the species composition of the avian community (Gutzwiller and Anderson 1992).

Interactions between landscape patterns and ecological processes are reciprocal. Ecological processes are regulated by landscape patterns, as we have just seen, but landscape patterns can in turn be caused by ecological processes, as where large grazing mammals shape the landscapes they in-

habit. The effects of moose (*Alces alces*) on Isle Royale in Lake Superior have been studied through the use of exclosures that have been in place since the 1940s. High rates of browsing by moose depress net primary production for the ecosystem, not just directly through removal of biomass, but also indirectly through alteration of nitrogen mineralization rates and litter decomposition rates. Moose browsing also shifts the composition of tree species found across the landscape toward spruce, and the predominance of spruce in turn feeds back to determine rates of biogeochemical processes (Pastor et al. 1988). The moose are thus both responding to and shaping the landscape. At a broader scale, landscape patterns interact with larger-scale disturbances, as we will see next.

Disturbance both creates and responds to landscape heterogeneity

Landscapes are naturally dynamic. In nature, change sometimes comes to ecosystems catastrophically—forests and prairies burn over large areas, and floods bring sudden inputs of sediment into river ecosystems—and at other times slowly, as a result of shifting climates and moving continents. We saw in Chapter 16 that disturbance is an important factor in determining community composition. Landscape ecologists have asked whether particular landscape patterns slow or accelerate the spread of disturbances or increase or decrease an ecosystem's vulnerability to disturbances (Turner 2005).

In 1988, forest fires burned nearly one-third of the 2.2 million acres of Yellowstone National Park. These fires followed decades of anthropogenic fire suppression and came in a summer of extreme drought and high winds. Such extensive fires are thought to have occurred in the northern Rockies at 100–500-year intervals over the past 10,000 years. The fires burned through forest stands of different ages and species compositions, leaving a complex mosaic of patches that were burned at different intensities (**Figure 23.8**). This large-scale disturbance was dictated in part by the forest conditions that existed at the time, and it was fueled by the wind and drought of a particular summer. But the fires of 1988 will probably dictate the composition of the landscape for decades, if not centuries, to come (Turner et al. 2003). Here, the disturbance—fire—was a primary force shaping the landscape pattern of the future. At the same time, it was also responding to the existing landscape structure. This reciprocal interaction between landscape and disturbance is a common one.

With the increase in humanity's ecological footprint, the nature and extent of landscape-level disturbance has been altered. Some places have been more subject to human disturbance than others. People first settled and cleared the areas with the most fertile soils for agriculture, and they naturally gravitated to good port locations, thus subjecting ecosystems in those places to the earliest anthropogenic disturbance. Areas close to human settlements were converted to agriculture or subjected to logging and hunting earlier than outlying areas.

FIGURE 23.8 Disturbances Can Shape Landscape Patterns   The fires that burned through nearly one-third of Yellowstone National Park in the summer of 1988 resulted in a complex mosaic of burned and unburned patches, which will shape the composition of the landscape for years to come. Areas that appear black in this aerial view of Madison Canyon were burned by intense crown fires, and brown patches were burned by severe ground fires, both of which killed most or all of the vegetation.

These disturbance patterns remain contributing factors to current biodiversity, even after people have left the land and it has reverted to forest (Butzer 1992).

Such *landscape legacies* shape plant communities in ways that are just starting to be understood. The effects of anthropogenic disturbance can be detected in ecological communities even centuries after the disturbance. In central France, Etienne Dambrine and his colleagues found that forest plant communities on the sites of Roman farming settlements still bore the mark of these disturbances 1,600 years later. The mapped forest area had not changed substantially since 1665, and was probably maintained as forest for centuries before that. The researchers studied plant diversity in the forest at various distances from sites of recently uncovered Roman ruins, and they found that plant species richness increased in the vicinity of the ruins. Examination of soil properties revealed that this increase was primarily a consequence of higher soil pH, which was thought to result from remnants of the lime mortar used in Roman buildings and from Roman agricultural practices. Soil phosphorus levels were also higher closer to the settlement sites (Dambrine et al. 2007) (**Figure 23.9**). How many other places on Earth might display the signatures of

FIGURE 23.9 Landscape Legacies   In central France, the legacy of Roman farming settlements, abandoned for nearly two millennia, is still reflected in plant species richness in the forest that replaced them. (A) More species were found closer to the center of settlement sites, including more species that prefer a higher soil pH. (B) Soil pH and (C) soil phosphorus were also elevated closer to the settlement sites. (After Dambrine et al. 2007.)

human activities long since abandoned in their current community structure?

Disturbance, whether natural or anthropogenic, is an important factor shaping the landscape. Some current anthropogenic disturbances are having far-reaching ecological effects, as we'll see in the next section.

Habitat Fragmentation

CONCEPT 23.2
Habitat fragmentation decreases habitat area, isolates populations, and alters conditions at habitat edges.

In 1986, a massive hydroelectric project in the Caroni River valley of Venezuela inundated 4,300 square kilometers of uneven terrain to create a reservoir known as Lago Guri (**Figure 23.10**). The result was the formation of scores of islands of tropical forest surrounded by water in what had formerly been a relatively intact forest. This change in the landscape presented an opportunity for John Terborgh and his students and colleagues to study the effects of fragmentation in a tropical dry forest ecosystem. What they found was disquieting. The smaller and medium-sized islands were lacking the top

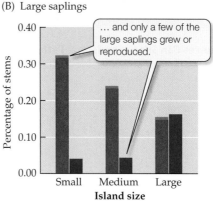

FIGURE 23.11 Effects of Habitat Fragmentation by Lago Guri
The very high abundances of herbivores on small and medium-sized islands in Lago Guri caused a dramatic decline in sapling establishment and survival. The bars show the percentages of (A) small saplings and (B) large saplings in study plots that left their size class through mortality or growth to a larger size, as well as the number of saplings recruited to each size class. (After Terborgh et al. 2006.)

FIGURE 23.10 The Islands of Lago Guri   An aerial view of Lago Guri, Venezuela. This lake formed when 4,300 km² of land were inundated by a hydroelectric dam, leaving isolated islands of tropical forest. This change in the landscape provided an ideal experimental system for studying the effects of forest fragmentation on a tropical biota.

predators found on the mainland, primarily wild cats (ocelots, jaguars, and pumas), raptors, and large snakes. As a result, populations of generalist herbivores, seed predators, and predators of invertebrates were 10 to 100 times more abundant on the islands than in intact forests. Species that increased in abundance included leaf-cutter ants, birds, rodents, frogs, spiders, howler monkeys, porcupines, tortoises, and lizards. The increased abundance of these species had a dramatic effect on the vegetation of these islands as tree recruitment decreased and tree mortality increased due to high rates of herbivory, primarily by leaf-cutter ants (**Figure 23.11**). Terborgh pointed to these effects as an indication of top-down regulation of vegetation communities (Terborgh et al. 2006). What lessons can we take from this "experiment" that apply to other fragmented ecosystems?

Ecologists recognize habitat fragmentation as one of the most prevalent and significant changes occurring in Earth's landscapes. When large blocks of habitat are flooded by dam construction, divided by roads, cleared of forests, or converted to human land uses, there are several consequences for the landscape and the species living there. The first is the simple loss of viable habitat area. Reductions in the area of suitable habitat available have contributed to the declines of thousands of species, including the red-cockaded woodpecker (see the Case Study in Chapter 22). Second, as the remaining habitat becomes divided into smaller and smaller patches, it is increasingly degraded and influenced by edge effects, as we saw in the case of the Biological Dynamics of Forest Fragments Project in Chapter 17. Third, fragmentation results in the spatial isolation of populations, making them vulnerable to the problems of small populations (see Chapter 10).

The process of habitat fragmentation may take place over many decades. A typical pattern begins with a clearing in a forest, which is then widened bit by bit until only isolated habitat fragments remain (**Figure 23.12**). Roads are typically catalysts of habitat conversion, though human access along rivers can also serve to accelerate deforestation. The principal drivers of habitat fragmentation are conversion of land for agriculture and urban expansion. Habitat fragmentation is a reversible process—the forests of northeastern United States, for example, are much more extensive than they were a century ago. The global trend, however, is toward net loss of forests (FAO, 2005) and increasingly fragmented forest, grassland, and riverine ecosystems. What are the ecological and evolutionary consequences of this fragmentation?

Fragmented habitats are biologically impoverished relative to intact habitats

When habitat is fragmented, some species go locally extinct within many of the fragments. There are a host of reasons why this occurs. There may be inadequate food resources, shelter, or nesting sites in the fragments, and animals may need to forage over larger areas than their conspecifics in intact habitat, using multiple fragments. Mutualisms may be disrupted as pollinators go missing or as mycorrhizal fungi fail to persist in a particular fragment. Breeding biology may be altered. Some fragments may lack the microenvironments needed for plant germination. The elimination of top predators, as we saw at Lago Guri, can result in excessive herbivory that prevents the plant community from regenerating. But local extinction is not inevitable; some species flourish under the changed conditions that follow fragmentation.

Fragmentation often leads to losses of top predators, giving rise to cascading effects, sometimes with dramatic consequences for the remaining community, as we saw at Lago Guri. An example with implications for human health is the growing risk of Lyme disease as a result of forest fragmentation in the northeastern United States. Brian Allan, Felicia Keesing, and Richard Ostfeld found that in the Hudson River valley of New York, forest fragments of less than 2 hectares contained very high populations of white-footed

(A)

(B)

(C)

(D)

FIGURE 23.12 The Process of Habitat Fragmentation
Historically intact habitats are gradually reduced with increased human presence. These contemporaneous photographs illustrate a process that typically takes decades to complete. (A) An intact eucalyptus forest in Western Australia. (B) Areas within the forest have been cleared for grazing. (C) The forest has become further fragmented over time. (D) Only a few remnants of forest remain.

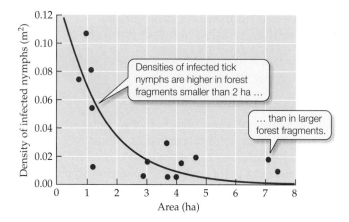

Densities of infected tick nymphs are higher in forest fragments smaller than 2 ha …

… than in larger forest fragments.

mice (*Peromyscus leucopus*), since those fragments do not support significant predator populations and the mice have few competitors there. White-footed mice are the most important reservoir of *Borrelia burgdorferi*, the spirochete bacterium that causes Lyme disease. Ticks are the vector of this disease. Tick nymphs collected in these small fragments are significantly more likely to carry the disease, and also occur at higher densities, than in larger forest fragments (**Figure 23.13**). The outcome is an increased risk of human infection with Lyme disease, ultimately a result of the biological impoverishment of habitat fragments (Allan et al. 2003).

FIGURE 23.13 Habitat Fragmentation Can Have Consequences for Human Health   The loss of predators from small forest fragments in New York State has led to elevated populations of white-footed mice in those fragments. As a result, densities of tick nymphs infected with the spirochete bacterium that causes Lyme disease are higher than in larger forest areas. (After Allan et al. 2003.)

The matrix between habitat fragments varies in permeability

Models for fragmented landscapes were initially derived from island biogeography and depicted habitat fragments as islands isolated in a "sea" of unsuitable matrix, just as the islands of Lago Guri literally are. But do those models truly fit? In Western Australia, Arnold and his colleagues used radiotelemetry to study the movements of *Macropus robustus*, the eastern wallaroo, between habitat fragments in a matrix of wheat fields

FIGURE 23.14 Habitat Islands   The wheat belt of Western Australia underwent a pronounced reduction in the area of native vegetation between 1920 and the present, leaving only scattered islands of habitat available to the eastern wallaroo (*Macropus robustus*). Wallaroos occasionally cross the matrix between habitat islands, but for the most part reside within the islands. (A) Aerial photograph of a region in Western Australia. (B) Representation of the habitat types shown in the photograph. (From Arnold et al. 1993.)

(A)

(B)

Agriculture
Open woodland/shrubland
Heath
Woodland/shrubland with ground vegetation

Where there are clusters of smaller fragments, the animals move more readily between fragments.

Wallaroos living in large habitat fragments tend to remain there.

(Arnold et al. 1993) (**Figure 23.14**). They found that wallaroos living in large habitat fragments tended to remain there, although young males would disperse to smaller fragments at some distance from their place of birth. Where the landscape was characterized by clusters of smaller fragments, the animals moved more readily between fragments, relying on corridors of natural vegetation along roads or small stepping stones of habitat for their movements. Thus, in eastern wallaroos, these habitat fragments are functioning islands surrounded by a matrix that they will occasionally cross. But do habitat patches always function like islands?

There has been increased recognition that fragmented landscapes are more complex than island models would suggest, at least for some of the species that live there. The matrix may be *permeable* to some extent, and may form a mosaic of different patch types, of which some are more permeable than others. In an example from South America, Traci Castellón and Kathryn Sieving studied the dispersal of a small insectivorous understory bird, the chucao tapaculo (*Scelorchilus rubecula*), after translocating individual birds to habitat fragments located in different landscape contexts. They found that birds translocated to fragments surrounded by pasture were much more reluctant to leave the fragments to get to larger forest blocks than were birds that either had a shrubby habitat to cross or that were in fragments connected to larger forest blocks by a forested corridor (Castellón and Sieving 2006). Similar observations were made in a study of rodents in the Atlantic Forest of Brazil, in which some species moved readily through the matrix, while other

species were hesitant to cross into unfamiliar patch types (Pardini 2004). As this study showed, the permeability of the matrix is species-dependent.

Edge effects change the physical environment and species abundances

As intact habitat is fragmented, an abrupt boundary between two dissimilar patch types is created. The length of this habitat boundary, or *edge*, increases as fragmentation increases. **Edge effects** are the diverse biotic and abiotic changes that are associated with such a boundary. The effect of edge formation is a change in the physical environment over a certain distance into the remaining fragment, as a result of which biological interactions and ecological processes can change as well (**Figure 23.15**). The course of these changes plays out over time, so we can separate the primary responses to fragmentation and edge formation from the secondary responses that develop later (see Figure 17.24).

We learned in Chapter 17 about edge effects seen in large-scale experiments in Brazil. The effects of abiotic changes at a forest edge are also illustrated by a study of microclimates 10–15 years after the clear-cutting of an old-growth Douglas fir forest in the Pacific Northwest (Chen et al. 1995). Edges were generally characterized by higher temperatures, higher wind speeds, and more light penetration. Diurnal temperature extremes were also greater at the edges as more heat was radiated from the forest edge at night. These abiotic edge effects varied in how far they extended into the forest interior. They were also more pronounced on south- and southwest-facing edges. The biotic consequences of these abiotic changes included higher rates of decomposition, more wind-thrown trees and thus more woody debris on the forest floor, and differential seedling survival of some tree species (Pacific fir) over others (Douglas fir and western hemlock).

Habitat edges can serve either as barriers to or facilitators of dispersal. Novel species interactions may take place at the junctions of two ecosystems. Some species may benefit from foraging in one habitat and reproducing in another. Invasive species are commonly more abundant in habitat edges, and demographic dynamics may be altered for many native species (Fagan et al. 1999).

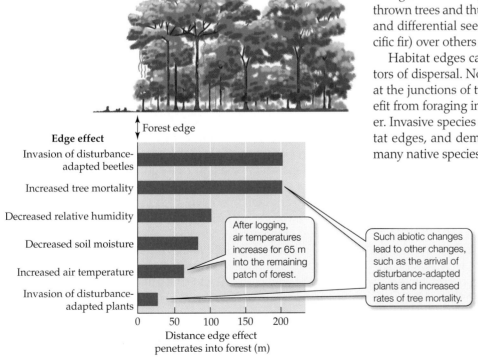

FIGURE 23.15 Edge Effects
When an intact forest is first fragmented, abiotic conditions of the environment change near the edge of the patch of forest that remains, giving rise to biotic changes. (After Laurance et al. 2002.)

Lower breeding success is common for birds nesting closer to habitat edges due to higher rates of egg predation by medium-sized predators such as raccoons, crows, weasels, and foxes as well as higher rates of nest parasitism, especially by cowbirds. In the tallgrass prairie of Wisconsin, Johnson and Temple (1990) studied the reproductive success of five species of ground-nesting birds. They found that proximity to a wooded edge of the prairie habitat significantly increased the probability of nest predation by medium-sized predators and of nest parasitism by cowbirds, significantly lowering reproductive success (Johnson and Temple 1990). Similar patterns have been observed in other prairies, in Scandinavian forests, in eastern deciduous forests, and in the tropics (Paton 1994). Some biologists have characterized edges as "biological traps" as a result of the increased risks that some species face there (Battin 2004).

Species living near habitat edges also face increased risks from human activity. Species that are vulnerable to hunting, to selective logging, or to harvesting for other purposes have been found to decline following the creation of an edge. Domesticated grazing animals may wander into habitat fragments, causing degradation, and other domesticated species, especially cats and dogs, may negatively affect populations of wild species. The net outcome of these altered biological dynamics in fragment edges is a shift in the biota of the fragment, with potential long-term consequences for species perseverance.

Fragmentation alters evolutionary processes

In the time since G. Evelyn Hutchinson's 1965 depiction of "the ecological theater and the evolutionary play," the stage set has been substantially rearranged by the footprint of humanity. The "evolutionary play" will indeed go on, but in altered ways that we are only now trying to understand. What are the evolutionary consequences when populations of all species are split into smaller and more isolated populations and thrown together in new communities that lack historical precedent?

You have already read in Chapters 10 and 22 about the genetic and demographic problems of small, isolated populations. Marcel Goverde and his colleagues studied the consequences of fragmentation by watching bumblebee behavior in the Jura Mountains of Switzerland (Goverde et al. 2002). Their experimental plots included meadow fragments of different sizes separated by closely mown matrix and control plots in unfragmented meadow habitat. The researchers studied the foraging behavior of bumblebees as they visited the flowers of wood betony (*Stachys officinalis*), which were common in both experimental fragments and control plots. The bees visited fragments less frequently than they visited control plots, and once there, they tended to stay longer in the fragments. Ultimately, these two changes in bumblebee behavior in response to fragmentation resulted in a lower probability of pollination, and an increased likelihood of inbreeding, for the wood betony in the fragments—resulting in an altered evolutionary trajectory for those plants.

In many cases, habitat fragmentation increases rates of inbreeding and genetic drift and alters selection regimes for those species confined to fragments. Keller and Largiadèr found significant genetic differentiation between populations of a flightless ground beetle (*Carabus violaceus*) that had been isolated by roads, particularly by a highway (Keller and Largiadèr 2003). Where plant populations become small and isolated, their chances of encountering their pollinators, their pathogens, their herbivores, their dispersers, and their competitors all may be reduced, with subsequent evolutionary consequences. Animal breeding systems and survival patterns may also be altered in small fragments. In New Hampshire, New England cottontails in small habitat fragments had male-skewed sex ratios and higher mortality rates than rabbits in large habitat blocks, both factors that would influence selection (Barbour and Litvaitis 1993).

We have only begun to study the evolutionary implications of habitat fragmentation, and we still have much to learn. Such evolutionary information is only one part of what must be considered in designing reserves that will work well to maintain biodiversity in landscapes increasingly modified by humans.

Designing Nature Reserves

CONCEPT 23.3
Biodiversity can best be sustained by large reserves connected across the landscape and buffered from areas of intense human use.

You may have a favorite national park, such as Everglades National Park in Florida, Grand Canyon National Park in Arizona, Bialowieski National Park in Poland, or Torres del Paine National Park in Chile. How did these places get to be national parks? What were they before they were parks? Are they the best possible sites for maintaining biodiversity in their regions? Now consider how well the land around you is functioning to sustain native species. Your view is undoubtedly shaped by where you are right now, by what the human history of your area is, and by how effective past conservation work there has been. We turn now to an examination of the ways in which people can work to improve the likelihood of the persistence of species native to their region.

To counteract habitat loss, conservation planners worldwide are working to locate and design protected areas where species can persist. The identification and preservation of core natural areas, buffer areas surrounding them, and habitat corridors connecting them is key to maintaining and allowing the growth of populations. In some cases, degraded lands can be restored as viable habitat for wild species.

Core natural areas should be large and compact

The principles of landscape ecology and conservation biology have come together to guide biologists in selecting the most vital lands for conservation. In the design of new nature reserves, **core natural areas** are those where the conservation of biodiversity and ecological integrity takes precedence over other values or uses, and "where nature can operate in its own way in its own time" (Noss et al. 1999). Populations that are able to maintain themselves in core areas may serve as sources of individuals for populations outside the protected area. Ideally, core areas also provide enough land to meet the large habitat area requirements of top predators.

Madagascar is a large island that is a global priority for conservation. It has a rich biota and many endemic species, including more than 70 species of lemurs, a group of primates found only on Madagascar. The biota of Madagascar is seriously imperiled, as only 15% of the island's original forest remains. Efforts are under way to put more land into conservation. In designing a new national park in northeastern Madagascar, Claire Kremen and her colleagues examined both the biological and the socioeconomic circumstances of

the region. Their design (**Figure 23.16**) was based on a core natural area that extended across several elevational and precipitation zones, encompassing a range of vegetation types. The proposed core area encompassed habitat for all of the region's rare species of butterflies, birds, and primates, and had as yet been little affected by deforestation. The researchers excluded areas close to villages that had already been fragmented and where hunting had negatively affected animal populations (Kremen et al. 1999). The Masoala National Park, which opened in 1997, is now the largest national park in Madagascar at 211,230 hectares (over 521,000 acres). With proper management, the park will give the unique biodiversity of this region an improved chance of being maintained in perpetuity.

Ideally, core natural areas must be large and uncut by roads, or even by trails. Thus, not all protected areas qualify as core natural areas. Many do not fully serve the purpose of protecting the whole biota from human interference. Most national parks in the United States were not established with the conservation of biodiversity as their primary mission, but rather to preserve scenery, often on land that was not useful for anything else. Conservation planners recognize that many countries do not have the luxury of carving out thousands

FIGURE 23.16 Designing Masoala National Park Masoala National Park in Madagascar was formed after careful planning that took both ecological and socioeconomic concerns into account. It preserves habitat for many threatened species, including the red ruffed lemur, which is endemic to this region of Madagascar. This map was simplified from more complex maps generated by using GIS techniques to analyze satellite imagery.

of hectares of land to be solely dedicated to biodiversity conservation, and thus frequently the cores of reserves do not meet all criteria to serve as core natural areas.

In the design of nature reserves, some spatial configurations are better than others for fostering the persistence of biodiversity (**Figure 23.17**). Overall, large, compact, and connected reserves are ideal, but there may be times when smaller or disconnected reserves may be more desirable; for example, diseases may spread less easily between isolated smaller reserves than within a large reserve. The primary biological objectives of reserve configuration are the maintenance of the largest possible populations of organisms, the provision of habitat for species throughout their area of distribution, and the provision of adequate area for maintenance of natural disturbance regimes.

Realistically, in many settings where conservation is being accomplished, either the landscape or the social context does not permit adhering to these principles (Williams et al. 2005). There are many smaller reserves that have been established with the conservation of a single species or ecological community as the main conservation objective. Such **biological reserves**, which may be only a few hectares in size, are nevertheless an important part of our conservation efforts. Particularly where human population density is high and large reserves are unfeasible, critically situated smaller reserves may be the best available option.

Core natural areas should be buffered by compatible land uses

Due to many constraints, only relatively small areas of land can be designated as core natural areas. However, if we are to conserve the majority of the world's species, much larger areas of land will have to be able to provide adequate habitat for biodiversity persistence (Soulé and Sanjayan 1998). We can augment the effectiveness of protected areas by surrounding them with **buffer zones**, large areas with less stringent controls on land use, yet which are at least partially compatible with many species' resource requirements. Such lands can be managed in ways that permit the production of needed human resources, such as timber, fiber, wild fruits, nuts, and medicines, but still maintain some habitat value. Activities that may be compatible with the conservation function of buffer zones include selective logging, grazing, agriculture, tourism, and limited residential development (Groom et al. 1999).

In the plan for Masoala National Park, Kremen and her colleagues included a buffer zone on the eastern side of the park, which consisted of more than 71,000 hectares of forest land designated for sustainable timber harvesting (see Figure 23.16). Areas were first identified that were at high risk of deforestation due to their proximity to villages. Researchers established how much wood each family, and thereby each village, consumed, and calculated how much area would be required to meet this need on a sustainable basis. The buffer zone augments the effective area of the park for many lowland species, even though they may be subjected to some level of hunting or collection (Kremen et al. 1999).

Elsewhere, as at Masoala, ecologists have found that buffer zones may serve as *population sinks* for some species, as animals that stray out of core areas and into buffer zones become vulnerable to hunting, roadkill, or other sources of mortality. In Peru, where slash-and-burn agriculture is commonly practiced just outside nature reserves, wild animals such as agoutis, armadillos, or tapirs commonly come to farmers' plots to feed on crops, causing considerable damage. These animals are then targeted by hunters (Naughton-Treves et al. 2003). Even in the sparsely inhabited region of Peru that was subject to study, such hunting has altered the relative abundances of mammals in the forest, which ultimately may affect forest composition in the core area.

FIGURE 23.17 The Best Spatial Configurations for a Core Natural Area Some spatial configurations are usually better than others for fostering biodiversity. (After Diamond 1975 and Williams et al. 2005.)

If we can succeed in establishing core areas for protection surrounded by sparsely inhabited buffer zones, have we done all that is necessary for conservation? Recall that landscape connectivity is another important consideration in reserve design.

Corridors can help maintain biodiversity in a fragmented landscape

Habitat corridors—linear patches that connect blocks of habitat—have become a staple of urban, suburban, and rural planning. It was clear to many ecologists early on that if habitat fragmentation was the problem, then connectivity might be the solution: if populations are at risk of isolation, then we should ensure that there are corridors of habitat that link them together. It made intuitive sense.

When designing Masoala National Park (see Figure 23.16), Kremen and her colleagues looked at the larger landscape and anticipated connections that would be important in the future. Many of Masoala's target species are also found in areas northwest of the park, in lands that lie between Masoala and two important protected areas to the north. The park plan included three narrow corridors to provide connections to those protected areas. The researchers developed this part of the plan by examining maps, but out of expediency, they did not actually do studies of animal movements (Kremen et al. 1999).

The intended function of habitat corridors is to prevent the isolation of populations in fragments. But do we know that corridors actually help to overcome this isolation? And do corridors work at all scales, for beetles as well as for wolves? Is a stream corridor in the suburbs providing necessary landscape connectivity for some species? At the continental scale, could we link Yellowstone to the Yukon through habitat corridors, as some have proposed? Experimental and observational studies of corridors' utility have shown mixed results.

Nick Haddad and his colleagues established a test of the utility of corridors at the Savannah River Ecological Laboratory in South Carolina. They set up patches of early successional habitat in a matrix of pine forest, some of them connected by corridors, and observed the movements of organisms between patches (**Figure 23.18**). Their results showed that the corridors did indeed serve to facilitate the movement of butterflies, pollen, and bird-dispersed fruits (Tewksbury et al. 2002).

Other studies, however, have found no benefits of corridors, and still others have found negative effects. For example, in the same experimental system at the Savannah River Ecological Laboratory, indigo bunting (*Passerina cyanea*) nest predation was higher in the patches connected by corridors (Weldon 2006). There are also concerns that corridors could facilitate movement of pathogens (Hess 1994) and of invasive species (Simberloff and Cox 1987).

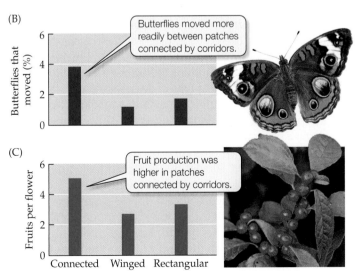

FIGURE 23.18 How Effective Are Habitat Corridors?
(A) Nick Haddad and his colleagues tested the effectiveness of habitat corridors by cutting experimental patches out of a pine forest, but leaving corridors between some of the patches. They then observed (B) the movement of the common buckeye butterfly (*Junonia coenia*) between patches, and (C) fruit production (which provides evidence of pollination) in winterberry (*Ilex verticillata*). (After Tewksbury et al. 2002.)

Ecological restoration can increase biodiversity in degraded landscapes

What if needed migration corridors are lacking and organisms' ability to migrate is impaired by a hostile matrix? This was just the case in Guanacaste Province on the Pacific coast of Costa Rica, where Santa Rosa National Park, in a lowland area of tropical dry forest, was largely separated by 35 km of cattle pasture and forest fragments from the upland forest habitat of the nearby mountains. Tropical ecologist Dan Janzen understood the necessity of elevational migration for many tropical insects, birds, and mammals, and also saw that the tropical dry forest that he had spent his career studying was fast disappearing. Janzen's effort to reverse this trend became one of the largest and most ambitious ecological restoration projects ever undertaken in the Neotropics. Now covering some 120,000 hectares of land and 70,000 acres of marine reserve, the Area de Conservación Guanacaste (ACG) includes protected areas, including three national parks, a protected corridor linking them, and the surrounding agricultural areas. The region is home to some 230,000 species, or 65% of the species in Costa Rica (Daily and Ellison 2002).

Cattle ranches have occupied much of the land between the parks for decades. Janzen has launched an effort to restore these pasturelands to native tropical dry forest, transitioning to moist tropical forest and cloud forest at higher elevations. His strategies, all guided by ecological research, include planting trees, suppressing fires, and limiting hunting (to maintain mammalian and avian seed dispersers). Fire suppression is necessary to halt fires that burn readily in pastures covered in Jaragua grass (*Hyparremia rufa*), an invasive introduced from Africa. Grazing will be maintained for some time in some areas to suppress the Jaragua grass; cows and horses have also been found to help in tree seed dispersal. Ultimately, the plan for the ACG calls for the restoration of about 75,000 hectares to tropical dry forest and cloud forest (Allen 1988).

Ecological restoration is being applied in many ecosystems with varying degrees of success. To be successful, restoration ecologists must first correctly diagnose the ecological state of an area, decide what should be the goals of the restoration, and then apply their understanding of ecological processes to recreating the desired type of ecosystem on the land. Anthony Bradshaw, a founder of restoration ecology, referred to this process as the "acid test" of ecology: "Each time we undertake restoration we are seeing whether, in the light of our knowledge, we can recreate ecosystems that function, and function properly" (Bradshaw 1987).

Our ecological knowledge is being put to this "acid test" in Guanacaste. Janzen is already seeing some signs that the project is working, but the process is a very long and slow one. His efforts to restore not only the landscape, but also the people's relationship with the land and the organisms that live there, reflects a profound shift in how we relate to and manage land. In the next section, we will look more closely at how ecology gets applied when we make decisions about how to use the land and manage natural resources sustainably.

Ecosystem Management

CONCEPT 23.4
Ecosystem management is a collaborative process with the maintenance of long-term ecological integrity as its core value.

The year was 1989. The setting was a packed public hearing in Olympia, Washington. A convoy of logging trucks and about 300 loggers had made the journey to the capital to defend their jobs. There was talk that the northern spotted owl could be listed as a threatened species under the U.S. Endangered Species Act, which would place its old-growth forest habitat off-limits to logging. Tempers were flaring among loggers and others supported by the timber industry. "When it comes to choosing between owls and the welfare of families, the hell with the owl as far as I'm concerned," said a state politician. At times, some of the testimony was drowned out by the honking of truck horns. The contentious debate about the logging of forests in the Pacific Northwest was reduced to "owls versus jobs," and resulted in bumper stickers and T-shirts, vandalism by both sides, and lots of angry words exchanged. This was the state of the "art" of public discourse over resource use at that time.

Some people recognized that there might be a better way to make decisions about the use of public lands. The conflict in the Pacific Northwest was in part the outcome of a long history of top-down management of natural resources, based on resource production and extraction. In 1995, a federal Interagency Ecosystem Management Task Force was formed to develop alternatives to this approach (DellaSala and Williams 2006).

Approaches to managing natural resources have become more collaborative over time

Through most of the twentieth century, management of natural resources on public lands focused on maintaining individual resources of economic interest, whether timber, deer, ducks, or scenery for visitors. This focus remained at the core of many land management policies until the 1980s. Eventually, natural resource agencies expanded their missions to include "multiple use," in recognition that different people had different interests in public lands and that it was possible to manage public lands to meet diverse and at times competing demands. This was frequently done through spatial compartmentalization of uses, in which different blocks of land were designated as timber extraction zones, recreation zones, or wilderness areas (Meffe et al. 2002).

TABLE 23.1

Differences between Traditional Natural Resource Management and Ecosystem Management

Traditional management	Ecosystem management
Emphasis on commodities and natural resource extraction	Emphasis on balance between commodities, amenities, and ecological integrity
Equilibrium perspective; stability; climax communities	Nonequilibrium perspective; dynamics and resiliency; shifting mosaics
Reductionism; site-specific	Holism; view of lands in landscape context
Predictability and control	Uncertainty and flexibility
Solutions developed by resource management agencies	Solutions developed through discussions among all stakeholders
Confrontation; single-issue polarization; public as adversary	Consensus building; multiple issues; partnerships

Source: Meffe et al. 2002.

Since the 1980s, with our greater awareness of the necessity of preserving biodiversity, our goals for land management have shifted again. Ecosystem management has emerged as a way to expand the scope of management to include protection of all native species and ecosystems on the landscape and to focus not only on the sustainability of a resource of interest, but the sustainability of the whole system. Yet, because the past history of more utilitarian management approaches is still culturally intact, ecosystem management today takes many guises and is interpreted in myriad ways (Franklin 1996).

What is ecosystem management? Most simply stated, it is "managing ecosystems so as to assure their sustainability" (Franklin 1996). A committee of the Ecological Society of America arrived at a less simple but more comprehensive consensus definition in 1996: "**Ecosystem management** is management driven by explicit goals, executed by policies, protocols, and practices, and made adaptable by monitoring and research based on our best understanding of the ecological interactions and processes necessary to sustain ecosystem structure and function" (Christensen et al. 1996) (**Table 23.1**). This definition emphasizes sustainability, but also recognizes the need for setting goals and using science to evaluate and adjust management practices over time.

The conflict in the late 1980s over old-growth forests in the Pacific Northwest was a stimulus to ecologists, forest service staff, industry, and citizens to seek a less confrontational way to make decisions. Since that time, more collaborative decision making has been combined with better incorporation of science to arrive at management plans that sustain both biodiversity and people's livelihoods and are responsive to changing conditions. In ecosystem management, the focus is on a particular biophysical ecosystem, or *ecoregion*, delineated by natural boundaries rather than political boundaries: a watershed, a mountain range, a stretch of coastline. The full range of *stakeholders*—people with some interest in the project—becomes involved in decision making for the ecoregion, joined together by the common goal of maintaining its ecological integrity and economic viability.

Ecosystem management sets sustainable goals, implements policies, monitors effectiveness, and adjusts as necessary

Ecosystem management is a process, one that may be implemented in different ways for different projects. Most ecosystem management projects begin with the gathering of scientific data to define the nature of the problems in the ecosystem and the use of this information to set sustainable goals. To meet those goals, a set of actions is needed, many of which may require adapting new policies. Once a new policy is implemented, the ecosystem is monitored to gauge whether the action brings about the desired result. Adjustments to the policies are then made as needed in an iterative process known as **adaptive management** (**Figure 23.19**),

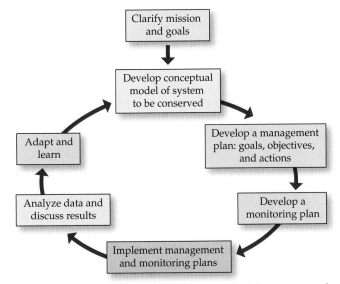

FIGURE 23.19 Adaptive Management Is a Vital Component of Ecosystem Management Adaptive management is a systematic way of learning from past management actions and adjusting future decisions accordingly. (After Margoluis and Salafsky 1998.)

in which management actions are seen as experiments and future management decisions are determined by the outcome of present decisions.

Monitoring is a vital component of adaptive management. For example, Mark Boyce developed a model predicting elk and wolf populations in Yellowstone following the reintroduction of wolves. He and his colleague Nathan Varley have now taken an adaptive management approach by adjusting this model based on demographic data for the first 10 years of wolf presence. Since their original model estimated elk numbers well, but underestimated wolf numbers, they knew that some of the model's assumptions needed adjustment (Varley and Boyce 2006). This approach will be important for determining acceptable hunting levels for elk and for future adjustments in response to changing circumstances.

Humans are an integral part of ecosystems

Human actions affect natural ecosystems, and human economies are affected by natural resource levels. Ecosystem managers must not only manage natural resources and biodiversity across large landscapes, but also devise plans that protect both natural ecosystems and human economies. Ecosystem management incorporates human social and economic factors as a fundamental part of the decision-making process, along with legal requirements and, of course, ecological integrity (**Figure 23.20**). The integration of these different components is seen as necessary to achieve a successful management outcome.

As we have seen, people need nature for many reasons, ranging from the economic to the spiritual. Ecosystem management incorporates education of the public about their reliance on ecosystem services as part of its mission. It also engages the public in helping to solve those problems that degrade the ecosystem services that they rely on.

Any conservation plan that excludes the human component will not be accepted, ultimately, by the stakeholders. The plan for Masoala National Park took the needs of the people living around the park into consideration. Conservation planners not only calculated their wood needs and provided for them in a buffer zone designated for managed forestry, but also surveyed the region for tree species that would have value in an export market and included them in an economic plan for future use. The idea was to remove economic pressure from park resources by identifying ways that people could support themselves and increase their incomes using resources outside the park. In addition, Kremen's team worked in conjunction with local people and with the Malagasy government to develop the plan, recognizing the importance of local acceptance of any proposal they made. In the end, the park plan provided for the economic needs of the people, identifying forest resources that could be used to enrich the region, as well as for the habitat requirements of all the taxa included in the planners' analysis. Whether the park can achieve these goals will be ascertained only over time (Kremen et al. 1999).

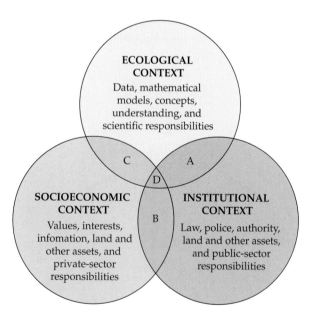

FIGURE 23.20 Humans Are an Integral Part of Ecosystem Management Ecosystem management integrates interests derived from ecological, institutional, and socioeconomic contexts. The letters represent the overlap of the three contexts: A, zone of regulatory or management authority; B, zone of social obligations; C, zone of informal decisions (as opposed to legal requirements); D, zone of win–win–win partnerships.

Wolves in the Yellowstone Landscape

The reintroduction of wolves to the northern Rockies in 1995 reflected the shift to an ecosystem management approach to decision making. It was a bold step that followed years of study and preparation. That it happened at all reflects a quantum shift in human attitudes toward nature over the last century. In the late 1800s and early 1900s, wolves were feared and reviled. They were perceived as a threat to people and livestock—a perception in accordance with reality as far as livestock were concerned. They were hunted to extinction in the area of Yellowstone National Park by the late 1930s and throughout virtually all of the contiguous United States not long thereafter.

The removal of a top predator can alter the landscape substantially, in part because herbivores that were once controlled by the predator may increase in number and negatively affect vegetation dynamics, as we saw in the case of the Lago Guri islands. With the removal of wolves from Yellowstone came declines in the growth and reproduction of riparian tree species. While aspen trees had reproduced successfully along the banks of rivers and streams throughout the 1800s and into the 1900s, they failed to reproduce at all for a 70-year period following the removal of wolves (Ripple and Beschta 2007). The likely reason was heavy browsing by herbivores such as elk, which roamed freely along rivers and streams once the wolves were gone. Support for this explanation comes from the ways in which elk behavior and aspen growth and reproduction changed after wolves were reintroduced.

The reintroduction began in the winter of 1995–1996, when 31 wolves captured in Canada were released into the park. Their numbers increased rapidly; by 2004, there were about 250 wolves in the park. Following the reintroduction, other changes occurred as well. Elk, the wolves' principal prey, were initially naive and very vulnerable to wolf predation. The elk have since modified their behavior, changing their movement patterns and showing a preference for foraging in places that provide high visibility. As a consequence, cottonwoods, aspens, and willows have begun to recover in some areas. The early signs of recovery appear to be concentrated in areas where elk face a high risk of predation, such as locations where visibility is poor, escape routes are lacking, or ambush sites are common. Thus, elk

may be avoiding areas where they are most vulnerable to attack by wolves, allowing trees in those areas to recover (Ripple and Beschta 2004). Merely the elks' fear of wolves is enough to change their behavior, and the implications of this change are reflected at the landscape level.

The recovery of riparian vegetation has had significant consequences for stream hydrology, as more willows lining stream banks have both slowed stream flow and increased sedimentation (Beschta and Ripple 2006). In addition, the increased growth of riparian tree species is expected to provide shade and habitat for trout, which prefer shade-cooled waters, and for migratory birds. More riparian bird species have been observed in a similarly recuperating ecosystem in Alberta (Hebblewhite et al. 2005). As populations of willow—a preferred food for beavers—have increased, new beaver colonies have appeared. The dams built by the beavers have changed patterns of water flow, creating marshlands that may favor the return of otters, ducks, muskrats, and mink (**Figure 23.21**). Clearly, the presence of wolves has the potential to change Yellowstone ecosystems. This makes the return of the wolf a grand ecological experiment, one whose results will be monitored closely in years to come.

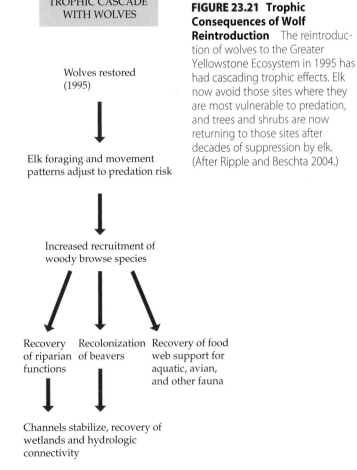

FIGURE 23.21 Trophic Consequences of Wolf Reintroduction The reintroduction of wolves to the Greater Yellowstone Ecosystem in 1995 has had cascading trophic effects. Elk now avoid those sites where they are most vulnerable to predation, and trees and shrubs are now returning to those sites after decades of suppression by elk. (After Ripple and Beschta 2004.)

Future Changes in the Yellowstone Landscape

The addition of just one key species to the Yellowstone ecosystem has had broad and cascading ecological and landscape effects. But other, even more fundamental changes are taking place in the Yellowstone ecosystem. Recall from Chapters 2–4 that climate is the single most important determinant of where species live. With rising concentrations of greenhouse gases in the atmosphere, climatic warming is anticipated in the coming century. Will Yellowstone be able to maintain its current biological diversity in the face of global climate change?

A modeling study showed what the vegetation of the region surrounding Yellowstone National Park might look like under a doubling of current atmospheric CO_2 concentrations, which could happen within a century (**Figure 23.22**). Generally, the projections are for higher temperatures, no increase in precipitation, more frequent fires, and upslope and northward migrations of many species. The model projects shifts in forest communities, with some species declining within the park and others increasing their range to include the park. Species currently rare or absent from the GYE that may increase substantially include gambel oak, western red cedar, and ponderosa pine. A near elimination of whitebark pine is predicted as that species moves north (Bartlein et al. 1997).

The loss of the whitebark pine would trigger a number of other ecological shifts. This tree is a keystone species that produces large, fatty, and nutritious nuts, a primary food source for the Clark's nutcracker as well as for both black and grizzly bears. Clark's nutcracker, in turn, is an important disperser of the whitebark pine's seeds (Siepielski and Benkman 2007). If the whitebark pine is displaced from the Yellowstone ecosystem by global warming, it is likely that these associated species will also move north.

What this model cannot tell us is whether the movement of organisms can keep pace with climate change. The challenge posed to conservation planners is to conserve sufficient landscape connectivity to allow this biological mass migration to take place. The Northern Rockies may have a better chance than many parts of the world to

weather these changes due to the presence of extensive conservation lands there. But how will species of the Great Plains or of Europe fare?

(A) Gambel oak (*Quercus gambellii*)

Yellowstone National Park

(B) Western red cedar (*Thuga plicata*)

Under a doubling of atmospheric CO_2 concentrations, western red cedar, which does not now exist in the park, is predicted to extend its range there.

(C) Ponderosa pine (*Pinus ponderosa*)

(D) Whitebark pine (*Pinus albicaulis*)

Whitebark pine is predicted to disappear from much of the park and other regions where it now lives.

■ No change: present now and at $2 \times CO_2$
■ Range contraction: present now; absent at $2 \times CO_2$
■ Range expansion: not present now; present at $2 \times CO_2$

FIGURE 23.22 Projected Effects of Climate Change in the Northern Rockies Shifts in the distributions of some principal tree species in the northern Rocky Mountains are projected by a model of a future climate driven by twice the current atmospheric CO_2 concentrations. Note the introduction of some species currently uncommon in the region, (A) gambel oak, (B) western red cedar, and (C) ponderosa pine, and the near disappearance of (D) whitebark pine. (After Bartlein et al. 1997.)

Conservation biologists have turned to landscape-level tools and climate models to predict the coming changes. We are already beginning to see the effect of climate change in shifts in plant and animal distributions, plant phenology, and animal behavior; we know that these changes are just the start. In Canada, it is predicted that 75%–80% of national parks will see shifts in their dominant vegetation. In South Africa, some critical ecosystems for plant diversity are projected to lose substantial area. In the Great Basin of North America, between 9% and 62% of mammal species could be lost as a result of a 3°C warming (Hannah et al. 2002). How do we plan for such changes? Is there a way to avert these losses?

The tools described in this chapter will be key to preparing for protection of biodiversity in the future. Landscape ecology and the use of remote sensing tools are critical to understanding current patterns of biodiversity and can help us to predict future ones. While over the past 30 years we have put enormous effort into selecting, establishing, and undertaking management of new protected areas, we need to ask how well these areas will harbor their species in a warmer world. If losses are projected under climate change, are there steps we can take now that can improve landscape connectivity, that can create or improve buffer zones around core natural areas, that can restore degraded areas to greater ecological integrity? And can people with different interests come together to dedicate lands to conservation?

The challenges will be considerable with a growing human population and growing demands on the land. Ecologists will have the critical role of providing the scientific information needed to make decisions about how we proceed as a society. The future of untold numbers of species relies on how effective we can be at this task.

SUMMARY

CONCEPT 23.1 Landscape ecology examines spatial patterns and their relationship to ecological processes and changes.

- A landscape is a heterogeneous area of land made up of a dynamic mosaic of different components that interact through the exchange of materials, energy, and organisms.

- Landscapes are characterized by their composition—the elements that constitute them—as well as by their structure—how those elements are arranged on the landscape.

- Landscape patterns regulate ecological processes by determining how easily organisms can move among elements as well as by regulating biogeochemical cycling.

- Landscape patterns both shape and are shaped by disturbances.

CONCEPT 23.2 Habitat fragmentation decreases habitat area, isolates populations, and alters conditions at habitat edges.

- Habitat fragments are generally biologically impoverished compared with the intact habitat from which they were derived.

- Once a habitat fragment is isolated from surrounding habitat, organisms reliant on that habitat may be isolated or may be able to cross the intervening matrix to some extent.

- The edges of habitat fragments have a different abiotic environment and different demographic dynamics than interior habitats.

- The isolation of populations and the shifts in ecological communities that result from habitat fragmentation alter the evolutionary process.

CONCEPT 23.3 Biodiversity can best be sustained by large reserves connected across the landscape and buffered from areas of intense human use.

- The ideal spatial configuration for a core natural area is large, compact, and connected to or close to other protected natural areas.

- Core natural areas should be surrounded by buffer zones predominated by uses that are compatible with biodiversity conservation but which still allow human economic uses.

- Habitat corridors are instrumental in facilitating the movement of organisms between natural areas.

- Ecological restoration allows areas that have been degraded in the past to support native species and ecosystem processes once again.

CONCEPT 23.4 Ecosystem management is a collaborative process with the maintenance of long-term ecological integrity as its core value.

- Collaboration among all stakeholders is key to arriving at effective management plans.

- Ecosystem management is a process of setting sustainable goals, developing and implementing land use management policies, monitoring the effectiveness of prior decisions, and adapting plans accordingly.

- Humans are an integral part of ecosystems, and conservation plans that include the economic and social well-being of local human populations are more likely to succeed over the long term.

REVIEW QUESTIONS

1. How are islands of terrestrial habitat that result from fragmentation like actual islands surrounded by water? How are they different? How do the principles of island biogeography apply to habitat islands in a fragmented landscape?

2. Are corridors just long, skinny habitat patches? Describe how a corridor is like the habitat blocks it is meant to join and how it is different, and outline the implications for organisms using the corridor. Do you think corridors are beneficial? Do you think they are necessary? Why?

3. The western boundary of Yellowstone National Park, where it borders a national forest, is visible from space (check it out with Google Earth at 44°21 45.19 N 111°05 50.08 W). Explain this in terms of the contrasting missions of a national park and that of a national forest. What are the ecological implications of these contrasting institutional functions for (a) biodiversity and (b) ecosystem management?

4. Describe the protected areas in the vicinity of your ecology class. Rank them by size, describe who has jurisdiction over each of them, and evaluate how well you think they may be doing to protect native biodiversity. Consider them as a network. How well connected are they? Is there any way landscape connectivity could be improved in your region? What is the potential and the need for increasing the area of conserved land in your region?

5. Give an example of a conservation effort in your region that has an ecosystem management approach. What is the ecosystem that is being protected? What are some of the target species or biological communities this effort seeks to protect? Who are the stakeholders involved in decision making? How well does this effort adhere to the core value of seeking to maintain ecological integrity?

HANDS-ON PROBLEM SOLVING

Patch Movement: Crickets versus Cybercrickets

This Web exercise (www.sinauer.com/ecology) explores how organisms move across patches in the landscape. You will interpret data from a recent paper that simulated the movement of virtual organisms between patches across landscapes with different levels of connnectivity, and then compared the simulation results to results from manipulation studies. Do real crickets move like cybercrickets?

SUGGESTED READINGS

Dale, V. H. and 8 others. 2000. Ecological principles and guidelines for managing the use of land. *Ecological Applications* 10(3): 639–670. *This report was written by a committee of the Ecological Society of America to provide guiding principles for land managers. It is an excellent and readable synopsis of these critical ideas.*

Lindenmayer, D. and J. Fischer. 2006. *Habitat Fragmentation and Landscape Change: An Ecological and Conservation Synthesis.* Island Press, Washington, DC. *Provides thorough coverage of the causes and consequences of habitat fragmentation. The presentation of changes in ecosystems, particularly anthropogenic changes, and how they affect organisms is clear and comprehensive.*

Meffe, G., L. Nielsen, R. L. Knight and D. Schenborn. 2002. *Ecosystem Management: Adaptive, Community-Based Conservation.* Island Press, Washington, DC. *This book teaches about ecosystem management through fictional case studies designed to reflect the actual complexity of problem solving with multiple stakeholders. It gives weight to the biological, economic, and sociological considerations necessary to manage ecosystems sustainably.*

Soulé, M. E. and J. Terborgh. 1999. *Continental Conservation: Scientific Foundations of Regional Reserve Networks.* Island Press, Washington, DC. *This book makes a case for considering conservation on the scale of continents. The authors present a case for restoring top predators to ecosystems and for developing core natural areas surrounded by buffers and connected by wildlife corridors.*

Turner, M., R. H. Gardner and R.V. O'Neill. 2001. *Landscape Ecology in Theory and Practice: Pattern and Process.* Springer Verlag, New York. *A clear and general overview of the science and methods of landscape ecology. The CD that accompanies the book allows close looks at all the figures.*

Wiens, JA. 1997. Metapopulation dynamics and landscape ecology. In *Metapopulation Biology: Ecology, Genetics, and Evolution,* eds. IA Hanski and ME Gilpin, Academic Press, San Diego. *This book chapter explores the similarities and differences of two sub-disciplines of ecology that focus on how spatial heterogeneity affects ecological processes: landscape ecology and the study of metapopulations (see Chapter 10 of our textbook, Ecology).*

24 Global Ecology

CONCEPT 24.1 Elements move among geological, atmospheric, and biological pools at a global scale.

CONCEPT 24.2 Earth is warming at an unprecedented rate due to anthropogenic emissions of greenhouse gases.

CONCEPT 24.3 Anthropogenic emissions of sulfur and nitrogen cause acid deposition, alter soil chemistry and affect the health of ecosystems.

CONCEPT 24.4 Losses of ozone in the stratosphere and increases in ozone in the troposphere each pose risks to organisms.

FIGURE 24.1 A Decline in Coral Reefs
The incidence of coral bleaching is increasing worldwide.

CASE STUDY

Dust in the Wind and the Decline of Coral

Coral reefs are unique among marine ecosystems for their high productivity and the diversity of organisms associated with them. Millions of species occupy the reefs, using the unique habitat of calcium carbonate formations created by the coral, and the high energy supply based on high net rates of primary production (see Chapter 3 for more thorough description of coral reefs). In addition to their importance as hot spots of marine biodiversity, coral reefs provide critical habitat for economically important fish and recreation for humans. However, coral reefs throughout the world are experiencing increasing rates of disease, loss of symbiotic algae (bleaching), and mortality (**Figure 24.1**). Corals are particularly sensitive to changes in their physical environment, requiring a narrow range of temperatures, clear water, and high salinity (see Chapter 16). Although a long-term general decline in coral reefs has been occurring for centuries (Pandolfi et al. 2003), the rate of decline has increased substantially since the 1970s (**Figure 24.2**).

Ecologists studying coral reefs have suggested that the increased rate of reef decline is associated with human activities. Factors that have been implicated include greater transport of erosional sediments from terrestrial ecosystems, water pollution, and overharvesting of biotic resources, including keystone predators and dynamite fishing. While these and many other factors may contribute to local coral reef decline, they don't explain the widespread nature of the decline: it is occurring both near and far from sources of human perturbations. What could explain the losses of healthy corals over their entire range?

A potential explanation for the widespread decline in the health of coral reefs was proposed by Virginia Garrison and colleagues (Garrison et al. 2003). A hint came from examination of some of the more common coral diseases. Of the four pathogenic diseases that have been described in corals, one of the most common, aspergillosis, is caused by a widespread terrestrial fungus, *Aspergillus sydowii*. Aspergillosis occurs in humans as well, infecting cavities in the lungs, and causing breathing difficulty and fever. How did a fungus usually found

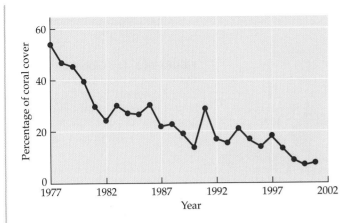

FIGURE 24.2 A Record of Coral Reef Decline Changes in the percent cover of coral at monitoring sites throughout the Caribbean Sea show greater than an 80% decrease over a 25-year period. (After Gardner et al. 2003.)

in soil become a major disease in marine corals? In the immortal words of Bob Dylan, "The answer my friend, is blowing in the wind." Garrison and colleagues have proposed that windblown dust carries fungal spores from distant terrestrial ecosystems to reef ecosystems, where they germinate and infect large numbers of corals, debilitating and eventually killing many.

The hypothesis proposed by Garrison and her colleagues, however, raises many questions. Where is the dust coming from, and why would it have become more important after the 1970s? How can the fungus survive transport in the atmosphere, experiencing extreme temperatures and exposure to high levels of ultraviolet radiation? Is aspergillosis the primary cause of disease, or is it a secondary infection that follows infection by a marine fungus? What other roles might dust deposition play in the recent decline of corals?

n Chapter 21, we reviewed the cycling of nutrients within ecosystems associated with biological uptake and decomposition. The movements of these biologically important elements, however, are linked at a global scale that transcends ecological boundaries at the ecosystem and biome level. Ecological processes at the ecosystem scale (e.g., net primary production, decomposition) influence global phenomena (e.g., greenhouse gases emissions and uptake). In addition, the realization that humans are increasingly changing the physical and chemical environment at a global scale has fostered a greater awareness of ecology at these larger spatial scales. Atmospheric emissions of pollutants, dust, and greenhouse gases have caused widespread environmental problems, including climate change, acid precipitation, eutrophication, and loss of stratospheric ozone. A major focus of global ecology is therefore the study of the extensive environmental effects of human activities.

The first part of this chapter will cover the global-level cycles of carbon, nitrogen, phosphorus, and sulfur, which are related to, but distinct from, the ecosystem-level cycles covered in Chapter 21. Knowledge of these cycles is important for understanding global environmental change. Humans have had profound effects on these element cycles, and the associated environmental changes will be discussed in the final sections.

Global Biogeochemical Cycles

CONCEPT 24.1
Elements move among geological, atmospheric, and biological pools at a global scale.

In this section we will follow the cycling of carbon, nitrogen, phosphorus, and sulfur at the global scale, incorporating the geological and atmospheric components of those cycles. These particular elements are emphasized both because of

their importance to biological activity and because of their roles in human alteration of the global environment. The cycles are discussed in terms of the *pools*, or reservoirs, where the elements occur and the *fluxes*, or rates of movement, between pools. For example, terrestrial plants would be a pool for carbon, while photosynthesis would represent a flux—in this case, the movement of carbon from the atmospheric pool to the terrestrial plant pool.

The global carbon cycle is closely associated with energy

Carbon (C) is critically important for biological activity because of its role in energy transfer and biological structures. At a global scale, C is a relatively dynamic element, moving between atmospheric, terrestrial, and oceanic pools relatively quickly (over weeks to decades). It is important that we understand this global C cycle because changes in the fluxes of C among these pools are influencing Earth's climate system. As we saw in Chapter 2, C in the atmosphere occurs primarily as carbon dioxide (CO_2) and methane (CH_4), greenhouse gases that influence atmospheric absorption and reradiation of infrared radiation from Earth's surface. Thus, any changes in the atmospheric concentrations of these gases can have profound effects on the global climate, as we will see later in this chapter.

There are four major global pools of C: atmosphere, oceans, land surface (including soils and vegetation), and sediments and rock (**Figure 24.3**) (Schlesinger 1997). The largest of these pools is the combination of sediments and rock, which contain 99% of the global C. The C in the sediment and rock pool is found primarily in the form of carbonate minerals and organic compounds. It is the most stable of the major pools, taking up and releasing C on geological time scales. The oceanic pool consists of two main components, surface waters (to depths of 75–200 m), where most marine biological activity occurs, and deeper, colder waters. There is

FIGURE 24.3 The Global Carbon Cycle The numbers are the major pools (boxes) and annual fluxes (arrows) of carbon (pools in Pg, 10^{15}g; fluxes in Pg/year; GPP, total ecosystem photosynthesis). (Additional data from IPCC 2007.)

agricultural development, and the burning of fossil fuels. Prior to the mid-nineteenth century, forest clearing was the largest contributor to anthropogenic C release to the atmosphere. Removing the forest canopy warms the soil surface, increasing rates of decomposition and heterotrophic respiration. Burning of trees also releases CO_2, as well as small amounts carbon monoxide (CO) and methane (CH_4), to the atmosphere.

The anthropogenic flux of C to the atmosphere from the terrestrial pool has continued to increase in recent decades. In 1970, anthropogenic CO_2 emissions to the atmosphere were 4.1 Gt (10^{15} g) of C per year; by 2005, this rate had doubled (IPCC 2007). Burning of fossil fuels accounts for approximately 80% of the anthropogenic C flux to the atmosphere; the remaining 20% is associated with deforestation. During the last half of the twentieth century deforestation for agricultural development shifted from occurring predominantly in mid-latitudes of the northern hemisphere to the tropics. Approximately half of these anthropogenic CO_2 emissions are taken up by the oceans and terrestrial biota. However, this proportion will decrease with time as the fluxes of terrestrial ecosystem and ocean uptake will not keep pace with the rate of atmospheric CO_2 increase (IPCC 2007).

Photosynthesis is sensitive to the concentration of CO_2 in the atmosphere. As a result, we might expect increases in rates of photosynthesis as anthropogenic CO_2 emissions increase, primarily in plants with the C_3 photosynthetic pathway (see Chapter 5). Experiments have shown, however, that for some herbaceous plants, these increases may be short-term, and that acclimatization to elevated CO_2 concentrations may occur with time. For other plants, such as forest trees, increases in photosynthetic rates may be more sustained, resulting in a CO_2 fertilization effect. It is extremely important that we understand the response of forest ecosystems to elevated CO_2 concentrations. Because so much of terrestrial NPP, and thus C uptake, occurs in these ecosystems, their response will have a profound effect on the fate of anthropogenic CO_2 emissions. However, it is difficult to manipulate atmospheric CO_2 concentrations experimentally in an intact forest. In one successful approach, called free-

a net uptake of C from the atmosphere by the ocean surface as CO_2 dissolves in the water due to a concentration gradient between the atmosphere (higher concentration) and the ocean (lower concentration). There is relatively little mixing between the ocean surface and the deeper waters, although C is transferred between them by the sinking of detritus and carbonate shells of marine organisms, and the flow of polar ocean currents to deeper layers. Most of the oceanic C (> 90%) is in the deeper waters. Some release of this deep ocean pool of C occurs when upwelling brings C-rich water to the surface, releasing CO_2 to the atmosphere.

The terrestrial pool, which includes vegetated and non-vegetated land surfaces and their associated soils, is the largest pool of biologically active C. Soils contain approximately twice as much C as the plant pool. The terrestrial pool exchanges C with the atmosphere primarily by photosynthetic uptake of CO_2 by plants and respiratory CO_2 release by plants and heterotrophs. Prior to the Industrial Revolution that began in the early nineteenth century, CO_2 uptake and release from the terrestrial pool were roughly equal, with no net change in atmospheric CO_2.

As a result of a rapidly growing human population over the past two centuries and associated industrial and agricultural development, there has been an increase in the release of C to the atmosphere from the terrestrial pool. This **anthropogenic** (human-associated) release of C is the result of *land-use-change*, which is mainly the clearing of forests for air CO_2 enrichment, or FACE, researchers inject CO_2 into the air through vertical pipes surrounding stands of trees while monitoring the atmospheric concentration of CO_2 within the experimental stands. The rate of CO_2 injection is controlled to maintain a relatively constant elevated level.

A FACE experiment was used to investigate the ecosystem effects of elevated CO_2 in a young loblolly pine (*Pinus taeda*) forest in North Carolina by Evan DeLucia and his colleagues (DeLucia et al. 1999) (**Figure 24.4**). The experiment was initiated in 1997, when three plots exposed to elevated CO_2 concentrations (200 ppm above ambient levels) and three control plots exposed to ambient CO_2 levels were established. The researchers used measurements of tree basal area to estimate aboveground NPP and repeated collections of soil cores to estimate fine root growth and belowground NPP. DeLucia and his colleagues found that the elevated CO_2 concentrations increased the overall NPP of the forest by 25%. Input of C into the soil, both from aboveground litter and from belowground fine root turnover, also increased. The results of this experiment indicate that forests may be an important sink (reservoir) for anthropogenic CO_2. However, DeLucia et al. suggested that their forest stand represents the upper limit of potential CO_2 uptake, and that older forests,

and forests with lower water and nutrient supplies, may not have as great a capacity to take up CO_2. Results from other FACE experiments in forest ecosystems have found similar responses to elevated CO_2 (an average increase in NPP of 23%; Norby et al. 2005).

Atmospheric CO_2 directly affects the acidity (pH) of the oceans by diffusing into seawater and forming carbonic acid:

$$CO_2 + H_2O \leftrightarrows H_2CO_3 \text{ (carbonic acid)} \leftrightarrows H^+ + HCO_3^-$$
$$\text{(bicarbonate)} \leftrightarrows 2\,H^+ + CO_3^{2-} \text{ (carbonate)}$$

Significant increases in ocean acidity have been forecast using model simulations incorporating the expected increases in anthropogenic CO_2 emissions over the twenty-first century (Caleira and Wickett 2003). These predicted changes will have two main effects on the multitude of marine organisms that form their protective external shells from calcium carbonate, including corals and plankton. First, the increase in acidity will dissolve the existing shells of the organisms. Second, lower concentrations of carbonate in seawater will decrease the organisms' ability to synthesize shells (Feely et al. 2004; Orr et al. 2005). Both effects will increase their mortality and lower their abundances. Anthropogenic emissions of CO_2 therefore have potential to tremendously alter the diversity and function of marine ecosystems.

Emissions of CH_4 to the atmosphere from the terrestrial pool have also increased as a result of human activities. Although atmospheric concentrations CH_4, at 500 ppb, are much lower than those of CO_2, the potential effects of even small increases in CH_4 on climate change are significant because it is a more effective greenhouse gas per molecule than CO_2. CH_4 is emitted naturally by anaerobic methanogenic bacteria that live in wetlands and shallow marine sediments. Methanogenic bacteria in the rumens of ruminant animals are also a source of atmospheric CH_4. Anthropogenic emissions of CH_4 have doubled since the early nineteenth century as a result of the processing and burning of fossil fuels, agricultural development (primarily that of rice, which is grown in flooded fields), burning of forests and crops, and livestock production (IPCC 2007). As a result, atmospheric CH_4 concentrations have more than doubled over the past two centuries, but this increase slowed considerably following 1995.

Atmospheric concentrations of C have changed dynamically throughout Earth's history in association with geological and climatic changes. Concentrations of CO_2 have ranged from greater than 3,000 parts per million (ppm) 60 million years ago to

FIGURE 24.4 A FACE Experiment The circles visible in this aerial photo are free-air CO_2 enrichment (FACE) treatment rings in a loblolly pine (*Pinus taeda*) forest at the Duke Experimental Forest. CO_2 is released from plastic pipes surrounding treatment plots at a rate calculated to raise the CO_2 concentration to 200 parts per million above ambient atmospheric CO_2 concentrations.

FIGURE 24.5 Temporal Changes in Atmospheric CO₂ and CH₄ Atmospheric CO_2 and CH_4 concentrations have varied with temperature over the past 400,000 years. Gas concentrations were measured in bubbles trapped in Antarctic ice; temperatures were estimated using oxygen isotopic analyses. (Data from IPCC 2007.)

less than 200 ppm 140,000 years ago. Over the past 400,000 years, variations in the concentrations of CO_2 and CH_4, measured in tiny bubbles preserved in polar ice, have followed glacial–interglacial cycles (see Chapter 2) (**Figure 24.5**). The lowest CO_2 concentrations during this period were associated with glacial periods. Over most of the past 12,000 years, atmospheric CO_2 concentrations remained relatively stable, varying between 260 and 280 ppm. Since the mid-nineteenth century, however, concentrations have increased at a rate faster than at any other time over the past 400,000 years, reaching values of about 380 ppm in 2005 (IPCC 2007). Even if anthropogenic CO_2 emissions were decreased dramatically, atmospheric CO_2 concentrations would remain elevated due to a time lag (decades to centuries) in oceanic uptake. Thus, elevated CO_2 concentrations will have effects for a long time to come, with or without rapid mitigation of anthropogenic emissions. The influence of CO_2 and CH_4 on climate change will be discussed later in this chapter.

Biological fluxes dominate the global nitrogen cycle

Nitrogen (N) plays a key role in biological processes as a constituent of proteins and enzymes, and it is one of the resources that most commonly limits primary production (see Chapter 19). Thus, cycles of N and C are tightly coupled through the processes of photosynthesis and decomposition.

The largest pool of atmospheric N (> 90%) is dinitrogen gas (N_2) (**Figure 24.6**). As we saw in Chapter 21, N_2 is very stable chemically and cannot be used by most organisms, with the important exception of nitrogen-fixing bacteria, which are able to convert it to a more chemically usable form. Terrestrial N_2 fixation by these bacteria provides approximately 128 teragrams (Tg; 10^{12} g) of biologically available N per year (Cleveland et al. 1999, Galloway et al. 2004)

and supplies 12% of the annual biological demand (Schlesinger 1997). The remaining 88% is derived from uptake of N from the soil in chemically available forms released by decomposition. Oceanic N_2 fixation contributes another 120 Tg annually to the global budget. Geological deposits of N associated with sediments containing organic matter, represent a much smaller fraction of the global pool than in the case of C.

Although the pools of N at land and ocean surfaces are relatively small, they are very active biologically, and they are held tightly by internal ecosystem cycling processes. Fluxes from these biologically active oceanic and terrestrial pools are small relative to the rates of internal cycling, usually less than 10% (Chapin et al. 2002). The natural flux of N that occurs between terrestrial and oceanic pools via movement of N in rivers is tiny, but it plays an important biological role by enhancing primary production in estuaries and salt marshes. Denitrification, a microbial process that occurs in anoxic soils and in the ocean (see Chapter 21), results in movement of N (as N_2) from terrestrial and marine ecosystems to the atmosphere. These ecosystems also lose N through burial of organic matter in sediments and burning of biomass.

Human activities have altered the global N cycle tremendously. Anthropogenic fluxes are now the dominant components of the N cycle (Galloway et al. 2004) (**Figure 24.7**). The rate of fixation of atmospheric N_2 by humans now exceeds natural terrestrial biological rates, and emissions of N associated with industrial and agricultural activities are causing widespread environmental changes, including acid precipitation, discussed later in this chapter. Three major processes account for these anthropogenic effects on the global N cycle.

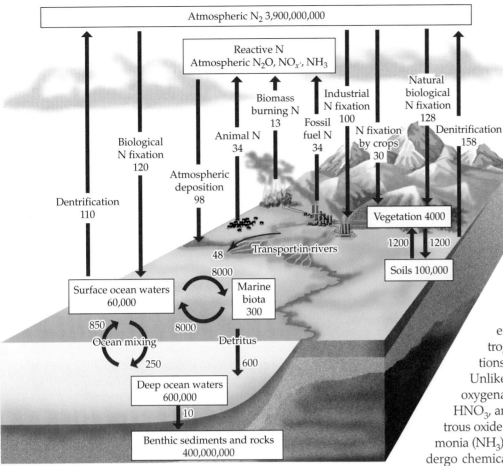

FIGURE 24.6 The Global Nitrogen Cycle
The numbers are the major pools (boxes) and annual fluxes (arrows) of nitrogen (pools in Tg, 10^{12}g; fluxes are Tg/year). (Data from Cleveland et al. 1999 and Galloway et al. 2004.)

biological N_2 fixation. In addition, flooding of agricultural fields for some crops, such as rice, has increased N_2 fixation associated with anaerobic cyanobacteria. Finally, anthropogenic emissions of certain gaseous forms of nitrogen have greatly increased the concentrations of these compounds in the atmosphere. Unlike N_2, these compounds, which include oxygenated nitrogen compounds (NO, NO_2, HNO_3, and NO_3^-; known collectively as NO_x), nitrous oxide (N_2O, also known as laughing gas), ammonia (NH_3), and peroxyacetyl nitrate (PAN), can undergo chemical reactions in the atmosphere and are potentially available for biological uptake. Fossil fuel combustion is the primary source of anthropogenic nitrogenous gas emissions. Other contributors include biomass burning associated with deforestation, denitrification and volatilization (conversion to gaseous form) of fertilizers, and emissions from livestock feedlots and human sewage treatment plants. These reactive forms of N are returned to terrestrial and marine ecosystems through the process of atmospheric deposition (see Chapter 21).

First, agricultural fertilizer is manufactured using the Haber–Bosch process, which involves combining air (78% N_2) and hydrogen gas under high pressures and temperatures to form ammonium. The manufacture of fertilizer was responsible for major advances in agricultural food supply during the twentieth century. Second, the growing of N_2-fixing crops such as soybeans, alfalfa, and peas has significantly increased

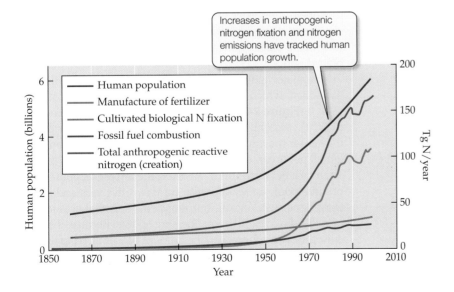

FIGURE 24.7 Changes in Anthropogenic Fluxes in the Global N Cycle   Increases in fertilizer production through the Haber–Bosch process, use of N_2-fixing crops (e.g., soybeans), and combustion of fossil fuels have all contributed to the tremendous increase in biologically available (reactive) N. (After Galloway et al. 2004.)

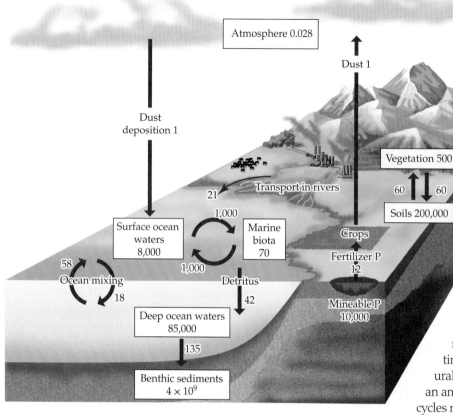

Atmosphere 0.028

Dust 1

Dust deposition 1

Vegetation 500

Transport in rivers

21

1,000

60   60

Surface ocean waters 8,000

Marine biota 70

Soils 200,000

Crops

58

1,000

Fertilizer P 12

Ocean mixing

Detritus

18

42

Mineable P 10,000

Deep ocean waters 85,000

135

Benthic sediments 4×10^9

FIGURE 24.8 The Global Phosphorus Cycle The numbers are the major pools (boxes) and annual fluxes (arrows) of phosphorus (pools in Tg; fluxes in Tg/year).

The global phosphorus cycle is dominated by geochemical fluxes

Phosphorus (P) limits primary production in some terrestrial ecosystems—particularly those with old, well-weathered soils, such as tropical lowland forests—and in some aquatic and marine ecosystems. Phosphorus availability can also control the rate of biological N_2 fixation because that process has a relatively high metabolic demand for P. Consequently, the C, N, and P cycles are linked to one another through photosynthesis and NPP, decomposition, and N_2 fixation.

Unlike C and N, P has essentially no atmospheric pool, with the exception of dust (**Figure 24.8**). Gaseous forms of P are extremely rare. The largest pools of P are in terrestrial soils and marine sediments. Phosphorus is released from sedimentary rocks in biologically available forms by weathering. The largest fluxes of P occur in internal ecosystem cycles, which form a tight recycling loop between biological uptake by plants and microorganisms and release by decomposition. Typically, very little of the P cycling through terrestrial and aquatic ecosystems is lost. In terrestrial ecosystems, most P loss is associated with occlusion of P (i.e., chemical conversion from soluble to insoluble, biologically unavailable forms). Movement of P from terrestrial to aquatic ecosystems occurs primarily through erosion and movement of particulate organic matter, mainly bits and pieces of plants, into streams. Much of the P transported from terrestrial to marine ecosystems (about 90%) is lost when it is deposited in deep ocean

sediments. Ultimately, P in sediments in both marine and terrestrial ecosystems is cycled again in association with tectonic uplift and weathering of rocks, which occurs on a scale of hundreds of millions of years. Anthropogenic effects on the global P cycle are associated with application of agricultural fertilizers, discharges of sewage and industrial wastes, and increases in terrestrial surface erosion. Phosphorus fertilizers are usually derived from the mining of uplifted, ancient marine sediment deposits. Mining releases four times more P annually than is liberated through natural rock weathering. Globally, P fertilizer is applied in an amount equivalent to approximately 20% of the P that cycles naturally through terrestrial ecosystems (Schlesinger 1997). While occlusion of P in the soil minimizes the flux of anthropogenic P from terrestrial to aquatic ecosystems, that flux still has significant potential for negative environmental effects. One such effect is eutrophication in lakes, as described in the case of Lake Washington (see p. 470).

Biological and geochemical fluxes both determine the global sulfur cycle

Sulfur (S) is a constituent of some amino acids and ribonucleic acids (DNA, RNA), but it is probably never in short supply for organismal growth. Sulfur plays important roles in atmospheric chemistry, and as with C, N, and P, anthropogenic changes to the global S cycle have important negative environmental consequences, primarily through the generation of acid rain.

The major global pools of S are in rocks, sediments, and the ocean, which contains a large pool of dissolved sulfate (SO_4^{2-}) (**Figure 24.9**). Weathering of S-containing minerals, mainly sedimentary pyrite, releases more soluble forms that may enter the atmosphere or oceans. Fluxes of S among the global pools can occur in gaseous, dissolved, or solid forms. There is a net movement of S from the terrestrial pool to the oceanic pool, associated with transport in rivers and in atmospheric dust. Volcanic eruptions emit substantial amounts of sulfur dioxide (SO_2) into the atmosphere. Because they are episodic events, however, the amount of S emitted to the atmosphere by volcanic eruptions is approximately the same as the amount blown into the atmosphere as dust from bare

FIGURE 24.9 The Global Sulfur Cycle The numbers are the major pools (boxes) and annual fluxes (arrows) of sulfur (pools in Tg; fluxes in Tg/year).

soils on a time scale of centuries. Oceans emit S to the atmosphere as small particles of windborne ocean spray and as gaseous emissions associated with microbial activity. Bacteria and archaea in anaerobic soils also emit sulfur-containing gases such as hydrogen sulfide (biogenic emissions). Most gaseous S compounds in the atmosphere undergo transformation to SO_4^{2-} and H_2SO_4 (sulfuric acid), which are removed relatively quickly by precipitation.

Anthropogenic emissions of S to the atmosphere have quadrupled since the Industrial Revolution. Most of these anthropogenic emissions are associated with the burning of S-containing coal and oil and the refining of ore from mining. What goes up comes down in the form of atmospheric deposition, usually within the same region from which it was emitted, but not always. Long-distance transport of fine dust is an episodic event, associated with droughts and major wind events. Increases in erosion associated with clearing of vegetation and overgrazing have contributed to anthropogenic input of S into the atmosphere as dust. Transport of S in rivers has doubled over the past 200 years (Schlesinger 1997).

Human activities have resulted in changes in all four of the global biogeochemical cycles we have just described, and as we have noted, some of those changes have had important environmental effects. Let's turn our attention to those effects now.

Global Climate Change

CONCEPT 24.2
Earth is warming at an unprecedented rate due to anthropogenic emissions of greenhouse gases.

Throughout this book, we have emphasized the role that climate plays in ecological processes, including the distributions and physiological performance of organisms, rates of resource supply, and the outcomes of biological interactions such as competition. Thus, changes in climate, particularly changes in the frequency of extreme events such as extensive droughts, violent storms, or extreme high and low temperatures, will have profound effects on ecological patterns and processes. Because they are disturbances that result in significant mortality within populations, these extreme events are often critical in determining species' geographic ranges.

In Chapter 2, we learned the difference between weather and climate. *Weather* refers to the current state of the atmosphere around us at any given time, including temperature, humidity, cloudiness, and precipitation. *Climate* is the long-term description of weather, including both average conditions and the full range of variation. *Climatic variation* occurs at a multitude of time scales, from the daily changes associated with daytime solar heating and nighttime cooling, to seasonal changes associated with the tilt of Earth's axis, to decadal changes associated with interactions between ocean currents and the atmosphere (such as the Pacific Decadal Oscillation, described in Chapter 2). **Climate change**, on the other hand, refers to directional change in climate over a period of several decades. Earth is currently undergoing climate change.

Evidence of climate change is substantial

Climate change is distinguished from climatic variation by significant trends lasting at least three decades. Based on analy-

FIGURE 24.10 Changes in Global Temperature and Precipitation (A) Mean annual global temperature anomalies between 1880 and 2005, averaged from air and sea surface temperature records from a number of meteorological stations and normalized to sea level. (B) Regional trends in mean annual global temperatures for 2001–2005. The values for parts A and B are expressed as differences (anomalies) from the average global temperature between 1951 and 1980. (C) Trends in global precipitation from 1901 to 2005, expressed as percentage change over that time period. (A, B after Hansen et al. 2006; C after IPCC 2007.)

(A)

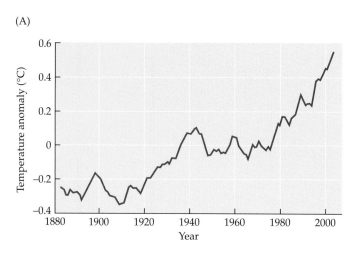

(B) 2001–2005 mean surface temperature anomaly (°C)

ses of records from numerous climate monitoring stations, atmospheric scientists have determined that Earth is currently experiencing a significant change in climate (IPCC 2007) (**Figure 24.10A**). During the twentieth century, the average global surface temperature increased 0.6°C (± 0.2°C), with the greatest change occurring in the past 50 years (**Figure 24.10B**). This rapid rise in global temperature is unprecedented in the past 10,000 years, although temperature changes at similar rates may have occurred at the onset and end of some glacial cycles. The 1990s was the warmest decade of the previous 1,000 years, and 2005 was the warmest year in over a century (IPCC 2007). In association with this warming trend, there has been a widespread retreat of mountain glaciers, thinning of the polar ice caps and melting of permafrost, and a rise in sea level of 15 cm since 1900.

The warming trend has been heterogeneous across the globe, with some regions warming, others not changing significantly, and some even cooling. The warming trend has been most significant in the mid- to high latitudes of the Northern Hemisphere. Changes in precipitation have also occurred, with more precipitation in terrestrial portions of the high latitudes of the Northern Hemisphere and drier weather in the subtropics and tropics (**Figure 24.10C**). There has also been a tendency toward greater frequencies of some extreme weather events, such as hurricanes (including massive storms such as Hurricane Katrina in 2005), droughts, and high temperature extremes (IPCC 2007).

What are the causes of the observed climate change?

As we saw in Chapter 2, climate change may result from changes in the amount of solar radiation absorbed by Earth's surface or in the amount of absorption and reradiation of infrared radiation by gases in the atmosphere. Changes in ab-

sorption of solar radiation by Earth may be associated with the amount of radiation emitted by the sun, variation in Earth's position relative to the sun, or in the reflection of solar radiation by clouds or surfaces with high reflectivity (albedo), such as snow and ice.

The warming of Earth by atmospheric absorption and reradiation of infrared radiation emitted by Earth's surface is known as the **greenhouse effect** (see Figure 2.4). This phenomenon is associated with radiatively active **greenhouse gases** in the atmosphere, primarily water vapor, CO_2, CH_4, and N_2O. The effectiveness of these gases in absorbing radiation is associated with their concentrations and chemical properties. Water vapor contributes the most to greenhouse warming, but its atmospheric concentration varies greatly from region to region, and changes in its average concentration have been small. Of the remaining greenhouse gases, which tend to be better mixed into the atmosphere, CO_2 con-

(C)

Change in precipitation per century, 1901–2005 (%)

<-100   -80   -60   -40   -20   -5   5   20   40   60   No data

tributes the most to greenhouse warming, followed by CH_4 (which has about 30% of the effect of CO_2; although its effect per molecule is greater, its atmospheric concentration is much lower) and N_2O (with about 10% of the effect of CO_2).

As we saw earlier in our discussion of the global biogeochemical cycles of C and N, atmospheric concentrations of CO_2, CH_4, and N_2O are increasing substantially, primarily as a result of fossil fuel combustion and land use change (**Figure 24.11**). Are increases in anthropogenic emissions of these greenhouse gases responsible for global climate change? To evaluate the underlying causes of climate change, its potential effect on ecological and socioeconomic systems, and our options for limiting climate change associated with human activities, the World Meteorological Organization and the United Nations Environment Programme established the Intergovernmental Panel on Climate Change (IPCC) in 1988. The IPCC convenes panels of experts in atmospheric and climatic science to evaluate climatic trends and the probable causes for any such changes. These experts use a combination of sophisticated modeling and analysis of data from the scientific literature to evaluate potential underlying causes of observed climate change, and scenarios for climate change in the future. The IPCC releases assessment reports periodically to enhance the understanding of climate change among scientists, policymakers, and the general public. In recognition of their efforts to spread "knowl-

edge about man-made climate change," the IPCC was awarded the Nobel Peace Prize in 2007.

In its third assessment report, released in 2001, the IPCC concluded that the majority of the observed global warming is attributable to human activities (**Figure 24.12**). While this conclusion is debated in the political arena, it is backed by the majority of the world's leading atmospheric scientists.

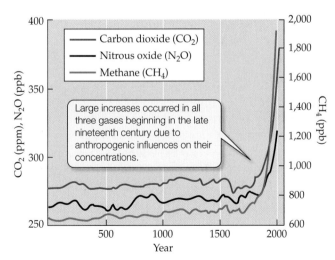

FIGURE 24.11 Increases in Greenhouse Gases Atmospheric concentrations of three greenhouse gases—CO_2, CH_4, and N_2O—over the past 2,000 years have been determined from ice cores and direct atmospheric measurements since 1958. (After IPCC 2007.)

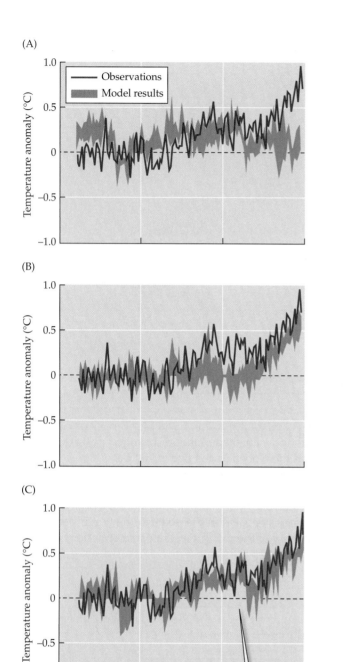

(A)

(B)

(C)

The results of the model that used both natural climatic and anthropogenic factors were the best match to the observed global temperature changes.

FIGURE 24.12 Contributors to Global Temperature Change
IPCC scientists compared observed global temperature changes between 1850 and 2000 with the results of computer models. The models predicted the temperature changes that would be expected in that period based on (A) natural climatological factors only, including variation in solar radiation and atmospheric concentrations of aerosols from volcanic eruptions, (B) anthropogenic factors only, including concentrations of greenhouse gases and sulfate aerosols, and (C) a combination of natural and anthropogenic factors. The values are expressed as anomalies relative to the mean temperature between 1901 and 1950. The results indicate that anthropogenic factors have played a large role in the observed warming. (After IPCC 2007.)

When did humans begin to affect climate?

There is widespread agreement that human effects on the global climate system began around 1900, when greenhouse gases produced during the Industrial Revolution began to accumulate in appreciable amounts in the atmosphere. This conclusion assumes that the climate over the previous several millennia was relatively stable under natural conditions. However, a recent hypothesis proposed by William Ruddiman suggests that human influences on the climate system began about 5,000 years ago with the advent of agricultural development in Europe, Asia, and the Americas (Ruddiman 2003).

Ruddiman based his hypothesis on an evaluation of climate change during the glacial–interglacial cycles of the past 400,000 years and its association with trends in atmospheric concentrations of CO_2 and CH_4 (see Figure 24.5). Through multiple glacial–interglacial cycles, atmospheric CO_2 and CH_4 concentrations peaked just before the onset of climatic cooling associated with decreases in the intensity of solar radiation due to changes in the shape of Earth's orbit. Concentrations then declined as the cooling progressed. Peaks in CO_2 and CH_4 concentrations occurred about 10,500 years ago, but the expected decreases following the peak were reversed about 8,000 and 5,000 years ago, respectively, when small increases in CO_2 and CH_4 concentrations occurred (**Figure 24.13**). During this period, humans were developing organized agricultural techniques, including forest clearing and irrigation. Ruddiman suggested that increases in atmospheric CO_2 and CH_4 concentrations associated with these agricultural practices prevented Earth from entering a new glacial cycle about 8,000 years ago. He also noted that reductions in atmospheric CO_2 concentrations during the past 5,000 years have coincided with human disease epidemics and abandonment of agriculture. While Ruddiman's hypothesis is controversial, it challenges us to consider what is "natural" climate change and what is "anthropogenic" climate change.

Will the climate continue to grow warmer? The IPCC's models project an additional increase in average global temperature of 1.8°C–4.0°C over the twenty-first century (IPCC

Paul Crutzen, a Nobel Prize winner for his work on stratospheric ozone loss, has suggested that we have entered a new geological period, which he called the *Anthropocene* (*anthropo,* "human;" *cene,* "recent") *epoch* (geological age) to indicate the extensive impact of humans on our environment, particularly the climate system (Crutzen and Stoermer 2000).

(A) CO_2

(B) CH_4

Instead of decreasing, as they did at about the same point in previous glacial–interglacial cycles, greenhouse gas concentrations began to increase.

Observed trend
Natural trend

Years before present

Years before present

FIGURE 24.13 A Delayed Ice Age? William Ruddiman compared observed trends in atmospheric concentrations of (A) CO_2 and (B) CH_4 with those expected based on repeated glacial–interglacial patterns over the past 400,000 years. Ruddiman suggested that the deviation from the expected decreases resulted from the development of agriculture about 5,000 years ago. (After Ruddiman 2003.)

2007). The range of variation in this estimate is associated with uncertainties about future rates of anthropogenic greenhouse gas emissions and about the behavior of the terrestrial–atmosphere–ocean system. Model simulations incorporating different economic development scenarios have predicted vastly different future rates of emissions. Aerosols in the atmosphere represent another source of uncertainty in the models' predictions. Aerosols, which reflect solar radiation, have a cooling effect on global temperatures; for example, emissions of large amounts of aerosols associated with major volcanic eruptions have had notable cooling effects at a global scale (**Box 24.1**). While some aerosols have been increasing in the atmosphere (e.g., dust in association with land use change and desertification), others have decreased (e.g., SO_4^{2-}, due to decreasing anthropogenic SO_2 emissions). Water in the atmosphere may play contrasting roles: clouds may have a cooling effect, while water vapor, which may increase due to greater evapotranspiration, may increase greenhouse warming. Despite these uncertainties in predicting the magnitude of future climatic warming, further ecologically meaningful increases in global temperatures are a certainty.

The ecological consequences of climate change are substantial

What does a 1.8°C–4.0°C change in temperature mean for biological communities? We can get a sense of what such a temperature change might mean by comparing it with the climatic variation associated with elevation in mountains. A median value for the projected temperature change (2.9°C) would correspond to a 500 m (1,600-foot) shift in elevation. In the Rocky Mountains, this change in climate would correspond approximately to a full change in vegetation zone, from subalpine forest (dominated by spruce and fir) to montane forest (dominated by Ponderosa pine). Thus, if we assume perfect tracking of climate change by the current vegetation, climate change during the twenty-first century would result in an elevational shift in vegetation zones of 200–860 m. Similar predictions for latitudinal climatic shifts suggest movement of biological communities of 500–1,000 km toward the poles.

Climate–vegetation correlations, such as those described in Chapter 4, are useful as a demonstration of what could happen with climate change, but it would be naive to use them to predict what will actually happen to biological communities. We know that communities are structured by a multitude of factors, including climate, particularly climatic extremes, as well as species interactions, the dynamics of succession, dispersal ability, and barriers to dispersal (as described in Unit 4). Because the coming climate change will be rapid relative to the climate changes that have shaped current biological communities, it is unlikely that the same assemblages of organisms will form the communities of the future.

Paleoecological records of past communities reinforce the suggestion that novel communities may emerge with climate change by showing that some plant communities of the past were quite different from modern plant communities. Jonathan Overpeck and colleagues used pollen records to reconstruct large-scale vegetation change since the most recent glacial maximum in eastern North America (18,000 years ago) (Overpeck et al. 1992). They found not only that community types had made latitudinal shifts as the climate

BOX 24.1

Models, Volcanoes, and Climate Change

When atmospheric scientists realized in the 1970s that greenhouse gas concentrations in the atmosphere were increasing, they began to evaluate the effects of these gases on future climate change using computer models of Earth's climate system. These climate models, which simulate atmospheric and terrestrial physics associated with heat exchange at numerous points on Earth, can be incredibly complex. How well do they predict the effects of changes in atmospheric concentrations of greenhouse gases and aerosols on global temperatures? Two examples demonstrate that climate models have provided reasonably accurate predictions, giving us some confidence in their forecasts.

In 1992, NASA scientist James Hansen used a climate model to make a bold prediction: he announced that a recent volcanic eruption in the Philippines would cool Earth's average temperature by 0.5°C (1°F) over a 15-month period (Hansen et al. 1992). Earth would then begin to warm, returning to the temperatures observed before the eruption by 1995. The volcano in question was Mount Pinatubo, which in June of 1991 exploded in one of the largest eruptions of the twentieth century (**Figure A**). Five cubic kilometers (about 1 cubic mile) of

volcanic material blasted out of the mountain, sending a cloud of gas and ash to a height of 35 kilometers (22 miles). Simultaneously, huge avalanches of hot gas and ash roared down the sides of the mountain, killing hundreds of people and filling valleys with up to 200 m (660 feet) of volcanic deposits.

The eruption of Mount Pinatubo sent 17 million tons of sulfur dioxide into the stratosphere. Sulfur dioxide is invisible, but it reacts with water vapor to form aerosols of sulfuric acid that partially block incoming solar radiation. In the stratosphere, sulfuric acid aerosols can persist for several years and be spread by wind around the globe, potentially cooling the entire planet.

Hansen's predictions were right on target, down to the month and tenth of degree: average global temperatures dropped 0.5°C over a 15-month period, then gradually rose again, reaching their previous levels by 1995. Mount Pinatubo provided a unique opportunity to model a short-term effect of changes in aerosol concentrations on global tempera-

tures. How well do climate models perform with changes in greenhouse gases?

The scientists who prepared the IPCC's fourth report (2007) compared projections made for its first three reports (1990, 1996, and 2001) with observed changes in climate to see how well climate models had predicted climate change over the period from 1990 to 2005. The predictions for mean annual global temperatures made in the earlier reports (an increase of 0.1°C–0.3°C per decade) were very close to the temperatures actually recorded (an increase of 0.2°C per decade, **Figure B**). Even when the uncertainties associated with future greenhouse gas emissions are taken into account, the models have been fairly accurate in their predictions of climate change. These results give us some confidence that the models can provide reasonable estimates of future climatic scenarios and take into account the most important factors influencing the relationships between greenhouse gas emissions and climate change. Nevertheless, improvements are continually being made to climate models to improve their forecasting ability.

FIGURE A   Eruption of Mount Pinatubo, 1991

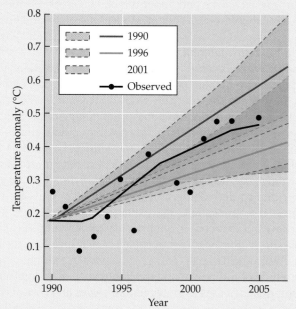

FIGURE B   Can Climate Models Forecast Climate Change?   Best estimates for changes in mean annual global temperatures relative to mean temperature from 1961 to 1990 from the first three IPCC reports matched the actual changes between 1990 and 2005 reasonably well. The shaded regions indicate variation in the estimates due to uncertainties about future greenhouse gas emissions. (Note that there was no "best estimate" line for the 2001 report.) (After IPCC 2007.)

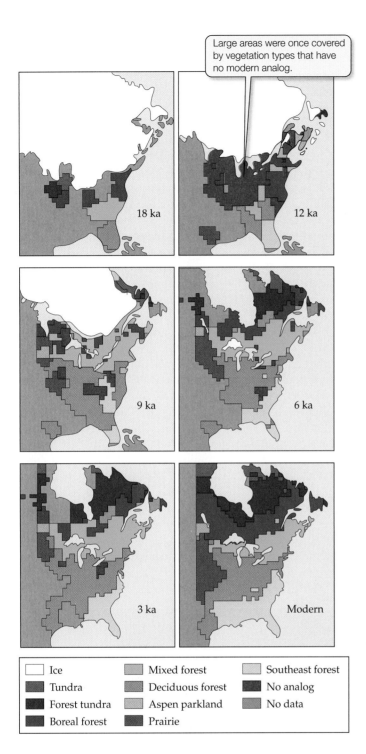

Large areas were once covered by vegetation types that have no modern analog.

18 ka

12 ka

9 ka

6 ka

3 ka

Modern

☐ Ice	▨ Mixed forest	▨ Southeast forest
■ Tundra	▨ Deciduous forest	■ No analog
■ Forest tundra	▨ Aspen parkland	▨ No data
■ Boreal forest	▨ Prairie	

FIGURE 24.14 Past Vegetation Change Vegetation types in eastern North America have changed since the last glacial maximum, 18,000 years ago (dates are years before present). Vegetation composition was determined from preserved pollen deposits. (After Overpeck et al. 1992.)

be the only way for them to avoid extinction. Their rates of dispersal, and barriers to their dispersal associated with anthropogenic habitat fragmentation, will be important constraints on their responses to climate change. For most animals, mobility is not a problem, but their habitat and food requirements are intimately associated with the presence of specific vegetation types.

Plant dispersal rates are, on average, much slower than the predicted rate of climate change. In order to track the projected change in climate over the next century, plant species populations will need to move 5–10 km per year. Plant species that have animal-dispersed seeds, and which can establish viable populations and grow to reproductive maturity in a relatively short time, may be able to disperse rapidly enough to keep pace with climate change. However, this kind of dispersal strategy is common mainly in ruderal (weedy) herbaceous plants. Shrubs and trees have much slower rates of dispersal; as a result, there may be significant time lags in their response to climate change. In addition, barriers to dispersal may prevent organisms of all kinds from migrating in response to climate change. Dams, for example, may prevent fish from moving to water with more suitable temperatures. Fragmentation of habitat by human development may pose significant barriers to dispersal for some species (see Chapter 23). Without dispersal corridors, species face a greater probability of local population extinction in the face of climate change.

In addition to affecting species' geographic ranges, climate change will affect ecosystem functions, such as net primary production (NPP), decomposition, and nutrient cycling and retention. Both photosynthesis and respiration are sensitive to temperature, and because their balance determines NPP, the direct effects of climatic warming on NPP may be relatively minor. As indicated in Chapter 19, however, variation in NPP is related to water and nutrient availability and vegetation type, all of which may be affected by climate change. Changes in precipitation regimes and evapotranspiration rates resulting from climatic change may strongly influence both water and nutrient availability. Because of the regional nature of climate change and the resulting changes in vegetation types, both increases and decreases in NPP may occur. Thus, the effect of climate change on NPP will probably not be uniform. The effect of warming on nutrient supplies will be most pronounced in mid- to high-latitude terrestrial ecosystems, where low temperatures constrain rates of nutrient cycling and soils have large pools of nutrients. As a result, climate change may lead to increases in NPP in some temperate and boreal forest ecosystems.

warmed, but also that community types without modern analogs (**Figure 24.14**) existed under climatic regimes that were unique and no longer present. Overpeck and his colleagues concluded that future vegetation assemblages would follow similar trends, given the predicted rapid rate of global warming and the potential for the emergence of unique climatic patterns with no current analogs.

Because climate change will be continue to be relatively rapid, it is likely that evolutionary responses will not be possible for most plants and animals, and thus dispersal may

Ecological responses to climate change are apparent

As noted earlier, global warming of 0.6°C has occurred over the past century. Several physical environmental changes have occurred concurrently, including the retreat of glaciers, enhanced melting of sea ice, and a rise in sea level. Have biological systems responded to climate changes? Numerous reports of biological changes are consistent with recent global warming (Walther et al. 2002). These changes include earlier migration of birds and earlier spring greening of vegetation.

Although they are more difficult to link directly to climate, there have been changes in the geographic ranges of species that are consistent with warming. For example, Georg Grabherr and colleagues investigated the flora of vascular plants found on summits of mountains in the European Alps and compared the current species composition with records dating back to the eighteenth and early nineteenth centuries (Grabherr et al. 1994). They found a consistent trend of upward movement of species from lower elevations onto the summits of the mountains (**Figure 24.15**). Similarly, Camille Parmesan and colleagues recorded a northward shift in the ranges of European nonmigratory butterfly species (Parmesan et al. 1999). Of the 35 species examined, 63% had shifted their ranges northward, while only 3% had shifted their ranges southward. An estimated 279 plant and animal species have shown geographic range shifts that are consistent with recent climate changes (Parmesan and Yohe 2003).

Migratory animals may also be adversely affected by climate change (Root et al. 2003). For example, migrating marine species, including whales and fish, may be forced into longer journeys due to substantial changes in the distribution of their prey species as ocean temperatures warm. Some migratory birds in England and North America have been arriving at nest sites as much as 3 weeks earlier than they did 30 years ago due to warmer spring temperatures and faster snow melt. However, plants and invertebrate prey species have responded faster to climate change than the migrating birds, resulting in a mismatch of bird arrival and prey availability. On the other hand, longer breeding seasons may increase the number of offspring produced by some bird species, particularly in high latitude ecosystems.

Changes in community composition, as well as local extinctions of species, may also be indicators of climate change. These effects may be particularly apparent in some nonmotile marine species. As indicated in the Case Study, corals are sensitive to changes in water temperature. Warm temperatures can kill the algal symbionts of corals, resulting in bleaching of coral reefs. Since 1980, several bleaching events have occurred around the world in association with warmer-than-average sea surface temperatures. In 1998, one of the warmest years ever recorded, 16% of coral reefs across the globe experienced bleaching (Walther et al. 2002). Changes in the abundances of marine foraminiferans, a type of zooplankton, also reflect trends in global warming during the last century (Field et al. 2006). Foraminiferans have characteristic shells that allow them to be identified in marine sediments. Cores collected from benthic sediments can be examined to determine changes in the species composition of foraminiferans over time. Because the environmental tolerances of different species are known, changes in the abundances of foraminiferan species provide a means of reconstructing ma-

FIGURE 24.15 Plants Are Moving Up the Alps   Grabherr and colleagues compared historical records of vascular plant species richness on the summits of mountains in the European Alps with censuses taken in the early 1990s. The green curve indicates the pattern of species richness with change in the elevation of the summit for historical records, while the red curve indicates the present relationship. (After Grabherr et al. 1994.)

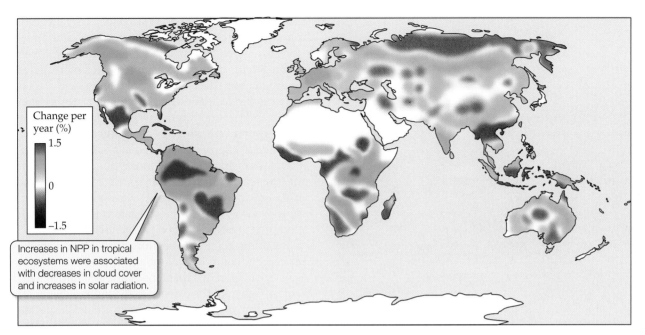

Change per year (%)
1.5
0
−1.5

Increases in NPP in tropical ecosystems were associated with decreases in cloud cover and increases in solar radiation.

FIGURE 24.16 Changes in Terrestrial NPP Changes in net primary production (NPP) observed between 1982 and 1999, expressed as percentage change per year, as determined using remote sensing. (After Nemani et al. 2003.)

rine environments of the past. Following the mid-1970s, an increase in tropical and subtropical foraminiferan species, and a decrease in temperate and polar species, occurred in the eastern North Pacific Ocean, indicating a warming of ocean waters there.

Finally, changes in global NPP and in the balance of C exchange in the Arctic also indicate biological responses to climate change. Ramakrishna Nemani and colleagues used remote sensing data to examine global patterns of NPP over an 18-year period (1982–1999) (Nemani et al. 2003). In addition, they evaluated changes in the potential climate-associated constraints on plant growth (sunlight, temperature, and water availability) during the study period. They found that global NPP increased 6% during the study period, or 0.3% per year (**Figure 24.16**). Tropical ecosystems exhibited the largest increase in NPP, which was associated with increases in solar radiation due to less cloud cover during the study period. There was a notable decrease in NPP in northern latitudes, which coincided with research indicating that some areas of the Arctic switched from a net uptake of CO_2 from the atmosphere (acting as a *sink*) to a net export of CO_2 (acting as a *source*) (Oechel et al. 1993). Large amounts of C are stored in the soils of boreal and Arctic ecosystems as a result of strong temperature constraints on decomposition and the long-term buildup of carbon in the soils since the last glacial maximum. However, warming during the twentieth century increased the rate of CO_2 loss from Arctic soils, such that losses now exceed gains from NPP. Warming of high-latitude

terrestrial ecosystems could provide a positive feedback to climate change by enhancing their losses of CO_2 and CH_4. However, the rates of CO_2 loss from Arctic ecosystems have decreased since the early 1990s, possibly due to changes in rates of nutrient cycling and physiological and compositional changes in the plants (Oechel et al. 2000).

Biological indicators of global climate change are diverse, and they are increasing with time. Experiments, modeling, and comparisons with historical and paleoecological records have all given us clues to how Earth's biota will respond to climate change. However, substantial uncertainties in predicting biological responses to climate change still exist, many of which are associated with other environmental changes that are occurring at the same time. In the next section, we'll look at two such anthropogenic changes that are having profound effects on ecosystems: emissions of sulfur and nitrogen into the atmosphere.

Acid and Nitrogen Deposition

CONCEPT 24.3
Anthropogenic emissions of sulfur and nitrogen cause acid deposition, alter soil chemistry, and affect the health of ecosystems.

The negative effects of air pollution have been known since at least the time of the ancient Greeks, when laws protected the quality of air, as indicated by its odor (Jacobson 2002). Since the Industrial Revolution, air pollution has mainly been associated with urban industrial centers, power plants, and oil and gas refineries. These stationary sources of atmospher-

FIGURE 24.17 Air Quality Monitoring in Grand Canyon National Park
Visibility in national parks and wilderness areas, such as the Grand Canyon, has been compromised by emissions of anthropogenic pollutants, including sulfate aerosols. These pollutants not only lower visibility, but also pose a health hazard to the organisms that come into contact with them, including humans.

ic pollutants mainly affect the areas immediately adjacent to them and are usually considered regional rather than global problems. However, effective emissions dispersal strategies (e.g., tall smokestacks), widespread industrial development during the twentieth century, and greater emissions of pollutants from mobile sources, such as automobiles, have increased the spatial extent of air pollution tremendously.

Fossil fuel combustion, agriculture, and urban and suburban development have influenced fluxes of N and S to an even greater degree than fluxes of C. Emissions of N and S into the atmosphere have resulted in two related environmental issues: acid precipitation and N deposition. Emissions of N and S are only a subset of the multiple types of air pollution, but they are among the most far-reaching. Sites affected by acid precipitation and N deposition now include national parks and wilderness areas (**Figure 24.17**).

Acid precipitation causes nutrient imbalances and aluminum toxicity

The detrimental effects of air pollution on nearby vegetation, buildings, and human health have been known for several centuries, although the causes were not well understood. In England during the mid-nineteenth century, industrial processes that released acidic compounds into the atmosphere, primarily hydrochloric acid, were implicated as a major

source of harmful pollution (Jacobson 2002). Legislation was enacted in 1863 to reduce these acidic emissions. The term "acid rain" was first coined by Robert Angus Smith, an inspector charged with enforcing the new regulations. Despite such legislation, acid precipitation continued to be a problem throughout the nineteenth and twentieth centuries in large industrialized urban centers. During the 1960s, awareness of the widespread effects of acid precipitation, including its effects on nearby "pristine" ecosystems and agriculture, increased. In particular, damage to forests and mortality among aquatic organisms in northern Europe, parts of Asia, and northeastern North America prompted greater attention to acid precipitation.

Sulfuric acid (H_2SO_4) and nitric acid (HNO_3) are the main acidic compounds found in the atmosphere. As we saw earlier in this chapter, sulfuric acid forms in the atmosphere from the oxidation of gaseous sulfur compounds emitted by the burning of fossil fuels, especially some forms of coal and oil, and the smelting of metal-containing ores. Nitric acid originates from the oxidation of NO_x compounds, which are also emitted into the atmosphere by fossil fuel combustion. Sulfuric and nitric acids can dissolve in cloud water, and fall from the sky with precipitation (wet deposition). Naturally occurring precipitation has a pH of 5.0 to 5.6, due to the natural dissolution of carbon dioxide and formation of carbonic acid. Acid precipitation has a pH range from 5.0 to 2.0. Acidic compounds may also be deposited to Earth's surface when they form aerosols too large to be suspended, or when they attach to the surfaces of dust particles. This process is dry deposition.

Research has focused on determining the causes of environmental degradation associated with acid precipitation. Initially, the acidity was considered the main culprit. However, in most cases, rainfall and surface waters did not have a low enough pH to cause the observed biological responses. An exception occurs in regions at high latitudes or high elevations that develop a seasonal snowpack. During winter, acidic compounds accumulate in the snow. When temperatures increase in spring, water percolates through the snowpack, leaching out all of the accumulated soluble compounds. The first meltwater of spring is therefore more acidic than the precipitation that fell during the winter. This *acid pulse* has the potential to be toxic to sensitive organisms in soils and streams, including microorganisms, invertebrates, amphibians, and fish.

The vulnerability of organisms in soils, streams, and lakes to inputs of acid precipitation is determined by the ability of their chemical environment to counteract the acidity, known as its **acid neutralizing capacity**. The acid neutralizing capacity of soils and water is usually associated with concentrations of base cations, including Ca^{2+}, Mg^{2+}, and K^+. Soils derived from parent material with high concentrations of

FIGURE 24.18 Air Pollution Has Damaged European Forests The high tree mortality seen in this forest in southeastern Germany is associated with acid rain and the resulting nutrient imbalance, particularly losses of nutrient cations. Extensive forest decline occurred in Germany and northern Czechoslovakia (now part of the Czech Republic) in the 1970s and 1980s.

these cations, such as limestone, are better able to neutralize acid precipitation than those from more acidic parent material, such as granites.

The detrimental effects of acid precipitation on plants and aquatic organisms are associated with biogeochemical reactions in the soil. As H^+ percolates through the soil, it replaces Ca^{2+}, Mg^{2+}, and K^+ at cation exchange sites on the surfaces of clay particles (see Chapter 21). These cations are released into the soil solution and can then leach out of the rooting zone of plants. The loss of base cations leads to a decrease in soil pH, or soil acidification. Deficiencies in Ca and Mg, sometimes in combination with other stresses, were associated with large-scale mortality of trees in European forests during the 1970s and 1980s (**Figure 24.18**). In advanced stages of soil acidification, aluminum (Al^{3+}) is released into the soil from cation exchange sites. Aluminum is toxic to plant roots, soil invertebrates, and aquatic organisms, including fish. The combination of increasing acidity in precipitation and increasing aluminum concentrations in terrestrial runoff has been linked to fish die-offs in lakes and streams in northern Europe and eastern North America.

The realization that acid precipitation was negatively affecting the biota of forest and lake ecosystems prompted enhanced monitoring of atmospheric deposition and, eventually, laws to limit acidic emissions. Restrictions on emissions of S in North America and Europe have resulted in significant reductions in the acidity of precipitation (**Figure 24.19**). Forests are recovering from the effects of acid precipitation in central Europe, thanks to legislation limiting S emissions as well as decreased industrial activity in the former Soviet Union. Stream chemistry measurements also reflect reduced acidity of precipitation and recovery of aquatic ecosystems. Acid precipitation re-

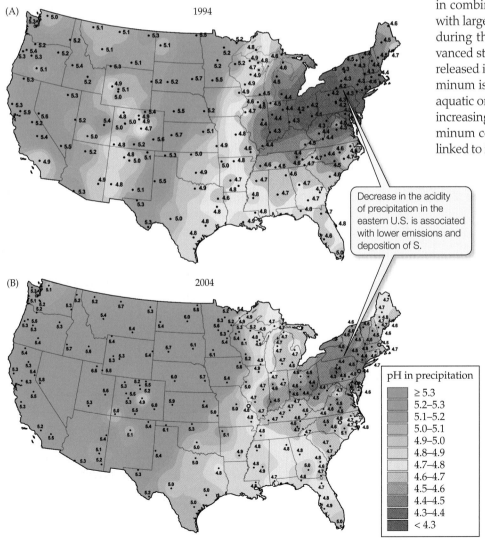

Decrease in the acidity of precipitation in the eastern U.S. is associated with lower emissions and deposition of S.

pH in precipitation

	≥ 5.3
	5.2–5.3
	5.1–5.2
	5.0–5.1
	4.9–5.0
	4.8–4.9
	4.7–4.8
	4.6–4.7
	4.5–4.6
	4.4–4.5
	4.3–4.4
	< 4.3

FIGURE 24.19 Decreases in Acid Precipitation The acidity of precipitation in the United States as measured in (A) 1994 and (B) 2004.

(A) 1860

(B) Early 1990s

N deposition varies regionally, but overall, it has increased more than threefold since 1860 ...

(C) 2050

... and is expected to double again by 2050.

Deposition of inorganic N (mg N/m^2/year)

5,000
2,000
1,000
750
500
250
100
50
25
5

mains a problem, however, in some countries that have experienced rapid industrial development, such as China and India.

Too much of a good thing can be bad: N deposition

Anthropogenic emissions of nitrogenous compounds into the atmosphere have greatly altered global N cycles. Many of these compounds are considered *reactive N*, because unlike nitrogen gas (N_2), they can participate in chemical reactions in the atmosphere, soils, and water, and are therefore biologically reactive. Reactive N can fall back to Earth (via dry and wet deposition) after being transported some distance in the atmosphere. Globally, anthropogenic emissions and deposition of reactive N compounds are more than three times greater now than they were in 1860 (Galloway et al. 2004) (**Figure 24.20**). Emissions and deposition of reactive N are expected to double between 2000 and 2050 as industrial development increases to keep pace with the human population. Greater deposition of N will increase the supply of N for biological activity, but this abundance will come with an environmental cost.

The role of N as a determinant of rates of primary production in ecosystems ranging from tundra to the open ocean was described in Chapter 19. Nitrogen plays an important role in photosynthesis, which forms the base of the food webs providing energy to all other organisms. Considerable benefit to humanity has accrued from the widespread application of N fertilizers to crops since the early twentieth century. We might expect, therefore, that an increased supply of N would facilitate plant growth and greater overall production in a N-limited ecosystem. Primary production has indeed increased in some ecosystems as a result of increased N deposition (e.g., forests in Scandinavia; Binkley and Högberg 1997). Nitrogen deposition may be partly responsible for a greater uptake of atmospheric CO_2 by terrestrial ecosystems in the Northern Hemisphere (IPCC 2007).

FIGURE 24.20 Historic and Projected Changes in Nitrogen Deposition
(A) Estimated deposition of inorganic N in 1860. (B) Measured rates for the early 1990s. (C) Projected rates for 2050. (After Galloway et al. 2004.)

Legend:
- N mineralization
- Leaf N concentration
- NPP
- Ca:Al ratios
- Nitrification
- NO_3^- leaching and soil acidification

Greater N availability initially enhances plant growth (NPP), and greater N concentrations increase the rates of N cycling processes (mineralization, nitrification).

Eventually, losses of calcium and magnesium and greater aluminum concentrations lead to lower plant growth and a decrease in NPP.

At high rates of N input, the ecosystem reaches the final stage of N saturation: the system cannot retain N from both inputs and mineralization, and leaching of N into groundwater occurs.

y-axis: Relative units (0, 50, 100, 150, 200)
x-axis: Increasing N input

FIGURE 24.21 Effects of Nitrogen Saturation Aber and colleagues provided a conceptual model of forest ecosystem responses to increasing inputs of inorganic N. (After Aber et al. 1998.)

Although primary production is increasing in some ecosystems due to N deposition, there is also strong evidence that N deposition is associated with environmental degradation, loss of diversity, and acidification. While N limits primary production in many terrestrial ecosystems, the capacity of vegetation, soils, and soil microbes to take up greater N inputs can be exceeded; this condition is known as "nitrogen saturation" (Aber et al. 1998) (**Figure 24.21**). Greater concentrations of inorganic N compounds (NH_4^+ and NO_3^-) in the soil lead to enhanced rates of microbial processes (nitrification and denitrification; see Chapter 21) that release nitrous oxide (N_2O), a potent greenhouse gas. NO_3^- is easily leached from soils and can move into groundwater, eventually entering aquatic ecosystems. When NO_3^- moves through the soil, it carries cations, including K^+, Ca^{2+}, Mg^{2+}, in solution to maintain a charge balance. As in the case of acid precipitation, losses of these cations can lead to nutrient deficiencies and eventually to acidification of soils.

Most aquatic ecosystems are limited by P, so the biological uptake of anthropogenic NO_3^- that enters them from terrestrial ecosystems may be relatively small (although there is greater processing of N than expected; see Figure 21.16). Riverine transport of N to nearshore marine ecosystems has increased as rates of N deposition have increased (Howarth et al. 1996). Primary production in estuarine and marsh communities is often limited by N, and thus the influx of N from terrestrial sources into these ecosystems has resulted in eutrophication (see Chapter 21) resulting in heavy algal growth. Eutrophication can create anoxic conditions in the bottom waters of nearshore ecosystems. High inputs of organic matter from the increased algal production lead to high rates of decomposition by microorganisms, which consume most of the available oxygen. The resulting anoxic conditions are lethal for most marine life, including fish. Anoxic conditions may occur over large areas, creating "dead zones." Dead zones of up to 18,000 km^2 form annually in the Gulf of Mexico, and similar dead zones form in the Baltic Sea, the Black Sea, and Chesapeake Bay.

In nutrient-poor ecosystems, many plants have adaptations that lower their nutrient requirements, which lowers their capacity to take up additional inputs of N. As a result, faster-growing species may outcompete the species adapted to low-nutrient conditions. Eventually, this competition can lead to losses of diversity and alteration of community composition. In Holland, species-rich heath communities adapted to low-nutrient conditions have been replaced by species-poor grassland communities as a result of very high rates of N deposition (Berendse et al. 1993). In Great Britain, Carly Stevens and colleagues surveyed grassland communities across the country with a range of N deposition rates (**Figure 24.22A**). At 68 sites, they measured the mean plant species richness in multiple study plots, along with several environmental variables, to try to explain the variation in plant diversity among the sites. The environmental variables included nine soil chemical factors, nine physical environmental variables, grazing intensity, and the presence or absence of grazing enclosures (Stevens et al. 2004). Of the 20 possible factors that may have influenced differences in species richness among the study sites, the amount of N deposition explained the greatest amount of variation (55%, **Figure 24.22B**): higher inputs of N were associated with lower species richness. While this study was correlative in nature, its results are supported by a multitude of experimental studies in which adding N to experimental plots decreased species richness, often resulting in the loss of rare species (Suding et al. 2005). Stevens and colleagues suggest that a general trend of decreasing plant diversity has occurred in the British Isles, and will continue to occur with increasing anthropogenic N deposition. High rates of N deposition also facilitate the successful spread of some invasive plant species at the expense of native species (Dukes and Mooney 1999).

(A)

(B)

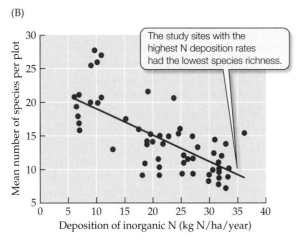

FIGURE 24.22 Nitrogen Deposition Lowers Species Diversity
(A) Inorganic N deposition in the United Kingdom. Dots on the map indicate the study sites where plant species richness in grassland ecosystems was measured. (B) Correlation between rates of inorganic N deposition and plant species richness. (After Stevens et al. 2004.)

The ecological effects of sulfur and nitrogen result when atmospheric deposition returns anthropogenic emissions to Earth's surface. In the next section we'll describe some anthropogenic compounds that exert negative effects while remaining in the atmosphere.

Atmospheric Ozone

CONCEPT 24.4
Losses of ozone in the stratosphere and increases in ozone in the troposphere each pose risks to organisms.

Ozone is good for biological systems, but only when it isn't in close contact with them. In the upper atmosphere (the stratosphere), ozone provides a shield that protects Earth from harmful radiation. When in contact with organisms in the lower atmosphere (the troposphere), however, ozone can harm them. Detrimental changes in ozone concentrations have occurred in both the stratosphere and the troposphere as a result of anthropogenic emissions of air pollutants.

Loss of stratospheric ozone increases transmission of harmful radiation

About 2.3 billion years ago, prokaryotes evolved the capacity to carry out photosynthesis, initiating a significant alteration of Earth's atmosphere. Oxygen released by photosynthesis began to accumulate in the atmosphere, leading to a series of changes that facilitated the evolution of greater physiological and biological diversity, including the appearance of aerobic respiration. The increase oxygen also led to the formation of a layer of ozone (O_3) in the stratosphere (10–50 km altitude). Ozone acts as a shield protecting Earth's surface from high-energy ultraviolet-B (UVB) radiation (0.25–0.32 μm). UVB radiation is harmful to all organisms, causing damage to DNA, damage to photosynthetic pigments in plants and photosynthetic bacteria, impairment of immune responses, and cancerous skin tumors in animals, including humans.

Stratospheric ozone concentrations change seasonally as a result of changes in atmospheric circulation patterns, particularly in the polar zones, where they decline in spring. (Jacobson 2002). British scientists measuring ozone concentrations in the Antarctic were the first to record an unusually large decrease in springtime stratospheric ozone concentrations starting in 1980. Since then, there has been a progressively larger decrease in stratospheric ozone concentrations in the Antarctic each spring, lasting about 2 months. Ozone levels have decreased by as much as 70% relative to levels recorded in the same season prior to 1980 (**Figure 24.23**). There has also been an increase in the spatial extent of the Antarctic region experiencing a decrease in ozone, called the **ozone hole**. An ozone hole is defined as an area with less than 220 Dobson units (= 2.7 × 10^16 molecules of ozone) per square centimeter; prior to 1980, ozone concentrations had never been recorded below this level. Ozone decreases have been recorded from 25° S to the South Pole. Similar reductions in ozone have been recorded in the Arctic (from 50° N to the North Pole), although the magnitude of the decrease has not been as great (thus conferring the name **Arctic ozone**

dent, since ozone concentrations have not dropped below 220 Dobson units).

The decrease in stratospheric ozone was predicted in the mid-1970s by Mario Molina and Sherwood Rowland, who discovered that certain chlorinated compounds, particularly chlorofluorocarbons (CFCs), could destroy ozone molecules. CFCs were developed in the 1930s for use as refrigerants and were later found to be useful as propellants in spray cans dispensing hair spray, paint, deodorants, and many other products. By the 1970s, as much as a million metric tons of CFCs were being produced every year. Molina and Rowland (1974) found that CFCs did not degrade in the troposphere and could remain there for a very long time. In the stratosphere, however, CFCs react with other compounds, particularly in the polar regions during winter, to produce reactive chlorine molecules that destroy ozone. Other anthropogenic compounds with the same effect include carbon tetrachloride, used as a solvent and to fumigate grain, and methyl chloroform, used as an industrial solvent and degreaser. A single free chlorine atom has the potential to destroy 10^5 ozone molecules. Thus, the potential danger posed by chlorinated compounds to the stratospheric ozone layer was clear to Molina and Rowland.

Increases in UVB radiation at Earth's surface have been recorded at the same time decreasing concentrations of stratospheric ozone have been observed (Madronich et al. 1998). The increases have been most striking in the Antarctic region, with as much as a 130% increase in UVB radiation during the spring. Increases in the Northern Hemisphere also have been recorded, including a 22% increase at mid-latitudes during the spring.

These increases in UVB radiation at Earth's surface have coincided with a higher incidence of skin cancer in humans. The occurrence of skin cancer is now approximately ten times higher than it was in the 1950s. UVB radiation had an important role in the evolution of pigmentation in humans (Jablonski 2004). The production of melanin, a protective skin pigment, was selected for in humans living at low latitudes, where ozone levels are naturally lowest and the highest levels of UVB radiation reach Earth's surface. As humans migrated away from equatorial Africa into colder climates with less sunlight, high amounts of melanin in the skin limited production of vitamin D, resulting in selection for lower melanin production in peoples of higher latitudes. As lighter-skinned humans have subsequently migrated into environments with higher UVB

(A)

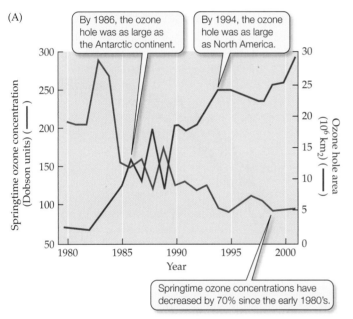

By 1986, the ozone hole was as large as the Antarctic continent.

By 1994, the ozone hole was as large as North America.

Springtime ozone concentrations have decreased by 70% since the early 1980's.

FIGURE 24.23 The Antarctic Ozone Hole (A) Since 1980, there have been dramatic decreases in springtime ozone concentrations over the Antarctic region, with concentrations dropping below the threshold for ozone hole status (220 Dobson units) for a large proportion of the area after 1984. (B) Average ozone levels over Antarctica for the months of September 1979 and September 2000, with the lowest ozone levels shown in purple, demonstrate the dramatic decrease and its spatial extent that occurred during this period.

(B) 1979   2000

radiation, to which their complexions are not adapted, they have increased their risk of skin cancers. This has become particularly true for populations at high latitudes in the Southern Hemisphere, including Australia, New Zealand, Chile, Argentina, and South Africa, where exposure to UVB is enhanced by stratospheric ozone loss. Concern is particularly great in Australia, where nearly 30% of the population is diagnosed with some form of skin cancer.

Substantial evidence exists to indicate that increasing UVB radiation has important ecological effects (Caldwell et al. 1998; Paul and Gwynn-Jones 2003). Amphibians appear to be particularly susceptible to the harmful effects of increased UVB radiation, primarily at the larval stage (see the Case Study in Chapter 1). Sensitivity to UVB radiation varies among the coexisting species within a community, and as a result, changes in community composition are likely to result from increased UVB radiation. The potential for detrimental UVB effects due to stratospheric ozone loss is greatest at high latitudes and at high elevations (> 3,000 m), due to the already lower atmospheric filtering of UV radiation there.

The realization of the rapid changes in stratospheric ozone concentrations, and of their probable anthropogenic cause, resulted in several international conferences on ozone destruction in the 1980s. At these conferences, the Montreal Protocol, an international agreement calling for a reduction and eventual end to the production and use of CFCs and other ozone-degrading chemicals, was developed. The Montreal Protocol has been signed by more than 150 countries. Concentrations of CFCs have remained the same or, in most cases, declined since the Montreal Protocol went into effect in 1989 (**Figure 24.24**). A progressive recovery of the ozone

FIGURE 24.24 Progress against the Ozone Killers
Measurements of atmospheric concentrations of ozone-destroying compounds at five monitoring locations across the globe show that several of them have declined since the signing of the Montreal Protocol in 1989.

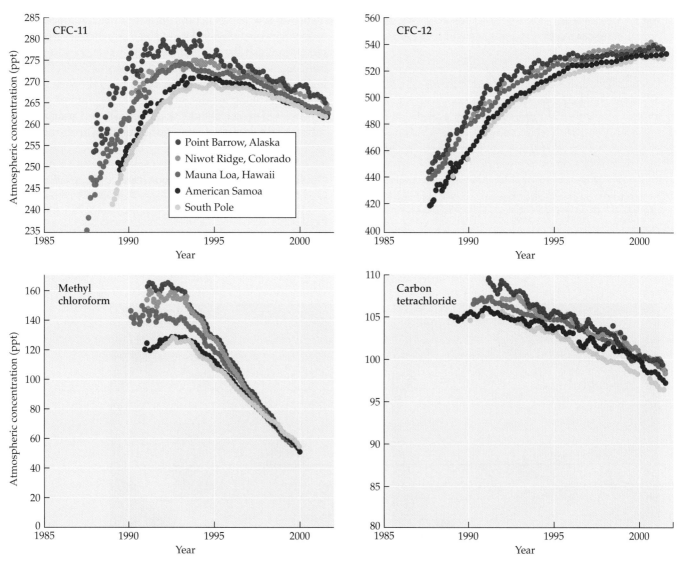

layer is expected to occur over several decades, since slow mixing of the troposphere and stratosphere, with the long-lived CFCs they still contain, will result in a time lag before stratospheric ozone concentrations rise. Full recovery of the ozone layer is expected by 2050.

Tropospheric ozone is harmful to organisms

Ninety percent of Earth's ozone is found in the stratosphere. The remaining 10% occurs in the troposphere. Tropospheric ozone is generated by a series of reactions involving sunlight, NO_x, and volatile organic compounds such as hydrocarbons, carbon monoxide, and methane. In some regions, natural vegetation can be an important source of these volatile organic compounds, which include terpenes (which give pines their characteristic odor) and isoprene. Under clean atmospheric conditions, the amount of ozone produced in the troposphere is very small, but anthropogenic emissions of ozone precursor molecules have greatly increased its production. Air pollutants that produce ozone can travel long distances and thus tropospheric ozone production is a widespread concern.

Tropospheric ozone is environmentally damaging for two main reasons. First, ozone is a strong oxidant; that is, the oxygen in it reacts easily with other compounds. Ozone causes respiratory damage and is an eye irritant in humans and other animals. Increases in the incidence of childhood asthma have been linked to exposure to ozone. Ozone damages the membranes of plants and can significantly decrease their photosynthetic rates and growth. Ozone also increases the susceptibility of plants to other stresses, such as low water availability. Decreases in crop yields have been associated with exposure to ozone. Characteristic symptoms of ozone pollution have been found in native plants near urban areas since the 1940s and 1950s (e.g., in the San Gabriel Mountains near Los Angeles and in the northern Alps in Italy), but more recently, symptoms have been noted in national parks and wilderness areas farther from sources of pollution. For example, plants in the Sierra Nevada of California are negatively affected by ozone generated in the Central Valley and the San Francisco and Los Angeles urban areas (Bytnerowicz et al. 2003). Growth rates of trees in forests of the eastern United States are as much as 10% lower than they would be in the absence of ozone (Chappelka and Samuelson 1998).

Second, ozone is a greenhouse gas that can contribute to global climate change. Ozone has a short life span in the atmosphere relative to other greenhouse gases, however, and its concentration can vary greatly from place to place. Thus, the effect of anthropogenic ozone on climate change is difficult to estimate.

Strategies to limit tropospheric ozone production have focused on lowering anthropogenic emissions of NO_x and volatile organic compounds. In most developed countries, efforts to lower emissions of ozone-producing compounds have met with success. In the United States, for example,

emissions of volatile organic compounds dropped by 50% between 1970 and 2004, emissions of NO_x dropped by 30% (U.S. EPA 2005), and tropospheric ozone concentrations are decreasing near large urban areas (Fiore et al. 1998). Regulation of emissions of ozone-producing compounds has not been as strict in some developing countries, however. Ozone is a serious air pollutant in urban and agricultural regions of China and India, but stricter environmental regulations are now being put in place.

CASE STUDY REVISITED
Dust in the Wind and the Decline of Coral

We've seen throughout this chapter that many aspects of global ecology involve some interaction with the atmosphere: greenhouse gases and climate change, emissions and deposition of N and S, and stratospheric destruction and tropospheric production of ozone. Could the widespread decline of coral reefs also have an atmospheric component?

Virginia Garrison and her colleagues have suggested that atmospheric dust deposition has contributed to the widespread nature of coral reef decline (Garrison et al. 2003). The dust influences coral health in several ways. First, as noted earlier, fungal spores may be carried with the dust. These spores may be carried within cracks on the surfaces of mineral particles in the dust, where they would be protected from the damaging effects of UVB radiation. Dust samples collected from the atmosphere have been found to contain viable spores of some of the fungi associated with coral diseases, including pathogenic forms of *Aspergillus sydowii* (Garrison et al. 2003). *Aspergillus* causes the initial infection in the coral disease aspergillosis, but that is followed by secondary infections caused by other marine fungi that do not normally attack healthy corals.

Second, the dust itself is a source of nutrients for marine biota—including pathogens. Nutrient enrichment in general significantly increases the severity of coral diseases (Bruno et al. 2003). Deposition of iron (Fe) can trigger greater incidence of coral diseases by enhancing the growth of pathogenic sulfide-oxidizing bacteria. Finally, anthropogenic pollutants, such as pesticides and volatile chemicals released by industrial processes and trash burning, are adsorbed onto the surfaces of dust particles. These pollutants are then transported through the atmosphere with the dust and deposited in distant regions, such as tropical coral reefs. This pollutant-laden dust may also be contributing to the decline in coral health.

Where is the dust coming from, and why have its effects been more apparent since the 1970s? The origin of atmospheric dust can be determined by measuring its particle size and chemistry, particularly the ratios of certain key elements, which indicate the parent material. Another ap-

FIGURE 24.25 Global Dust Storms (A) This satellite image shows a massive plume of dust from China spreading across the Korean Peninsula and out into the western Pacific Ocean in April 2006. The image was acquired by a satellite-based Moderate resolution Imaging Spectroradiometer (MODIS). Similar dust storms have ejected dust into the atmosphere that has traveled as far as Europe, two-thirds of the way around the globe. (B) Sources of the dust deposited in the Caribbean region include the deserts of North Africa and Asia. (B after Garrison et al. 2003.)

(A)

(B)

proach involves reconstruction of the route of the air mass carrying the dust using weather maps, a technique known as *back-trajectory analysis*. In addition, visible remote sensing data from satellites can track the movement of dust storms (**Figure 24.25A**). Results from these techniques indicate that the dust deposited in the tropical Atlantic, Indian, and Pacific oceans has two main sources: the Sahara and Sahel regions of North Africa and the Gobi and Takla Makan deserts of Asia (**Figure 24.25B**). These regions have experienced significant recent **desertification**, a process that results in the loss of plant cover and acceleration of the loss of soil. A major cause of desertification is the development of agriculture in marginal sites; such agricultural development in these regions has increased substantially since the 1970s. This pattern of desertification, and subsequent increases in emissions of dust to the atmosphere, is therefore consistent with the timing of coral reef decline.

Multiple factors are contributing to the decline of coral reefs. While many of these factors have local origins, such as increased sediment and pollutant inputs from nearby rivers, dust deposition is a global phenomenon, affecting coral reefs regardless of their proximity to the sources of the pathogens, nutrients, and pollutants carried by the dust.

CONNECTIONS IN NATURE

A Historical Perspective on Dust and Ecology

Recent increases in desertification have accelerated dust emissions, transport, and deposition across the globe. Asian dust has been detected in the European Alps, traveling two-thirds of the way around the globe in approximately a week (Grousset et al. 2003). The effects of dust deposition on ecological processes are not fully under-

stood, but they include several processes mentioned above, such as nutrient and microorganismal movement between continents and from continents to oceans. Dust deposition of nutrients can have important consequences for primary production and the global carbon cycle. The supply of iron (Fe) from dust deposition may be important for oceanic primary production (Mahowald et al. 2005) as we saw in Chapter 19, and inputs of cations from dust are important to primary production in tropical forests in the Amazon (Okin et al. 2004). The ecological effects of imported pathogens have been largely unexplored.

From a historical point of view, the current magnitude of dust movement is not unprecedented. During glacial periods, large amounts of loess (rock ground to fine powder by glaciers) have been transported across the globe. Has dust deposition affected coral reefs in the past? Coral reefs are one of the most ancient ecosystems on Earth, probably first appearing in the Cambrian period, over 500 million years ago (Wood 1998), predating the appearance of angiosperms (in the early Cretaceous, over 130 million years ago). During Earth's history, coral reefs have occupied from less than a hundredth to as much as ten times the area they currently occupy.

The role of dust deposition in the natural waxing and waning of coral reefs is uncertain, and most of the historical variation in reef extent may be associated with changes in water temperature and sea level related to climate change.

However, the potential for dust to negatively affect coral reefs exemplifies the importance of global phenomena that influence local ecological patterns and processes. Dust clearly has large influences on nutrient inputs and enhancement of primary production in both marine and terrestrial ecosystems. Some of the other environmental problems we have discussed in this chapter share these features of global transport and local effects: N deposition and tropospheric ozone formation, for example, are widespread forces that influence ecosystems in remote locations far from the source of the pollutants. Global ecology is a field that will continue to grow as we realize the widespread importance of global biogeochemical cycles in local ecological phenomena and the role of humans in intensifying their effects.

SUMMARY

CONCEPT 24.1 Elements move among geological, atmospheric, and biological pools at a global scale.

- The global carbon cycle includes large exchanges of CO_2 between the atmosphere and Earth's surface associated with photosynthesis, respiration, and diffusion into the oceans, and within the last 150 years, anthropogenic emissions of CO_2 and CH_4.

- Atmospheric concentrations of CO_2 and CH_4 are increasing due to burning of fossil fuels, deforestation, and agricultural development.

- Elevated atmospheric CO_2 may increase terrestrial plant growth and the acidity of the oceans, causing significant ecological changes.

- Global fluxes of nitrogen are associated with biological uptake and transformations, but human fixation and emissions now dominate the global nitrogen cycle.

- The global cycles of phosphorus and sulfur include both geochemical and biological fluxes.

- Anthropogenic fluxes of phosphorus associated with mining and industrial emissions of sulfur far exceed natural fluxes associated with weathering.

CONCEPT 24.2 Earth is warming at an unprecedented rate due to anthropogenic emissions of greenhouse gases.

- Elevated levels of CO_2, CH_4, N_2O, and other greenhouse gases in the atmosphere have warmed Earth, particularly since the 1950s, and this warming trend is expected to continue throughout the twenty-first century.

- Large changes in species distributions, community composition, and ecosystem function are expected as a result of global climate change.

- Recent changes in species' geographic ranges and in ecosystem carbon exchange have been attributed to climate change.

CONCEPT 24.3 Anthropogenic emissions of sulfur and nitrogen cause acid deposition, alter soil chemistry, and affect the health of ecosystems.

- Sulfuric and nitric acids form in the atmosphere from compounds emitted mainly by the burning of fossil fuels, and are subsequently deposited on Earth's surface as acid precipitation.

- Acid precipitation causes nutrient imbalances and aluminum toxicity in soils, altering community composition and ecosystem function.

- Atmospheric deposition of reactive nitrogen compounds emitted from anthropogenic sources can increase productivity in some ecosystems, but may also lead to soil acidification, dead zones in nearshore aquatic ecosystems, losses of species diversity, and increases in invasive species.

CONCEPT 24.4 Losses of ozone in the stratosphere and increases in ozone in the troposphere each pose risks to organisms.

- Anthropogenic emissions of chlorinated compounds have led to a loss of stratospheric ozone since the 1980s, particularly at high latitudes, and thus to an increase in the levels of harmful ultraviolet-B radiation reaching Earth's surface.

- Reactions involving volatile organic compounds, many of which are anthropogenic air pollutants, generate ozone in the troposphere, where it can harm organisms.

REVIEW QUESTIONS

1. What are the major biological influences on the global carbon cycle? How have human influences during the past two centuries affected the fluxes of CO_2 associated with these biological influences (i.e., other than fossil fuel burning) and subsequently increased atmospheric CO_2 concentrations?

2. Global climate change has occurred at an accelerated rate since the latter part of the twentieth century, and this trend is expected to continue into the twenty-first century. As the climate changes, terrestrial animals are capable of migrating to regions where the climate is optimal for their function. Despite their mobility, ecologists are still predicting that many animal species will experience local extinctions. Explain why animal responses to climate change will involve factors other than physiological functioning and dispersal rates.

3. Describe how acid precipitation and nitrogen deposition are linked, and how they differ.

4. How can ozone in the atmosphere be both good and bad for organisms?

HANDS-ON PROBLEM SOLVING

Climate Change in the Arctic | This Web exercise (www.sinauer.com/ecology) explores regional variation in global warming. Different regions of Earth and the associated troposphere are warming at different rates. You will interpret data from a recent paper showing pronounced warming in the Arctic region at different levels in the atmosphere.

SUGGESTED READINGS

The following two articles focus on human alterations of nitrogen cycles at ecosystem and global scales. The Aber et al. paper provides a framework for evaluating ecological change during progressive stages of nitrogen saturation in forest ecosystems. The Galloway et al. paper provides a context for evaluating human effects on nitrogen cycles based on past, current, and future alterations of nitrogen biogeochemistry.

Aber, J., W. and 9 others. 1998. Nitrogen saturation in temperate forest ecosystems: Hypotheses revisited. *BioScience* 48: 921–934.

Galloway, J. N. and 14 others. 2004. Nitrogen cycles: Past, present, and future. *Biogeochemistry* 70: 153–226.

Intergovernmental Panel on Climate Change. 2007. *Climate Change 2007: The Physical Science Basis; Impacts, Adaptation and Vulnerability; Mitigation of Climate Change.* Cambridge University Press, Cambridge; also available on the IPCC Web site: http://www.ipcc.ch/. *These three reports are authored by the leading scientists in climate change research from atmospheric science, Earth systems science, ecology, economics, and sociology. The reports represent a consensus on the current status of climate change, its causes, its effects on ecological and economic systems, and strategies to minimize those effects.*

Jacobson, M. Z. 2002. *Atmospheric Pollution: History, Science, and Regulation.* Cambridge University Press, Cambridge. *A fascinating treatment of air pollution science from a multitude of perspectives, including the history, chemistry, atmospheric physics, ecology, and mitigation of air pollution problems.*

Ruddiman, W. F. 2003. The anthropogenic greenhouse era began thousands of years ago. *Climatic Change* 61: 261–293. *William Ruddiman's suggestion that humans began modifying global climate with the advent of organized agriculture, delaying the onset of another Ice Age, remains very controversial, but makes fascinating reading.*

Schlesinger, W. H. 1997. *Biogeochemistry: An Analysis of Global Change.* 2nd ed. Academic Press, San Diego, CA. *An excellent introduction to biogeochemistry at the global and cosmic scales. Clear descriptions of the major global cycles of biologically important elements, and human influences on those cycles, are included.*

The following two articles clearly express the magnitude of global environmental change and its importance to the present and future of status of Earth's ecosystems. They emphasize the ascension of humans to dominance as the major force of change on planet Earth.

Vitousek, P. M. 1994. Beyond global warming: Ecology and global change. *Ecology* 75: 1861–1876.

Vitousek, P. M., H. A. Mooney, J. Lubchenco and J. M. Melillo. 1997. Human domination of Earth's ecosystems. *Science* 277: 494–499.

APPENDIX
Some Metric Measurements Used in Ecology

MEASURES OF	UNIT	EQUIVALENTS	METRIC → ENGLISH CONVERSION
Length	meter (m)	base unit	1 m = 39.37 inches = 3.28 feet
	kilometer (km)	1 km = 1000 (10^3) m	1 km = 0.62 miles
	centimeter (cm)	1 cm = 0.01 (10^{-2}) m	1 cm = 0.39 inches
	millimeter (mm)	1 mm = 0.1 cm = 10^{-3} m	1 mm = 0.039 inches
	micrometer (μm)	1 μm = 0.001 mm = 10^{-6} m	
	nanometer (nm)	1 nm = 0.001 μm = 10^{-9} m	
Area	square meter (m^2)	base unit	1 m^2 = 1.196 square yards
	hectare (ha)	1 ha = 10,000 m^2	1 ha = 2.47 acres
Volume	liter (l)	base unit	1 l = 1.06 quarts
	milliliter (ml)	1 mL = 0.001 l = 10^{-3} l	1 ml = 0.034 fluid ounces
	microliter (μl)	1 μL = 0.001 ml = 10^{-6} l	
Mass	gram (g)	base unit	1 g = 0.035 ounces
	kilogram (kg)	1 kg = 10^3 g	1 kg = 2.20 pounds
	teragram (Tg)	1 Tg = 10^{12} g	
	petagram (Pg)	1 Pg = 10^{15} g	
	milligram (mg)	1 mg = 10^{-3} g	
	microgram (μg)	1 μg = 10^{-6} g	
	picogram (pg)	1 pg = 10^{-12} g	
Temperature	degree Celsius (°C)	base unit	°C = $\frac{5}{9}$(°F − 32)
			0°C = 32°F (water freezes)
			100°C = 212°F (water boils)
			20°C = 68°F ("room temperature")
Pressure	Megapascal (MPa)		1 MPa = 145 PSI (pounds per square inch)
Energy	joule (J)		1 J ≈ 0.24 calorie = 0.00024 kilocalorie*

*A *calorie* is the amount of heat necessary to raise the temperature of 1 gram of water 1°C. The *kilocalorie*, or nutritionist's calorie, is what we commonly think of as a calorie in terms of food.

Answers to Review Questions

Chapter 1

1. The phrase "connections in nature" is meant to evoke the fact that interactions among organisms and their environment can cause events in nature to be interconnected. As a result of such connections, an action that directly affects one part of an ecological community may cause unanticipated side effects in another part of a community. Various examples related to amphibian deformities and population declines illustrate such connections and their side effects. For example, it appears that the addition of fertilizers to ponds has led to the following chain of events: the fertilizer stimulates increased algal growth, which then leads to increased snail abundance, increased *Ribeiroia* abundance, and hence, more frequent amphibian deformities.

2. Ecology is the scientific study of interactions between organisms and their environment. The scope of ecology is broad, and may address virtually any level of biological organization (from molecules to the biosphere). Most ecological studies, however, include emphasis on one or more of the following levels: individuals, populations, communities, or ecosystems. Thus, if an ecologist studied the effects of a particular gene, they probably would emphasize how the gene affected interactions in nature—they might, for example, study how a gene affected the ability of an organism to cope with its environment, or how a gene affected interactions among species. Compared with a geneticist or cell biologist, an ecologist would be less likely to emphasize either the gene itself or its effects on the workings of a cell, and more likely to study how the gene affected interactions in nature that occur at the individual, population, community, or ecosystem levels.

3. The scientific method summarizes how the process of scientific inquiry typically occurs. The four key steps in this inquiry process are: (1) observe nature and ask a question about those observations; (2) use previous knowledge or intuition to develop hypotheses (possible answers) to those questions; (3) evaluate different hypotheses by performing experiments, collecting new observations, and/or analyzing results from quantitative models; and (4) use the results from the approaches taken in (3) to modify the hypotheses, pose new questions, or draw conclusions about the natural world. An essential feature of many scientific investigations is a *controlled experiment* in which results from an experi-

mental group (that has the factor being tested) are compared with results from a control group (that lacks the factor being tested).

Chapter 2

1. Extreme environmental conditions, such as high and low temperatures or droughts, are important determinants of mortality in organisms. As a result, species' geographic distributions often reflect extreme environmental conditions as well as average conditions. The timing of physical changes is also important, as exemplified by the response of vegetation to the timing of precipitation, which is not indicated by average annual conditions.

2. Differences in the intensity of solar radiation at Earth's surface establish latitudinal gradients of surface heating. Greater heating in the tropics results in rising air currents, which establish large-scale atmospheric circulation cells, called Hadley cells. Polar cells form where cold, dense air descends at the poles. Between the Hadley and Polar cells are the Ferrell cells, driven by the movement of the Hadley and Polar cells and the exchange of energy between equatorial and polar air masses. The temperate zone is found at mid-latitudes in association with the Ferrell cells.

3. a. Mountains influence climate by steering the movements of large air masses. Mountains also force air masses passing over them to rise, cooling the air. Precipitation increases on their windward slopes, since the cooler air has a lower capacity to hold water vapor. When the air mass descends on the leeward slopes, it is drier and warmer, and thus the climate there is also drier and warmer, a phenomenon known as the "rain-shadow effect." Mountains can also create local upslope and downslope winds due to differential heating and cooling of their slopes.

   b. Ocean water can absorb more energy without a change in its own temperature than land can. As a result, land areas adjacent to oceans experience smaller seasonal temperature changes than inland areas. Oceans may also provide moisture to air masses passing over adjacent land surfaces, enhancing precipitation.

4. Salinization is a progressive increase in soil salinity due to surface evapotranspiration of water. Desert areas have high rates of evapotranspiration and little precipi-

tation to leach salts to deeper soil layers. Some desert soils also have impervious soil layers underlying the surface layer that impede leaching, increasing the potential for salinization.

Chapter 3

1. Plant growth forms are good indicators of the physical environment, particularly climatic and soil conditions. Because they cannot move to a new environment, plants have evolved morphological features that allow them to cope with their physical environment, including its extremes. Leaf life span (evergreen versus deciduous leaves), for example, reflects the fertility of the soil. Some biomes, such as grasslands, can also be indicators of disturbances such as predation or fire. Animals can be important features of and controls on biome distribution, but their mobility renders them less useful as indicators of biomes.

2. Those biomes containing resources that most benefit human economies have been the most transformed. Those resources include fertile soils, forage for livestock, and a climate conducive to growing crops. Therefore, the most heavily transformed biomes include grasslands, which often have fertile soils and forage for livestock, tropical seasonal forests and savannas, which have forage for livestock and a climate conducive to growing crops, and deciduous forests, which have fertile soils and a climate conducive to crops. Tropical rainforests are one of the most rapidly changing biomes today due to deforestation for cropland and pastures.

3. According to the river continuum concept, water velocity, stream bed particle size, and shading by riparian vegetation all decrease as rivers move downstream. As a result, the importance of the surrounding terrestrial ecosystems as sources of energy for stream organisms tends to decrease downstream. Stream insects include more shredders near the source of a stream and more collectors in the lower portions. Attached plants and free-floating algae become more abundant downstream.

4. Light penetration varies according to the depth and clarity of the water. Where there is enough light for photosynthesis (the photic zone), photosynthetic organisms provide food for consumers, increasing the abundance of organisms. The stability of the substratum determines whether organisms can anchor themselves or will be covered with sand. Nearshore zones with rocky substrata tend to have the most abundant and diverse communities. Photosynthetic organisms are more sparse in shallow ocean zones with sandy bottoms and below the photic zone in the open ocean.

Chapter 4

1. Plants as a group exhibit very slightly greater tolerances of temperature extremes than ectotherms (see Figure 4.7), and both of these groups have tolerances much greater than those of endothermic animals. Plants and ectotherms, which cannot generate heat internally, are more reliant on tolerance as a strategy for adapting to body temperature variation, while endotherms rely on avoidance of body temperature extremes through internal heat generation and behavior, such as migration. Plants can exhibit avoidance of temperature extremes through leaf deciduousness.

2. The seasonal change in the thickness of the arctic fox's fur represents acclimatization to cold winter temperatures. Arctic foxes grow thicker fur in winter, increasing their insulation and decreasing their lower critical temperature. This seasonal acclimatization allows the arctic fox to stay warmer in winter and invest less energy in metabolic heat production than would be needed without the thicker fur.

   The thick fur of the arctic fox, and its ability to alter its fur thickness in response to environmental temperatures, represent adaptations that its relative, the bat-eared fox, does not share. The bat-eared fox has not been exposed to cold conditions that would favor the growth of thicker fur. In the distant past, species intermediate between the arctic and bat-eared foxes may have moved to colder, more seasonal climates where natural selection favored individuals with the capacity to grow thicker fur in winter.

3. a. Transpiration is an evaporative cooling mechanism that allows the plant to lower its leaf temperature below the air temperature. However, transpiration also results in water loss from the plant. If the water is not replaced, because the soil is too dry or the water loss is too rapid, the plant will experience water stress, and the rates of its physiological processes, such as photosynthesis, will decrease.

   b. Dark animals may be able to warm themselves more effectively, but they may also be more visible to their predators or prey. In many cases, it appears that camouflage is more important than the ability to absorb sunlight effectively.

4. The principal ways in which plants determine their resistance to water loss are by adjusting the degree of opening of their stomates and by the thickness of the outer cuticle. Arthropods have cuticles that are extremely resistant to water loss. Similarly, skin thickness in amphibians, birds, and mammals affects their resistance to water loss. Reptiles have particularly thick skin, often overlain by scales, that provides a very effective barrier to water movement into the atmosphere. Note, howev-

er that increasing the resistance of a barrier to water loss requires trade-offs with evaporative cooling as well as gas exchange.

Chapter 5

1. Autotrophy is the use of sunlight (photosynthesis) or inorganic chemicals (chemosynthesis) to fix CO_2 and synthesize energy storage compounds containing carbon–carbon bonds. Photosynthesis occurs in archaea, bacteria, protists, algae, and plants. Heterotrophy is the consumption of organic matter to obtain energy. The organic matter includes both living and dead organisms. Living organisms vary in their mobility, and their consumers (predators) have adapted ways to improve their efficiency in capturing their food (prey). Dead organic matter can be eaten and digested internally by multicellular heterotrophs or externally broken down by enzymes excreted into the environment and then absorbed by archaea, bacteria, and fungi.

2. In the short term, plants can alter the locations of their chloroplasts to acclimatize to changing light levels. Over longer time frames, plants can grow leaves with different thicknesses, vary the numbers of chloroplasts in their leaves, and change the density of the light-harvesting pigments on their photosynthetic membranes. Too much light can damage the photosynthetic membranes. Plants can lower the light energy absorbed through leaf and chloroplast movement and dissipate excess energy using the xanthophyll cycle. Photorespiration may also use up excess energy that occurs under intense light conditions.

3. C_4 plants spatially separate CO_2 fixation and the Calvin cycle. They have a biochemical pump that takes up CO_2 in the mesophyll tissue and concentrates it at the site of the Calvin cycle, in the bundle sheath tissue. The bundle sheath is surrounded by a waxy coating that keeps CO_2 from diffusing out. The higher CO_2 concentrations in the bundle sheath essentially eliminate the energy-robbing process of photorespiration in C_4 plants. Although additional energy is required to operate the C_4 biochemical pump, a higher rate of photosynthesis more than compensates for that energy expenditure, and the photosynthetic rates of C_4 plants are often higher than those of C_3 plants. C_4 plants also maintain greater water use efficiency than C_3 plants.

4. CAM plants open their stomates to take up CO_2 at night, when the humidity of the air is higher than it is during the day. They store CO_2 in the form of a four-carbon organic acid, then release it to the Calvin cycle during the day. The storage of CO_2 allows the stomates to be closed during the day, when the potential for transpirational water loss is greater.

5. Live animals are a higher-quality food source, but they are rarer and thus harder to find, and they may have defense mechanisms that require expenditure of energy overcome. Plant detritus is abundant in many ecosystems, so little energy needs to be expended in locating it, but its food quality is low.

6. The greater expenditure of energy required by species B to fly between patches would dictate that it needs to spend longer in each patch in order to meet the assumptions of the marginal value theorem. Because its overall rate of energy gain in the habitat is lower, due to greater amount of energy it expends in traveling between patches, species B should deplete each patch to a greater degree before leaving it than species A.

Chapter 6

1. Natural selection acts as a sorting process, favoring individuals with some heritable traits over individuals with other heritable traits. As a result, the frequency of the favored traits in a population may increase over time. When this occurs, the frequencies of alleles that determine the favored traits also increase over time, causing the population to evolve. But the individuals in the population do not evolve—each individual either has the trait favored by selection or it does not.

2. By consistently favoring individuals with one heritable trait over individuals with other heritable traits, natural selection can lead to a steady increase in the frequency of alleles that determine the favored trait. Although gene flow and genetic drift can also cause the frequency of alleles that determine an advantageous trait to increase over time, each of these processes can also do the reverse—that is, they can promote an increase in the frequency of disadvantageous alleles. Gene flow, for example, can transfer disadvantageous alleles to a population, thereby impeding adaptive evolution. Similarly, the random fluctuations in allele frequencies that result from genetic drift can promote an increase in the frequency of a disadvantageous allele. Hence, natural selection is the only evolutionary mechanism that consistently causes adaptive evolution.

3. Patterns of evolution over long time scales result from large-scale processes such as speciation, mass extinction, and adaptive radiation. The fossil record shows us that life on Earth has changed greatly over time, as seen in the rise and fall of different groups of organisms (for example, the rise of the amphibians and their later fall as reptiles became the dominant group of terrestrial vertebrates). Such changes in the diversity of life are due in part to speciation, the process by which one species splits to form two or more species. The rise and fall of different groups of organisms is also determined by

mass extinctions and adaptive radiations. By removing large proportions of the species on Earth, a mass extinction forever changes the course of evolution and hence alters the patterns of evolution observed after the extinction event. Similarly, by promoting an increase in the number of species in a group of organisms, an adaptive radiation also shapes the patterns of evolution observed over long time scales.

4. Evolution occurs as organisms interact with one another and with their environment. Hence, evolution occurs partly in response to ecological interactions, and those interactions help to determine the course of evolution. The reverse is also true: as the species in a biological community evolve, the ecological interactions among those species change. Thus, ecology and evolution have joint effects because they both depend on how organisms interact with one another and their environments.

Chapter 7

1. Any organisms that are intermediate in the characteristics represented by the r–K continuum or Grime's triangular model would be appropriate answers to this question. The triangular model in particular is well suited to intermediate strategies. In his original paper, Grime described many intermediate types of plant strategies and gave examples.

    In many plants and marine invertebrate animals, dispersal is negatively correlated with propagule size: smaller propagules can disperse farther than larger ones. In invertebrate animals, smaller egg size is also correlated with longer development times and increased reliance on exogenous food (rather than yolk provided in the egg) to complete development. In vertebrates, however, smaller egg sizes actually lead to more rapid development to hatching. In both cases, the correlation between egg size and development time is striking, and the pattern that is favored will vary with environmental conditions (e.g., temperature, rates of predation on larvae, etc.). An important reason why species that live in the same habitats may still exhibit different reproductive patterns is that in different years different strategies may be favored depending on the particular environmental conditions. For example, in years with abundant food availability, a small-egg strategy may be favored, as offspring can acquire resources readily from the environment. However, in years when food is limited, a large-egg strategy may be advantageous due to its decreased reliance on external energy sources.|

2. Asexual reproduction allows even a single individual to quickly increase the population size and allows a single highly successful genotype to dominate the population. The primary benefit of sex is the recombination of genetic material through the merging of unique genotypes. This recombination allows deleterious mutations to be purged from the population and allows potentially beneficial new combinations of genes to be introduced. The maintenance of both sexual and asexual reproduction allows rotifers to (1) quickly increase the size of the reproductive population under beneficial environmental conditions while (2) retaining the capacity to purge any deleterious mutations that may build up after prolonged periods of asexual reproduction.

3. Removal of small to medium-sized fish might produce selection for rapid growth through the size ranges that are favored by the fishery. This might lead to reproduction at older ages and larger sizes if there is a trade-off between growth and reproduction. Fish that are selected to grow quickly would allocate fewer resources to reproduction at smaller sizes so that they could allocate more resources to growth. Unfortunately, this is not the only effect of the Nassau grouper fishery. Because of heavy overfishing for both small and large fish and methods that target fish when they come together in large groups to spawn, Nassau grouper populations have declined precipitously.

Chapter 8

1. Complicating factors discussed in the text include (1) limited knowledge about the dispersal capabilities of the organism under study, (2) the fact that populations may have a patchy structure, and (3) the fact that individuals may be hard to define. The first two factors—limited information about dispersal and patchy populations—can make it difficult to determine the area within which individuals interact, hence constituting a population. The third factor—difficulty in defining individuals—applies to the many organisms that reproduce asexually to form clones. In such organisms, it can be hard to determine what an individual is, thus making it difficult to estimate abundance.

2. The simplest reason that no species is found everywhere is that much of Earth does not provide suitable habitat. There can, in turn, be many reasons why portions of Earth are not suitable for a particular species. For example, the abiotic or biotic conditions of an environment may limit the growth, survival, or reproduction of the species, as may disturbance or the interaction between abiotic and biotic conditions. Furthermore, a species may be absent from environments where we would expect it to thrive because of limited dispersal and historical factors (including evolutionary history and continental drift).

3. A niche model is a predictive tool that models the environmental conditions occupied by a species based on

the conditions at known localities. Niche models can be used to predict the future distribution of an invasive species by collecting as much data as possible about environments where the invasive species currently is found. Those data are then used to construct a niche model, which in turn is used to identify currently unoccupied locations that are likely to provide suitable habitat for the species. For such predictions to accurately reflect the future spread of the organism, information also must be gathered about its dispersal capabilities.

4. For a conservative estimate, assume there are 20 otters per square kilometer, each of which eats 20% of its body weight in food each day. Since urchins, on average, weigh 0.55 kg each, a kilogram of urchins consists of roughly $1/0.55 = 1.82$ urchins. Thus, the number of urchins per square kilometer that an otter population would be expected to eat each year is

(20 otters/km^2) × (0.2 × 23 kg/otter per day) × (365 days/year) × (1.82 urchins/kg) = 61,116 urchins/km^2/year

Chapter 9

1. a.

Age (x)	N_x	S_x	l_x
0	100	0.4	1.0
1	40	0.375	0.4
2	15	0.333	0.15
3	5	0	0.05
4	0		0

b. In a cohort life table, the fate of a group of individuals born during the same time period (a cohort) is followed from birth to death. This type of life table is often used for sessile or relatively immobile organisms that do not have long life spans. For organisms that are highly mobile or long-lived, a static life table may be used in which the survival and fecundity of individuals of different ages are observed during a single time period.

2. a.

Year	N_0	N_1	N_2
0	50	50	50
1	100	16.7	25
2	116.7	33.3	8.3
3	150	38.9	16.7
4	188.9	50	19.4
5	238.9	63	25
6	301.9	79.6	31.5

From these results we estimate:

$\lambda = 31.5/25 = 1.26$

The stable age distribution is 73% N_0; 19% N_1; 8% N_2.

b.

Year	N_0	N_1	N_2
0	80	50	20
1	130	26.7	25
2	156.7	43.3	13.3
3	200	52.2	21.7
4	252.2	66.7	26.1
5	318.9	84.1	33.3
6	403	106.3	42

From these results we estimate:

$\lambda = 31.5/25 = 1.26$

The stable age distribution is 73% N_0; 19% N_1; 8% N_2.

c. Based on the results in parts (a) and (b), we estimate:

$\lambda = 1.26$

The stable age distribution is 73% N_0; 19% N_1; 8% N_2.

3. a. 3240.

b. Substituting the values $N_0 = 40$, $\lambda = 3$, and $t = 27$, we have

$N_t = N_0\lambda^t = 40 \times 3^{27}$

c. In this case, we have the values $N_0 = 100$, $\lambda = 0.75$, and $t = 3$, which we plug into the relation

$N_t = N_0\lambda^t = 100 \times (0.75)^3 = 42.19$

4. Factors that regulate populations are density-dependent: when N (the number of individuals in a population) is below some level, there is a tendency for the population size to increase, whereas when N goes above some level, there is a tendency for the population size to decrease. This may be true even if density-independent factors, such as year-to-year variations in temperature or rainfall, are the primary cause of year-to-year changes in abundance.

Chapter 10

1. There are many built-in time lags in the responses of populations to changes in density. For example, the amount of available food may increase or decrease between the time the parent generation feeds and the time its offspring are born. In such a situation, the number of offspring produced may be more closely related to the previous conditions than to the conditions at the time of their birth. As a result of such time lags, the population may experience delayed density dependence, which may cause it to fluctuate in abundance over time.

2. Small populations can be threatened by chance events associated with genetic factors, demographic stochasticity, environmental stochasticity, and natural catastrophes. Genetic factors that increase the risk of extinction in small populations include genetic drift and inbreeding, both of which can increase the frequencies of harm-

ful alleles. Demographic stochasticity results from chance events related to the reproduction and survival of individuals; such chance events can cause population growth rates to drop, as might occur if a storm happened to kill most of the females in a small population, leaving few females to produce the next generation of offspring. Environmental stochasticity refers to unpredictable variation in environmental conditions; such variation can cause population growth rates to vary dramatically from year to year, increasing the chance of extinction in small populations. Finally, natural catastrophes can cause sudden reductions in population size, subjecting a population to increased risks from genetic factors, demographic stochasticity, and environmental stochasticity.

3. a. Yes, as illustrated by the two generations of parents and offspring in this diagram:

KEY: FC = female child; MC = male child; FG = female grandchild, MG = male grandchild

  b. No, all of the individuals in the second generation of offspring are related to one another. As illustrated by this example, inbreeding is likely to be common in small populations.

4. a. As the proportion of the habitat that is suitable for the species drops, the colonization rate may decrease (because the distance between remaining populations increases) and the extinction rate may increase (because the loss of habitat may cause the remaining populations to become smaller, making them more prone to extinction). Once the extinction rate exceeds the colonization rate, the metapopulation will decline to extinction because existing populations will become extinct more rapidly than new populations are established.

  b. A large habitat patch is likely to have a larger population than a small habitat patch. As a result, the populations in the two large habitat patches are less likely to go extinct than the others. Thus, the large habitat patches could serve as source populations from which individuals disperse to small patches, thereby

reducing the extinction rate of the small populations (the rescue effect). Positioning the large habitat patches far from each other and far from any of the small habitat patches would make it less likely that they could have a rescue effect, thus making it more likely that the metapopulation would not persist.

Chapter 11

1. a. The key difference is that the levels of a resource can be depleted, but those of a physical factor cannot.

  b. Immediately after application of the fertilizer, it is likely that the intensity of competition for nitrogen will drop because soil nitrogen levels will not be as limited in supply. However, as the added nitrogen is used up by the plants (and leached from the soil by rainfall), soil nitrogen levels will decrease, and the intensity of belowground competition for nitrogen will increase.

2. Four general features of competition and an example of each: (1) Competition can be direct (as in allelopathy) or indirect (as in both pitcher plants and bedstraws). (2) Competition is often asymmetrical (as in Tilman's diatoms). (3) Competition can occur between closely (as in bedstraws) or distantly (as in ants and rodents) related species. (4) Competition can determine the distribution and abundance of the competing species (as in Connell's barnacles or *Neotamias* chipmunks).

3. a. Results from laboratory experiments, field observations, and mathematical models all suggest that competing species are more likely to coexist when they use resources in different ways. For example, in Gause's experiments with *Paramecium*, *P. caudatum* coexisted with *P. bursaria*, most likely because one species fed primarily on bacteria, the other on yeast. Likewise, in the case of four species of *Anolis* lizards that lived together on Jamaica and ate similar food, Schoener's field observations indicated that these species used resources in different ways (an example of resource partitioning). Finally, graphical analyses of the Lotka–Volterra competition model indicates that competing species can coexist when the inequality shown in Equation 11.2 holds; that inequality is more likely to hold when competing species use resources in very different ways (e.g., when α and β are not close to 1).

  b. Because $\beta = 1.6$ and there are 140 individuals of species 1, it would take $1.6 \times 140 = 224$ individuals of species 2 to reduce its own per capita growth rate by the same amount that the 140 individuals of species 1 do. Therefore, because there are 230 individuals of species 2 present, species 2 is having a slightly greater effect on its own per capita growth rate than is species 1.

c. The statement is not correct. For example, if $\alpha = 0.5$ and $\beta = 1$, Equation 11.2 predicts that both species will persist when $0.5 < K_1/K_2 < 1$. Thus, for example, if $K_1 = 100$ and $K_2 = 150$, both species should persist when $\alpha = 0.5$ and $\beta = 1$. (The statement can be shown to be false in many other ways; for example, in Figure 11.13B, values for α, β, K_1, and K_2 can be selected such that species 2 always drives species 1 to extinction, even though $\alpha < \beta$.)

4. a. Possible reasons why these meadows might harbor one or the other (or both) of these two plant species: (1) both species could persist at all locations, but one species (or the other) has yet to disperse to some meadows; (2) the physical conditions of the meadows differ such that in some meadows species 1 is favored, while in others species 2 is favored, and in still others, the species can partition resources such that both persist; (3) the abundances of herbivores or pathogens that feed on species 1 or 2 may vary between the meadows, causing the outcome of competition to differ from meadow to meadow; (4) the rates of a periodic disturbance such as fire may differ among meadows (if one of the species is an inferior competitor but is more tolerant of fire); and (5) evolutionary change may have occurred in some meadows but not others, causing a range of possible outcomes, such as a competitive reversal or character displacement.

b. Addition and removal experiments would help us to evaluate these possible explanations for the observed distributions of the species. For example, in meadows where only species 1 was found, individuals of species 2 could be planted next to some individuals of species 1, but not others. Similarly, in meadows where both species are found, removal experiments could be performed in which individuals of species 1 could be removed from the vicinity of some species 2 individuals, but not others (and vice versa).

Chapter 12

1. Most predators have a broad diet in that they eat a wide range of prey species. Although there are a substantial number of herbivores that can eat many different plant species, the majority of herbivores are insects, most of which feed on just one or a few plant species.

2. A prey individual that cannot evade a predator is killed and eaten. While herbivores do not typically kill their food plants, they do have powerful negative effects on the plants on which they feed. As a result of this strong selection pressure that predators and herbivores exert on their food organisms, prey species and plants have evolved a wide range of defensive mechanisms that increase the chance that they will not be eaten.

Predators and herbivores must eat if they are to survive, so there is also strong selection on predators and herbivores to overcome the defenses of their food organisms. These effects are pervasive because all organisms must obtain food—setting in motion the conflicts described in the previous sentences. The effects are pronounced because there is such strong selection for both defensive and counterdefensive mechanisms.

3. In this chapter, we claim that exploitative interactions such as predation and herbivory can have pronounced effects on ecological communities.

a. Evidence described in this and preceding chapters indicates that exploitative interactions such as predation and herbivory can have a powerful effect on the abundances and distributions of populations of the interacting species. Ecological communities are composed of sets of interacting populations. Thus, this claim is likely to be true because if predation or herbivory affect the populations involved in those interactions, then they will also have an effect on other populations that interact with those populations.

b. The scientific evidence strongly supports this claim. In this chapter, we described several cases in which the effects of predation or herbivory were so pronounced that they altered ecological communities greatly, in some cases causing a shift from one community type to another. For example, such effects were seen for lesser snow geese feeding on marsh grasses, artic foxes feeding on seabirds, and aquatic snails feeding on large aquatic plants.

4. The prey population changes in size according to the equation

$$\frac{dN}{dt} = rN - aNP$$

where N is the prey density and P is the predator density. The prey population does not change in size when $dN/dt = 0$, or in this case, when

$$\frac{dN}{dt} = rN - aNP = 0$$

From this we can calculate that the prey population does not change in size when $rN = aNP$, or $P = r/a$. This implies that if there are exactly $P = r/a$ predators in the population, the prey population does not change in size. Thus, we can conclude that if there were more than r/a predators, the prey population should decrease in size (and if there were fewer than r/a predators, the prey population should increase in size).

Chapter 13

1. Ectoparasites live on the surface of the host, while endoparasites live inside the body of the host. Examples of ectoparasites include plants such as dodder and fungi

such as rusts or smuts; examples of endoparasites include tapeworms and bacterial pathogens such as *Mycobacterium tuberculosis*. Ectoparasites can disperse more easily from one host individual to the next than can endoparasites; however, ectoparasites are at greater risk from natural enemies than are endoparasites.

2. a. Host organisms have a wide range of defensive mechanisms that include a protective outer covering, an immune system that kills or limits the effectiveness of the parasite, and biochemical conditions inside the host's body that reduce the ability of the parasite to grow or reproduce.

   b. The statement could be true if the plant populations in Australia possessed specific defensive features that limited the ability of the parasite to grow or reproduce, yet the populations in Europe lacked such adaptations. Among many other possible examples, plants in the Australian populations might possess a specific allele that enabled them to kill or disable the parasite—hence causing the parasite to have mild effects there—whereas plants in the European populations might lack this allele, making them vulnerable to parasite attack.

3. Parasites can greatly reduce the growth, survival, or reproduction of host individuals, thereby reducing the growth rate of host populations. As a result, we would expect that parasites could also alter both the outcomes of species interactions and the composition of ecological communities. For example, if two plant species compete for resources and one typically outcompetes the other, a parasite that reduces the performance of the superior competitor may cause a competitive reversal in which the inferior competitor becomes the superior competitor. Such changes in the outcome of species interactions will cause changes in the relative abundances of the interacting species, thus altering the ecological community.

4. a. A given disease will become established and spread in a given host population only if the density of susceptible hosts exceeds a critical threshold density (S_T). The concept of a threshold density has considerable medical and ecological importance because it indicates that a disease will *not* spread if the density of susceptible hosts can be held below the threshold density.

   b. A disease can become established and spread only if the density of infected individuals (I) increases over time. This means that a disease will become established and spread if the rate of change of the density of infected individuals is greater than zero, which can be written mathematically as $dI/dt > 0$. Based on Equation 13.1, we can therefore calculate that a disease will become established and spread if $dI/dt = \beta SI - dI > 0$.

This equation implies that the disease will spread if $\beta SI > dI$, which can be rewritten as $\beta S > d$ or $S > d/\beta$.

Thus, we can conclude that the disease will become established and spread if the density of susceptible individuals (S) exceeds d/β; hence, $S_T = d/\beta$ is the critical, threshold density.

Chapter 14

1 Commensalism and mutualism share a number of characteristics: they are both very common, they can evolve in many ways, and they can cease to be beneficial if conditions change such that the costs of the interaction exceed its benefits. In addition, some evidence indicates that positive interactions may be particularly common in harsh environments. Positive interactions can also differ from one another in that they can range from species-specific, obligate, and coevolved relationships to those that show none of these three characteristics.

2. When a species in a mutualistic interaction provides its partner with a benefit, that action comes at a cost to the species providing the benefit. If circumstances change such that the costs of the interaction are greater than the benefits, a species may cease to provide benefits to its partner, or it may penalize its partner. The fact that mutualists may stop providing benefits to their partners when it is not advantageous for them to do so has convinced researchers that mutualism is not an altruistic interaction.

3. Initially, we could expect a decrease in the growth or reproduction of the coral species that are most sensitive to high water temperatures. If high temperatures continued long enough to cause repeated bleaching, these more sensitive species would begin to suffer heavy mortality. As a result of the decreased growth, reproduction, and survival of the these sensitive species, the species composition of the reef would change: those coral species that were better able to tolerate high water temperatures would constitute an increasingly high percentage of the corals found in the reef. Such changes in the composition of the coral reef community might also affect other species; for example, a fish that depended on an increasingly rare coral for shelter or food might also decline in abundance. As water temperatures continued to rise, other, less sensitive corals would also experience negative effects. Eventually, if temperatures continued to rise, the abundance of all corals in the reef would decline, as would the abundance of the many species that depend on the reef.

Chapter 15

1. A community is a group of interacting species that exist together at the same place and time. The interactions among species within communities equal more than the sum of all the individual species. They act in a synergistic fashion to shape community structure through time. In practical terms, communities can be characterized by the physical habitat that they occupy (e.g., deserts, streams, or sand dunes) or by the dominant biological habitat (e.g., trees, corals, or kelp).

2. Species richness tells us the number of species in a community, but not the relative abundances of those species. If two communities had similar numbers of species, but great differences in species evenness, species richness would not reflect this very important difference. Species diversity measurements would. Rank abundance curves show us how abundance varies among species and whether, for example, some species are much more abundant than others. Knowing the relative abundances of species allows hypotheses to be generated about how those species are interacting in the community.

3. Dominant species have a large effect on other species due to their high abundance and biomass. For example, kelp or trees have a large influence on species diversity by virtue of providing habitat, food, and other services that are directly related to their size. Keystone species have a strong effect, not because of their abundance or biomass, but because of the role they play in the community. For example, sea otters and wolves can have strong effects on their respective communities by preying on herbivores that, in turn, eat primary producers. This indirect interaction allows primary producers, which can have their own strong effects, to have higher abundances. Finally, ecosystem engineer species are able to create, modify, or maintain physical habitat for themselves and other species. Trees and kelp are examples of ecosystem engineers that are dominant species and beavers are an example of a keystone species ecosystem engineer.

4. a. Direct effect 1. Removing seagulls will increase the abundance of gooseneck barnacles due to decreased predation by seagulls.

    Direct effect 2. Removing seagulls will increase the abundance of the limpet *Lottia digitalis* due to decreased predation by seagulls.

   b. Indirect effect 1. Removing seagulls will decrease or not change the abundance of mussels due to increased competition with gooseneck barnacles.

    Indirect effect 2. Removing seagulls will decrease or not change the abundance of the limpet *Lottia strigatella* due to increased competition with the limpet *Lottia digitalis*.

    Indirect effect 3. Removing seagulls will decrease or not change the abundance of phytoplankton due to increased predation by gooseneck barnacles.

    Indirect effect 4. Removing seagulls will decrease or not change the abundance of seaweeds due to increased predation by the limpet *Lottia digitalis*.

Chapter 16

1. Abiotic and biotic agents of change include those listed in Table 16.1. Intense disturbances such as hurricanes, tsunamis, fires, and volcanic eruptions can cause major damage, but are relatively infrequent. Other agents of change, such as sea level rise, competition, or parasitism may not cause major damage, but may be frequent or constant. Still others, such as predation or logging, may be intense and relatively frequent, but not very extensive, thus forming patches of available resources. Abiotic and biotic agents of change can interact in two ways. Species such as ecosystem engineers can change abiotic conditions in ways that affect other species. Similarly, abiotic conditions can change the outcome of species interactions, thus influencing the trajectory of succession.

2. Primary succession involves the colonization of habitats devoid of life. Species colonizing these habitats must deal with stressful conditions and act as ecosystem engineers to create soils, nutrients, and food. Secondary succession involves the reestablishment of a community in which most of the organisms have been destroyed. Under these conditions, colonizing species are able to utilize the legacy effects of the preexisting species, but are likely to face more competition for resources.

3. A hypothetical community might be a newly cleared vacant lot in an unnamed city. The facilitation model would be supported if the first species to arrive were stress-tolerant and had the ability to modify their habitat in positive ways. In this case, these early species would facilitate the growth of later species, which are better competitors but less stress-tolerant. Over time, these later species would dominate as they outcompeted the facilitating species. The tolerance model assumes that the earliest species modify the environment, but in ways that neither help nor hinder later species. Later species are merely those that live longer and tolerate stressful conditions longer than early species. Finally, the inhibition model would be supported if the early species created conditions that benefited themselves, but inhibited later species. Only through the removal of those inhibitory early species—for example, via disturbance or stress— would later species be able to displace them.

4. It is hard to know whether a community is stable because stability depends on the spatial and temporal

scale at which the community is observed. All communities fluctuate and change over time, but how long must we wait for the community to return to some original state before we assume it is stable? There is no single answer. Although Sutherland did observe the formation of alternative communities on his tiles when predators were manipulated, did he follow the communities long enough, and at a large enough spatial scale, to show stability? Again, it depends on how you define "stability," leaving us with an unresolved question.

Chapter 17

1. The largest spatial scale is the global scale, which covers the entire world, over which there are major differences in species diversity and composition with latitude and longitude. At the global scale, these patterns are controlled by speciation, extinction, and immigration. The next scale down is the regional scale, defined by uniform climate and species that are bound by dispersal to the region. Within a region, species diversity and composition depend on immigration and extinction rates across the topographic and environmental landscape. The regional species pool (also called gamma diversity) has an important influence on the species present at the next scale down, the local scale (also called alpha diversity). The relationship between regional and local species diversity can help determine whether regional processes or the local effects of species interactions and physical conditions determine the membership of a local community.

2. Six biogeographic regions, which represent distinct groups of biota that vary in species diversity and composition, have been identified on land. These biogeographic regions reflect the evolutionary isolation of species due to continental drift. Thus, the ancestors of many modern species may have occurred together in the evolutionary past, but since Pangaea began breaking up into the continents we know today, they have evolved in very separate ways. There are significant impediments to dispersal within oceans, such as currents, thermal gradients, differences in water depth, and continents themselves, so it is conceivable that significant biogeographic regionalism will be found in the oceans as well as on land, but that effort has received considerably less attention.

3. The three main hypotheses focus on (1) geographic area and temperature, (2) evolutionary history, and (3) productivity. The first hypothesis proposes that both the geographic area and thermal stability of the tropics might promote both higher speciation rates and lower extinction rates, thereby increasing species' population sizes and geographic ranges. Speciation rates should increase because larger geographic ranges should lead to greater genetic isolation. Extinction rates should decrease because

larger population sizes should lower the risk of extinction due to chance events and larger species ranges should spread risk over a larger area.

The second hypothesis suggests that the tropics have had a longer evolutionary history than the temperate or polar zones because of the greater climatic stability there. This stability should have allowed more species to evolve without the interruption of severe climatic conditions that would have hindered speciation and increased extinctions.

The third hypothesis suggests that productivity increases species diversity by promoting larger population sizes, which should lead to lower extinction rates and overall higher species richness.

4. Design 3 would be the best choice because it has the largest and closest fragments of rare habitat. The equilibrium theory of island biogeography predicts that the largest islands will have the lowest extinction rates because species population sizes will be larger than on smaller islands. In addition, the theory predicts that islands closest to sources of species (e.g., mainland areas, or in this case, other, larger fragments of the rare habitat) will have the highest species diversity because immigration rates will be higher than on more distant islands.

Chapter 18

1. Yes. Community membership is dependent on dispersal, abiotic factors, and biological interactions. Given all the introductions of non-native species that have occurred worldwide, it is clear that "getting there" has been an important constraint on species entrance into communities. In this particular case, the seeds on the ecologist's shoes are physically and biologically adapted to prairie grassland communities and are thus prime candidates for successful introduction into New Zealand grassland communities.

2. Equilibrium theory relies on ecological and evolutionary compromises that result in resource partitioning as the mechanism of relatively stable coexistence. Resource partitioning is the idea that coexistence among species is possible if the species in a community use its resources in slightly different ways. In contrast, nonequilibrium theory relies on population fluctuations due to disturbance, stress, or predation as the mechanism of coexistence. As long as species' populations never reach their carrying capacities, competitive exclusion will not occur, and coexistence will be possible.

3. The lottery model best supports the tropical rainforest data set. The lottery model assumes that resources made available by the mortality of individuals are captured at random by recruits from a larger pool of colonists such

that no one species has an advantage, and that species diversity is maintained as a result.

4. Species diversity–community function relationships can vary depending on two variables: the degree of overlap in the ecological functions of species and variation in the strength of ecological functions of species. Graph A is best described by the complementarity hypothesis, which proposes that as species richness increases, there will be a linear increase in community function. This linear relationship occurs because each additional species added to the community has a unique and equally incremental effect on community function. Graph B is best described by the redundancy hypothesis, in which there is an upper limit on the effect of species richness on community function. This curvilinear relationship occurs because the unique functional contributions of species reach a threshold due to the overlap of species' functions. Graph C best describes the driver and passenger hypothesis, which suggests that the strengths of the effects of species' functions vary dramatically. "Driver" species have a large effect on community function, while "passenger" species have a minimal effect. The addition of driver and passenger species to a community has unequal effects on community function, producing a "stair step" curve. Graph D best describes a variation on the driver and passenger hypothesis in which the curve levels off at high species richness as the functions of the species become redundant.

Chapter 19

1. Primary production is the source of the energy entering an ecosystem, and therefore determines the amount of energy available to support that ecosystem. Primary production is also an important part of the exchange of carbon between the atmosphere and the biosphere, and thus is important in determining the atmospheric concentration of CO_2, an important greenhouse gas. Finally, primary production is a measure of the functioning of an ecosystem and provides a biological indicator of the ecosystem's response to stress.

2. As NPP increased in terrestrial ecosystems, the leaf area index would increase along with overall plant biomass. The amount of shading would increase as the leaf area index increases, and light would become increasingly limiting to growth. To compensate, plants would allocate more growth to stems and less to roots, so as to increase their height and overtop neighbors in order to acquire more light.

3. The researchers found a correlation between NPP and soil temperature, and assumed that the causal link was through the effect of soil temperature on root growth. While this may be correct, the researchers failed to show the causal link conclusively, which would require careful experimentation, or at least more thorough measurements of the effect of soil temperature on the factors that can influence plant growth. For example, soil temperature can have a pronounced influence by affecting the rate of decomposition of organic matter in the soil, and thus the availability of nutrients, which may influence growth rates.

4. Harvest techniques are simple and don't require high-tech equipment. However, harvesting can be very labor-intensive, may fail to account for production that is lost to herbivores or decomposition, and is impractical for large areas. Remote sensing provides estimates of NPP at larger spatial scales and can be used at frequent intervals. However, remote sensing is expensive and requires handling of massive amounts of data. Since it is based on absorption of light by chlorophyll, remote sensing can potentially overestimate NPP if a plant canopy is physiologically inactive.

5. The ^{15}N composition in the tissues of a heterotroph is reflective of the sources of food it is eating. If an animal population relies on a single food source, its ^{15}N composition would tend to have little variation among individuals, particularly if its food source was plants. An omnivore, by definition, consumes both plants and animals. A population of omnivores should therefore show greater variation among individuals in ^{15}N tissue composition as a result of their varying degrees of reliance on plant and animal foods.

Chapter 20

1. Population B should have a higher assimilation efficiency due to the higher food quality of its diet. The garbage and plant component of population A's diet is higher in materials that are difficult to digest, and its C:N ratio is also lower than that of population A's rodent diet. Thus, the amount of food assimilated would be greater in population B.

2. The seasonal and diurnal temperature variations in these animals' environments are different, and should result in different production efficiencies. The marine environment is more thermally stable, and thus the marine mammals should need to invest less energy in coping with temperature changes than the mammals in the terrestrial ecosystem. As a result, the marine mammals should be able to invest more energy in growth and reproduction.

3. The forest would have a greater amount of total energy flowing through its different trophic levels because a greater amount of energy would enter that ecosystem at the first trophic level. However, a larger proportion of the energy entering the lake ecosystem would pass

through its higher trophic levels due to its higher consumption and production efficiencies.

4. The effect of the trophic cascade on primary production (plant growth) would be less if the herbivore were a specialist. Increases in the abundance of a specialist herbivore would result in greater consumption of one or a few plant species, potentially releasing other plant species from the effects of competition with the species being consumed.

Chapter 21

1. The transformation of minerals in rock involves both the physical breakdown (mechanical weathering) and chemical alteration (chemical weathering) of the minerals. Mechanical weathering occurs through expansion and contraction of solid materials due to freezing–thawing or drying–rewetting cycles, gravitational forces such as landslides, and pressure exerted by plant roots. Mechanical weathering exposes the surfaces of mineral particles to chemical weathering. Weathering is a soil-building process, leading to the development of ever finer mineral particles and greater release of the nutrients in the minerals. The release of CO_2 and organic acids into the soil from organisms and detritus enhances the rate of chemical weathering.

2. The original source of nitrogen for plants is dinitrogen gas (N_2) in the atmosphere, but they cannot use it unless it is converted to other forms by the process of nitrogen fixation. Only prokaryotes can carry out nitrogen fixation, which is an energetically expensive process. Some plants, such as legumes, have symbiotic relationships with nitrogen-fixing prokaryotes. As ecosystems develop, nitrogen builds up in the pool of detritus and is converted into soluble organic and inorganic forms through decomposition. Some of the nitrogen released by decomposition is consumed by microorganisms, lowering the supply available to plants.

3. While both primary production and decomposition influence the buildup of organic matter and associated nutrients in the soil, decomposition is more sensitive to climatic controls than primary production. The mean residence time (or rate of turnover) is therefore more strongly controlled by decomposition. Low soil temperatures in boreal forests result in very long mean residence times. High rates of decomposition limit the buildup of soil organic matter in tropical forests, and the mean residence times of nutrients such as nitrogen and phosphorus are two orders of magnitude lower (i.e., turnover rates are 200 times faster) than those in boreal forests.

4. Nutrient cycling in the water column is efficient in both tropical and temperate-zone lakes, but organic matter is progressively lost from the surface layers in both sys-

tems, falling into the sediments in the benthic zone, where oxygen concentrations, and thus decomposition rates, are low. In the temperate zone, some of these nutrient-rich sediments are brought back to the surface layers during seasonal turnover of water, where they are decomposed, providing nutrients to support production. Turnover is largely absent in tropical lakes, which are therefore more dependent on external inputs of nutrients from streams and terrestrial ecosystems.

Chapter 22

1. This is a thought exercise, and answers will be personal, but should reflect that conservation biologists are scientists, motivated by a personal sense that loss of biodiversity is a problem—in part an ethical problem—that society needs to address. Answers should reflect that science must be rigorous and can be applied to solving conservation problems even if there are values behind the science.

2. Biotic homogenization is occurring at a number of levels through the spread of invasive species and through the increasing dominance of generalist species in concert with a loss of more specialized species. The three components of biodiversity undergoing increased homogenization are species diversity, genetic diversity, and ecosystem diversity. Range expansions of invasive species, the increase in generalist species, and the loss of more specialized species through range contractions or extinction are leading to taxonomic homogenization. Hybridization between native and introduced species contributes to genetic homogenization. Finally, increasing taxonomic homogenization, particularly the loss of specialized species, may result in losses of functional groups and cause ecosystems to become more functionally similar.

3. The principal threats to biodiversity are habitat loss, degradation, and fragmentation; the spread of invasive species; overharvesting; and climate change. For some species, disease poses a threat, and for others, particularly aquatic species, pollution is a significant threat. Many freshwater mussel species of North America are threatened by both by pollution and by the invasion of the zebra mussel. The Pyrenean ibex was driven extinct by hunting, climate change, disease, and competition with domesticated and non-native species. Many other examples are possible.

4. The classification system set up by Natural Heritage/NatureServe is a documentation of each species' conservation status from a biological perspective, while a listing under the U.S. Endangered Species Act is a legal designation. While federally endangered species would generally also be considered globally rare by Natural Heritage/NatureServe, the reverse does not necessarily hold true; many extremely rare or threatened species are

not on the federal endangered species list. The Endangered Species Act (ESA) provides legal protection for listed species, and it requires the designation of critical habitat and the development and implementation of a recovery plan for those species. In contrast, Natural Heritage/NatureServe can only recommend the protection of species.

5. Answers to this question will depend on where students are located and what species they identify. The object of this question is to make students aware of species of conservation concern, threats to biodiversity, and efforts that are under way to protect species in their own region. It also invites them to identify research needs and to think about scientific approaches to conservation.

Chapter 23

1. Habitat islands resemble actual islands in the way that they spatially isolate populations of some species from one another, with potential demographic and genetic consequences. They differ from islands, however, in that the matrix between habitat fragments may be more or less permeable to some species, so that movement between habitat fragments may be constrained, but may still occur with some frequency. As we saw in Chapter 17, the principles of island biogeography apply to habitat islands in that there is immigration to fragments, extinction within fragments, and some equilibrium level of species diversity. Larger habitat islands can sustain greater species diversity than smaller fragments.

2. In a sense, corridors are long, skinny habitat patches. Animals may nest in them, plants will germinate in them if conditions are right, and predation and competition occur in them. But they are likely to be biologically impoverished relative to larger habitat blocks because of the effects of their narrow dimensions on their biotic and abiotic properties. They are likely to resemble edge habitat in experiencing more light, more rapid nutrient cycling, and more predation than larger habitat blocks. They may be more vulnerable to invasive species, and they may permit movement of diseases between habitat blocks. Nevertheless, they are generally beneficial, at least for some species, in allowing movement of organisms across a fragmented landscape.

3. National forests and national parks have different management objectives and the difference in the resulting land uses is visible from space, in the form of a clear line separating clear-cut patches of the Targhee National Forest from the uncut forests of Yellowstone National Park. National forests permit the harvesting of timber, which is generally not permitted in national parks. Timber harvesting makes for a patchy forest of different-aged stands, which may support a different group of species than is found in a national park, and may favor early successional species over old-growth-associated species. While both national parks and national forests have a mandate to protect biodiversity, national parks must balance these aims with recreation and visitor needs, while national forests must include timber production needs in their mission as well. Under an ecosystem management approach, the emphasis would be regional, and so the national forest and national park administrations would be working together to achieve conservation goals set by consensus.

4. Answers to this question will be place-specific and will depend on where students are located. The object of this question is to get students to evaluate the size, connectivity, and effectiveness of protected areas in their region. They may also find themselves wondering what kinds of lands serve as protected areas (i.e., state and federal lands, local public lands, lands held by land trusts and private conservation organizations such as the Audubon Society or The Nature Conservancy) and see the challenge of keeping track of where all these lands are. They should consider the role that private lands play in harboring local species, and they should consider how lands change over time depending on their ownership. In considering connectivity between protected areas, they should evaluate the protection status of corridors and whether those corridors are vulnerable to future disruption. They may consider possibilities for habitat restoration that could improve the landscape-level picture.

5. Again, the response here will be place-specific. This question invites students to identify and examine regional ecosystem management projects. They may choose a watershed protection project, a regional effort to preserve a forest, or a coordinated effort to protect habitat for a particular species. The key will be to identify a diverse partnership of stakeholders coming together to address land use in a region in a manner that serves both human needs and regional biodiversity. Students are asked to identify the successes of the program, and they may also identify the societal compromises and actions that slow progress in land protection or in resolving problems causing habitat degradation.

Chapter 24

1. The two major biological influences on the global carbon cycle are photosynthesis, which takes up CO_2 from the atmosphere, and respiration, which releases CO_2 back to the atmosphere. Prior to the Industrial Revolution, uptake by photosynthesis and release by respiration were roughly equal at a global scale, and thus there was no net flux associated with Earth's biota. However, increasing human population growth rates resulted in increasing deforestation and agricultural development, resulting in greater

decomposition and heterotrophic respiration due to warming of the soil surface. As a result, atmospheric CO_2 concentrations increased. Deforestation was the primary reason for increasing atmospheric CO_2 concentrations until the early part of the twentieth century.

2. While animals can respond to climate change by moving, their habitats cannot. Animals are dependent on plants to provide their food (or food for their prey). Climate change will be so rapid that evolutionary responses will not be possible for most species of plants, and the dispersal rates of most species are too slow to track the predicted climate changes. Loss of habitat will therefore result in decreased population growth for some animals as a result of climate change. Additionally, migrating animals may respond to climate change more slowly than non-migrating species. As a result, prey species may be less abundant or absent when these animals arrive at their destination.

3. One of the components of acid precipitation is nitric acid (HNO_3), which is a reactive form of nitrogen (i.e., it is biologically available). Both acid deposition and nitrogen deposition can cause acidification of soils and surface waters, loss of nutrient cations, and increases in aluminum. Nitrogen deposition can increase plant growth when it occurs at low rates. Over long periods, or at high rates of N deposition, N saturation occurs, leading to increased plant susceptibility to stress and, as noted above, soil acidification.

4. The effect of atmospheric ozone on organisms depends on where in the atmosphere it is found. Ozone in the stratosphere acts as a shield against high-energy ultraviolet-B radiation, which is harmful to organisms. In contrast, ozone in the troposphere damages organisms that come in direct contact with it. Ozone in the troposphere also acts as a greenhouse gas, contributing to global climate change.

Glossary

Numbers in brackets refer to the chapter(s) where the term is introduced.

absolute population size   The actual abundance of a population or species. [8]

abundance   The number of individuals in a species that are found in a given area; abundance is often measured by population size or population density. [8]

acclimatization   An organism's adjustment of its physiology, morphology, or behavior to lessen the effect of an environmental change and minimize the associated stress. [4]

acid neutralizing capacity   The ability of the chemical environment to counteract acidity, associated with concentrations of base cations, including Ca^{2+}, Mg^{2+}, and K^+. [24]

acidity   A measure of the ability of a solution to behave as an acid, a compound that releases protons (H^+) to the water in which it is dissolved. *Compare* alkalinity. [2]

adaptation   A physiological, morphological, or behavioral trait with an underlying genetic basis that enhances the survival and reproduction of its bearers in their environment. [1, 4, 6]

adaptive evolution   A process of evolutionary change in which adaptive traits tend to increase in frequency in a population over time. [6]

adaptive management   A component of ecosystem management in which management actions are seen as experiments and future management decisions are determined by the outcome of present decisions. [23]

adaptive radiation   An event in which a group of organisms gives rise to many new species that expand into new habitats or new ecological roles in a relatively short time. [6]

aerosols   Solid, liquid, or gaseous particles suspended in the atmosphere. [21]

age structure   The proportions of a population in each age class. [9]

albedo   The amount of solar radiation reflected by a surface, usually expressed as a percent of the incident solar radiation. [2]

alkalinity   A measure of the ability of a solution to behave as a base, a compound that takes up protons (H^+) or releases hydroxide ions (OH^-). *Compare* acidity. [2]

Allee effect   A decrease in the population growth rate (r or λ) as the population density decreases. [10]

allele   One of two or more forms of a gene that result in the production of different versions of the protein that the gene encodes. [6]

allelopathy   A form of interference competition in which individuals of one species release toxins that directly harm individuals of other species. [11]

allochthonous   The input of energy and nutrients from outside the ecosystem. *Compare* autochthonous. [20]

allometry   Differential growth of body parts that results in a change of shape or proportion with size. [7]

alpha diversity   Species diversity at the local or community scale. *Compare* beta diversity, gamma diversity. [17]

alternation of generations   A complex life cycle, found in many algae and all plants, in which there is both a multicellular diploid form, the sporophyte, and a multicellular haploid form, the gametophyte. [7]

alternative stable states   Different community development scenarios that are possible at the same location under similar environmental conditions. [16]

anisogamy   Production of two types of gametes of different sizes. *Compare* isogamy. [7]

anoxic   Having a low oxygen concentration. [2]

anthropogenic   Of, relating to, or caused by humans or their activities. [24]

aposematic coloration   *See* warning coloration. [12]

arbuscular mycorrhizae   Mycorrhizae in which the fungal partner grows into the soil, extending some distance away from the plant root, and also grows between some root cells while penetrating others. *Compare* ectomycorrhizae. [14]

Arctic ozone dent   An area of the stratosphere over the Arctic region where ozone concentrations are low, but have not dropped below 220 Dobson units. [24]

assimilation efficiency   The proportion of ingested food that is assimilated by an organism. [20]

atmospheric deposition   The movement of particulate and dissolved matter from the atmosphere into ecosystems by gravity or in precipitation. [21]

atmospheric pressure   The pressure the exerted on a surface due to the mass of the atmosphere above it. [2]

autochthonous   Energy and nutrients produced within the ecosystem. *Compare* allochthonous. [20]

autotroph   An organism that converts energy from sunlight or from inorganic chemical compounds in the environment into chemical energy stored in the carbon–carbon bonds of organic compounds. *Compare* heterotroph. [5]

avoidance   A response of an organism to stressful environmental conditions that lessens their effect through some behavior or physiological activity that minimizes exposure to the stress. *Compare* tolerance. [4]

balance of nature   The idea that natural systems tend to return to an original, preferred state after a disturbance. [1]

benthic   Of or relating to the bottom of a body of water. [3]

beta diversity   The change in species number and composition, or turnover of species, as one moves from one community to another. *Compare* alpha diversity, gamma diversity. [17]

bioaccumulation   The progressive concentration of a substance in an organism's body over its lifetime. [20]

biodiversity   A term used to describe the diversity of important ecological entities that span multiple spatial scales, from genes to species to communities. [15]

biofuel   A liquid or gas fuel made from plant material (biomass). [18]

biogeochemistry   The study of the physical, chemical, and biological factors that influence the movements and transformations of chemical elements. [21]

biogeography   The study of variation in species composition and diversity among geographic locations. [17]

biological crust   A soil crust composed of a mix of species of cyanobacteria, lichens, and mosses; also called a cryptobiotic or cryptogamic crust. [21]

biological reserve An often small nature reserve established with the conservation of a single species or ecological community as the main conservation objective. [23]

biomagnification An increase in the tissue concentrations of a substance at higher trophic levels that results as animals at each trophic level consume prey with increasing concentrations of the substance. [20]

biome A terrestrial biological community shaped by the regional climate, soil, and disturbance patterns where it is found, usually classified by the growth form of its most abundant plants. [3]

biosphere The highest level of biological organization, consisting of all living organisms on Earth plus the environments in which they live; the biosphere is located between the lithosphere and the troposphere. [1, 3]

biotic resistance Interactions of the native species in a community with non-native species that exclude or slow the growth of those non-native species. [18]

bottom-up control Limitation of the abundance of a population by nutrient supply or by the availability of food. *Compare* top-down control. [10]

boundary layer A zone close to a surface where a flow of fluid, usually air, encounters resistance and becomes turbulent. [4]

buffer zone A portion of a nature reserve surrounding a core natural area where controls on land use are less stringent than in the core natural area, yet land uses are at least partially compatible with many species' resource requirements. *Compare* core natural area. [23]

C₃ photosynthetic pathway A biochemical pathway involving the uptake of CO_2 by the enzyme ribulose 1,5 bisphosphate carboxylase/oxygenase (rubisco) and synthesis of sugars by the Calvin cycle. *Compare* C₄ photosynthetic pathway, crassulacean acid metabolism. [5]

C₄ photosynthetic pathway A biochemical pathway involving the daytime uptake of CO_2 by the enzyme phosphoenol pyruvate carboxylase (PEPcase) in mesophyll cells; the carbon is then transferred as a four carbon acid to the bundle sheath cells, where CO_2 is released to the Calvin cycle for sugar synthesis. *Compare* C₃ photosynthetic pathway, crassulacean acid metabolism. [5]

Calvin cycle The biochemical pathway used by photosynthetic and chemosynthetic organisms to fix CO_2 and synthesize sugars. [5]

carbon neutral Of or relating to fuels that produce an amount of CO_2 when burned that is equal to or less than the amount taken up by the plants from which they are made. [18]

carrying capacity The maximum population size that can be supported indefinitely by the environment; the concept of a carrying capacity is represented by the term K in the logistic equation (Equation 9.7). [9]

catchment The area in a terrestrial ecosystem that is drained by a single stream; a common unit of study in terrestrial ecosystem studies; also called a watershed. [21]

cation exchange capacity A soil's ability to hold nutrient cations such as Ca^{2+}, K^+, and Mg^{2+} and exchange them with the soil solution, determined by the clay content of the soil. [21]

cavitation A disruption of the water column in the xylem of a plant that can occur under dry conditions when high tension (negative pressure potential) on the water in the xylem pulls air into it, or following freezing of water in the xylem, and coalescence of air bubbles when the ice melts. [4]

character displacement A process in which competition causes the forms of competing species to evolve to become more different over time, thereby causing species to become more different where they live together than where they live apart. [11]

cheater In a mutualism, an individual that increases its production of offspring by overexploiting its mutualistic partner. [14]

chemical weathering The chemical breakdown of soil minerals leading to the release of soluble forms of nutrients and other elements. *Compare* mechanical weathering. [21]

chemolithotrophy *See* chemosynthesis.

chemosynthesis The use of energy from inorganic chemical compounds to produce carbohydrates; also known as *chemolithotrophy*. [5]

clays Fine soil particles (<2 μm) that have a semicrystalline structure and weak negative charges on their surfaces that can hold onto cations and exchange them with the soil solution. [21]

climate The long-term description of weather, based on averages and variation measured over decades. *Compare* weather. [2]

climate change Directional change in climate over a period of several decades or longer. [24]

climax stage The final stage of succession, viewed as a stable end point that experiences little change. [16]

cline A pattern of gradual change in a characteristic of an organism over a geographic region. [6]

clone A genetically identical copy of an individual. [8]

clumped dispersion A dispersion pattern in which individuals are grouped together. *Compare* random dispersion, regular dispersion. [8]

coarse-filter Of or referring to an approach to conservation biology that focuses on habitats, landscapes, and ecosystems. *Compare* fine-filter. [22]

coevolution The evolution of two interacting species, each in response to selection imposed by the other. [13]

cohort life table A life table in which the fate of a group of individuals born during the same time period (a cohort) is followed from birth to death. [9]

commensalism An interaction between two species in which individuals of one species benefit while individuals of the other species do not benefit and are not harmed. [14]

community A group of interacting species that occur together at the same place and time. [1, 15]

community structure The set of characteristics that shape a community, including the number, composition, and abundance of species. [15]

compensation An adaptive growth response of plants to herbivory in which removal of plant tissues stimulates the plant to produce new tissues. [12]

competition An interaction between two species in which individuals of both species are harmed when both species use a resource that limits their ability to grow and reproduce. [11]

competition coefficient A constant used in the Lotka–Volterra competition model to describe the effect of one competing species on the other. [11]

competitive exclusion principle The principle that two species that use a limiting resource in the same way cannot coexist indefinitely. [11]

competitive plants In Grime's plant life history model, plants that are superior competitors under conditions of low stress and low disturbance. *Compare* ruderals, stress-tolerant plants. [7]

complementarity hypothesis A hypothesis proposing that as the species richness of a community increases, there is a linear increase in the positive effects of those species on community function. *Compare* redundancy hypothesis. [18]

complex life cycle   A life cycle in which there are at least two distinct stages that differ in their habitat, physiology, or morphology. [7]

conduction   The transfer of sensible heat through the exchange of kinetic energy by molecules in direct contact with one another. *Compare* convection. [2]

conservation biology   The scientific study of phenomena that affect the maintenance, loss, and restoration of biodiversity. [22]

consumer   An organism that obtains its energy by eating other organisms or their remains. *Compare* producer. [1]

consumption efficiency   The proportion of the biomass available in an ecosystem that is ingested. [20]

context-dependent   A view of species interactions as variable in strength and direction under different environmental conditions. [15]

continental climate   The climate typical of terrestrial areas in the middle of large continental land masses at high latitudes, characterized by high variation in seasonal temperatures. *Compare* maritime climate. [2]

continental drift   The slow movement of tectonic plates (sections of Earth's crust) across Earth's surface. [17]

controlled experiment   A standard scientific approach in which an experimental group (that has the factor being tested) is compared with a control group (that lacks the factor being tested). [1]

convection   The transfer of sensible heat through the exchange of air and water molecules as they move from one area to another. *Compare* conduction. [2]

convergence   The evolution of similar growth forms among distantly related species in response to similar selection pressures. [3]

cooperative breeding   A behavioral pattern in which young animals postpone breeding and instead help their parents raise offspring. [8]

core natural area   A portion of a nature reserve where the conservation of biodiversity and ecological integrity takes precedence over other values or uses. *Compare* buffer zone. [23]

Coriolis effect   The influence of the frictional forces associated with the spinning of Earth on its axis on wind and ocean current direction. [2]

crassulacean acid metabolism (CAM)   A photosynthetic pathway in which CO_2 is fixed and stored as an organic acid at night, and then released to the Calvin cycle during the day. *Compare* C_3 photosynthetic pathway, C_4 photosynthetic pathway. [5]

crypsis   A defense against predators in which prey species have a shape or coloration that provides camouflage and allows them to avoid detection. [12]

cryptobiotic crust   *See* biological crust.

cryptogamic crust   *See* biological crust.

damped oscillations   A pattern of population fluctuations where the extent to which the population rises and falls in abundance gradually become smaller over time. [10]

decomposition   The physical and chemical breakdown of detritus by detritivores for energy capture, leading to the release of nutrients as simple, soluble organic and inorganic compounds that can be taken up by other organisms. [21]

delayed density dependence   Delays in the effect of population density on population size that can contribute to population fluctuations. [10]

demographic stochasticity   Chance events associated with whether individuals survive or reproduce. [10]

denitrification   A process by which certain bacteria convert nitrate (NO_3^-) into nitrogen gas (N_2) and nitrous oxide (N_2O) under anoxic conditions. [21]

density-dependent   Of or referring to a factor that causes birth rates, death rates, or dispersal rates to change as the density of the population changes. *Compare* density-independent. [9]

density-independent   Of or referring to a factor whose effects on birth and death rates are independent of population density. *Compare* density-dependent. [9]

desertification   Degradation of formerly productive land in arid regions resulting in loss of plant cover and acceleration of soil erosion. [3, 24]

detritus   Freshly dead or partially decomposed remains of organisms. [8]

diapause   A state of suspended animation or dormancy in which an organism can survive unfavorable conditions. [7]

direct development   A simple life cycle that goes directly from fertilized egg to juvenile without passing through a free-living larval stage. [7]

direct interactions   Interactions between two species, including traditional relationships such as competition, predation, and positive interactions. *Compare* indirect interactions. [15]

directional selection   Selection that favors individuals with one extreme of a heritable phenotypic trait. *Compare* disruptive selection, stabilizing selection. [6]

dispersal limitation   A situation in which a species' limited capability for dispersal prevents it from reaching areas of suitable habitat. [8]

dispersion   The spatial arrangement of individuals within a population. [8]

disruptive selection   Selection that favors individuals with a phenotype at either extreme over those with an intermediate phenotype. *Compare* directional selection, stabilizing selection. [6]

distribution   The geographic area where individuals of a species are present. [8]

disturbance   An abiotic event that kills or damages some organisms and thereby creates opportunities for other organisms to grow and reproduce. [7, 8, 16]

dominant species   A species that has large, community-wide effects by virtue of its size or abundance, its strong competitive ability, or its provision of habitat or food for other species, also called a foundation species. [15]

doubling time (td)   The number of years it takes a population to double in size. [9]

driver and passenger hypothesis   A hypothesis proposing that the strengths of the effects of species' ecological functions on their communities vary dramatically, such that "driver" species have a large effect on community function, while "passenger" species have a minimal effect. [18]

dynamic equilibrium model   An elaboration of the intermediate disturbance hypothesis proposing that species diversity is maximized when the level of disturbance and the rate of competitive displacement are roughly equivalent. [18]

ecological footprint   The total area of productive ecosystems required to support a population. [9]

ecological niche   The physical and biological conditions that a species needs to grow, survive, and reproduce. [8]

ecology   The scientific study of interactions between organisms and their environment. [1]

ecosystem   A community of interacting organisms plus the physical environment in which they live. [1, 19]

ecosystem engineer A dominant species that influences its community by creating, modifying, or maintaining physical habitat for itself and other species. [15]

ecosystem management An approach to habitat management in which scientifically-based policies and practices guide management decisions on how best to meet an overarching goal of sustaining ecosystem structure and function for long periods of time. [23]

ecosystem services Natural processes that sustain human life that depend on the functional integrity of natural communities and ecosystems. [22]

ecotype A population with adaptations to a unique environment. [4]

ectomycorrhizae Mycorrhizae in which the fungal partner typically grows between plant root cells and forms a mantle around the exterior of the root. *Compare* arbuscular mycorrhizae. [14]

ectoparasite A parasite that lives on the surface of another organism. *Compare* endoparasite. [13]

ectotherm An animal that regulates its body temperature primarily through energy exchange with its external environment. *Compare* endotherm. [4]

edge effects Biotic and abiotic changes that are associated with an abrupt habitat boundary such as that created by habitat fragmentation. [17, 23]

El Niño Southern Oscillation (ENSO) A climatic oscillation in the Pacific Ocean causing widespread climate variation and changes in upwelling currents, characterized by a switch in the positions of high-pressure and low-pressure systems over the equatorial Pacific, leading to a weakening of the easterly trade winds that normally push warm water toward Southeast Asia. *See also* La Niña. [2]

endemic Occurring in a particular geographic location and nowhere else on Earth. [17]

endoparasite A parasite that lives inside the body of its host organism. *Compare* ectoparasite. [13]

endosperm Nutrient-rich material in a seed that sustains the developing embryo and often the young seedling. [7]

endotherm An animal that regulates its body temperature primarily through internal metabolic heat generation. *Compare* ectotherm. [4]

environmental science An interdisciplinary field that incorporates concepts from the natural sciences (including ecology) and the social sciences (e.g., politics, economics, ethics), focused on how people affect the environment and how we can address environmental problems. [1]

environmental stochasticity Erratic or unpredictable changes in the environment. [10]

epilimnion The warm surface layer of water in a lake, lying above the thermocline, that forms during the summer in some lakes of temperate and polar regions. *Compare* hypolimnion. [2]

equilibrium theory A proposed explanation for the coexistence of species in a community that relies on relatively stable coexistence due to ecological and evolutionary "compromises" that result in resource partitioning. *Compare* nonequilibrium theory. [18]

equilibrium theory of island biogeography A theory proposing that the number of species on an island or in an island-like habitat results from a dynamic balance between immigration rates and extinction rates. [17]

eutrophic Nutrient-rich; characterized by high primary productivity. *Compare* oligotrophic. [21]

eutrophication A change in the nutrient status of an ecosystem from nutrient poor to nutrient rich; such a change may occur naturally in some lakes due to the accumulation of sediments, but it may also be caused by anthropogenic nutrient inputs. [21]

evapotranspiration The sum of water loss through evaporation and transpiration . [2]

evolution (1) Change in allele frequencies in a population over time. (2) Descent with modification; the process by which organisms gradually accumulate differences from their ancestors. [1, 6]

evolutionarily significant unit An appropriate target of management within a species (e.g., subspecies, populations) as determined by genetic analyses; also called a *management unit*. [22]

exploitation A relationship in which one organism benefits by feeding on, and thus directly harming, another. [12]

exploitation competition An interaction in which species compete indirectly through their mutual effects on the availability of a shared resource. *Compare* interference competition. [11]

exponential growth Change in the size of a population of a species with continuous reproduction by a constant proportion at each instant in time. *Compare* geometric growth. [9]

exponential population growth rate (r) A constant proportion by which a population of a species with continuous reproduction changes in size at each instant in time; also called the intrinsic rate of increase. *Compare* geometric population growth rate. [9]

extent In landscape ecology, the area or time period over which observations are collected for a region under study; together with grain, extent characterizes the scale at which a landscape is studied. *Compare* grain. [23]

extinction vortex A pattern in which a small population that drops below a certain size may decrease even further in size, perhaps spiraling towards extinction; an extinction vortex results from the vulnerability of small populations to chance genetic, demographic, and environmental events. [22]

facultative CAM The ability to switch from C_3 to CAM photosynthesis when water availability decreases. [5]

fecundity The average number of offspring produced by a female while she is of age x (denoted F_x in a life table). [9]

Ferrell cell A large-scale, three-dimensional pattern of atmospheric circulation in each hemisphere, located at mid-latitudes between the Hadley and polar cells. [2]

fine-filter Of or referring to an approach to conservation biology that focuses on genes, populations, and species. *Compare* coarse-filter. [22]

finite rate of increase *See* geometric population growth rate. [9]

fixation (1) With respect to the genetic composition of a population, an allele frequency of 100%. [6] (2) The uptake of the gaseous form of a compound, including CO_2 in photosynthesis and N_2 in nitrogen fixation, by organisms for use in metabolic functions. [5]

flagship species A charismatic species that may be emphasized in conservation efforts because it helps to garner public support for a conservation project. [22]

focal species One of a group of species selected as a priority for conservation efforts and chosen because its ecological requirements differ from other species in the group, thereby helping to ensure that as many different species as possible receive protection. [22]

food web A diagram showing the connections between organisms and the food they consume. [20]

foundation species *See* dominant species.

fugitive species A species whose persistence depends on its being able to disperse from one place to another as environmental conditions change. [11]

functional group   A subset of the species in a community that function in similar ways, but do not necessarily use the same resources. *Compare* guild. [15]

gametophyte   A multicellular haploid life cycle stage, found in many algae and all plants, that produces haploid gametes. *Compare* sporophyte, alternation of generations. [7]

gamma diversity   Species diversity at the regional scale; the regional species pool. *Compare* beta diversity, gamma diversity. [17]

gene flow   The transfer of alleles from one population to another via the movement of individuals or gametes. [6]

gene-for-gene interaction   A specific defensive response that makes particular plant genotypes resistant to particular parasite genotypes. [13]

genet   In plants and animals that reproduce asexually, the entire genetic individual; a genet may consist of multiple, genetically-identical parts, each of which has the potential to function as an independent physiological unit. *Compare* ramet. [8]

genetic drift   A process in which chance events determine which alleles are passed from one generation to the next, thereby causing allele frequencies to fluctuate randomly over time; the effects of genetic drift are most pronounced in small populations. [6, 10]

genotype   The genetic makeup of an individual. [6]

geographic range   The entire geographic region over which a species is found. [4, 8]

geometric growth   Change in the size of a population of a species with discrete reproduction by a constant proportion from one discrete time period to the next. *Compare* exponential growth. [9]

geometric population growth rate (λ)   A constant proportion by which a population of a species with discrete reproduction changes in size from one discrete time period to the next; also called the finite rate of increase. *Compare* exponential population growth rate. [9]

grain   In landscape ecology, the size of the smallest homogeneous unit of study (such as a pixel in a digital image), which determines the resolution at which a landscape is observed; together with extent, grain characterizes the scale at which a landscape is studied. *Compare* extent. [23]

gravitational potential   The energy associated with gravity. [4]

greenhouse effect   The warming of Earth by gases in the atmosphere that absorb and reradiate infrared energy emitted by Earth's surface. [24]

greenhouse gases   Atmospheric gases that absorb and reradiate the infrared radiation emitted by Earth's surface, including water vapor (H_2O), carbon dioxide (CO_2), methane (CH_4), and nitrous oxide (N_2O). [2]

gross primary production (GPP)   The amount of energy that autotrophs capture by photosynthesis and chemosynthesis. [19]

guild   A subset of the species in a community that use the same resources, whether or not they are taxonomically related. *Compare* functional group. [15]

habitat corridor   A relatively narrow patch that connects blocks of habitat and often facilitates the movement of species between those blocks. [23]

habitat degradation   Anthropogenic changes that reduce the quality of habitat for many, but not all, species. [22]

habitat fragmentation   The breaking up of once continuous habitat into a complex matrix of spatially isolated habitat patches amid a human-dominated landscape. [10, 22]

habitat loss   The outright conversion of an ecosystem to another use by human activities. [22]

habitat mutualism   A mutualism in which one partner provides the other with shelter, a place to live, or favorable habitat. [14]

Hadley cell   A large-scale, three-dimensional pattern of atmospheric circulation in each hemisphere in which air is uplifted at the equator and subsides at about 30° N and S. [2]

heat capacity   The amount of energy required to raise the temperature of a substance. [2]

herbivore   An organism that eats the tissues or internal fluids of living plants or algae. [12]

heterogeneity   The quality or state of consisting of dissimilar elements. [23]

heterotroph   An organism that obtains energy by consuming energy-rich organic compounds from other organisms. Compare *autotroph*. [5]

horizons   Layers of soil distinguished by their color, texture, and permeability. [21]

horizontal interactions   Non-trophic interactions, such as competition and some positive interactions, that occur within a trophic level. [15]

host   An organism on or within which a parasite or other symbiont lives. *Compare* host, symbiosis. [13]

hypolimnion   The densest, coldest water layer in a lake, lying below the thermocline. *Compare* epilimnion. [2]

hyporheic zone   The portion of the substratum below and adjacent to a stream bed where water movement still occurs, either from the stream or from groundwater moving into the stream. [3]

hypothesis   A possible answer to a question developed using previous knowledge or intuition. *See* scientific method. [1]

hysteresis   An inability to shift back to the original community type, even when the original conditions are restored. [16]

inbreeding   Mating between related individuals. [10]

indirect interactions   Interactions in which the relationship between two species is mediated by a third (or more) species. *Compare* direct interactions. [15]

induced defense   In plant–herbivore interactions, a defense against herbivory, such as production of a secondary compound, that is stimulated by herbivore attack. [12]

interaction strength   A measure of the effect of one species' population on the size of another species' population. [15, 20]

interaction web   A concept that has been introduced to describe both the trophic (vertical) and non-trophic (horizontal) interactions in a traditional food web. [15]

interference competition   An interaction in which species compete directly by performing antagonistic actions that interfere with the ability of their competitors to use a resource that both require, such as food or space. *Compare* exploitation competition. [11]

intermediate disturbance hypothesis   A hypothesis proposing that species diversity in communities should be greatest at intermediate levels of disturbance because competitive exclusion at low levels of disturbance and mortality at high levels of disturbance should reduce species diversity. [18]

interspecific competition   Competition between members of different species. *Compare* intraspecific competition. [11]

Intertropical Convergence Zone (ITCZ)   The zone of maximum solar radiation, atmospheric uplift, and precipitation within the tropical zone. [2]

intraspecific competition   Competition between individuals of a single species. *Compare* interspecific competition. [11]

intrinsic rate of increase   *See* exponential population growth rate. [9]

invasive species   Introduced species that survive and reproduce in their new environment, sustain a growing population, and have large effects on the native community. [22]

island biogeography   *See* equilibrium theory of island biogeography.

isogamy   The production of equal-sized gametes. *Compare* anisogamy. [7]

isolation by distance   A pattern in which patches located far away from occupied patches are less likely to be colonized than are nearby patches. [10]

iteroparous   Having the capacity to reproduce multiple times in a lifetime. *Compare* semelparous. [7]

jump dispersal   A long-distance dispersal event by which a species colonizes a new geographic region. [10]

K   *See* carrying capacity. [7]

***K*-selection**   In the *r*- and *K*-selection approach to classifying life history strategies, the selection pressures faced by organisms that live in environments where population densities are high (at or near the carrying capacity, *K*). *Compare* *r*-selection. [7]

keystone species   A strongly interacting species that has a large effect on energy flow and on community structure and composition disproportionate to its abundance or biomass. [15, 20]

La Niña   A stage of the El Niño Southern Oscillation consisting of a stronger-than-average phase of the normal climatic pattern, with high atmospheric pressure off the coast of South America and low pressure in the western Pacific. [2]

land use change   The conversion of land from natural to human-disturbed communities for resource extraction, such as agriculture or pastoral land use. [3]

landscape   An area that is spatially heterogeneous either in its elements (such as the vegetative cover types found at different locations) or in how those elements are arranged. [23]

landscape composition   In landscape ecology, the kinds of elements or patches comprised by a landscape and how much of each kind is present. [23]

landscape ecology   The study of landscape patterns and the effects of those patterns on ecological processes. [23]

landscape structure   In landscape ecology, the physical configuration of the different compositional elements of a landscape. [23]

lapse rate   The rate at which atmospheric temperature decreases with increasing distance from the ground. [2]

latent heat flux   Heat transfer associated with the phase change of water, such as evaporation, sublimation, or condensation. [2]

latent period   A period in which an individual is infected with a disease but cannot spread the disease. [13]

leaf area index   The area of leaves per unit of ground area (a dimensionless number, since it is an area divided by an area). [19]

lentic   Of or referring to still water. *Compare* lotic. [3]

life history   A record of major events relating to an organism's growth, development, reproduction, and survival. [7]

life history strategy   The overall pattern in the timing and nature of life history events, averaged across all the individuals of a species. [7]

life table   A summary of how survival and reproductive rates in a population vary with the age of individuals; in species for which age is not informative or is difficult to measure, life tables are often based on the size or life history stage of individuals. [9]

lignin   A structural compound that strengthens plant tissues. [21]

lithosphere   Earth's surface crust and upper mantle; the layer of Earth that includes the tectonic plates. [3]

litter   Fresh, undecomposed organic matter. [21]

littoral zone   The nearshore zone of a lake where the photic zone reaches to the bottom. [3]

local scale   A spatial scale that is essentially equivalent to a community. [17]

loess   Thick layers of sediment deposited by wind. [21]

logistic growth   Change in the size of a population that is rapid at first, then decreases as the population approaches the carrying capacity of its environment. [9]

lotic   Of or relating to flowing water. *Compare* lentic. [3]

Lotka–Volterra competition model   A modified form of the logistic equation used to model competition (see Equation 11.1). [11]

lottery model   A nonequilibrium hypothesis proposing that species diversity in communities is maintained by a "lottery" in which resources made available by the effects of disturbance, stress, or predation are captured at random by recruits from a larger pool of potential colonists. [18]

lower critical temperature   The environmental temperature at which the heat loss of an endotherm is greater than its basal metabolic heat production, so that its body temperature begins to drop, triggering an increase in metabolic heat generation. [4]

macroparasites   Relatively large parasite species, such as arthropods and worms. *Compare* microparasites. [13]

macrophyte   A rooted aquatic plant. [3]

marginal value theorem   A conceptual optimal foraging model proposing that an animal should stay in a food patch until the rate of energy gain in that patch has declined to the average rate for the habitat, then depart for another patch. [5]

maritime climate   The climate typical of coastal terrestrial regions that are influenced by an adjacent ocean, characterized by low daily and seasonal variation in temperature. *Compare* continental climate. [2]

mark–recapture   An approach to estimating population size in which a subset of the individuals in a population are captured, marked, and released; after a suitable period of time, individuals are then captured again, and the proportion of marked individuals found in the second capture is used to estimate the population size. [8]

mass extinction   An event in which a large proportion of Earth's species are driven to extinction worldwide in a relatively short time. [6]

matric potential   The energy associated with attractive forces on the surfaces of large molecules inside cells or on the surfaces of soil particles. [4]

mean residence time   The amount of time an average molecule of an element spends in a pool, such as a plant or the soil, before leaving it. [21]

mechanical weathering   The physical breakdown of rocks into progressively smaller particles without a chemical change. *Compare* chemical weathering. [21]

mesotrophic   Having a nutrient status that is intermediate between oligotrophic and eutrophic, usually used in reference to lakes. [21]

metamorphosis   An abrupt transition in form from the larval to the juvenile life cycle stage that is sometimes accompanied by a change in habitat. [7]

metapopulation   A set of spatially isolated populations linked to one another by dispersal. [10]

microparasites   Parasite species too small to be seen with the naked eye, such as bacteria, protozoans, and fungi. *Compare* macroparasites. [13]

Milankovitch cycles   Cycles of regular change over thousands of years in the shape of Earth's orbit, in the angle of tilt of its axis, and in its orientation toward other celestial bodies that change the intensity of solar radiation primarily at high latitudes. [2]

mimicry   A defense against predators in which prey species resemble less palatable organisms or physical features of their environment, causing potential predators to mistake them for something less desirable to eat. [12]

mineralization   The chemical conversion of organic matter into inorganic nutrients. [21]

morphs   Discrete phenotypes with few or no intermediate forms. [7]

mosaic   A composite of heterogeneous elements. [23]

mutation   Change in the DNA of a gene. [6]

mutualism   A mutually beneficial interaction between individuals of two species (a +/+ relationship). [14]

mycorrhizae   Symbiotic associations between plant roots and various types of fungi that are usually mutualistic. [14]

natural catastrophe   An extreme environmental event such as a flood, severe windstorm, or outbreak of disease that can eliminate or drastically reduce the sizes of populations. [10]

natural selection   The process by which individuals with certain heritable characteristics tend to survive and reproduce more successfully than individuals with other heritable characteristics. [1, 6]

nekton   Swimming organisms capable of overcoming water currents. *Compare* plankton. [3]

net ecosystem exchange   *See* net ecosystem production. [19]

net ecosystem production   The combined ecosystem fluxes of CO_2, including net primary production and heterotrophic respiration. [19]

net primary production (NPP)   The amount of energy (per unit time) that autotrophs capture by photosynthesis and chemosynthesis, minus the amount used in cellular respiration. [1, 19]

net reproductive rate (R_0)   The mean number of offspring produced by an individual in a population during its lifetime. [9]

net secondary production   The balance between energy gain through ingestion, and energy losses by cellular respiration and egestion in heterotrophs. [19]

niche model   A predictive tool that models the ecological niche occupied by a species based on the conditions at localities the species is known to occupy. [8]

nitrification   A process by which certain chemoautotrophic bacteria, known as nitrifying bacteria, convert ammonia (NH_3) and ammonium (NH_4^+) to nitrate (NO_3^-) under aerobic conditions. [21]

nitrogen fixation   The process of taking up nitrogen gas (N_2) and converting it into chemical forms that are more chemically available to organisms. [21]

nitrogenase   A nitrogen-fixing enzyme found only in certain bacteria. [21]

nonequilibrium theory   A proposed explanation for the coexistence of species in a community that relies on fluctuating conditions that maintain their coexistence by keeping dominant species from monopolizing resources. *Compare* equilibrium theory. [18]

North Atlantic Oscillation   An oscillation in atmospheric pressures and ocean currents in the North Atlantic Ocean that affects climatic variation in Europe, in northern Asia, and on the eastern coast of North America. [2]

nurse plant   An adult plant that provides shade or other protection allowing plants of its own or other species to germinate and survive. [14]

nutrient   A chemical element required by an organism for its metabolism and growth. [21]

nutrient cycling   The cyclic movement of nutrients between organisms and the physical environment. [1, 21]

occlusion   A process by which soluble phosphorus combines with iron, calcium, and aluminum to form insoluble compounds that are unavailable as nutrients. [21]

oligotrophic   Nutrient-poor, characterized by low primary productivity. *Compare* eutrophic. [21]

omnivore   An organism that feeds on more than one trophic level and/or feeds on both plants and animals. [15]

ontogenetic niche   A size- or life cycle stage-specific ecological role. [7]

optimal   Resulting in the maximization of fitness. [7]

optimal foraging   A theory proposing that animals will maximize the amount of energy acquired per unit of time, energy, and risk involved in finding food. [5]

osmotic adjustment   An acclimatization response to changing water availability in terrestrial and aquatic environments that involves changing the solute concentration, and thus the osmotic potential, of the cell. [4]

osmotic potential   The energy associated with dissolved solutes. [4]

ozone hole   An area of the stratosphere with an ozone concentration of less than 220 Dobson units (= 2.7×10^{16} molecules of ozone) per square centimeter; found primarily over the Antarctic region. [24]

Pacific Decadal Oscillation (PDO)   A long-term oscillation in sea surface temperatures and atmospheric pressures in the North Pacific Ocean that affects climates primarily in northwestern North America. [2]

paedomorphic   Resulting from a delay of a developmental event relative to sexual maturation. [7]

parasite   An organism that lives in or on a host organism and feeds on its tissues or body fluids. [12, 13]

parasitoid   An insect that lays one or a few eggs on or in a host organism (itself usually an insect), which the resulting larvae remain with, eat, and almost always kill. [12]

parent material   The rock or mineral material broken down by weathering to form a soil; usually the bedrock underlying the soil. [21]

pathogen   A parasite that causes disease. [13]

pelagic zone   The open water zone of a lake or ocean at some distance from the shore. [3]

permafrost   A subsurface soil layer that remains frozen year-round for at least 3 years. [3]

phenotype   The observable characteristics of an organism. [6]

phenotypic plasticity   The ability of a single genotype to produce different phenotypes under different environmental conditions. [7]

photic zone   The surface layer of a lake or ocean where enough light penetrates to allow photosynthesis. [3]

photorespiration   A chemical reaction in photosynthetic organisms in which the enzyme rubisco takes up O_2, leading to the breakdown of sugars, the release of CO_2, and a net loss of energy. [5]

photosynthesis   A process that uses sunlight to provide the energy needed to take up CO_2 and synthesize sugars. [5]

physical factor   A feature of the environment that affects organism function and population growth rates but is not consumed or depleted. Compare *resource*. [11]

phytoplankton   Photosynthetic plankton. *Compare* zooplankton. [3]

pioneer stage   The first stage of primary succession. [16]

plankton   Small, often microscopic organisms that live suspended in water; although many plankton are mobile, none can swim strongly enough to overcome water currents. *Compare* nekton. [3]

polar cell   A large-scale, three-dimensional pattern of atmospheric circulation in which air subsides at the poles and moves toward the equator when it reaches Earth's surface, and is replaced by air moving through the upper atmosphere from lower latitudes. [2]

polar zone   The major climatic zone above 60° N and S. [2]

polyphenism   A type of phenotypic plasticity in which a single genotype produces multiple, discrete morphological forms. [7]

pool   (1) The total amount of a nutrient or other element found within a component of an ecosystem. [21] (2) A relatively deep and slowly flowing portion of a stream. *Compare* riffle. [3]

population   A group of individuals of the same species that live within a particular area and interact with one another. [1, 8]

population cycles   A pattern of population fluctuations in which alternating periods of high and low abundance occur after nearly constant intervals of time. [10]

population density   The number of individuals per unit of area. [8]

population fluctuations   The most common pattern of population growth, in which population size rises and falls over time. [10]

population outbreak   An extremely rapid increase in the number of individuals in a population. [10]

population regulation   A pattern of population growth in which one or more density-dependent factors increase population

size when numbers are low and decrease population size when numbers are high. [9]

population size   The number of individuals in a population. [8]

population viability analysis (PVA)
Projection of the potential future status of a population through use of demographic models; a PVA approach is often used to estimate the likelihood that a population will persist for a certain amount of time in different habitats or under different management scenarios. [22]

positive interactions   Interactions between species in which one or both species benefit and neither is harmed. [14]

predator   An organism that kills and eats other organisms, referred to as its prey. [12]

pressure potential   The energy associated with the exertion of pressure. [4]

primary producer   *See* producer.

primary production   The chemical energy in an ecosystem generated by autotrophs, derived from the fixation of carbon during photosynthesis and chemosynthesis. *See also* gross primary production, net primary production. [19]

primary succession   Succession that involves the colonization of habitats devoid of life. *Compare* secondary succession. [16]

producer   An organism that can produce its own food by photosynthesis or chemosynthesis; also called a *primary producer* or *autotroph*. *Compare* consumer.[1]

production efficiency   The proportion of assimilated food that is used to produce new consumer biomass. [20]

pubescence   The presence of hairs on the surface of an organism. [4]

quadrat   A sampling area (or volume) of any size or shape. [8]

r-selection   In the *r*- and *K*-selection approach to classifying life history strategies, the selection pressures faced by organisms that live in environments where population densities are usually low. *Compare K-selection. [7]

radiatively active gases   See greenhouse gases. [2]

rain-shadow effect   The effect a mountain range has on climate by forcing moving air upward, causing it to cool and release precipitation on the windward slopes, resulting in lower levels of precipitation and soil moisture on the leeward slope. [2]

ramet   An actually or potentially physiologically independent member of a genet

that may compete with other members for resources. *Compare* genet. [8]

random dispersion   A dispersion pattern that is similar to what would occur if individuals were positioned at locations selected at random. *Compare* clumped dispersion, regular dispersion. [8]

rank abundance curve   A graph that plots the proportional abundance of each species in a community relative to the others in rank order, from most abundant to least abundant. [15]

redundancy hypothesis   A hypothesis that assumes an upper limit on the positive effect of species richness on community function because once species richness reaches some threshold, the functions of species in the community will overlap. *Compare* complementarity hypothesis. [18]

regional scale   A spatial scale that encompasses a geographic area where the climate is roughly uniform and the species contained therein are often restricted to that region by their dispersal capabilities. [17]

regional species pool   All the species contained within a region; sometimes called *gamma diversity*. [17]

regular dispersion   A dispersion pattern in which individuals are relatively evenly spaced throughout their habitat. *Compare* clumped dispersion, random dispersion. [8]

relative population size   An estimate of population size based on data that are related in an unknown way to the absolute population size, but can be compared from one time period or place to another. [8]

replicate   To perform each treatment of a controlled experiment, including the control, more than once. [1]

reproductive effort   The quantity of energy and resources an organism devotes to reproduction. [7]

rescue effect   A tendency for high rates of immigration to protect a population from extinction. [10]

resilience   The rate of recovery after a disturbance. [18]

resistance   The influence of a barrier on the movement of compounds such as water or CO_2. [4]

resource   A feature of the environment that is required for growth, survival, or reproduction and which can be consumed or otherwise used to the point of depletion. Compare *physical factor*. [11]

resource partitioning   The use of limiting resources by different species in a community in different ways. [11, 18]

resource ratio hypothesis   A hypothesis that posits that species can coexist in a community by using resources the same resources but in different proportions. [18]

riffle A fast-moving portion of a stream. *Compare* pool. [3]

ruderals In Grime's plant life history model, plants that are adapted to environments with high levels of disturbance and low levels of stress. *Compare* competitive plants, stress-tolerant plants. [7]

salinity The concentration of dissolved salts in water. [2]

salinization A process by which high rates of evapotranspiration in arid regions result in a progressive buildup of salts at the soil surface. [2]

sand The coarsest soil particles (0.05–2 mm). [21]

savanna A type of community dominated by grasses with intermixed trees and shrubs. [3]

scale The spatial or temporal dimension at which ecological observations are collected. [1]

scientific method An iterative and self-correcting process by which scientists learn about the natural world, consisting of four steps: (1) observe nature and ask a question about those observations; (2) develop possible answers to that question (hypotheses); (3) evaluate competing hypotheses with experiments, observations, or quantitative models; (4) use the results of those experiments, observations, or models to modify the hypotheses, pose new questions, or draw conclusions. [1]

secondary compound One of a wide variety of chemicals produced by plants and used in functions such as defense against herbivory, protection from harmful radiation, and soil nutrient cycling; secondary compounds are so-named because they are synthesized on side branches off of the primary metabolic pathways that produce the carbohydrates, amino acids, lipids, and other compounds common to all organisms. [12]

secondary production Energy in an ecosystem that is derived from the consumption of organic compounds produced by other organisms. *Compare* primary production. [19]

secondary succession Succession that involves the reestablishment of a community in which most, but not all, of the organisms have been destroyed. *Compare* primary succession. [16]

semelparous Reproducing only once in a lifetime. *Compare* iteroparous. [7]

senescence A decline in the fitness of an organism with age as a result of physiological deterioration. [7]

sensible heat flux The transfer of heat through the exchange of energy by conduction or convection. [2]

sequential hemaphroditism A change or changes in the sex of an organism during the course of its life cycle. [7]

service mutualism A mutualism in which one partner performs an ecological service for the other. [14]

Shannon index The index most commonly used to describe species diversity quantitatively. [15]

sit-and-wait predator A predator that hunts by remaining in one place and attacking prey that move within striking distance. [12]

soil A mix of mineral particles, detritus, dissolved organic matter, water containing dissolved minerals and gases (the soil solution), and organisms that develops in terrestrial ecosystems. [21]

speciation The process by which one species splits into two or more species. [6]

species A population or group of populations whose members have similar characteristics and can interbreed with one another, but not with members of other such groups. [6]

species accumulation curve A graph that plots species richness as a function of the total number of individuals that have accumulated with each additional sample. [15]

species–area relationship The relationship between species richness and area sampled. [17]

species composition The identity of the species present in a community. [15]

species diversity A measure that combines both the number of species (species richness) in a community and their relative abundances compared with one another (species evenness). [15]

species evenness The relative abundances of species in a community compared with one another. [15]

species richness The number of species in a community. [15]

sporophyte A multicellular diploid life cycle stage, found in many algae and all plants, that produces haploid spores. *Compare* gametophyte, alternation of generations. [7]

stability The tendency of a community to remain the same in structure and function. [18]

stabilizing selection Selection that favors individuals with an intermediate phenotype. *Compare* directional selection, disruptive selection. [6]

stable Returning to the original state after some perturbation. [16]

stable age distribution A population age structure that does not change from one year to the next. [9]

stable limit cycle A pattern of population fluctuations in which abundance cycles indefinitely. [10]

static life table A life table that records the survival and reproduction of individuals of different ages during a single time period. [9]

stomate A pore in plant tissues, usually leaves, surrounded by specialized guard cells that control its opening and closing. [4]

stratification The layering of water in oceans and lakes due to differences in water density and temperature with depth. [2]

stress An abiotic factor that results in a decrease in the rate of an important physiological process, thereby lowering the potential for an organism's survival, growth, or reproduction. [4, 7, 16]

stress-tolerant plants In Grime's plant life history model, plants that are adapted to conditions of high stress and low disturbance. *Compare* competitive plants, ruderals. [7]

subsidence In meteorology, a sinking (downward) motion in the atmosphere, usually over a broad area, leading to the development of a high pressure cell. *Compare* uplift. [2]

succession The change in species composition over time as a result of abiotic and biotic agents of change. [16]

surrogate species A species selected as a priority for conservation with the assumption that its conservation will serve to protect many other species with overlapping habitat requirements. [22]

survival rate The proportion of individuals of age x that survive to be age $x + 1$ (denoted S_x in a life table). [9]

survivorship The proportion of individuals that survive from birth (age 0) to age x (denoted l_x in a life table). [9]

survivorship curve A graph that is based on survivorship data (l_x) and that plots the number of individuals from a hypothetical cohort (typically, of 1,000 individuals) that will survive to reach different ages. [9]

symbiont As the smaller member of a symbiosis, an organism that lives in or on an organism of another species, its host. *Compare* host, symbiosis. [13]

symbiosis A relationship in which two species live in close physiological contact with each other. *Compare* host, symbiont. [14]

taxonomic homogenization A worldwide reduction of biodiversity resulting from the spread of non-native and native generalists coupled with declining abundances and distributions of native specialists and endemics. [22]

temperate zone The major climatic zone between 30° and 60° N and S. [2]

thermocline The zone of rapid temperature change in a lake beneath the epilimnion and above the hypolimnion. [2]

thermoneutral zone The range of environmental temperatures over which endotherms maintain a constant resting metabolic rate. [4]

threshold density The number of individuals in a population susceptible to a disease at (or above) which the disease will become established and spread. [13]

tides Patterns of rising and falling ocean water generated by the gravitational attraction between Earth and the moon and sun. [3]

till Thick layers of sediment deposited by glaciers. [21]

tolerance The ability to survive stressful environmental conditions. *Compare* avoidance. [4]

top-down control Limitation of the abundance of a population by consumers. *Compare* bottom-up control. [10]

torpor A state of dormancy in which endotherms drop their lower critical temperature and associated metabolic rate. [4]

trade-off An organism's allocation of its limited energy or other resources to one structure or function at the expense of another. [7]

transpiration The evaporation of water from the inside of a plant, typically through stomatal pores. [2]

trophic cascade A change in the rate of consumption at one trophic level that results in a series of changes in species abundances and species compositions at lower trophic levels. [15, 20]

trophic efficiency A measure of the transfer of energy between trophic levels, consisting of the amount of energy at one trophic level divided by the amount of energy at the trophic level immediately below it. [20]

trophic facilitation An interaction in which a consumer is indirectly facilitated by a positive interaction between its prey and another species. [15]

trophic level A group of species in a community that obtain energy in similar ways, classified by the number of feeding steps by which the group is removed from autotrophs, which are the first trophic level. [15, 20]

trophic mutualism A mutualism in which one or both of the mutualists receives energy or nutrients from its partner. [14]

trophic pyramid A common approach to conceptualizing trophic relationships in an ecosystem in which a stack of rectangles is constructed, each of which represents the amount of energy or biomass within one trophic level. [20]

tropical zone The major climatic zone between 25° N and S, encompassing the equator. [2]

turgor pressure Hydrostatic pressure that develops in a plant cell when water moves into it, following a gradient in water potential. [4]

turnover (1) The mixing of the entire water column in a stratified lake when all the layers of water reach the same temperature and density. [2] (2) The replacement of one species with another over time or space. [17]

turnover rate *See* mean residence time. [21]

type I survivorship curve A survivorship curve in which newborns, juveniles, and young adults all have high survival rates and death rates do not begin to increase greatly until old age. [9]

type II survivorship curve A survivorship curve in which individuals experience a constant chance of surviving from one age to the next throughout their lives. [9]

type III survivorship curve A survivorship curve in which individuals die at very high rates when they are young, but those that reach adulthood survive well later in life. [9]

umbrella species A surrogate species selected with the assumption that protection of its habitat will serve as an "umbrella" to protect many other species; often a species with large or specialized habitat requirements or which is easy to count. [22]

uplift (meteorology) The rising of warm, less dense air in the atmospheredue to surface heating. *Compare* subsidence. [2]

upwelling The rising of deep ocean waters to the surface where prevailing wind patterns cause surface waters to flow away from a coast. [2]

variability The range of abiotic conditions in an environment. [2]

vertical transmission The spread of a disease from mother to newborn. [13]

vicariance An approach to biogeography that explains the geographic distribution of species in terms of events such as continental drift that result in the geographic isolation of populations that once were connected to one another. [17]

warning coloration A defense against predators in which prey species that contain powerful toxins advertise those toxins with bright coloration; also referred to as *aposematic coloration*. [12]

water potential The overall energy status of water in a system; the sum of osmotic potential, gravitational potential, and matric potential. [4]

watershed *See* catchment. [21]

weather The current temperature, humidity, precipitation, wind, and cloudiness. *Compare* climate. [2]

weathering The physical and chemical processes by which rock minerals are broken down, eventually releasing soluble nutrients and other elements. [21]

zero population growth isoclines Lines derived from the Lotka–Volterra competition model marking the variables at which population does not increase or decrease in size. [11]

zooplankton Nonphotosynthetic plankton. *Compare* phytoplankton. [3]

Illustration Credits

The following figures use elements originally rendered for Sadava et al., *Life: The Science of Biology* (Sinauer Associates and W.H. Freeman, 2008): Figures 2.24, 5.11, 7.10, 13.3, 13.9, and 17.10.

The following figures use elements originally rendered for Cain et al., *Discover Biology* (Sinauer Associates and W.W. Norton, 2002): Figures 1.8, 1.11, 8.9, 8.16, 11.8, and 11.18.

Cover photo: The community of organisms inhabiting Botswana's savanna includes African elephants (*Loxodonta africana*), impala (*Aepyceros melampus*), mourning doves (*Streptopelia decipiens*), and acacia trees (*Acacia* sp.). © Frans Lanting/Minden Pictures.

UNIT OPENERS

Unit 1: American bison (*Bison bison*) in Yellowstone National Park. © M. and J. Miller/AGE Fotostock.

Unit 2: A population of king penguins (*Aptenodytes patagonicus*), South Georgia Island, Antarctica. © Alexander Hafemann/istockphoto.com.

Unit 3: A great blue heron (*Ardea herodias*) and its catfish prey, Everglades National Park, Florida. © Paul Tessier/istockphoto.com.

Unit 4: A coral reef community in the Red Sea, Egypt. © Edwin van Wier/ShutterStock.

Unit 5: Primary production and decomposition in a temperate forest ecosystem. © Aleksander Bolbot/ShutterStock.

Unit 6: Satellite image of the Sundarbans region of India and Bangladesh, which encompasses the largest remaining tract of mangrove forest (deep green) in the world. NASA image created by Jesse Allen, Earth Observatory, using data obtained from the University of Maryland's Global Land Cover Facility.

Chapter 1 1.1: © Pieter Johnson. 1.2: Data from IUCN 2000, AmphibiaWeb (www.amphibiaweb.org), and Hero and Shoo 2003. 1.5: © Joseph M. Kiesecker. 1.6A: © Phototake Inc./Alamy. 1.6: © Joseph M. Kiesecker. 1.6 *organism*: © Uwe Ohse/Shutterstock. 1.6 *population*: © blickwinkel/Alamy. 1.6 *community*: © Georgie Holland/AGE Fotostock. 1.6 *biosphere*: Courtesy of NASA. 1.7: After www.cdc.gov/ncidod/dvbid/westnile/Mapsactivity/surv&control106Maps.htm. 1.9A: © LMR Group/Alamy. 1.9B: Courtesy of Andrew Sinauer. 1.9C: © Jeremy Woodhouse/

Photodisc Green/Getty Images. 1.9D: © Visual&Written SL/Alamy. 1.12A: Courtesy of Tim Cooper and Richard Lenski, Michigan State University. 1.12B: Courtesy of Simone DesRoches and Dolph Schluter. 1.12C: Courtesy of the U.S. Forest Service. 1.13: Courtesy of Walter Carson.

Chapter 2 2.1: © Steve Bloom Images/Alamy. 2.3: Courtesy of Craig Allen. 2.5, 2.6, 2.8, 2.10, 2.11, 2.12: After Ahrens, 2005. 2.13: After IPCC 2005. 2.14A: Courtesy of the Center for Sustainability and the Global Environment (SAGE) through their *Atlas of the Biosphere*, www.sage.wisc.edu/atlas. 2.14B: Courtesy of the NOAA National Oceanographic Data Center. 2.15: After Strahler and Strahler 2005. 2.16: Courtesy of the Center for Sustainability and the Global Environment (SAGE) through their *Atlas of the Biosphere*, www.sage.wisc.edu/atlas. 2.17: Data from www.globalbioclimatics.org. 2.18B *west*: © Jill Stephenson/Alamy. 2.18B *east*: © Danita Delimont/Alamy. 2.18B *bottom*: Courtesy of NASA. 2.20: After Ahrens, 2005. 2.22: After Dodson 2005. 2.23: Data from the NOAA Tropical Atmosphere Ocean Project. 2.25: Courtesy of NOAA. 2.26: After Lomolino et al. 2006. 2.27: Courtesy of NOAA.

Chapter 3 3.1: © Frans Lanting/Minden Pictures. 3.2: © Peter Snowball/NHMPL. 3.3 *sclerophyll*: © Marion Brenner/Botanica/Jupiter Images. 3.3 *deciduous*: © Arco Images/Alamy. 3.3 *cactus*: © Dan Eckert/istockphoto.com. 3.3 *needle-leaved*: © imagebroker/Alamy. 3.3 *grass*: © maurice joseph/Alamy. 3.3 *broad-leaved*: © Frans Lanting/Minden Pictures. 3.3 *forb*: Courtesy of William Bowman. 3.4: After Whittaker 1975. 3.5B, C: Courtesy of the Center for Sustainability and the Global Environment (SAGE) through their *Atlas of the Biosphere*, www.sage.wisc.edu/atlas. *Tropical rainforest, left*: © Frans Lanting/Minden Pictures. *Tropical rainforest, right*: © Murray Cooper/Minden Pictures. 3.6: Courtesy of NASA/Goddard Space Flight Center Scientific Visualization Studio. *Tropical seasonal forest, left*: © Nigel Dennis/AGE Fotostock. *Tropical seasonal forest, right*: © Pete Oxford/naturepl.com. *Hot subtropical desert, left*: © Carr Clifton/Minden Pictures. *Hot subtropical desert, right*: © Gerry Ellis/Minden Pictures. 3.7: David McIntyre. *Temperate grasslands, left*: © Tim Fitzharris/Minden Pictures. *Temperate grasslands, right*: © Pavel Filatov/Alamy. *Temperate shrublands, left*: © AfriPics.com/

Alamy. *Temperate shrublands, right*: © blickwinkel/Alamy. *Temperate deciduous forests, left*: © Tim Fitzharris/Minden Pictures. *Temperate deciduous forests, right*: © Digital Archive Japan/Alamy. *Temperate evergreen forests, left*: © John N.A. Lott/Biological Photo Service. *Temperate evergreen forests, right*: © James Brunker/Alamy. 3.8: © David Noton/naturepl.com. *Boreal forests, left*: © Jan Hodder & Michael Graybill/Biological Photo Service. *Boreal forests, right*: © Jose Schell/naturepl.com. 3.9: Courtesy of U.S. Forest Service Research. *Tundra, left*: © All Canada Photos/Alamy. *Tundra, right*: Courtesy of William Bowman. 3.10: © FLPA/Alamy. 3.11 *grassland, montane, alpine*: Courtesy of William Bowman. 3.11 *subalpine*: © Tim Fitzharris/Minden Pictures. 3.12: © Pete Oxford/Minden Pictures. 3.15A: © Phototake Inc./Alamy. 3.15B: © Laguna Design/Science Photo Library/Photo Researchers, Inc. 3.17: © Mark Brownlow/naturepl.com. 3.18: © Eric Horan/AGE Fotostock. 3.19: © Pete Oxford/Minden Pictures. 3.20: © Suzanne Long/Alamy. 3.22: © Georgette Douwma/naturepl.com. 3.23: Courtesy of Earth Sciences and Image Analysis Laboratory, NASA Johnson Space Center. 3.24: © Ralph A Clevenger/Flirt Collection/photolibrary.com. 3.25A: © D. P. Wilson/Minden Pictures. 3.25B: © Harold Taylor/OSF/photolibrary.com. 3.26: © Norbert Wu/Minden Pictures. 3.27: © North Wind Picture Archives/Alamy. 3.29A: Courtesy of the Konza Prairie LTER.

Chapter 4 4.1: Courtesy of J. M. Storey. 4.2 *sylvatica*: David McIntyre. 4.2 *maculata*: © Robin Arnold/istockphoto.com. 4.4: © Craig Lovell/Eagle Visions Photography/Alamy. 4.7: After Willmer et al. 2005. 4.8: After Nobel 1983. 4.9A: © Brian Sullivan/Visuals Unlimited. 4.9B: Courtesy of G. H. Holroyd and A. M. Hetherington. 4.10: Courtesy of James Ehleringer. 4.12: © Dinodia Images/Alamy. 4.13A: © Alastair MacEwen/OSF/photolibrary.com. 4.13B: © Takahashi, via Creative Commons. 4.14: After Willmer et al. 2005 and Schmidt-Nelson 1997. 4.15: Courtesy of William Bowman. 4.17: After Armitage et al. 2003. 4.17 *inset*: Courtesy of Zdeno Kostka. 4.18: After Kramer 1983. 4.21: After Slatyer 1967. 4.23: After Willmer et al. 2005 and Edney 1980. 4.24, 4.25: After Schmidt-Nelson 1997. 4.27A, B: © blickwinkel/Alamy. 4.27C: © Dennis Kunkel Microscopy, Inc.

Crisafulli et al. 2005, courtesy of Charlie M. Crisafulli. 16.21A: © Jonathon Titus, courtesy of John Bishop.

Chapter 17 17.1: © David R. Frazier Photolibrary, Inc./Alamy. 17.2B: © R. O. Bierregaard. 17.3A: © Tui de Roy/Minden Pictures. 17.3B: © Carr Clifton/Minden Pictures. 17.3C: © Tim Fitzharris/Minden Pictures. 17.3D: © Staffan Widstrand/naturepl.com. 17.4: Courtesy of Sally Hacker. 17.8B: © The Natural History Museum, London. 17.11: After Pielou 1979. 17.14B1: © Jerome Whittingham/istockphoto.com. 17.14B2: Courtesy of Andrew Sinauer. 17.14B3: © Wayne Tam/istockphoto.com. 17.22: From Symons 1888. 17.23A *top*: Courtesy of DigitalGlobe. 17.23A *middle and bottom*: Courtesy of E. O. Wilson.

Chapter 18 18.1: © Larry Michael/naturepl.com. 18.2: Courtesy of G. David Tilman. 18.3: © Jim Havey/Alamy. 18.5A: © Karen Gowlett-Holmes/OSF/photolibrary.com. 18.5B: © Willem Kolvoort/naturepl.com. 18.5B *inset*: Courtesy of the U.S. Geological Survey. 18.6: © Arco Images/Alamy. 18.7: After Ricklefs 2001. 18.13: © Phototake Inc./Alamy. 18.20: © Jurgen Freund/naturepl.com.

Chapter 19 19.1: © Mark Spear, Woods Hole Oceanographic Institution. 19.2: © Verena J. Tunnicliffe. 19.4: Courtesy of Andrew Sinauer. 19.7A: Courtesy of Stephen Ausmus/USDA ARS. 19.7B: Courtesy of Dylan Fischer. 19.9A: Courtesy of Sean Burns. 19.10: Image by Robert Simmon, NASA GSFC Earth Observatory, based on data provided by Watson Gregg, NASA GSFC. 19.13: Courtesy of William Bowman. 19.15: Courtesy of Wayne Wurtsbaugh and the American Association of Limnology and Oceanography. 19.16: Courtesy of Fisheries and Oceans Canada, Experimental Lakes Area. Reproduced with the permission of Her Majesty the Queen in Right of Canada, 2007. 19.17B: © Ken Johnson/Monterey Bay Aquarium Research Institute. 19.21: From Shank et al. 1998.

Chapter 20 20.1: © Norbert Rosing/OSF/photolibrary.com. 20.2: © Bryan & Cherry Alexander Photography/Alamy. 20.4A *left*: © WoodyStock/Alamy. 20.4A *right*: David McIntyre. 20.8: © Flip Nicklin/Minden Pictures. 20.16: Courtesy of Sally Hacker. 20.21: © Darryl Leniuk/AGE Fotostock.

Chapter 21 21.1, 21.2: Courtesy of Jayne Belnap/USGS. 21.4: After Brady and Weil 2001. 21.5A: © Ken Wagner/Visuals Unlimited. 21.5B: © E. H. Newcomb and S. R. Tandon/Biological Photo Service. 21.7: After Paul and Clark 1996. 21.12: Courtesy of DigitalGlobe. 21.15: © Louise Heusinkveld/Alamy. 21.19B: After Dodson 2004. 21.20: Courtesy of Jason C. Neff. 21.21: © Chris Evans/River to River CWMA/Bugwood.org.

Chapter 22 22.1: © Barry Mansell/naturepl.com. 22.2B: © David Sieren/Visuals Unlimited. 22.3: © Bristol City Museum/naturepl.com. 22.11: © Franz Pagot/Alamy. 22.15: © Digital Vision Ltd./AGE Fotostock. 22.16A-C: Courtesy of the U.S. Fish and Wildlife Service. 22.16D: © Yva Momatiuk & John Eastcott/Minden Pictures. 22.18: © Gerry Ellis/DigitalVision/PictureQuest. 22.19, 22.20: Courtesy of the U.S. Fish and Wildlife Service.

Chapter 23 23.1: © Brakefield Photo/AGE Fotostock. 23.3: Courtesy of Brad Stratton and Kirstin Seleen/New York State Chapter of The Nature Conservancy. 23.4: Courtesy of NASA. 23.7: © imagebroker/Alamy. 23.8: Courtesy of Jim Peaco/NPS Photo. 23.10: © Porky Pies Photography/Alamy. 23.12: From McIntyre and Hobbs 1999, courtesy of Richard Hobbs. 23.14A: Courtesy of DigitalGlobe. 23.14A *inset*: © Jean-Paul Ferrero/Auscape/Minden Pictures. 23.16: © Rainer Raffalski/Alamy. 23.18A: Courtesy of DigitalGlobe. 23.18 *butterfly*: © Millard H. Sharp/Photo Researchers, Inc. 23.18 *berries*: © Bruce Coleman, Inc./Alamy. Box 23.1: After Turner et al. 2001.

Chapter 24 24.1: © Georgette Douwma/naturepl.com. 24.3: After Chapin et al. 2002. 24.4: Courtesy of Evan DeLucia. 24.6, 24.8, 24.9: After Chapin et al. 2002. 24.15: Courtesy of William Bowman. 24.17: Courtesy of Julie Winchester/Cooperative Institute for Research in the Atmosphere. 24.18: © vario images GmbH & Co. KG/Alamy. 24.19: Courtesy of the National Atmospheric Deposition Program (NRSP-3). Figure 24.23A After Jacobson 2002. 24.23B: Courtesy of NASA. 24.24: Courtesy of NOAA. 24.25A: Courtesy of MODIS Rapid Response Project at NASA/GSFC. Box 24.1 (A): Courtesy of Dave Harlow/U.S. Geological Survey.

Literature Cited

A

Aber, J., W. and 9 others. 1998. Nitrogen saturation in temperate forest ecosystems: Hypotheses revisited. *BioScience* 48: 921–934.

Adey, W. H. and R. S. Steneck. 2001. Thermogeography over time creates biogeographic regions: A temperature/space/time-integrated model and an abundance-weighted test for benthic marine algae. *Journal of Phycology* 37: 677–698.

Agrawal, A. A. and M. Fishbein. 2006. Plant defense syndromes. *Ecology* 87: S132–S149.

Ahrens, C. D. 2005. *Essentials of Meteorology*. 4th ed. Thomson Brooks/Cole, Belmont, CA.

Allan, B. F., F. Keesing and R. S. Ostfeld. 2003. Effect of forest fragmentation on Lyme disease risk. *Conservation Biology* 17 (1): 267–272.

Allee, W. C. 1931. *Animal Aggregations, A Study in General Sociology*. University of Chicago Press, Chicago.

Allen, E. B., P. J. Temple, A. Bytnerowicz, M. J. Arbaugh, A. G. Sirulnik and L. E. Rao. 2007. Patterns of understory diversity in mixed coniferous forests of Southern California impacted by air pollution. *Scientific World Journal* 7: 247–263.

Allen, W. H. 1988. Biocultural restoration of a tropical forest. *BioScience* 38 (3): 156–161.

Allin, E. F. and J. A. Hopson. 1992. Evolution of the auditory system in Synapsida ("mammal-like reptiles" and primitive mammals) as seen in the fossil record. In *The Evolutionary Biology of Hearing*, D. B. Webster, R. R. Fay and A. N. Popper (eds.), 587–614. Springer, New York.

Alliende, M. C. and J. L. Harper. 1989. Demographic studies of a dioecious tree. I. Colonization, sex, and age structure of a population of *Salix cinerea*. *Journal of Ecology* 77: 1029–1047.

Allison, G. 2004. The influence of species diversity and stress intensity of community resistance and resilience. *Ecological Monographs* 74: 117–134.

Alpert, P. 2006. Constraints of tolerance: Why are desiccation tolerant organisms so small or rare? *Journal of Experimental Biology* 209: 1575–1584.

Amos, W. and A. Balmford. 2001. When does conservation genetics matter? *Heredity* 87: 257–265.

Andelman, S. J. and W. F. Fagan. 2000. Umbrellas and flagships: Efficient conservation surrogates or expensive mistakes? *Proceedings of the National Academy of Sciences USA* 97 (11): 5954–5959.

Andrewartha, H. G. and L. C. Birch. 1954. *The Distribution and Abundance of Animals*. University of Chicago Press, Chicago.

Anthony, P. A., J. A. M. Holtum and B. R. Jackes. 2002. Shade acclimation of rainforest leaves to colonization by lichens. *Functional Ecology* 16: 808–816.

Arcese, P. and J. N. M Smith. 1988. Effects of population density and supplemental food on reproduction in song sparrows. *Journal of Animal Ecology* 57: 119–136.

Arias, E. 2004. *United States Life Tables, 2001*. National Vital Statistics Reports, vol. 52, no. 14. National Center for Heath Statistics, Hyattsville, MD. http://www.cdc.gov/nchs/data/nvsr/nvsr52/nvsr52_14.pdf.

Armitage, K. B., B. C. Woods and C. M. Salsbury. 2000. Energetics of hibernation in woodchucks (*Marmota monax*). In *Life in the Cold*, G. Heldmaier and M. Klingenspor (eds.), 73–80. Springer, New York.

Armitage, K. B., D. T. Blumstein and B. C. Woods. 2003. Energetics in hibernating yellow belly marmots (*Marmota flaviventris*). *Comparative Biochemistry and Physiology* A 134: 101–114.

Arnold, G. W., D. E. Steven, J. R. Weeldenburg and E. A. Smith. 1993. Influences of remnant size, spacing pattern and connectivity on population boundaries and demography in euros *Macropus robustus* living in a fragmented landscape. *Biological Conservation* 64: 219–230.

Attum, O., P. Eason, G. Cobbs and S. M. B. El Din. 2006. Response of a desert lizard community to habitat degradation: Do ideas about habitat specialists/generalists hold? *Biological Conservation* 133 (1): 52–62.

B

Bais, H. P., R. Vepachedu, S. Gilroy, R. M. Callaway and J. M. Vivanco. 2003. Allelopathy and exotic plant invasion: From molecules and genes to species interactions. *Science* 301: 1377–1380.

Baker, C. S., M. L. Dalebout, G. M. Lento and N. Funahashi. 2002. Gray whale products sold in commercial markets along the Pacific coast of Japan. *Marine Mammal Science* 18 (1): 295–300.

Balmford, A., R. E. Green and M. Jenkins. 2003. Measuring the changing state of nature. *Trends in Ecology and Evolution* 18 (7): 326–330.

Barbour, M. S. and J. A. Litvaitis. 1993. Niche dimensions of New England cottontails in relation to habitat patch size. *Oecologia* 95: 321–327.

Barnosky, A. D., P. L. Koch, R. S. Feranec, S. L. Wing and A. B. Shabel. 2004. Assessing the causes of Late Pleistocene extinctions on the continents. *Science* 306: 70–75.

Bartlein, P., J. C. Whitlock and S. L. Shafer. 1997. Future climate in the Yellowstone National Park region and its potential impact on vegetation. *Conservation Biology* 11 (3): 782–792.

Battin, J. 2004. When good animals love bad habitats: Ecological traps and the conservation of animal populations. *Conservation Biology* 18 (6): 1482–1491.

Beall, C. M. 2006. Andean, Tibetan, and Ethiopian patterns of adaptation to high-altitude hypoxia. *Integrative and Comparative Biology* 46: 18–24.

Becerra, J. X. 2003. Synchronous coadaptation in an ancient case of herbivory. *Proceedings of the National Academy of Sciences USA* 100: 12804–12807.

Becerra, J. X. 2007. The impact of herbivore–plant coevolution on plant community structure. *Proceedings of the National Academy of Sciences USA* 104: 7483–7488.

Beck, B. B. 1980. *Animal Tool Behavior: The Use and Manufacture of Tools by Animals.* Garland STPM Press, New York.

Begon, M., M. Mortimer and D. J. Thompson. 1996. *Population Ecology: A Unified Study of Animals and Plants.* 3rd ed. Blackwell Science, Ltd. Oxford.

Beerling, D. J., C. P. Osborne and W. G. Chaloner. 2001. Evolution of leaf-form in land plants linked to atmospheric CO_2 decline in the Late Palaeozoic era. *Nature* 410: 352–354.

Begon, M., M. Mortimer and D. J. Thompson. 1996. *Population Ecology: A Unified Study of Animals and Plants.* 3rd ed., Blackwell Science Ltd., Oxford.

Beisner, B. E., D. T. Haydon and K. Cuddington. 2003. Alternative stable states in ecology. *Frontiers in Ecology and the Environment* 1: 376–382.

Beissinger, S. R. and M. I. Westphal. 1998. On the use of demographic models of population viability in endangered species management. *Journal of Wildlife Management* 62 (3): 821–841.

Bell, M. A., M. P. Travis and D. M. Blouw. 2006. Inferring natural selection in a fossil threespine stickleback. *Paleobiology* 32: 562–577.

Bellows, T. S. Jr. 1981. The descriptive properties of some models for density dependence. *Journal of Animal Ecology* 50: 139–156.

Belnap, J. 2003. The world at your feet: Desert biological crusts. *Frontiers in Ecology and Environment* 1: 181–189.

Belnap, J. and D. Eldridge. 2001. Disturbance and recovery of biological soil crusts. In *Biological Soil Crusts: Structure, Function, and Management,* J. Belnap and O. L. Lange (eds.), 363–383. Berlin: Springer-Verlag.

Belt, T. 1874. *The Naturalist in Nicaragua.* Murray, London.

Benkman, C. W. 1993. Adaptation to single resources and the evolution of crossbill (*Loxia*) diversity. *Ecological Monographs* 63: 305–325.

Benkman, C. W. 2003. Divergent selection drives the adaptive radiation of crossbills. *Evolution* 57: 1176–1181.

Benson, A. J. and D. Raikow. 2007. *Dreissena polymorpha.* USGS Nonindigenous Aquatic Species Database, Gainesville, FL. http://nas.er.usgs.gov/queries/FactSheet.asp?speciesID=5

Benton, M. J. and B. C. Emerson. 2007. How did life become so diverse? The dynamics of diversification according to the fossil record and molecular phylogenetics. *Palaeontology* 50: 23–40.

Benton, M. J. and R. J. Twitchett. 2003. How to kill (almost) all life: The end-Permian extinction event. *Trends in Ecology and Evolution* 18: 358–365.

Berendse, F., R. Aerts and R. Bobbink. 1993. Atmospheric nitrogen deposition and its impact on terrestrial ecosystems. In *Landscape Ecology of a Stressed Environment,* C. C. Vos and P. Opdam (eds.), 104–121. Chapman and Hall, London.

Berlow, E. L. 1997. From canalization to contingency: Historical effects in a successional rocky intertidal. *Ecological Monographs* 676: 435–460.

Berlow, E. L. 1999. Strong effects of weak interactions in ecological communities. *Nature* 398: 330–334.

Bernhardt, E. S. and 13 others. 2005. Can't see the forest for the stream? In-stream processing and terrestrial nitrogen exports. *BioScience* 55: 219–230.

Bertness, M. D. 1989. Interspecific competition and facilitation in a northern acorn barnacle population. *Ecology* 70: 257–268.

Bertness, M. D. 2007. *Atlantic Shorelines.* Princeton University Press, Princeton, NJ.

Bertness, M. D. and R. Callaway. 1994. Positive interactions in communities. *Trends in Ecology and Evolution* 9: 191–193.

Bertness, M. D. and S. D. Hacker. 1994. Physical stress and positive associations among marsh plants. *American Naturalist* 144: 363–372.

Bertness, M. D. and G. H. Leonard. 1997. The role of positive interactions in communities: Lessons from intertidal habitats. *Ecology* 78: 1976–1989.

Bertness, M. D. and S. W. Shumway. 1993. Competition and facilitation in marsh plants. *American Naturalist* 142: 718–724.

Beschta, R. L. and W. J. Ripple. 2006. River channel dynamics following extirpation of wolves in northwestern Yellowstone National Park, USA. *Earth Surface Processes and Landforms* 31: 1525–1539.

Bierregaard, R. O. Jr., C. Gascon, T. E. Lovejoy and R. C. G. Mesquita. 2001. *Lessons from Amazonia: The Ecology and Conservation of a Fragmented Forest.* Yale University Press, New Haven, CT.

Binkley, D. and P. Högberg. 1997. Does atmospheric deposition of acidity and nitrogen threaten Swedish forests? *Forest Ecology and Management* 92: 119–152.

Birkeland, C. 1997. Symbiosis, fisheries and economic development on coral reefs. *Trends in Ecology and Evolution* 12: 364–367.

Bishop, J., W. F. Fagan, J. D. Schade and C. M. Crisafulli. 2005. Causes and consequences of herbivory on Prairie lupine (*Lupinus lepidus*) in early primary succession. In *Ecological Responses to the 1980 Eruption of Mount St. Helens,* V. H. Dale, F. J. Swanson and C. M. Crisafulli (eds.), 151–161. Springer, New York.

Bjorkman, O. 1981. Responses to different quantum flux densities. In *Physiological Plant Ecology* I: *Encyclopedia of Plant Physiology,* O. L. Lange, P. S. Nobel, C. B. Osmond and H. Ziegler (eds.), 57–101. Springer-Verlag, Berlin.

Blais, J. M., D. W. Schindler, D. C. G. Muir, L. E. Kimpe, D. B. Donald and B. Rosenberg. 1998. Accumulation of persistent organochlorine compounds in mountains of western Canada. *Nature* 395: 585–588.

Blais, J. M., L. E. Kimpe, D. McMahon, B. E. Keatley, M. L. Mattory, M. S. V. Douglas and J. P. Smol. 2005. Arctic seabirds transport marine-derived contaminants. *Science* 309: 445–445.

Blaustein, A. R. and P. T. J. Johnson. 2003. Explaining frog deformities. *Scientific American* 288 (2): 60–65.

Boonstra, R., D. Hik, G. R. Singleton and A. Tinnikov. 1998. The impact of predator-induced stress on the snowshoe hare cycle. *Ecological Monographs* 79: 371–394.

Borland, A. M., H. Griffiths, C. Maxwell, M. S. J. Broadmeadow, N. M. Griffiths and J. D. Barnes. 1992. On the ecophysiology of the Clusiaceae in Trinidad: expression of CAM in *Clusia minor* L. during the transition from wet to dry season and characterization of three endemic species. *New Phytologist* 122: 349–357.

Bouzat, J. L., H. A. Lewin and K. N. Paige. 1998. The ghost of genetic diversity past: Historical DNA analysis of the greater prairie chicken. *American Naturalist* 152: 1–6.

Bowman, W. D. and C. J. Bilbrough. 2001. Influence of a pulsed nitrogen supply on growth and nitrogen uptake in alpine graminoids. *Plant and Soil* 233: 283–290.

Bowman, W. D. and T. R. Seastedt. 2001. *Structure and Function of an Alpine Ecosystem, Niwot Ridge, Colorado.* Oxford University Press, New York.

Bowman, W. D., T. A. Theodose, J. C. Schardt and R. T. Conant. 1993. Constraints of nutrient availability on primary production in two alpine communities. *Ecology* 74: 2085–2098.

Braby, M. F. 2002. Life history strategies and habitat templets of tropical butterflies in north-eastern Australia. *Evolutionary Ecology* 16: 399–413.

Bradshaw, A. D. 1987. Restoration: An acid test for ecology. In *Restoration Ecology—a Synthetic Approach to Ecological Restoration*, W. R. Jordan, M. E. Gilpin and J. D. Aber (eds.), 23–30. Cambridge University Press, Cambridge.

Brady, N. C. and R. R. Weil. 2001. *The Nature and Property of Soils.* Prentice Hall, Upper Saddle River, NJ.

Brewer, J. S. 2003. Why don't carnivorous pitcher plants compete with non-carnivorous plants for nutrients? *Ecology* 84: 451–462.

Bronstein, J. L. 1992. Seed predators as mutualists: Ecology and evolution of the fig/pollinator interaction. In *Insect-Plant Interactions* (Vol. 4), E. A. Bernays (ed.), p. 1–44. CRC Press, Boca Raton, FL.

Bronstein, J. L. 1994. Conditional outcomes in mutualistic interactions. *Trends in Ecology and Evolution* 9: 214–217.

Brooks, T., J. Tobias and A. Balmford. 1999. Deforestation and bird extinctions in the Atlantic forest. *Animal Conservation* 2: 211–222.

Brower, L. P. 1996. Monarch butterfly orientation: missing pieces of a magnificent puzzle. *The Journal of Experimental Biology* 199: 93–103.

Brown, J. H. and D. W. Davidson. 1977. Competition between seed-eating rodents and ants in desert ecosystems. *Science* 196: 880–882.

Brown, J. H. and A. Kodric-Brown. 1977. Turnover rates in insular biogeography: Effect of immigration on extinction. *Ecology* 58: 445–449.

Bruno, J. F., L. E. Petes, C. D. Harvell and A. Hettinger. 2003. Nutrient enrichment can increase the severity of coral diseases. *Ecology Letters* 6: 1056–1071.

Bullock, S., H. Mooney and E. Medina, eds. 1995. Seasonally dry tropical forests. Cambridge University Press, Cambridge.

Burgess, R. L. 1977. The Ecological Society of America: Historical data and preliminary analyses. In *History of American Ecology*, F. N. Egerton (ed.), 1–24. Arno Press, New York.

Bush, M. B. 2003. *Ecology of a Changing Planet.* 3rd ed. Prentice Hall, Upper Saddle River, NJ.

Buss, L. W. and J. B. C. Jackson. 1979. Competitive networks: Nontransitive competitive relationships in cryptic coral reef environments. *American Naturalist* 113: 223–234.

Buston, P. M. 2003a. Forcible eviction and prevention of recruitment in the clown anemonefish. *Behavioral Ecology* 14 (4): 576–582.

Buston, P. M. 2003b. Size and growth modification in clownfish. *Nature* 424: 145–146.

Buston, P. M. 2004. Territory inheritance in clownfish. *Proceedings of the Royal Society of London* B (suppl.) 271: S252–S254.

Butzer, K. W. 1992. The Americas before and after 1492: An introduction to current geographical research. *Annals of the Association of American Geographers* 82 (3): 345–368.

Bytnerowicz, A., M. J. Arbaugh and R. Alonso. 2003. *Ozone Pollution in the Sierra Nevada: Distribution and Effecs on Forests.* Elsevier, Amsterdam.

C

Cabin, R. J., S. G. Weller, D. H. Lorence, T. W. Flynn, A. K. Sakai, D. Sandquist and L. J. Hadway. 2000. Effects of long-term ungulate exclusion and recent alien species control on the preservation and restoration of a Hawaiian tropical dry forest. *Conservation Biology* 14 (2): 439–453.

Cain, M. L., H. Damman, R. A. Lue and C. K. Yoon. 2002. *Discover Biology.* 2nd ed. Sinauer Associates/W. W. Norton, Sunderland/New York.

Caldwell, M. M., L. O. Bjorn, J. F. Bornman, S. D. Flint, G. Kulandaivelu, A. H. Teramura and M. Tevini. 1998. Effects of increased solar ultraviolet radiation on terrestrial ecosystems. *Journal of Photochemistry and Photobiology* B: *Biology* 46: 40–52.

Caleira, K. and M. E. Wickett. 2003. Anthropogenic carbon and ocean pH. *Nature* 425: 365.

Callaway, R. M. and L. King. 1996. Temperature-driven variation in substrate oxygenation and the balance of competition and facilitation. *Ecology* 77: 1189–1195.

Callaway, R. M., E. H. DeLucia and W. H. Schlesinger. 1994. Biomass allocation of montane and desert ponderosa pine: An analog for response to climate change. *Ecology* 75 (5): 1474–1481.

Callaway, R. M. and 12 others. 2002. Positive interactions among alpine plants increase with stress. *Nature* 417: 844–848.

Carlsson, J. O. L., C. Brönmark and L.-A. Hansson. 2004. Invading herbivory: The golden apple snail alters ecosystem functioning in Asian wetlands. *Ecology* 85: 1575–1580.

Carlton, J. T. and J. B. Geller. 1993. Ecological roulette: The global transport and invasion of non-indigenous marine organisms. *Science* 261: 78–82.

Carroll, S. P. and C. Boyd. 1992. Host race radiation in the soapberry bug: Natural history with the history. *Evolution* 46: 1052–1069.

Carroll, S. P., H. Dingle and S. P. Klassen. 1997. Genetic differentiation of fitness-associated traits among rapidly evolving populations of the soapberry bug. *Evolution* 51: 1182–1188.

Carson, R. 1962. *Silent Spring.* Houghton Mifflin, Boston.

Cary, J. R and L. B. Keith. 1979. Reproductive change in the ten-year cycle of snowshoe hares. *Canadian Journal of Zoology* 57: 375–390.

Castellón, T. D. and K. E. Sieving. 2006. An experimental test of matrix permeability and corridor use by an endemic understory bird. *Conservation Biology* 20 (1): 135–145.

Caswell, H. 2001. *Matrix Population Models: Construction, Analysis, and Interpretation.* Sinauer Associates, Sunderland, MA.

Caughley, G. 1994. Directions in conservation biology. *Journal of Animal Ecology* 63 (2): 215–244.

Caughley, G., N. Shepherd and J. Short. 1987. *Kangaroos: Their Ecology and Management in the Sheep Rangelands of Australia.* Cambridge University Press, Cambridge.

Ceballos, G. and P. R. Ehrlich. 2002. Mammal population losses and the extinction crisis. *Science* 296 (5569): 904–907.

Cebrian, J. and J. Lartigue. 2004. Patterns of herbivory and decomposition in aquatic and terrestrial ecosystems. *Ecological Monographs* 74: 237–259.

Cerling, T. E., J. M. Harris, B. J. MacFadden, M. G. Leakey, J. Quade, V. Eisenmann and J. R. Ehleringer. 1997. Global vegetation change through the Miocene/Pliocene boundary. *Nature* 389: 153–158.

Černý, J. and T. Pačes. 1995. Acidification in the Black Triangle Region. In Acid Reign '95? Fifth International Conference on Acidic Deposition Science and Policy, Prague.

Channell, R. and M. V. Lomolino. 2000. Trajectories to extinction: Spatial dynamics of the contraction of geographical ranges. *Journal of Biogeography* 27 (1): 169–179.

Chapin, F. S. III., L. Moilanen and K. Kielland. 1993. Preferential use of organic nitrogen for growth by a non-mycorrhizal Arctic sedge. *Nature* 361: 150–153.

Chapin, F. S., L. R. Walker, C. L. Fastie and L. C. Sharman. 1994. Mechanisms of primary succession following deglaciation at Glacier Bay, Alaska. *Ecological Monographs* 64: 149–175.

Chapin, F. S. III, P. A. Matson and H. A. Mooney. 2002. *Principles of Terrestrial Ecosystem Ecology.* Springer, New York.

Chappelka, A. and L. Samuelson. 1998. Ambient ozone effects on forest trees of the eastern United States: A review. *New Phytologist* 139: 91–108.

Charnov, E. L. 1976. Optimal foraging, the marginal value theorem. *Theoretical Population Biology* 9: 129–136.

Charnov, E. L. 2002. Reproductive effort, offspring size and benefit–cost ratios in the classification of life histories. *Evolutionary Ecology Research* 4: 749–758.

Chen, J. Q., J. F. Franklin and T. A. Spies. 1995. Growing-season microclimatic gradients from clear-cut edges into old-growth douglas-fir forests. *Ecological Applications* 5 (1): 74–86.

Chen, J.-Y. and 8 others. 2004. Small bilaterian fossils from 40 to 55 million years before the Cambrian. *Science* 305: 218–222.

Christensen, N. L. and 12 others. 1996. The report of the Ecological Society of America Committee on the Scientific Basis for Ecosystem Management. *Ecological Applications* 6 (3): 665–691.

Christianou, M. and B. Ebenman. 2005. Keystone species and vulnerable species in ecological communities: Strong or weak interactors? *Journal of Theoretical Biology* 235 (1): 95–103.

Clements, F. E. 1916. *Plant Succession: An Analysis of the Development of Vegetation.* Publication 42. Carnegie Institution of Washington, Washington, DC.

Cleveland, C. C. and 10 others. 1999. Global patterns of terrestrial biological nitrogen (N_2) fixation in natural ecosystems. *Global Biogeochemical Cycles* 13: 623–645.

Clutton-Brock, T. H., F. E. Guinness and S. D. Albon. 1983. The costs of reproduction to red deer hinds. *Journal of Animal Ecology* 52: 367–383.

Cohen, J. E. 1995. *How Many People Can the Earth Support?* W. W. Norton, New York.

Colinvaux, P. 1978. *Why Big Fierce Animals Are Rare: An Ecologist's Perspective.* Princeton University Press, Princeton, NJ.

Coltman, D. W., P. O'Donoghue, J. T. Jorgenson, J. T. Hogg, C. Strobeck and M. Festa-Bianchet. 2003. Undesirable evolutionary consequences of trophy hunting. *Nature* 426: 655–658.

Connell, J. H. 1961a. The effects of competition, predation by *Thais lapillus*, and other factors on natural populations of the barnacle, *Balanus balanoides*. *Ecological Monographs* 31: 61–104.

Connell, J. H. 1961b. The influence of interspecific competition and other factors on the distribution of the barnacle *Chthamalus stellatus*. *Ecology* 42: 710–723.

Connell, J. H. 1978. Diversity in tropical rain forests and coral reefs. *Science* 199: 1302–1310.

Connell, J. H. 1983. On the prevalence and relative importance of interspecific competition: Evidence from field experiments. *American Naturalist* 122: 661–696.

Connell, J. H. and R. O. Slatyer. 1977. Mechanisms of succession in natural communities and their role in community stability and organization. *American Naturalist* 111: 1119–1144.

Connell, J. H. and W. P. Sousa. 1983. On the evidence needed to judge ecological stability or persistence. *American Naturalist* 121: 789–824.

Cooper, W. S. 1923a. The recent ecological history of Glacier Bay, Alaska: The interglacial forests of Glacier Bay. *Ecology* 4: 93–128.

Cooper, W. S. 1923b. The recent ecological history of Glacier Bay, Alaska: Permanent quadrats at Glacier Bay: An initial report upon a long-period study. *Ecology* 4: 355–365.

Copeyon, C. K., J. R. Walters and J. H. Carter III. 1991. Induction of red-cockaded woodpecker group formation by artificial cavity construction. *Journal of Wildlife Management* 55 (4): 549–556.

Costanzo, J. P., R. E. Lee Jr., A. L. DeVries, T. Wang and J. R. Layne Jr. 1995. Survival mechanisms of vertebrate ectotherms at subfreezing temperatures: Applications in cryomedicine. *FASEB Journal* 9: 351–358.

Cotton, P. A., S. D. Rundle and K. E. Smith. 2004. Trait compensation in marine gastropods: Shell shape, avoidance behavior, and susceptibility to predation. *Ecology* 85: 1581–1584.

Courchamp, F., T. Clutton-Brock and B. Grenfell. 1999. Inverse density dependence and the Allee effect. *Trends in Ecology and Evolution* 14: 405–410.

Courchamp, F., E. Angulo, P. Rivalan, R. J. Hall, L. Signoret, L. Bull and Y. Meinard. 2006. Rarity value and species extinction: The anthropogenic Allee effect. *PLoS Biology* 4 (12): 2405–2410.

Cowie, R. H. 2001. Decline and homogenization of Pacific faunas: The land snails of American Samoa. *Biological Conservation* 99 (2): 207–222.

Cowie, R. J. 1977. Optimal foraging in great tits (*Parus major*). *Nature* 268: 137–139.

Cowles, H. C. 1899. The ecological relations of the vegetation on the sand dunes of Lake Michigan. *Botanical Gazette* 27: 95–117, 167–202, 281–308, 361–391.

Crews, T. E., K. Kitayama, J. H. Fownes, D. A. Herbert, R. H. Riley, D. Mueller-Dombois and P. M. Vitousek. 1995. Changes in soil phosphorus fractions and ecosystem dynamics across a long chronosequence in Hawaii. *Ecology* 76: 1407–1424.

Crisafulli, C. M., J. A. MacMahon and R. R. Parmenter. 2005. Small-mammal survival and colonization on the Mount St. Helens volcano: 1980–2005. In *Ecological Responses to the 1980 Eruption of Mount St. Helens,* V. H. Dale, F. J. Swanson and C. M. Crisafulli (eds.), 183–197. Springer, New York.

Croll, D. A., J. L. Maron, J. A. Estes, E. M. Danner and G. V. Byrd. 2005. Introduced predators transform subarctic islands from grassland to tundra. *Science* 307: 1959–1961.

Crouse, D. T., L. B. Crowder and H. Caswell. 1987. A stage-based population model for loggerhead sea turtles and implications for conservation. *Ecology* 68: 1412–1423.

Crowcroft, P. 1991. *Elton's Ecologists.* University of Chicago Press, Chicago.

Crowder, L. B., D. T. Crouse, S. S. Heppell and T. H. Martin. 1994. Predicting the impact of Turtle Excluder Devices on logger-head sea turtle populations. *Ecological Applications* 4: 437–445.

Crutzen, P. J. and E. F. Stoermer. 2000. The "Anthropocene." *Global Change Newsletter* 41: 12–13.

Currie, C. R. and A. E. Stuart. 2001. Weeding and grooming of pathogens in agriculture by ants. *Proceedings of the Royal Society of London* B 268: 1033–1039.

Currie, C. R., J. A. Scott, R. C. Summerbell and D. Malloch. 1999. Fungus-growing ants use antibiotic-producing bacteria to control garden parasites. *Nature* 398: 701–704.

Currie, C. R., M. Poulsen, J. Mendenhall, J. J. Boomsma and J. Billen. 2006. Coevolved crypts and exocrine glands support mutualis-tic bacteria in fungus-growing ants. *Science* 311: 81–83.

D

Dahlgren, C. P. and D. B. Eggleston. 2000. Ecological processes underlying ontogenetic habitat shifts in a coral reef. *Ecology* 81 (8): 2227–2240.

Daily, G. C. and K. Ellison. 2002. *The New Economy of Nature: The Quest to Make Conservation Profitable.* Island Press, Washington, DC.

Dale, V. H., F. J. Swanson and C. M. Crisafulli, eds. 2005. *Ecological Responses to the 1980 Eruption of Mount St. Helens.* Springer, New York.

Dambrine, E., J. L. Dupouey, L. Laut, L. Humbert, M. Thinon, M. T. Beaufils and H. Richard. 2007. Present forest biodiversity pat-terns in France related to former Roman agriculture. *Ecology* 88 (6): 1430–1439.

Damman, H. and M. L. Cain. 1998. Population growth and viability analyses of the clonal woodland herb, *Asarum canadense. Journal of Ecology* 86: 13–26.

Darwin, C. 1859. *On the Origin of Species.* J. Murray, London.

Darwin, C. 1875. *Insectivorous Plants.* Appleton and Company, New York.

Daszak, P., A. A. Cunningham and A. D. Hyatt. 2000. Emerging infectious diseases of wildlife—Threats to biodiversity and human health. *Science* 287 (5452): 443–449.

Davidson, D. W., S. C. Cook, R. R. Snelling and T. H. Chua. 2002. Explaining the abundance of ants in lowland tropical rainfor-est canopies. *Science* 300: 969–972.

Davidson, J. 1938. On the growth of the sheep population in Tasmania. *Transactions of the Royal Society of South Australia* 62: 342–346.

Davidson, J. and H. G. Andrewartha. 1948. The influence of rainfall, evaporation and atmospheric temperature of fluctuations in the size of a natural population of *Thrips imaginis* (Thysanoptera). *Journal of Animal Ecology* 17: 200–222.

Davis, C. C. 1964. Evidence for the eutrophication of Lake Erie from phytoplankton records. *Limnology and Oceanography* 9: 275–283.

Davis, M. B., J. Ford and R. E. Moeller. 1985. Paleolimnology. In *An Ecosystem Approach to Aquatic Ecology: Mirror Lake and Its Environment,* G. E. Likens (ed.), 345–366. Springer-Verlag, New York.

Dawson, J. and R. Lucas. 2000. *Nature Guide to the New Zealand Forest.* A Godwit Book. Random House New Zealand, Auckland, NZ. 312 pp.

Dayton, P. K. 1971. Competition, disturbance and community organization: The provision and subsequent utilization of space in a rocky intertidal community. *Ecological Monographs* 41: 351–389.

Deevey, E. S. 1947. Life tables for natural populations. *Quarterly Review of Biology* 22: 283–314.

de Grammont, P. C. and A. D. Cuarón. 2006. An evaluation of threatened species categorization systems used on the American continent. *Conservation Biology* 20 (1): 14–27.

Delcourt, H. R. 2002. Creating landscape pattern. In *Learning Landscape Ecology,* S. E. Gergel and M. G. Turner (eds.), 62–82. Springer Verlag, New York.

Delcourt, P. A., H. R. Delcourt, C. R. Ison, W. E. Sharp and K. J. Gremillion. 1998. Prehistoric human use of fire, the eastern agricultural complex, and Appalachian oak–chestnut forests: Paleoecology of Cliff Palace Pond, Kentucky. *American Antiquity* 63: 263–278.

DellaSala, D. A. and J. E. Williams. 2006. The northwest forest plan: A global model of forest management in contentious times. *Conservation Biology* 20 (2): 274–276.

Del Moral, R., D. M. Wood and J. H. Titus. 2005. Proximity, microsites, and biotic interactions during early succession. In *Ecological Responses to the 1980 Eruption of Mount St. Helens,* V. H. Dale, F. J. Swanson and C. M. Crisafulli (eds.), 93–107. Springer, New York.

DeLucia, E. H. and 10 others. 1999. Net primary production of a forest ecosystem under experimental CO_2 enrichment. *Science* 284: 1177–1179.

DeMaster, D. P., A. W. Trites, P. Clapham, S. Mizroch, P. Wade, R. J. Small and J. Ver Hoef. 2006. The sequential megafaunal col-lapse hypothesis: Testing with existing data. *Progress in Oceanography* 68: 329–342.

Demmig-Adams, B. and W. W. Adams III. 1996. The role of xantho-phyll cycle carotenoids in the protection of photosynthesis. *Trends in Plant Science* 1: 21–26.

Denslow, J. S. 1987. Tropical rainforest gaps and tree species diver-sity. *Annual Review of Ecology and Systematics* 18: 431–451.

Department of Economic and Social Affairs, Population Division. 2007. *World Population Prospects: The 2006 Revision. Highlights.* United Nations Secretariat, New York.

Dewailly, E., P. Ayotte, S. Bruneau, C. Laliberte, D. C. G. Muir and R. J. Norstrom. 1993. Inuit exposure to organochlorines through the aquatic food-chain in Arctic Quebec. *Environmental Health Perspectives* 101: 618–620.

Diamond, J. M. 1975. The island dilemma: Lessons of modern bio-geographic studies for the design of natural reserves. *Biological Conservation* 7: 129–146.

Diamond, J. 1997. *Guns, Germs, and Steel: The Fates of Human Societies*. W. W. Norton, New York.

Dietl, G. P., G. S. Herbert and G. J. Vermeij. 2004. Reduced competition and altered feeding behavior among marine snails after a mass extinction. *Science* 306: 2229–2231.

Dirzo, R. and P. H. Raven. 2003. Global state of biodiversity and loss. *Annual Review of Environment and Resources* 28: 137–167.

Dixon, A. F. G. 1971. The role of aphids in wood formation. II. The effect of the lime aphid, *Eucallipterus tiliae* L. (Aphididae), on the growth of lime *Tilia × vulgaris* Hayne. *Journal of Applied Ecology* 8: 393–399.

Dobson, A. and M. Meagher. 1996. The population dynamics of brucellosis in the Yellowstone National Park. *Ecology* 77: 1026–1036.

Dodd, D. M. B. 1989. Reproductive isolation as a consequence of adaptive divergence in *Drosophila pseudoobscura*. *Evolution* 43: 1308–1311.

Dodson, C. H. and A. H. Gentry. 1991. Biological extinction in Western Ecuador. *Annals of the Missouri Botanical Garden* 78 (2): 273–295.

Dodson, S. 2004. *Introduction to Limnology*. McGraw Hill, New York.

Duffy, J. E., B. J. Cardinale, K. E. France, P. B. McIntyre, E. Thebault and M. Loreau. 2007. The functional role of biodiversity in ecosystems: Incorporating trophic complexity. *Ecology Letters* 10: 522–538.

Duggins, D. O. 1980. Kelp beds and sea otters: An experimental approach. *Ecology* 61: 447–453.

Duggins, D. O., C. A. Simenstad and J. A. Estes. 1989. Magnification of secondary production by kelp detritus in coastal marine ecosystems. *Science* 245: 170–173.

Dukes, J. S. and H. A. Mooney. 1999. Does global change increase the success of biological invaders? *Trends in Ecology and Evolution* 14: 135–139.

Dybdahl, M. F. and C. M. Lively. 1998. Host–parasite coevolution: Evidence for rare advantage and time-lagged selection in a natural population. *Evolution* 52: 1057–1066.

Dyer, L. A. and K. K. Letourneau. 1999a. Trophic cascades in a complex terrestrial community. *Proceedings of the National Academy of Sciences USA* 96: 5072–5076.

Dyer, L. A. and K. K. Letourneau. 1999b. Relative strengths of top-down and bottom-up forces in a tropical forest community. *Oecologia* 119: 265–274.

E

Eberhard, W. G. 2001. Under the influence: Webs and building behavior of *Plesiometa argyra* (Araneae, Tetragnathidae) when parasitized by *Hymenoepimecis argyraphaga* (Hymenoptera, Ichneumonidae). *Journal of Arachnology* 29: 354–366.

Edmondson, W. T. and A. H. Litt. 1982. *Daphnia* in Lake Washington. *Limnology and Oceanography* 27: 272–293.

Edney, E. B. 1980. The components of water balance. In *Insect Biology in the Future*, M. Locke and D. A. Smith, 39–58. Academic Press, New York.

Ehleringer, J. R. and C. S. Cook. 1990. Characteristics of *Encelia* species differing in leaf reflectance and transpiration rate under common garden conditions. *Oecologia* 82: 484–489.

Ehleringer, J. R., T. E. Cerling and B. R. Helliker. 1997. C_4 photosynthesis, atmospheric CO_2, and climate. *Oecologia* 112: 285–299.

Ehrlich, P. R. and E. O. Wilson. 1991. Biodiversity studies—Science and policy. *Science* 253 (5021): 758–762.

Elser, J. J. and 11 others. 2000. Nutritional constraints in terrestrial and freshwater food webs. *Nature* 408: 578–580.

Elton, C. 1927. *Animal Ecology*. Sidgwick and Jackson, London.

Elton, C. S. 1958. *The Ecology of Invasions by Animals and Plants*. Chapman and Hall, London.

Endler, J. A. 1986. *Natural Selection in the Wild*. Princeton University Press, Princeton, NJ.

Endler, J. A. 1995. Multiple-trait coevolution and environmental gradients in guppies. *Trends in Ecology and Evolution* 10: 22–29.

Ensign, S. H. and M. W. Doyle. 2006. Nutrient spiraling in streams and river networks. *Journal of Geophysical Research* 111: G04009, doi:10.1029/2005JG000114.

Epstein, E. and A. J. Bloom. 2005. *Mineral Nutrition of Plants: Principles and Perspectives*. 2nd ed. Sinauer Associates, Sunderland, MA.

Erickson, R. O. 1945. The *Clematis fremontii* var. *riehlii* population in the Ozarks. *Annals of the Missouri Botanical Garden* 32: 413–460.

Estes, J. A. and D. O. Duggins. 1995. Sea otters and kelp forests in Alaska: Generality and variation in a community ecology paradigm. *Ecological Monographs* 65: 75–100.

Estes, J. A., M. T. Tinker, T. M. Williams and D. F. Doak. 1998. Killer whale predation on sea otters linking oceanic and nearshore ecosystems. *Science* 282: 473–476.

Evans, R. D., R. Rimer, L. Sperry and J. Belnap. 2001. Exotic plant invasion alters nitrogen dynamics in an arid grassland. *Ecological Applications* 11: 1301–1310.

F

Fagan, W. F., R. S. Cantrell and C. Cosner. 1999. How habitat edges change species interactions. *American Naturalist* 153 (2): 165–182.

Faithfull, I. 1997. *Etiella behrii* (Zeller) (Lepidoptera: Pyralidae) bred from pods of gorse, *Ulex europaeus* L. Fabaceae. *Victorian Entomology* 27: 34.

FAO. Global Forest Resources Assessment 2005. Food and Agriculture Organization of the United Nations. www.fao.org/forestry/site/24690/en.

Farrell, T. M. 1991. Models and mechanisms of succession: An example from a rocky intertidal community. *Ecological Monographs* 61: 95–113.

Favreau, J. M., C. A. Drew, G. R. Hess, M. J. Rubino, F. H. Koch and K. A. Eschelbach. 2006. Recommendations for assessing the effectiveness of surrogate species approaches. *Biodiversity and Conservation* 15 (12): 3949–3969.

Feely, R. A., C. L. Sabine, K. Lee, W. Berrelson, J. Kleypas, V. J. Fabry and F. J. Millero. 2004. Impact of anthropogenic CO_2 on the $CaCO_3$ system in the oceans. *Science* 305: 362–366.

Fenner, F. and F. N. Ratcliffe. 1965. *Myxomatosis*. Cambridge University Press, Cambridge.

Ferraz, G., G. J. Russell, P. C. Stouffer, R. O. Bierregaard Jr., S. L. Pimm and T. E. Lovejoy. 2003. Rates of species loss from Amazonian forest fragments. *Proceedings of the National Academy of Sciences USA* 100: 14069–14073.

Field, C. B., M. J. Behrenfeld, J. T. Randerson and P. Falkowski. 1998. Primary productivity of the biosphere: Integrating terrestrial and oceanic components. *Science* 281: 237–240.

Field, D. B., T. R. Baumgartner, C. D. Charles, V. Ferreira-Bartrina and M. D. Ohman. 2006. Planktonic foraminifera of the California current reflect 20th-century warming. *Science* 311: 63–66.

Fiore, A. M., D. J. Jacob, J. A. Logan and J. H. Yin. 1998. Long-term trends in ground level ozone over the contiguous United States, 1980–1995. *Journal of Geophysical Research* 103 D1: 1471–1480.

Fischlin, A. and 9 others. 2007. Ecosystems, their properties, goods, and services. In *Climate Change 2007: Impacts, Adaptation and Vulnerability: Contribution of Working Group II to the Fourth Assessment Report of the Intergovernmental Panel on Climate Change*, M. L. Parry, O. F. Canziani, J. P. Palutikof, P. J. van der Linden and C. E. Hanson (eds.), 211–272. Cambridge University Press, Cambridge.

Fisher, S. G. and G. E. Likens. 1973. Energy budget in Bear Brook, New Hampshire: An integrative approach to stream ecosystem metabolism. *Ecological Monographs* 43: 421–439.

Fitter, A. H. 2005. Darkness visible: Reflections on underground ecology. *Journal of Ecology* 93: 231–243.

Flecker, A. S. and C. R. Townsend. 1994. Community-wide consequences of trout introduction in New Zealand streams. *Ecological Applications* 4: 798–807.

Fleishman, E., D. D. Murphy and P. E. Brussard. 2000. A new method for selection of umbrella species for conservation planning. *Ecological Applications* 10 (2): 569–579.

Flessa, K. W. 1975. Area, continental drift and mammalian diversity. *Paleobiology* 1: 189–194.

Foley, J. A., M. H. Costa, C. Delire, N. Ramankutty and P. Snyder. 2003. Green surprise? How terrestrial ecosystems could affect Earth's climate. *Frontiers in Ecology and Environment* 1: 38–44.

Fonseca, G. A. B. da. 1985. The vanishing Brazilian Atlantic forest. *Biological Conservation* 34: 17–34.

Forman, R. T. T. 1995. Some general principles of landscape and regional ecology. *Landscape Ecology* 10 (3): 133–142.

Food and Agriculture Organization of the United Nations. *See* FAO.

Frank, P. W., C. D. Boll and R. W. Kelly. 1957. Vital statistics of laboratory cultures of Daphnia pulex De Geer as related to density. *Physiological Zoology* 30: 287–305.

Franklin, J. F. 1996. Ecosystem management: An overview. In *Ecosystem Management: Applications for Sustainable Forest and Wildlife Resources*, M. S. Boyce and A. Haney (eds.), 21–53. Yale University Press, New Haven, CT.

Franklin, J. F. and C. T. Dyrness. 1988. *Natural Vegetation of Oregon and Washington*. Oregon State University Press, Corvallis.

Freeman, S. and J. C. Herron. 2007. *Evolutionary Analysis*. 4th ed. Pearson Prentice Hall, Upper Saddle River, NJ.

Frid, C. L. J. and E. Fordham. 1994. The morphology of the sub-littoral gastropod *Gibbula cineraria* along a gradient of wave action. *Ophelia* 40: 135–146.

Funk, D. J., P. Nosil and W. J. Etges. 2006. Ecological divergence exhibits consistently positive associations with reproductive isolation across disparate taxa. *Proceedings of the National Academy of Sciences USA* 103: 3209–3213.

Futuyma, D. J. 2005. *Evolution*. Sinauer Associates, Sunderland, MA.

G

Galloway, J. N., J. D. Aber, J. W. Erisman, S. P. Seitzinger, R. W. Howarth, E. B. Cowling and B. J. Cosby. 2003. The nitrogen cascade. *BioScience* 53: 341–356.

Galloway, J. N. and 14 others. 2004. Nitrogen cycles: Past, present, and future. *Biogeochemistry* 70: 153–226.

Gardner, T. A., I. M. Cote, J. A. Gill, A. Grant and A. R. Watkinson. 2003. Long-term region-wide declines in Caribbean Corals. *Science* 301:958–960.

Garrison, V. H. and 9 others. 2003. African and Asian dust: From desert soils to coral reefs. *BioScience* 53: 469–480.

Gascon, C., G. B. Williamson and G. A. B. da Fonseca. 2000. Receding forest edges and vanishing reserves. *Science* 288: 1356–1358.

Gasol, J. M., P. A. del Giorio and C. M. Duarte. 1997. Biomass distribution in marine planktonic communities. *Limnology and Oceanography* 42: 1353–1363.

Gaston, K. J., P. H. Williams, P. Eggleton and C. J. Humphries. 1995. Large scale patterns of biodiversity: Spatial variation in family richness. *Proceedings of the Royal Society of London* B 260: 149–154.

Gause, G. F. 1934a. Experimental analysis of Vito Volterra's mathematical theory of the struggle for existence. *Science* 79: 16–17.

Gause, G. F. 1934b. *The Struggle for Existence*. Williams & Wilkins, Baltimore, MD.

Geffeney, S. L., E. Fujimoto, E. D. Brodie III, E. D. Brodie Jr. and P. C. Ruben. 2005. Evolutionary diversification of TTX-resistant sodium channels in a predator–prey interaction. *Nature* 434: 759–763.

Gilg, O., I. Hanski and B. Sittler. 2003. Cyclic dynamics in a simple vertebrate predator–prey community. *Science* 302: 866–868.

Gilliam, F. S. and W. J. Platt. 2006. Conservation and restoration of the *Pinus palustris* ecosystem. *Applied Vegetation Science* 9 (1): 7–10.

Gleason, H. A. 1917. The structure and development of the plant association. *Bulletin of the Torrey Botanical Club* 44: 463–481.

Gotelli, N. J. and A. M. Ellison. 2004. *A Primer of Ecological Genetics*. Sinauer Associates, Sunderland, MA.

Gough, L., G. R. Shaver, J. Carroll, D. L. Royer and J. A. Laundre. 2000. Vascular plant species richness in Alaskan arctic tundra: The importance of soil pH. *Journal of Ecology* 88: 54–66.

Goverde, M., K. Schweizer, B. Baur and A. Erhardt. 2002. Small-scale habitat fragmentation effects on pollinator behaviour: Experimental evidence from the bumblebee *Bombus veteranus* on calcareous grasslands. *Biological Conservation* 104 (3): 293–299.

Grabherr, G., M. Gottfried and H. Pauli. 1994. Climate effects on mountain plants. *Nature* 369: 448.

Graham, I. M. and X. Lambin. 2002. The impact of weasel predation on cyclic field-vole survival: The specialist predator hypothesis contradicted. *Journal of Animal Ecology* 71: 946–956.

Grant, B. R. and P. R. Grant. 2003. What Darwin's finches can teach us about the evolutionary origin and regulation of biodiversity. *BioScience* 53: 965–975.

Grant, P. R. 1986. *Ecology and Evolution of Darwin's Finches*. Princeton University Press, Princeton, NJ.

Grime, J. P. 1977. Evidence for the existence of three primary strategies in plants and its relevance to ecological and evolutionary theory. *American Naturalist* 111 (982): 1169–1194.

Grime, J. P. 1997. Biodiversity and ecosystem function: The debate deepens. *Science* 277: 1260–1261.

Groom, M. J. and 7 others. 1999. Buffer zones: Benefits and dangers of compatible stewardship. In *Continental Conservation: Scientific Foundations of Regional Reserve Networks*, M. E. Soulé and J. Terborgh (eds.), 171–197. Island Press, Washington, DC.

Groom, M. J., G. K. Meffe and C. R. Carroll. 2006. *Principles of Conservation Biology*. 3rd ed. Sinauer Associates, Sunderland, MA.

Grosberg, R. K. and R. R. Strathmann. 1998. One cell, two cell, red cell, blue cell: The persistence of a unicellular stage in multicellular life histories. *Trends in Ecology and Evolution* 13 (3): 112–116.

Grousset, F. E., P. Ginoux, A. Bory and P. E. Biscaye. 2003. Case study of a Chinese dust plume reaching the French Alps. *Geophysical Research Letters* 30: 1277, doi:10.1029/2002GL016833.

Grutter, A. S., J. M. Murphy and J. H. Choat. 2003. Cleaner fish drives local fish diversity on coral reefs. *Current Biology* 13: 64–67.

Gurevitch, J., L. L. Morrow, A. Wallace and J. S. Walsh. 1992. A meta-analysis of competition in field experiments. *American Naturalist* 140: 539–572.

Gurevitch, J., S. M. Scheiner and G. A. Fox. 2006. *The Ecology of Plants.* 2nd ed. Sinauer Associates, Sunderland, MA.

Gutzwiller, K. J. and S. H. Anderson. 1992. Interception of moving organisms: Influences of patch shape, size, and orientation on community structure. *Landscape Ecology* 6 (4): 293–303.

H

Hacker, S. D. and M. D. Bertness. 1996. Trophic consequences of a positive plant interaction. *American Naturalist* 148: 559–575.

Hacker, S. D. and M. D. Bertness. 1999. Experimental evidence for the factors maintaining plant species diversity in a New England salt marsh. *Ecology* 80: 2064–2073.

Hacker, S. D. and S. D. Gaines. 1997. Some implications of direct positive interactions for community species diversity. *Ecology* 78: 1990–2003.

Haddrath, O. and A. J. Baker. 2001. Complete mitochondrial DNA genome sequences of extinct birds: Ratite phylogenetics and the vicariance biogeography hypothesis. *Proceedings of the Royal Society of London* B 268: 939–945.

Haines, B. L. 1978. Element and energy flows through colonies of the leaf-cutting ants, *Atta colombica*, in Panama. *Biotropica* 10: 270–277.

Hairston, N., F. Smith and L. Slobodkin. 1960. Community structure, population control and competition. *American Naturalist* 94: 421–425.

Hannah, L. and 7 others. 2002. Conservation of biodiversity in a changing climate. *Conservation Biology* 16 (1): 264–268.

Hansen, J., A. Lacis, R. Ruedy and Mki. Sato. 1992. Potential climate impact of Mount Pinatubo eruption. *Geophysical Research Letters* 19: 215–218, doi:10.1029/91GL02788.

Hansen, J., Mki. Sato, R. Ruedy, K. Lo, D. W. Lea and M. Medina-Elizade. 2006. Global temperature change. *Proceedings of the National Academy of Sciences USA* 103: 14288–14293,

Hansen, T. A. 1978. Larval dispersal and species longevity in lower Tertiary gastropods. *Science* 199: 885–887.

Hanski, I. 1999. *Metapopulation Ecology*. Oxford University Press, New York.

Harding, S. R. and J. R. Walters. 2002. Processes regulating the population dynamics of red-cockaded woodpecker cavities. *Journal of Wildlife Management* 66 (4): 1083–1095.

Hare, S. R. and R. C. Francis. 1994. Climate change and salmon production in the Northeast Pacific Ocean. In *Climate Change and Northern Fish Populations*, R. J. Beamish (ed.), 357–372. Canadian Special Publication of Fisheries and Aquatic Sciences, 121.

Harper, K. A. and 9 others. 2005. Edge influence on forest structure and composition in fragmented landscapes. *Conservation Biology* 19 (3): 768–782.

Harrison, P. 1987. *A Field Guide to Seabirds of the World*. Stephen Greene Press, Lexington, MA.

Harrison, P. and F. Pearce. 2001. *AAAS Atlas of Population and the Environment*. University of California Press, Berkeley.

Harrison, R. G. 1980. Dispersal polymorphisms in insects. *Annual Review of Ecology and Systematics* 11: 95–118.

Harshbarger, J. C., M. J. Coffey and M. Y. Young. 2000. Intersexes in Mississippi River shovelnose sturgeon sampled below Saint Louis, Missouri, USA. *Marine Environmental Research* 50 (1–5): 247–250.

Hart, M. W. and R. R. Strathmann. 1994. Functional consequences of phenotypic plasticity in echinoid larvae. *Biological Bulletin* 186: 291–299.

Hartl, D. L. and A. G. Clark. 1997. *Principles of Population Genetics*. 3rd ed. Sinauer Associates, Sunderland MA.

Hartnett, D. C. and G. W. T. Wilson. 1999. Mycorrhizae influence plant community structure and diversity in tallgrass prairie. *Ecology* 80: 1187–1195.

Hatch, A. C. and A. R. Blaustein. 2003. Combined effects of UV-B radiation and nitrate fertilizer on larval amphibians. *Ecological Applications* 13: 1083–1093.

Hawkins, C. E. and 11 others. 2006. Emerging disease and population decline of an island endemic, the Tasmanian devil *Sarcophilus harrisii*. *Biological Conservation* 131: 307–324.

Hays, J. D., J. Imbrie and N. J. Shackleton. 1976. Variations in Earths orbit—Pacemaker of ice ages. *Science* 194: 1121–1132.

Hearne, S. 1911. *A Journey from Prince of Wales's Fort in Hudson's Bay to the Northern Ocean in the years 1769, 1770, 1771 and 1772*. The Champlain Society, Toronto.

Hebblewhite, M. and 7 others. 2005. Human activity mediates a trophic cascade caused by wolves. *Ecology* 86 (8): 2135–2144.

Hechtel, L. J., C. L. Johnson and S. A. Juliano. 1993. Modification of antipredator behavior of *Caecidotea intermedius* by its parasite *Acanthocephalus dirus*. *Ecology* 74: 710–713.

Hector, A. and 33 others. 1999. Plant diversity and productivity experiments in European grasslands. *Science* 286: 1123–1127.

Hegazy, A. K. 1990. Population ecology and implications for conservation of *Cleome droserifolia*: A threatened xerophyte. *Journal of Arid Environments* 19: 269–282.

Henderson, S., P. Hattersley, S. von Caemmerer and C. B. Osmond. 1995. Are C_4 pathway plants threatened by global climatic change? In *Ecophysiology of Photosynthesis*, E.-D. Schulze and M. M. Caldwell (eds.), 529–549. Springer, New York.

Henry, M., J. M. Pons and J. F. Cosson. 2007. Foraging behaviour of a frugivorous bat helps bridge landscape connectivity and ecological processes in a fragmented rainforest. *Journal of Animal Ecology* 76 (4): 801–813.

Hero, J.-M. and L. P. Shoo. 2003. Conservation of amphibians in the old world tropics: defining unique problems associated with a regional fauna. In *Amphibian Conservation*, R. D. Semlitsch (ed.), 70-84. Smithsonian Books, Washington.

Hersperger, A. M. 2006. Spatial adjacencies and interactions: Neighborhood mosaics for landscape ecological planning. *Landscape and Urban Planning* 77 (3): 227–239.

Hess, G. R. 1994. Conservation corridors and contagious disease—A cautionary note. *Conservation Biology* 8 (1): 256–262.

Hetrick, B. A. D., G. W. T. Wilson and D. C. Hartnett. 1989. Relationship between mycorrhizal dependence and competitive ability of two tallgrass prairie grasses. *Canadian Journal of Botany* 67: 2608–2615.

Higgins, S. I., R. Nathan and M. L. Cain. 2003. Are long-distance dispersal events in plants usually caused by nonstandard means of dispersal? *Ecology* 84: 1945–1956.

Hik, D. S. and R. L. Jefferies. 1990. Increases in the net above-ground primary production of a salt-marsh forage grass: A test of the predictions of the herbivore-optimization model. *Journal of Ecology* 78: 180–195.

Hiller, N. L., S. Bhattacharjee, C. van Ooij, K. Liolios, T. Harrison, C. Lopez-Estraño and K. Haldar. 2004. A host-targeting signal in virulence proteins reveals a secretome in malarial infection. *Science* 306: 1934–1937.

Himmel, M. E., S.-Y. Ding, D. K. Johnson, W. S. Adney, M. R. Nimlos, J. W. Brady and T. D. Foust. 2007. Biomass recalcitrance: Engineering plants and enzymes for biofuels production. *Science* 315: 804–807.

Hochochka, P. W. and G. N. Somero. 2002. *Biochemical Adaptation, Mechanism and Process in Physiological Evolution*. Oxford University Press, Oxford.

Hoffmeister, T. S., L. E. M. Vet, A. Biere, K. Holsinger and J. Filser. 2005. Ecological and evolutionary consequences of biological invasion and habitat fragmentation. *Ecosystems* 8: 657–667.

Hollinger, D.Y. and 14 others. 1998. Forest–atmosphere carbon dioxide exchange in eastern Siberia. *Agricultural and Forest Meteorology* 90: 291–306.

Holmes, J. C. 1996. Parasites as threats to biodiversity in shrinking ecosystems. *Biodiversity and Conservation* 5 (8): 975–983.

Hooper, R. G., W. E. Taylor and S. C. Loeb. 2004. Long-term efficacy of artificial cavities for red-cockaded woodpeckers: Lessons learned from Hurricane Hugo. In *Red-Cockaded Woodpecker: Road to Recovery*, R. Costa and S. J. Daniels (eds.), 430–438. Hancock House Publishers, Blaine, WA.

Houserová, P., V. KubáÚ, S. Krácmar and J. Sitko. 2007. Total mercury and mercury species in birds and fish in an aquatic ecosystem in the Czech Republic. *Environmental Pollution* 145: 185–194.

Howarth, R. W. and 14 others. 1996. Regional nitrogen budgets and riverine N & P fluxes for the drainages to the North Atlantic Ocean: Natural and human influences. *Biogeochemistry* 35: 75–139.

Hudson, P. J., A. P. Dobson and D. Newborn. 1998. Prevention of population cycles by parasite removal. *Science* 282: 2256–2258.

Huffaker, C. B. 1958. Experimental studies on predation: Dispersion factors and predator–prey oscillations. *Hilgardia* 27: 343–383.

Huffaker, C. B. and C. E. Kennett. 1957. A ten-year study of vegetational changes associated with biological control of Klamath weed. *Journal of Range Management* 12: 69–82.

Hughes, J. B., J. J. Hellman, T. H. Ricketts and B. J. M. Bohannan. 2001. Counting the uncountable: Statistical approaches to estimating microbial diversity. *Applied and Environmental Microbiology* 67: 4399–4406.

Humphreys, W. F. 1979. Production and respiration in animal populations. *Journal of Animal Ecology* 48: 427–454.

Hunt, G. R. 1996. Manufacture and use of hook-tools by New Caledonian crows. *Nature* 379: 249–251.

Hunt, G. R. and R. D. Gray. 2003. Diversification and cumulative evolution in New Caledonian crow tool manufacture. *Proceedings of the Royal Society of London* B 270: 867–874.

Hunziker, J. H., R. A. Palacios, A. G. de Valesi and L. Poggio. 1972. Species disjunctions in *Larrea*: Evidence from morphology, cytogenetics, phenolic compounds and seed albumins. *Annals of Missouri Botanical Garden* 59: 224–233.

Huston, M. 1979. A general hypothesis of species diversity. *American Naturalist* 113: 81–101.

Hutchinson, G. E. 1959. Homage to Santa Rosalia or why are there so many kinds of animals? *American Naturalist* 93: 145–159.

Hutchinson, G. E. 1961. The paradox of the plankton. *American Naturalist* 95: 137–145.

Hutchinson, G. E. 1965. *The Ecological Theater and the Evolutionary Play.* Yale University Press, New Haven, CT.

I

IPCC. 2007. *Climate Change 2007*. Cambridge University Press, Cambridge; also available at the IPCC Web site: http://www.ipcc.ch/.

IPCC. 2007. *Climate Change 2007: The Physical Science Basis: Contribution of Working Group I to the Fourth Assessment Report of the Intergovernmental Panel on Climate Change*, S. Solomon, D. Qin, M. Manning, Z. Chen, M. Marquis, K. B. Averyt, M. Tignor and H. L. Miller (eds.). Cambridge University Press, Cambridge.

IUCN Global Amphibian Assessment. 2004. http://www.globalamphibians.org.

J

Jablonski, D. 1995. Extinctions in the fossil record. In *Extinction Rates*, J. H. Lawton and R. M. May (eds.), 25–44. Oxford University Press, Oxford.

Jablonski, D., K. Roy and J. W. Valentine. 2006. Out of the tropics: Evolutionary dynamics of the latitudinal diversity gradient. *Science* 314: 102–106.

Jablonski, N. G. 2004. The evolution of human skin and skin color. *Annual Review of Anthropology* 33: 585–623.

Jablonski, N. G. 2006. *Skin: A Natural History*. University of California Press, Berkeley.

Jackson, M. E. and R. D. Semlitsch. 1993. Paedomorphosis in the salamander *Ambystoma talpoideum*: Effects of a fish predator. *Ecology* 74 (2): 342–350.

Jacobson, M. Z. 2002. *Atmospheric Pollution, History, Science, and Regulation*. Cambridge University Press, Cambridge.

Jagersten, G. 1972. *Evolution of the Metazoan Life Cycle: A Comprehensive Theory*. Academic Press, London, New York.

Janzen, D. H. 1966. Coevolution of mutualism between ants and acacias in Central America. *Evolution* 20: 249–275.

Janzen, D. H. 1986. The future of tropical ecology. *Annual Review of Ecology and Systematics* 17: 305–324.

Jefferies, R. L., R. F. Rockwell and K. F. Abraham. 2003. The embarrassment of riches: Agricultural food subsidies, high goose numbers, and loss of Arctic wetlands—a continuing saga. *Environmental Reviews* 11: 193–232.

Jeon, K. W. 1972. Development of cellular dependence on infective organisms: Microsurgical studies in amoebas. *Science* 176: 1122–1123.

Jeschke, J. M. 2007. When carnivores are "full and lazy." *Oecologia* 152: 357–364.

Johnson, P. T. J., K. B. Lunde, E. G. Ritchie and A. E. Launer. 1999. The effect of trematode infection on amphibian limb development and survivorship. *Science* 284: 802–804.

Johnson, R. G. and S. A. Temple. 1990. Nest predation and brood parasitism of tallgrass prairie birds. *Journal of Wildlife Management* 54 (1) 106–111.

Jones, C. G., J. H. Lawton and M. Shachak. 1994. Organisms as ecosystem engineers. *Oikos* 69: 373–386.

Jones, C. G., J. H. Lawton and M. Shachak. 1997. Positive and negative effects of organisms as physical ecosystem engineers. *Ecology* 78: 1946–1957.

Jones, H. L. and J. M. Diamond. 1976. Short-time-base studies of turnover in breeding bird populations on the California Channel Islands. *Condor* 78: 526–549.

Jordan, M. J. and G. E. Likens. 1975. An organic carbon budget for an oligotrophic lake in New Hampshire. *Verhandlungen Internationale Vereinigung fur Theoretische und Angewandte Limnologie* 19: 994–1003.

Jousson, O. and 8 others. 2000. Invasive alga reaches California. *Nature* 408: 157–158.

K

Keever, K. 1953. Present composition of some stands of the former oak–chestnut forest in the southern Blue Ridge mountains. *Ecology* 34: 44–54.

Keller, I. and C. R. Largiadèr. 2003. Recent habitat fragmentation caused by major roads leads to reduction of gene flow and loss of genetic variability in ground beetles. *Proceedings of the Royal Society of London* B 270 (1513): 417–423.

Kenward, B., A. A. S. Weir, C. Rutz and A. Kacelnik. 2005. Tool manufacture by naive juvenile crows. *Nature* 433: 121.

Kessler, A. and I. T. Baldwin. 2001. Defensive function of herbivore-induced plant volatile emissions in nature. *Science* 291: 2141–2144.

Kessler, A., R. Halitschke and I. T. Baldwin. 2004. Silencing the jasmonate cascade: Induced plant defenses and insect populations. *Science* 305: 665–668.

Kideys, A. E. 2002. Fall and rise of the Black Sea ecosystem. *Nature* 297: 1482–1484.

Kiehl, J. T. and K. E. Trenberth. 1997. Earth's annual global mean energy budget. *Bulletin of the American Meteorological Society* 78: 197–208.

Kiesecker, J. M. 2002. Synergism between trematode infection and pesticide exposure: A link to amphibian limb deformities in nature? *Proceedings of the National Academy of Sciences USA* 99: 9900–9904.

Kinchy, A. J. 2006. On the borders of post-war ecology: Struggles over the Ecological Society of America's preservation committee, 1917–1946. *Science as Culture* 15 (1): 23–44.

King, A. A. and W. M. Schaffer. 2001. The geometry of a population cycle: A mechanistic model of snowshoe hare demography. *Ecology* 82: 814–830.

Kingsland, S. E. 1991. Defining ecology as a science. In *Foundations of Ecology: Classic Papers with Commentaries*, L. A. Real and J. H. Brown (eds.), 1–13. University of Chicago Press, Chicago.

Kinzig, A. P. and J. Harte. 2000. Implications of endemics–area relationships for estimates of species extinctions. *Ecology* 81 (12): 3305–3311.

Knapp, A. K. and 9 others. 2002. Rainfall variability, carbon cycling and plant species diversity in a mesic grassland. *Science* 298: 2202–2205.

Kodric-Brown, A. and J. M. Brown. 1993. Highly structured fish communities in Australian desert springs. *Ecology* 74: 1847–1855.

Koenig, W. D., F. A. Pitelka, W. J. Carmen, R. L. Mumme and M. T. Stanback. 1992. The evolution of delayed dispersal in cooperative breeders. *Quarterly Review of Biology* 67 (2): 111–150.

Kohler, S. L. and M. J. Wiley. 1997. Pathogen outbreaks reveal large-scale effects of competition in stream communities. *Ecology* 78: 2164–2176.

Köhler, W. 1927. *The Mentality of Apes*. Routledge & Kegan Paul, London.

Komdeur, J. 1992. Importance of habitat saturation and territory quality for evolution of cooperative breeding in the Seychelles warbler. *Nature* 358: 493–495.

Kondoh, M. 2003. Foraging adaptation and the relationship between food-web complexity and stability. *Science* 299: 1388–1391.

Korpimäki, E. and K. Norrdahl. 1998. Experimental reduction of predators reverses the crash phase of small-rodent cycles. *Ecology* 79: 2448–2455.

Koskela, T., S. Puustinen, V. Salonen and P. Mutikainen. 2002. Resistance and tolerance in a host plant–holoparasitic plant interaction: Genetic variation and costs. *Evolution* 56: 899–908.

Kozaki, A. and G. Takeba. 1996. Photorespiration protects C_3 plants from photooxidation. *Nature* 384: 557–580.

Kraaijeveld, A. R., K. A. Hutcheson, E. C. Limentani and H. C. J. Godfray. 2001, Costs of counterdefenses to host resistance in a parasitoid of *Drosophila*. *Evolution* 55: 1815–1821.

Kramer, P. J. 1983. *Water Relations of Plants*. Academic Press, New York, NY.

Krebs, C. J. and 7 others. 1995. Impact of food and predation on the snowshoe hare cycle. *Science* 269: 1112–1115.

Kremen, C., V. Razafimahatratra, R. P. Guillery, J. Rakotomalala, A. Weiss and J. S. Ratsisompatrarivo. 1999. Designing the

Masoala National Park in Madagascar based on biological and socioeconomic data. *Conservation Biology* 13 (5): 1055–1068.

Kricher, J. 1998. *A Field Guide to Eastern Forests.* Illustrated by G. Morrison. Houghton Mifflin, New York.

Krümmel, E. M. and 7 others. 2003. Delivery of pollutants by spawning salmon. *Nature* 425: 255.

Krützen, M., J. Mann, M. R. Heithaus, R. C. Connor, L. Bejder and W. B. Sherwin. 2005. Cultural transmission of tool use in bottlenose dolphins. *Proceedings of the National Academy of Sciences USA* 102: 8939–8943.

Kuhnlein, H. V., O. Receveur, D. C. G. Muir, H. M. Chan and R. Soueida. 1995. Arctic indigenous women consume greater than acceptable levels of organochlorines. *Journal of Nutrition* 125: 2501–2510.

L

Lack, D. 1947. The significance of clutch size. *Ibis* 89: 302–352.

Lambeck, R. J. 1997. Focal species: A multi-species umbrella for nature conservation. *Conservation Biology* 11 (4): 849–856.

Lambers, H., F. S. Chapin III and T. L. Pons. 1998. *Plant Physiological Ecology.* Springer, New York.

Lande, R. 1988. Demographic models of the northern spotted owl (*Strix occidentalis caurina*). *Oecologia* 75: 601–607.

Langerhans, R. B., M. E. Gifford and E. O. Joseph. 2007. Ecological speciation in *Gambusia* fishes. *Evolution* 61: 2056–2074.

Larcher, W. 1980. *Physiological Plant Ecology.* Springer-Verlag, Berlin.

Lauenroth, W. K. and O. E. Sala. 1992. Long-term forage production of North American shortgrass steppe. *Ecological Applications* 2: 397–403.

Laurance, W. F. 2001. The hyper-diverse flora of the Central Amazon. In *Lessons from Amazonia: The Ecology and Conservation of a Fragmented Forest,* R. O. Bierregaard Jr., C. Gascon, T. E. Lovejoy and R. C. G. Mesquita (eds.), 47–53. Yale University Press, New Haven, CT.

Laurance, W. F. and 10 others. 2002. Ecosystem decay of Amazonian forest fragments: A 22-year investigation. *Conservation Biology* 16: 605–618.

Law, R. 1975. Colonization and the evolution of life histories in *Poa annua.* Unpublished PhD thesis, University of Liverpool.

Law, R., A. D. Bradshaw and P. D. Putwain. 1977. Life-history variation in *Poa annua. Evolution* 31 (2): 233–246.

Lawler, S. P. 1993. Species richness, species composition and population dynamics of protists in experimental microcosms. *Journal of Animal Ecology* 62: 711–719.

Lawton, J. H. 2000. *Community Ecology in a Changing World.* Excellence in Ecology, O. Kinne (ed.), no. 11. Ecology Institute, Luhe, Germany.

Layne, J. R. and R. E. Lee. 1995. Adaptations of frogs to survive freezing. *Climate Research* 5: 53–59.

LeBlanc, G. A. 1995. Trophic-level differences in the bioconcentration of chemicals: Implications in assessing biomagnification. *Environmental Science and Technology* 29: 154–160.

Lennartsson, T., P. Nilsson and J. Tuomi. 1998. Induction of overcompensation in the field gentian, *Gentianella campestris. Ecology* 79: 1061–1072.

Levin, S. A. 1992. The problem of pattern and scale in ecology. *Ecology* 73 (6): 1943–1967.

Levine, J. M. 2000. Species diversity and biological invasions: Relating local process to community pattern. *Science* 288: 852–854.

Levins, R. 1969. Some demographic and genetic consequences of environmental heterogeneity for biological control. *Bulletin of the Entomological Society of America* 15: 237–240.

Levins, R. 1970. Extinction. In *Some Mathematical Problems in Biology,* M. Gerstenhaber (ed.), 75–107. American Mathematical Society, Providence, RI.

Levinton, J. S. 2001. *Marine Biology: Function, Biodiversity, Ecology.* 2nd ed. Oxford University Press, New York.

Lewison, R. L., L. B. Crowder, A. J. Read and S. A. Freeman. 2004. Understanding impacts of fisheries bycatch on marine megafauna. *Trends in Ecology and Evolution* 19 (11): 598–604.

Lewontin, R. C. 1969. The meaning of stability. In *Diversity and Stability in Ecological Systems,* 13–24. Brookhaven Symposia in Biology, no. 22. Brookhaven National Laboratories, Brookhaven, NY.

Likens, G. E. and F. H. Bormann. 1995. *Biogeochemistry of a Forested Ecosystem.* Springer-Verlag, New York.

Lindeman, R. L. 1942. The trophic–dynamic aspect of ecology. *Ecology* 23: 399–418.

Lindenmayer, D. B. and M. A. McCarthy. 2006. Evaluation of PVA models of arboreal marsupials: Coupling models with long-term monitoring data. *Biodiversity And Conservation* 15 (13): 4079–4096.

Lippincott, C. L. 2000. Effects of *Imperata cylindrica* (L.) Beauv. (cogongrass) invasion on fire regime in Florida sandhill. *Natural Areas Journal* 20: 140–149.

Little, D. P. and D. W. Stevenson. 2007. A comparison of algorithms for the identification of specimens using DNA barcodes: Examples from gymnosperms. *Cladistics* 23: 1–21.

Liu, H., E. S. Menges and P. F. Quintana-Ascencio. 2005. Population viability analyses of *Chamaecrista keyensis*: Effects of fire season and frequency. *Ecological Applications* 15 (1): 210–221.

Lively, C. M. 1989. Adaptation by a parasitic trematode to local populations of its snail host. *Evolution* 43: 1663–1671.

Lomolino, M. V., J. H. Brown and R. Davis. 1989. Island biogeography of montane forest mammals in the American Southwest. *Ecology* 70: 180–194.

Lomolino, M. V., B. R. Riddle and J. H. Brown. 2006. *Biogeography.* 3rd ed. Sinauer Associates, Sunderland, MA.

Loreau, M. and 11 others. 2001. Biodiversity and ecosystem functioning: Current knowledge and future challenges. *Science* 294: 804–808.

Lotka, A. J. 1932. The growth of mixed populations: Two species competing for a common food supply. *Journal of the Washington Academy of Sciences* 22: 461–469.

Louda, S. M., D. Kendall, J. Conner and D. Simberloff. 1997. Ecological effects of an insect introduced for the biological control of weeds. *Science* 277: 1088–1090.

Lovell, H. B. 1958. Baiting of fish by a green heron. *Wilson Bulletin* 70: 280–281.

Lubchenco, J. 1978. Plant species diversity in a marine intertidal community: Importance of herbivore food preference and algal competitive abilities. *American Naturalist* 112: 23–39.

M

MacArthur, R. H. 1955. Fluctuations in animal populations and a measure of community stability. *Ecology* 36: 533–536.

MacArthur, R. H. 1958. Population ecology of some warblers of Northeastern coniferous forests. *Ecology* 39: 599–619.

MacArthur, R. H. 1969. Patterns of communities in the tropics. *Biological Journal of the Linnaean Society* 1: 19–30.

MacArthur, R. H. and J. W. MacArthur. 1961. On bird species diversity. *Ecology* 42: 594–598.

MacArthur, R. H. and E. O. Wilson. 1963. An equilibrium theory of insular zoogeography. *Evolution* 17: 373–387.

MacArthur, R. H. and E. O. Wilson. 1967. *The Theory of Island Biogeography.* Princeton University Press, Princeton, NJ.

MacDonald, I. A. W., L. L. Loope, M. B. Usher and O. Hamann. 1989. Wildlife conservation and the invasion of nature reserves by introduced species: A global perspective. In *Biological Invasions: A Global Perspective*, J. A. Drake, H. A. Mooney, F. di Castri, R. H. Groves, F. J. Kruger, M. Rejmanek and M. Williamson (eds.), 215–256. SCOPE (Scientific Committee on Problems of the Environment), 37. John Wiley & Sons, Chichester.

Mack, R. N. and J. N. Thompson. 1982. Evolution in steppe with few large, hooved mammals. *American Naturalist* 119: 757–773.

Macnair, M. R. and P. Christie. 1983. Reproductive isolation as a pleiotropic effect of copper tolerance in *Mimulus guttatus*. *Heredity* 50: 295–302.

MacPhee, R. D. E. and P. A. Marx. 1997. Humans, hyperdisease, and first-contact extinctions. In *Natural Change and Human Impact in Madagascar*, S. M. Goodman and B. D. Patterson (eds.), 169–217. Smithsonian Institution Press, Washington, DC.

Madigan, M. T. and J. M. Martinko. 2005. *Brock Biology of Microorganisms.* Prentice Hall, Upper Saddle River, NJ.

Madronich, S., R. L. McKenzie, L. O. Bjorn and M. M. Caldwell. 1998. Changes in biologically active ultraviolet radiation reaching the Earth's surface. *Journal of Photochemistry and Photobiology B: Biology* 46: 5–19.

Magurran, A. E. 2004. *Measuring Biological Diversity.* Blackwell Science Publishing, Malden, MA.

Mahowald, N. and 8 others. 2005. The atmospheric global dust cycle and iron inputs to the ocean. *Global Biogeochemical Cycles* 19: GB4025 10.1029/2004GB002402.

Maitland, D. P. 1994. A parasitic fungus infecting yellow dungflies manipulates host perching behaviour. *Proceedings of the Royal Society of London* B 258: 187–193.

Mantua, N. J. 2001. The Pacific Decadal Oscillation. In *The Encyclopedia of Global Environmental Change*, vol. 1, *The Earth System: Physical and Chemical Dimensions of Global Environmental Change*, M. C. McCracken and J. S. Perry (eds.), 592–594. Wiley, New York.

Mantua, N. J. and S. R. Hare. 2002. The Pacific Decadal Oscillation. *Journal of Oceanography* 58: 35–44.

Mantua, N. J., S. R. Hare, Y. Zhang, J. M. Wallace and R. C. Francis. 1997. A Pacific interdecadal climate oscillation with impacts on salmon production. *Bulletin of the American Meteorological Society* 78: 1069–1079.

Margoluis, R. and N. Salafsky. 1998. *Measures of Success: Designing, Managing, and Monitoring Conservation and Development Projects.* Island Press, Washington, DC.

Margulis, L. and K. V. Schwartz. 1988. *Five Kingdoms: An Illustrated Guide to the Phyla of Life on Earth.* 3rd ed. W. H. Freeman, New York.

Maron, J. L. and M. Vila. 2001. When do herbivores affect plant invasion? Evidence for the natural enemies and biotic resistance hypotheses. *Oikos* 95: 361–373.

Marr, J. W. 1967. *Ecosystems of the East Slope of the Front Range of Colorado.* University of Colorado Studies, Series in Biology, no. 8. University of Colorado Press, Boulder.

Marschner, H. 1995. *Mineral Nutrition of Higher Plants.* Academic Press, San Diego, CA.

Marteinsdottir, G. and G. A. Begg. 2002. Essential relationships incorporating the influence of age, size and condition on variables required for estimation of reproductive potential in Atlantic cod *Gadus morhua*. *Marine Ecology Progress Series* 235: 235–256.

Marti, M., R. T. Good, M. Rug, E. Knuepfer and A. F. Cowman. 2004. Targeting malaria virulence and remodeling proteins to the host erythrocyte. *Science* 306: 1930–1933.

Martin, J. A., B. E. Hamilton, P. D. Sutton, S. J. Ventura, F. Menacker and M. L. Munson. 2003. Births: Final data for 2002. National Vital Statistics Reports, Vol. 52 no. 10. Hyattsville, Maryland: National Center for Heath Statistics. (Available at: http://www.cdc.gov/nchs/data/nvsr/nvsr52/nvsr52_10.pdf)

Martin, J. H. and 43 others. 1994. Testing the iron hypothesis in ecosystems of the equatorial Pacific Ocean. *Nature* 371: 123–129.

Martin, P. S. 1984. Prehistoric overkill: The global model. In *Quaternary Extinctions: A Prehistoric Revolution*, P. S. Martin and R. G. Klein (eds.), 354–403. University of Arizona Press, Tucson.

Martin, P. S. 2005. *Twilight of the Mammoths: Ice Age Extinctions and the Rewilding of America.* University of California Press, Berkeley.

Mattson, W. J. Jr. 1980. Herbivory in relation to plant nitrogen content. *Annual Review of Ecology and Systematics* 11: 119–161.

Mauseth, J. D. 1988. *Plant Anatomy.* Benjamin/Cummings, Menlo Park, CA.

May, R. M. 1973. *Stability and Complexity in Model Ecosystems.* Princeton University Press, Princeton, NJ.

May, R. M. 1976. Models for single populations. In *Theoretical Ecology: Principles and Applications*, R. M. May (ed.), 4–25. W. B. Saunders, Philadelphia.

May, R. M. 2006. Threats to tomorrow's world. *Notes and Records of the Royal Society* 60: 109–130.

May, R. M. and R. M. Anderson. 1983. Parasite-host coevolution. In *Coevolution*, D. J. Futuyma and M. Slatkin (eds.), pp. 186–206. Sinauer Associates, Sunderland, MA.

May, R. M., J. H. Lawton and N. E. Stork. 1995. Assessing extinction rates. In *Extinction Rates*, J. H. Lawton and R. M. May (eds.), 1–24. Oxford University Press, Oxford.

Mayr, E. 1956. Geographical character gradients and climatic adaptation. *Evolution* 10: 105–108.

McCaig, A. E., L. A. Glover and J. I. Prosser. 1999. Molecular analysis of bacterial community structure and diversity in unimproved and improved upland grass pastures. *Applied and Environmental Microbiology* 65: 1721–1730.

McCallum, H. and A. Dobson. 1995. Detecting disease and parasite threats to endangered species and ecosystems *Trends in Ecology and Evolution* 10 (5): 190–194.

McClain, M. E. and 11 others. 2003. Biogeochemical hot spots and hot moments at the interface of terrestrial and aquatic ecosystems. *Ecosystems* 6: 301–312.

McEvoy, P., C. Cox and E. Coombs. 1991. Successful biological control of ragwort, *Senecio jacobaea,* by introduced insects in Oregon. *Ecological Applications* 1: 430–442.

McIntyre, S. and R. Hobbs. 1999. A framework for conceptualizing human effects on landscapes and its relevance to management and research models. *Conservation Biology* 13 (6): 1282–1292.

McKane, R. B. and 10 others. 2002. Resource-based niches provide a basis for plant species diversity and dominance in Arctic tundra. *Nature* 415: 68–71.

McMahon, T. E. and J. C. Tash. 1988. Experimental analysis of the role of emigration in population regulation of desert pupfish. *Ecology* 69: 1871–1883.

McNeill, W. H. 1976. *Plagues and Peoples.* Anchor Press/Doubleday, Garden City, New York.

McNeilly, T. 1968. Evolution in closely adjacent plant populations. III. *Agrostis tenuis* on a small copper mine. *Heredity* 23: 99–108.

Medawar, P. B., 1952. *An Unsolved Problem of Biology.* H. K. Lewis, London.

Meffe, G., L. Nielsen, R. L. Knight and D. Schenborn. 2002. *Ecosystem Management: Adaptive, Community-Based Conservation.* Island Press, Washington, DC.

Meine, C., M. Soulé and R. F. Noss. 2006. "A mission-driven discipline": The growth of conservation biology. *Conservation Biology* 20 (3): 631–651.

Meinesz, A. 2001. *Killer Algae.* University of Chicago Press, Chicago.

Meire, P. M. and A. Ervynck. 1986. Are oystercatchers (*Haematopus ostralegus*) selecting the most profitable mussels (*Mytilus edulis*)? *Animal Behaviour* 34: 1427–1435.

Melillo, J. M., J. D. Aber and J. F. Muratore. 1982. Nitrogen and lignin control of hardwood leaf litter decomposition dynamics. *Ecology* 63: 621–626.

Mellars, P. 1989. Major issues in the emergence of modern humans. *Current Anthropology* 30: 349–385.

Memmott, J., N. M. Waser and M. V. Price. 2004. Tolerance of pollination networks to species extinctions. *Proceedings of the Royal Society of London* B 271 (1557): 2605–2611.

Menge, B. A. 1995. Indirect effects in marine rocky intertidal interaction webs: Patterns and importance. *Ecological Monographs* 65: 21–74.

Menge, B. A., E. L. Berlow, C. A. Blanchette, S. A. Navarette and S. B. Yamada. 1994. The keystone concept—Variation in interaction strength in a rocky intertidal habitat. *Ecological Monographs* 64: 249–286.

Menge, B. A., B. Daley and P. A. Wheeler. 1996. Control of interaction strength in marine benthic communities. In *Food Webs: Integration of Patterns and Dynamics,* G. A. Polis and K. O. Winemiller (eds.), 258–274. Chapman and Hall, New York.

Miller, G. T. 2007. *Living in the Environment: Principles, Connections, and Solutions.* Thomson Brooks/Cole, Belmont, CA.

Millennium Ecosystem Assessment. 2005. *Ecosystems and Human Well-being: Biodiversity Synthesis.* World Resources Institute, Washington, DC.

Mokany, K., R. J. Raison and A. S. Prokushkin. 2006. Critical analysis of root:shoot ratios in terrestrial biomes. *Global Change Biology* 12: 84–96.

Molina, M. J. and F. S. Rowland. 1974. Stratospheric sink for chlorofluoromethanes: Chlorine atom-catalysed destruction of ozone. *Nature* 249: 810–812.

Molino, J.-F. and D. Sabatier. 2001. Tree diversity in tropical rain forests: A validation of the intermediate disturbance hypothesis. *Science* 294: 1702–1704.

Moore, J. A. 1957. An embryologist's view of the species concept. In *The Species Problem,* E. Mayr (ed.), 325–388. American Association for the Advancement of Science, Washington DC.

Moore, S. E., T. J. Cole, E. M. E. Poskitt, B. J. Sonko, R. G. Whitehead, I. A. McGregor and A. M. Prentice. 1997. Season of birth predicts mortality in rural Gambia. *Nature* 388: 434.

Morellato, L. P. C. and C. F. B. Haddad. 2000. Introduction: The Brazilian Atlantic forest. *Biotropica* 32: 786–792.

Morin, X., C. Augspurger and I. Chuine. 2007. Process-based modeling of species' distributions: What limits temperate tree species' range boundaries? *Ecology* 88: 2280–2291.

Morris, W. F. and D. F. Doak. 2002. *Quantitative Conservation Biology: Theory and Practice of Population Viability.* Sinauer Associates, Sunderland, MA.

Mouritsen, K. N. and R. Poulin. 2002. Parasitism, community structure and biodiversity in intertidal ecosystems. *Parasitology* 124: S101–S117.

Mouritsen, K. N., L. T. Mouritsen and K. T. Jensen. 1998. Changes of topography and sediment characteristics on an intertidal mud-flat following mass-mortality of the amphipod *Corophium volutator. Journal of the Marine Biological Association of the United Kingdom* 78: 1167–1180.

Moutinho, P., C. D. Napstad and E. A. Davidson. 2003. Influence of leaf-cutting ant nests on secondary forest growth and soil properties in Amazonia. *Ecology* 84: 1265–1276.

Muir, J. 1915. *Travels in Alaska.* Houghton Mifflin, Boston.

Munger, J. 1984. Optimal foraging? Patch use by horned lizards (Iguanidae: Phrynosoma). *American Naturalist* 123: 654–680.

Murdoch, W. W., S. Avery and M. E. B Smith. 1975. Switching in a predatory fish. *Ecology* 56: 1094–1105.

Murphy, D. D. and C. M. Knopp. 2000. *Lake Tahoe Watershed Assessment.* Vol. 1. General Technical Report PSW-GTR-175. Pacific Southwest Research Station, Forest Service, U.S. Department of Agriculture, Albany, CA.

Myers, J. H. and D. Bazely. 1991. Thorns, spines, prickles, and hairs: Are they stimulated by herbivory and do they deter herbivores? In *Phytochemical Induction by Herbivores,* D. W. Tallamy and M. J. Raupp (eds.), 325–344. Wiley, New York.

Myers, R. A. and B. Worm. 2003. Rapid worldwide depletion of predatory fish communities. *Nature* 423 (6937): 280–283.

N

Nabhan, G. P. and A. R. Holdsworth. 1998. *State of the Sonoran Desert Biome: Uniqueness, Biodiversity, Threats and the Adequacy of Protection in the Sonoran Bioregion.* The Wildlands Project, Tucson, AZ.

Nachman, M. W., H. E. Hoekstra and S. L. D'Agostino. 2003. The genetic basis of adaptive melanism in pocket mice. *Proceedings of the National Academy of Sciences USA* 100: 5268–5273.

Naeem, S. L., J. Thompson, S. P. Lawler, J. H. Lawton and R. M. Woodfin. 1994. Declining biodiversity can alter the performance of ecosystems. *Nature* 368: 734–736.

Nager, R. G., P. Monaghan and D. Houston. 2000. Within-clutch trade-offs between the number and quality of eggs: Experimental manipulations in gulls. *Ecology* 81 (5): 1339–1350.

Naiman, R. J., C. A. Johnston and J. C. Kelley. 1988. Alterations of North American streams by beaver. *BioScience* 38: 753–762.

National Research Council. 2002. Effects of trawling and dredging on seafloor habitat. Committee on Ecosystem Effects of Fishing: Phase I—Effects of Bottom Trawling on Seafloor Habitats, Ocean Studies Board, Division on Earth and Life Studies, National Research Council. National Academy of Sciences Press, Washington, DC.

Naughton-Treves, L., J. L. Mena, A. Treves, N. Alvarez and V. C. Radeloff. 2003. Wildlife survival beyond park boundaries: The impact of slash-and-burn agriculture and hunting on mammals in Tambopata, Peru. *Conservation Biology* 17 (4): 1106–1117.

Nee, S., N. Colegrave, S. A. West and A. Grafen. 2005. The illusion of invariant quantities in life histories. *Nature* 309: 1236–1239.

Neff, J. C., R. L. Reynolds, J. Belnap and P. Lamothe. 2005. Multi-decadal impacts of grazing on soil physical and biogeochemical properties in southeastern Utah. *Ecological Applications* 15: 87–95.

Nemani, R. R. and 7 others. 2003. Climate-driven increases in global terrestrial net primary production from 1982 to 1999. *Science* 300: 1560–1563.

Neutel, A.-M., J. A. P. Heesterbeek and P. C. de Ruiter. 2002. Stability in real food webs: Weak links in long loops. *Science* 296: 1120–1123.

Newbold, J. D., J. W. Elwood, R. V. O'Neill and A. L. Sheldon. 1983. Phosphorus dynamics in a woodland stream ecosystem: A study of nutrient spiralling. *Ecology* 64: 1249–1265.

Nicholson, A. J. 1933. The balance of animal populations. *Journal of Animal Ecology* 2: 132–178.

Nicholson, A. J. 1957. The self-adjustment of populations to change. *Cold Spring Harbor Symposia on Quantitative Biology* 22: 153–173.

Nobel, P. S. 1983. *Biophysical Plant Physiology and Ecology*. W. H. Freeman, San Francisco.

Norby, R. J. and 13 others. 2005. Forest response to elevated CO_2 is conserved across a broad range of productivity. *Proceedings of the National Academy of Sciences USA* 102: 18052–18056.

Norris, K. 2004. Managing threatened species: The ecological toolbox, evolutionary theory and declining-population paradigm. *Journal of Applied Ecology* 41 (3): 413–426.

Noss, R., E. Dinerstein, B. Gilbert, M. Gilpin, B. J. Miller, J. Terborgh and S. Trombulak. 1999. Core areas: Where Nature reigns. In *Continental Conservation: Scientific Foundations of Regional Reserve Networks*, M. E. Soulé and J. Terborgh (eds.), 99–128. Island Press, Washington, DC.

O

Oates, J. F., M. Abedi-Lartey, W. S. McGraw, T. T. Struhsaker; and G. H. Whitesides. 2000. Extinction of a West African red colobus monkey. *Conservation Biology* 14 (5): 1526–1532.

O'Brien, E. L., A. E. Burger and R. D. Dawson. 2005. Foraging decision rules and prey species preferences of Northwestern Crows (*Corvus caurinus*). *Ethology* 111: 77–87.

O'Brien, S. J. and W. E. Johnson. 2005. Big cat genomics. *Annual Review of Genomics and Human Genetics* 6: 407–429.

Odum, H. T. and C. F. Jordan. 1970. Metabolism and evapotranspiration of the lower forest in a giant plastic cylinder. In *A Tropical Rain Forest*, H. T. Odum and R. F. Pigeon (eds.), 165–189. Division of Technical Information, U.S. Atomic Energy Commission.

Oechel, W. C., S. J. Hastings, G. Vourlitis, M. Jenkins, G. Riechers and N. Grulke. 1993. Recent change of Arctic tundra ecosystems from a net carbon dioxide sink to a source. *Nature* 361: 520–523.

Oechel, W. C., G. L. Vourlitis, S. J. Hastings, R. C. Zulueta, L. Hinzman and D. Kane. 2000. Acclimation of ecosystem CO_2 exchange in the Alaskan Arctic in response to decadal climate warming. *Nature* 406: 978–981.

Ogren, W. L. 1984. Photorespiration: Pathways, regulation, and modification. *Annual Review of Plant Physiology* 35: 415–442.

Ogutu-Ohwayo, R. 1993. The impact of predation by Nile perch, *Lates niloticus* L., on the fishes of Lake Nabugabo, with suggestions for conservation of endangered endemic cichlids. *Conservation Biology* 7: 701–711.

Okin, G. S., N. Mahowald, O. A. Chadwick and P. Artaxo. 2004. Impact of desert dust on the biogeochemistry of phosphorus in terrestrial ecosystems. *Global Biogeochemical Cycles* 18: GB2005.

Olden, J. D., N. L. Poff, M. R. Douglas, M. E. Douglas and K. D. Fausch. 2004. Ecological and evolutionary consequences of biotic homogenization. *Trends in Ecology and Evolution* 19 (1): 18–24.

Ono, M., T. Igarashi, E. Ohno and M. Sasaki. 1995. Unusual thermal defense by a honeybee against mass attack by hornets. *Nature* 377: 334–336.

Orr, J. C. and 26 others. 2005. Anthropogenic ocean acidification over the twenty-first century and its impact on calcifying organisms. *Nature* 437: 681–686.

Osmond, C. B., K. Winter and H. Ziegler. 1982. Functional significance of different pathways of photosynthesis. In *Physiological plant ecology*, vol. 2, O. L. Lange, P. S. Nobel, C. B. Osmond and H. Ziegler (eds.), 479–547. Encyclopedia of Plant Physiology, new series, vol. 12B. Springer Verlag, New York.

Overpeck, J. T., R. S. Webb and T. Webb. 1992. Mapping Eastern North American vegetation change of the past 18,000 years: No-analogs and the future. *Geology* 20: 1071–1074.

Owen-Smith, N. 1987. Pleistocene extinctions: The pivotal role of megaherbivores. *Paleobiology* 13: 351–362.

P

Pace, M. L., J. J. Cole, S. R. Carpenter and J. F. Kitchell. 1999. Trophic cascades revealed in diverse ecosystems. *Trends in Ecology and Evolution* 14: 483–488.

Packer, C., A. E. Pusey, H. Rowley, D. A. Gilbert, J. Martenson and S. J. O'Brien. 1991. Case study of a population bottleneck: Lions of the Ngorongoro Crater. *Conservation Biology* 5: 219–230.

Paine, R. T. 1966. Food web complexity and species diversity. *American Naturalist* 100: 65–75.

Paine, R. T. 1969. A note on trophic complexity and community stability. *American Naturalist* 103: 91–93.

Paine, R. T. 1974. Intertidal community structure: Experimental studies on the relations between a dominant competitor and its principal predator. *Oecologia* 15: 93–120.

Paine, R. T. 1979. Disaster, catastrophe, and local persistence of the sea palm, *Postelsia palmaeformis*. *Science* 205: 685–687.

Paine, R. T., J. C. Castillo and J. Cancino. 1985. Perturbation and recovery patterns of starfish-dominated intertidal assemblages in Chile, New Zealand, and Washington State. *American Naturalist* 125: 679–691.

Palumbi, S. R. 2001. Humans as the world's greatest evolutionary force. *Science* 293: 1786–1790.

Pandolfi, J. M. and 11 others. 2003. Global trajectories of the long-term decline of coral reef ecosystems. *Science* 301: 955–958.

Pardini, R. 2004. Effects of forest fragmentation on small mammals in an Atlantic Forest landscape. *Biodiversity and Conservation* 13: 2567–2586.

Park, T. 1948. Experimental studies of interspecies competition. I. Competition between populations of the flour beetles, *Tribolium confusum* Duvall and *Tribolium castaneum* Herbst. *Ecological Monographs* 18: 267–307.

Parmenter, A. W. and 8 others. 2003. Land use and land cover change in the Greater Yellowstone Ecosystem: 1975–1995. *Ecological Applications* 13 (3): 687–703.

Parmesan, C. and G. Yohe. 2003. A globally coherent fingerprint of climate change impacts across natural systems. *Nature* 421: 37–42.

Parmesan, C. and 12 others. 1999. Poleward shifts in geographical ranges of butterfly species associated with regional warming. *Nature* 399: 579–583.

Pastor, J., R. J. Naiman, B. Dewey and P. Mcinnes. 1988. Moose, microbes, and the boreal forest. *BioScience* 38 (11): 770–777.

Paton, P. W. C. 1994. The effect of edge on avian nest success: How strong is the evidence? *Conservation Biology* 8: 17–26.

Patterson, B. D. 1980. Montane mammalian biogeography in New Mexico. *Southwestern Naturalist* 25: 33–40.

Patterson, B. D. 1981. Morphological shifts of some isolated populations of *Eutamias* (Rodentia: Sciuridae) in different congeneric assemblages. *Evolution* 35: 53–66.

Paul, E. A. and F. E. Clark. 1996. *Soil Microbiology and Biochemistry*. Academic Press, San Diego, CA.

Paul, N. D. and D. Gwynn-Jones. 2003. Ecological roles of solar UV radiation: Towards an integrated approach. *Trends in Ecology and Evolution* 18: 48–55.

Paulay, G. 1997. Diversity and distribution of reef organisms. In *Life and Death of Coral Reefs*, C. Birkeland (ed.), 298–353. Chapman & Hall, New York.

Pearce, F. 1997. Why is the apparently pristine Arctic full of toxic chemicals that started off thousands of kilometres away? *New Scientist* 154: 24–27.

Pearcy, R. W. 1977. Acclimation of photosynthetic and respiratory carbon dioxide exchange to growth temperature in *Atriplex lentiformis* (Torr) Wats. *Plant Physiology* 59: 795–799.

Pearl, R. and L. J. Reed. 1920. On the rate of growth of the population of the United States since 1790 and its mathematical representation. *Proceedings of the National Academy of Sciences USA* 6: 275–288.

Peek, A. S., R. A. Feldman, R. A. Lutz and R. C. Vrijenhoek. 1998. Cospeciation of chemoautotrophic bacteria and deep sea clams. *Proceedings of the National Academy of Sciences USA* 95: 9962–9966.

Pellmyr, O. and C. J. Huth. 1994. Evolutionary stability of mutualism between yuccas and yucca moths. *Nature* 372: 257–260.

Pellmyr, O., C. C. Labandeira and C. M. Herrera, eds. 2002. *Plant–Animal Interactions: An Evolutionary Approach*. Blackwell Science Publishing, Malden, MA.

Perez, J. M., J. E. Granados, R. C. Soriguer, P. Fandos, F. J. Marquez and J. P. Crampe. 2002. Distribution, status and conservation problems of the Spanish Ibex, *Capra pyrenaica* (Mammalia: Artiodactyla). *Mammal Review* 32 (1): 26–39.

Peterson, B. J. and B. Fry. 1987. Stable isotopes in ecological studies. *Annual Review of Ecology and Systematics* 18: 293–320.

Peterson, G., C. R. Allen and C. S. Holling. 1998. Ecological resilience, biodiversity, and scale. *Ecosystems* 1: 6–8.

Pfennig, D. W. 1992. Proximate and functional causes of polyphenism in an anuran tadpole. *Functional Ecology* 6 (2): 167–174.

Pickett, S. T. A. 1989. Space-for-time substitution as an alternative to long-term studies. In *Long-term Studies in Ecology: Approaches and Alternatives*, G. E. Likens (ed.), 110–135. Springer-Verlag, New York.

Pielou, E. C. 1979. *Biogeography*. Wiley, New York.

Pimentel, D., E. H. Feinberg, P. W. Wood and J. T. Hayes. 1965. Selection, spatial distribution, and the coexistence of competing fly species. *American Naturalist* 99: 97–109.

Pimentel, D., L. Lach, R. Zuniga and D. Morrison. 2000. Environmental and economic costs of nonindigenous species in the United States. *BioScience* 50: 53–65.

Pimm, S. L. 2002. *Food Webs*. University of Chicago Press, Chicago.

Pimm, S. L., H. L. Jones and J. Diamond. 1988. On the risk of extinction. *American Naturalist* 132: 757–785.

Pinder, A. W., K. B. Storey and G. R. Ultsch. 1992. Estivation and hibernation. In *Environmental Physiology of the Amphibia*, M. E. Feder and W. W. Burggren (eds.), 250–274. University of Chicago Press, Chicago.

Polis, G. A. 1991. Complex trophic interactions in deserts: An empirical critique of food web theory. *American Naturalist* 138: 123–155.

Pollock, M. M., R. J. Naiman and T. A. Hanley. 1998. Plant species richness in riparian wetlands–a test of biodiversity theory. *Ecology* 79: 94–105.

Post, D. M. 2002. The long and short of food-chain length. *Trends in Ecology and Evolution* 17: 269–277.

Poulsen, M. and J. J. Boomsma. 2005. Mutualistic fungi control crop diversity in fungus-growing ants. *Science* 307: 741–744.

Pounds, J. A., M. L. P. Fogden and J. H. Campbell. 1999. Biological response to climate change on a tropical mountain. *Nature* 398: 611–615.

Power, M. 1995. Floods, food chains, and ecosystem processes in rivers. In *Linking Species and Ecosystems*, C. G. Jones and J. H. Lawton (eds.), 52–60. Chapman and Hall, New York.

Power, M. E., M. S. Parker and J. T. Wootton. 1996a. Disturbance and food chain length in rivers. In *Food Webs: Integration of Patterns and Dynamics*, G. A. Polis and K. O. Winemiller (eds.), 286–297. Chapman and Hall, New York.

Power, M. E. and 9 others. 1996b. Challenges in the quest for keystones. *BioScience* 46: 609–620.

Primack, R. B. and S. L. Miao. 1992. Dispersal can limit local plant distribution. *Conservation Biology* 6: 513–519.

R

Raab, T. K., D. A. Lipson and R. K. Monson. 1996. Non-mycorrhizal uptake of amino acids by roots of the alpine sedge *Kobresia myosuroides*: Implications for the alpine nitrogen cycle. *Oecologia* 108: 488–494.

Rabinowitz, D., S. Cairns and T. Dillon. 1986. Seven forms of rarity and their frequency in the flora of the British Isles. In *Conservation Biology: The Science of Scarcity and Diversity*, M. E. Soulé (ed.), 182–204. Sinauer Associates, Sunderland, MA.

Rahel, F. J. 2000. Homogenization of fish faunas across the United States. *Science* 288 (5467): 854–856.

Ralls, K. and J. D. Ballou. 2004. Genetic status and management of California Condors. *Condor* 106 (2): 215–228.

Ranta, P., T. Blom, J. Niemela, E. Joensuu and M. Siitonen. 1998. The fragmented Atlantic rain forest of Brazil: Size, shape and distribution of forest fragments. *Biodiversity and Conservation* 7 (3): 385–403.

Raphael, D. O. and P. S. Nobel. 1986. Growth and survivorship of *Agave deserti*: Influences of parent–ramet connections. *Botanical Gazette* 147: 78–83.

Raxworthy, C. J., E. Martinez-Meyer, N. Horning, R. A. Nussbaum, G. E. Schneider, M. A. Ortega-Huerta and A. T. Peterson. 2003. Predicting distributions of known and unknown reptile species in Madagascar. *Nature* 426: 837–841.

Raymond, M., C. Chevillon, T. Guillemaud, T. Lenormand and N. Pasteur. 1998. An overview of the evolution of overproduced esterases in the mosquito *Culex pipiens*. *Philosophical Transactions of the Royal Society of London* B 353: 1707–1711.

Redford, K. H. 1992. The empty forest. *BioScience* 42 (6): 412–422.

Redman, R. S., K. B. Sheehan, R. G. Stout, R. J. Rodriguez and J. M. Henson. 2002. Thermotolerance generated by plant/fungal symbiosis. *Science* 298: 1581.

Rees, W. E. 1992. Ecological footprints and appropriated carrying capacity: What urban economics leaves out. *Environment and Urbanization* 4: 121–130.

Reich, P. B., M. B. Walters and D. S. Ellsworth. 1997. From tropics to tundra: Global convergence in plant functioning. *Proceedings of the National Academy of Sciences USA* 94: 13730–13734.

Reiners, W. A., I. A. Worley and D. B. Lawrence. 1971. Plant diversity in a chronosequence at Glacier Bay, Alaska. *Ecology* 52: 55–69.

Relyea, R. A. 2003. Predator cues and pesticides: A double dose of danger for amphibians. *Ecological Applications* 13: 1515–1521.

Reusch, T. B. H., M. A. Häberli, P. B. Aeschlimann and M. Milinski. 2001. Female sticklebacks count alleles in a strategy of sexual selection explaining MHC polymorphism. *Nature* 414: 300–302.

Reznick, D. N., M. J. Bryant, D. Roff, C. K. Ghalambor and D. E. Ghalambor. 2004. Effect of mortality on the evolution of senescence in guppies. *Nature* 431: 1095–1099.

Ricciardi, A., R. J. Neves and J. B. Rasmussen. 1998. Impending extinctions of North American freshwater mussels (Unionoida) following the zebra mussel (*Dreissena polymorpha*) invasion. *Journal of Animal Ecology* 67 (4): 613–619.

Richardson, H. and N. A. M. Verbeek. 1986. Diet selection and optimization by northwestern crows feeding on Japanese littleneck clams. *Ecology* 67: 1219–1226.

Richter, B. D., D. P. Braun, M. A. Mendelson and L. L. Master. 1997. Threats to imperiled freshwater fauna. *Conservation Biology* 11 (5): 1081–1093.

Ricklefs, R. E. 2001. *The Economy of Nature*. 5th ed. W. H. Freeman, New York.

Rieseberg, L. H., A. Widmer, A. M. Arntz and J. M. Burke. 2002. Directional selection is the primary cause of phenotypic diversification. *Proceedings of the National Academy of Sciences USA* 99: 12242–12245.

Rieseberg, L. H. and 9 others. 2003. Major ecological transitions in wild sunflowers facilitated by hybridization. *Science* 301: 1211–1216.

Riley, S. P. D., H. B. Shaffer, S. R. Voss and B. M. Fitzpatrick. 2003. Hybridization between a rare, native tiger salamander and its introduced congener. *Ecological Applications* 13: 1263–1275.

Ripple, W. J. and R. L. Beschta. 2004. Wolves and the ecology of fear: Can predation risk structure ecosystems? *BioScience* 54 (8): 755–766.

Ripple, W. J. and R. L. Beschta. 2007. Restoring Yellowstone's aspen with wolves. *Biological Conservation* 138 (3–4): 514–519.

Rizki, R. M. and T. M. Rizki. 1990. Parasitoid virus-like particles destroy *Drosophila* cellular immunity. *Proceedings of the National Academy of Sciences USA* 87: 8388–8392.

Robertson, G. P., M. A. Huston, F. C. Evans and J. M. Tiedje. 1988. Spatial variability in a successional plant community: Patterns of nitrogen availability. *Ecology* 69: 1517–1524.

Rohde, K. 2005. *Nonequilibrium Ecology*. Cambridge University Press, Cambridge.

Rojstaczer, S., S. M. Sterling and N. J. Moore. 2001. Human appropriation of photosynthesis products. *Science* 294 (5551): 2549–2552.

Root, R. B. and N. Cappuccino. 1992. Patterns in population change and the organization of the insect community associated with goldenrod. *Ecological Monographs* 62: 393–420.

Root, T. L., J. T. Price, K. R. Hall, S. H. Schneider, C. Rosenzweig and J. A. Pounds. 2003. Fingerprints of global warming on wild animals and plants. *Nature* 421: 57–60.

Rosenzweig, M. L. 1992. Species diversity gradients: We know more and less than we thought. *Journal of Mammology* 73: 715–730.

Ross, P. S. 2006. Fireproof killer whales (*Orcinus orca*): Flame-retardant chemicals and the conservation imperative in the charismatic icon of British Columbia, Canada. *Canadian Journal of Fisheries and Aquatic Sciences* 63 (1): 224–234.

Rovira, A. D., C. D. Bowen and R. C. Foster. 1983. The significance of rhizosphere microflora and mycorrhizas in plant nutrition. In *Inorganic Plant Nutrition* (*Encyclopedia of Plant Physiology*, new series, Vol. 15B), A. Läuchli and R. L. Bieleski (eds.), 61–93. Springer-Verlag, Berlin.

Ruddiman, W. F. 2003. The anthropogenic greenhouse era began thousands of years ago. *Climatic Change* 61: 261–293.

Ryan, M. G., D. Binkley and J. H. Fownes. 1997. Age-related decline in forest productivity: Pattern and process. *Advances in Ecological Research* 27: 213–262.

S

Sale, P. F. 1977. Maintenance of high diversity in coral reef fish communities. *American Naturalist* 111: 337–359.

Sale, P. F. 1979. Recruitment, loss and coexistence in a guild of territorial coral reef fishes. *Oecologia* 42: 159–177.

Salisbury, F. B. and C. Ross. 1992. *Plant Physiology*. 4th ed. Wadsworth, Belmont, CA.

Sanderson, E. W., M. Jaiteh, M. A. Levy, K. H. Redford, A. V. Wannebo and G. Woolmer. 2002. The human footprint and the last of the wild. *BioScience* 52 (10): 891–904.

Sandquist, D. R. and J. R. Ehleringer. 2003. Population- and family-level variation of brittlebush (*Encelia farinosa*, Asteraceae) pubescence: Its relation to drought and implications for selection in variable environments. *American Journal of Botany* 90: 1481–1486.

Sargent, R. D. 2004. Floral symmetry affects speciation rates in angiosperms. *Proceedings of the Royal Society of London* B 271: 603–608.

Saunders, D. A., R. J. Hobbs and G. W. Arnold. 1993. The Kellerberrin project on fragmented landscapes: A review of current information. *Biological Conservation* 64: 185–192.

Sax, D. F. and S. D. Gaines. 2003. Species diversity: From global decreases to local increases. *Trends in Ecology and Evolution* 18 (11): 561–566.

Schall, J. J. 1992. Parasite-mediated competition in *Anolis* lizards. *Oecologia* 92: 58–64.

Schimel, J. P. and J. Bennett. 2004. Nitrogen mineralization: Challenges of a changing paradigm. *Ecology* 85: 591–602.

Schindler, D. W. 1974. Eutrophication and recovery in experimental lakes: Implications for lake management. *Science* 184: 897–899.

Schlesinger, W. H. 1997. *Biogeochemistry: An Analysis of Global Change*. 2nd ed. Acadmic Press, San Diego, CA.

Schluter, D. 1994. Experimental evidence that competition promotes divergence in adaptive radiation. *Science* 266: 798–801.

Schluter, D. 1998. Ecological causes of speciation. In *Endless Forms: Species and Speciation*, D. J. Howard and S. H. Berlocher (eds.), 114–129. Oxford University Press, Oxford.

Schluter, D., T. D. Price and P. R. Grant. 1985. Ecological character displacement in Darwin's finches. *Science* 227: 1056–1059.

Schmidt-Nielsen, B. and K. Schmidt-Nielsen. 1951. A complete account of the water metabolism in kangaroo rats and an experimental verification. *Journal of Cellular and Comparative Physiology* 38: 165–182.

Schmidt-Nielsen, K. 1964. *Desert Animals: Physiological Problems of Heat and Water*. Clarendon Press, Oxford.

Schmidt-Nielsen, K. 1997. *Animal Physiology, Adaptation and Environment*. Cambridge University Press, Cambridge.

Schnee, C., T. G. Kollner, M. Held, T. C. J. Turlings, J. Gershenzon and J. Degenhardt. 2006. The products of a single maize sesquiterpene synthase form a volatile defense signal that attracts natural enemies of maize herbivores. *Proceedings of the National Academy of Sciences USA* 103: 1129–1134.

Schoener, T. W. 1974. Resource partitioning in ecological communities. *Science* 185: 27–39.

Schoener, T. W. 1983. Field experiments on interspecific competition. *American Naturalist* 122: 240–285.

Schoener, T. W. and D. A. Spiller. 1996. Devastation of prey diversity by experimentally introduced predators in the field. *Nature* 381: 691–694.

Scholander, P. F., V. Walters, R. Hock and L. Irving. 1950. Body insulation of some arctic and tropical mammals and birds. *Biological Bulletin* 99: 225–236.

Schtickzelle, N. and M. Baguette. 2003. Behavioural responses to habitat patch boundaries restrict dispersal and generate emigration–patch area relationships in fragmented landscapes. *Journal of Animal Ecology* 72 (4): 533–545.

Schuur, E. A. G. 2003. Productivity and global climate revisited: The sensitivity of tropical forest growth to precipitation. *Ecology* 84: 1165–1170.

Schwartz, M. W. 1999. Choosing the appropriate scale of reserves for conservation. *Annual Review of Ecology and Systematics* 30: 83–108.

Schwarzkopf, L. and R. Shine. 1992. Costs of reproduction in lizards: Escape tactics and susceptibility to predation. *Behavioral Ecology and Sociobiology* 31: 17–25.

Seehausen, O., J. J. M. van Alphen and F. Witte. 1997. Cichlid fish diversity threatened by eutrophication that curbs sexual selection. *Science* 277 (5333): 1808–1811.

Selosse, M.-A. and F. Le Tacon. 1998. The land flora: A phototroph–fungus partnership? *Trends in Ecology and Evolution* 13: 15–20.

Sessions, S. K. and S. B. Ruth. 1990. Explanation for naturally occurring supernumary limbs in amphibians. *Journal of Experimental Zoology* 254: 38–47.

Shank, T. M., D. J. Fornari, K. L. Von Damm, M. D. Lilley, R. M. Haymon and R. A. Lutz. 1998. Temporal and spatial patterns of biological community development at nascent deep-sea hydrothermal vents (9°N, East Pacific Rise). *Deep-Sea Research* II 45: 465–516.

Shmida, A. and M. V. Wilson. 1985. Biological determinants of species diversity. *Journal of Biogeography* 12: 1–20.

Sidor, C. A. 2003. Evolutionary trends and the origin of the mammalian lower jaw. *Paleobiology* 29: 605–640.

Siepielski, A. M. and C. W. Benkman. 2007. Convergent patterns in the selection mosaic for two North American bird-dispersed pines. *Ecological Monographs* 77 (2): 203–220.

Sih, A. and B. Christensen. 2001. Optimal diet theory: When does it work, and when and why does it fail? *Animal Behavior* 61: 379–390.

Simberloff, D. 2003. Eradication: Preventing invasions at the outset. *Weed Science* 51 (2): 247–253.

Simberloff, D. and J. Cox. 1987. Consequences and costs of conservation corridors. *Conservation Biology* 1 (1): 63–71.

Simberloff, D. S. and E. O. Wilson. 1969. Experimental zoogeography of islands: The colonization of empty islands. *Ecology* 50: 278–296.

Simenstad, C. A., J. A. Estes and K. W. Kenyon. 1978. Aleuts, sea otters, and alternative communities. *Science* 200: 403–411.

Sinervo, B. 1990. The evolution of maternal investment in lizards: An experimental and comparative analysis of egg size and its effects on offspring performance. *Evolution* 44: 279–294.

Sjögren-Gulve, P. 1994. Distribution and extinction patterns within a northern metapopulation of the pool frog, *Rana lessonae*. *Ecology* 75: 1357–1367.

Slatyer, R. D. 1967. *Plant–Water Relationships*. Academic Press, New York.

Smil, V. 2002. *The Earth's Biosphere: Evolution, Dynamics and Change*. MIT Press, Cambridge, MA.

Smith, A. T. 1974. The distribution and dispersal of pikas: Consequences of insular population structure. *Ecology* 55:1112–1119.

Smith, F. D. M., R. M. May, R. Pellew, T. H. Johnson and K. R. Walter. 1993. How much do we know about the current extinction rate? *Trends in Ecology and Evolution* 8 (10): 375–378.

Smith, F. E. 1961. Density-dependence in the Australian thrips. *Ecology* 42: 403–407.

Smith, L. J., A. T. Holycross, C. W. Painter and M. E. Douglas. 2001. Montane rattlesnakes and prescribed fire. *Southwestern Naturalist* 46: 54–61.

Smith, S. M. 1975. Innate recognition of coral snake pattern by a possible avian predator. *Science* 187: 759–760.

Smith, T. B. 1993. Disruptive selection and the genetic basis of bill size polymorphism in the African finch *Pyrenestes*. *Nature* 363: 618–620.

Snyder, N. F. R., S. R. Derrickson, S. R. Beissinger, J. W. Wiley, T. B. Smith, W. D. Toone and B. Miller. 1996. Limitations of captive breeding in endangered species recovery. *Conservation Biology* 10 (2): 338–348.

Sokal, R. R. and F. J. Rohlf. 1995. *Biometry*. 3rd ed. W. H. Freeman, New York.

Solé, R. V. and J. Montoya. 2001. Complexity and fragility in ecological networks. *Proceedings of the Royal Society of London* B 268: 2039–2045.

Soulé, M. E. 1985. What is conservation biology? *BioScience* 35 (11): 727–734.

Soulé, M. E. and M. A. Sanjayan. 1998. Conservation targets: Do they help? *Science* 279 (5359): 2060–2061.

Soulé, M. E. and B. A. Wilcox (eds.). 1980. *Conservation Biology: An Evolutionary–Ecological Perspective*. Sinauer Associates, Sunderland, MA.

Sousa, W. P. 1979a. Disturbance in marine intertidal boulder fields: The non-equilibrium maintenance of species diversity. *Ecology* 60: 1225–1239.

Sousa, W. P. 1979b. Experimental investigations of disturbance and ecological succession in a rocky intertidal algal community. *Ecological Monographs* 49: 227–254.

Sousa, W. P. 2001. Natural disturbance and the dynamics of marine benthic communities. In *Marine Community Ecology*, M. D. Bertness, S. D. Gaines and M. E. Hay (eds.), 85–130. Sinauer Associates, Sunderland, MA.

Spencer, K. A. 1972. Handbooks for the identification of British insects, *Diptera, Agromyzidae*. (Vol. X, Part 5 g). Royal Entomological Society, London.

Spielman, D., B. W. Brook and R. Frankham. 2004. Most species are not driven to extinction before genetic factors impact them. *Proceedings of the National Academy of Sciences USA* 101 (42): 15261–15264.

Sprent, J. I. 1979. *The Biology of Nitrogen Fixing Organisms*. McGraw Hill, London.

Springer, A. M. and 7 others. 2003. Sequential megafaunal collapse in the North Pacific Ocean: An ongoing legacy of industrial whaling? *Proceedings of the National Academy of Sciences USA* 100: 12223–12228.

Steadman, D. W. 1995. Prehistoric extinctions of Pacific island birds—Biodiversity meets zooarchaeology. *Science* 267 (5201): 1123–1131.

Stebbins, G. L. 1974. *Flowering Plants: Evolution Above The Species Level*. Belknap, Cambridge, MA.

Stein, B. A., L. S. Kutner and J. S. Adams. 2000. *Precious Heritage*. Oxford University Press, Oxford.

Steneck, R. S., S. D. Hacker and M. N. Dethier. 1991. Mechanisms of competitive dominance between crustose coralline algae: An herbivore-mediated competitive reversal. *Ecology* 72: 938–950.

Stevens, C. J., N. B. Dise, J. O. Mountford and D. J. Gowling. 2004. Impact of nitrogen deposition on the species richness of grasslands. *Science* 303: 1876–1879.

Stevens, O. A. 1932. The number and weight of seeds produced by weeds. *American Journal of Botany* 19: 784–794.

Stiling, P. D. 2002. *Ecology: Theories and Applications*. 4th ed. Prentice-Hall, Upper Saddle River, NJ.

Stinner, J. N. and V. H. Shoemaker. 1987. Cutaneous gas exchange and low evaporative water loss in the frogs *Phyllomedusa sauvagei* and *Chiromantis xerampelina*. *Journal of Comparative Physiology* 157: 423–427.

Storey, K. B. 1990. Life in a frozen state: Adaptive strategies for natural freeze tolerance in amphibians and reptiles. *American Journal of Physiology* 258: R559–R568.

Strahler, A. and A. Strahler. 2005. *Physical Geography*. 3rd ed. John Wiley & Sons, Hoboken, NJ.

Strayer, D. L. and H. M. Malcom. 2007. Effects of zebra mussels (*Dreissena polymorpha*) on native bivalves: The beginning of the end or the end of the beginning? *Journal of the North American Benthological Society* 26 (1): 111–122.

Strebel, P. M. and S. L. Cochi. 2001. Waving goodbye to measles. *Nature* 414: 695–696.

Suding, K. and 7 others. 2005. Functional and abundance-based mechanisms explain diversity loss due to nitrogen fertilization. *Proceedings of the National Academy of Sciences USA* 102: 4387–4392.

Sutherland, J. P. 1974. Multiple stable points in natural communities. *American Naturalist* 108: 859–873.

Symons, G., 1888. *The Eruption of Krakatau and Subsequent phenomena: Reports of the Krakatau Committee of the Royal Society*. Trubner & Co., London.

T

Taiz, L. and E. Zeiger. 2006. *Plant Physiology*. 4th ed. Sinauer Associates, Sunderland, MA.

Tanabe, S. 2002. Contamination and toxic effects of persistent endocrine disrupters in marine mammals and birds. *Marine Pollution Bulletin* 45 (1–12): 69–77.

Tansley, A. G. 1917. On competition between *Galium saxatile* L. (*G. hercynicum* Weig.) and *Galium sylvestre* Poll. (*G. asperum* Schreb.) on different types of soil. *Journal of Ecology* 5: 173–179.

Tansley, A. G. 1935. The use and abuse of vegetational concepts and terms. *Ecology* 16: 284–307.

Tebbich, S., M. Taborsky, B. Fessl and D. Blomqvist. 2001. Do wood-pecker finches acquire tool-use by social learning? *Proceedings of the Royal Society of London* B 268: 2189–2193.

Terborgh, J. 1973. On the notion of favorableness in plant ecology. *American Naturalist* 107: 481–501.

Terborgh, J., K. Feeley, M. Silman, P. Nunez and B. Balukjian. 2006. Vegetation dynamics of predator-free land-bridge islands. *Journal of Ecology* 94 (2): 253–263.

Tewksbury, J. J. and 9 others. 2002. Corridors affect plants, animals, and their interactions in fragmented landscapes. *Proceedings of the National Academy of Sciences USA* 99 (20): 12923–12926.

Thomas, C. D. and T. M. Jones. 1993. Partial recovery of a skipper butterfly (*Hesperia comma*) from population refuges: Lessons for conservation in a fragmented landscape. *Journal of Animal Ecology* 62: 472–481.

Thomas, F., A. Schmidt-Rhaesa, G. Martin, C. Manu, P. Durand and F. Renaud. 2002. Do hairworms (Nematomorpha) manipulate the water seeking behaviour of their terrestrial hosts? *Journal of Evolutionary Biology* 15: 356–361.

Thomas, F. and 7 others. 2003. Biochemical and histological changes in the brain of the cricket *Nemobius sylvestris* infected by the manipulative parasite *Paragordius tricuspidatus* (Nematomorpha). *International Journal for Parasitology* 33: 435–443.

Thompson, J. N. 1998. Rapid evolution as an ecological process. *Trends in Ecology and Evolution* 13: 329–332.

Thrall, P. H. and J. J. Burdon. 2003. Evolution of virulence in a plant host–pathogen metapopulation. *Science* 299: 1735–1737.

Tilman, D. 1977. Resource competition between plankton algae: An experimental and theoretical approach. *Ecology* 58: 338–348.

Tilman, D. 1997. Community invisibility, recruitment limitation, and grassland biodiversity. *Ecology* 78: 81–92.

Tilman, D. and J. A. Downing. 1994. Biodiversity and stability in grasslands. *Nature* 367: 363–365.

Tilman, D., M. Mattson and S. Langer. 1981. Competition and nutrient kinetics along a temperature gradient: An experimental test of a mechanistic approach to niche theory. *Limnology and Oceanography* 26: 1020–1033.

Tilman, D., D. Wedin and J. Knops. 1996. Productivity and sustain-ability influenced by biodiversity in grassland ecosystems. *Nature* 379: 718–720.

Tilman, D., J. Hill and C. Lehman. 2006. Carbon-negative biofuels from low-input high-diversity grassland biomass. *Science* 314: 1598.

Tinker, D. B., W. H. Romme and D. G. Despain. 2003. Historic range of variability in landscape structure in subalpine forests of the Greater Yellowstone Area, USA. *Landscape Ecology* 18: 427–439.

Townsend, C. R., M. R. Scarsbrook and S. Doledec. 1997. The inter-mediate disturbance hypothesis, refugia, and biodiversity in streams. *Limnology and Oceanography* 42: 938–949.

Tracy, R. L. and G. E. Walsberg. 2002. Kangaroo rats revisited: Re-evaluating a classic case of desert survival. *Oecologia* 133: 449–457.

Trites, A. W. and C. P. Donnelly. 2003. The decline of Steller seal lions *Eumetopias jubatus* in Alaska: A review of the nutritional stress hypothesis. *Mammal Review* 33: 3–28.

Tsukaya, H., K. Fujikawa and S. Wu. 2002. Thermal insulation and accumulation of heat in the downy inflorescences of *Saussurea medusa* (Asteraceae) at high elevation in Yunnan, China. *Journal Plant Research* 115: 263–268.

Turchin, P. 1990. Rarity of density dependence or population regu-lation with lags? *Nature* 344: 660–663.

Turchin, P. 2003. *Complex Population Dynamics: A Theoretical/Empirical Synthesis.* Princeton University Press, Princeton, NJ.

Turner, M. G. 2005. Landscape ecology in North America: Past, present and future. *Ecology* 86 (8): 1967–1974.

Turner, M. G., R. H. Gardner and R. V. O'Neill. 2001. *Landscape Ecology in Theory and Practice: Pattern and Process.* Springer Verlag, New York.

Turner, M. G., W. H. Romme and D. B. Tinker. 2003. Surprises and lessons from the 1988 Yellowstone fires. *Frontiers in Ecology and the Environment* 1 (7): 351–358.

U

Uddenberg, N., T. Fagerström and A. Jeffner. 1995. Kontakten med urkällan: Naturen, människan och religionen. In *Det Stora Sammanhanget*, N. Uddenberg (ed.), 141–186. Doxa, Nora, Sweden.

Umina, P. A., A. R. Weeks, M. R. Kearney, S. W. McKechnie and A. A. Hoffmann. 2005. A rapid shift in a classic clinal pattern in *Drosophila* reflecting climate change. *Science* 308: 691–693.

U.S. Congress, Office of Technology Assessment. 1993. *Harmful Non-Indigenous Species in the United States.* OTA-F-565. Washington, DC: U.S. Government Printing Office.

U.S. Environmental Protection Agency. 2005. *Evaluating Ozone Control Programs in the Eastern United States: Focus on the NO_x Budget Trading Program, 2004.* EPA454-K-05–001. http://www.epa.gov/airtrends/2005/ozonenbp.pdf.

V

Van Breemen, N. and A. C. Finzi. 1998. Plant-soil interactions: Ecological aspects and evolutionary implications. *Biogeochemistry* 42: 1–19.

van den Bosch, F., R. Hengeveld and J. A. J. Metz. 1992. Analysing the velocity of animal range expansion. *Journal of Biogeography* 19: 135–150.

van der Heijden, M. G. A. and 7 others. 1998. Mycorrhizal fungal diversity determines plant biodiversity, ecosystem variability and productivity. *Nature* 396: 69–72.

Vander Zanden, M. J. and J. B. Rasmussen. 1996. A trophic position model of pelagic food webs: Impact on contaminant bioaccu-mulation in lake trout. *Ecological Monographs* 66 (4): 451–471.

Van Dover, C. L. 2000. *The Ecology of Deep-Sea Hydrothermal Vents.* Princeton University Press, Princeton, NJ.

Van Lear, D. H., W. D. Carroll, P. R. Kapeluck and R. Johnson. 2005. History and restoration of the longleaf pine–grassland ecosys-tem: Implications for species at risk. *Forest Ecology and Management* 211: 150–165.

Vannote, R. L., G. W. Minshall, K. W. Cummins, K. R. Sedell and C. E. Cushing. 1980. The River Continuum Concept. *Canadian Journal of Fisheries and Aquatic Sciences* 37: 130–137.

Van Tyne, J. and A. J. Berger. 1959. *Fundamentals of Ornithology*. Wiley, New York.

Varley, N. and M. S. Boyce. 2006. Adaptive management for reintroductions: Updating a wolf recovery model for Yellowstone National Park. *Ecological Modelling* 193: 315–339.

Veblen, T. T., T. Kitzberger and J. Donnegan. 2000. Climatic and human influences on fire regimes in ponderosa pine forests in the Colorado Front Range. *Ecological Applications* 10: 1178–1195.

Vellend, M., J. A. Myers, S. Gardescu and P. L. Marks. 2003. Dispersal of *Trillium* seeds by deer: Implications for long-distance migration of forest herbs. *Ecology* 84: 1067–1072.

Vermeij, G. J. 1991. When biotas meet: Understanding biotic interchange. *Science* 253 (5024): 1099–1104.

Vermeij, G. J. 1993. *A Natural History of Shells*. Princeton University Press, Princeton, NJ.

Veron, J. E. N. 2000. *Corals of the World*. Australian Institute of Marine Science, Townsville, AU.

Vitousek, P. M. 1977. The regulation of element concentrations in mountain streams in the northeastern United States. *Ecological Monographs* 47: 65–87.

Vitousek, P. M. and H. Farrington. 1997. Nutrient limitation and soil development: Experimental test of a biogeochemical theory. *Biogeochemistry* 37: 63–75.

Vitousek, P. M. and C. B. Field. 1999. Ecosystem constraints to symbiotic nitrogen fixers: A simple model and its implications. *Biogeochemistry* 46: 179–202.

Vitousek, P. M., H. A. Mooney, J. Lubchenco and J. M. Melillo. 1997. Human domination of the Earth's ecosystems. *Science* 277: 494–499.

Volterra, V. 1926. Variazioni e fluttuazioni del numero d'individui in specie animali conviventi. *Memorie della R. Accademia dei Lincei* 2: 31–113. Reprinted 1931 as "Variations and fluctuations of the number of individuals in animal species living together," in R. N. Chapman, *Animal Ecology*, McGraw-Hill, New York.

W

Wackernagel, M. and 10 others. 2002. Tracking the ecological overshoot of the human economy. *Proceedings of the National Academy of Sciences USA* 99: 9266–9271.

Wagner, W. L., M. M. Bruegmann, D. R. Herbst and J. Q. C. Lau. 1999. Hawaiian vascular plants at risk: 1999. *Bishop Museum Occasional Papers* 60: 1–58.

Walker, T. W. and J. K. Syers. 1976. The fate of phosphorus during pedogenesis. *Geoderma* 15: 1–19.

Wallace, A. R. 1855. On the law which has regulated the introduction of new species. *Annals and Magazine of Natural History* 16.

Wallace, A. R. 1858. On the tendency of varieties to depart indefinitely from the original type. *Journal of the Proceedings of the Linnean Society, Zoology* 3: 53–62.

Wallace, A. R. 1860. On the zoological geography of the Malay Archipelago. *Journal of the Linnaean Society of London* 4: 172–184.

Wallace, A. R. 1869a. *The Malay Archipelago, The Land of The Orangutan And The Bird of Paradise: A Narrative of Travel With Studies of Man And Nature*. Macmillan, London.

Wallace, A. R. 1869b. On the physical geography of the Malay Archipelago. *Proceedings of the Royal Geographical Society of London* 7 (5): 206–213.

Wallace, A. R. 1876. *The Geographic Distribution of Animals*. 2 vols. Harper and Brothers, New York.

Wallace, A. R. 1878. *Tropical Nature and Other Essays*. Macmillan, London.

Walters, C. 1995. *Fish on the Line: The Future of Pacific Fisheries*. A report to the David Suzuki Foundation, Fisheries Project Phase I, 219–2211 West Fourth Avenue, Vancouver, BC, V6K 4S2, Canada.

Walters, J. R. 1991. Application of ecological principles to the management of endangered species: The case of the red-cockaded woodpecker. *Annual Review of Ecology and Systematics* 22: 505–523.

Walters, J. R., C. K. Copeyon and J. H. Carter. 1992. Test of the ecological basis of cooperative breeding in red-cockaded woodpeckers. *Auk* 109 (1): 90–97.

Walther, G.-R. and 8 others. 2002. Ecological responses to recent climate change. *Nature* 416: 389–395.

Waples, R. S. 1998. Evolutionarily significant units, distinct population segments, and the Endangered Species Act: Reply to Pennock and Dimmick. *Conservation Biology* 12 (3): 718–721.

Ward, M. P., J. S. Milledge and J. B. West. 1995. *High Altitude Medicine and Physiology*. Chapman and Hall, London.

Weathers, K. C., M. L. Cadenasso and S. T. A. Pickett. 2001. Forest edges as nutrient and pollutant concentrators. *Conservation Biology* 15: 1506–1514.

Webb, N. R. and L. E. Haskins. 1980. An ecological survey of heathlands in the Poole Basin, Dorset, England, in 1978. *Biological Conservation* 17: 281–296.

Webberley, K. M. and 8 others. 2004. Host reproduction and a sexually transmitted disease: Causes and consequences of *Coccipolipus hippodamiae* distribution on coccinellid beetles. *Journal of Animal Ecology* 73: 1–10.

Weber, N. A. 1966. Fungus-growing ants. *Science* 153: 587–604.

Weber, R. E. 2007. High-altitude adaptations in vertebrate hemoglobins. *Respiratory Physiology and Neurobiology* 158: 132–142.

Weir, A. A. S., J. Chappell and A. Kacelnik. 2002. Shaping of hooks in New Caledonian crows. *Science* 297: 981.

Weis, A. E. and W. G. Abrahamson. 1986. Evolution of host-plant manipulation by gall makers: Ecology and genetic factors in the *Solidago–Eurosta* system. *American Naturalist* 127: 681–695.

Weiss, R. A. and A. J. McMichael. 2004. Social and environmental risk factors in the emergence of infectious disease. *Nature Medicine* 10: S70–S76.

Weldon, A. J. 2006. How corridors reduce Indigo Bunting nest success. *Conservation Biology* 20 (4): 1300–1305.

Werner, E. E. 1988. Size scaling and the evolution of complex life cycles. In *Size-Structured Populations*, B. Ebenman and L. Persson (eds.), 61–81. Springer-Verlag, Berlin.

Werner, E. E. and J. F. Gilliam. 1984. The ontogenetic niche and species interactions in size-structured populations. *Annual Review of Ecology and Systematics* 15: 393–425.

Westemeier, R. L. and 8 others. 1998. Tracking the long-term decline and recovery of an isolated population. *Science* 282: 1695–1698.

Westoby, M. 1997. What does "ecology" mean? *Trends in Ecology and Evolution* 12: 166.

White, G. 1789. *The Natural History of Selborne*. Reprinted 1977, Penguin, New York.

White, J., J. Wilson and K. Horskins. 1997. The role of adjacent habitats in rodent damage levels in Australian macadamia orchard systems. *Crop Protection* 16 (8): 727–732.

Whiteman, C. D. 2000. *Mountain Meteorology: Fundamentals and Applications*. Oxford University Press, New York.

Whittaker, R. H. 1975. *Communities and Ecosystems*. Macmillan, New York.

Whittaker, R. H., K. J. Willis and R. Field. 2001. Towards a general, hierarchical theory of species diversity. *Journal of Biogeography* 28: 453–470.

Wilbur, H. M. 1997. Experimental ecology of food webs: Complex systems in temporary ponds. The Robert H. MacArthur Award Lecture. *Ecology* 78: 2279–2302.

Wilcove, D. S., D. Rothstein, J. Dubow, A. Phillips and E. Losos. 1998. Quantifying threats to imperiled species in the United States. *BioScience* 48 (8): 607–615.

Wilkie, D. S. and J. F. Carpenter. 1999. Bushmeat hunting in the Congo Basin: An assessment of impacts and options for mitigation. *Biodiversity and Conservation* 8 (7): 927–955.

Williams, C. B. 1943. Area and number of species. *Nature* 152: 264–267.

Williams, C. B. 1964. *Patterns in the Balance of Nature*. Academic Press, London.

Williams, J. C., C. S. ReVelle and S. A. Levin. 2005. Spatial attributes and reserve design models: A review. *Environmental Modeling and Assessment* 10 (3): 163–181.

Willig, M. R., D. M. Kaufman and R. D. Stevens. 2003. Latitudinal gradients of biodiversity: Pattern, process, scale, and synthesis. *Annual Review of Ecology and Systematics* 34: 273–309.

Willmer, P., G. Stone and I. Johnston. 2005. *Environmental Physiology of Animals*. Blackwell Publishing, Malden, MA.

Wilson, E. O. 1994. *Naturalist*. Island Press, Washington, DC.

Wilson, E. O. and D. S. Simberloff. 1969. Experimental zoogeography of islands: Defaunation and monitoring techniques. *Ecology* 50: 267–278.

Wilson, S. D. and D. Tilman. 1993. Plant competition and resource availability in response to disturbance and fertilization. *Ecology* 74: 599–611.

Windsor, D. A. 1998. Most of the species on Earth are parasites. *International Journal for Parasitology* 28: 1939–1941.

Winterhalder, B. P. 1980. Canadian fur bearer cycles and Cree-Ojibwa hunting and trapping practices. *American Naturalist* 115: 870–879.

Wirth, R., H. Herz, R. J. Ryel, W. Beyschlag and B. Hölldobler. 2003. *Herbivory of Leaf-Cutting Ants: A Case Study on Atta columbica in the Tropical Rainforest of Panama*. Springer, Berlin.

Witman, J. D., R. J. Etter and F. Smith. 2004. The relationship between regional and local species diversity in marine benthic communities: A global perspective. *Proceedings of the National Academy of Sciences USA* 101: 15664–15669.

Wood, R. 1998. *Reef Evolution*. Oxford University Press, Oxford.

Woodroffe, R. 1999. Managing disease threats to wild mammals. *Animal Conservation* 2: 185–193.

Wright, S. J. 1981. Intra-archipelago vertebrate distributions: The slope of the species–area relation. *American Naturalist* 118: 726–748.

Wright, S. J. 2005. Tropical forests in a changing environment. *Trends in Ecology and Evolution* 10: 553–560.

Y

Yates, T. L. and 15 others. 2002. The ecology and evolutionary history of an emergent disease: Hantavirus pulmonary syndrome. *BioScience* 52: 989–998.

Yoda, J. A., T. Kira, H. Ogawa and K. Hozumi. 1963. Self-thinning in overcrowded pure stands under cultivated and natural conditions. *Journal of Biology, Osaka City University* 14: 107–120.

Yoshida, T., L. E. Jones, S. P. Ellner, G. F. Fussmann and N. G. Hairston Jr. 2003. Rapid evolution drives ecological dynamics in a predator–prey system. *Nature* 424: 303–306.

Z

Zar, J. H. 2006. *Biostatistical Analysis*. 5th ed. Prentice Hall, Upper Saddle River, NJ.

Zhao, M., S. W. Running and R. R. Nemani. 2006. Sensitivity of Moderate Resolution Imaging Spectroradiometer (MODIS) terrestrial primary production to the accuracy of meteorological reanalyses. *Journal of Geophysical Research–Biogeosciences* 111: G01002.

Index

Page numbers followed by the letter *f* denote figures; those followed by the letter *t* denote tables